Advances in Soft Computing 50

Editor-in-Chief: J. Kacprzyk

Advances in Soft Computing

Further volumes of this series can be found on our homepage: springer.com

Mieczyslaw A. Klopotek, Slawomir T. Wierzchon, Kryzysztof Trojanowski (Eds.)
Intelligent Information Processing and Web Mining, 2006
ISBN 978-3-540-33520-7

Ashutosh Tiwari, Joshua Knowles, Erel Avineri, Keshav Dahal, Rajkumar Roy (Eds.)
Applications and Soft Computing, 2006
ISBN 978-3-540-29123-7

Bernd Reusch, (Ed.)
Computational Intelligence, Theory and Applications, 2006
ISBN 978-3-540-34780-4

Jonathan Lawry, Enrique Miranda, Alberto Bugarín Shoumei Li, María Á. Gil, Przemysław Grzegorzewski, Olgierd Hryniewicz,
Soft Methods for Integrated Uncertainty Modelling, 2006
ISBN 978-3-540-34776-7

Ashraf Saad, Erel Avineri, Keshav Dahal, Muhammad Sarfraz, Rajkumar Roy (Eds.)
Soft Computing in Industrial Applications, 2007
ISBN 978-3-540-70704-2

Bing-Yuan Cao (Ed.)
Fuzzy Information and Engineering, 2007
ISBN 978-3-540-71440-8

Patricia Melin, Oscar Castillo, Eduardo Gómez Ramírez, Janusz Kacprzyk, Witold Pedrycz (Eds.)
Analysis and Design of Intelligent Systems Using Soft Computing Techniques, 2007
ISBN 978-3-540-72431-5

Oscar Castillo, Patricia Melin, Oscar Montiel Ross, Roberto Sepúlveda Cruz, Witold Pedrycz, Janusz Kacprzyk (Eds.)
Theoretical Advances and Applications of Fuzzy Logic and Soft Computing, 2007
ISBN 978-3-540-72433-9

Katarzyna M. Węgrzyn-Wolska, Piotr S. Szczepaniak (Eds.)
Advances in Intelligent Web Mastering, 2007
ISBN 978-3-540-72574-9

Emilio Corchado, Juan M. Corchado, Ajith Abraham (Eds.)
Innovations in Hybrid Intelligent Systems, 2007
ISBN 978-3-540-74971-4

Marek Kurzynski, Edward Puchala, Michal Wozniak, Andrzej Zolnierek (Eds.)
Computer Recognition Systems 2, 2007
ISBN 978-3-540-75174-8

Van-Nam Huynh, Yoshiteru Nakamori, Hiroakira Ono, Jonathan Lawry, Vladik Kreinovich, Hung T. Nguyen (Eds.)
Interval / Probabilistic Uncertainty and Non-classical Logics, 2008
ISBN 978-3-540-77663-5

Ewa Pietka, Jacek Kawa (Eds.)
Information Technologies in Biomedicine, 2008
ISBN 978-3-540-68167-0

Didier Dubois, M. Asunción Lubiano, Henri Prade, María Ángeles Gil, Przemysław Grzegorzewski, Olgierd Hryniewicz (Eds.)
Soft Methods for Handling Variability and Imprecision, 2008
ISBN 978-3-540-85026-7

Juan M. Corchado, Francisco de Paz, Miguel P. Rocha, Florentino Fernández Riverola (Eds.)
2nd International Workshop on Practical Applications of Computational Biology and Bioinformatics (IWPACBB 2008), 2009
ISBN 978-3-540-85860-7

Juan M. Corchado, Sara Rodríguez, James Llinas, José M. Molina (Eds.)
International Symposium on Distributed Computing and Artificial Intelligence 2008 (DCAI 2008), 2009
ISBN 978-3-540-85862-1

Juan M. Corchado, Sara Rodríguez,
James Llinas, José M. Molina (Eds.)

International Symposium on Distributed Computing and Artificial Intelligence 2008 (DCAI 2008)

Springer

Editors

Juan M. Corchado
Departamento de Informática y Automática
Facultad de Ciencias
Universidad de Salamanca
Plaza de la Merced S/N
37008, Salamanca
Spain
E-mail: corchado@usal.es

Sara Rodríguez
Departamento de Informática y Automática
Facultad de Ciencias
Universidad de Salamanca
Plaza de la Merced S/N
37008, Salamanca
Spain
E-mail: srg@usal.es

James Llinas
Department of Industrial and Systems Engineering
University at Buffalo
315 Bell Hall
Buffalo, NY 14260-2050
U.S.A.
E-mail: llinas@acsu.buffalo.edu

José M. Molina
EPS Universidad Carlos III de Madrid
Departamento de Informática
Avenida de la Universidad Carlos III, 22
Colmenarejo, 28270 Madrid
Spain
E-mail: molina@ia.uc3m.es

ISBN 978-3-540-85862-1 e-ISBN 978-3-540-85863-8

DOI 10.1007/978-3-540-85863-8

Advances in Soft Computing ISSN 1615-3871

Library of Congress Control Number: 2008933602

Typeset & Cover Design: Scientific Publishing Services Pvt. Ltd., Chennai, India.

Printed in acid-free paper

5 4 3 2 1 0

springer.com

Preface

The International Symposium on Distributed Computing and Artificial Intelligence is an annual forum that brings together ideas, projects, lessons, etc. associated with distributed computing, artificial intelligence and its applications in different themes. This meeting has been held at the University of Salamanca from the 22th to the 24th of October 2008. This symposium has be organized by the Biomedicine, Intelligent System and Educational Technology Research Group (http://bisite.usal.es/) of the University of Salamanca. The technology transfer in this field is still a challenge and for that reason this type of contributions has been specially considered in this edition. This conference is the forum in which to present application of innovative techniques to complex problems. The artificial intelligence is changing our society. Its application in distributed environments, such as the Internet, electronic commerce, mobile communications, wireless devices, distributed computing, and so on is increasing and is becoming an element of high added value and economic potential, both industrial and research. These technologies are changing constantly as a result of the large research and technical effort being undertaken in both universities and businesses. The exchange of ideas between scientists and technicians from both academic and business areas is essential to facilitate the development of systems that meet the demands of today's society.

This symposium has evolved from the Iberoamcrican Symposium on Distributed Computing and continues to grow and prosper in its role as one of the premier conferences devoted to the quickly changing landscape of distributed computing, artificial intelligence and the application of AI to distributed systems. This year's technical program is extremely strong and diverse, with contributions in both established and evolving areas of research. Submitted papers came from over 16 different countries, representing a truly "wide area network" of research activity. The DCAI technical program includes 88 papers (74 long papers, 12 short papers and 2 doctoral consortium) selected from a submission pool of 142 papers, from 16 different countries. In addition to the main DCAI program, DCAI'08 is host to several other events: Third Symposium of Ubiquitous Computing and Ambient Intelligence 2008, Second International Workshop on Practical Applications of Computational Biology & Bioinformatics, and Second International Workshop on User-Centric Technologies and applications.

We thank the excellent work of the local organization members and also from the members of the Program Committee for their excellent reviewing work.

October 2008

Juan M. Corchado
Sara Rodríguez
James Llinas
José M. Molina

Organization

General Co-chairs

Juan M. Corchado – University of Salamanca
James Llinas (Chairman) – State University of New York
José. M. Molina (Cochairman) – University of Carlos III de Madrid
Sara Rodríguez – University of Salamanca

Program Committee

James Llinas (Chairman) – State University of New York
José. M. Molina (Cochairman) – University of Carlos III de Madrid
Adriana Giret – Universidad Politécnica de Valencia
Alberto Fernández – Universidad Rey Juan Carlos
Alicia Troncoso Lora – Universidad Pablo de Olavide, Sevilla
Álvaro Herrero – Universidad de Burgos
Ana Cristina García Bicharra – Universidad Federal Fluminense
Ángel Alonso – Universidad de León
Antonio Berlanga de Jesús – Universidad Carlos III de Madrid
Antonio Moreno – Universidad Rovira y Virgili
Araceli Sanchís – Universidad Carlos III de Madrid
B. Cristina Pelayo García–Bustelo – Universidad de Oviedo
Beatriz López – Universitat de Girona
Bogdan Gabrys – Bournemouth University
Bruno Baruque – Universidad de Burgos
Carina González – Universidad de la Laguna
Carlos Carrascosa – Universidad Politécnica de Valencia
Carmen Benavides – Universidad de León
Daniel Gayo Avello – Universidad de Oviedo
Daniel Glez-Peña – Universidad de Vigo
David de Francisco – Telefónica I+D
Eladio Sanz – Universidad de Salamanca
Eleni Mangina – University College Dublin
Emilio Corchado – Universidad de Burgos
Estefanía Argente – Universidad Politécnica de Valencia
Eugénio Oliveira– Universidade do Porto
Evelio J. González – Universidad de la Laguna

Faraón Llorens Largo – Universidad de Alicante
Fernando Díaz – Universidad de Valladolid
Fidel Aznar Gregori – Universidad de Alicante
Florentino Fdez-Riverola – Universidad de Vigo
Francisco Pujol López – Universidad de Alicante
Helder Coelho – Universidade de Lisboa
Javier Carbó – Universidad Carlos III de Madrid
Javier de Andrés Suárez – Universidad de Oviedo
Javier Martínez Elicegui– Telefónica I+D
Jesús García Herrero – Universidad Carlos III de Madrid
José M. Molina – Universidad Carlos III de Madrid
José R. Méndez – Universidad de Vigo
José R. Villar – Universidad de Oviedo
José V. Álvarez-Bravo – Universidad de Valladolid
Juan A. Botia – Universidad de Murcia
Juan J. Álvarez-Sánchez – Universidad de Valladolid
Juan M. Serrano – Universidad Rey Juan Carlos
Juan Manuel Cueva Lovelle – Universidad de Oviedo
Juan Pavón – Universidad Complutense de Madrid
Lourdes Borrajo – Universidad de Vigo
Luis Alonso – Universidad de Salamanca
Luis Correia – Universidad de Lisboa
Luis F. Castillo – Universidad Autónoma de Manizales
Manuel González-Bedia – Universidad de Zaragoza
Manuel Resinas – Universidad de Sevilla
Maria del Mar Pujol López – Universidad de Alicante
María H. Mejía-Salazar – Universidad de Caldas
Miguel Angel Patricio – Universidad Carlos III de Madrid
Miguel Rebollo – Universidad Politécnica de Valencia
Oscar Sanjuan Martínez – Universidad de Oviedo
Rafael Corchuelo – Universidad de Sevilla
Ramón Rizo – Universidad de Alicante
Rubén Fuentes – Universidad Complutense de Madrid
Tzai-Der Wang – Cheng Shiu University
Vicente Botti – Universidad Politécnica de Valencia
Vicente Julian – Universidad Politécnica de Valencia

Organising Committee

Juan M. Corchado (Chairman) – University of Salamanca
Sara Rodríguez (Cochairman) – University of Salamanca
Dante I. Tapia – University of Salamanca
Juan F. De Paz – University of Salamanca
Javier Bajo – University of Salamanca
Cristian Pinzón – University of Salamanca
Rosa Cano – University of Salamanca
Aitor Mata – University of Salamanca

Contents

Paralell and Evolutionary Algorithms

Multiagent Systems III

Genetic Algorithms

New Algorithms and Applications

Ambient Intelligence and Context-Aware

Applications III

A Simulated Annealing Method to Cover Dynamic Load Balancing in Grid Environment

Mauricio Paletta[1] and Pilar Herrero[2]

[1] Departamento de Ciencia y Tecnología, Universidad Nacional Experimental de Guayana, Av. Atlántico, Ciudad Guayana, Venezuela
mpaletta@uneg.edu.ve

[2] Facultad de Informática. Universidad Politécnica de Madrid, Campus de Montegancedo S/N, 28.660 Boadilla del Monte, Madrid, Spain
pherrero@fi.upm.es

Abstract. High-performance scheduling is critical to the achievement of application performance on the computational grid. New scheduling algorithms are in demand for addressing new concerns arising in the grid environment. One of the main phases of scheduling on a grid is related to the load balancing problem therefore having a high-performance method to deal with the load balancing problem is essential to obtain a satisfactory high-performance scheduling. This paper presents SAGE, a new high-performance method to cover the dynamic load balancing problem by means of a simulated annealing algorithm. Even though this problem has been addressed with several different approaches only one of these methods is related with simulated annealing algorithm. Preliminary results show that SAGE not only makes it possible to find a good solution to the problem (effectiveness) but also in a reasonable amount of time (efficiency).

Keywords: Grid Computing, load-balancing, Simulated Annealing.

1 Introduction

Grid Computing (GC) [5, 11] emerged in the mid 1990s as an alternative to design distributed computing infrastructure for sharing resources. With the evolution of the computational grid, new scheduling algorithms are in demand for addressing new concerns arising in the grid environment. In this environment the scheduling problem is to schedule a stream of applications from different users to a set of computing resources to maximize system utilization [10].

But, the complexity of scheduling problem increases with the size of the grid and becomes highly difficult to solve effectively, so that the developed of techniques that provide an optimal or near optimal solution for large grids has been required [9]. In this sense, high-performance schedulers become a critical part of the programming environment for the computational grid, and therefore an area of very active research and development is evolving [3]

In the same order of ideas, there are three main phases of scheduling on a grid [8]: 1) resource discovery, which generates a list of potential resources; 2) gathering information about those resources and choosing the best set to match the application requirements (load-balancing problem); 3) task execution, which includes file staging and cleanup.

A load-balancing strategy, which relates with the second phase of scheduling on a grid, deals with partitioning a program into smaller tasks that can be executed concurrently and mapping each of these tasks to a computational resource (this problem is

J.M. Corchado et al. (Eds.): DCAI 2008, ASC 50, pp. 1–10, 2009.
springerlink.com

also known as task allocation or mapping problem). An efficient load balancing strategy avoids the situation where some resources are idle while others have multiple jobs queued up. Therefore, by developing strategies that can map these tasks to resources in a way that balances out the load, the total processing time will be reduced with improving the utilization of these resources.

In this sense, having a high-performance method to deal with the load-balancing problem related with the scheduling on a grid is an important part the programming environment for the computational grid should take into consideration.

On the other hand, and due to the fact that load-balancing is a combinatorial optimization problem the use of heuristics is useful to cope in practice with its difficulty [9]. Most of the research on load-balancing strategy in particular, and scheduling in general, focused on static scenarios that, in most of the cases, employ heuristic methods (see next section for details). Although static techniques have proved effectiveness, the dynamic nature of the grid requires effective and efficient dynamic allocation of available resources techniques, even though there is often a trade-off between solution quality and speed in achieving a solution [18].

This paper presents a new heuristic method designed to solve the dynamic load-balancing problem in grid environments. This method, called SAGE (Simulated Annealing to cover dynamic load balancing in Grid Environment), is a dynamic load-balancing method based on Simulated Annealing (SA) algorithm that is properly configured and developed whereby optimal or near-optimal task allocations can "evolve" during the operation of the grid system.

As part of the SA configuration, a valid representation of the current state of the grid environment must be defined. In this case the specifications given by Herrero et al in [13, 14] are used to do this. Thus, the results of preliminary experiments obtained using a test application and presented in this paper may be associated with possible results to achieve in real or simulated grid environments. These results show that SAGE not only makes it possible to find a good solution to the problem but also in a reasonable amount of time, obtaining the high-performance feature it is looking for.

SAGE could be integrated in any grid model as CAM [14] (Collaborative/Cooperative Awareness Management), specifically in the scheduling management component, with the aim to give an alternative way to deal with the load-balancing problem in grid environments. So that, users from existing grid models, in general, may benefit with another way to deal the load-balancing problem and decide which algorithm to use according to the results obtained.

The paper is organized as follow. Section 2 presents the existing related work in the area, a briefly description of the SA algorithm is presented in section 3, section 4 presents SAGE, results of some experimental tests are presented in section 5. Finally, section 6 exposes the paper conclusions as well as the future work related with this research.

2 Related Work

Intensive research has been done in the area of scheduling algorithms for grid computing and many results have been widely accepted. An interesting reading on the state of the art about this area is given by Fangpeng et al in [7].

Some heuristic approaches for this problem include local search [20], Tabu Search [1], genetic algorithms [1, 12, 22, 24], Ant Colony Optimization (ACO) algorithm [10, 17], intelligent agents [6, 13], particle swarm optimization [2], fuzzy based scheduling [16], economic-based approaches [5], Multi-Objective Evolutionary Algorithm (MOEA), and Great Deluge (GD) algorithm [18]. Regard to SA who is the subject in which the proposal presented in this paper was designed, some approaches can be reviewed in [1, 4, 21, 23]. The main difference between these previous SA-based works and SAGE is that they are offline methods (tasks are collected in a set and after that they are scheduled) and SAGE is designed to be an online method (every time a new task has to be executed it is scheduled based on current conditions).

Next section presents a briefly description of the SA method including the elements that are needed to define for solving combinatorial optimization problems using this method.

3 SA Method

SA [15, 19] is a random-search technique (a generalization of a Monte Carlo method) which exploits an analogy between the way in which a metal cools and freezes into a minimum energy crystalline structure (the annealing process) and the search for a minimum in a more general system; it forms the basis of an optimization technique for combinatorial and other problems. The concept originates from the way in which crystalline structures are brought to more ordered states by an "annealing process" of repeated heating and slowly cooling the structures.

In SA, a system is initialized at a temperature T with some configuration whose energy is evaluated to be E. A new configuration is constructed by applying a random change, and the change in energy dE is computed. The new configuration is unconditionally accepted if it lowers the energy of the system. If the energy of the system is increased by the change, the new configuration is accepted depending on the probability distribution of Boltzmann [19]; this process is repeated sufficient times at the current temperature to sample the search space; then the temperature is decreased; the process is repeated at the successively lower temperatures until a frozen state is achieved.

This procedure allows the system to move to lower energy states, while still jumping out of local minima (especially at higher temperatures) due to the probabilistic acceptance of some upward moves. SA theory states that if temperature is lowered sufficiently slowly (by carefully controlling the rate of cooling), the solid will reach thermal equilibrium, which is an optimal state (global optimum).

To solve a problem through this strategy it is necessary to define the following elements:

1) The representation of a valid solution to the problem.
2) The cost function (energy) that measures the quality of each solution.
3) The mechanism of transition of the space of solutions from t to t+1 (dynamic of the model).
4) The parameters that control the rate of cooling.

Next section presents the way in which these four elements are defined according to the load-balancing problem in grid environments to have a high-performance method to cover this problem.

4 SAGE: A SA Based Method to Cover Load Balancing in Grid Environments

4.1 The Grid Model

The grid specifications considered in this study can be reviewed in detail in [13, 14]. The distributed environment *DE* is composed of n hosts h_i ($1 \leq i \leq n$) each composed of several computational resources r_j each of a specific type (CPU power, memory, disk space, knowledge, etc.). If the following statements are considered:

1) There is a maximum of m types or classes of resources ($1 \leq j \leq m$).
2) A particular host h_i may have more than one resource of the same type.
3) A particular host h_i may not have a resource for a specific type.

$$\overline{DE} = \left| \begin{array}{cccc} & r_1 & \cdots & r_m \\ h_1 : & f_{11} & \cdots & f_{1m} \\ \vdots & \vdots & \ddots & \vdots \\ h_n : & f_{n1} & \cdots & f_{nm} \end{array} \right. \tag{1}$$

$$f_{ij} = \begin{cases} \textit{Maximum}, \text{entire resource is available} \\ \textit{Medium}, \text{part of the resource is available} \\ \textit{Null}, \text{the resource is saturated or is not available} \end{cases} \tag{2}$$

Then and as can be seen in (1), *DE* can be represented using a matrix with n rows and m columns; each row represents a host with the corresponding vector of resources; each column is associated with a specific type of resource. In this sense a particular item f_{ij} of the matrix corresponds to the degree in which the resource type r_j of the host h_i is being used at a given time. As can be seen in (2) there are three possible values for setting each f_{ij}: *Maximum* when h_i has at the disposal all the resource associated to the type r_j; *Medium* when h_i has at the disposal only a part of the resource associated to the type r_j, and *Null* when h_i has not resources r_j at the disposal either because it is saturated or because it is not available in h_i.

A task T who needs resources from this grid-based system is composed by a collection of processes p_k each associated with the resource type necessary to complete the specific process. So that, T can be seen as a vector of p elements each associated to a valid value that represents a resource type:

$$\overline{T} = (p_1, \ldots, p_p)\,;\ \forall k, 1 \leq k \leq p, 1 \leq p_k \leq m \tag{3}$$

In this sense, the load-balancing problem related to this system is as follows: for any task T, looks a valid relationship (given by the resource type) between all the processes p_k of T according to the current state of *DE*, and considering the following statements:

1) It must be done dynamically, so in a reasonable time.
2) It must be done considering the load-balancing of each host.
3) It must be done considering the load-balancing of each resource type.

In other words, it is necessary to find, in a efficient, effective and dynamical way, what $f_{ij}(t)$ must be changed for each p_k in order to obtain the new state of *DE* or *DE*(t+1).

4.2 Representing a Valid Solution

A solution *S* for this problem is represented with a vector of size *p*; each element s_l ($1 \leq l \leq p$) has the host in which the process p_l of *T* has been assigned. So that, the value of s_l is the row index *i* and the value of p_l is the column index *j* for the element f_{ij} of *DE* in which the process *l* has been assigned.

This way, having the task $T = (p_1,\ldots,p_p)$ and the solution $S = (s_1,\ldots,s_p)$ it is possible to know all the f_{ij} involved in the distribution of the *p* processes of *T* among the *n* hosts of *DE*.

4.3 Measuring a Solution

As was previously read in Section 3, one of the elements to define for using the SA strategy is the cost function or energy that measures the quality of a specific solution. In this sense the following statements are considered:

1) While lower energy (system temperature) is better quality in the solution.
2) Energy is directly proportional to the current state of the resources.
3) There should be a restriction to balance the type of resources of each host.
4) There should be a restriction to balance the hosts of each type of resource.

First, it is necessary to define values for measuring the current state of the resources. Due to the fact that energy or solution quality is directly proportional to the current state of the resources, these values are defined according to the following guidelines:

1) Reward (low or no energy) to those resources that are full available.
2) Punish (high energy) those that are saturated.
3) Severely punish (very high energy) invalid situations (a solution which assigns a process to a saturated or not available resource).

If "τ(x)" is the function that transforms the current state f_{ij} of a resource then:

$$\tau(f_{ij}) = \begin{cases} 0, f_{ij} = Maximum \\ 1, f_{ij} = Medium \\ 2, f_{ij} = Null \\ 5, \text{other case} \end{cases} \tag{4}$$

Therefore, minimum energy ε_m (top quality) from the current state of *DE* depends on the accumulated resource use:

$$\varepsilon_m(DE) = \sum_{i,j}^{n,m} \tau(f_{ij}) \tag{5}$$

To define restrictions for balancing the type of resources of each host and the hosts of each type of resource, Th_i is the current resource accumulated for the host h_i (sum of the elements of the row i in DE) and Tr_j is the current host accumulated for the resource type r_j (sum of the elements of the column j in DE):

$$Th_i = \sum_{j=1}^{m} \tau(f_{ij}) \tag{6}$$

$$Tr_j = \sum_{i=1}^{n} \tau(f_{ij}) \tag{7}$$

Therefore, for balancing the resources from the same host h_i their values are in the order of Th_i / n, and for balancing the hosts from the same resource type r_j their values are in the order of Tr_j / m. Any difference to these values must be considered in the calculation of the restrictions. Using: 1) squares to handle the negative differences, 2) adjustment factors μ and δ to determine the degree of importance which we want to consider each restriction, and 3) divided between 2 to facilitate the derivation of the term, the restrictions Ψ_1 and Ψ_2 are defined as follow:

$$\Psi_1 = \frac{\mu}{2}\left(\sum_{j=1}^{m}\left(\sum_{i=1}^{n}(Tr_j/n - \tau(f_{ij}))^2\right)\right) \tag{8}$$

$$\Psi_2 = \frac{\delta}{2}\left(\sum_{i=1}^{n}\left(\sum_{j=1}^{m}(Th_i/m - \tau(f_{ij}))^2\right)\right) \tag{9}$$

Finally, for a current situation of DE at time t and a specific solution S the energy of the system with these conditions is calculated as follow:

$$\varepsilon(DE(t), S) = \frac{1}{2}\varepsilon_m(DE) + \Psi_1 + \Psi_2 \tag{10}$$

4.4 The Mechanism of Transition and Cooling Control Parameters

The rules governing change of DE between time t to t+1 in the dynamic of the system affect only those resources f_{ij} that have been assigned with any process in the optimal solution S obtained once a SA iteration was executed. In this sense, the new values are calculated according to the following statements:

1) $f_{ij}(t) = Maximum \Rightarrow f_{ij}(t+1) = Medium$.
2) $f_{ij}(t) = Medium \Rightarrow f_{ij}(t+1) = Null$.
3) $f_{ij}(t) =$ other case $\Rightarrow f_{ij}(t+1) =$ other case.

The first mapping or tentative solution S is generated from a uniform random distribution. The mapping is mutated randomly changing the host of a specific process, and the new solution is evaluated. If the new solution is better (less energy), it replaces the old one. If the new solution is worse, it is accepted depending on the probability distribution of Boltzmann. After each mutation, the system temperature is reduced to 95% of its current value. The heuristic stops when there is no change in the solution for 5 iterations or the system temperature approaches a value closes to zero or the total of iterations reaches the value of n x m x p. The initial value of the temperature is equal to n x m.

Next section presents some experimental results based on the efficiency and effectiveness of the method previously defined.

5 Experimental Results

The results presented in this section are taken from a testing experimental application which is possible to indicate the number of hosts n, the number of resources types m, the maximum number of processes p for a task T, the initial state (null or random) of DE, and the values for the adjustment factors μ and δ for both restrictions. The hardware where the tests were done is 2.16GHz and 2 GB RAM.

Fig. 1 shows a screen shot of the testing application. At the bottom left is the matrix *DE* in the current state (6 hosts and 10 resource types in the example); at the bottom right are the current results of balancing including run time in the process execution (1 sec.). As it can be seen in the results all processes were assigned to resources that were in a state equal to *Maximum* (value of 0) which is the optimal situation for this problem.

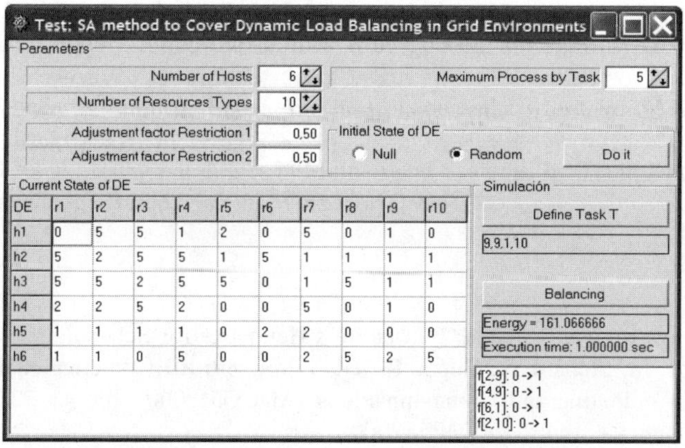

Fig. 1. A screen shot of the application for testing the method

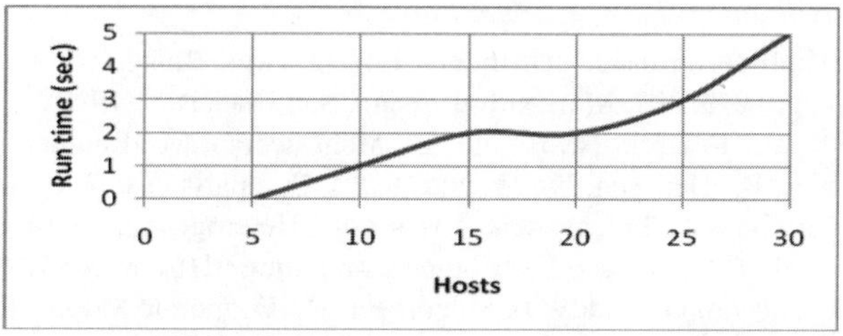

Fig. 2. Execution times with $m = 10$, $p = 5$ and $n \in \{5,10,15,20,25,30\}$

It is important to mention that all parameters can be changed dynamically previous to execute the balancing algorithm, as for example to increment the number of hosts in the system. Fig. 2 shows the execution times (in seconds) to obtain an optimal solution or close to the optimum of a grid with 10 different types of resources, tasks composed by 5 processes and varying the number of hosts from 5 to 30 (5 in 5).

6 Conclusions and Future Work

In this paper we present SAGE, a SA based method to cover dynamic load-balancing problem in Grid Environments. The results show that the heuristic strategy proposed finds good solutions (closer to the optimum) in a reasonable time (a few seconds). Therefore, this algorithms guarantees efficient and effective load-balancing of any task requires resources from a grid-based system and it also can be applied in dynamic way.

Even though this method has not yet been tested neither in a real grid environment nor a grid environment simulation, the manner in which the current state of a grid environment is represented is suitable for real grid environments. Therefore it is expected that the results of preliminary experiments are also achieved in real scenarios.

We are working on the term for calculating the energy that can be adapted to meet other needs in the scheduling problem of a grid. As a future work we will handle a large scale of possibilities for the degree in which the resource types are being used at a given time, as for example using fuzzy logic techniques. We are also working on integrating this algorithm in simulated applications schedulers by using grid modeling and simulating toolkit to obtain more accurate results in experiments and compare them with other similar work.

References

1. Abraham, A., Buyya, R., Nath, B.: Nature's Heuristics for Scheduling Jobs in Computational Grids. In: Sinha, P.S., Gupta, R. (eds.) Proc. 8th IEEE International Conference on Advanced Computing and Communications (ADCOM2000), pp. 45–52. Tata McGraw-Hill Publishing Co. Ltd, New Delhi (2000)
2. Abraham, A., Liu, H., Zhang, W., Chang, T.G.: Scheduling Jobs on Computational Grids Using Fuzzy Particle Swarm Algorithm. In: Proc. of 10th International Conference on Knowledge-Based & Intelligent Information & Engineering Systems, England, pp. 500–507 (2006)
3. Berman, F.: High-performance schedulers. In: The Grid: Blueprint for a New Computing Infrastructure, pp. 279–309. Morgan Kaufmann, San Francisco (1999)
4. Braun, R., Siegel, H., Beck, N., Boloni, L., Maheswaran, M., Reuther, A., Robertson, J., Theys, M., Yao, B., Hensgen, D., Freund, R.: A Comparison of Eleven Static Heuristics for Mapping a Class of Independent Tasks onto Heterogeneous Distributed Computing Systems. Journal of Parallel and Distributed Computing 61(6), 810–837 (2001)
5. Buyya, R., Abramson, D., Giddy, J., Stockinger, H.: Economic Models for Resource Management and Scheduling in Grid Computing. Journal of Concurrency and Computation: Practice and Experience 14(13-15), 1507–1542 (2002)

6. Cao, J., Spooner, D.P., Jarvis, S.A., Nudd, G.R.: Grid load balancing using intelligent agents. Future Generation Computer Systems 21(1), 135–149 (2005)
7. Fangpeng, D., Selim, G.A.: Scheduling Algorithms for Grid Computing: State of the Art and Open Problems Technical Report No. 2006-504, Queen's University, Canada, 55 pages (2006), http://www.cs.queensu.ca/TechReports/Reports/2006-504.pdf
8. Fernandez-Baca, D.: Allocating Modules to Processors in a Distributed System. IEEE Transactions on Software Engineering 15(11), 1427–1436 (1989)
9. Fidanova, S.: Simulated Annealing for Grid Scheduling Problem. In: Proc. IEEE John Vincent Atanasoff International Symposium on Modern Computing (JVA 06), 10.1109/JVA.2006.44, pp. 41–44 (2006)
10. Fidanova, S., Durchova, M.: Ant Algorithm for Grid Scheduling Problem. In: Lirkov, I., Margenov, S., Waśniewski, J. (eds.) LSSC 2005. LNCS, vol. 3743, pp. 405–412. Springer, Heidelberg (2006)
11. Foster, I., Kesselman, C., Tuecke, S.: The Anatomy of the Grid: Enabling Scalable Virtual Organizations. International Journal of Supercomputer applications and High Performance Computing 15(3), 200–222 (2001)
12. Grosan, C., Abraham, A., Helvik, B.: Multiobjective Evolutionary Algorithms for Scheduling Jobs on Computational Grids. In: Guimares, N., Isaias, P. (eds.) IADIS International Conference, Applied Computing 2007, Salamanca, Spain, pp. 459–463 (2007) ISBN: 978-972-8924-30-0
13. Herrero, P., Bosque, J.L., Pérez, M.S.: An Agents-Based Cooperative Awareness Model to Cover Load Balancing Delivery in Grid Environments. In: Meersman, R., Tari, Z., Herrero, P. (eds.) OTM-WS 2007, Part I. LNCS, vol. 4805, pp. 64–74. Springer, Heidelberg (2007)
14. Herrero, P., Bosque, J.L., Pérez, M.S.: Managing Dynamic Virtual Organizations to get Effective Cooperation in Collaborative Grid Environments. In: Meersman, R., Tari, Z. (eds.) OTM 2007, Part II. LNCS, vol. 4804, pp. 1435–1452. Springer, Heidelberg (2007)
15. Kirkpatrick, S.: Optimization by Simulated Annealing: Quantitative Studies. Journal of Statistical Physics 34(5-6), 975–986 (1984)
16. Kumar, K.P., Agarwal, A., Krishnan, R.: Fuzzy based resource management framework for high throughput computing. In: Proc. of the 2004 IEEE International Symposium on Cluster Computing and the Grid, pp. 555–562 (2004)
17. Lorpunmanee, S., Sap, M.N., Abdullah, A.H., Chompoo-inwai, C.: An Ant Colony Optimization for Dynamic Job Scheduling in Grid Environment. International Journal of Computer and Information Science and Engineering 1(4), 207–214 (2007)
18. McMullan, P., McCollum, B.: Dynamic Job Scheduling on the Grid Environment Using the Great Deluge Algorithm. In: Malyshkin, V.E. (ed.) PaCT 2007. LNCS, vol. 4671, pp. 283–292. Springer, Heidelberg (2007)
19. Metropolis, N., Rosenbluth, A.W., Rosenbluth, M.N., Teller, A.H., Teller, E.: Equations of state calculations by fast computing machines. Journal of Chemical Physics 21(6), 1087–1091 (1953)
20. Ritchie, G., Levine, J.: A fast, effective local search for scheduling independent jobs in heterogeneous computing environments, Technical report, Centre for Intelligent Systems and their Applications, School of Informatics, University of Edinburgh (2003)
21. Yarkhan, A., Dongarra, J.: Experiments with Scheduling Using Simulated Annealing in a Grid Environment. In: Parashar, M. (ed.) GRID 2002. LNCS, vol. 2536, pp. 232–242. Springer, Heidelberg (2002)

22. Ye, G., Rao, R., Li, M.: A Multiobjective Resources Scheduling Approach Based on Genetic Algorithms in Grid Environment. In: Fifth International Conference on Grid and Cooperative Computing Workshops, pp. 504–509 (2006)
23. Young, L., McGough, S., Newhouse, S., Darlington, J.: Scheduling Architecture and Algorithms within the ICENI Grid Middleware. In: Proc. of UK e-Science All Hands Meeting, Nottingham, pp. 5–12 (2003)
24. Zomaya, A.Y.: The YH (2001) Observations on using genetic algorithms for dynamic load-balancing. IEEE Transactions On Parallel and Distributed Systems 12(9), 899–911 (2001)

Research and Design of a Dynamic Forest Growing Simulation System Based on HLA

Fan Jing, Dong Tianyang, and Sun Siang

College of Software Engineering
Zhejiang University of Technology
Hangzhou, 310014, China
fangjing@zjut.edu.cn, dty@zjut.edu.cn, ssahz@126.com

Abstract. In order to address the issues of high coupling of tree shape modeling with communication modeling, as well as limited simulation scale in traditional forest simulation, this paper presents a novel dynamic forest growing simulation system after analyzing the application of High Level Architecture (HLA) and Open Grid Services Architecture (OGSA) technology. In this forest growing simulation system based on HLA, the communication model is constructed by HLA specification and the real-time communication between tree models is carried out by the Run-Time Infrastructure (RTI). The experiments show that the simulation system can avail of all available resources in the grid, and resolve the problem of the simulation results of visual reality and accuracy that can't be satisfied at the same time. Grid services can make the traditional simulation application from single computer distribute to multiple grid nodes. It will improve the speed of virtual scene rendering and the utilization efficiency of grid resources.

Keywords: Grid, HLA, OGSA, Virtual forest.

1 Introduction

The simulation of forest growing from the perspective of the ecosystem is a new cross-research project, which uses virtual reality technology to simulate the evolution of the forest ecological landscape and integrates various disciplines of technology and knowledge such as computer graphics, forestry, geography, ecology and so on [1]. The L system provided by biologist Lindenmayer is widely used in plants modeling. In addition, the plants modeling methods also include AMAP model, the function iteration system (IFS), fractal methods, branch matrix and Dual-Scale Automaton Model. For large-scale forest scene rendering, many mature technologies have been presented, such as the bulletin board, similar examples technology, and view-dependent adaptive level of details. But all these technologies are coupled with communication model during the process of modeling the plants shape model. When we need to model a new species, we have to completely rewrite the grammar. So the modification and maintenance are very difficult. At the same time, the simulation model based on these technologies can't use mathematical formula to quantify communication model. Therefore, it can't be applied in the agriculture, forestry and industry, because the simulation results can't satisfy the demand of reality and science.

HLA resolves the problem of reusability and interoperability [2]. HLA is an open and object-oriented architecture. Its most notable feature is to separate from the function implementation, operation management and communications by using generic

J.M. Corchado et al. (Eds.): DCAI 2008, ASC 50, pp. 11–18, 2009.
springerlink.com

and relatively independent support-service procedures that can hide the details [3]. HLA was first used in the military field to simulate the offensive and defensive weapons systems, thus the strategy and tactics could be researched. By now, HLA technology hasn't been applied in the forest growing simulation. This research will use HLA technology to simulate the evolution of the forest ecological landscape. The different trees will be the federal members. These tree federal members establish the communication models in accordance with the forestry and communicate through RTI between the entities.

Usually the virtual forest simulation needs enormous computing resources. If the simulation carried out only on one computer, it is required to reach the balance between the visual effects and the complexity of algorithm. Therefore, it is difficult to give attention to each other at the same time. Grid computing by using the joint and distributed computing model, can maximize the use of computing power in existing networks, improve the utilization efficiency of computing resources and achieve the aim of the various grid resource sharing.

In order to address the issues of high coupling of tree shape modeling with communication modeling, as well as limited simulation scale in traditional forest simulation, this paper presents a novel dynamic forest growing simulation system after analyzing the application of HLA and OGSA technology.

This research will encapsulate the MDS service of Globus to realize dynamic discovering of the grid resources. The GRAM and GRSS services of the Globus are applied to achieve the security and dynamic allocation for the grid resources. The communication of tree model is built by HLA specification, the RTI is used to simulate the real-time communication between tree models, and the possibility of distributed RTI communication through OGSA resource management service provided by GT3 is discussed.

2 The Dynamic Forest Growing Simulation System Based on HLA

This paper presents a dynamic forest growing simulation system based on HLA. This system is divided into grid resources support layer, tree entities communication layer and forest simulation application layer. Its structure is shown in Figure 1.

Grid resources support layer uses the MDS and GRAM components of Globus to find the simulation nodes and dynamically allocate resources, which solves the problem of static allocation of limited resources. These resources include various grid resources, such as computing resources, storage resources, equipment resources and so on. The resource management interface of every type resources need to be customized in the realization.

The tree entities communication layer is composed of two parts: tree model module and communication unit. The tree model is built with L-system. The communication unit includes RTIExec module, FedExec module and LibRTI module. The RTIExec module is a global process that is in charge of the management of the federation. The FedExec module manages the members' behaviors of joining or exiting the federation,

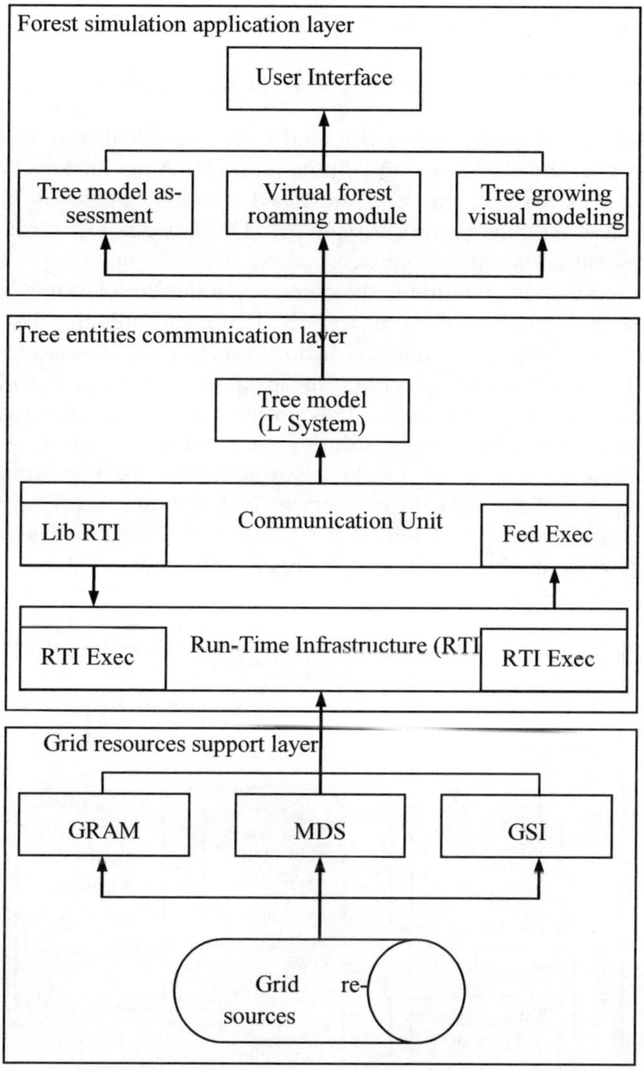

Fig. 1. The structure of dynamic forest growing simulation system

and provides a mechanism to cooperate with these members. The LibRTI module is the function library that provides the interface of HLA.

Forest simulation application layer is responsible for the conversion from a two-dimensional entity to a three-dimensional entity. Tree growing visual modeling tool can change the simulation entities into a three dimensional shape to the user by using L grammar, texture mapping and other technologies. The users can use the virtual real-time roaming tools to browse the forest scenes for the aim of assessment, then modify the defects of the simulation entities through tree growth model.

3 Key Issues in the Dynamic Forest Growing Simulation System

3.1 Mechanism of Resource Discovery in Grid Environment

In order to discover the grid resource efficiently, the resource discovery module of the dynamic forest growing simulation system based on HLA is composed of the simulation resources index service, the semantic-based resources aggregation service, RTI factory service and federate factory service [6]. The flowchart of resource discovery in grid environment is shown in Figure 2.

The RTI factory service module is the core part of the Simulation system, which is in charge of the management of the process RTIExec and building the RTI ambassador. By creating the instance of the RTI factory service, the simulation grid environment is built. The federate factory service encapsulates all the tree models that may be provided by different organization [5]. This service provides the federate ambassador and creates the instance of tree models during the simulation. The resources aggregation service module discovers all available resources for the tree simulation, by requesting the service of the RTI factory service and federate factory service [7]. The discovered resources will be stored in the simulation database. The simulation resources index service module can search the resources in the simulation database.

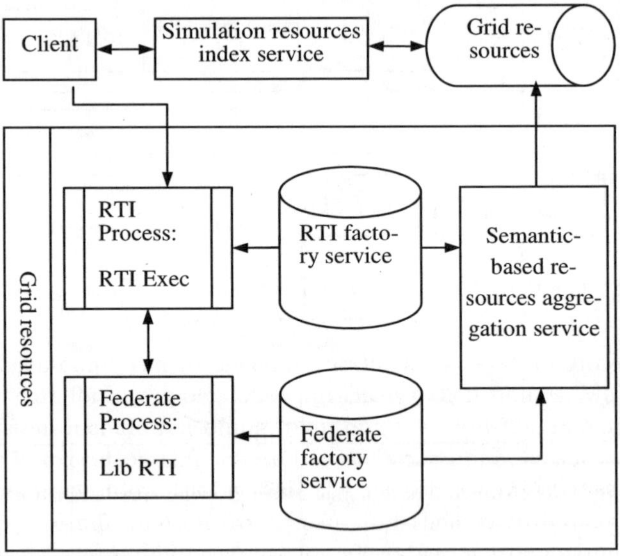

Fig. 2. The Mechanism of Resource Discovery in Grid Environment

3.2 The Design of Simulation Entities

Simulation entities will be divided into two layers: presentation layer and interactive layer. The dynamic forest growing simulation system based on HLA has two categories simulation entities: one is the tree simulation entity, and the other is topography simulation entity.

In the presentation layer, the established trees simulation entity has the following advantages: because of the separation between interactive layer and the presentation layer, it is easy to maintain; the method of random number modeling is utilized in the tree models by using the topology and geometry parameters; more entities of the same species that have different shapes can be created by using the same grammar.

As shown in Figure 3, tree simulation entities include various types of trees. Firstly, the user can select the three dimensional model of tree organs, tree branches structure model and the corresponding texture according to botanical knowledge. Then, the influence of environment on trees entities and the impact of competition between tree entities are controlled by the rules and parameters. At last, the tree models are constructed by using the L-system. That utilizes the iterative and recursive rules to achieve dynamic tree growing simulation [4]. The topography entities that don't involve dynamic changes can be constructed by reading the DEM data and mapping the textures. Finally, the stand model combines the tree model with the topography model to build the forest model.

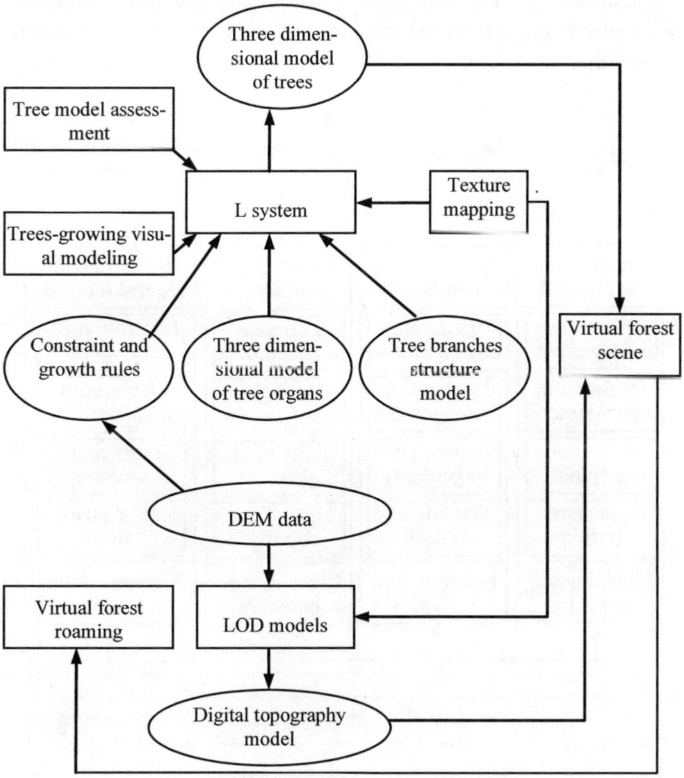

Fig. 3. The flowchart of virtual forest scene rendering

In the interactive layer, the federal members are divided into two groups: the tree federal members and the environment federal members. There are six type interfaces defined in HLA: federation management, declaration management, object management, ownership, time management and data distribution management. Four type interfaces are used in this research: federation management, declaration management, time management and data distribution management. In this research, the federation management module is used to start up the federation; The declaration management module is used to define the interactive interface for the federal members; The time management module defines the simulation time step and synchronizes the simulation process; The data distribution management module is in charge of the parameter's passing between the federal members.

When rendering a large-scale forest scene, the tree federal members and the environment federal members will communicate through the RTI ambassador module and the federate ambassador module to create the virtual objects. The interaction between the federal members is shown in Figure 4. The federal members can't communicate with each other directly. The interaction between the federal members must be carried out by the RTI ambassador module and the federate ambassador module.

The competition between trees leads to the parameters changes of light, soil and temperature and humidity. The tree federal members are only interested in the parameters of light intensity, soil moisture, soil temperature, air temperature, air humidity, location and other attributes. So every federal member publishes these attributes

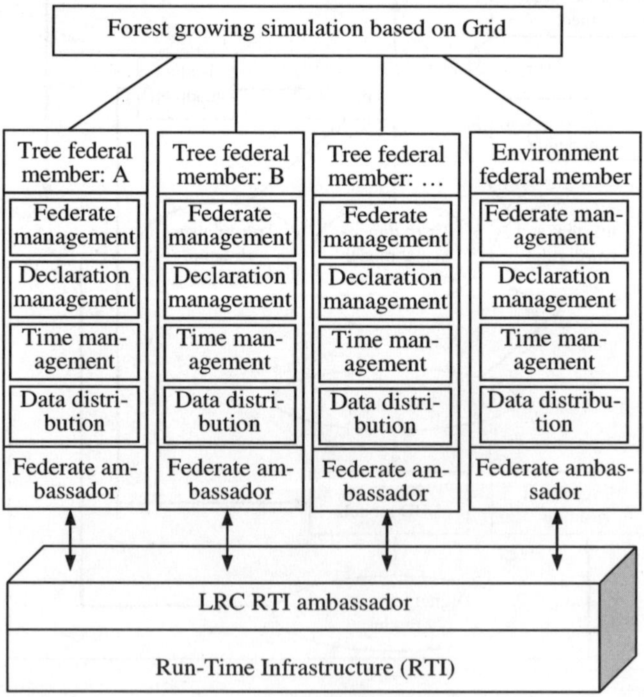

Fig. 4. The interaction between the federal members

of the trees through a statement management service. The environment federal members order these attributes and reflect the changes according to the interactive rules and the parameters of other tree federal members. So this competition between the trees can be expressed by the interaction between the tree federal members and the environment federal members. The environment federal members will pass the changed property value to the tree federal members through the data distribution management services. The tree federal members will translate these parameters to the L grammar. L system will reflect the parameters in the three dimensional shape of trees, thus completing the real-time influences between the trees.

For the simulation applications, each federal member needs to be coordinated. Therefore, the time management service is used to define the simulation step (such as one year, one month), in order to ensure consistent time and dimension in various simulation entities and reach the aim of the whole forest growing simulation. In this way, the presentation layer and interactive layer will be separate. Interactive layer can construct different communication models according to all kinds of knowledge. Thus, the simulation system is easier to maintain and has the maximum scalability.

4 Result and Conclusion

The simulation system is implemented based on the Windows platform. A simple experiment was carried out in two computers having CPU of Pentium IV 3.06GHZ and memory of 1024M bytes. Figure 5 shows the screenshot of virtual forest scene. In this virtual forest scene, there are 1600 five-year trigeminal trees. It took the dynamic forest growing simulation system based on HLA 6500ms to render the virtual scene. If the same virtual scene is rendered in one computer, it need spend 9600ms. It indicates that the simulation system has a good performance in the field of dynamic forest growth simulation.

Fig. 5. The scene of virtual forest

In this research, a novel forest simulation architecture: a dynamic forest growing simulation system based on HLA is presented. The grid technology, HLA technology and 3D plants modeling technology are used in this forest simulation system. This system can avail of all available resources in the grid, and resolve the problem of the simulation results of visual reality and accuracy that can't be satisfied at the same time. The advantage of the system using the HLA/RTI can simulate the interaction between the trees and improve the scientific degree of the simulation results. Grid services can make the traditional simulation application from single computer distribute to multiple grid nodes. Therefore, it will improve the speed of virtual scene rendering and the utilization efficiency of grid resources.

The future work of this research will be carried out in the followings: (1) research on the dynamic loading of grid to enhance the fault-tolerant of this system, (2) research on the modeling of tree organs, such as the swing branches in the wind, to improve the reality of virtual forest.

References

[1] Pan, Z.G.: Distributed graphics and its application, pp. 133–179. Posts & Telecommunications Press, Beijing (1977) (in chinese)

[2] Karlsson, M., Olsson, L.: PRTI 1516-rationale and design. In: Fall Simulation Interoperability Workshop, pp. 338–350 (2001)

[3] Douglas, D.W.: Rationale and design of MAK real-time RTI. In: Spring Simulation Interoperability Workshop, pp. 132–141 (2001)

[4] Prusinkiewicz, P., Lindermayer, A., Hanan, J.: Developmental models of herbaceous plants for computer imagery purposes. Computer Graphics 22, 141–150 (1988)

[5] Foster, I., Kesselman, C.: Globus: a metacomputing infrastructure toolkit. Intl J. Supercomputer Applications 11, 115–128 (1997)

[6] Jianquan, T., Weiqing, T., Jingbo, D., et al.: Message-oriented middleware on grid environment based on OGSA. In: International Conference on Computer Networks and Mobile Computing, vol. 10, pp. 424–427 (2003)

[7] Talia, D.: The open grid services architecture: where the grid meets the Web. IEEE Internet Computing 6, 67–71 (2002)

A User Management Web System Based on Portlets for a Grid Environment Integrating Shibboleth, PURSe, PERMIS and Gridsphere

David Mera, José M. Cotos, José R.R. Viqueira, and José Varela

Systems Laboratory, Electronics and Computer Science Department. Technological Research Institute, University of Santiago de Compostela, Campus Sur, 15782 Santiago de Compostela, Spain
david.mera@usc.es, manel.cotos@usc.es, joserios@usc.es, eljpet@usc.es

Summary. We propose in this project the development of a distributed collaborative environment, which will constitute a virtual laboratory for multidisciplinary research projects related to oceanographic remote sensing. We will give an overview and the current state of this project, and concretely we will show the security access management module. We propose a well balanced solution between security and simplicity based on the integration of several technologies, where a user can either be registered through a web portal using a portlets system or access directly via Shibboleth. Grid access and job execution are controlled by a Role Base Access Control system (RBAC) that makes use of attribute certificates for the users and Public Key Infrastructure (PKI).

Keywords: Grid Computing, Shibboleth, PERMIS, Portlets, RBAC, PURSe, PKI.

1 Introduction

The increasing number of research projects for Earth Observation due to the launching of new missions every year, with higher spatial and radiometric resolution, makes the data analysis a very hard and tedious job, even unfeasible. Thus, it would be desirable that the research community had a simple and efficient access to the available datasets in order to get the best results.

GRID technology allows the research community to undertake collaborative computation, sharing heterogeneous resources such as hardware (clusters, PCs, sensors...), software (Operative Systems, scientific software,) and datasets, and all of them connected through a network, i.e. Internet.

RETELAB is being developed as a distributed collaborative working environment which will constitute a virtual laboratory for multidisciplinary research projects related to oceanographic remote sensing [2].

Once the web portal and the GRID infrastructure are completed, we will undertake the development of various testbed applications. These testbeds will enable on the one hand, to test the project and on the other hand, to get useful software that the oceanographers would be able to use.

J.M. Corchado et al. (Eds.): DCAI 2008, ASC 50, pp. 19–23, 2009.
springerlink.com

2 Approach Overview

This paper is focused on the management and security control of RETELAB users. The aim of this phase was the development of a simple and secure access approach to the web portal and GRID system.

Current solutions have common problems such as the fact that they use a command line interface and it is neither comfortable nor attractive for users. A user friendly interface is basic to make users feel comfortable with a tool. Another problem is the computer science knowledge that the users are supposed to have. They usually need to spend a lot of time learning, installing and configuring tools instead of working on their projects.

The low security level in the final hosts is another issue to take into account. Security highly depends on the final users and their hosts and on the way they preserve their private keys and passwords.

3 Approach Description

We decided to develop an access system based on a web portal with a comfortable and user friendly interface, which does not demand computer skills from the users. As we cannot forget the security, the system will be based on a public key infrastructure (PKI). Moreover, we should find a solution to get and to store user credentials. Thus, we should develop a well balanced solution between security and simplicity. The access to the resources should be managed by roles which simplify user management and enable a more specific control of user rights. Hence, we decided to use the RBAC model [8]. Finally, it would be desirable to have a Single Sign-On (SSO) network where users would be able to use their credentials to access our system without the need to register. That is, a reliable partner network in which we can trust our partners and their access systems. Thus, the users of the partner organizations would be able to access our system through their own access systems.

4 System Architecture

Facing the problem, we came up with a web portal based on portlets technology [1]. We have used GridSphere [5] as the portlet container to develop our portal. The Globus Toolkit [4], specifically its version 4 (GT4), was used to develop the Grid System. In addition, we have also integrated some portlets developed by the Open Grid Computing Environment (OGCE). The register module is based on the PURSe [7] system that was improved to enable the management of user attribute certificates (ACs), and supports the role paradigm. Public Key Certificates (PKCs) got by PURSe are stored in the online credential repository MyProxy [6]. We have used PERMIS [3] software to develop the RBAC model. Finally, we have also integrated Shibboleth [9] to obtain a SSO network.

A typical scenario to register a user in RETELAB is shown in Fig.1. First, the root user uses the Registration Portlet to register a new user. Next, the

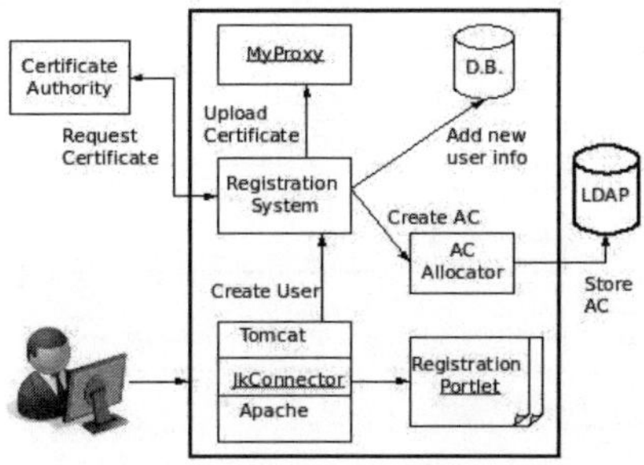

Fig. 1. User Register

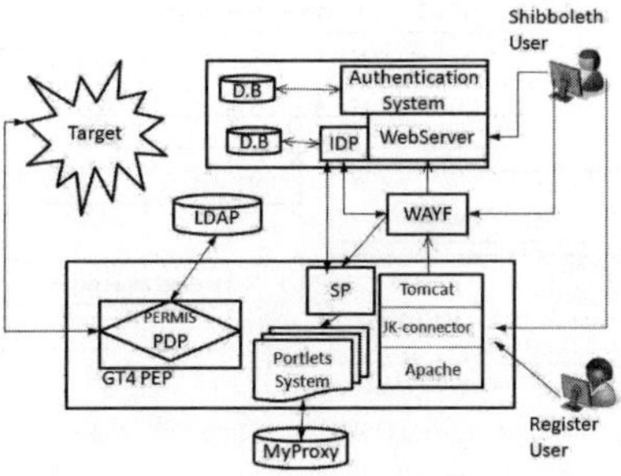

Fig. 2. User Access

Registration System is invoked with the user data and it would get a PKC for the user, via the Certificate Authority (CA). Once the Register System obtains the certificate, it constructs a proxy with the user PKC that is stored in MyProxy. After that, the Registration System gets a user AC through the AC Allocator using the user roles. Such AC is stored into the LDAP directory. Finally, the remainder user data is saved in a database (DB).

The user access and control architecture is shown in Fig.2. The process is described as follows: First, the user is enabled to access the System either by the use of a local account or by a Shibboleth based access using a partner account. After that, either the system or the user retrieves the user proxy from MyProxy. Such a proxy enables the running of jobs in the Grid system using the jobManagerPortlet. When a job is sent, the Policy Decision Point (PDP) will allow or deny the request. We can have different PDPs working together but one of them must be fixed. This is the PERMIS PDP that provides the RBAC model. This PDP gets the user AC with the user roles, stored in the LDAP. The PDP allows or denies the action according to the roles, the action and the target where the user wishes to access.

5 Implementation Details

As it has been mentioned, RETELAB is being implemented with portlets that are deployed in GridSphere. The user registration system is supported by the PURSe system. PURSe allows to add new users to the web portal and to store their data into a DB. It can also obtain user PKCs via a CA and store them into the MyProxy server. Thus, we only had to improve the PURSe code to enable the management of user ACs, which are needed to assign user roles.

The Fig.3 shows the classes used to register users. We developed a system to associate roles with users using ACs certificates and the AC system generator of PERMIS as a basis. We tried to develop the system with the slightest possible impact and interference with the PURSe code. Thus, we developed several RETELAB classes which can be added, such a jar file. It is also necessary to add a new register module in the PURSe web.xml file and to create a properties file

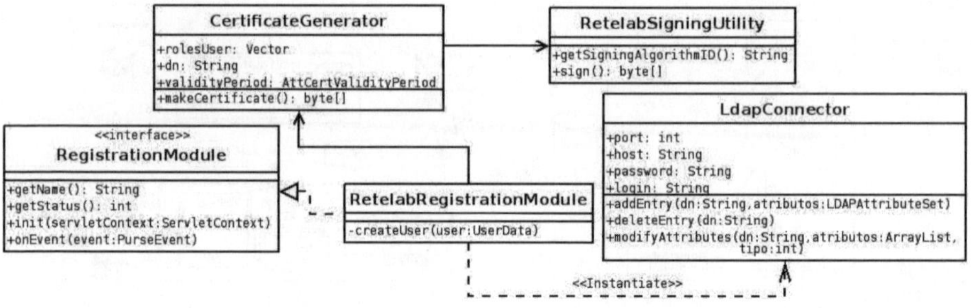

Fig. 3. Class Diagram of User Register

with the LDAP access information. A brief description of the relevant steps is given next: the RetelabRegistrationModule is listening the PURSe events. When a new user is created the createUser method is invoked with user personal information (distinguished name, roles, etc.). It gets a user AC in a X.509 format by the use of the CertificateGenerator class. Finally, the AC would be stored in a LDAP by means of LDAPConnector class.

Once we have new users registered, their certificates stored in MyProxy and their ACs in the LDAP we need to manage the user access, the roles and either allow or deny actions according to their roles.

The users can access the system through RETELAB either once they have been registered with PURSe or directly using Shibboleth, which is integrated with GridSphere as an authentication module.

The RBAC model is supported by PERMIS and it manages the user actions through user roles that are defined into user ACs. Most GT4 services are web services, and PERMIS has developed software for integrating PERMIS with GT4 as a PDP that controls the access to Grid Services. The PERMIS authorization service needs to be configured accordingly to the LDAP, which has the ACs and the system policy.

The PERMIS configuration explains how to control the access to a Grid service, but the way to manage the execution of the jobs through the WS-GRAM services is not intuitive. The desired results were obtained modifying the authorization method used to allow the access to the createManagedJob operation of the WS-GRAM Webservice called ManagedJobFactoryService.

6 Conclusions and Future Work

In this paper we have presented the user access module that will be used in the RETELAB Grid system. Our main aim is to develop an easy to use and accesible Grid system for the researchers. The user access module was the first step to get it and often it is under consideration.

To achieve this goal, several technologies were integrated to find a good solution that covered the expectations. Finally we got an open source and based in standars system, composed with backed technologies.

The next task in the RETELAB project will be to focus our work in the integration of the Grid Technology with OGC web Services Standars.

Acknowledgement. The authors wish to thank the Ministerio de Educación y Ciencia of Spain (ESP2006-13778-C04) for financial support.

References

1. Abdelnur, A., Hepper, S.: Java TM Portlet Specification. Version 1.0 (2003) (retrieved March 2008), `http://www.jcp.org/en/jsr/detail?id=168`
2. Mera, D., Cotos, J.M., Saco, P., Gómez, A.: An integrated Solution to the Security User Access in the RETELAB Grid Project, using a Web System based on Portlets and a RBAC Model by means of User Attribute Certificates and PKI. In: Silva, F., Barreira, G., Ribeiro, L. (eds.) Proceedings of 2nd Iberian Grid Infrastructure Conference, pp. 296–307 (2008) ISBN 978-84-9745-288-5
3. Chadwick, D.W., Otenko, A.: The PERMIS X.509 role based privilege management infrastructure. Future Generation Computer Systems 19(2), 277–289 (2003)
4. Foster, I., Kesselman: The grid: blueprint for a new computing infrastructure. Morgan Kaufmann Publishers Inc., San Francisco (1999)
5. Novotny, J., Russell, M., Wehrens, O.: GridSphere: a portal framework for building collaborations. Concurrency and Computation: Practice & Experience 16(5), 503–513 (2004)
6. Novotny, J., Tuecke, S., Welch, V.: An Online Credential Repository for the Grid: MyProxy. In: 10th IEEE International Symposium on High Performance Distributed Computing (2001)
7. PURSE: Portal-based User Registration Service (Retrieved March 2008), `http://www.Gridscenter.org/solutions/purse`
8. Sandhu, R.S., Coyne, E.J., Feinstein, H.L., Youman, C.E.: Role-based access control models. Computer 29(2), 38–47 (1996)
9. Shibboleth, Web Single Sign-On and Federating Software (retrieved April 2008), `http://shibboleth.internet2.edu/`

Speeding Up in Distributed SystemC Simulations

V. Galiano[1], H. Migallón[1], D. Pérez-Caparrós[1], J.A. Palomino[2], and M. Martínez[3]

[1] Miguel Hernández University, Av. de la Universidad s/n, 03202, Elche, Spain
vgaliano@umh.es, hmigallon@umh.es, davidperez@ieee.org
[2] University of Alicante, Ctra. San Vicente del Raspeig s/n, 03071, Alicante, Spain
japb4@alu.ua.es
[3] Design of Systems on Silicon (DS2), C.R. Darwin, 2, 46980, Paterna, Spain
marcos.martinez@ds2.es

Abstract. SystemC is becoming a popular framework for System on Chip (SoC) design and verification. Innovation in SoC design does not come easily. Smaller features, faster speeds, and intricate digital logic require more simulation resources. SystemC is designed to run the entire simulation solely on one processor. Though designers can model concurrent threads, those threads are executed sequentially as one process. In this paper, we analyze the efficiency of the main approaches to distribute the simulation of SystemC models and we present our own approach which is based on the analyzed ones. We compare the effectiveness of our approach with previous ones using a model with *n* RISC processors and a single *n*-port RAM.

Keywords: SystemC, Distributed Simulation, RTL, SoC, MPI.

1 Introduction

SystemC is becoming more and more used for the SoC design. Built on top of C/C++, it allows the full object-oriented power of the language, while providing constructs to easily describe concurrent hardware behavior. Moreover, SystemC provides, in front of other traditional digital modeling systems languages (as VHDL or Verilog), a separation between communication and functionality and different levels of abstraction depending on the needs of each simulation. In addition to these aspects, the nature of an open source standard makes SystemC a dynamic and powerful language with a wide variety of tools and libraries.

On the other hand, as the complexity of SystemC models increases they require computation resources that single computer cannot provide. There is a need for tools that allow distributed simulation of SoC described in SystemC across multiple computers. We propose to apply principles of Parallel Discrete Event Simulation (PDES) [5] to speed simulation results. The development of such a tool is the main goal of our current work.

2 Related Work

There have been several SystemC kernel parallelization attempts, but it seems that no one is definitive or used as standard. All of them, to the best of our knowledge,

J.M. Corchado et al. (Eds.): DCAI 2008, ASC 50, pp. 24–28, 2009.
springerlink.com

use conservative algorithms to maintain the consistency of the simulated model. Optimistic algorithms are harder to implement and require too many resources during simulation [5].

There are basically two strategies to distribute SystemC simulations. Some authors have tried to parallelize/distribute SystemC by directly modifying its simulation kernel. The other strategy is based on wrapping communications between LPs (Logical Processes) using a self-developed library, which extends SystemC communication library.

A major drawback of modifying the SystemC kernel is the need to provide a continuous support to follow future versions of the standard. On the other hand, there is a great advantage as long as it is a more customizable approach.

Cox [4] and Combes et al. [2] propose a SystemC kernel modification approach. Both proposals use MPI [11] for communication between LPs, which are wrapped in a top-level module. This strategy obtains reasonable performance results for well-balanced coarse-grained models. Those LPs are manually defined and distributed by the simulated model designer. Cox approach avoid the explicit lookahead by choosing a robust synchronization algorithm [1][3].

Other approaches, following the second strategy mentioned above, include their own communication library that bridges LPs and synchronizes shared signals between them using explicit lookahead.

Trams [8][9][10] and Hamabe [6] present solutions that avoid modifying SystemC source code. Hamabe implementation uses MPI for communications and synchronization, while Trams propose the use of TCP/IP sockets. In Trams proposal, each LP is defined in separated executable pieces due to its communication technique.

There are other authors working on geographically distributed SystemC simulations [12]. Communications are made over Internet protocols and middleware such as SOAP, RMI or CORBA. However, their goal is not to obtain a better performance as well as our work aims.

We have developed an approach which is based on Trams and Hamabe proposals It consists of a simulation wrapper module that allows communications between manually distributed LPs using MPI. For synchronization it makes use of explicit lookahead. We show the benefits of our approach experimentally in the following sections.

3 Comparison between Non-distributed and Distributed SystemC Approaches

In order to compare our distributed SystemC approach with the non-distributed, we have developed a configurable hardware model that allows setting different simulation parameters.

3.1 Simulated Model

For performance evaluation, it has been used a model with n RISC processors, placed into separated simulation kernels, connected to a single RAM, which is accessible by *n* ports. The shared data RAM is in the same simulation kernel than the first CPU, but this imbalance can be omitted; the RAM model is insignificant compared with the RISC CPU models.

Each CPU reads in a small assembly program and executes it and writes the result back to data memory for some loop iterations.

The simulated model computation load is insignificant to get a performance gain in a distributed environment. It has been added a function that implements an algorithm with complexity $O(n^2)$ which simulates a certain level of computation load in each simulation cycle. The parameter that controls load level is CPU_LOAD.

3.2 Simulation Platform

Two dedicated clusters of machines have been used for gathering practical measurements:

- **Picolo:** CentOS GNU/Linux cluster with 24 nodes with two Dual-Core AMD Opteron(tm) Processor 2214, 8GB RAM and connected with Gigabit Ethernet. The software used is composed by a C++ compiler (gcc-4.0.2), the SystemC library (v2.2.0) and the MPI library (mpich-1.2.7p1).
- **-Dende:** Ubuntu GNU/Linux cluster with 5 nodes with a Intel(R) Core(TM)2 Duo CPU E6750 at 2.66GHz, 2GB RAM and connected with Gigabit Ethernet. The software used is composed by a C++ compiler (gcc-4.1.3), the SystemC library (v2.2.0) and the MPI library (mpich-1.2.7p1).

3.3 Simulated Cases and Measurement Conditions

Two simulation cases have been developed for performance comparison between non-distributed and distributed SystemC implementation:

- **non-distributed:** non-distributed version of the simulated model, this test case is used as the reference model and consists of a single kernel.
- **distributed:** distributed version of the simulated model that has been implemented using the MPI approach.

The simulation run times for all the performance tests has been obtained for 249678 simulated clock cycles (2496780ns with a 10ns clock cycle). The results shown include only the time that has been spent by the simulation process, initialization and elaboration times are not considered. Every simulation case has been run three times, the smallest simulation time has been used for further analysis.

3.4 Speedup Depending on Computation Load

As shown in Fig. 1 the higher CPU Load parameter the better speedup value. As example, in Picolo, with 16 processors, we get a speedup greater than 4. Moreover, we must consider that memory requirements are lower in a distributed simulation because

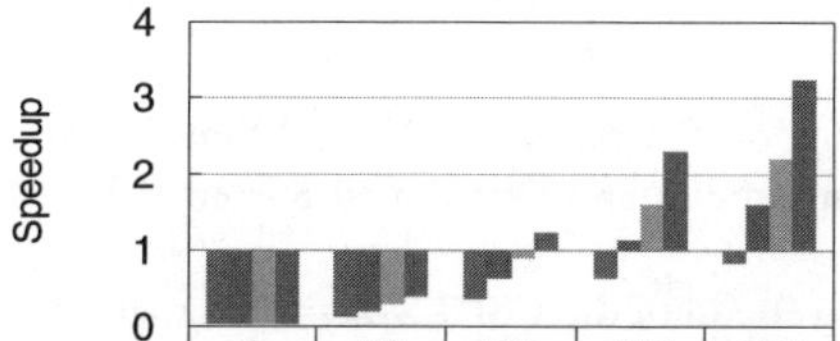

	0	50	100	150	200
2 CPU	0,04	0,13	0,38	0,65	0,84
4 CPU	0,03	0,21	0,64	1,13	1,60
8 CPU	0,04	0,29	0,90	1,59	2,20
16 CPU	0,05	0,39	1,22	2,31	3,23

Load Level

(a) Dende

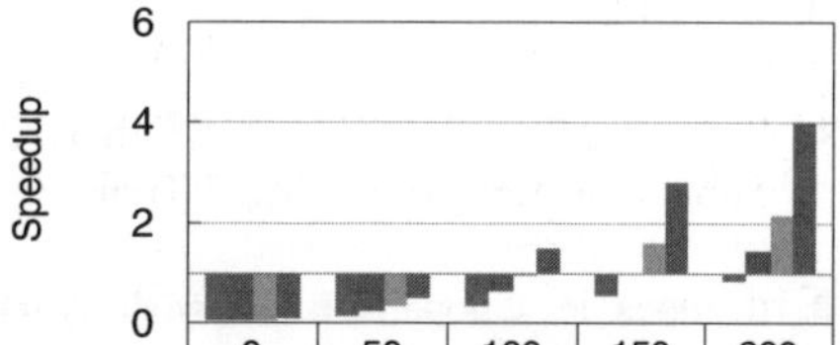

	0	50	100	150	200
2 CPU	0,03	0,15	0,35	0,54	0,84
4 CPU	0,06	0,26	0,65	0,98	1,43
8 CPU	0,06	0,35	0,97	1,59	2,13
16 CPU	0,09	0,49	1,49	2,78	4,02

Load Level

(b) Picolo

Fig. 1. Behaviour of the speedup depending of the CPU load parameter and the number of simulated CPUs on *Dende* (a) and *Picolo* (b) clusters

the global model is divided in different simulation kernels. In both systems, scalability is possible and is not architecture-machine dependent.

4 Conclusion

This paper presents a state of the art and performance comparison between distributed and non-distributed simulations of SystemC models. Not many researchers have dealt with the distribution of SystemC simulations.

Among the several projects to distribute SystemC simulations there are two methods to distribute SystemC models, some authors tried to modify the SystemC kernel, the other method consists of using standard communication libraries to synchronize signals between LPs. All the performance tests have been executed using an approach that follows the second method. In this paper, a configurable reference SystemC model has been developed to evaluate the main benefits and drawbacks of the distributed solution. This model has *n* RISC processors sharing a single *n*-port data RAM. Its *n* simulation kernels have been spread across different computing nodes in two different dedicated clusters. For the communication and synchronization between the n kernels, it has been used the MPI approach following Hamabe's solution.

The simulation results show that a distributed SystemC model can achieve a considerable performance gain. This gain occurs when the SystemC model reaches a certain level of computation load per signal synchronization cycle.

The results obtained encourage us to follow working with MPI implementation. Our future work will be focused on implement a new communication library that could be used in a wider range of SystemC models.

Acknowledgements. This work has been partially funded by the Ministry of Industry, Commerce and Tourism under project number FIT-330100-2007-98.

References

1. Bagrodia, R.L., Takai, M.: Performance Evaluation of Conservative Algorithms in Parallel Simulation Languages. IEEE Transactions on Parallel and Distributed Systems, 395–411 (2000)
2. Combes, P., Chopard, B., Zory, J.: A parallel version of the OSCI SystemC kernel. In: Alexandrov, V.N., van Albada, G.D., Sloot, P.M.A., Dongarra, J. (eds.) ICCS 2006. LNCS, vol. 3994, pp. 653–660. Springer, Heidelberg (2006)
3. Chandy, K.M., Sherman, B.: The Conditional Event Approach to Distributed Simulation. In: Proceedings of the SCS Multiconference on Distributed Simulation, Society for Computer Simulation International, vol. 21, pp. 93–99 (1989)
4. Cox, D.R.: RITSim: Distributed SystemC Simulation. Master Thesis. Rochester Institute of Technology (2005)
5. Fujimoto, R.M.: Parallel and Distributed Simulation Systems. Wiley-Interscience, Chichester (2000)
6. Hamabe, M.: SystemC with MPI for Clustering Simulation, `http://www5a.biglobe.ne.jp/~hamabe/SystemC`
7. Open SystemC Initiative (OSCI): IEEE Standard SystemC Language Reference Manual (2006)
8. Trams, M.: Conservative Distributed Discrete Event Simulation with SystemC using Explicit Lookahead. Digital Force White Paper (2004)
9. Trams, M.: A First Mature Revision of a Synchronization Library for Distributed RTL Simulation in SystemC. Digital Force White Paper (2004)
10. Trams, M.: User Manual for Distributed SystemC Synchronization Library Rev.1.1.1. Digital Force White Paper (2005)
11. Dongarra, J., Huss-Lederman, S., Otto, S., Snir, M., Walkel, D. The Message Passing Interface (MPI) Standard (1998), `http://www-unix.mcs.anl.gov/mpi`
12. Meftali, S., Dziri, A., Charest, L., Marquet, P., Dekeyser, J.L.: SOAP Based Distributed Simulation Environment for System-on-Chip (SoC) Design. In: Forum on Specification and Design Languages, FDL 2005 (2005)

Multiagent Approach for Supply Chain Integration by Distributed Production Planning, Scheduling and Control System

Pawel Pawlewski[1], Paulina Golinska[1], Marek Fertsch[1], Jesus A. Trujillo[2], and Zbigniew J. Pasek[3]

[1] Institute of Management Engineering., Poznan University of Technology, 60-695 Poznan, Poland
pawel.pawlewski@put.poznan.pl, paulina.golinska@put.poznan.pl, marek.fertsch@put.poznan.p

[2] Dep. Automation and Systems, University of Valladolid, E47008, Valladolid, Spain
jestru@eis.uva.es

[3] Dep. Industrial Syst. Engineering, University of Windsor, 401 Sunset Avenue Windsor, Canada
zjpasek@uwindsor.ca

Abstract. The changing business environment in which manufacturers are acting creates the need for more effective production processes planning, scheduling and control methods that are able to deal with uncertainties inherent in internal processes and external deliveries. The aim of the paper is to introduce the multiagent approach method for production planning, scheduling and control applicable in conditions of supply chain (SC) able to overcome the limitation of standard MRP/ERP systems in changing environment. Traditional approaches very often do not consider the influence of uncertainty inherent in production processes and supplies. Therefore, there is a need for the integration of manufacturing process planning, scheduling and control systems for generating more realistic and effective plans. Conceptual framework for the multi-agent approach method involves the hybrid solutions combining the advantages of MRP simple logic and theory of constrains (TOC) ability to synchronize all production and material flow in supply chain. Authors discuss how application of TOC buffers monitoring procedures can help to improve the control of synchronized production and material flow in supply chain.

Keywords: distributed manufacturing system, multiagent systems, supply chain, distributed process planning and scheduling.

1 Introduction

In number of industries where high value products are made on basis of make-to-order strategy the manufacturer has sufficient power over the parties involved in supply chain and is able to act as a leader. In following paper we refer by all assumptions to such a situation. Supply chain (SC) is defined as a flexible and cooperative business network of suppliers, manufacturer and distributors through which raw material are acquired, transformed within value adding (VA) processes and delivered as final goods to customers. The supply chain performance is defined in following paper as an ability to fulfill clients orders on-time, so the perspective of product delivery process (PDP) is taken in consideration by all assumptions. The SC performance is measured

J.M. Corchado et al. (Eds.): DCAI 2008, ASC 50, pp. 29–37, 2009.
springerlink.com

as a ratio regarding customers orders delivered due-to-date to overall customers orders being executed in supply chain in defined period. In order to optimize the performance, supply chain functions especially production planning must operate in an integrated manner. In order increase the SC performance the production processes planning, scheduling and control methods have to be able to deal with uncertainties inherent in internal processes and external deliveries. Uncertainty can be defined as the conditions within supply chain when probability of particular events/disturbances appearance cannot be counted. There is already a variety of information systems to facilitate the flow of materials, information and funds from MRP (Materials Requirements Planning) and ERP (Enterprise Resource Planning), to newly developed Supply Chain Management (SCM) systems. The production planning, scheduling and control based mainly on MRP logic allows integration of the functional areas within enterprise at the operational level. Main limitations of such systems is intra-enterprise focus. Existing SCM systems provide analytical tools for advanced planning but they lack the integration with MRP/ERP system.

2 Approaches to Production Planning, Scheduling and Control

Traditional approaches to production planning and scheduling in MRP based logic do not consider real-time machine workloads and shop floor dynamics. Therefore, there is a need for the integration of manufacturing process planning and control systems for generating more realistic and effective plans. The overview of approaches to production planning and scheduling can be found in [11]. Traditional approaches are to production planning and scheduling are based on:

- Centralized Optimization Algorithms: in number of researches process planning and scheduling is combined as a single scheduling optimization problem. In order to reduce the computation complexity of a big central optimization algorithm, some researchers try to split the optimization problem into several steps, each with a distinguishing objective [e.g. 15].
- Close Loop Optimization: some researchers argued that NLPP-oriented approaches for integrating process planning and scheduling are still in an offline mode. All the acceptable schedules made at the predictive scheduling stage are almost immediately subject to changes on the shop floor owing to the rescheduling factors, such as machine breakdowns, materials shortage, order cancellation, due date changes [e.g. 9].
- Distributed Process-Planning (DPP) Approaches: in order to integrate with scheduling, DPP advocates a more flexible plan scenario by separating the whole process to several steps [e.g.7].

Agent-Based Approaches provide a distributed intelligent solution by multiagent negotiation, coordination, and cooperation. The following researches refer to application of multiagent systems for production planning purpose [11]:

- bidding based approach - the process routes and schedules of a part are accomplished through the contract net bids. No centralized process planner is applied although same planning techniques are applied to every machine. The task allocation and process alternative selection are achieved through the hierarchical bidding processes between machine agents and shop floor manager, between upper level machine agents and lower level machine agents, and between machine agents and tool agents.

- a multiagent architecture based on separation of the rough process-planning task as a centralized shop floor planner from the detailed process planning conducted through agent negotiations.
- based on cascading auction protocol provides a framework for integrating process planning and hierarchical shop floor control. The integration of the real-time online process planning and shop floor control is achieved through a recursive auction process carried out in parallel among part management agent and multiple resource management agents.

The application of multiagent can be extended to whole supply chain due to following potential advantages of distributed manufacturing scheduling [12] logic :

- usage of parallel computation through a large number of processors, which may provide scheduling systems with high efficiency and robustness.
- ability to integrate manufacturing process planning and scheduling.
- possibility for individual resources to trade off local performance to improve global performance, leading to cooperative scheduling.
- possibility of connection directly to physical devices and execution of real-time dynamic rescheduling with respect to system stability.
- it provides the manufacturing system with higher reliability and device fault tolerance.
- usage of mechanisms similar to those being used in manufacturing supply chains mainly negotiation.
- the manufacturing capabilities of manufacturers can be directly connected to each other and optimization is possible at the supply chain level, in addition to the shop floor level and the enterprise level.
- possibility of application of other techniques may be adopted at certain levels for decision-making, for example: simulated annealing [5], genetic algorithm ect. [2].

There is a research gap regarding integration of process planning, manufacturing scheduling and control [11]. Agent-based approaches provide a natural way to integrate planning and control activates and makes possible simultaneously the optimization of these functions. However it increases the complexity of the problem. Proposed by authors conceptual framework for the multiagent approach method involves the hybrid solutions combining the advantages of MRP simple logic and theory of constrains (TOC) ability to synchronize all production and material flow in supply chain. The applications of TOC as synchronization mechanism allows to reduce a number of parameters to be control so it allows to simplify the complexity of integration problem.

3 Production Planning, Scheduling and Control in Supply Chain

In following research authors refers to production planning strategy based on the Planed Order Release schedule and MRP concept where Master Production Schedule is the main planning and control schedule. It states what kind of end products should be produced and it helps preliminary verify whether customers' orders can be fulfilled due-to-date. MPS is main driver and information source for further material requirements planning and accompanying calls or supplies and allows making details production schedules for production system resources in planning horizon [14] .

Due uncertainty inherent in SC in manufacturing systems things rarely go as expected. Researches examined a variety of buffering or dampening techniques to minimize the effect of uncertainty in manufacturing systems based on MRP logic. Comprehensive literature review can be found in [3]. In situation when the dampening or buffering techniques are not efficient the modification of Master Production Schedule (MPS) is needed. The high replanning frequency in order to overcome the uncertainty induces the system nervousness when a minor change in MPS creates significant changes in material requirements planning (MRP). The availability of materials is often limited due the fact that suppliers have similar bottlenecks and schedules variations transmitted from sub-tier suppliers. In supply chain where manufacturer acts as a leader the manufacturing system nervousness has negative impact on overall SC performance, so it can be assumed that changes in MPS should be reduced.

Due the fact that MPS is the main driver and information source for further material requirements planning and details production schedules for production system resources it can be assumed that in supply chain where exists a single manufacturer and a limited number of 1-tier suppliers and the manufacturer has sufficient power over the others parties involved in supply chain a global MPS can be elaborated. Taking in consideration following the distributed planning can be described as a situation where in centralized process of preparing plans the distributed sub plans for each supply chain's participant are elaborated. The sub-plans for supplier side and distribution side are dependent on manufacturer capacities, so it can be treated as main constrain in planning process.

4 Multiagent Approach for Supply Chain Integration

4.1 Model

Agent-based system is defined in following paper as a multi-agent system that's acts as a support tool and utilized the databases of main system (ERP system). Multi-agent

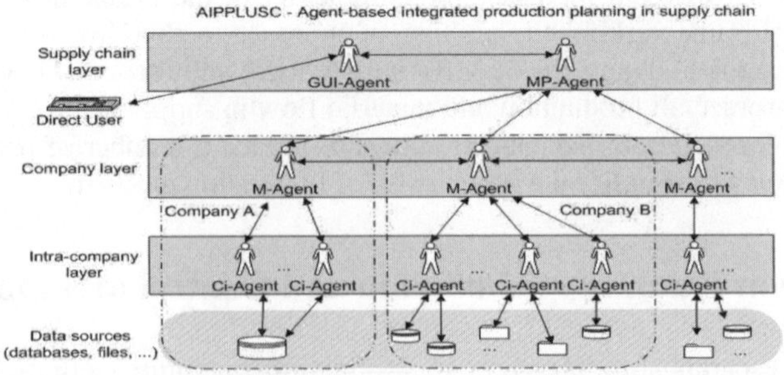

Fig. 1. AIPLUSC model [1]

system is a collection of heterogeneous, encapsulated applications (agents) that participate in the decision making process [8]. The architecture of proposed tool (AIPPLUSC) is based on the assumption that system will support the MPS creation in ERP system and will be plug in to ERP system database by for example java connector. Figure 1 illustrates structure and the amount of agents which can be found in each layer. The proposed model is based of assumption that agent-based system will act as "glue" integration existing information systems within each company at global supply chain level. The planning problem in following paper is described at three layers reflecting to:

1) supply chain perspective so called global planning;

the entity level where global plan is divided to sub-plans which are executed by each company and being transform for individual production schedule at company level and where local re-planning activities takes place

intra-company sub-layer where production control activities are executed and information about disturbances are gathered and passed to upper levels.

The graphical user interface agent creates a graphical user interface (GUI) for the interaction of the MAS to production manager (direct users). The GUI-Agent is able to initialize and sent behavior parameters and messages to the Master Planning Agent (MP-Agent). The MP-Agent is exactly one in the system because the data from all the managing agents (M-Agent) is fused at this agent to generate re-planning schedules for the production, distributors or supplier. The MP-Agent is responsible for control of the logic of all agents and creates the plans for the M-Agent. The detailed description of the distributed planning algorithm can be found in [6]. The planning process can be presented as following algorithm:

1. Definition of chain goal and set of performance indicators
2. Generate an initial MPS for manufacturer taking in consideration customer orders assign to planning horizon plan and capacities constrains
3. Negotiate the initial plan with supply and distribution side and find a plan with lower number of constrains among them (so called feasible MPS for supply chain)
4. Decompose the feasible MPS for sub-plans for supply, manufacturer and distribution side
5. Insert synchronization among sub-plans based on TOC concept for time buffers and Drum-Buffer-Line concept where manufacturer sub-MPS is giving pace for supplies and distribution planning activities
6. Allocate sub-plans to agents using task-passing mechanism, if failure come back to previous step or generate new global MPS (step 2)
7. Initiate plan monitoring when plans are executed in TOC green buffer no additional re-planning needed when plans are executed in yellow TOC buffer the re-planning at local level, if plans are executed in TOC red buffer go to step 2.

M-Agents are initialized by the MP-Agent and they are responsible for translation of the global plan into detail schedules. The agent is allowed to prepare the number of alternative(contingency) local plans as long as there are not conflicting with global

MPS. The local replanning activities are allowed as long as they don't influence the global MPS. When replanning activity affects the global MPS it has to be passed to MP-Agent. The CI-Agent is responsible for control of plans execution within one company based on given performance indicators. It reports to M-Agent in upper layer whether production plans are executed according to given MPS.

4.2 Buffer Control Mechanism

In order to put synchronization between entities involved in supply chain and at the same time to limit the number of data being send among agents authors have decided to refer to Theory of Constrains control mechanism Drum-Buffer-Rope (DBR) [10]. This concept can be applied in conditions of supply chain due the fact that DBR allows to synchronize the use of resources without having to actively control each resource. The purpose of drum is to exploit the constraint of the closed loop supply chain. Buffers are time windows protecting the Supply Chain global constrain and bounded to it the critical path of product delivery process (PDP). The DBR buffers refers to time needed for deliveries of materials and component.

The size of buffer in stable manufacturing system is as three times the average lead time to the constraint from the raw material release point [4]. In SC where uncertainty in process and deliveries lead-times appears the buffer can be counted as defined as the multiplied minimum cumulative processing time for the individual part

$$BS_i = MULTI \cdot \sum_{j=1} PT_{ij} \quad (1)$$

Where BS_i is the size of buffer for part i=(1,2,3,..,n), PT_{ij} is the minimum processing time for operation j on part i MULTI is a constant multiplier. The mentioned above buffer is divided in three area: green (no action needed), yellow (prepare for actions) and red (react). The size of each part of buffer can be equal, but it is recommended to revise the performance of buffer's green, yellow and read zone on regular basis in order to tune their size to statistical fluctuations appearing in material flows in particular manufacturing environment.

4.3 Potential Implementation

The implementation process has to take in consideration the following elements: the environment for creation of intelligent agent and the possibility to interact with existing systems. Authors refer to FIPA[17] specifications standards and JADE[16] framework due its simplicity and potential for high level of interoperability between agents. Figure 2 presents the potential application of proposed solution.

Supply chain layer allows to defined all global planning constrains like size of the buffers. The definitions of constrains is based on TOC theory. The constrains are search for optimization of the supply chain throughput. Due the fact that all processes are set to the process of most constrained conditions namely bottleneck, it can be assume that the improvement of productivity of all processes at the same speed pace as the improvement of constrained conditions can be expanded to the entire supply

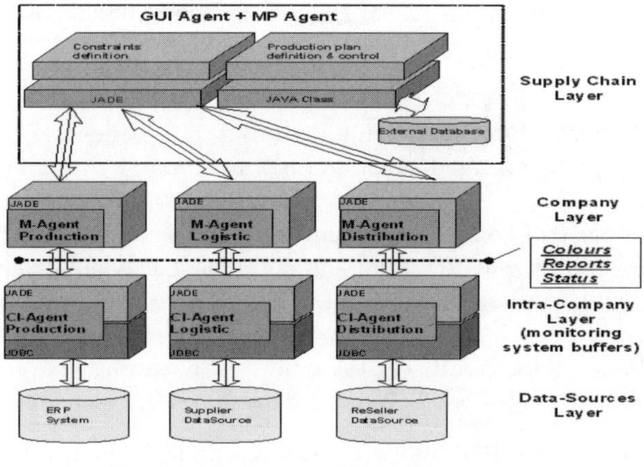

Fig. 2. Potential implementation of AIPLUSC

chain. Due manufacturer's MPS is main information driver for whole SC authors assumed that in order to define the initial plan for supply chain first the manufacturer MPS taking in consideration customer orders assign to planning horizon plan and "potential bottlenecks". The initial MPS is then negotiated by MP agent with M-Agents in area of raw materials supplies and distribution in order to find a plan with lower number of constrains among them (so called feasible MPS for supply chain), that is then passed for execution to agents located at company layer. The color reports status refer to time window in PDP. Main responsible for monitoring of system buffer are Ci-agents, that check every defined period of time the size of time buffer on bases of data for ERP. According to achieved results it reports to upper level (appropriate M-Agent) the following status: green (everything ok), yellow (there is a risk of delay) or red (alarm). The buffer control logic is following (figure 3.):

1. when green- it sends a message to layer 2 (M-agent) and don't do anything until next check period;
2. when it is yellow Ci-agent reports to its own M-Agent due is not allow to contact M-agent at other company. The M-agent begins the reactive replanning procedure according to schema presented in Figure 4. The schema can be simplified as:
   - when there is the contingency plans plan allowing the local replanning without changes of due-dates to other sub-plans in production of planning process it make the replanning.
   - when there is contingency plan but there is a risk of delay it negotiate with next M-agent
   - when it is no contingency plan it informs MP-Agent, who makes replanning
3. when red- it sends a message to alarm the M-agent. M-agent send back message to recalculate the buffer. When the answer from Ci-agent is confirming (again red) then M-Agent reports it to MP-agent, who sends the message to M-agent how big is the delay (time is main characteristics in this approach due performance is measure by ability to deliver product to customer on-time).

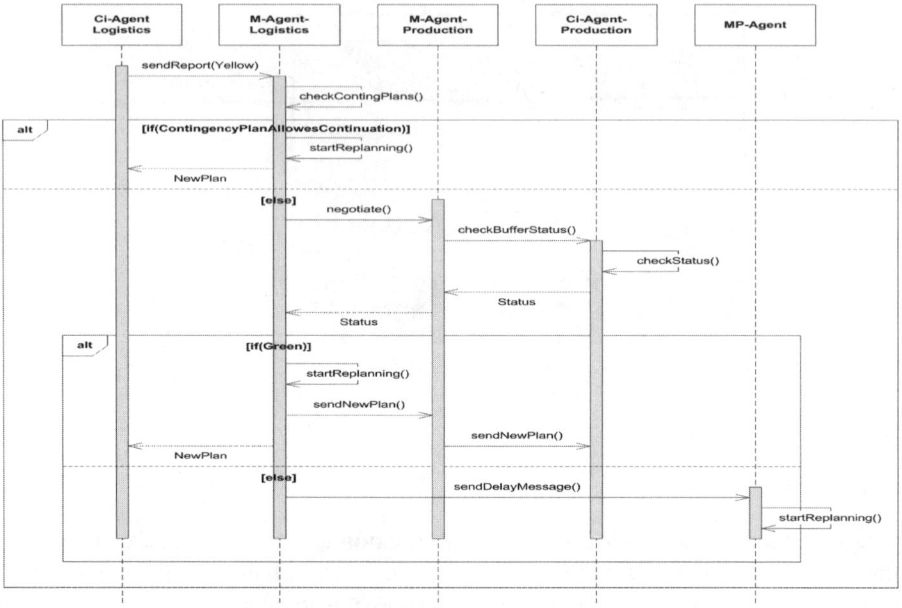

Fig. 3. Buffer control mechanism – yellow flag [1]

5 Conclusions

In following paper authors present the concept of an agent based system for integrated production planning in supply chain AIPPLUSC, the multi-agents distributed system that reflects the needs for distributed decision making in multi-entities environment. The system design requirements refer to increase of integration of information and material flow between entities that are involved in network. The issues related to distributed planning process were discussed and a new algorithm for hierarchical production planning by multi-agent system based on concept of centralized planning for distributed plans was proposed. Perspectives for further research should reflect to elaboration of communication algorithm among agent and structure of standardized input and output data structure for all companies involved in supply chain.

Acknowledgements

The origin of following paper refers to the research project for Paulina Golinska at Department of Business Informatics, Carl von Ossietzky University of Oldenburg financially supported by DAAD. The continuation of the research was conducted thanks to collaboration between Poznan University of Technology (Poland), University of Valladolid (Spain) and University of Windsor (Canada).

References

1. Golinska, P., Brehm, N., Fertsch, M., Marx Gomez, J., Oleskow, J., Pawlewski, P.: The proposal of production planning and control system applicable by supply chain integration by through agent-based solutions. In: The proceedings of the 19th ICPR, 19th International Conference on Production Research, Valparaiso, Chile, 27.07-02.08 (2007)
2. Goldberg, D.: Genetic Algorithms in Search, Optimization, and Machine learning. Addison-Wesley, Reading (1989)
3. Guide, V.D.R., Shiverasta, R.: A review of techniques for buffering against uncertainty with MRP systems. Prod., Planning and Control 11, 223–233 (2000)
4. Guide Jr., V.D.R.: Scheduling using DBR in a remanufacturing environment. Int. Journal of Production Research 34, 1081–1091 (1996)
5. Kirkpatrick, S., Gelatt, D., Vecchi, M.P.: Optimization by simulating annealing. Science 220, 671–680 (1983)
6. Jennings, N.R., Wooldridge, M.J.: Applications of Intelligent Agents. Agent Technology: Foundations, Applications, and Markets, pp. 3–28. Springer, Heidelberg (1998)
7. Chang, F.T.S., Zhang, J., Li, P.: Modelling of integrated distributed and co-operative process planning system using an agent-based approach. Proc. Inst. Mech. Eng., Part B-J. Eng. Manuf. 215(B10), 1437–1451 (2001)
8. Pechoucek, M., Říha, A., Vokrínek, J., Marík, V., Prazma, V.: ExPlanTech: Applying Multi-agent Systems in Production Planning. Production Planning and Control 3(3), 116–125 (2003)
9. Saygin, C., Kilic, S.: Integrating flexible process plans with scheduling in flexible manufacturing systems. Int. J. Adv. Manufacturing Tech. 15(4), 268–280
10. Schragenheim, E., Ronen, B.: Drum–buffer–rope shop floor control. Productions and Inventory Management Journal 31(3), 18–22 (1996)
11. Shen, W., Wang, L., Hao, Q.: Agent-Based Distributed Manufacturing Process Planning and Scheduling: A state-of-art survey. IEEE Transactions on Systems, Man and Cybernetics-Part C: Applications and Reviews 36(4) (2006)
12. Shen, W.: Distributed manufacturing scheduling using intelligent agent. IEEE Expert /Intell.Syst. 17(1), 88–94 (2002)
13. Towill, D.R., Childerhouse, P., Disney, S.M.: Integrating the Automotive Supply Chain: Where are we Now? International Journal of Physical Distribution and Logistics Management 32(2), 79–95 (2002)
14. Viera, G.E., Favetto, F.: Understanding the complexity of Master Production Scheduling Optimization. In: Proceeding of the 18th ICPR, Salerno, Italy (2005)
15. Zijm, W.H.M.: The integration of process planning and shop floor scheduling in small batch part manufacturing. Ann CIRP 44(1), 429–432 (1995)
16. `http://jade.tilab.com`
17. `http://www.fipa.org`

Multiagent System Implementation for Network Management Based on SNMP Protocol

Néstor D. Duque M.[1], María Helena Mejía S.[1,2], Gustavo Isaza[2], and Adriana Morales[2]

[1] Universidad Nacional de Colombia, Sede Manizales, Colombia
[2] Universidad de Caldas, Colombia
ndduqueme@unal.edu.co, mhmejiasa@unal.edu.co,
gustavo.isaza@ucaldas.edu.co, adriana.morales@ucaldas.edu.co
Semillero de investigación GAIA(Grupo Ambientes Inteligentes Adaptativos): David Rubio J. Jaime E Alvarez S, Angelo Quintero R

Abstract. The multiagent system for the network management that is presented in this article consists of several agents whose main aim is to facilitate the effort of network management according to established policies. These agents are based on Java and are constructed with the support of JADE platform. The agents use SNMP protocol to execute their tasks on the network. An agent based architecture increase the integration, adaptability, cooperation, autonomy and the efficient operation in heterogeneous environment in the network supervision.

Keywords: SNMP protocol, Multiagent System, Multiagent platform, Network management.

1 Introduction

The network management can be difficult when the network have a considerable number of network devices, because verify the correct functioning of all of them at the same time involves to have the enough human resource for its monitoring; is for that reason that necessarily certain tasks must be automatized and that these tasks generate results that at the moment of be consolidated can be consulted in any computer inside the network. The Simple Network Management Protocol (SNMP), allows manipulating certain variables of the network devices that support it, it is used to monitor and to control the behavior of those devices [1].

The intelligent agents are software entities with special properties that allow them to execute in an autonomous way and make decisions in agreement with predefined conditions of the environment in which they live. This article presents the development of a multiagent system (MAS) for the network management, supported with four agents that perform particular actions and that implement their communication with the messages passing. These messages are contextualized within a specific ontology for this system.

The agents involved in this system have the ability to extract information from network devices through the use of SNMP, analyze it and verify potential abnormalities that can occur; when these cases happens, the system notifies it through alarms or traps to the network managers so they can respond accordingly.

The construction of MAS allows shaping the environment for managing network distributed and autonomous, capable of adapting and transforming environmental conditions on which operates with minimal human intervention.

J.M. Corchado et al. (Eds.): DCAI 2008, ASC 50, pp. 38–46, 2009.
springerlink.com

The rest of the material is organized as follows: The following paragraph contains some concepts of the protocol of network administration SNMP, then makes a specific reference to some features of the development methodologies of MAS that were promptly used in the project, the numeral 4 details the proposal phases in the analysis, design and construction, while paragraph 5 presents the results obtained, to finish with conclusions and future work.

2 SNMP Protocol

The Simple Network Management Protocol (SNMP) is an application layer protocol that facilitates the exchange of management information between network devices. It is part of the TCP/IP protocol suite. SNMP enables network administrators to manage network performance, find and solve network problems, and plan for network growth. An SNMP managed network includes three main components: managed devices, network management systems (NMS) and agents. [2][3]

A managed device is a network node that contains an SNMP agent and resides on a managed network. Managed devices collect and store management information and make this information available using SNMP. Managed devices, can be servers, switches, hosts, routers and printers, etc. [2]

An agent is a software module that resides in a managed device. An agent has local knowledge of management information and translates that information in a compatible way with SNMP. [2][3]

The Network Management System executes applications that monitor and control managed devices. NMS manage the processing and memory resources required for network management.

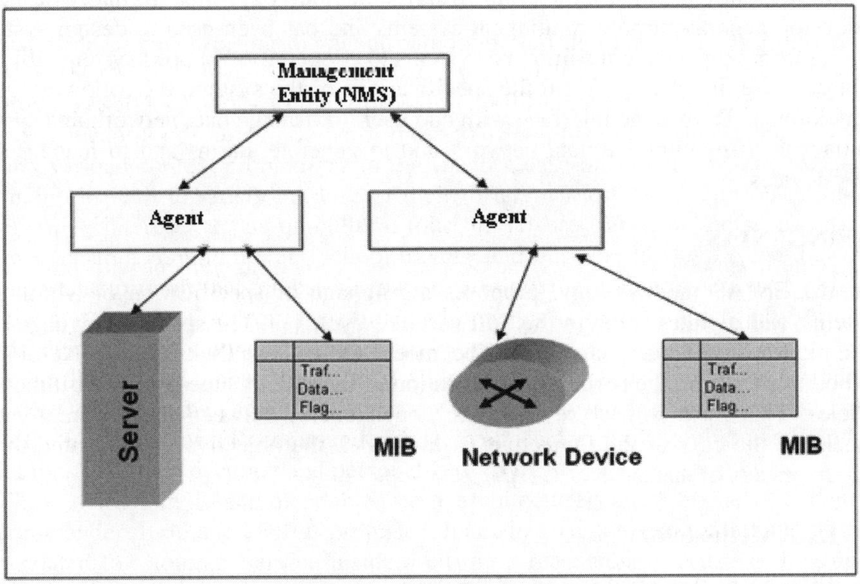

Fig. 1. SNMP Architecture

Management Information Base (MIB) is an information collection that is organized hierarchically. The MIB are accessed using a network management protocol such as SNMP. SNMP uses these structures to define the interface for a resource. A MIB consists of a set of object definitions, each of which exposes some property of the resource to be managed. These definitions must be extremely flexible, since SNMP is designed to support any resource type.

The next figure presents an SNMP model and architecture using generic agents.

3 Multi-Agent System (MAS) Methodology

In [10] is described the use of different methodologies to analyze the problem in discussion. In our case, we used GAIA, INGENIAS and MASE to compare the agent software engineer life cycle until the implementation.

A MAS is a system composed of multiple interacting intelligent agents, oriented to solve problems which are difficult for a single agent.

3.1 GAIA

GAIA is a methodology for agent-oriented analysis and design. The GAIA methodology is both general, in that it is applicable to a wide range of multi-agent systems, and comprehensive, in that it deals with both the macro-level (societal) and the micro-level (agent) aspects of systems. GAIA is founded on the view of a multi-agent system as a computational organization consisting of various interacting roles [4], [5].

3.2 MaSE

The Multiagent Systems Engineering (MaSE) methodology, was originally designed to develop general-purpose multiagent systems and has been used to design systems ranging from computer virus immune systems to cooperative robotics systems [6].

In our case, in order to fulfill the specifications of the system, the following goals were defined: To manage interface with end user, to manage data network and system multiagent, to monitor the data network and to generate alarms and to manage network devices.

3.3 INGENIAS

The INGENIAS methodology, proposes a language of specification of Multiagent Systems, and its integration in the Software life cycle [7]. The specific language uses meta-models and natural language. The model Objectives/Tasks in the INGENIAS methodology allows the partial documentation of the task of supervising fulfillment of policies. This methodology reviews the dependencies between the different meta-models. In the case of the Objective/Tasks model, the objective is associated to the tasks by means of instances of *GTAfecta*.

3.4 Used Methodology

The different methodologies allowed making a comparative analysis of the problem and their possible process of software engineer based on agents studying different sequences and life cycles. In this case, we worked with INGENIAS, GAIA and MASE to determine the best solution to integrate Multiagent systems in networking management with SNMP. [10]

4 Architecture and Construction of Multiagent System

The multiagent platform was build on Java using the JADE framework [8],[9] advantages that simplify the MAS construction and provides a set of tools for platform debug.

The MAS consists of *Agente Interfaz, Agente Gestor, Agente de Información* and *Agente de Traps* [9]. The agent's names are used in Spanish because in the design and implementation are defined in this language. The Figure 2 is the diagram of components, where agents are represented as Java Classes and the tools that support them.

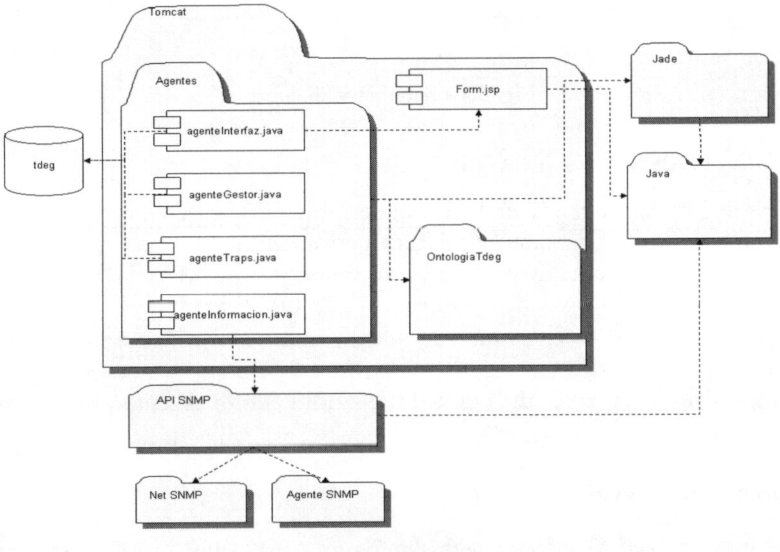

Fig. 2. Components diagram

The Figure 3 is the System Collaboration diagram where agents and interchanged messages with their own performative and the action implemented in the ontology is represented.

The system can be managed via the Internet using Apache Tomcat as Web server, being necessary to integrate the multiagent platform and the server.

4.1 Interfaz Agent

This agent is dedicated to receive IP directions and user specified OID, this information is needed to know the device and the SNMP characteristic subject of study, besides, this agent will present the information related to abnormal actions during the operation of the network, this information is show in traps and alarms.

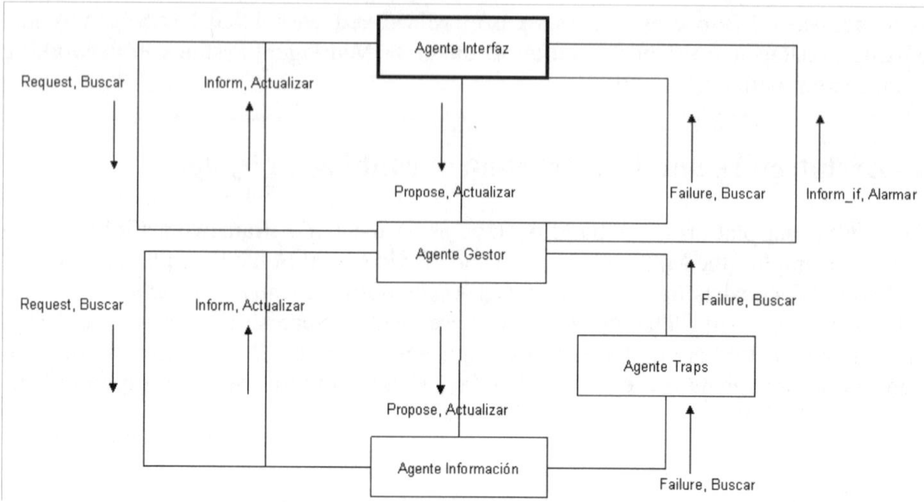

Fig. 3. Collaboration diagram

The *Interfaz* Agent has 4 behaviors:

- *getValores*: It receive the IP and OID values previously entered by the user through the Web interface. It uses the REQUEST performative.
- *getReportes*: It receives the result of the user query to the MAS using the INFORM performative.
- *getAlarmaGestor*: It receives emitted alarms from *Gestor* Agent be shown to the user.
- *getTrapGestor*: It receives the emitted traps from *Gestor* Agent to be shown to the user.

4.2 Información Agent

This agent is devoted to interact with the *Traps* Agent, with the *Gestor* Agent and whit the SNMP API used for communication with network devices. This agent acts as intermediary between the MAS and the monitored network.

Información Agent has 4 behaviors to execute different tasks on the network. This agent rewrites the JADE Agent class methods *setup*() and *takedown*(). The behaviors are:

- *getRequest*: It receives an ACL message with Request performative and with the ontology *Buscar* action retrieves a MIB value from an specific IP. The response is sent to the *Gestor* Agent with a message that includes an *Actualizar* action with the Inform performative.
- *getRequest1*: It is the same to the pervious one, but it uses the Query_If and Inform_If performatives to receive and send the messages.
- *setRequest*: It is devoted to update a MIB value in an specific IP, it receives a message with the *Actualizar* action and Propose performative, but it doesn't respond with a message.

- *porcesaTrap*: Its function is to gather all the generated traps in the network and to send it to the *Gestor* Agent within a message with the Failure performative that consists of the ontology's *Buscar* action.

4.3 Traps Agent

This agent is devoted to receive the traps that are sent from *Información* Agent and to store the data (Date, source IP and Trap OID) into the database. It notifies to the *Gestor* Agent about the trap.

The only one behavior of this agent performs 3 tasks:

- It receives the messages originated form *Information* Agent with the Failure performative that has into its content the ontology's *Buscar* action.
- It stores in the database the information related to the trap.
- It sends to the *Gestor* Agent the same message received from the *Información* Agent.

4.4 Management Agent

It is the intermediary between the *Interfaz* Agent and the *Información* Agent. It receives the request of OIDs and sends it to the *Información* Agent; furthermore it receives the results and responds to the source agent (*Interfaz* Agent). This agent tracks several OIDs in some devices to know the behavior of the network to generate the appropriated alarms according to the defined user's threshold. The alarm will be stored in the database and a message will be sent to the *Interfaz* Agent informing about the problem, describing the failure, the device that generates the alarm and the altered OID.

This agent has 5 behaviors that define the tasks that must be accomplished depending on the received message.

- *getAgentInterfaz*. This CyclicBehaviour type behavior receives a message of Request type with the *Buscar* action and accomplish the forward function with the IP and the OID received from *Information* Agent with a Request type message.
- *getAgentInformacion*: This CyclicBehaviour type behavior receives an Inform type message with the device IP, and the requested OID and its value, it creates an Inform type message to be sent to *Interfaz* Agent that performed the original request.
- *getTrapTraps*: It is a CyclicBehaviour type behavior that receives a message from the *Traps* Agent indicating which was the device that generates the trap, it creates a new message with the Failure performative and send it to the *Interfaz* Agent to notify the received trap.
- *getTestRed*: It is a behavior that check several OIDs in specific devices, within a period of 10s to identify whether the status of the network is operating according to the established policy.
- *getAgentInformacionTestRed*: As a result of the query performed in the previous behavior about OIDs, the *Information* Agent sends an Inform_if type message including the OID's value. This value is confronted with the values defined in the policy, previously stored in a database to produce an alarm that will be stored in a database for historical purposes.

4.5 Ontology

The developed ontology [11] consists of the concepts *IP*, *OID*, *Valor*, *Descripción* and of the actions *Buscar*, *Actualizar* and *Alarmar*.

The *IP* concept is a string that contains the IP address of a specific device. The *OID* concept is a string that contains the number of references for an OID variable. The *Valor* concept is a string that represents the MIB's variable value in a specific device in a specific time. The *Descripción* concept is a string that describes the cause of the generated alarm.

The *Buscar* action consists of concepts *IP* and *OID*, this action has several purposes: it is used to indicate to the *Información* Agent to search for a specific OID variable in a specific IP address, when the *Información* Agent receives a trap, sends it to the *Traps* Agent including the *Buscar* action whit the trap OID and the source IP. The *Buscar* action is used to pass the IP and OID values between the *Interfaz* Agent and the *Gestor* Agent.

The *Actualizar* action consists of concepts *IP*, *OID* and *Valor*; inside the ontology it has the function of communicate to the *Gestor* Agent about a specific value to a MIB's variable represented by the OID in the device with the corresponding IP to be assigned. The data is sent to the *Información* Agent.

The *Actualizar* action is used when the *Información* Agent responds to a query about a variable, then it sends inside of the message an *Actualizar* action that informs the actual OID value, related to the requested IP: it accomplish the same function of the message that goes from *Gestor* Agent to *Interfaz* Agent.

The *Alarmar* action consists of *OID* and *Descripción* concepts, this action is used to identify some abnormal situation presented during the network operation, it is, when a value defined by the policy is altered generating the abnormal situation.

5 Results

The system execution in a controlled environment gave the awaited results according to the limitations of the established policies and the defined thresholds. The effect of the execution of the Multiagent system to monitor the network can detect if the policy associated to the location of the device is being violated. The agents Behaviour and the detection among them through the message exchange allows to make the network scan; later the *Management* Agent generates an alarm that is stored in the database and is sent to the *Interfaz* Agent that is the one in charge of showing it to the users.

6 Conclusions

To specify completely a MAS in the analysis and design phases and their implementation, it is necessary to use different models using agent software engineering.

In this phase, the project increased the complete functionality of network supervision and reduced tuning the policies. It allows introducing other learning schemes that turn the platform in a solid network management system that becomes autonomous.

The Ontology integration for the MAS in SNMP is an important data representation to solve the heterogeneous problems in distributed network management systems and in a scalable semantic model find inferences using reasoning tools.

The emerging intelligent agent paradigm is an optimal solution for the distributed network management problem.

The proposed system could be compared against the traditional management technique in terms of response time and speedup using the prototype implemented using the performance management as the case study.

Also conflicts at the moment appeared at the same time for making the execution of all the agents, since these and all the used classes must be compiled with the same version of the Java Virtual Machine.

The agents have been proposed as a solution to the problem of the management of increasingly heterogeneous networks. This research extends an agent framework targeted at network management with an architecture and design for integration with an SNMP agent. The future network management solutions need to be adaptable, flexible, and intelligent without increasing the burden on network resources, the project addressed this requirement with a new management platform architecture based on multi-agents.

Acknowledgments. This work has been partially supported by joint call Universidad Nacional de Colombia and Universidad de Caldas. 2008.

References

1. RFC: 1155, 1157, 3584, 2571, 1212
2. Subramanyan, R., Miguel-Alonso, J., Fortes, J.A.B.: Design and Evaluation of a SNMP-based Monitoring System for Heterogeneous, Distributed Computing, Technical Report, TR-ECE 00-11, School of Electrical and Computer Eng., Purdue University (2000)
3. Case, J., Fedor, M. RFC.: 1157: Simple Network Management Protocol (SNMP). MIT Laboratory for Computer Science (May 1990) Consulted March 2008, `http://www.faqs.org/rfcs/rfc1157.html`
4. Wooldridge, M., Jennings, N., Kinny, D.: The Gaia Methodology for Agent-Oriented Analysis and Design. In: Autonomous Agents and Multi-Agent Systems. Kluwer Academic Publishers, Netherlands (2000)
5. Zambonelly, F., Jennings, N.R., Wooldridge, M.: Developing Multiagent Systems: The Gaia Methodology. ACM Transactions on Software Engineering and Methodology 12(3), 317–370 (2003)
6. DeLoach, S.A.: Engineering Organization-Based Multiagent Systems. In: Garcia, A., Choren, R., Lucena, C., Giorgini, P., Holvoet, T., Romanovsky, A. (eds.) SELMAS 2005. LNCS, vol. 3914, pp. 109–125. Springer, Heidelberg (2006)
7. Pavón, J., Gómez-Sanz, J.: Agent Oriented Software Engineering with INGENIAS. In: Mařík, V., Müller, J.P., Pěchouček, M. (eds.) CEEMAS 2003. LNCS (LNAI), vol. 2691, pp. 394–403. Springer, Heidelberg (2003)
8. JADE: Java Agent Development Framework. Ingeniería de la Información de la Universidad de Parma, `http://jade.tilab.com/`

9. Bellifemine, F., Caire, G., Trucco, T., Giovanni, R.: Jade Programmer's Guide, Jade Tutorial, Tilab (2003)
10. Mejía, M., Duque, N.D., Morales, A.: Metodologías para Sistema Multiagente en el caso de estudio: Gestión de Redes basado en el Protocolo SNMP. In: Tendencias en Ingeniería de Software e Inteligencia Artificial, Medellín (2007)
11. Obitko, M., Snášel, V.: Ontology Repository in Multi-Agent System. From Proceeding Artificial intelligence and applications (Austria, 2004)

A Multiagent Architecture Applied to Dynamic Generation of CV Documents

Evelio J. González, Alberto Hamilton, Lorenzo Moreno, Jonatán Felipe, and Vanesa Muñoz

Departamento de Ingeniería de Sistemas y Automática y ATC
Universidad de La Laguna
Av. Astrofísico Fco. Sánchez s/n, 38207, La Laguna, Tenerife, Spain
ejgonzal@ull.es

Abstract. The aim of this paper is to present a dynamic system for the automatic and dynamic generation of CV documents in an academic and research environment. For that purpose, the authors have integrated Multiagent Systems (MAS) with XML and Apache Cocoon, designing a web portal where the users can manage their CV data. Regarding to the use of Apache Cocoon and apart from showing its great potential, one of the main contributions of the work presented in this paper consists of the dynamic generation of the web environment.

Keywords: Multiagent System, CV Generation, Apache Cocoon.

1 Introduction

Every Universitary model usually requires its members to manage a big amount of personal data, such as publications in journals and attended conferences. Different official institutions often require researchers and students for these data in order to different purposes such as awarding a contract or research fellowship, annual reports, etc. Unfortunately, it is usual that each institution has its own template to fill, so researchers are often condemned to waste their time typing the same data in different documents. It has been calculated that the generation of a CV takes an average 3-hour period and that an automated system could save at least $25,000 per 100 generated CV's. It would be desirable a user-friendly web environment in which researchers and students could manage their personal data and generate their updated CV with only a click. For that purpose, the authors have decided to use Apache Cocoon [1,2] as base of the designed web environment. An interesting tool provided by Apache Cocoon is its forms (Cocoon forms or CForms), a XML way to build forms that can be filled by the users.

This system for the dynamic generation of CV documents has been integrated in a Multiagent System, originally developed for the automatic management of agendas in a Universitary Scenario taking the advantage of the use of ontologies [3] expressed in a highly expressive language, OWL.

Why do the authors use agents in the generation of CV documents when it seems that a simple database for each user could be sufficient? The answer lies in the human behaviour. Firstly, it has been observed that sometimes the corresponding author of an article forgets to communicate the acceptance of that paper to the rest of the authors.

J.M. Corchado et al. (Eds.): DCAI 2008, ASC 50, pp. 47–51, 2009.
springerlink.com

Even if the corresponding author send, e.g., an e-mail to a co-author communicating the good news, this co-author is usually so busy that he/she prefers to update his/her CV database later, taking the risk of 'losing' the paper in his/her CV. This behaviour implies that each user sends an e-mail to every related colleague, looking for that 'lost' publications, whenever he/she needs to present a CV document, wasting a lot of time in this way. Thus, a general database – an initial attempt was initially implemented in Apache Cocoon, covering all the users' merits, could be a good solution. However, once more the authors have bumped into the human behaviour. In spite of the security offered by the database manager, a significant number of members of the Universitary Scenario were reluctant to insert their data, claiming that they did not want leave their data in a centralized system where they could be accessed by malicious people. In this context, the features provided by the multiagent systems – distribution, reliability, proactivity, autonomous and reactive behaviour, etc.- seem to be especially useful.

2 Multiagent Architecture

The application of MAS to this problem is justified by the following reasons.

- The environment is dynamic. For instance, the number of users and their preferences can change in an unpredictable way.
- The agents form a distributed system and it is not necessary a permanent connection. The agents are who interact, not the users.
- Extensibility. Using both MAS and ontologies, new types of agents (or new instances of the same agents, even implemented by different developers) can be added easily to the system, making its functionality grow in a dynamic way. In general, this easiness cannot be reached by centralized systems, for example, a central server that every user interacts with via their Web browser.

The authors have implemented a MAS for planning and scheduling in a University Research Group. Originally the system was composed of 6 different types of agents.

User Agent (UA): This agent is an end-user interface, which shows the schedule to its related user and allows it to ask for a meeting or a resource. When it occurs, this agent tries to locate its negotiator agent and communicates what user needs. Once the negotiator has finished its work, the user agent receives the result and shows it to the user.

Negotiator Agent (NA): The implementation of the meeting and resource negotiation algorithm is applied via this agent. When it is asked by its related user agent for a meeting negotiation, it looks in the Directory Facilitator (DF) for the negotiator agents of the rest of the intended attendees. Then, the negotiation process begins. Alternatively, in the case of a resource negotiation, it looks for the resource agent.

Ontology Agent (OA): It provides ontology services to an agent community, so that the identification of a shared ontology for communication between two agents is facilitated. The definition of an external ontology, managed by an OA, provides numerous general advantages: it permits consultation with regard to concepts, the updating and use of ontologies and it eliminates the need to program the entire ontology in every agent, hence reducing required resources.

Resource Agent (RA): This agent is invoked when a resource negotiation occurs. Firstly, it asks the Ontology Agent for the instances of the selected resource type.

Mail Agent (MA): When an agenda change is confirmed, the Mail Agent is requested by the respective negotiator agent to send an email to the user via the mail software. For this purpose, it asks to OA for the email address of the user, as these data are stored in the ontology.

Rule Agent (RuA): This agent provides the system with the ability of learning from the users based on the previous behavior of that user or ad hoc preferences. The RuA is consulted whenever there is an agenda change in order to organize a meeting. The NA will consult with this agent in order to determine whether or not the user is supposed to agree to a possible agenda change.

In this scenario, several mobile agents have been integrated. Each user owes its own mobile agent, called CVSearchAgent (CVSA) that is periodically migrating in the network looking for new merits in which its user is involved. It is clear from the nature of the scenario that the CVSA's do not need to be always active as users do not need to be continuously submitting their CV. Thus, their activity is reduced to a few hours each 7-15 days. This fact and the characteristic of mobility make that the network is not overloaded, as the interaction with other agents and the search of new merits can be done off-line.

Whenever a new merit is found, the CVSA asks the MASplan MA for sending an e-mail to the corresponding user, informing of the result of the search. This way, the user can obtain a copy of the found data for other purposes different to the generation of CVs. Apart from this interaction, the CVSA send these data to the UA for its inclusion in the local user CV database. The implementation of these local databases does not affect the dynamic generation of the forms as the XSD's are accessed from a remote server.

Each user owes its own database, stored in the local system and accessed through the Cocoon web portal environment with an authorization and verification process. That database should be managed by an agent in the system. In order to reuse code and resources, the authors have implemented that management functions in the UA code. Thus, the CVSA interacts with the corresponding UA when it is necessary. One of the actions to be carried out is to avoid redundancy in the merits.

Regarding to the use of ontologies by the agents, the original MASplan one has complemented with the inclusion of new concepts related to CV activities. These definitions make easier the interaction among the CVSA's and the UA's. When both agents start a conversation, they compare the research areas of their corresponding users. In case of disjointness, there is no need of search in the XML database, avoiding its computational cost, and maybe more important, avoiding future unproductive conversations. A more refined version, currently in progress, will consist in using the DF functionality as a yellow pages service of research activities. Another open line in this project consists of providing more intelligence to the system. The purpose would be that the CVSA's themselves were able to extract the rules of interaction among users regarding analogue research areas- for example, if two areas are compatible or not – from the list of keywords in the titles of publications or common authors in the merits.

For that purpose, several well-known techniques can be used, e.g., Dempster-Shafer method. After that deduction, the CVSA would interact with the RuA and OA for the inclusion of the deduced axiom in the ontology.

3 Cocoon Web Portal

In this section, the Cocoon web portal will be described. It is noted, as stated above, that the interaction of the web portal with the MAS is done via the UA's.

The data involved in the structure of the system have been divided into four categories, declared in their respective schemas.

- Courses (lectures, seminaries): name of the course, targets, etc.
- Personal data: name, address, spoken languages, etc.
- Merits: list of the merits to be included in a CV, such as publications in journals, conferences, books, chapters, research activities, etc.
- Groups: list of research groups, members, address, director, etc. – accessed only by the administrators of the system.

It is noted that for avoiding repetition in this structure, each document includes a reference field, a kind of main key in a database. This key is the number of identity card (DNI) of the user.

As stated above, the web environment is generated in a dynamic way. It is interesting that the information could change following the structure of the XML files, in other words, following the changes produced in the XML schemas. This purpose has been reached through the design of a set of XSL transformations that allow generating Cocoon forms. As expected, a change in the XML schema involves a change in the corresponding CForm. The mentioned files collect the XSD structure and extract recursively the data and turn it into a CForm structure. All of them have a similar structure, based on recursive invocations with attribute inheritance and recollection in groups of information. Thus, the differences are not in their structure but in their content, due to their specific syntax using their respective namespaces. Mainly, the idea consists in reading the schema from the root element to its children elements, at the same time as each child element inherits the attributes of its parent element. This structure will be used whenever the system needs to extract information from a XSD document. The first phase of the project was only focused on this web environment. When a user desires to generate an updated CV, a XSL-FO transformation is applied in order to get that CV which is accessed in the browser.

4 Conclusion

In this paper, the authors have presented a multi-agent system for the automatic generation of CV's for a Universitary scenario. The agents related to this purpose have been integrated with a previously designed system for planning and scheduling in a University Research Group Scenario. The included agents, called CVSA, are mobile-one for each user in the system- and their task consists of the search of merits related to their corresponding user. This search is carried out interacting with other agents, called UA, in charge of the management of local merits databases. The main reason for this distribution, and thus for the use of a multiagent system, lies in the human behaviour.

The original system included an ontology for the automatic meeting negotiation in an intelligent way amongst several researchers, among other several features. This ontology was implemented in OWL, a highly expressive markup language. In the context of this work, some new concepts have been introduced, for example axioms related to the disjointness of different research areas, facilitating the interaction among the agents involved. Currently, the authors are working on the addition of new interesting features in the system, as the ability of the agents of deducing axioms online. Other interesting aspect of this work is the implementation of a dynamic web environment for the management of the CV generation. The development of this environment has been carried out using Apache Cocoon. The web environment is generated in a dynamic way.

The system has been used successfully for the generation of the documentation nearly 100 professors, so the use of this web environment has saved a lot of time in the data treatment.

References

1. Mazzocchi, S.: Adding XML Capabilities with Cocoon. In: ApacheCON 2000, Orlando (2000)
2. Noels, S.: Standards Applied: Using Apache Cocoon and Forrest. In: XML Europe 2003, London, England (2003)
3. Falasconi, S., et al.: Using ontologies in Multi-Agent Systems. In: Tenth Knowledge Acquisition for Knowledge-Based Systems Workshop (1996)

HoCa Home Care Multi-agent Architecture

Juan A. Fraile[1], Javier Bajo[1], Belén Pérez Lancho[2], and Eladio Sanz[2]

[1] Universidad Pontificia de Salamanca, Compañía 5, 37002, Salamanca, Spain
{jafraileni,jbajope}@upsa.es
[2] Departamento de Informática y Automática, Universidad de Salamanca, Plaza de la Merced s/n, 37008, Salamanca, Spain
{lancho,esanz}@usal.es

Summary. This paper presents a Hybrid Multi-Agent Architecture for the control and supervision of dependent environments, based on an Ambient Intelligence model. HoCa architecture incorporates a management system of alerts based on SMS and MMS technologies, and an automated identification, localization, and movement control system based on Java Card and RFID technologies. HoCa is independent from the programming language and operating system in that it is executable. The core of the architecture is formed by both deliberative agents and reactive agents that interact to offer efficient services. The architecture has been tested in a real environment and the results obtained are presented in this paper.

Keywords: Dependent environments, Ambient Intelligence, Multiagent Systems, Home Care.

1 Introduction

There is currently considerable growth in the development of automation technologies, such as home automation and Ambient Intelligence (AmI). One of their principal objectives is to look after the user's well-being and obtain a more friendly, rational, productive, sustainable and secure relationship for users within their environments. Several architectures based on agent utilization have emerged thanks to the appearance of intelligent spaces and the integration of devices that are programmable via computer networks [11]. These have stimulated the development of ubiquitous computation, which is the most promising technological approximation for resolving the challenge of developing strategies that allow the early detection and prevention of problems in an automated environment.

The main objective of this paper is to define a hybrid Multi-Agent Architecture for the control and the supervision of open environments. It involves developing an architecture that allows automated identification, localization, alarms management and control of movement. The users who utilize the system in which this architecture is applied will be able to gain wireless access to all the information that they need to perform their work. The novel innovation at the core of the architecture is a real time communication protocol that allows

J.M. Corchado et al. (Eds.): DCAI 2008, ASC 50, pp. 52–61, 2009.
springerlink.com

secure and rapid communication between the reactive agents and the system sensors. These reactive agents, whose response time is critical, are influenced by deliberative BDI agents, which are located inside the platform given that a very fluid communication already exists between them. Additionally, the architecture manages an alert or alarm system across the agents' platform specially designed to work with mobile devices. The alert system contains different levels of urgency. The alert level is determined by the deliberative agent who, depending on the alert level, then emits the alert to either a reactive agent or a deliberative agent.

The paper is organized as follows: The second section presents the problem that prompted this work. The third section presents the proposed architecture, and the fourth section gives the results and conclusions obtained after applying the proposed architecture to a real case in an environment of dependence.

2 General Description of the Problem

The use of intelligent agents is an essential component for analyzing information on distributed sensors [12] [14]. These agents must be capable of both independent reasoning and joint analysis of complex situations in order to be able to achieve a high level of interaction with humans [3]. Although multi-agent systems already exist and are capable of gathering information within a given environment in order to provide medical care [9] [5], there is still much work to be done. It is necessary to continue developing systems and technology that focus on the improvement of services in general. After the development of the internet there has been continual progress in new wireless communication networks and mobile devices such as mobile telephones and PDAs. This technology can help to construct more efficient distributed systems capable of addressing new problems [6].

Hybrid architectures try to combine deliberative and reactive aspects, by combining reactive and deliberative modules [5]. The reactive modules are in charge of processing stimuli that do not need deliberation, whereas the deliberative modules determine which actions to take in order to satisfy the local and cooperative aims of the agents. The aim of modern architectures like Service Oriented Architecture (SOA) is to be able to interact among different systems by distributing resources or services without needing to consider which system they are designed for. An alternative to these architectures are the multi-agent systems, which can help to distribute resources and to reduce the centralization of tasks. Unfortunately the complexity of designing multi-agent architecture is great since there are not tools to either help programme needs or develop agents.

Multi-agent systems combine aspects of both classic and modern architectures. The integration of multi-agent systems with SOA and web services has been recently investigated [1]. Some investigators focus on the communication among these models, whereas others focus on the integration of distributed services, especially web services, in the agents' structure [3] [10] [13].

These works provide a good base for the development of multi-agent systems. Because the majority of them are in the development stage, their full potential in a real environment is not known. HoCa has been implemented in a real environment and not only does it provide communication and integration among distributed agents, services and applications, but it also provides a new method for facilitating the development of multi-agent systems, thus allowing the agents and systems to function as services.

HoCa implements an alert and alarm system across the agent's platform, specially designed to be used by mobile devices. The platform agents manage this service and determine the level of alert at every moment so that they can decide who will receive the alert and when. In order to identify each user, HoCa implements a system based on Java Card [15] and RFID (Radio Frequency IDentification) microchip technology in which there will be a series of distributed sensors that provide the necessary services to the user.

3 Proposed Architecture

The HoCa model architecture uses a series of components to offer a solution that includes all levels of service for various systems. It accomplishes this by incorporating intelligent agents, identification and localization technology, wireless networks and mobile devices. Additionally, it provides access mechanisms to multi-agent system services, through mobile devices, such as mobiles phones or PDAs. Access is provided via wi-fi wireless networks, a notification and alarm management module based on SMS and MMS technologies, and user identification and localization system based on Java Card and RFID technologies. This system is dynamic, flexible, robust and very adaptable to changes of context.

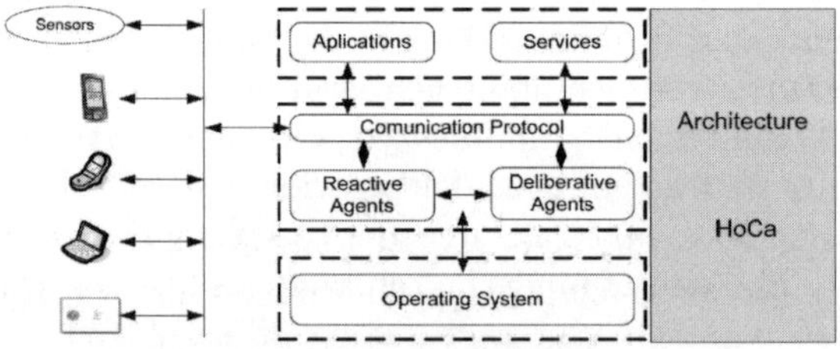

Fig. 1. HoCa Framework

HoCa architecture describes four basic blocks that can be seen in Figure 1: Applications, Services, Agents Platform and Communication Protocol. These blocks constitute the whole functionality of the architecture.

3.1 Agents Platform in HoCa

This platform is the core of the architecture and integrates two types of agents, each of which behaves differently for specific tasks, as shown in Figure 2.

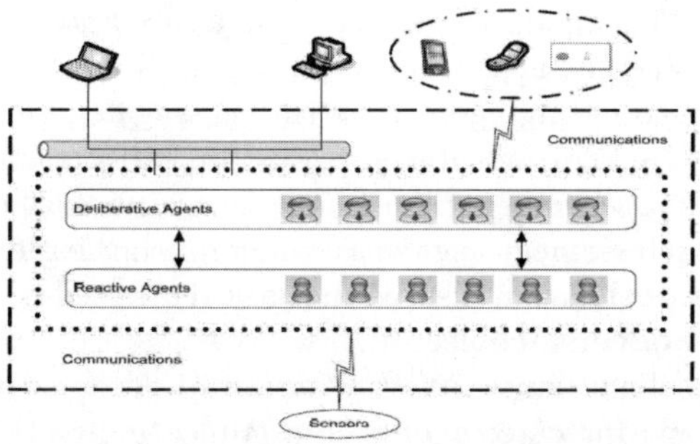

Fig. 2. Agents platform structure in the HoCa Architecture

The first group of agents is made up of deliberative BDI agents, which are in charge of the management and coordination of all system applications and services. These agents are able to modify their behaviour according to the preferences and knowledge acquired in previous experiences, thus making them capable of choosing the best solution. Deliberative agents constantly deal with information and knowledge. Because they can be executed on mobile devices, they are always available and they provide ubiquitous access for the users. There are different kinds of agents in the architecture, each one with specific roles, capabilities and characteristics. This fact facilitates the flexibility of the architecture to incorporate new agents. However, there are pre-defined agents which provide the basic functionalities of the architecture:

- CoAp Agent: This agent is responsible for all communications between applications and the platform. Manages the incoming requests from the applications to be processed by services. It also manages responses from services to applications. CoAp Agent is always on "listening mode". Applications send XML messages to the agent requesting for a service, then the agent creates a new thread to start communication using sockets.
- CoSe Agent: It is responsible for all communications between services and the platform. The functionalities are similar to CommApp Agent but backwards. This agent is always on "listening mode" waiting for responses of services. Manager Agent indicates CommServ Agent the service that must be invoked. Then, CommServ Agent creates a new thread with its respective socket and sends an XML message to the service.
- Directory Agent. Manages the list of services that can be used by the system. For security reasons, the list of services is static and can only be modified manually, however services can be added, erased or modified dynamically. The list contains the information of all trusted available services.
- Supervisor Agent. This agent supervises the correct functioning of the agents in the system. Supervisor Agent verifies periodically the status of all agents registered in the architecture by means of sending ping messages. If there is no response, the agent kills the agent and creates another instance of that agent.

- Security Agent. This agent analyzes the structure and syntax of all incoming and outgoing XML messages. If a message is not correct, the Security Agent informs the corresponding agent that the message cannot be delivered.
- Manager Agent. Decides which agent must be called taking into account the users preferences. Users can explicitly invoke a service, or can let the Manager Agent decide which service is better to accomplish the requested task. If there are several services that can resolve the task requested by an application, the agent selects the optimal choice.
- Interface Agent. This kind of agent has been designed to be embedded in users' applications. Interface agents communicate directly with the agents in HoCa so there is no need to employ the communication protocol, but FIPA ACL specification. The requests are sent directly to the Security Agent, which analyzes the requests and sends them to the Manager Agent.

The second group is made up of reactive agents. Most of the research conducted within the field of multi-agent systems focuses on designing architectures that incorporate complicated negotiation schemes as well as high level task resolution, but don't focus on temporal restrictions. In general, the multi-agent architectures assume a reliable channel of communication and, while some establish deadlines for the interaction processes, they don't provide solutions for limiting the time the system may take to react to events.

It is possible to define a real-time agent as an agent with temporal restrictions for some of its responsibilities or tasks [11]. From this definition, we can define a real-time multi-agent system (Real Time Multi-Agent System, RT-MAS) as a multi-agent system in which at least one of the agents is a real-time agent. The use of RT-MAS makes sense within an environment of critical temporal restrictions, where the system can be controlled by autonomous agents that need to communicate among themselves in order to improve the degree of system task completion. In this kind of environments every agent requires autonomy as well as certain cooperation skills to achieve a common goal.

3.2 HoCa Communication Protocol

Communication protocol allows applications, services and sensors to be connected directly to the platform agents. The protocol presented in this work is open and independent of programming languages. It is based on the SOAP standard and allows messages to be exchanged between applications and services as shown in Figure 3.

However, interaction with environmental sensors requires Real-time Transport Protocol (RTP) [4] which provides transport functions that are adapted for applications that need to transmit real-time data such as audio, video or simulation data, over multicast or unicast network services. The RTCP protocol is added to RTP, allowing a scaleable form of data supervision. Both RTP and RTCP are designed to work independently from the transport and lower network services. They are in charge of transporting data with real-time characteristics, and of supervising the quality of service, managing the information for all the entities taking part in the current session.

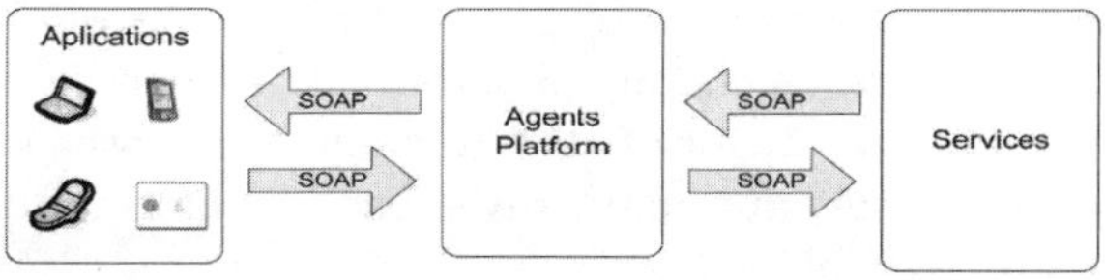

Fig. 3. Communication using SOAP messages in HoCa

The communications between agents within the platforms follows the FIPA ACL (Agent Communication Language) standard. This way, the applications can use the platform to communicate directly with the agents.

3.3 Location and Identification System in HoCa

This system incorporates Java Card [15] and RFID [7] technologies. The primary purpose of the system is to convey the identity of an object or person, as with a unique serial number, using radio waves. Java Card is a technology that permits small Java applications (applets) to be run safely in microchip smart cards and similar embedded devices. Java Card gives the user the ability to program applications that can be run off a card so that it has a practical function in a specific application domain. The main features of Java Card are portability and security; it is described in ISO 7816. The data are stored in the application and the Java Card applications are executed in an isolated environment, separate from the operating system and from computer that reads the card. The most commonly used algorithms, such as DES, 3DES, AES, and RSA, are cryptographically implemented in Java Card. Other services such as electronic signature or key generation are also supported.

RFID technology is grouped into the so-called automatic identification technologies. But RFID provides more information than other auto-identification technologies, speeds up processes without losing reliability, and requires no human intervention.

The combination of these two technologies allows us to both identify the user or identifiable element, and to locate it, by means of sensors and actuators, within the environment, at which time we can act on it and provide services. The microchip, which contains the identification data of the object to which it is adhered, generates a radio frequency signal with this data. The signal can be picked up by an RFID reader, which is responsible for reading the information and sending it, in digital format, to the specific application.

3.4 Using HoCa to Development a Multi-Agent System for Dependent Environment

The alert system is integrated into the HoCa architecture and uses mobile technology to inform users about alerts, warnings and information specific to the daily routine of the application environment. This is a very configurable system that allows users to select the type of information they are interested, and to receive it immediately on their mobile phone or PDA. It places the information to be sent into information categories. The users determine the information they are interested in. The system automatically sends the information to each of the users as soon as it is available.

4 Proposed Architecture

Ambient Intelligence based systems aim to improve people quality of life, offering more efficient and easy to use services and communication tools to interact with other people, systems and environments. One of the most benefited segments of population with the development of these systems is elderly and dependent people. Agents and multi-agent systems in dependency environments are becoming a reality, especially on health care. Most agents-based applications are related to the use of this technology in patients monitoring, treatment supervision and data mining.

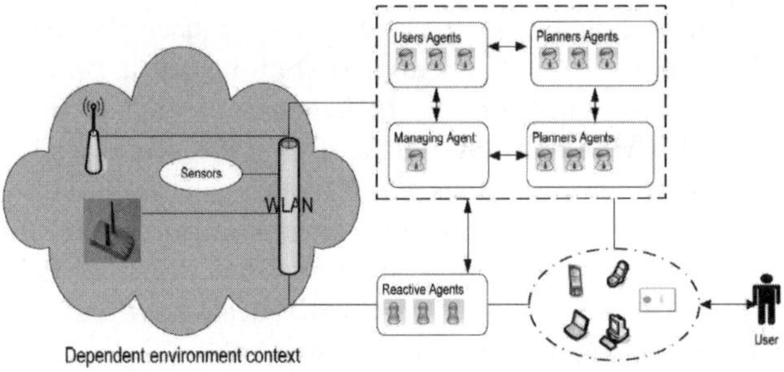

Fig. 4. HoCa structure in a dependent environment

HoCa has been employed to develop a multi-agent system aimed to enhance assistance and care for low dependence patients at their homes. Main functionalities in the system include reasoning, planning mechanisms, management alerts and responses in execution time offered to certain stimuli, as shown in Figure 4. These functionalities allow the system the use of several context-aware technologies to acquire information from users and their environment. Among the technologies used are mobile systems for alerts service managing across PDA and mobile phones, Java Card elements for identification and presence detectors and access control.

Each agent in the system has its own functionalities. If an agent needs to develop a task in collaboration with other agent a request form is send. There are priority tasks that a set of agents can perform. This ensures that the priority tasks are always available. There are four types of agents:

- User Agents manage user personal data and their behavior. They are responsible through the system, identify and locate implemented by the architecture. They determine the status of the user and offering services in the environment as a correct temperature, automatic lighting, access blocking or opening, etc.
- SuperUser Agent runs on mobile devices and inserts new tasks into the Manager Agent to be processed by a reasoning mechanism. It also needs to interact with the User Agents to impose new tasks and receive periodic reports, and with the ScheduleUser Agents to ascertain plans' evolution.
- SheduleUser Agent schedules the users' daily activities obtaining dynamic plans depending on the tasks needed for each user. It manages scheduled-users profiles, tasks, available time and resources. Every agent generates personalized plans depending on the scheduled-user profile.
- Manager Agent runs on a Workstation and plays two roles: the security role that monitors the users and the manager role that handle the databases and the tasks assignment. It must provide security for the users and ensure the tasks assignments are efficient.

On the other hand there are a number of reactive agents that work in collaboration with the deliberative agents. These agents are in change of control devices interacting with sensors (access points, lights, temperature, alarms detection, etc.). They receive information, monitor cnvironment services and also check the devices status connected to the system. All information is treated by the reactive agent and it is sent to the manager agent to be processed.

5 Results and Conclusions

HoCa has been used to develop a multi-agent system for monitoring dependent patients at home. The main features of this system include reasoning and planning mechanisms, and alert and response management. Most of these responses are reactions in real time to certain stimuli, and represent the abilities that the reactive agents have in the HoCa architecture based platform. To offer all these features the system uses various technologies and acquires information from users and the surrounding environment. Some of the technologies used to test the system include mobile technology for managing service alerts through PDAs and mobile phones, and Java Card technology for identification and access control.

One of the main contributions of the HoCa architecture is the alert system. We implemented several test cases to evaluate the management of alerts integrated into the system. This allowed us to determine the response time for warnings generated by the users, for which the results were very satisfactory, with response times shorter than those obtained prior to the implementation of HoCa. The system studies the information collected, and applies a reasoning process which allows alerts to be automatically generated. For these alerts, the system does not only take response time into account, but also the time elapsed between alerts, and the user's profile and reliability, in order to generalize reactions to common

Table 1. Comparison between the HoCa and the ALZ-MAS architectures

Factor	HoCa	ALZ-MAS
Average response time to incidents (min.)	8	14
Assisted incidents	12	17
Average number of daily planned tasks	12	10
Average number of daily services completed	46	32
Time employed to attend and alert (min.)	75	90

situations. The results show that HoCa fits perfectly within complex systems by correctly exploiting services and planning mechanisms.

Table 1 presents the results obtained after comparing the HoCa architecture to the previously developed ALZ-MAS architecture [6] in a case study on medical care for patients at home. The ALZ-MAS architecture allows the monitoring of patients in geriatric residences, but home care is carried out through traditional methods. The case study presented in this work consisted of analysing the functioning of both architectures in a test environment. The HoCa architecture was implemented in the home of 5 patients and was tested for 30 days. The results were very promising. The data shown in Table 1are the results obtained from the test cases. They show that the alert system improved the communication between the user and the dependent care services providers, whose work performance improved, allowing them to avoid unnecessary movement such as travels and visits simply oriented to control or supervise the patient. The user identification and location system in conjunction with the alert system has helped to notably reduce the percentage of incidents in the environment under study. Moreover, in addition to a reduction in the number of incidents, the time elapsed between the generation of a warning and solution decreased significantly. Finally, due to the many improvements, the level of user satisfaction increased with the introduction of HoCa architecture since patients can live in their own homes with the same level of care as those offered at the residence.

References

1. Ardissono, L., Petrone, G., Segnan, M.: A conversational approach to the interaction with Web Services. In: Computational Intelligence, vol. 20, pp. 693–709. Blackwell Publishing, Malden (2004)
2. Bahadori, S., Cesta, A., Grisetti, G., Iocchi, L., Leone, R., Nardi, D., Oddi, A., Pecora, F., Rasconi, R.: RoboCare: Pervasive Intelligence for the Domestic Care of the Elderly. Artificial Intelligence 1(1), 16–21 (2003)
3. Bonino da Silva, L.O., Ramparany, F., Dockhorn, P., Vink, P., Etter, R., Broens, T.: A Service Architecture for Context Awareness and Reaction Provisioning. In: IEEE Congress on Services (Services 2007), pp. 25–32 (2007)
4. Carrascosa, C., Bajo, J., Julian, V., Corchado, J.M., Botti, V.: Hybrid multi-agent architecture as a real-time problem-solving model. Expert Systems With Applications 34(1), 2–17 (2008)

5. Corchado, J.M., Bajo, J., de Paz, Y., Tapia, D.: Intelligent Environment for Monitoring Alzheimer Patients, Agent Technology for Health Care. Decision Support Systems 34(2), 382–396 (2008)
6. Corchado, J.M., Bajo, J., Abraham, A.: GERAmI: Improving the delivery of health care. IEEE Intelligent Systems 23(2), 19–25 (2008)
7. ITAA. Radio Frequency Identification. RFID...coming of age. In: White paper, Information Technology Association of America (2004), `http://www.itaa.org/rfid/docs/rfid.pdf`
8. Julian, V., Botti, V.: Developing real-time multi-agent systems. Integrated Computer-Aided Engineering 11(2), 135–149 (2004)
9. Mengual, L., Bobadilla, J., Trivio, G.: A fuzzy multi-agent system for secure remote control of a mobile guard robot. In: Favela, J., Menasalvas, E., Chávez, E. (eds.) AWIC 2004. LNCS (LNAI), vol. 3034, pp. 44–53. Springer, Heidelberg (2004)
10. Ricci, A., Buda, C., Zaghini, N.: An agent-oriented programming model for SOA and web services. In: 5th IEEE International Conference on Industrial Informatics, pp. 1059–1064 (2007)
11. Rigole, P., Holvoet, T., Berbers, Y.: Using Jini to integrate home automation in a distributed software-system. In: Plaice, J., Kropf, P.G., Schulthess, P., Slonim, J. (eds.) DCW 2002. LNCS, vol. 2468, pp. 291–303. Springer, Heidelberg (2002)
12. Tapia, D.I., Bajo, J., De Paz, F., Corchado, J.M.: Hybrid Multiagent System for Alzheimer Health Care. In: Bajo, J., Corchado, E.S., Herrero, l., Corchado, J.M. (eds.) Hybrid Artificial Intelligence Systems. HAIS 2006, Universidad de Salamanca, pp. 1–18 (2006)
13. Walton, C.: Agency and the Semantic Web. Oxford University Press, Inc., Oxford (2006)
14. Wooldridge, M.: An Introduction to MultiAgent Systems. John Wiley and Sons, Chichester (2002)
15. Chen, Z.: Java Card Technology for Smart Cards. Addison Wesley Longman (2000) ISBN 0201703297

Social Identity Management in Social Networks

Diego Blanco, Jorge G. Sanz, and Juan Pavón

Dep. Ingeniería del Software e Inteligencia Artificial
Universidad Complutense Madrid
Ciudad Universitaria s/n, 28040 Madrid, Spain
diego.blanco@fdi.ucm.es, jjgomez@sip.ucm.es, jpavon@fdi.ucm.es

Abstract. There is a lot of hype about social networks and related software within the concept Web 2.0 in the business and technical worlds. Given the fact that information is being widely spread in an exponential way with rooted dependencies among networks of social networks, we could find our digital identity exposed in ways that we could consider as inappropriate. Further than censorship, there is an arising need of knowing what is happening with the information we are delivering somewhere in the net, managing the distribution and the quality of detailed information. This paper discusses on which ways social identity management could be approached and further analyzed.

Keywords: Social networks, social network identity, privacy.

1 Introduction

Although social networks have been analyzed comprehensively in the last decade [1,2], in the last three or four years the outburst of social sites has gained momentum. Not only Facebook, Flickr, LinkedIn or tuenti have achieved a notorious impact in web community or media, but a hundred of new sites have appeared asking for its piece of cake [3]. Although its popularity started up in the visitor volume, the time or traffic spent in these sites turned out to be different of what up to the date used to be the standard.

The revolution that this 'new old brand' channel has started up offers a differential approach to knowledge management about users. People are connected by social links (friendship, colleagues, family, students, and so on) so that they can share a common set of interests. And those links have a strong dependency with confidence on the individual. People that everyone decides to relate to, are those who they want in their 'inner circle'. This pattern iterates up to the whole humankind (if they share the same kind of tools and exist in the same network) [12].

Groups are created in the social network around common interests as a means of widening the social connections. Those circles are exploited in different ways:

- Social activities
- Traditional o new wave marketing
- Targeted cross or up-selling

The game starts when analysis is done within the network. The extent to which our presence in the social network can reach is unprecedented and uncontrollable. Who and how are people accessing our social network identity and data? How to establish

J.M. Corchado et al. (Eds.): DCAI 2008, ASC 50, pp. 62–70, 2009.
springerlink.com

suitable filters to what can and cannot be seen or accessed? Is there any way of establishing behavior management for our social identity? How to manage cross social networks bounds and links? How to harness the social engineering applied to social networks? In other words, future contractor should not be allowed to find photos of last private party in the Canary Islands. Which are the risks of disorganized social growth? Will we reach a saturation point?

In the next sections a framework is established for analyzing the ways in which social identity should be managed.

This paper is divided in 5 sections. This section introduces the social networks and defines the context of following analysis. Section 2, establishes the basic structure of information necessary to manage social identity within a social network. Section 3 reviews the paths of evolution of social identity in social networks from existing or inferred information. Section 4 analyzes the way in which information in the social networks affects social identity and a structure for managing it. In section 5 conclusions and future work are described.

2 Social Network Identity

There are social network identity initiatives such as OpenID[13] where focus is placed in a unique login for different web sites, also known as Internet Single Sign-On. This initiative lacks all the semantics necessary to discover what is known about any user or managing that information. In order to cover this need, it is introduced the concept of Social Network Identity.

Social Network Identity (denoted as SNI from now on) describes the potential information –what is known and what eventually could be guessed- about an individual within a social network. It is characterized by two different sets of attributes:

- *Explicit attributes:* All the information placed in the network on purpose about an individual. This is denoted as EA = $\{ea_1, ea_2, \ldots, ea_n\}$, the set of explicit attributes provided in a certain Social Network
- *Implicit attributes:* All the information that could be inferred from the explicit attributes or from our behavior within the network. This is denoted as IA= $\{ia_1, ia_2, \ldots, ia_n\}$, the set of attributes discovered.

The SNI can be established as a function on these two sets. Nonetheless, social networks have a great grouping component. Being so, the SNI is partitioned for different *g* groups an individual belongs to. Explicit attributes imply what an individual wants other individuals in the social network know about him/her. Filters can be defined so that access to his/her explicit attributes is limited [14], for instance, to the closer environment, any specific groups or concrete persons.

Accessible information, denoted as AI, for different groups *g*, can be defined in terms of explicit attributes. The knowledge individuals in a certain group should get from us can be expressed as:

$$AI_g = \pi_g(EA) \tag{1}$$

SNI can now be expressed as:

$$SNI = \bigcup_{g=1}^{n} f(AI_g, IA_g) \tag{2}$$

EA could be expressed now as:

$$EA = \bigcup_{g=1}^{n} AI_g \tag{3}$$

Whereas EA can be dynamically modified and depends on what kind of information users want to expose, the implicit attributes are not affected by filters but are inferred from what user really does; this involves IA affects every AI_g.

There is an important privacy subject to be considered here. Privacy breaches will be found using the following function:

$$fPriv(EA) = \left(\# \, ea_y \in EA \,\middle|\, \exists \, ia_x \in IA, sem(ia_x) \equiv sem(ea_y)\right) \tag{4}$$

being *sem* a function that analyses the semantics behind an attribute.

3 Social Network Management: Scenarios

Social Identity Management becomes complex when value of (4) increases, thinning the line that separates what an individual wants the social network to know about him/her and what the social network effectively knows.

For instance, an individual could be state explicitly that he does not like reading about sports (explicit attribute) but he could be spending all day long connected to see what is happening at The Football League (implicit attribute). The information that has been effectively gathered is useless here if it cannot be managed properly, avoiding privacy issues.

SNI quality will depend on several premises:

- The more suitable the filters are for different groups, the better our digital identity will be managed. Function seen in (4) could be constrained to every AI (1), rewritten as $fPriv(AI_g)$.
- The more disjoint and non-interfering be EA and IA, the easier will be to manage SNI without violating privacy subjects

SNI privacy considerations can be quantified using the following indicator.

$$SNIIndicator(EA, IA) = {}^{\#\,(EA \cap_{semantic} IA)}\!/_{\#\,IA} \tag{5}$$

By analyzing different scenarios derived from (5), the following graphic depicts the relation between cardinality of both sets EA and IA, and the relationship between their semantic intersections. The analyzed situation is based on both sets cardinality and its semantic commonalities.

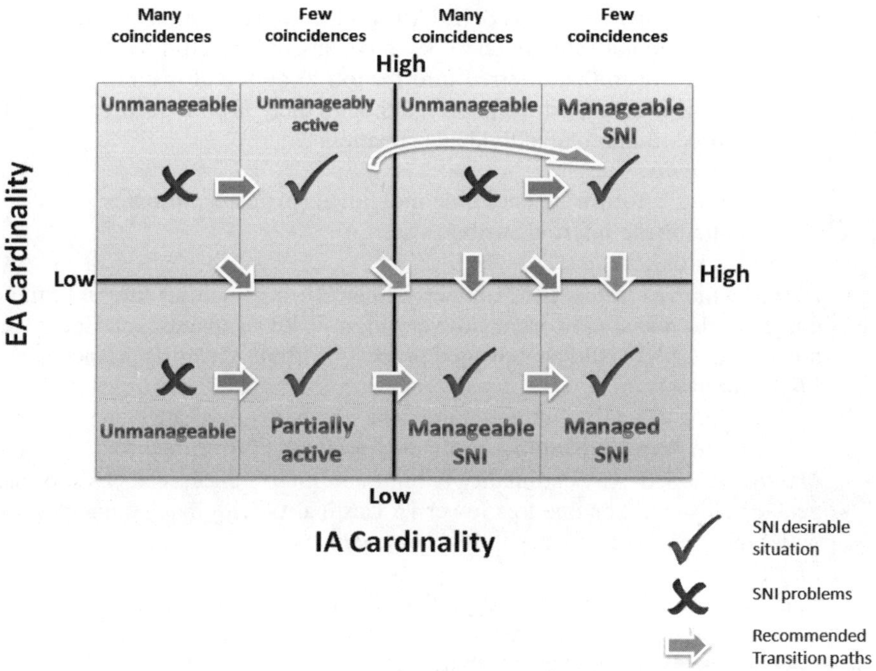

Fig. 1. SNI indicator based on EA, IA and its semantic intersection

There will be Social Network Management desirable situations:

- Higher EA cardinality: implies a better controlled information placed in the social network
- Lower #(EA$\cap_{\text{semantic}}$IA), where an individual exposed identity weights at least as much as the inferred one: implies a margin of SNI management capability without violating the expressed privacy.

And also avoidable situations:

- Higher #(EA$\cap_{\text{semantic}}$IA), where an individual exposed identity weights much less than the inferred one: implies a high risk of violating the expressed privacy.

Under this viewpoint, digital identity would fall within one of the following groups:

- *Unmanageable:* EA are very similar to IA and the cardinality of semantic intersection is high. These SNIs present serious problems in order to manage relationships with them, having to manipulate a double morale.

  Migration paths: In order to stop having unmanageable SNI, several alternatives could be defined:
  - o Behave as one has stated so that inferred attributed reduce its cardinality and thus reducing the semantic intersection cardinality.
  - o Reducing the inferred attributes set.
  - o A mix of both.

- *Unmanageably active:* In this SNI, IA is high compared with EA; this involves that someone do not want much to be known about himself directly but he participates actively so that inferred information to be found is high. These SNIs force to manage relationships poorly, although there is a lot of information that could be used to improve individuals situation.
  Migration paths:
  - Making EA grow up enriching individuals exposed attributes
  - Reducing the inferred attributes set.
  - A mix of both
- *Partially active:* In this SNI, neither IA nor EA cardinalities are high; the exposure in the social network is not very high. With a semantic intersection cardinality low, SNI could be managed properly without violating privacy.
  Migration paths:
  - Making EA grow up enriching individuals exposed attributes and keeping the IA set with a low cardinality to avoid SNI degradation.
- *Manageable SNI:* EA cardinality is high and interferences with IA are manageable. This could be due to a lower IA cardinality with many coincidences or with higher IA cardinality with few coincidences.
  Migration paths:
  - Behave as one has stated so that inferred attributed reduce and thus reducing the semantic intersection cardinality.
  - Reducing the inferred attributes set.
  - A mix of both
- *Managed SNI:* EA cardinality is high, IA cardinality is low with few coincidences between both sets. This is the ideal situation which minimizes the privacy risks and which simplifies the SNI management.

SNI evolution under the described circumstances can be described using the following diagram:

Fig. 2. Social Network Management: SNI migration path

4 Managing SNI

Evolving EA is simple as far as people are interested in sharing more knowledge about them. But what happens when they are not interested in sharing? Their behavior will speak for themselves although they do not want to. Information shared will do for them.

How much does shared information affect the IA for an individual? This introduces another concept, the ***network impact value for information*** (denoted as NIVI from now on). NIVI is affected by:

- *Information control (IC)*: Information is controllable or not. A friend could be asked to remove a photo without problems; but asking an aggressive opponent not to speak ill of us could be quite difficult.
- *Information Relevance (IR)*: Information generated in a closer circle about us seems more relevant that thirds gossiping. IR depends on network topology and connections architecture. The higher the relevance is, the bigger will be the NIVI.
- *Deviation from exposed attributes (DEA)*: Information which contains a semantics which is different from the exposed one will have a higher DEA value.

The following formula describes the NIVI within SNI for a piece of information *i* published in some Node within the network:

$$NIVI(i, Node) = p * \left({(IR(i, Node) * DEA(i))} \big/ {IC(i, Node)} \right) \tag{6}$$

NIVI describes the impact that information has on IA. Relevance and Deviation from stated behavior will increase the NIVI while a higher control about the information will allow us to minimize it.

4.1 DEA Analysis

DEA function handles the gap between EAs and IAs. Let us imagine a photo where someone is drunk with a couple of friends and one of them has uploaded it to a social network where all of them belong to or not. How many of that individual EAs are violated o new IAs defined? Information has to be decorated to get to this kind of guessing. Decorating can take different ways. For instance, pattern recognition could be used to guess whether his face is in the photo. Or someone could tag the photo including his name along with terms as his college, year, and so on. Inferred data generated could be overwhelming [15, 16].

Given the fact that DEA will depend on the semantic richness of different pieces of information, the volume of attributes that could be discovered should be managed to get a controlled domain of information to work with.

Let us denote *SR(i)* as the semantic relevance of a piece of information. From the SNI point of view, all the information in the network should maintain a low SR(i) value, this is, be as semantically irrelevant as possible so that the impact on IA be minimized. A photo from somebody taken 20 years ago without additional tagging will probably have a low SR. The same photo, tagged with college, year of the photo and class name will have a semantic value because relationship of somebody with the photo could be guessed and further analyzed, so its SR will be higher. Imagine the scenario. Whether that person lives in a city with only one college, he/she could have kept in secret his/her college name, but the second photo could help infer a good amount of IAs: College, age (approximated), even a possible set of known people (other persons in the same photo).

SR could be expressed as a function of all the relevant attributes that could be discovered.

$$SR(i) = f(a_{i1}, a_{i2}, \ldots, a_{in}) \quad (7)$$

Denoting SK as a function reflecting the knowledge about an individual in the social network in terms of EA, DEA could be defined in terms of SK and SR as:

$$DEA(i) = |\, SK - SR(i)| \quad (8)$$

Or more generally for n attributes.

$$DEA = |\, n * SK - \sum_{i=1}^{n} SR(i)| \quad (9)$$

4.2 Information Control

This parameter has to be analyzed in terms of social network metrics [2, 3]. IC depends on anyone's role in the network and the capability he is able to show so that information is managed. Lower values of ***Betweeness*** (people that can be reached indirectly from direct links) would increase our control of information, with a direct path to the information holders. Same would happen with ***Closeness***, which establishes a measure of minimum paths to reach there were the information is relevant. ***Eigenvector centrality (EC)*** measures the importance of anybody as nodes in the network by means of assignment of relative scores to all nodes depending on connections.

$$IC(i, Node) = \frac{f\big(Node, Betweeness(self), Closeness(self)\big)}{EC(self)} \quad (10)$$

IC will yield a higher value as our influence in the network increases, in reach, expanding our network value and in shortening the jumps to get where we wanted to. Finally, our node value in the network will normalize the control that can be exerted.

4.3 Information Relevance

This function ponders the effect of network architecture on the information. If the information is created by the most active and popular member in the network, its diffusion will have a greater impact about any other people than information generated on the limits of the net. For instance, it is much more relevant information communicated by Jesus in the Bible that the one coming from Jonah. Information relevance has to be analyzed in line with ***user network value*** (UNV) to establish the attention to be paid to it.

$$IR(i, Node) = f(i, UNV(Node)) \quad (11)$$

User network value (denoted as UNV) is a function which reflects an absolute value of a user in a network based on the participation and relevance in the same way as PageRank [17, 18] assigns a value to pages.

5 Conclusions

In this paper a framework has been introduced for analyzing social identity from two main standpoints: evolution of social network identity, to align what is effectively known about individuals and what it is known but should not be used in interactions, and network information management, this is, try to harness the impact derived from information existence in the social networks and the insights that could be achieved. Computer Social network are bringing new mechanisms for information spreading due to its natural correspondence to human social networks that are somehow affecting the knowledge or even the control of what is out there about anybody. This framework intends to help in tracking and managing the social network identity, our digital blueprint, without limiting the inherent capabilities of the media.

Future works will include the development of User Network Value function, so that information can be aligned with the people managing it, simulation of social network identity value generation and migration paths and a deeper analysis of network impact value for information.

Acknowledgments

This work has been performed in the context of the research project INGENIAS 2, TIN2005-08501-C03, which has been funded by Spanish Council on Research and Technology (with FEDER support).

References

1. Scott, J.P.: Social Network Analysis: A Handbook, 2nd edn. Sage Publications Ltd., Thousand Oaks (2000)
2. Wasserman, S., Faust, K.: Social Network Analysis: Methods and Applications. Cambridge University Press, Cambridge (1994)
3. De Jonghe, A.: Social Networks Around The World: How is Web 2.0 Changing Your Daily Life? BookSurge Publishing (2008)
4. Javier, G., et al.: Personalización: más allá del CRM y el marketing relacional. Pearson, London (2004)
5. Krebs, V.: The Social Life of Routers. Internet Protocol Journal 3, 14–25 (2000)
6. Valente, T.W.: Social network thresholds in the diffusion of innovations. Social Networks 18, 69–89 (1996)
7. Valente, T.W.: How to search a social network. Social Networks 27, 187–203 (2005)
8. Boyd, D.M., Ellison, N.B.: Social Network Sites: Definition, History, and Scholarship. Journal of Computer-Mediated Communication 13, 210–230 (2007)
9. Guimera, R., Sales-Pardo, M., Amaral, L.A.N.: Classes of complex networks defined by role-to-role connectivity profiles. Nature physics, 63–69 (January 2007)
10. Bygge, R.: The cost of anti-social networks: Identity, Agents and neo-luddites (2006), `http://www.firstmonday.org/issues/issue11_12/bigge/index.html`
11. Boyd, D.: Identity Production in a Networked Culture: Why Youth Heart MySpace. American Association for the Advancement of Science (2006), `http://www.danah.org/papers/AAAS2006.html`

12. Travers, Jeffrey, Milgram, S.: An Experimental Study of the Small World Problem. Sociometry 32(4), 425–443 (1969)
13. Proyecto OpenID.net (2008), `http://openid.net`
14. Ackerman, M.S., et al.: The Do-I-Care Agent: Effective Social Discovery and Filtering on the Web
15. Li, W.-S., Hara, Y., Fix, N., Candan, S., Hirata, K., Mukherjea, S.: Brokerage architecture for stock photo industry. In: 7th International Workshop on Research Issues in Data Engineering (RIDE 1997) High Performance Database Management for Large-Scale Applications, p. 91 (1997)
16. Pastra, K., Saggion, H., Wilks, Y.: Intelligent Indexing of Crime Scene Photographs. IEEE Intelligent Systems 18(1), 55–61 (2003)
17. Page, L., Brin, S., Motwani, R., Winograd, T.: The PageRank citation ranking: Bringing order to the web. Technical report, Stanford University, Stanford, CA (1998)
18. Brin, S., Page, L.: The anatomy of a large-scale hypertextual Web search engine. In: Proceedings of the Seventh International World Wide Web Conference (1998)

STRS: Social Network Based Recommender System for Tourism Enhanced with Trust

Fabian Bustos, Juan López, Vicente Julián, and Miguel Rebollo

Department of Information Systems and Computation
Technical University of Valencia
Camino de Vera s/n, 46022, Valencia, Spain
fbustos@dsic.upv.es, jlopez@dsic.upv.es, vinglada@dsic.upv.es, mrebollo@dsic.upv.es

Abstract. Nowadays, the development and use of Recommender Systems has grown significantly as a tool to counteract the overload information problem, by selecting a subset of personalized items from a huge set of information based on user preferences. On the other hand, Multi-Agent System applications have shown to be an important area where the Recommender System theory can be applied. This paper presents a tourism Recommender System architecture integrating Multi-Agent Technology and Social Network Analysis, applying trust concepts to create relevant and good quality personalized recommendations trying to solve the tourist Recommender Systems issues.

Keywords: Multi-agent Systems, Recommender Systems, Trust and Reputation.

1 Introduction

Software development using the agent paradigm, and more specifically Multi-Agent systems (MAS), is growing every day because of their reliability and great potential to provide useful solutions to daily problems in different areas, like information retrieval, education or tourism. Some of these MAS applications, can be often affected by an overload content problem, as in the specific case of tourism, where user needs are more difficult to satisfy because of their different lifestyles (for instance long or short trips), interests (focus on cultural tourism, leisure travel, or sacred travel) and motives (such as family travels, school travels, and business travels). Nevertheless, it is important to point that almost all tourists are looking for an experience, trying to get the most satisfaction from their time and money. As stated, tourists have different needs, demanding quality and also diversity.

According to this, those kinds of MAS applications need to integrate techniques that help to reduce the gap between users' expectations and their experiences. Therefore, the necessity for a Recommender System appears to fulfill tourist needs and expectations in a more accurate and personalized way.

Thus, a tourism Recommender System architecture called Social-Net Tourism Recommender System (STRS) is proposed. Its main goal is to process all the information related to a tourism service inside a city in a reasonable time, providing up-to-date, comprehensive and accurate information, all personalized according to the user profile. Moreover, this recommender has been improved including a model of trust

J.M. Corchado et al. (Eds.): DCAI 2008, ASC 50, pp. 71–79, 2009.
springerlink.com

and reputation to provide predictive and higher quality recommendations and, at the same time, acting as a filter to prevent spam content. This STRS is basically composed by the Multi-Agent Tourism System (MATS) [1], a previous work in the multi-agent systems and tourism information services area, enriched by means of the recommender systems technology.

The remainder of this paper is organized as follows: section 2 provides a general overview on recommender systems. Section 3 gives a general description of the multi-agent tourism system. Section 4 describes the recommender system and the trust model developed and applied inside the MATS. Section 5 describes the resulting Social-Net Tourism Recommender System; and finally section 6 shows some tests to measure the quality and efficiency of the Social-Net Tourism Recommender System.

2 Related Work

Recommender Systems emerged as a research field thanks to the development and posterior growing of the web. Experiences with the web have shown that there are many different types of users looking for information and entertainment. Because of this growing, users have faced with the need of some helpful mechanism to filter and customize all the information resources identifying the adequate content to satisfy their expectations from a potentially huge set of content.

There are some efforts done by different researchers and companies that cover the recommender system area extensively. Basically, a recommender system is defined as software that attempts to predict items that may fulfill user needs and expectations. In other words, recommender systems help users to identify items of interest. These recommendations are generally made in two ways: by calculating the similarity between items and recommending items related to those in which the user has expressed interest, or by calculating the similarity between users in the system and recommending items that are liked to similar users. Different examples of applications employing recommender system can be found in many different contexts as [2, 3], news [4, 5, 6], music [7, 8, 9, 10, 11], films [12], web sites [13] and e-commerce [14].

The success of the recommendation depends on two parameters: relevance and quality. Relevance refers to items recommended being similar or close to the user likes and needs. Quality is more difficult to define, but with quality we intent to not only provide the items relevant to the users but also the "best" items that are relevant to the user's preferences. The better the quality and relevance are, the greater the possibility of fulfilling user goal is. Here, it is important to count on mechanisms that help to avoid unnecessary content (accepting just high quality content) in terms of the recommendation. Including a trust mechanism within the recommender system will help to provide higher quality recommendations, obtained by getting the recommendation from the trustworthiness entities inside the system, making the recommendation more accurate and reliable, and improving the recommender system performance by filtering all the spam content. In addition, entities can share trust information about other entities inside the system, propagating their trust information and helping others to decide whose to accept recommendations and which content to accept.

Work has been reported to introduce trust into the Recommender Systems domain. In the FilmTrust website [15], the use of Semantic Web-base social networks augmented with trust is integrated into a movie recommender system, to create predictive movie recommendations. In the TREPPS [16] recommender system, authors present a recommender system based on the trust of social networks. They use the trust computing (done by means of two fuzzy logic applications - fuzzy inference system and fuzzy MCDM method) to appropriately assess the quality and the veracity of peer production services.

3 The Multi-Agent Tourism System

As commented before, The Multi-Agent Tourism System (MATS) is the result of a previous work done in the multi-agent systems and the tourism information services field. This is a MAS application that offers different services to users (tourist in a city), helping them to manage the time they will spend in the city in an efficient way. With this system, tourist can find information about places of personal interest like restaurants, movie theaters, museums, theatres and other places of personal interest like monuments, churches, beaches, parks, etc., according to their preferences using their handheld devices (like a mobile phone or a PDA). Once a specific place has been selected, tourist would be able to establish a process to make a reservation in a restaurant, buy tickets for a film, etc., in a given time period. The system also allows users to request a plan for a specific day in the city with a number of activities recommended according to their preferences, where the reservation of a restaurant(s) for lunch, dinner or both can be made. The implemented system has a reactive behavior responding to queries coming from the environment (users). Those queries correspond to some basic services as search about places of interest, reservation process or request a day planning service. However, these services are simple and can be improved with more sophisticated negotiation processes and the system can be extended including new services as a recommendation service.

4 Integrating Recommender System Technology Enhanced with Trust Aspect Inside the MATS

The aim of this work is to integrate Recommender System Technology to the Multi-Agent Tourism System commented below in order to improve the system with a new recommendation service. In addition to this service, a trust component has been added increasing the performance of the Recommender System in terms of accuracy, relevance and quality.

4.1 The Multi-Agent Social Recommender System

The proposed Recommender System approach integrates Multi-Agent technology with Social Network Analysis. This recommender system uses social networks analysis to model communities of users and attempts to identify all the relationships among them. Using these relationships between users, the recommender can determine which users are more related to the user who made the request for a recommendation and recommends the items that those related users like the most.

Five different types of agents basically form the Multi-Agent Social Recommender System (MARS) [17]: the User Agent, Data Agent, Recommender Agent, Social-Net Agent and the Group Agent.

- The User Agent is an interface between the user and the MARS. Through the User Agent, users ask for recommendations and provide the system with information related to their characteristics, likes and preferences. When new users join the system, they send a request to the recommender system to be accepted, with some information about them. This would be the basic profile.
- The Data Agent manages a database with user information, updating the database information and the profiles. The database has the cases and records of user's interaction within the MARS. The Data Agent is in charge of handling the database related to the items. Using each user's profile and the item database, the Data Agent finds the items that are relevant to them and stores that information on the database.
- The Recommender Agent receives all the queries of recommendation and user registration. If the Recommender Agent receives a new user registration query, it informs the Social-Net Agent and the Data Agent to include the user on the platform. The Recommender Agent provides the recommended item or items to the User Agent and receives the feedback from it with the evaluation of the recommendation made by the user. This feedback is used by the Data Agent and the Social-Net Agent to update their values.
- The Social-Net Agent adds a node onto the social network when a user joins the platform. Through the profile and user parameters, it determines the new user's similarity with regards to the other profiles. Each similarity value labels the link between two nodes in our non-directed valued graph. To do this, we need mathematical models that indicate how to compute similarity between user profiles.

  To estimate the value of the similarity between two users we use the Pearson Coefficient because it can determine negative correlations. The formula is given in formula (1). In this equation, *r(x,y)* represents the similarity between user *x* and user *y*, *n* is the number of overlapping tastes (i.e. kinds of restaurants in which user *x* and user *y* have common values).

$$r(x,y)=\frac{\sum_{i=1}^{n}(R_{x,i}-\bar{R}_x)(R_y-\bar{R}_y)}{\sum_{i=1}^{n}(R_{x,i}-\bar{R}_x)^2\sum_{i=1}^{n}(R_y-\bar{R}_y)^2} \qquad (1)$$

- The Group Agent is responsible for conducting an exhaustive analysis of the user data and establishes the creation of groups of users in our social network that have similar tastes and preferences. The main function of the Group Agent is to find the subgroup of nodes that are accessible (if there is at least one possible path between them) to the others, but not necessarily adjacent.

  Another important feature offered by this agent inside the recommender system, it's the possibility to make recommendations for group of users that have similar preferences.

Test have been made to this Recommender System showing that it can provide appropriate and good quality recommendations to users, but, at the same time, some weakness have been detected. For example, in order to determine users similarity and produce a good quality recommendation, The Group Agent need users involved in the recommendation process to have some items rated in common. So, when users do not have rated any similar items, the quality of the recommendation is low. The quality of the recommendation can also be affected by the 'cold-star' (new users entering the system with no items rated). However, we believe that adding a trust component inside the Recommender System will help to overcome all the problems commented above and also will improve the quality of the final recommendation. Next section will introduce a Trust and Reputation component and its integration inside the MARS is discussed.

4.2 Trust and Reputation Model

In order to overcome the weakness commented before, a Trust and Reputation component was designed to be integrated within the recommender system. This component will provide the Recommender System with some improvements as filtering unnecessary content or supplying higher quality recommendations. Moreover, entities can share trust information about other entities inside the system, propagating their trust information and helping others to decide who will accept recommendations and which content will accept. Also, this trust propagation will help to solve the 'cold-star' problem, because new users can infer trust relationships from others and can be easily integrated within the recommender system.

This section will discuss the Trust and Reputation component. As stated before, this component was designed using social network analysis. In general, the component has two basic but important modules: Social Network Design Module and Decision Module. Next subsection will give a detailed view of these two modules and their functionalities.

Social Network Design Module

This module is in charge to build and recalculate the network and its values each time after changes have been detected inside the system. For each node (new) that enters the system, this module is in charge to create it, initialize it inside the network and, at the same time, recalculate the values in the network. The module has a number of complex data set to save all the information related to each node inside the network, for example, name, items that likes the most, prestige level, general trust and trust by item (constantly recalculated inside this module). Here, it is important to clarify that the module create a different view of the social network for each single actor, and different subnetworks for each item or service that each actor provide inside the system.

Decision Module

Basically, the Decision Module is in charge to decide whom to interact with. It works based on the information obtained through direct experiences with others actors, as well as witness information (information obtained from the experiences of other actors). This values can be obtained from the data set contained inside the Social Network Design Module and are used for this module to find possible actors which are trustworthy enough to interact with. Also, this module is in charge to compute and find the prestigious agent, who will be asked first about the reputation of others. This prestigious agent is calculated using an important social network concept called prestige, defined by the formula (2):

$$P_D^{'}(n_i) = \frac{\sum_{i=1}^{g} d_I(n_i)}{g-1} \tag{2}$$

The higher prestige occurs when $P_D = 1$.

However, in our case, we need to consider also the weight of those links (representing the trust value from others about the agent who is being calculated its prestige) to find the real prestigious agent. A new concept called reputation agent prestige is introduced in formula (3) to ensure that the prestigious agent is the agent being considered trustworthiness by the others.

$$\mathrm{Re}\,p(n_k) = \frac{\sum_{i=1}^{g} d_I(n_i)}{g-1} \times \frac{w_{ik}}{(g-1)} = \frac{\sum_{i=1}^{g} d_I(n_i) \times w_{ik}}{(g-1)^2} \tag{3}$$

In the above equations, $d_I(n_i)$ correspond to the number of adjacent nodes to node n_i, w_{ik} is the trust value of adjacent nodes about the node being evaluated about its prestige and g is the total number of nodes. The *indegree* of the node n_i represent the numbers of links such as ($l_k = \langle n_i, n_j \rangle$), for all l_k in L, and all n_j in N.

This trust component is going to be included inside the User Agent in the Recommender System, so it can evaluate and state the trustworthiness of others and decide if the content provided by certain user is relevant to be accepted or not. This component will also help to predict and spread trust of other users giving more information and knowledge to the Recommender System.

5 Social-Net Tourism Recommender System

According to the models and features described before, the resulting system integrating those new functionalities has been called Social-Net Tourism Recommender System (STRS). The STRS is basically formed by two subsystems that collaborate to provide up-to-date and accurate tourism recommendation: the MATS and the MASR. The STRS uses social networks to model communities of users and attempts to identify the relations among them. Using the relation between the users on each different subject (restaurant likes, museum likes, theatre likes), the STRS identify which users are more similar to the user that wants to get a recommendation and recommend the items the users like the most.

The recommender process is described as follows. User Agent will be the interface agent between the user and the STRS. By means the User Agent, the user will send the requests of the recommendation and will give the information related to the user characteristics, likes and preferences.

The Recommender Agent receives the request for the recommendation from the User Agent, deals with it and resends it to the Social-Net Agent, asking for the identifiers of the users that are the most similar to the one that has requested the service. The Social-Net Agent receives the request with the identifier, locates the node corresponding to the user who asks for a recommendation on the social network and evaluates which nodes have the highest positive correlation value. The Social-Net Agent sends back a list to the Recommender Agent. After this, the Recommender Agent sends this list to the Data Agent. The Data Agent takes each one of the user identifiers and, for each one of them, recovers the items that have the highest likeness from the database, making a list and giving it back to the Recommender Agent, the Recommender Agent sends the list to the Broker Agent to check if the sights to recommend are available. The Recommender Agent sends the highest ranked available sights (this list also contains the id of the agent who made the recommendation about the site) from the definitive list to the User Agent as the result of the initial request. With this list, the user agent can use its trust values about others to decide which content to accept.

6 Test and Results

In order to evaluate the behavior of the STRS, it was necessary to determine some tests to establish it's scalability over time and the quality of the recommendations as the number of users grows, and as the user interacts with the system. The tests have been done using a single platform running on a dedicated computer, running the MATS (Broker Agent, Sight Agents and Plan Agent), and in another dedicated computer was executed the MASR (Recommender Agent, Data Agent and Social-Net Agent and several User Agents). A sight type has been selected to analyze the behavior of the recommender system wanting to evaluate the quality of the recommendation itself, but it has to be clear that user's likes and dislikes are different depending on the type of the sight. The sights selected for the test are the restaurants which can be considered a good representative because for a tourist is typically difficult to find a good restaurant due to their quantity and the variety of restaurants in a city.

The quality of the recommendation was measured using some tourist-agents to simulate different queries and evaluate the final recommendation. Each tourist-agent has a profile with different topics; each topic corresponds to a type of food, assigning a random value from 0 to 10 for each topic. When the tourist-agent receives the recommendation, it will evaluate the recommendation made by the STRS applying a standardized evaluation of each one of the items recommended against the values of each topic. The feedback will be the average value of the evaluation of each recommended item. Figure 1a shows the increase of the average quality value over time (iterations), with a number of fifty tourist-agents.

This Figure shows three lines, the dotted one corresponds to the quality response over time for the evaluation of the recommendations made by the MARS; the lighter one corresponds to the items recommended by the MATS without the MARS, and finally, the darker one corresponds to the evaluation of the recommendations made by the MATS adding the trust component. As can be seen, the quality of the recommendation without the MARS will be lower than the recommendation obtained with the MARS. Moreover, the quality of the recommendation improves over time (each iteration), due to the interaction of the tourist-agents with the system. According to this, when the user interacts with the application (asking for recommendation, providing feedback, etc.) provides more information and knowledge about his likes and dislikes and the system has a more accurate description of the user characteristics. Also, as the system evolves, the trust calculations are more accurate (because the trust model has more specific information about others to model their behavior and know their real tastes and expertise) to filter unnecessary content from the recommendation, improving the quality and relevance of the final recommendation accepted by the interested user.

Another important issue to be analyzed is the cost for the STRS to provide the recommendation. The variation of the average response time over the number of users is shown in Figure 1b. The average response time increases due to the increase in the number of users. Obviously, when the number of users increases, the data used for the recommendation also increases. Thus, the response time for a recommendation is acceptable because the user does not have to wait for a long time for a recommendation. The reason for this is that the similarity calculations for the social network are made offline and again, the trust knowledge about others helps to maintain the average response time in an acceptable range of values, causing prestigious agents be detected with a higher certainty degree.

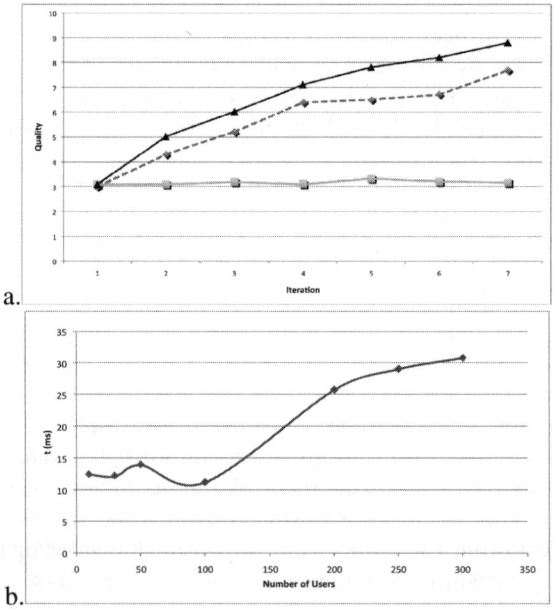

Fig. 1. (a) Recommendation Evaluation. (b) Average response time for recommendation.

References

1. Lopez, J.S., Bustos, F.A., Julian, V.: Tourism services using agent technology: A multi-agent approach. INFOCOMP Journal of Computer Science, 51–57 (2007) Special edn.
2. Amazon Auctions, `http://auctions.amazon.com`
3. Woodruff, A., Gossweiler, R., Pitkow, J., et al.: Enhancing a digital book with a reading recommender. In: Proceedings of the SIGCHI conference on Human factors in computing systems, pp. 153–160 (2000)

4. Bomhardt, C.: NewsRec, a SVM-driven Personal Recommendation System for News Websites. In: Proceedings of the IEEE/WIC/ACM international Conference on Web intelligence, pp. 545–548. IEEE Computer Society, Los Alamitos (2004)
5. Lee, H.J., Park, S.J.: MONERS: A news recommender for the mobile web. Expert Systems with Applications 32, 143–150 (2007)
6. Lang, K.: NewsWeeder: Learning to filter netnews. In: Machine Learning: Proceedings of the Twelfth International Conference (1995)
7. Resnick, P., et al.: GroupLens: An Open Architecture for Collaborative Filtering of Netnews. Internal Research Report, MIT Center for Coordination Science (1994)
8. CDNow, `http://www.cdnow.com`
9. MyStrands, `http://www.mystrands.com/`
10. LaunchCast, `http://music.yahoo.com/launchcast/`
11. Chen, H., Chen, A.L.: A music recommendation system based on music data grouping and user interests. In: Proceedings of the Tenth international Conference on information and Knowledge Management (2001)
12. Mediaunbound, `http://www.mediaunbound.com`
13. MovieLens, `http://www.movielens.umn.edu`
14. Pazzani, M., Muramatsu, J., Billsus, D.: Syskill & Webert: Identifying interesting Web Sites. In: Proceedings of Thirteenth National Conference on Artificial Intelligence, pp. 54–61 (1996)
15. Golbeck, J.: Generating predictive movie recommendations from trust in social networks. In: Proceedings of the fourth international conference on trust management (2006)
16. Li, Y.M., Kao, C.P.: TREPPS: A Trust-based Recommender System for Peer Production Services. Expert Systems with Applications (2008)
17. Lopez, J.S., Bustos, F.A., Rebollo, M., et al.: MultiAgent Recommender System: A Collaborative Social Network Model. In: IWPAAMS 2007, pp. 159–168 (2007)

An Agent-Based Simulation Tool to Support Work Teams Formation

Juan Martínez-Miranda and Juan Pavón

Dep. Ingeniería del Software e Inteligencia Artificial, Universidad Complutense de Madrid.
Ciudad Universitaria s/n, 28040 Madrid
jmartinez@microart.cat, jpavon@fdi.ucm.es

Abstract. The team configuration process is typically performed by a manager based on past experience and available (though frequently scarce, uncertain and dynamic) information about the cognitive, personal and social characteristics of the potential team members. To support this decision-making process we propose an agent-based model where a virtual team can be configured using the characteristics of the real candidates to form the team, and given a set of tasks, the model generates the possible performance of the team-members. The model is based on Fuzzy Set Theory and Fuzzy Logic and it has been validated with an industrial project involving 23 team members and 23 tasks. The architecture of the model and the initial results are presented in this paper.

Keywords: Agent-Based Simulation, Human Performance Modelling, Emotional Models, Social Behaviour, Work Team Formation.

1 Introduction

Modelling of human performance has been an area of research since early 70's [7], given its applicability to a wide range of areas such as testing and evaluation (e.g. of user interfaces [3]), training (e.g. safety-critical domains, where a wrong action in the real world may have catastrophic results [4]), and operation analysis. Although the achievement of a good model involving human behaviour is still a great challenge (due to the "instability", unpredictability and the ability to perform independent actions that characterise the human behaviour), in the last years some techniques and models are emerging mainly within the military and social science domains, which clearly indicate that some valid modelling of human behaviour is possible [19].

This kind of models can be usefully applied for the analysis and study of the influence of human behaviour within work teams. The correct configuration of a work team is not a trivial task because it should take into consideration not only the technical competence and the availability of the possible team members, but also personal, psychological and social aspects of the people. The influence of these human characteristics at work has been studied since different research areas such as Psychology, [1], Sociology [21], Human Resources Management [20], and Organisational Behaviour [14].

This paper presents ongoing research work towards the simulation of human behaviour within a work team. The model represents a set of human characteristics as the internal state of individual software agents. Each agent represents a real person

J.M. Corchado et al. (Eds.): DCAI 2008, ASC 50, pp. 80–89, 2009.
springerlink.com

with its particular internal state, and the resulting behaviour of the agent is a consequence of the simulated interaction with its team-mates and with its assigned tasks. The ultimate goal of this research is the development of a software simulation tool that can assist project managers during the decision-making process for the configuration of work teams.

2 Related Work

Computer simulations to analyse and understand complex phenomena have been well applied in several research domains including Political Sciences [22], Economics [16], Social Sciences [13] and Environmental Sciences [11]. The success in the use of simulations within the mentioned research areas relies on the feasibility to *play* with the behaviour of the modelled phenomenon under study by changing the conditions of its environment and/or its internal parameters to observe the consequences in a controlled experiment. Note that, most of the times, such task would generate a high cost or would be even impossible to realise in real life.

The study of human behaviour has been one of these complex phenomena that in the last years have taken advantage of the use of computer simulations and specifically of agent-based simulations. The agent-based approach enhances the potential of computer simulation as a tool for theorising about social scientific issues, since it facilitates the modelling of artificial societies of autonomous intelligent agents [5]. In [18] an agent-based model to represent human behaviour is presented where the agents can be characterised by a set of attributes and their behaviour can be described by a set of simple decision heuristics. The work presented in [9] uses intelligent agents to represent human behaviour of military teamwork by dividing the agents behaviour into three categories: *leadership*, *individual* and *team-enabling*.

When the model includes the representation of qualitative human characteristics (such as levels of motivation, emotional states, personality trends, among some others) it is difficult to accept that a numeric value could be used to measure any of these characteristics. One appropriate method of solution is the use of fuzzy logic. A fuzzy emotional model for decision-making applied to a mobile robot has been described in [8]. This model deals with three negative emotions: *fear*, *pain* and *anger*. Another related work is presented in [12], where fuzzy agents with dynamic personalities are proposed: fuzzy sets are defined for personality traits and facets and the concise representation of personality knowledge is processed through fuzzy rules.

3 Description of the TEAKS Model

The agent based model for work team analysis, which is called TEAKS (TEAm Knowledge-based Structuring) is described in the following subsections.

3.1 General Architecture

The general architecture of a TEAKS model is shown in Fig. 1. A TEAKS model consists of a set of agents, representing each of the persons involved in a project, and a project configuration that determines their tasks and relationships.

Work team initial configuration. In a first step, the internal state of each agent is configured using the personal characteristics of the people who may integrate the work team. Also, the characteristics of the project that the work team will develop are configured at this step by the Project Manager. The relationship between the agents and the project is set up by the assignment of each task of the project to one or more agents. This entire configuration will be the input to the next step where the simulations take place.

Simulation of work teams. This process is based on the simulation of the interaction among the agents and the project to get the estimation of the work team performance. The interaction among the team members and its assigned tasks generate the agent behaviour. This behaviour determines the performance of each team member. During

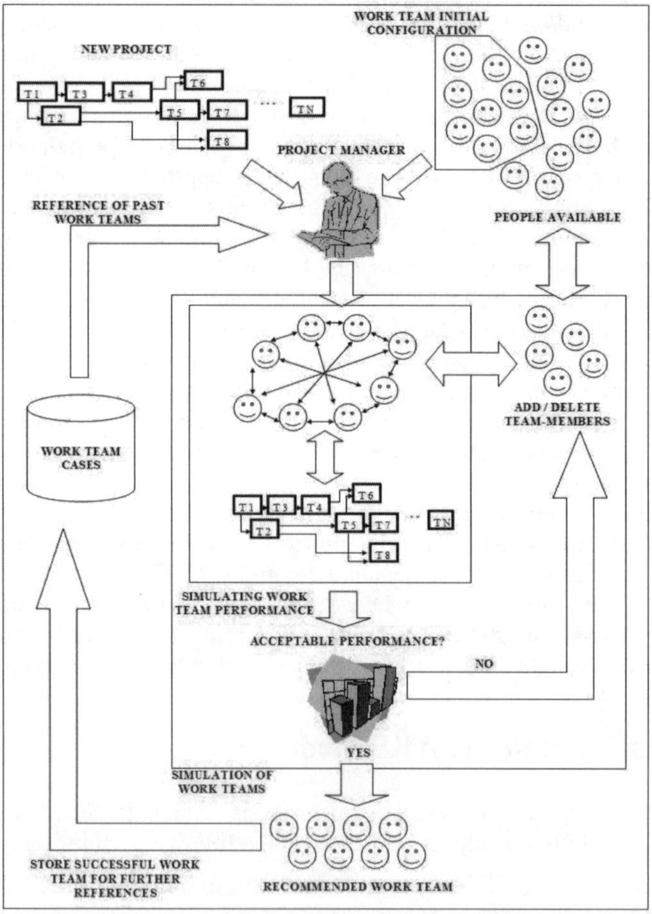

Fig. 1. TEAKS Architecture

this step, the configuration of the team can be changed to analyse if the introduced change(s) improves the global performance of the work team. That means that new agents with different internal state can be added to the team, some can be removed and/or new reassignments of the project's tasks can be made. Once these modifications are introduced, new simulations can be executed to analyse the consequences and observe which team presents a better performance.

Proposed work team final configuration. The team with the best performance is taken into account in order to get the information related to each team-member performance, which will be used by the project manager for the configuration of the real work team.

3.2 Internal Models of the Agents

The TEAKS model considers a set of human characteristics to observe how the possible individual behaviour within the work team can be obtained from the combined values in this agent's internal state, which is affected by the interaction with the internal state of the other agents and by the characteristics of the assigned task(s). In its current version, the model of the agent's internal state incorporates the following human characteristics: personality trends, a set of basic emotions, social characteristics and cognitive capabilities.

Personality trends. During the last years, several studies have shown the importance of personality traits in job performance [15]. We have selected a set of four personality traits to model as part of the internal state of the agents taken from [2]: ***amiable:*** steadiness, co-operating with others to carry out the tasks; ***expressive:*** influencing others, forming alliances to accomplish results; ***analytical:*** compliance, working with existing circumstances to promote quality in products and services; and ***driver:*** dominance, shaping the environment by overcoming opposition to accomplish the tasks.

Emotional State. Emotions have been proven to be an important factor at work place [10], and the critical decision is to select the set of basic emotions to include it into the internal state of the agents. From the extended psychology classification of the basic emotions, we have selected a set of four basic emotions to model the agents' emotional state at work. Two of them are positive emotions and the other two have a negative influence over people's performance: positive emotions: desire and interest of the person to develop a specific task in a given moment. Negative emotions: disgust and anxiety generated by a given task in specific moment.

We made this selection given the context of application and thinking in the most common emotions produced by the activities of a person at work. In an industrial project it is more common that one specific task produces the *interest* or *desire* (in developing that task) positive emotions in a worker than the *happiness* or *joy* emotions (most commonly identified during personal life situations). On the other hand, the negative emotions *disgust* and *anxiety* produced by specific work activities are more common than *fear*, *pain* or *sadness*, even though that there are some special circumstances through the development work activities that can produce these emotions (such as dangerous or high risk tasks), but currently we concentrate on projects where these type of tasks do not frequently appear.

Social Characteristics. In any group of people, human relations are important to achieve a good communication and co-ordination among the group members. The characteristics of human relations in groups and teams (such as competence, trust, co-operation, among others) are the main topic of research in areas such as social psychology, social sciences and organisational behaviour. Modelling all these characteristics is out of the scope of this research, but nevertheless, we considered a small set of social characteristics: ***introverted / extroverted*:** one person can be predominantly concerned with his/her own thoughts more that the social relations with the others (introverted) or he/she can be an outgoing, socially confident person (extroverted).

Cognitive Capabilities. Quite often cognitive capabilities and intelligence are mainly considered to select a person for a job and there are several tests to measure a person IQ. Human cognitive capabilities involve several brain processes such as learning, reasoning and memory among others. Modelling these brain processes and their interactions to generate an intelligent behaviour has been one of the main goals of Artificial Intelligence and it is, as well as human social characteristics, out of the scope of this research. The cognitive capabilities of a person were taken into account in the model by considering the personal degree of expertise in a particular domain. Thus, to represent the technical knowledge of a person, five parameters to describe the role of the person within the team were identified: ***Project Manager:*** a person in charge of managing the project and assigning tasks to the other team-members; ***Co-ordinator:*** the person in charge of specialised tasks, re-configuration of tasks and re-location of resources; ***Specialist:*** the person in charge of complex and specialised tasks; ***Technician:*** the person who can deal with technical and no-specialised tasks; and ***Assistant:*** the person in charge of not complex, routine and repetitive tasks. In addition of each person's role, each team-member has two other independent-role parameters to represent his/her cognitive capabilities: ***level of experience*** and ***level of creativity***.

3.3 Work Performance Parameters

The resulting team members' performance is calculated by taking into account its internal characteristics, the interaction with its assigned task and the interaction with the rest of team-members. We have considered five parameters to evaluate the expected final performance of each team member: ***goals achievement***, ***timeliness*** and ***quality*** of the performed task; ***level of collaboration***, and ***level of required supervision*** during each one of the assigned tasks.

The Project. The other important aspect to be considered in the TEAKS model is the project that the team has to work on. The PMBOK definition of a project is used to model the distinctive and common characteristics of the majority of industrial projects: *"a project is a temporary endeavour undertaken to create a unique product or service"* [6]. These kinds of projects are divided into *finite* and *unique* tasks assigned to one or more members of the work team. For the TEAKS model, a project is represented by its division into tasks and every task is represented by the following parameters: ***number of participants*** in the task, ***estimated duration*** (measured in days),

sequence (two or more tasks could be executed sequentially or in parallel), ***priority within the project, deadline, cost, quality, application domain, task description, level of difficulty*** and ***type of task*** (required specialisation level).

4 Getting the Agent's Behaviour

The behaviour of every agent modelled in the virtual team (i.e. the simulation of his/her performance within the project individually and as a team-member), is generated by the combination of above mentioned human characteristics. Given the difficulty to accept numerical values could be used to measure the intensity of either human emotion, experience or creativity level, we use fuzzy values to assess the parameters in the agent's internal state and two of the tasks parameters (difficulty and type). Using these fuzzy values and a set of pre-defined fuzzy rules, the simulated individual behaviour of each one of the virtual team members is obtained in a three-step cyclical process represented in the Fig. 2.

In the first step, the initial values in the emotional parameters of the agents are modified influenced by the characteristics of the task assigned to the agent, the personality of its team-mates and its own personality. In the second step, each emotional parameter of the agent is defuzzified to introduce random values in the model. This randomness represents the non-deterministic nature of human emotions: given the same person under similar circumstances, his/her reaction in front of these circumstances will not always be exactly the same. During the third step, the crisp values in the emotional state of the agent are fuzzified again to generate the fuzzy values in each one of the agent's performance parameters.

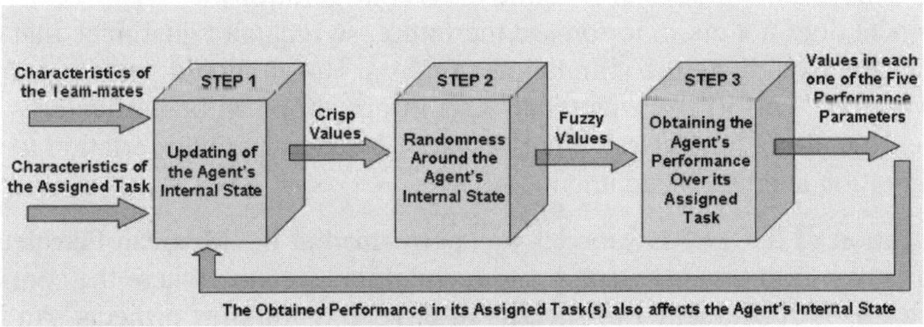

Fig. 2. Getting the agent's behaviour process

Finally, the obtained agent's performance over its assigned task, updates the emotional state of the agent again representing how this performance affects its emotional state. An example of the fuzzy rules used during this process is as follows (where T1 represents a task and A1 an agent):

IF T1 is assigned to A1 AND T1 currently has a *high delay* AND A1 has a *driver* personality with *high intensity* AND … THEN

The *desire* emotion of A1 has a *high increase*,
The *interest* emotion of A1 has a *high increase*,

The *disgust* emotion of A1 *maintains the same value*,
The *anxiety* emotion of A1 has a *low increase*

IF A1 has a *high creativity level*; A1 has a *driver* personality with *high intensity* AND T1 requires a *high specialisation level* THEN

The *goals satisfaction* of A1 in T1 is *normal*
The *timeliness* of T1 has a *medium advance*
The obtained *quality* on T1 has a *medium increase*
The *team collaboration level* of A1 is *normal*
The *individual contribution* of A1 to the task has a *medium increase*
The *required supervision level* of A1 is *normal.*

5 Validation of the Model

The TEAKS model has been implemented using the JADE Multi-Agent framework (http://jade.tilab.com/), JESS (http://herzberg.ca.sandia.gov/jess/), and FuzzyJess (http://www.iit.nrc.ca/IR_public/fuzzy/fuzzyJToolkit2.html) for the reasoning mechanism of every agent. This prototype was implemented with the following assumptions and limitations:

1. The software agents do not work to solve any real project but with the only purpose of simulating their interactions with other agents and with their tasks.
2. A plausible set of global behaviours of a team is obtained by averaging its behaviour over a statistically significant number of simulations.
3. The most suitable team configuration can be obtained by comparing the sets of global behaviours for several possible team configurations.
4. The model does not claim to foresee the future, so it cannot guarantee that the team will behave exactly as the simulation suggests. But it should provide information about possible performance patterns. This information can be particularly useful in the identification of undesirable performance patterns and their relation to the team configuration and task assignment.

A validation of the TEAKS model was performed at the Mexican Petroleum Institute (IMP, www.imp.mx), a research and technological centre where the configuration of work teams is fundamental to success in large and complex projects. An Information Technology project was selected for this validation due to the availability of the project manager, the accessibility to confidential documents that evaluate the performance of some of the people involved in the project and the recent finalisation of that project. In particular, this case study was a project on the development and implementation of a Geographical Information System (GIS) for internal use at the IMP.

This project was divided into 23 tasks and assigned (coincidentally) to 23 team-members (1 project manager, 1 coordinator, 15 specialists and 6 technicians). The results generated by our simulation tool were compared with the evaluation reports provided by the project manager. One of the main problems during the validation was that not all the people involved in the project permitted the access to his/her performance evaluation forms and only the final performance of 14 people could be compared with the results obtained from the simulation.

The comparison results between the real and the simulated work team was better when comparing the *goals achievement* and the *level of required supervision* performance parameters where the coincidence in the obtained values was present in 6 of the 14 people and in 5 of the 14 people respectively. In addition, for the rest of the people where the results were not coincident, the differences in the values did not exceed more than one unit of measure in for each team-member.

In the rest of the performance parameters (*quality of the developed tasks*, *timeliness of the developed tasks*, *level of collaboration during the duration of the project* plus an extra-parameter used in the IMP's projects and configured in the model: *level of individual contribution to the project*) the differences between the virtual and the real team exceeded the two units of measure for the 10% of the evaluated team-members and in one unit of measure for the 80% of the team-members.

Additionally, the project manager of the real work team performed an analysis of the results from the simulation. His evaluation and general comments were satisfactory and positive. In his own words: "*given the* (above-described) *limitations of the tool, the results are interesting and resemble a part of the reality*" and he also argued that this software could be useful for some specific type of projects, such as those projects where the project manager has never worked with the possible team-members before, and those projects where a strong personal interaction between the team-members is fundamental for the success of the project (e.g. projects of deep-water exploration or projects with tasks at the petroleum platforms, where a group of people works together for long periods of time without any contact to the outer world).

6 Conclusions and Future Work

Although the results from the IMP case study lead us to think that the proposed model can really be a useful tool for the work teams formation, we are conscious about the limitations of the model. Our next steps in this research are directed towards the enhancement of the model in two directions: by one hand, we are studying how to integrate into the model the effect of previous interaction (e.g. previous experience of working together) between the team-members in their performance on a new project and, on the other hand, we are also analysing different possibilities to make the model completely extensible to different types of projects thinking that in some projects the parameters related with the emotional states of the people are more determinant for the success of the project than e.g. the social characteristics and vice-versa.

The modelling of previous interaction between the team-members will be addressed by inserting the concept of trust and/or reputation as an additional parameter in the relationship between the team-members. In this direction, we are considering some model of trust such the one presented in [17] where the level of trust between people evolves depending on the good or bad obtained results in the task or activity in which the people is working together.

Acknowledgments. This work has been developed with support of the program "Grupos UCM-Comunidad de Madrid" with grant CCG07-UCM/TIC-2765, and the project TIN2005-08501-C03-01, funded by the Spanish Council for Science and Technology.

References

1. Bechtoldt, M.N., Welk, C., Hartig, J., Zapf, D.: Main and moderating effects of self-control, organizational justice, and emotional labour on counterproductive behaviour at work. European Journal of Work and Organizational Psychology 16(4), 479–500 (2007)
2. Biegler, L.T., Grossmann, I.E., Westerberg, A.: Systematic Methods of Chemical Process Design. The Physical and Chemical Engineering Sciences ch. I, 1–21 (1997)
3. Bonnie, J.E., Dario, S.D.: Multi-Purpose Prototypes for Assessing User Interfaces. Pervasive Computing Systems. IEEE Pervasive Computing 4(4), 27–34 (2005)
4. Bruzzone, A.G., Figini, F.: Modelling Human Behaviour in Chemical Facilities and Oil Platforms. In: Proceedings of Summer Computer Simulation Conference SCSC 2004, pp. 538–542 (2004)
5. Conte, R., Gilbert, N., Sichman, J.S.: MAS and Social Simulation: A Suitable Commitment. In: Sichman, J.S., Conte, R., Gilbert, N. (eds.) MABS 1998. LNCS (LNAI), vol. 1534, pp. 1–9. Springer, Heidelberg (1998)
6. Duncan, W.R.: A Guide to the Project Management Body of Knowledge (PMBOK). Project Management Institute (1996)
7. Dutton, J.M., Starbuck, W.H.: Computer Simulation of Human Behaviour. John Willey, New York (1971)
8. El-Nasr, M.S., Yen, J., Ioerger, T.R.: FLAME – Fuzzy Logic Adaptive Model of Emotions. Autonomous Agents and Multi-Agent Systems 3, 219–257 (2000)
9. Ellis, J.E., Martin, M.W.: Human Behavior Representation of Military Teamwork. Master's Thesis, Naval Postgraduate School (2006)
10. Fisher, C.D.: Mood and emotions while working: missing pieces of job satisfaction? Journal of Organisational Behaviour. 21(2), 185–202 (2000)
11. Gernaey, K.V., van Loosdrecht, M.C.M., Henze, M., Lind, M., Jørgensen, S.B.: Activated sludge wastewater treatment plant modeling and simulation: state of the art. Environmental Modelling and Software 9(9), 763–783 (2004)
12. Ghasem-Aghaee, N., Ören, T.I.: Towards Fuzzy Agents with Dynamic Personality for Human Behaviour Simulation. In: Proceedings of the 2003 Summer Computer Simulation Conference, pp. 3–10 (2003)
13. Gilbert, N., Troitzsch, K.G.: Simulation for the Social Scientist. Open University Press (2005)
14. Grant, A.M., Campbell, E.M., Chen, G., Cottone, K., Lapedis, D., Lee, K.: Impact and the art of motivation maintenance: The effects of contact with beneficiaries on persistence behavior. Organizational Behavior and Human Decision Processes 103(1), 53–67 (2007)
15. Heller, D., Judge, T.A., Watson, D.: The confounding role of personality and trait affectivity in the relationship between job and life satisfaction. Journal of Organisational Behaviour 23(7), 815–835 (2002)
16. Marks, R.E.: Validating Simulation Models: A General Framework and Four Applied Examples. Computational Economics 30(3), 265–290 (2007)
17. Martínez-Miranda, J., Jung, B., Payr, S., Petta, P.: The Intermediary Agent's Brain: Supporting Learning to Collaborate at the Inter-Personal Level. In: Proceedings of the 7th Conference on Autonomous Agents and Multiagent Systems AAMAS (to appear, 2008)
18. Pahl-Wost, C., Ebenhöh, E.: Heuristics to characterise human behaviour in agent based models. In: iEMSs 2004 International Congress: Complexity and Integrated Resources Management, Germany (June 2004)

19. Richard, W.P., Mavor, A.S.: Human Behaviour Representation, Military Requirements and Current Models. In: Modelling Human and Organisational Behaviours. National Academy Press (1998)
20. Sturges, J.: All in a day's work? Career self-management and the management of the boundary between work and non-work. Human Resource Management Journal 18(2), 118–134 (2008)
21. Syed, J.: From Transgression to Suppression: Implications of Moral Values and Societal Norms on Emotional Labour. Gender, Work & Organization 15(2), 182–201 (2008)
22. Yamakage, S., Hoshiro, H., Mitsutsuji, K., Sakamoto, T., Suzuki, K., Yamamoto, K.: Political Science and Multi-Agent Simulation: Affinities, Examples and Possibilities. In: Terano, T., et al. (eds.) Agent-Based Approaches in Economic and Social Complex Systems IV, pp. 165–182 (2007)

A Decentralized Model for Self-managed Web Services Applications

José Mª Fernández de Alba, Carlos Rodríguez, Damiano Spina, Juan Pavón[1], and Francisco J. Garijo[2]

[1] Dep. Ingeniería del Software e Inteligencia Artificial, Universidad Complutense de Madrid
Facultad de Informática, Ciudad Universitaria s/n, 28040, Madrid, Spain
jpavon@fdi.ucm.es
[2] Telefónica Investigación y Desarrollo
Emilio Vargas, 6, 28043, Madrid, Spain
fgarijo@tid.es

Abstract. Self-management in distributed systems is a way to cope with the growing complexity of these ones today, and its support in existing systems requires a transformation in their architectures. This work presents a decentralized model for the implementation of self-management capabilities, which also has the advantage of avoiding the single point of failure (SPOF) issue, providing more robustness to the management system. The proposed architecture has been validated in a real distributed application.

Keywords: self-management, autonomic computing, multi-agent system.

1 Introduction

Autonomic Computing is a concept initially developed by IBM [1], with the intention to cope with the increasing complexity of systems, as they grow in the number of elements and information. This solution intends to automate many of the functions associated with computing. In concrete, as [1] specifies, a computing system has the autonomic capability if it can manage itself only with the high-level objectives from administrators. The goal of such *self-management* ability is to free system administrators from the details of system operation and maintenance while providing users with a high-performance service.

The four main aspects of self-management are: self-configuration (automatic seamless configuration parameter adjustment), self-optimization (automatic performance tuning), self-healing (automatic detection and reparation of problems) and self-protection (automatic prevention from attacks and cascading errors).

IBM proposed a layered architecture [11] in which the upper layers contain the Autonomic Managers (AMs), and the lowest layer is populated by the managed resources. The management interfaces of these resources are encapsulated as service endpoints, so that they can be accessed via an enterprise communication technology, like Web Services. The recommended management model is Web Service Distributed Management (WSDM) standard [3]. The AMs in the control layer are cooperating agents [5], which achieve their management goals following high-level policies. They share a knowledge source, which provide a common domain model and the high-level information.

J.M. Corchado et al. (Eds.): DCAI 2008, ASC 50, pp. 90–98, 2009.
springerlink.com

The WSDM specification [3] enables management-related interoperability among components from different systems and facilitates integration of new ones, improving scalability. It also provides mechanisms for proactively analyzing different component properties such as quality of service, latency, availability, etc. In [14] is described an implementation example which is based on the IBM approach using a centralized architecture with a common Knowledge Repository.

Other self-management architectures like RISE [12] are domain-specific. They focus on particular system aspects such as: image management [12], workflow adaptation [13] and pervasive computing [15].

The work presented in this paper proposes a framework for incorporating self-management capabilities into Web Services applications based on WSDM model. It provides Web Services and Web Applications with autonomous features such as fault diagnosis, dynamic rebinding, file restoring, and resource substitution.

The approach consists on making each WS component of the system Self-Managed. Instead of having a common Knowledge Repository, which is often a Single Point Of Failure (SPOF), each self-managed Component has self-knowledge about its own dependences, and social knowledge about their dependent components. The aim of the paper is to describe the proposed approach, which is illustrated with a working example of self-healing. The validation framework is based on a website supporting a distributed social network for artists.

The paper begins with the architectural approach presented in section 2. A more detailed description of this architecture is shown in section 3, focusing the planning model in section 4. The case study and the validation of the proposed approach are in section 5. Finally, a summary of the work done and future work are discussed in the conclusions at section 6.

2 Approach for Enabling Self-management Features

The proposed approach focus on distributed systems based upon Web Services technology. These systems could be transformed into self-management systems by applying a self-management framework to each component. The basic idea is to make each system component (WS) self-managed, by enhancing them with new management components implementing self-management capabilities. Then make the self-managed components

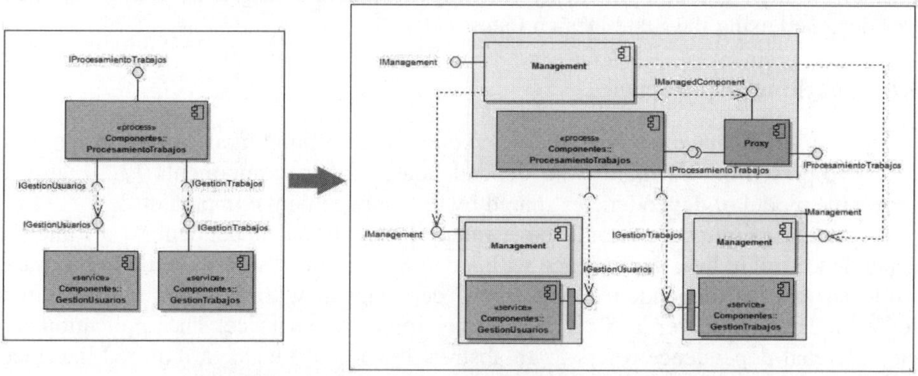

Fig. 1. Transforming a system into a self-managed system

cooperate in order to make the overall system self-managed. Figure 1, gives an example of transformation based on the studied case.

Each component has internal parts like files, libraries, etc., and possibly dependences with other components and servers. These components will be monitored, analyzed and controlled to provide self-management capabilities for each component and the whole system.

3 Self-managed Architecture

Figure 2 illustrates the internal structure of a service component. The "Component" is the original component that provides the logical operation of the service. The "Management" and "ProxyOfControl" components implement the management capabilities and are added to build the "NewComponent", which now has the original logical operation and self-management capability.

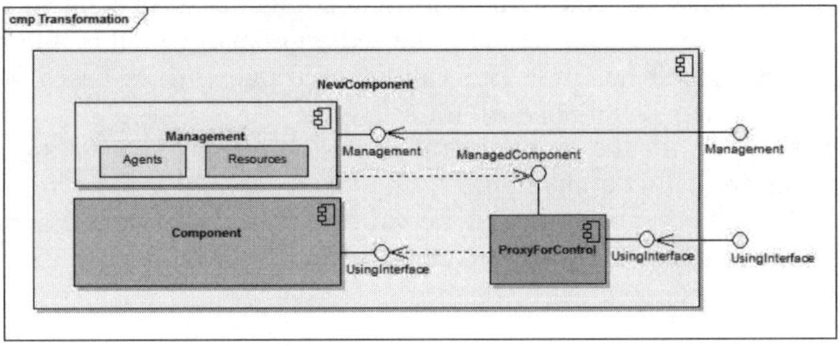

Fig. 2. Self-management component Architecture

The Management Component is made of packaged agents, resources and a model, which will be described later. The Management Interface offers operations to others self-managed components, which might use them to know its operational status.

The "ProxyOfControl" component controls the access to the managed component, avoiding possible misuses in inappropriate states, and providing information about the state of the managed component by catching Technical Exceptions. This component was designed using the State Design Pattern [10].

3.1 Modelling Dependencies

Achieving self-management capabilities require a conceptual model of the domain involved representing explicitly the dependencies among components [2]. Figure 3 shows the model of dependencies shared by the management components.

A managed component could have internal or external dependences: an internal dependence might be a dependence with a computing entity such as a file or a library, while an external dependence might be a dependence with a server, e.g. an email server, a database server, a file server or any application service. The application service external dependence refers to an abstract service, it means, a required interface, which is resolved at runtime.

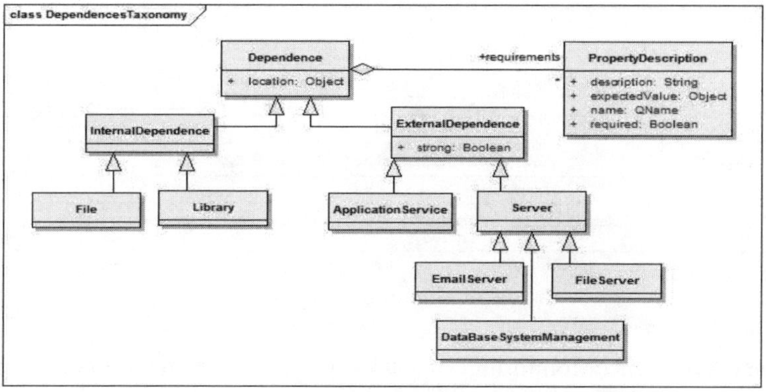

Fig. 3. Dependences

All dependences have a location. The location is an attribute indicating how to access the component that supplies the dependence, for usage or monitoring purpose. For application services, the location refers to the required interface, and the registry service location to find out a particular service to supply the dependence.

The dependence also has a set of requirements that define the properties to be satisfied by the related component.

A property description has the following attributes:

- Name: a full-qualified name.
- Description: a human-readable description.
- Required: if it is required or optional.
- Expected value: the expected value of the property.

Examples of properties are: can read, can write, XML well-formed, syntax of content validated, file names patterns, availability, time of response, etc.

3.2 Self-management Agents and Resources

The logical control was designed using the Multi-Agents paradigm [5], and it is implemented using component patterns based on the ICARO-T framework [6] [7]. There are four types of agents:

- **The Manager:** It is responsible for creating, terminating and management the rest of agents and resources in the Management Component.
- **The Installer:** It is responsible for verifying the internal dependences and to fix possible errors that may happen during the first execution of the managed component.
- **The Runtime agent:** It is responsible for verifying the right functioning of external dependences at runtime, passing control to the Repair agent when errors occur.
- **The Repair agent:** It has the responsibility of fixing errors send by the Runtime agent.

Resources perform different task required by the agents. Some of them are responsible for monitoring the properties of the managed component's dependences.

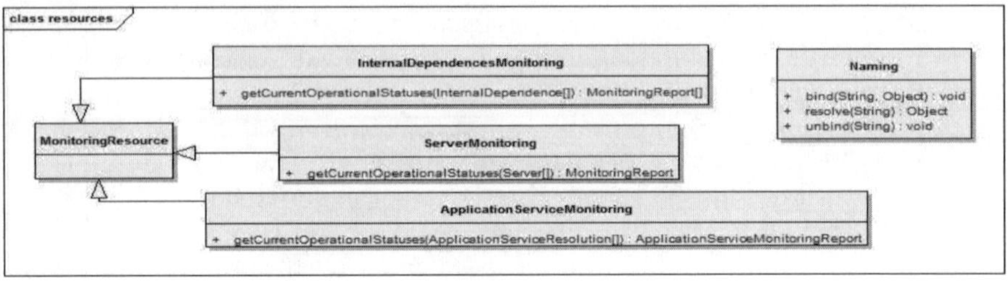

Fig. 4. Monitoring Resources

"InternalDependenceMonitoring" resource is in charge of getting the properties values of each managed component's internal dependence, and of inferring its operational status.

The "ServerMonitoring" resource is responsible for monitoring the servers as File servers, Database servers, etc, which are used by the managed components.

The "ApplicationServiceMonitoring" resource is responsible for monitoring application services used by the managed component. It monitors specific services instead of abstract services. It generates reports containing the service resolution of the abstract service dependence.

Monitoring resources generate reports that are read by agents to get the operational status of both internal dependencies of managed components, and external dependencies of those components. The Information about what to monitor is provided by the two XML description files: the Internal Dependence Description File (IDDF), and the External Dependence Description File (EDDF).

Agents use the Resources to monitor and analyze the internal structure of the managed component. The monitoring of external components is performed through queries and publish/subscribe mechanisms. Agents also gather information about the Managed Component state from the "ProxyOfControl". This information is used to achieve fault diagnosis.

3.3 The Behaviour of a Self-managed Component

The computing behaviour of a self-managed component will be illustrated with a working self-repair example taken from the Artist Community system, which has been used for validating the approach. The scenario is based on the failure of one running components –"GestionUsuarios"–, which affects the component "ProcesamientoTrabajos" depending on it. The system behaviour is depicted in figure 5. The Runtime Agent in "ProcesamientoTrabajos" detects the possible malfunction through its monitoring capability.

The Runtime Agent publishes the inferred status and stops the Managed Component "ProcesamientoTrabajos" because repair is needed. Then, it requests to the Manager to create an instance of the Repair Agent, which will be in charge of solving the problem. This agent first elaborates a repair plan in order to rebind an alternative of "GestionUsuarios", and then instantiates and executes the plan.

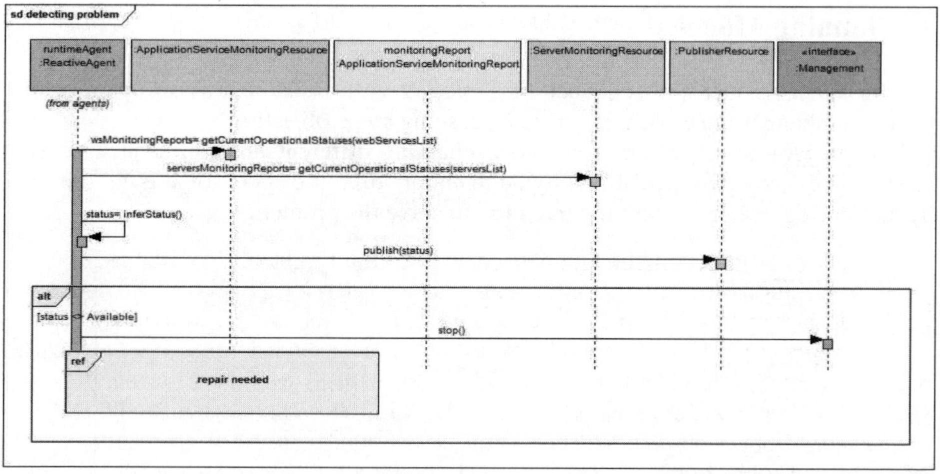

Fig. 5. A repair case

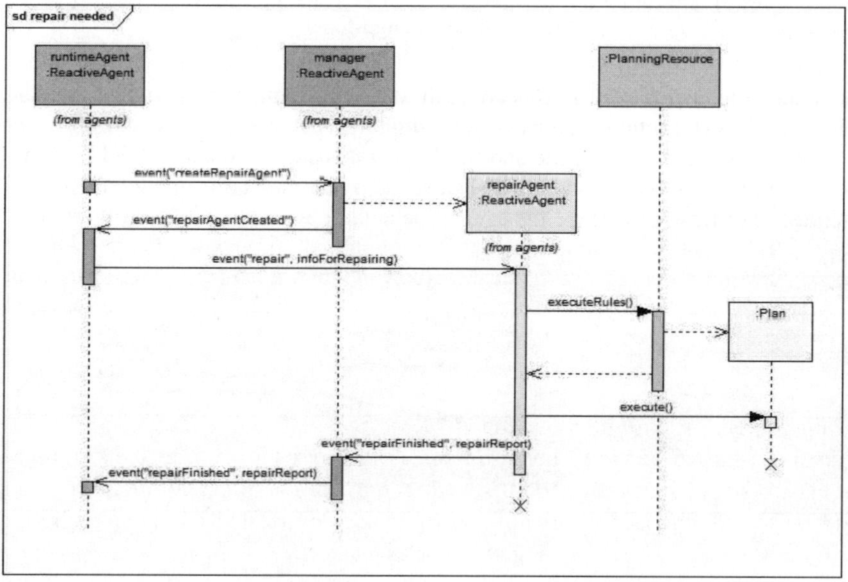

Fig. 6. Creation and execution of a plan by the Repair Agent

The repairing plan successes because an alternative service is available in the system. The new service rebound by the execution of the plan will be monitored in the next cycles of the Runtime Agents. If the new service's status is Available, the Runtime Agent will infer an Available status for the managed component, start it and publish the new status.

4 Planning Model

A Plan in this model is a sequence of Tasks. A Task is defined as an operator that somehow changes the environment state pursuing some objective.

The preparation of a plan consists in chaining different tasks in sequence. This process is dynamically performed by an agent anytime it detects some issue reported by monitoring resources with the intention to solve the problem.

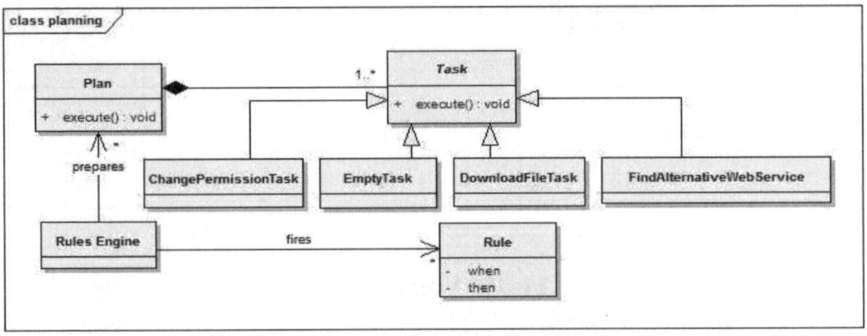

Fig. 7. The Planning Model

For an agent to decide which tasks are included in the plan, a set of "when-then" rules whose "when" part contains the possible symptoms detected for the possible issues. These rules are defined in a text file and fired by a rules engine based on RETE algorithm [9]. A rule example is given in figure 8. The set of predefined tasks and the rules file can be extended in order to customize the agents' behaviour against some issue.

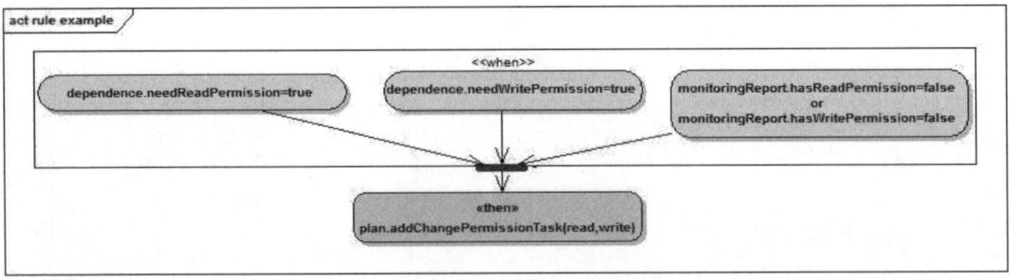

Fig. 8. A rule example

The preparation of the plan is finished when there are no more rules to fire. The plan is then ready to be executed, usually by the agent that prepared it.

5 Validation

The framework has been validated building a distributed system for assisting a Graphic Arts Community (the users) and then enhancing each system component with self-management capabilities.

The system is made of separated components that perform the different functions, some of them requiring others to their own functionality. The system is implemented using Java™ language and JAX-WS framework to support remote access via Web Services technology. Their interfaces and Access Points are registered in a central UDDI Registry. In addition, the system uses a SMTP Server, a Database Server and a UDDI Registry Server.

The transformation framework is made of a set of classes and file resources implemented with Java™, which are included together with business classes to generate a unique deployable component that runs on the same platform.

After framework application, the system has been successfully tested with a collection of significant scenarios including: restoration of missing files, XML validations, rebinding of services with replicated Web Services, etc.

Results showed that, although the computational overload is perceptibly increased, user-system interactions are not affected, while service continuity and stability are significantly improved.

6 Conclusions

The results obtained with the prototype show that the self-managed components perform successfully local monitoring, dynamic plan synthesis, and plan execution for component troubleshooting. Coordination among components is also achieved for fault diagnosis and self-healing. Compared to other approaches based on hierarchical management structures, making each component self-managed enforces their autonomy for failure detection and problem solving, and peer-to peer communication among components provides robustness and fault tolerance at system level. Decentralized control has also well known shortcomings with respect to centralized approaches, as the need of more sophisticated protocols for communication and cooperation. However, for large systems the advantages overcome the disadvantages, because this kind of architecture avoids bottlenecks, are more flexible and can be easily extended.

Future work should focus on self-optimization, self-configuration and self-protection. The last objective could be achieved following a similar approach consisting on enhancing each component with self-protection capabilities. This idea may not be applicable to the first two objectives. Achieving self-optimization and self-configuration would require system-wide parameter adjustment based on global system features that must be obtained seamlessly from system components. Therefore, individual components should "agree" on the proposed changes to achieve the global system behaviour. This might be done through the introduction of component's *choreographies* –group tasks carried out by agents in order to achieve common objectives, which are supported by some interaction protocols.

Another key issue is the automatic generation of Proxy classes and configuration files from code annotations made by developers in the business component code. This might be done by developing specific tools that will interpret the code annotations to detect component's usage of Web Services and other external components, as well as internal dependences. This annotation-oriented dependence declaration style, seems more intuitive and less error-prone than hardcoding dependency description files.

Finally, the self-management framework can be applied to itself since it is also a system. This can be useful to prevent errors in management tasks and to ensure that the machinery (configuration files and auxiliary classes) is ready.

Acknowledgements. We gratefully acknowledge Telefónica I+D, and the INGENIAS group for their help and support to carry out this work.

References

1. Kephart, J.O., Chess, D.M.: The Vision of Autonomic Computing. Computer Magazine on January, 41–50 (2003)
2. D'Souza, D.F., Wills, A.C.: Objects, Components and Frameworks With UML. Addison Wesley, Reading (1999)
3. OASIS WSDM Standards, An Introduction to WSDM, `http://docs.oasis-open.org/wsdm/wsdm-1.0-intro-primer-cd-01.pdf`
4. OASIS WSDM Standards, Management Using Web Services Part 2, `http://docs.oasis-open.org/wsdm/wsdm-muws2-1.1-spec-os-01.pdf`
5. Mas, A.: Agentes Software y Sistemas Multi-Agentes: Conceptos, Arquitecturas y Aplicaciones
6. Garijo, F.J., Bravo, S., Gonzalez, J., Bobadilla, E.: BOGAR_LN: An Agent Based Component Framework for Developing Multi-modal Services using Natural Language. In: Conejo, R., Urretavizcaya, M., Pérez-de-la-Cruz, J.-L. (eds.) CAEPIA/TTIA 2003. LNCS (LNAI), vol. 3040, p. 207. Springer, Heidelberg (2004)
7. The ICARO-T Framework. Internal report, Telefónica I+D (May 2008)
8. IBM, An architectural blueprint for autonomic computing, section 2.2
9. Forgy, C.: Rete: A Fast Algorithm for the Many Pattern/Many Object Pattern Match Problem. Artificial Intelligence 19, 17–37 (1982)
10. Gamma, E., Helm, R., Johnson, R., Vlissides, J.: Design Patterns: Elements of Reusable Object-Oriented Software
11. IBM, An architectural blueprint for autonomic computing
12. Lee, J., Jeong, K., Lee, H., Lee, I., Lee, S., Park, D., Lee, C., Yang, W.: RISE: A Grid-Based Self-Configuring and Self-Healing Remote System Image Management Environment. In: Proceedings of the Second IEEE International Conference on e-Science and Grid Computing (e-Science 2006) (2006)
13. Lee, K., Sakellariou, R., Paton, N.W., Fernandes, A.A.A.: Workflow Adaptation as an Autonomic Computing Problem. In: WORKS 2007 (2007)
14. Martin, P., Powley, W., Wilson2, K., Tian, W., Xu1, T., Zebedee, J.: The WSDM of Autonomic Computing: Experiences in Implementing Autonomic Web Services. In: International Workshop on Software Engineering for Adaptive and Self-Managing Systems (SEAMS 2007) (2007)
15. Ahmed, S., Ahamed, S.I., Sharmin, M., Haque, M.M.: Self-healing for Autonomic Pervasive Computing. In: Adams, C., Miri, A., Wiener, M. (eds.) SAC 2007. LNCS, vol. 4876. Springer, Heidelberg (2007)

FUSION@, A SOA-Based Multi-agent Architecture

Dante I. Tapia, Sara Rodríguez, Javier Bajo, and Juan M. Corchado

Departamento Informática y Automática
Universidad de Salamanca
Plaza de la Merced s/n, 37008, Salamanca, Spain
{dantetapia,srg,jbajope,corchado}@usal.es

Abstract. This paper presents a multi-agent architecture which facilitates the integration of distributed services and applications to optimize the construction of multi-agent systems. The architecture proposes a new and easier method to develop distributed multi-agent systems, where applications and services can communicate in a distributed way, even from mobile devices, independent of a specific programming language or operating system. The core of the architecture is a group of deliberative agents acting as controllers and administrators for all applications and services. The functionalities of the agents are not inside their structure, but modelled as services. This approach provides a higher ability to recover from errors and a better flexibility to change the agents' behaviour at execution time.

Keywords: Multi-Agent Systems, Services Oriented Architectures, Distributed Computing.

1 Introduction

The continuous development of software and systems requires creating increasingly complex and flexible applications, so there is a trend toward reusing resources and share compatible platforms or architectures. In some cases, applications require similar functionalities already implemented into other systems which are not always compatible. At this point, developers can face this problem through two options: reuse functionalities already implemented into other systems; or re-deploy the capabilities required, which means more time for development, although this is the easiest and safest option in most cases. While the first option is more adequate in the long run, the second one is most chosen by developers, which leads to have replicated functionalities as well as greater difficulty in migrating systems and applications. Moreover, the absence of a strategy for integrating applications generates multiple points of failure that can affect the systems' performance. This is a poorly scalable and flexible model with reduced response to change, in which applications are designed from the outset as independent software islands.

This paper describes a *Flexible User and ServIces Oriented multi-ageNt Architecture* (FUSION@). One of the most important characteristics is the use of intelligent agents as the main components in employing a service oriented approach, focusing on distributing the majority of the systems' functionalities into remote and local services and applications. The architecture proposes a new and easier method of building distributed multi-agent systems, where the functionalities of the systems are not integrated into the structure of the agents, rather they are modelled as distributed services and applications which are invoked by the agents acting as controllers and coordinators.

J.M. Corchado et al. (Eds.): DCAI 2008, ASC 50, pp. 99–107, 2009.
springerlink.com

Agents have a set of characteristics, such as autonomy, reasoning, reactivity, social abilities, pro-activity, mobility, organization, etc. which allow them to cover several needs for dynamic environments, especially ubiquitous communication and computing and adaptable interfaces. Agent and multi-agent systems have been successfully applied to several scenarios, such as education, culture, entertainment, medicine, robotics, etc. [6], [15]. The characteristics of the agents make them appropriate for developing dynamic and distributed systems as they possess the capability of adapting themselves to the users and environmental characteristics [8]. The continuous advancement in mobile computing makes it possible to obtain information about the context and also to react physically to it in more innovative ways [8]. The agents in this architecture are based on the deliberative Belief, Desire, Intention (BDI) model [9], [3], [12], where the agents' internal structure and capabilities are based on mental aptitudes, using beliefs, desires and intentions [7]. Nevertheless, complex systems need higher adaptation, learning and autonomy levels than pure BDI model [3]. This is achieved in FUSION@ by modelling the agents' characteristics to provide them with mechanisms that allow solving complex problems and autonomous learning.

FUSION@ has been designed to facilitate the development of distributed multi-agent systems. Agents have the ability to dynamically adapt their behaviour at execution time. FUSION@ provides an advanced flexibility and customization to easily add, modify or remove applications or services on demand, independently of the programming language. It also formalizes the integration of applications, services, communications and agents.

In the next section, the specific problem description that essentially motivated the development of FUSION@ is presented. Section 3 describes the main characteristics of this architecture and briefly explains some of its components. Section 4 presents the results and conclusions obtained.

2 Problem Description and Background

Excessive centralization of services negatively affects the systems' functionalities, overcharging or limiting their capabilities. Classical functional architectures are characterized by trying to find modularity and a structure oriented to the system itself. Modern functional architectures like Service-Oriented Architecture (SOA) consider integration and performance aspects that must be taken into account when functionalities are created outside the system. These architectures are aimed at the interoperability between different systems, distribution of resources, and the lack of dependency of programming languages [5]. Services are linked by means of standard communication protocols that must be used by applications in order to share resources in the services network [1]. The compatibility and management of messages that the services generate to provide their functionalities is an important and complex element in any of these approaches.

One of the most prevalent alternatives to these architectures is agents and multi-agent systems technology which can help to distribute resources and reduce the central unit tasks [1]. A distributed agents-based architecture provides more flexible ways to move functions to where actions are needed, thus obtaining better responses at execution time, autonomy, services continuity, and superior levels of flexibility and scalability than centralized architectures [4]. Additionally, the programming effort is reduced because it is only necessary to specify global objectives so that agents cooperate in solving problems and reaching specific goals, thus giving the systems the ability to generate knowledge and experience. Unfortunately, the difficulty in developing a multi-agent architecture is higher and because there are no specialized programming tools to develop agents, the programmer needs to type a lot of code to create services and clients [14]. It is also necessary to have a more complex system analysis and design, which implies more time to reach the implementation stage. Moreover, the system control is reduced because the agents need more autonomy to solve complex problems. The development of agents is an essential piece in the analysis of data from distributed sensors and gives those sensors the ability to work together and analyze complex situations, thus achieving high levels of interaction with humans [11].

Agent and multi-agent systems combine classical and modern functional architecture aspects. Multi-agent systems are structured by taking into account the modularity in the system, and by reuse, integration and performance. Nevertheless, integration is not always achieved because of the incompatibility among the agents' platforms. The integration and interoperability of agents and multi-agent systems with SOA and Web Services approaches has been recently explored [1]. Some developments are centred on communication between these models, while others are centred on the integration of distributed services, especially Web Services, into the structure of the agents. Bonino da Silva, et al. [2] propose merging multi-agent techniques with semantic web services to enable dynamic, context-aware service composition. They focus on SOA in designing a multi-agent service composition as an intelligent control layer, where agents discover services and adapt their behaviour and capabilities according to semantic service descriptions. Ricci et al. [13] have developed a java-based framework to create SOA and Web Services compliant applications, which are modelled as agents. Communication between agents and services is performed by using what they call "artifacts" and WSDL (Web Service Definition Language). Shafiq et al. [16] propose a gateway that allows interoperability between Web Services and multi-agent systems. This gateway is an agent that integrates Foundation for Intelligent Physical Agents (FIPA) and The World Wide Web Consortium (W3C) specifications, translating Agent Communication Language (ACL), SOAP and WSDL messages, and combines both directories from agents' platforms and web services. Li et al. [10] propose a similar approach, but focus on the representation of services. They use SOAP and WSDL messages to interact with agents. Walton [18] presents a technique to build multi-agent systems using Web Services, defining a language to represent the dialogs among agents. There are also frameworks, such as Sun's Jini and IBM's WebSphere, which provide several tools to develop SOA-based systems. Jini uses Java technology to develop distributed and adaptive systems over dynamic environments. Rigole et al. [14] have used Jini to create agents on demand into a home automation system, where each agent is defined as a service in the network. WebSphere provides tools for several operating systems and programming languages. However, the systems developed using these frameworks are not open at all because the framework is closed and services and applications must be programmed using a specific programming language that support their respective proprietary APIs.

Although these developments provide an adequate background for developing distributed multi-agent systems integrating a service oriented approach, most of them are in early stages of development, so it is not possible to actually know their potential in real scenarios. FUSION@ has an advantage regarding development because we have already implemented it into a real scenario. In addition, FUSION@ not only provides communication and integration between distributed agents, services and applications; it also proposes a new method to facilitate the development of distributed multi-agent systems by means of modelling the functionalities of the agents and the systems as services. Another feature in this architecture is security, which is managed by the agents. All communications must take place via the agents, so services cannot share their resources unless the agents allow it. Besides, services defined for each system must always be available, so they are not shared with other systems unless it is specified.

3 FUSION@, A SOA-Based Multi-agent Architecture

FUSION@ is a modular multi-agent architecture, where services and applications are managed and controlled by deliberative BDI (Belief, Desire, Intention) agents [9], [3], [12]. Deliberative BDI agents are able to cooperate, propose solutions on very dynamic environments, and face real problems, even when they have a limited description of the problem and few resources available. These agents depend on beliefs, desires, intentions and plan representations to solve problems [7]. Deliberative BDI agents are the core of FUSION@. There are different kinds of agents in the architecture, each one with specific roles, capabilities and characteristics. This fact facilitates the flexibility of the architecture in incorporating new agents. As can be seen on Figure 1, FUSION@ defines four basic blocks which provide all the functionalities of the architecture.

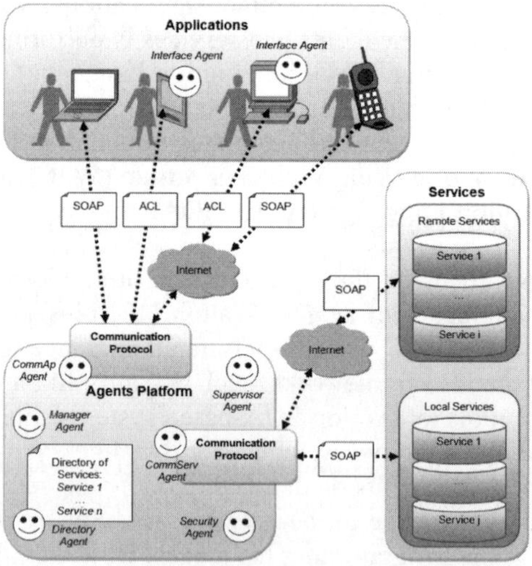

Fig. 1. FUSION@ model

- Applications. These represent all the programs that can be used to exploit the system functionalities. They can be executed locally or remotely, even on mobile devices with limited processing capabilities, because computing tasks are largely delegated to the agents and services.

- Agents Platform. This is the core of FUSION@, integrating a set of agents, each one with special characteristics and behaviour. An important feature in this architecture is that the agents act as controllers and administrators for all applications and services, managing the adequate functioning of the system, from services, applications, communication and performance to reasoning and decision-making.
- Services. They are the bulk of the functionalities of the system at the processing, delivery and information acquisition levels. Services are designed to be invoked locally or remotely. Services can be organized as local services, web services, or even as individual stand alone services.
- Communication Protocol. This allows applications and services to communicate directly with the agents platform. The protocol is completely open and independent of any programming language. This protocol is based on SOAP specification to capture all messages between the platform and the services and applications [5]. All external communications follow the same protocol, while the communication among agents in the platform follows the FIPA Agent Communication Language (ACL) specification. Applications can make use of agents platforms to communicate directly (using FIPA ACL specification) with the agents in FUSION@, so while the communication protocol is not needed in all instances, it is absolutely required for all services.

There are pre-defined agents which provide the basic functionalities of FUSION@:

- CommApp Agent. This agent is responsible for all communications between applications and the platform. It manages the incoming requests from the applications to be processed by services. It also manages responses from services (via the platform) to applications. All messages are sent to Security Agent for their structure and syntax to be analyzed.
- CommServ Agent. It is responsible for all communications between services and the platform. The functionalities are similar to CommApp Agent but backwards. Manager Agent signals to CommServ Agent which service must be invoked. All messages are sent to Security Agent for their structure and syntax to be analyzed. This agent also periodically checks the status of all services to know if they are idle, busy, or crashed.
- Directory Agent. It manages the list of services that can be used by the system. For security reasons [17], FUSION@ does not include a service discovery mechanism, so applications must use only the services listed in the platform. However, services can be added, erased or modified dynamically. There is information that is constantly being modified: the service performance (average time to respond to requests), the number of executions, and the quality of the service (QoS). This last data is very important, as it assigns a value between 0 and 1 to all services. All new services have a quality of service (QoS) value set to 1. This value decreases when the service fails (e.g. service crashes, no service found, etc.) or has a subpar performance compared to similar past executions. QoS is increased each time the service efficiently processes the tasks assigned.
- Supervisor Agent. This agent supervises the correct functioning of the other agents in the system. Supervisor Agent periodically verifies the status of all agents registered in the architecture by sending ping messages. If there is no response, the Supervisor agent kills the agent and creates another instance of that agent.

- Security Agent. This agent analyzes the structure and syntax of all incoming and outgoing messages. If a message is not correct, the Security Agent informs the corresponding agent (CommApp or CommServ) that the message cannot be delivered. This agent also directs the problem to the Directory Agent, which modifies the QoS of the service where the message was sent.
- Manager Agent. Decides which agent must be called by taking into account the QoS and users preferences. Users can explicitly invoke a service, or can let the Manager Agent decide which service is best to accomplish the requested task. This agent also checks if services are working properly. It requests the CommServ Agent to send ping messages to each service on a regular basis. If a service does not respond, CommServ informs Manager Agent, which tries to find an alternate service, and informs the Directory Agent to modify the respective QoS.
- Interface Agent. This kind of agent was designed to be embedded in users' applications. Interface agents communicate directly with the agents in FUSION@ so there is no need to employ the communication protocol, rather the FIPA ACL specification. The requests are sent directly to the Security Agent, which analyzes the requests and sends them to the Manager Agent. These agents must be simple enough to allow them to be executed on mobile devices, such as cell phones or PDAs. All high demand processes must be delegated to services.

FUSION@ is an open architecture that allows developers to modify the structure of these agents, so that agents are not defined in a static manner. Developers can add new agent types or extend the existing ones to conform to their projects needs. However, most of the agents' functionalities should be modelled as services, releasing them from tasks that could be performed by services. Services represent all functionalities that the architecture offers to users and uses itself. As previously mentioned, services can be invoked locally or remotely. All information related to services is stored into a directory which the platform uses in order to invoke them, i.e., the services. This directory is flexible and adaptable, so services can be modified, added or eliminated dynamically. Services are always on "listening mode" to receive any request from the platform. It is necessary to establish a permanent connection with the platform using sockets. Every service must have a permanent listening port open in order to receive requests from the platform. Services are requested by users through applications, but all requests are managed by the platform, not directly by applications. When the platform requests a service, the CommServ Agent sends an XML message to the specific service. The message is received by the service and creates a new thread to perform the task. The new thread has an associated socket which maintains communication open to the platform until the task is finished and the result is sent back to the platform. This method provides services the capability of managing multiple and simultaneous tasks, so services must be programmed to allow multi-threading. However, there could be situations where multi-tasks will not be permitted, for instance high demanding processes where multiple executions could significantly reduce the services performance. In these cases, the Manager Agent asks the CommServ Agent to consult the status of the service, which informs the platform that it is busy and cannot accept other requests until finished. The platform must then seek another service that can handle the request, or wait for the service to be idle. To add a new service, it is necessary to manually store its information into the directory list managed by the Directory Agent. Then, CommServ Agent sends a ping message to the service. The service responds to the ping message and the service is added to the platform. A service can be virtually any program that performs a specific task and shares its resources with the platform. These programs can provide methods to access data bases, manage connections, analyze data, get information from external devices (e.g. sensors, readers, screens, etc.), publish information, or even make use of other services. Developers have are free to use any programming language. The only requirement is that they must follow the communication protocol based on transactions of XML (SOAP) messages.

4 Results and Conclusions

Several tests have been done to demonstrate if the FUSION@ approach is appropriate to distribute resources and optimize the performance of multi-agent systems. Most of these tests basically consist on the comparison of two simple configurations (System A and System B) with the same functionalities. These systems are specifically designed to schedule a set of tasks using a planning mechanism [6]. System A integrates this mechanism into a deliberative BDI agent, while System B implements FUSION@, modelling the planning mechanism as a service.

A task is a java object that contains a set of parameters (TaskId, MinTime, MaxTime, ScheduleTime, UserId, etc.). ScheduleTime is the time in which a specific task must be accomplished, although the priority level of other tasks needing to be accomplished at the same time is factored in. The planning mechanism increases or decreases ScheduleTime and MaxTime according to the priority of the task: ScheduleTime= ScheduleTime-5min*TaskPriority and MaxTime=MaxTime+5min*TaskPriority.

To generate a new plan (i.e. scheduling), an automatic routine sends a request to the agent. In System A, the agent processes the request and executes the planning mechanism. On the other hand, System B makes use of FUSION@, so the request is processed by the Manager Agent which decides to use the planner service (i.e. the planning mechanism modelled as a service). The platform invokes the planner service which receives the message and starts to generate a new plan. Then, the solution is sent to the platform which delivers the new plan to the corresponding agent. Table 1 shows an example of the results delivered by the planning mechanism for both systems.

An agenda is a set of non organized tasks that must be scheduled by means of the planning mechanism or the planner service. There were 30 defined agendas each with

Table 1. Example of the results delivered by the planning mechanism

Time	Activity
19:21	Exercise
20:17	Walk
22:00	Dinner

50 tasks. Tasks had different priorities and orders on each agenda. Tests were carried out on 7 different test groups, with 1, 5, 10, 15, 20, 25 and 30 simultaneous agendas to be processed by the planning mechanism. 50 runs for each test group were performed, all of them on machines with equal characteristics. Several data have been obtained from these tests, focusing on the average time to accomplish the plans, the number of crashed agents, and the number of crashed services. For System B five planner services with exactly the same characteristics were replicated.

Figure 2 shows the average time needed by both systems to generate the paths for a fixed number of simultaneous agendas. System A was unable to handle 15 simultaneous agendas and time increased to infinite because it was impossible to perform those requests. However, System B had 5 replicated services available, so the workflow was distributed, and allowed the system to complete the plans for 30 simultaneous agendas. Another important data is that although the System A performed slightly faster when processing a single agenda, performance was constantly reduced when new simultaneous agendas were added. This fact demonstrates that the overall performance of System B is better when handling distributed and simultaneous tasks (e.g. agendas), instead of single tasks.

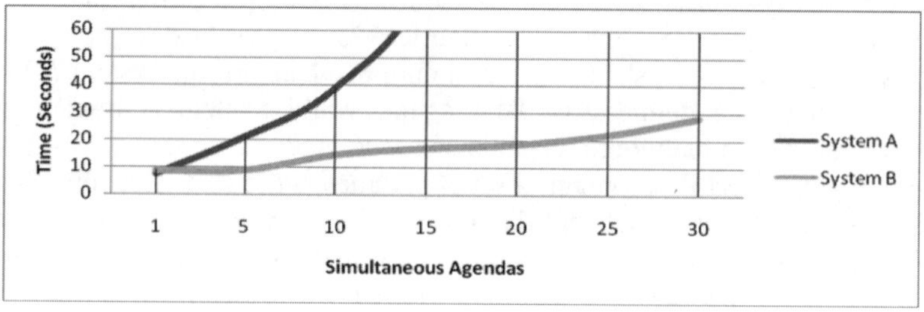

Fig. 2. Time needed for both systems to schedule simultaneous agendas

The architecture presented in this paper proposes an alternative where agents are based on the BDI (Belief, Desire, Intention) model and act as controllers and coordinators. FUSION@ exploits the agents' characteristics to provide a robust, flexible, modular and adaptable solution that covers most of the requirements of a wide diversity of projects. All functionalities, including those of the agents, are modelled as distributed services and applications. By means of the agents, the systems are able to modify their behaviour and functionalities at execution time. Developers can create their own functionalities with no dependency on any specific programming language or operating system.

Results demonstrate that FUSION@ is adequate for distributing composite services and optimizing performance for multi-agent systems. Future work consists on applying this architecture into composite multi-agent systems, as well as extending the experiments to obtain more decisive data.

Acknowledgements. This work has been partially supported by the TIN2006-14630-C03-03 and the IMSERSO 137/07 projects. Special thanks to Tulecom for the technology provided and the know-how supported.

References

1. Ardissono, L., Petrone, G., Segnan, M.: A conversational approach to the interaction with Web Services. Computational Intelligence 20, 693–709 (2004)
2. Bonino da Silva, L.O., Ramparany, F., Dockhorn, P., Vink, P., Etter, R., Broens, T.: A Service Architecture for Context Awareness and Reaction Provisioning. In: IEEE Congress on Services (Services 2007), pp. 25–32 (2007)
3. Bratman, M.E., Israel, D., Pollack, M.E.: Plans and resource-bounded practical reasoning. Computational Intelligence 4, 349–355 (1988)
4. Camarinha-Matos, L.M., Afsarmanesh, H.: A Comprehensive Modeling Framework for Collaborative Networked Organizations. Journal of Intelligent Manufacturing 18(5), 529–542 (2007)
5. Cerami, E.: Web Services Essentials Distributed Applications with XML-RPC, SOAP, UDDI & WSDL, 1st edn. O'Reilly & Associates, Inc., Sebastopol (2002)
6. Corchado, J.M., Bajo, J., De Paz, Y., Tapia, D.I.: Intelligent Environment for Monitoring Alzheimer Patients. In: Agent Technology for Health Care. Decision Support Systems. Eslevier, Amsterdam (in press, 2008)
7. Georgeff, M., Rao, A.: Rational software agents: from theory to practice. In: Jennings, N.R., Wooldridge, M.J. (eds.) Agent Technology: Foundations, Applications, and Markets. Springer, New York (1998)
8. Jayaputera, G.T., Zaslavsky, A.B., Loke, S.W.: Enabling run-time composition and support for heterogeneous pervasive multi-agent systems. Journal of Systems and Software 80(12), 2039–2062 (2007)
9. Jennings, N.R., Wooldridge, M.: Applying agent technology. Applied Artificial Intelligence 9(4), 351–361 (1995)
10. Li, Y., Shen, W., Ghenniwa, H.: Agent-Based Web Services Framework and Development Environment. Computational Intelligence 20(4), 678–692 (2004)
11. Pecora, F., Cesta, A.: Dcop for smart homes: A case study. Computational Intelligence 23(4), 395–419 (2007)
12. Pokahr, A., Braubach, L., Lamersdorf, W.: Jadex: Implementing a BDI-Infrastructure for JADE Agents. In: EXP - in search of innovation (Special Issue on JADE), Department of Informatics, University of Hamburg, Germany, pp. 76–85 (2003)
13. Ricci, A., Buda, C., Zaghini, N.: An agent-oriented programming model for SOA & web services. In: 5th IEEE International Conference on Industrial Informatics (INDIN 2007), Vienna, Austria, pp. 1059–1064 (2007)
14. Rigole, P., Holvoet, T., Berbers, Y.: Using Jini to Integrate Home Automation in a Distributed Software-System. In: Plaice, J., Kropf, P.G., Schulthess, P., Slonim, J. (eds.) DCW 2002. LNCS, vol. 2468, pp. 291–303. Springer, Heidelberg (2002)
15. Schön, B., O'Hare, G.M.P., Duffy, B.R., Martin, A.N., Bradley, J.F.: Agent Assistance for 3D World Navigation. LNCS, vol. 1, pp. 499–499. Springer, Heidelberg (1973)
16. Shafiq, M.O., Ding, Y., Fensel, D.: Bridging Multi Agent Systems and Web Services: towards interoperability between Software Agents and Semantic Web Services. In: Proceedings of the 10th IEEE International Enterprise Distributed Object Computing Conference (EDOC 2006), pp. 85–96. IEEE Computer Society, Washington (2006)
17. Snidaro, L., Foresti, G.L.: Knowledge representation for ambient security. Expert Systems 24(5), 321–333 (2007)
18. Walton, C.: Agency and the Semantic Web. Oxford University Press, Inc., Oxford (2006)

INGENIAS-SCRUM Development Process for Multi-Agent Development

Iván García-Magariño[1], Alma Gómez-Rodríguez[2],
Jorge Gómez-Sanz[1], and Juan C. González-Moreno[2]

[1] Universidad Complutense de Madrid
{ivan_gmg,jjgomez}@fdi.ucm.es
[2] Universidade de Vigo
{alma,jcmoreno}@uvigo.es

Summary. One of the key issues in development of Multi-agent System is following the more suitable development process for a particular problem. This means that development process models must be defined. Definition leads to better process understanding, facilitates process measurement and improvement, and constitutes the basis for automating the process itself. The main goal of this paper is the definition of a process for agent-based development. This definition is based on the Software Process Engineering Metamodel(SPEM) and applied in the definition of the software engineering process of INGENIAS MAS methodology following a SCRUM approach.

Keywords: Multiagent Systems(MAS), development process, SCRUM, INGENIAS, AOSE.

1 Introduction

Some authors [2, 3], indicate that the modeling of processes for agent-based development is an increasing demand due to many reasons. Some of the Agent-oriented Software Engineering(AOSE) methodologies [12, 4, 1, 11] may be followed using different process models depending on the system to be constructed. In these cases, the anticipated definition of the process, by means of its modeling, is very useful. Besides, the process models can be shared among Multi-agent System (MAS) designers, providing inexperienced designers with the process models defined by experienced ones. Moreover, defining the process models provide the basis for automating the process itself, in the same way as it is done in other engineering fields, and opens the possibility of adapting CASE tools for a particular process within a methodology.

This paper approaches the definition of a process model for agent-based development. When addressing this issue several perspectives can be followed. Some methodologies define a unique process of development, while others may follow different processes depending on the system, the team, etc. This paper incorporates this latest approach, applying a particular process to a well defined methodology. In fact, this paper focuses on a a particular methodology: INGENIAS, which has shown its suitability in AOSE, and on applying the SCRUM philosophy of management to INGENIAS process of development.

J.M. Corchado et al. (Eds.): DCAI 2008, ASC 50, pp. 108–117, 2009.
springerlink.com

The definition of processes is made easier using an editor which allows designers to create the suitable diagrams and descriptions in an automated way. In the last two years, an important number of tools have been proposed to model and specify Software Development Process [5, 3, 7], most of them based on the SPEM standard.

The structure of the remaining of the paper follows. Next two sections introduce briefly the SCRUM process of management and INGENIAS methodology of development. Section 4 provides a detailed definition of INGENIAS-SCRUM process, introducing several subsection where details about the lifecycle, the disciplines, etc. are explained. Finally, Section 5 presents the conclusions and future work.

2 SCRUM

A SCRUM is a mechanism in the sport of rugby for getting an out-of-play ball back into play. The term was adopted in 1987 by Ikujiro Nonaka and Hirotaka Takeuchi to describe hyper-productive development. Ken Schwaber formalized the process of SCRUM at OOPSLA 1995 [15].

As pointed in [14], SCRUM is an empirical Agile project management framework which is used to iteratively deliver to the customer software increments of high value. It is a simple framework used to organize teams and get work done more productively with higher quality. SCRUM relies on self organizing, empowered teams in order to deliver the product increments. It also relies on a customer, the Product Owner, which provides the development team with a list of desired features, using business value as the mechanism for prioritization.

From the previous paragraphs it must be clear that SCRUM is a model for management of the process of software development. It is not a methodology, but a framework which can fit with several methodologies. The SCRUM process is particularly suitable for Knowledge Engineering Developments based on the use of MAS, because of the agile development and the user implication.

3 INGENIAS Methodology

INGENIAS methodology covers analysis and design of MAS, and it is intended for general use, with no restrictions on application domain [13, 9]. It is based on UML diagrams and process (RUP), trying in this way to facilitate its use and apprenticeship. New models are added and the UML ones are enriched to introduced agent and organizational concepts. At the moment the metamodels proposed cover the whole life cycle and capture different views of the system. In the latest years, INGENIAS Metamodels have demonstrated their capability and maturity, as supporting specification for the development of MAS [13, 12].

Recently the INGENIAS Agent Framework (IAF) has been proposed taking into account the experience in application of INGENIAS methodology during several years enabling a full model driven development. This means that, following the guidelines of the IAF, an experienced developer can focus most of its effort in specifying the system, converting a great deal of the implementation in a matter of transforming automatically the specification into code. A MAS, in IAF, is constructed over the JADE platform. It can be distributed along one or several containers in one or many computers. To enable this feature, the IAF allows the declaration of different deployment configurations.

4 SCRUM for INGENIAS Methodology

The necessity of defining process models for agent-based development has been firstly established in [2]. After this work, many others have modeled different kinds of processes within a well known methodology. In this paper processes and methodologies are thought to be in two orthogonal dimensions that can be merged or interchanged during the MAS development process. This could be only be done if development resources and deliverables are previously defined and modeled into a well known process. Other alternative is to create from scratch a new process which matchs the necessities of the development team for fulfilling the user requirements. In this section, the guidelines used in [7] to specify and define a process model are followed for adapting the SCRUM Development Process for the INGENIAS Methodology. Moreover, SCRUM-INGENIAS process has proved his utility, because it has been used for developing a MAS system, based on delphi process for document relevance evaluation [6].

In the latest years an important number of tools for modeling and specifying Software Development Processes have been proposed. Most of them are based on the SPEM standard. One of these tools is instance Eclipse Process Framework (EPF) [5]; as part of the EPF documentation, two possible development processes are described: OpenUp and SCRUM. Moreover, in the EPF Composer Overview, authors propose as an example of the EPF capabilities, the model of a new process obtained by reusing several activities defined for OpenUp into a SCRUM process skeleton. On the other hand, in [2, 10] the approach is to find a set of common activities for different Agent Oriented Software Engineering Methodologies (AOSEM). The objective, in this case, is to reuse those activities across real Multi-Agent Software Development Life Cycle Processes.

In this paper, process activities can not be shared across different development processes in AOSEM because the intended meaning may be different and those activities may use different resources (human, software and hardware). Nevertheless, different processes for the same methodology which use the same resources can be used. This section addresses this goal; that is, to show how an agile project management technique, such as SCRUM, can be used within a methodology like INGENIAS, whose development process nature is basically those of the Rational original Unified Process Development (UPD) and which has been recently mapped to the OpenUP process (see [8]).

When trying to map a well established methodology/process into a new process, it is necessary to define the steps to be done for obtaining the desired result. In [8] several steps that must be followed in the definition of a new development process models for AOSEM are introduced, adopting SPEM as model for the specification. These steps are:

1. *Identify the process model with an existent process model* if possible, if not define from zero the new one taking as basis the next steps.
2. *Define the lifecycle view.* Identify the phases, iterations and sequence of application. In this step, other resources different from time are not considered.
3. *Define the disciplines.* Disciplines in SPEM determine process packages which take part in the process activities, related with a common *subject.* That is, disciplines represent a specialization of selected subactivities, where these new subactivities can not appear in other packages or disciplines. In this step, resources are the subject of the activities defined.
4. *Define the guidance and suggestion view.* The *Guidances* provide information about certain model elements. Each *Guidance* is associated to a *Guidance Kind.* This step is focused in exemplifying and documenting the activities previously defined.

After applying the previous steps to IAF and SCRUM, the results can be synthesized in the following points:

1. **Process Model:** There are several MAS that could be rapidly constructed with INGENIAS by reusing previous developments. Recently, the INGENIAS Agent Framework (IAF) for JADE has been proposed and documented as a successful approach in this context.
2. **Lifecycle:** In SCRUM, each release is produced in a number of brief iterations called Sprints (see Fig. 1). All the work is done in two basic phases: the *Preparation Phase* (before the first sprint) and the *Sprint Phases*(successive sprints leading to the release).
3. **Disciplines:** The team defines *tasks* required for developing the Sprint: Daily Scrum, Initiate Product Backlog, Manage problems, Plan Release, Sprint Planning, Sprint Retrospective, Sprint Review and Update product backlog.
4. **Guidance and suggestion view:** The guidance for implementing the process with INGENIAS could be found in the IAF documentation.

All the aforementioned technique steps are further explained in the following subsections.

4.1 Lifecycle

Although, SCRUM does not describe engineering activities required for product development, INGENIAS-SCRUM process must do it in order to adjust to IAF recommendations. IAF allows combining the classic approach of coding applications with modern techniques of automatic code generation. In this latest case, the generated MAS works over the JADE platform and there are additional tools available for this framework which can be applied. This requires a correct IAF specification, that is, determine the following aspects in a sound way:

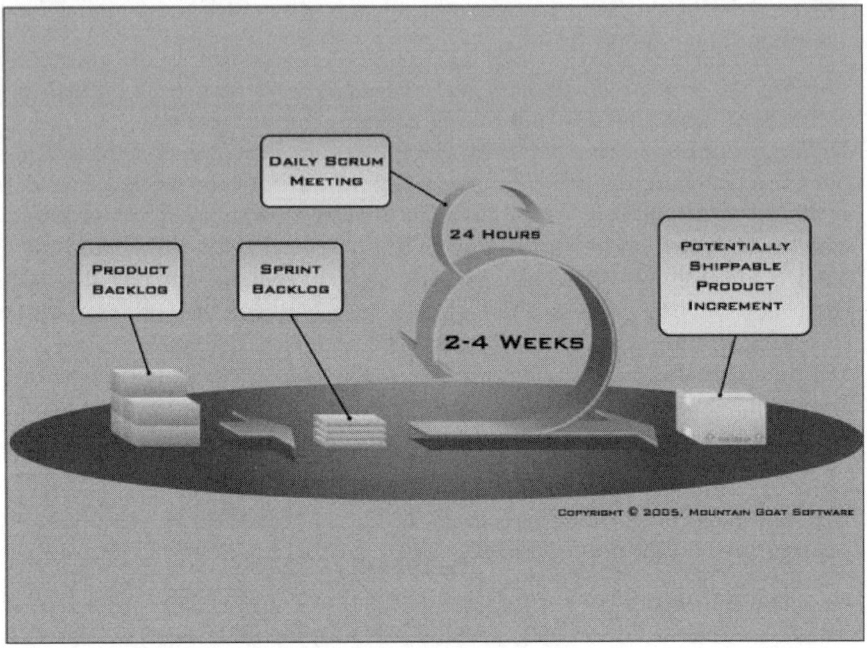

Fig. 1. SCRUM lifecycle

- **Interactions.** Interactions describe how agents do coordinate. Coordination is defined in terms of information transfer units.
- **Tasks.** Tasks are the basic units of agent's behavior. They modify the agent mental state and perform actions over applications.
- **Agents.** Agents are the system main building blocks. An agent is defined completely when its perception (main tasks) and its coordination means are described.
- **Deployment.** Deployment shows how agents are instantiated and initialized in the final system.
- **Tests.** Tests describe the testing units that the MAS should pass.
- **Goal**. The agents pursue certain goals. Goals guide agent's behavior.
- **Mental state**. The tasks performed by agents depend on their mental state and on their pursued goals.

IAF requires the use of the INGENIAS Development Kit (IDK). IDK contains a graphical editor for working with the specification models. Accordingly to IDK, the SCRUM definition for the *Preparation Phase* (see Fig. 2) comprises the tasks: *Initiate Product Backlog*, *Plan Release* and *Preparation Tasks*.

The *INGENIAS Product Backlog* contains the product requirements established with the IDK. This process can be done by adapting a model known from a previous project (i.e. IDK-IAF distribution comes with a complete cinema

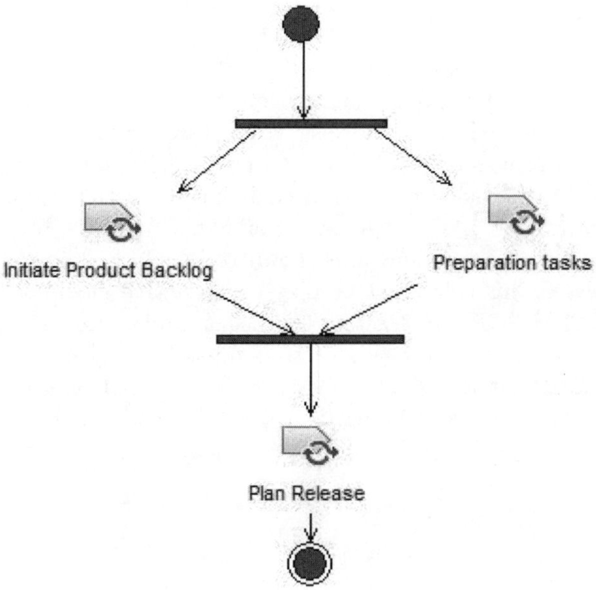

Fig. 2. Preparation Phase workflow

project which can be used for other distributed e-commerce developments) or by defining a completely new product backlog with the editor.

After this initial model is completed, in the *Preparation Tasks*, the *SCRUM Master* and the *Product Owner* establish the *Plan Release* which defines the needed *Sprints*.

From the SCRUM perspective and taking into account that IAF is based on the automatic generation of code, the project team must be completely involved in getting the release in the planned sprints. So, the INGENIAS specification must be established, using the IDK, as the core of the development. From this core, the different scripts and sources will be automatically produced. Nevertheless, at the first stage, the generated code may be incomplete and the programmer should add, if necessary, code in the tasks. This code must be incorporated in the specification by means of the IDK or the CodeUploader [1] application which will guarantee that next code generation does not overwrite the added code. Furthermore, for some MAS, external components may be necessary, and the IDK provides a mechanism to integrate these components into the MAS. In addition, a developer may find necessary to modify code already generated. In order to do that, in each *sprint* (see Fig. 3) the following tasks must be performed: *Plan Sprint*, *Update Product Backlog*, *Daily Works*, *Manage Problems*, *Conduct SCRUM Daily Meeting*, *Review Sprint*, and *Conduct Retrospective*.

[1] http://grasia.fdi.ucm.es/

4.2 Disciplines

Disciplines in SPEM determine process packages which take part in the process activities, related with a common *subject*. That is, disciplines represent a specialization of selected subactivities, where these new subactivities can not appear in other packages or disciplines. As previously pointed, in this approach the disciplines are the tasks required in each sprint, so the intended meaning of each task, according to IAF, must be explained. But first, the roles and products involved in the development must be introduced [8]. The roles, in this case, are:*Product Owner*, this role must be play by an active *customer* as in eXtreme Programming(XP); *SCRUM Master*, the *coach* and main of the development team; *SCRUM Team*, a *collective role* that must be played by the team members and *Stakeholder*, anyone that does not directly participate on the project but can influence the product being developed, that is, an *interested party*.

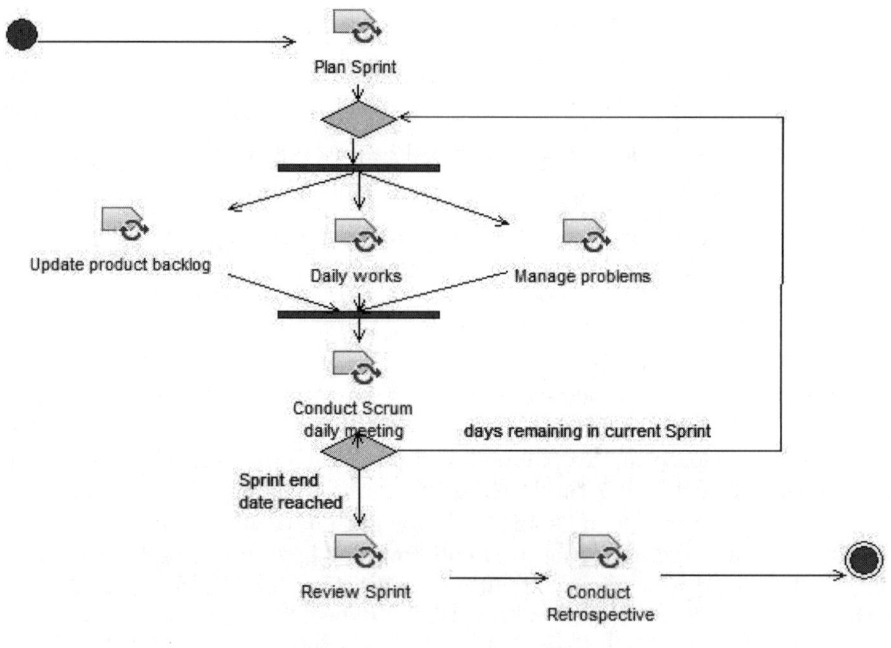

Fig. 3. Sprint Iteration

The products or artifacts involved in a SCRUM development process are: *Product backlog*, *Sprint backlog* and *Product increment*. The *product backlog* contains the product requirements and has the purpose of listing all the functionalities which must be implemented from a customer perspective. The *sprint backlog* is the list of things to do from the development team point of view. It could be understood as a fine-grained planning on particular tasks. Finally, the *product increment* is a partial product obtained at the end of each sprint, which can be deployed in the production environment or simply made available to users. From the INGENIAS-SCRUM perspective those artifacts are referred to the INGENIAS model and JADE code produced in each release. An INGENIAS model documented with the IDK can accomplish a low o high level of detail. Also, in the description of each model the *SCRUM Master* can fix the work to be done in the next release, where release can be identified with the *package* entity of the INGENIAS model.

The tasks that must be performed during the development lifecycle are:

- *Initiate Product Backlog.*- Explained in the previous subsection.
- *Preparation tasks.*- Explained in the previous subsection.
- *Plan Release.*- Explained in the previous subsection.
- *Sprint Planning.*- It involves the following steps: *Define the sprint's goal, Select the items, Identify tasks from items, Estimate tasks, Assigning tasks, Getting team commitment.*
- *Update product backlog.*- The update should take sprint scope changes into account for planning the next sprints. It must be performed in three steps: *Collecting changes, Reprioritizing items* and *Reestimating items.*
- *Daily SCRUM.*- The team performs backlog tasks to reach sprint goal in a free way, each SCRUM team selects the tasks to be performed or modified.
- *Sprint Retrospective.*- It allows the feedback and adaptation of the process.
- *Sprint Review.*- It shows what has been done and works.
- *Manage problems.*- This means to take into consideration the events that happen at every moment in a project and try to solve them.
- *Release work.*- This latest Sprint prepares product release.

4.3 Guidances

Developing a system with code generation facilities requires some guidance. In IAF documentation, several guidelines for development are proposed. In MAS, we recommend specially the use of two kinds of guidance: *Technique* and *Guideline*. The *technique* provides an algorithm to create a work product. The *guideline* is a set of rules and recommendations about a work product organization. As examples, the most relevant INGENIAS Guidances are detailed in [7].

5 Conclusions and Future Work

The definition of the software process engineering model of a MAS methodology makes it easier to learn and use the methodology. This paper provides a technique to define process models for agent-based development, using SPEM techniques. In particular, the technique is used for the redefinition of the SCRUM development process for the INGENIAS methodology. Furthermore, this work can guide a MAS team of developers through the steps of definition of a development process for MAS construction using an agent framework (the IAF) and the SCRUM process management. The description of the steps to follow, provided in the paper, can simplify the definition of processes for non expert engineers.

In addition, the description of how SCRUM can be used for managing a project for the INGENIAS methodology provides an example of how a methodology can incorporate several processes. That is, INGENIAS may follow a RUP process of development or can use other specific processes, such as SCRUM.

In the future, other methodologies and processes can be mixed in a particular process definition using the available tools for modeling. The models constructed in this way can assist the MAS designer in selecting the appropriate methodology and process model for a specific MAS, team of development or other circumstances. Moreover, the SCRUM process for the INGENIAS methodology can be compared with other MAS processes by means of certain metrics.

Acknowledgments

This work has been supported by the following projects: *Methods and tools for agent-based modeling* supported by Spanish Council for Science and Technology with grants TIN2005-08501-C03-01 and TIN2005-08501-C03-03 co-financed with FEDER funds and Grant for Research Group 910494 by the Region of Madrid (Comunidad de Madrid) and the Universidad Complutense Madrid. Some figures which facilitate de understanding of this paper have been obtained form EPF [5].

References

1. Bernon, C., Cossentino, M., Pavón, J.: Agent-oriented software engineering. Knowl. Eng. Rev. 20(2), 99–116 (2005)
2. Cernuzzi, L., Cossentino, M., Zambonelli, F.: Process models for agent-based development. Engineering Applications of Artificial Intelligence 18(2), 205–222 (2005)
3. Cossentino, M., Sabatucci, L., Seidita, V.: An Agent Oriented Tool for New Design Processes. In: Proceedings of the Fourth European Workshop on Multi-Agent Systems (2006)
4. Cuesta, P., Gómez, A., González, J.C., Rodríguez, F.J.: The MESMA methodology for agent-oriented software engineering. In: Proceedings of First International Workshop on Practical Applications of Agents and Multiagent Systems (IWPAAMS 2002), pp. 87–98 (2002)
5. Eclipse: Eclipse Process Framework (EPF), `http://www.eclipse.org/epf/`
6. García-Magariño, I., Pérez Agera, J.R., Gómez-Sanz, J.J.: Reaching consensus in a multi-agent system. In: International workshop on practical applications on agents and multi-agent systems, Salamanca, Spain (2007)
7. García-Magariño, I., Gómez-Rodríguez, A., González, J.C.: Modeling INGENIAS development process using EMF (In Spanish). In: 6th International Workshop on Practical Applications on Agents and Multi-agent Systems, IWPAAMS 2007, Salamanca, Spain, November 12/13, 2007, pp. 369–378 (2007)
8. García-Magariño, I., Gómez-Rodríguez, A., González, J.C.: Definition of Process Models for Agent-based Development. In: 9th International Workshop on AOSE, Lisbon, Portugal, May 12/13 (2008)
9. Gómez-Sanz, J.J., Fuentes, R.: Agent oriented software engineering with ingenias. In: Garijo, F.J., Riquelme, J.-C., Toro, M. (eds.) IBERAMIA 2002. LNCS (LNAI), vol. 2527. Springer, Heidelberg (2002)

10. Henderson-Sellers, B., Giorgini, P.: Agent-Oriented Methodologies. Idea Group Inc. (2005)
11. Mas, A.: Agentes Software y Sistemas Multi-Agentes. Pearson, Prentice Hall (2004)
12. Pavón, J., Gómez-Sanz, J.: Agent Oriented Software Engineering with INGENIAS. In: Mařík, V., Müller, J.P., Pěchouček, M. (eds.) CEEMAS 2003. LNCS (LNAI), vol. 2691, pp. 394–403. Springer, Heidelberg (2003)
13. Pavón, J., Gómez-Sanz, J.J., Fuentes, R.: Model Driven Development of Multi-Agent Systems. In: Rensink, A., Warmer, J. (eds.) ECMDA-FA 2006. LNCS, vol. 4066, pp. 284–298. Springer, Heidelberg (2006)
14. Schwaber, K., Beedle, M.: Agile Software Development with Scrum. Prentice Hall PTR, Upper Saddle River (2001)
15. Sutherland, J.: Business object design and implementation workshop. In: OOPSLA 1995: Addendum to the proceedings of the 10th annual conference on Object-oriented programming systems, languages, and applications (Addendum), pp. 170–175. ACM, New York (1995)

Using Agents for Long-Term Digital Reservation the PROTAGE Project

Josep Lluis de la Rosa[1], Johan E. Bengtsson[2], Raivo Ruusalepp[3], Ann Hägerfors[4], and Hugo Quisbert[4]

[1] EASY Innova & University of Girona
peplluis@eia.udg.edu
[2] InterNIT
johan@internit.se
[3] Estonian Business Archives
raivo@eba.ee
[4] Luleå University of Technology
ann.hagerfors@ltu.se, hugo.quisbert@ltu.se

Abstract. This is a summary of the PROTAGE project and a proposal for creation of agents useful for automation of the vast amount of human interactions and work otherwise needed in digital preservation. The potential of the technology is illustrated with an agent for appraisal of an individual's personal files.

Keywords: Agent technology, digital preservation, PROTAGE.

1 Introduction

PROTAGE is a 3 year project that is funded under the EU FP7 ICT programme as one of the “radically new approaches to digital preservation”. The mission of the project is to make long-term digital preservation easy enough for all computer users to preserve their own content, while reducing the cost and increasing the capacity of memory institutions to preserve digital information. The rapidly growing volume of digital information is causing data transfer from active IT systems to digital repositories at an increasing pace. This makes it necessary to find new levels of automation and self-reliance in digital archiving and preservation solutions. The increasing diversity in sizes and complexity among new digital resources imply that the repository systems must become highly automated and adaptable to various types of input, storage and access.

Until now no real-world deployment areas have been found, where agents can provide unique benefits or advantages compared with other already existing solutions [1]. So, in this situation, what should the agent research community do? The approach as suggested by [3] is threefold:

a) Look for killer applications, where agents are the natural or native solutions;
b) Develop further different pieces of agent technology;
c) Look for agents in the proper places, because in fact, they are already here.

The PROTAGE project will carry out research on the potential of software agent ecosystems to support the automation of digital preservation tasks. The objectives are to 1) to identify agent based technology suitable for the preservation field, and preserva-

J.M. Corchado et al. (Eds.): DCAI 2008, ASC 50, pp. 118–122, 2009.

tion models suitable for agent technology adaptation; 2) to further develop existing models/methods, taking into account stakeholder needs and requirements; 3) explore integration of PROTAGE solutions in other preservation environments.

2 Digital Preservation Issues

Digital preservation has become a pervasive and ubiquitous problem and concerns everyone who has digital information to keep. So far, only large memory institutions with expert knowledge and bespoke tools have been able to tackle it. A recent review of digital preservation research agendas noted that despite twenty years of active research there is still a lot of work to be done to solve the core issues. Therefore, radically new approaches to digital preservation are needed to support high volumes of data, dynamic and volatile digital content, keeping track of evolving meaning and usage context of digital content, safeguarding integrity, authenticity and accessibility over time, and models enabling automatic and self-organizing approaches to preservation [2].

Currently the level of automation in digital preservation solutions is low. The preservation process currently involves many manual stages but should be approached in a flexible and distributed way, combining intelligent automated methods with human intervention. The scalability of existing preservation solutions has been poorly demonstrated; and solutions have often not been properly tested against diverse digital resources or in heterogeneous environments.

Research in digital preservation domain has moved away from trying to find one ideal solution to the digital preservation problem, instead focusing on defining practical solutions for different preservation situations. These solutions have to utilise the expert know-how of memory institutions, be based on industry standards and above all, be scalable and adaptable to disparate environments.

3 A First Idea

Agent technology holds promise for digital preservation, using models for an automatic and self-organizing approach. The PROTAGE project proposes a fresh approach to overcome the fragmented expertise in digital preservation. We are using web services where people cooperate for learning how and when to preserve digital objects, where skilled specialists publish their expertise or answer queries of other people that use searches. This way, individuals and expert users (curators, preservation specialists, public and private agencies) will work openly to provide solutions, advice and web services, to enable long-term object preservation.

An open community sharing all types of resources (web services, wikis, blogs, etc) and recipes for appraisal, transfer, file format conversion and other digital preservation methods, is a new approach that will provide emergent value. However, the increasingly overload of experts while maintaining dedicated knowledge-bases and answering user queries, may hamper community problem-solving. Thus, the new cooperative approach will need tools for automating responses, to make recommendations and resources on their behalf, and exchange guidance in proactive ways: this is the mission of agent technology.

Agents will encapsulate a digital object with sufficient information with the mission of "surviving", that is "preserve your (object's) life as long as you can", making digital preservation an exciting native application of software agents.

4 An Agent Born for Digital Preservation - Giulia

This is in fact the simplest, example of a 'situated agent" that makes decisions considering environment (institutional), introspection, and social perspectives.

The agent will **appraise** whether a digital object is worth preserving after user has discarded it to the recycling bin. The agent will invoke web services [5] to decide whether other digital objects of the same type were selected for preservation by this user, and to identify whether this type of digital object belongs to a category that public authorities (national archives, national libraries, etc) have mandated for permanent preservation. The agent collaborates with other agents to find out if a copy of this object has already been retained elsewhere, and exchanges information to form a global view on the types of objects that should be preserved for posterity. Then the agent may warn the user about the need to keep the object, or even takes proactive action to ensure that the object is preserved.

Fig. 1. Deleting manuscript.rtf

The picture is showing how a digital object *manuscript.rtf* is deleted and the mechanisms that Giulia uses to appraise whether it should be retained instead or not. Let's follow the example and see the concrete tasks of the user agents:

1. Monitoring when the user (or the OS) deletes a file
   "an object, *manuscript.rtf*, has been deleted..."
2. Ask a web service about legal regulations that mandate selection for long-term preservation: "is it allowed to delete *manuscript.rtf*?"
3. Ask other agents for opinions about whether to preserve the deleted file:
   "Have other users deleted *manuscript.rtf*?"
4. Analyze the information of previously deleted files in this and other computers used by this user:
   "Has the user deleted similar objects before?"

A possible run is given in Figure 3. The answers are: "no", regarding institutional regulations, "no", regarding other users, and "yes" regarding this user. When applying a voting scheme as aggregation of all the answers, then "no" is the winner and the agent will make no more recommendations to the user. If the aggregated answer was "yes", then the user agent would warn the user not to delete this object and treat it accordingly the appropriate regulations set.

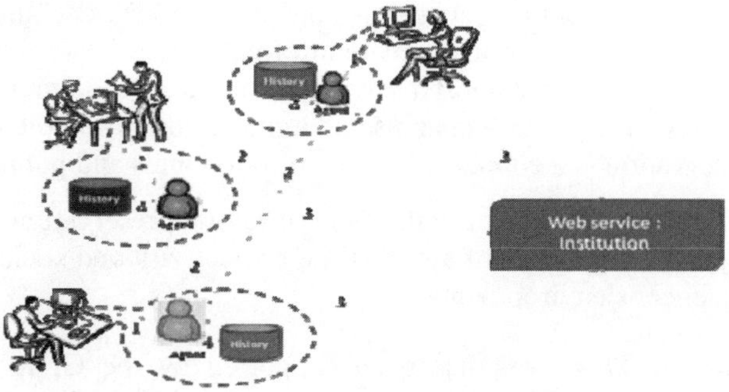

Fig. 2. Giulia is composed of both interacting agents and web services

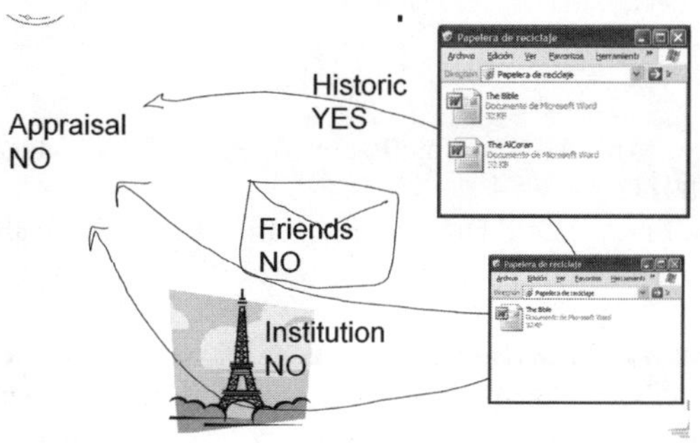

Fig. 3. Keep it or not? The situated agent decide

5 Discussion

The solution proposed by the PROTAGE project is to link digital objects to long-term digital preservation processes by using agent-based software technology. Based on the latest research on digital preservation strategies and on autonomous systems, the PROTAGE project will build and validate flexible and extensible software agents for long-term digital preservation and access that can cooperate with and be integrated in existing and new preservation systems. Intended application areas for prototypes produced by the project cover submission of digital material as well as monitoring of preservation systems and transfer to and between repositories. Targeted end users of the project results are curators, digital content creators, and individuals managing their own digital collections.

The PROTAGE project opens up a new and novel approach to digital preservation. The multi-agent preservation tools have the potential to enable integrated automated digital preservation operations in digital archives and libraries of different scales throughout Europe. It could also positively impact the preservation of file collections in the computers of individuals.

The PROTAGE project will apply agent technology to address critical long-term preservation issues, and find out whether digital preservation could be a killer application for software agents. The Giulia example illustrates that

a) agents automate interaction between people, and workflow within organisations;
b) agents make decisions with incomplete information;
c) agents are well suited to support a self-organising long-term archive;
d) web services are useful for human users, digital curators and their agents;
e) digital preservation is a common interest of individuals and public agencies.

The PROTAGE project will test in the field of digital preservation - at a very large scale - the autonomy, high level of automation, pro-activity and social capabilities of agents in heterogeneous environments.

Acknowledgements. This work has been supported by the Grant EU project Nº 216746 PReservation Organizations using Tools in AGent Environments (PROTAGE), FP7 and the PROTAGE consortium (www.protage.eu).

References

1. Hendler, J.: Where Are All the Intelligent Agents? IEEE Magazine Intelligent Systems 22(3), 2–3 (2007)
2. Digital Preservation Europe: DPE Digitial Preservation Research Roadmap (2007), http://www.digitalpreservationeurope.eu/publications/reports/dpe_research_roadmap_D72.pdf
3. Chen, W., Valckenaers, P., McBurney, P., Luck, M., Bobick, M., Dignum, F.: The Challenge of Finding Intelligent Agents. IEEE Magazine Intelligent Systems 22(4), 3–7 (2007)
4. Luck, M., McBurney, P., Shehory, O., Willmott, S.: Agent Technology Roadmap: A Roadmap for Agent Based Computing, AgentLink III (2005)
5. Payne, T.R.: Web Services from an Agent Perspective. IEEE Intelligent Systems 23(2), 12–14 (2008)

Management of Distributed and Redundant Storage in High Demand Web Servers for Heterogeneous Networks Access by Agents

Enrique Torres Franco[1], Oscar Sanjuán Martínez[2], José Daniel García Sánchez[3], Luis Joyanes Aguilar[1], Rubén González Crespo[1], and Sergio Rios Aguilar[1]

[1] Computer Science Faculty, Pontifical University of Salamanca, Madrid, Spain
{enrique.torres,luis.joyanes,ruben.gonzalez,
sergio.rios}@upsam.net
[2] Department of Computer Science, University of Oviedo, Oviedo, Asturias, Spain
osanjuan@uniovi.es
[3] Department of Computer Science, Carlos III University, Madrid, Spain
josedaniel.garcia@uc3m.es

Abstract. To improve the performance of a web server, the most usual solution is based on construcingt a distributed architecture, in which a set of nodes offer the web service. A web site is composed by a set of elements or resources, where each one of them can be of a specific type. The present article describes a new architecture proposed to add dynamic replication of the resources, supporting high demand file provisioning for web-based access arriving from heterogeneous (mobile & fixed) networks.

Keywords: software agents, emergent techniques, multi-agent, storage management, distributed storage, redundant storage, mobile services, mobile provisioning .

1 Introduction

In order to improve the performance of a web server, the most usual solution is based on constructing a distributed architecture, in which a set of nodes offers the web service. To improve the performance it is enough to add new nodes to the set. The server nodes get petitions and process them, but single file image is offered to the client, so does not have to know the names or the places of the nodes that forms the web system [1].

Those nodes usually do not have intelligence or advanced management mechanisms, and in this case it will be necessary, as we will see later, the introduction of agents that implement "emergent rules" [2][3] that assures the correct distribution of the web content.

A web site comprises a set of elements or resources, where each one of them can be of a specific type (HTML page, image, video, download file, music, etc.). A page is composed of a primary element that references a series of secondary elements (included on the page).

The web site receives petitions from the clients; each petition from a web page of a user is composed by several petitions to the server, one for each object that forms the page. The client establishes a connection to the server for each one of those petitions, and it is through this connection that it receives the response [4].

J.M. Corchado et al. (Eds.): DCAI 2008, ASC 50, pp. 123–131, 2009.
springerlink.com

The present article is structured as described: the second section presents the ideal use of agents and emergent systems to solve the problem. Then, in the third section the developed architecture for the web server distributed with static replication of content is described. In the fourth section it is explained the more common policies of petition distribution are explained. Next, in the fifth section, the underlying mathematic theory to the replication methods is explained briefly. On the sixth section the architecture proposed to add dynamic replication is show. Then on the seventh section the algorithm that is included on the proposed architecture is exposed, and finally on the eighth section there is the closing conclusion.

2 Use of Agents and Emergent Techniques

The technological advance of communications implies the approach of new scenarios in which terms like ubiquity, intelligence systems, distributed systems, high availability, secured networks, etc. necessarily make use of new techniques, new methodologies and new systems.

Traditional Artificial Intelligence systems are conceived individually as a monolithic agent whose behavior would be described as 'rational' by an external observer. In the early eighties, the system for distributed resolution of problems was characterized by a convergent acting way on the different nodes of a network, generally with centralized control. On this way, the several components are impassive in view of the acting of the rest of the network components.

According to [5], we are immersed on the 'agent age', and this is why many efforts to classify, define and normalize the paradigm exist. Also, they have a clear multidisciplinary character, joining such different areas like Psychology, Sociology, Artificial -intelligence and Software Engineering of course.

In the early nineties, multiagent systems with decentralized control and reusable modules appear. The agents of a multiagent system were conceived as independent from a specific problem and the system is provided with communication protocols generic enough. Recently, work is oriented to the study of the interoperability of heterogeneous systems or autonomous governing agents and their way to adapt themselves to dynamic environments.

The agents should have as application field unpredictable and uncertain environments. 'An Agent is a computer system that located on a specific environment, is able to act by autonomous way to reach its objectives' [6]. Approaches closer to software engineering tend to consider them as a simply object evolution [7].

That's the why we have choose an agent architecture to develop our system, we are using GAIA methodology and emergency. The emergency is that happens when a elements system with relative simplicity is organized spontaneously and without laws until a intelligence behavior appears. Lower level agents adopt behaviors that are the same that the high level ones: the ants made colonies; the urbanites neighborhood. They are based on evolutives techniques and in many cases on the interaction with the user [3]. This is that happens when 'the all is more intelligence that the sum of its parts'.

The emergent systems have the following characteristics:

- There is not hierarchical control from top to down (top-down) which tells system what to do (for example, 'mount a stack of Wood).

These techniques have been applied on similar problems as is shown on [8].

3 At the Beginning: Static Replication

The most extended distributed architecture is the Web System based on cluster [1]. In this architecture a distribution node between clients and servers is used, called Switch Web. The Switch Web receives all the petitions directed to the visible IP address and, using a distribution of petitions algorithm, decides what server must process each petition [9].

In this architecture routing depend on the content when the Switch analyzes the HTTP petition before choosing a destiny server. It is poorly efficient, but it allows to apply policies of distribution that considers the value of each petition.

The implementation trough TCP gateway allows assigning successive petitions from a client to different servers; because of this, on this solution it is possible to apply distribution policies more complex that allow a better balance of the charge between servers [10], so it is where it will be used.

4 Requests Distribution Policies

With the different architectures established, the requests distribution policy that the Switch Web will use shall be determined. The ideal result of a requests assignment algorithm should make all nodes have an equal quantity of tasks. Also, it has to demand a minimal computational effort, to prevent delays on the resend of the following petitions that would reduce the performance of the system. [11]

Applying the solution through a TCP gateway, the Switch Web is responsible for executing the Distribution Algorithm [1], because it is in charge of transferring the client's information to the nodes and vice versa. In the algorithm it can make use of the petition's content because the decision is performed at an application level.

Using policies based on server data allows taking care of the node state that forms the Web System, allowing selection of the less busy node, opening the possibility of dynamically choosing the more suitable node to resolve the petition. This selection can distributive equally the petitions on the servers or make petitions on the same server to obtain the maximum number of successes on the disks cache and the file system.

The assignment to the least busy node assigns the petitions based on the load of the nodes, for example, selecting the node with the least open connections. This policy applied to the algorithm of the Switch Web assigns the petitions to a node in relation of the number of petitions that this node is serving [12].

Partial replication of content allows reaching a big scalability on the volume of the data manipulated without spoiling the resultant system trustfulness. The architecture based on the Switch Web adapts perfectly to the situation of partial contents replication [13].

On a solution based on partial replication every element e_i is stored on a subset of the set of server nodes. This fact has impacts on the system architecture. The main reason is that this restriction takes out the freedom of distributing a petition to any server node. Also, it makes it necessary to establish a mechanism to determine in an efficient form which nodes keep a specific resource [13].

With the use of partial replica strategies, an element e_i only resides on a subset of server nodes. When a petition for an element comes, the assignation can only have as a result one of the nodes that content this element.

The use of partial replication imposes the need of performing the assignment of replicas of every element to the different nodes, including a mechanism of petition assignment to the nodes [14].

The simpler diagram of content assignation to the server nodes it is the static nature one. In this outlook, it is determined, in an initial moment, the number of replicas that is going to have from every element and the server node in which is going to be allocated every one of the replicas.

An more advanced diagram would be the one for dynamic content assignment to the server nodes. In this case, the assignment of content to the nodes is periodically assured to set the number of replicas of every resource and the node where they should be stored. The pursued objective is the adaptation to the needs of the service.

This objective is in line with the strict time and performance constraints usually found in servicing mobile requests. In fact, the radio part of the communication supporting the service requests involves high latency and its impact can be barely avoided, so it gains special importance to control the fixed communication service side. [15]. For example, a mobile location based service could clearly benefit from the use of dynamic content replication when serving static zonal maps, not only in synchronous service mode, but especially in the asynchronous/continuous service mode.

5 Mathematic Theory

The elements that compose a web site may be seen as a set of primary elements ($E^P = \{ e_1^p, e_2^p, \dots, e_n^p \}$) and a secondary elements set ($E^s = \{ e_1^s, e_2^s, \dots, e_m^s \}$) where each element has an associate size t_i and they should be distributed along a set of server nodes $S = \{ s_1, s_2, \dots, s_r \}$, where a node has a storage capacity c_i.

In this web server, the application of partial replication means that not all the elements e_i are in all servers s_i, an assignment mechanism which determines where each e_i is located shell be defined[14].

The simpler variants to resolve this type of replication assign the same replica number to each element and they distribute them in a uniform way along the nodes, by the initial cyclic assignment (element i is distributed from the node i) or by the final cyclic assignment (element i is distributed from the next node to the one than ended distribution i-1).

Those simple variants do not consider the nodes capacity limits, or the fact that different nodes may have different capacities. Because there is a storage limitation, it is needed to calculate the quantity of replicas of each element possible given the available storage space. If it is considered that the each node capacity is ci and each element size is ti, it may be declared that a way to calculate the replica number may be the shown in the equation (1).

$$r \leq \frac{\Sigma c_j}{\Sigma t_i} \tag{1}$$

This value isn´t correct at all, because it doesn´t consider that an element cannot be shared between two server nodes, but it establishes a maximum replica values that cannot be exceeded.

The adaptation of the algorithms that were mentioned before (initial cyclic assignment and final cyclic assignment) for that maximum value r is as simple as doing a storage iteration of all elements for each increase in the quantity of replicas until r is reached or an iteration happens that cannot store all the replicas.

A second alternative to resolve the problem is use the greedy algorithm, which sorts the elements from highest to lowest and it begins with the assignment of the bigger size elements. It begins trying the assignment of the r replicas and it reduces the number if the assignment is not possible. It is a simple algorithm and it balances better the distribution between servers.

The algorithm shown until now assigns the same number of replica to each element. This distribution is appropriate for environments in which the request are distributed in an uniform way along the set of elements, but in a web site some elements are often more popular than others, so they receive more requests. In this kind of environments, it is more convenient to assign a different number of replicas to each element of the web site, commonly more replicas the more popular an element is, that is, the greater the probability of a client asking for.

In this model each element has a number of replicas r_i, which can be obtained by a particularization of the previous inequality in which it considers the request probability pi, making as it appears in the equation (2), establishing as limits that there should be at least one copy of each element and as a maximum as much copies as a server nodes there are.

$$r_i = \frac{p_i \Sigma c_j}{t_i} \tag{2}$$

It seems evident that the element popularity change with time, so the previous equation suggests a new partial replication model called dynamic partial replication. If dynamic partial replication is used, the replicas number of each element change as the popularity changes, which can be calculate as a function of the number of request received for each element.

6 Proposed Architecture for the Dynamic Partial Replication

After studying the problem of distributed web storage with redundancy, and having analyzed the Agent Systems, we are going to propose an implementation of the architecture based on Agents and Emergent Systems.

The required implementation to develop a web site with partial replication needs to perform actions on the Switch Web and on the server node.

- The Switch Web should contain all the logic required to assign petitions to the servers (Request Distributor Agent), as well as counting them to the effect of modifying the popularity and determine if the popularity causes a change in the replication (Copies Control Agent). In this machine, the module that lets to the administrator add or remove elements (Extern Control Agent) must also be contained.
- The server node should include the required logic to obtain an element when the Switch Web considers that it should create an additional replica in this node, making a petition to another server node (Content Control Agent) and the component that resolves the petitions with disk access (Disks Access Agent).

To be able to create the prior model, it is necessary to specify the algorithm that will use each component, supposing that the Receiver and Sender modules only transfer petitions that arrive to each system.

7 Proposed Algorithm for Replica Number Control

After the architecture that will be used is established, we have to consider the proposed algorithm for the dynamic replication, which will be located in the replicas module referred in the previous point. This algorithm tries to optimize dynamically the copies number as a function of the number of requests that each file receives. In that way, in real time, the replicas number of those files with a large amount of requests will be increased in order to cover the demand. At the same time, the copies number of the files with fewer requests will be decreased in order to free storage space from the nodes.

Before delving into the proposed solution, it should be considered that the dynamic replication confronts three issues: number of copies each file, selection of the node where the copies will be stored in and the moment when the algorithm will be activated.

7.1 Number of Copies for Each File

Each file may have a different number of replicas. However, there is a problem, the initial situation, when there isn´t any information about the requests for the files. So, two different situations may arise at the moment of assigning the initial number of copies to each file: when there is a new file on the system and when the system has information about the file requests.

When there is a new file on the system there would not be any information about these file requests, so the copies number will have been calculated with another data. In this case, the file size is known and there are many studies that relate the files requests on internet with the Pareto distribution [16] which is shown in the equation (3). In this equation x is a random value which represents the file size and a is the Pareto shape. Several studies have been established [17][18][19] that the optimal value for a in this case is 1.2.

$$P(X = x) = \frac{a}{xa+1}\ \forall\, x \geq 1; F(X \leq x) = 1 - \frac{1}{xa}; F(X \geq x) = \frac{1}{xa} \tag{3}$$

Once the requests percentage of each file is known, the corresponding number of copies should be calculated. For this, both the probability of each file selection and the free space on the system nodes will be considered.

The algorithm will assign all the files once, causing that the nodes' free space to decrease by a determined percentage. This percentage will be subtracted from the probability that the files will be requested. As long as that value is higher than 0 a copy of that file will be introduced on the system. This will give the total number of copies of each file, so we should compare this number with the copies that there were on the system and add o delete copies according to the new number. If the new number is higher, the difference between the old and the new copies number will be added, and if the new number is lower, the difference between the new and old number of copies will be erased. The algorithm to obtain the number of copies is as follows:

```
insert=true;
while (insert)
{
 insert=false;
 for (int i=0;i<nFiles;i++)
 {
   if (pFile[8]>0)
   {
     aSpace-=size[8];
     copies[8]++;
     insert=true;
   }
 }
 for (int i=0;i<nFiles;i++)
 pFile[8]-= 1-aSpace/tSpace;
}
```

7.2 Nodes in Which the Copies Will Be Stored in

Before the file copies assignment among the different nodes, it has been considered that the system nodes have a finite storage space, so this aspect should be controlled.

A quite proved solution for several projects that approach the same problem [14][20] is the greedy algorithm with previous sorting of the replicas by their size. This algorithm obtains good results and its simplicity does not overload the system.

7.3 Moment When the Algorithm Will Be Activated

The proposed system is based on the solution given by [21]. The system will consider that the file replicas should be increased when the quality of service becomes compromised. This occurs when the response time for a file is higher than the previously established time. This situation is called timing failure.

According to the proposed model, each file request is accompanied whit the pair <response time, response time probability>. These values belong to each file and will be established by the administrator, who can change them at a moment to optimize the system.

The administrator will establish for each file the maximum value for the response time and the probability of the response time will be kept. The last value is considered because fewer timing failures do not put the quality of service at risk and while the fail percentage is under the limit, the copies number will be kept. In this way, the replicas module will count, with each request done from the users, the file that has been requested and whether there is a timing failure. When the requests percentage carried out with the established response time is under the established probability, the algorithm will be actived in order to optimize the files copies number.

Each time that a new file is added to the system, the number of copies for all the files should be recalculated. The new file probability is given by a Pareto distribution.

8 Resume

The agent systems, at least at the conceptual level facilitate the development of distributed systems, especially distributed Web server. This model is intended to maintain the same access time to a server, reducing the storage space.

References

[1] Cardellini, V., Casalicchio, E., Colajanni, M., Yu, P.S.: The state of the art in locally distributed Web-server systems. ACM Computing Surveys 34(2), 263–311 (2002)

[2] Johnson, S.: Emergence: The Connected Lives of Ants, Brains, Cities, and Software, Touchstone (2001)

[3] Resnick, M.: Turtles, Termites, and Traffic Jams: Explorations in Massively Parallel Microworlds. MIT Press, Boston (1997)

[4] Berners-Lee, T., Fielding, R., Frystyk, H.: Hypertext Transfer Protocol - HTTP/1.0. RFC 1945. Internet Engineering Task Force (May 1996)

[5] Corchado, J.M., Molina, J.M.: Introducción a la Teoría de Agentes y Sistemas Multiagente. Edite Publicaciones Científicas, Salamanca, España (2002)

[6] Wooldridge, M.: An Introduction to MultiAgent Systems. John Wiley and Sons, New York (2002)

[7] Odell, J.: Objects and agents compared. Journal of object technology 1, 41–53 (2002)

[8] Welicky, L., Sanjuán, O.: Improving Performance and Server Resource Usage with Page Fragment Caching in Distributed Web Servers. In: World Comp. Proceedings, Las Vegas, Nevada, USA (2007)

[9] Schroeder, T., Goddard, S., Ramamurthy, B.: Scalable web server clustering technologies. IEEE Network 14(3), 38–45 (2000)

[10] Andreolini, M., Colajanni, M., Nuccio, y.M.: Scalability of content aware server switches for cluster based Web information systems. In: Proc. of 12th International World Wide Web Conf (WWW 2003), Budapest, Hungary (2003)

[11] Colajanni, M.: Emerging internet-based services: new frontiers for performance models and applications. In: Proceedings of the First International Conference on the Quantitative Evaluation of Systems, pp. 122–123. IEEE Computer Society, Los Alamitos (2004)

[12] Teo, Y.M., Ayani, R.: Comparison of load balancing strategies on cluster-based web servers. Simulation 77(5–6), 185–195 (2001)

[13] Garcia, J.D., Carretero, J., Fernandez, J., Garcia, F., Singh, D.E., Calderon, A.: On the Reliability of Web Clusters with Partial Replication of Contents. In: First International Conference on Availability, Reliability and Security (ARES 2006), pp. 617–624. IEEE Computer Society, Los Alamitos (2006)

[14] Garcia, J.D.: Propuestas Arquitectónicas para servidores Web distribuidos con réplicas parciales, PhD Tesis, Universidad Carlos III de Madrid, Madrid (June 2005)

[15] Rios, S., et al.: Plataforma de Provisión de Servicios Móviles basados en localización con sistemas GNSS y de Inteligencia de Red para soporte de Agentes Inteligentes. In: IWPAAMS 2007. Sixth International Workshop on Practical Applications of Agents and Multiagent Systems, Salamanca, pp. 251–258 (2007)

[16] Fischer, M.J., Masi, D.M.B., Gross, D., Shortle, J., Brill, P.H.: Using quantile estimates in simulating internet queues with pareto service times. In: Proc. of the Winter Simulation Conference, vol. 1, pp. 477–485 (2001)

[17] Adya, A., Bahl, P., Padhye, J., Wolman, A., Zhou, L.: A Multi-Radio Unification Protocol for IEEE 802.11Wireless Networks. In: First International Conference on Broadband Networks, pp. 344–354 (2004)

[18] Leith, D.J., Shorten, R.N., McCullagh, G.: Experimental Evaluation of Cubic-TCP. In: Proc. Protocols for Fast Long Distance Networks, Los Angeles, USA (2007)

[19] VuBrugier, G., Stanojevi´c, R.S., Leith, D.J., Shorten, R.N.: A critique of recently proposed Buffer Sizing strategies. ACM Computer Communication Review 37(1) (January 2007)

[20] Tse, S.S.H.: Approximate Algorithms for Document Placement in Distributed Web Servers. IEEE Transactions on Parallel and Distributed Systems 16(6), 489–496 (2005)

[21] Krishnamurthy, S., Sanders, W.H., Cukier, M.: An Adaptive Quality of Service Aware Middleware for Replicated Services. IEEE Transactions on Parallel and Distributed Systems 15(11), 1112–1125 (2003)

On Accelerating the ss-Kalman Filter for High-Performance Computation

C. Pérez[1], L. Gracia[2], N. García[1], J.M. Sabater[1], J.M. Azorín[1], and J. de Gea[3]

[1] Miguel Hernández University, Spain
carlos.perez@umh.es
[2] Technical University of Valencia, Spain
luigraca@isa.upv.es
[3] Bremen University, Germany
jdegea@informatik.uni-bremen.de

Abstract. This paper presents the equations of the steady state Kalman Filter (ssKF) for both variable and constant sampling times, in order to state how important it is for the stability of this filter to have a constant sampling time. Under the condition of a constant sampling time (achieved here by using reconfigurable hardware), the steady-state Kalman Filter is then rewritten using a matrix property that will allow an efficient implementation in a parallel processor (although not in a sequential one), substantially improving the filter performance. This work also presents the solution to the particular cases for the propagation of the filter which can be found when implementing the algorithm, and demonstrates that the error introduced by using a fixed-point numerical implementation is stable with time.

Keywords: FPGA, Parallel implementation, steady state Kalman filter.

1 Introduction

In 1960, Dr. Kalman published his work called "A New Approach to Linear Filtering and Prediction Problems", presenting one of the most popular algorithms of modern science. Afterwards, several investigations have focussed on accelerating the calculation of this algorithm popularly called Kalman Filter (KF). These investigations obtained a sub-optimal case of the KF called steady-state Kalman filter or ssKF (see [2] for more information), with a lower computational cost than the original KF. This new filter was used mainly in aerial applications [13][2][11] (although nowadays it is used in many other fields [14]). Currently, the efforts to accelerate the ssKF algorithm are focussed almost exclusively on optimal hardware and software implementations.

On the other hand, an important issue has drawn little attention in most implementations: the precision of the ssKF is affected by the variability of the sampling time "T" (subsection 2.1 shows the equation with the influence of $T(n)$) [12], and that is the reason why the use of computers with high-level software (including the operating system - OS), can cause problems in some practical implementations due to the priority in application execution. In this paper, the high-speed implementation of a ssKF is presented and the effect of the variability in the sampling time is

J.M. Corchado et al. (Eds.): DCAI 2008, ASC 50, pp. 132–141, 2009.
springerlink.com

studied. The three possible cases that might be encountered when implementing the filter in a control scheme are presented. Finally, the effects caused to the state propagation and to the "processor" are studied.

A FPGA has been used for the real implementation of the filter which allows us to parallelize the filter calculation and to obtain a considerable advantage towards sequential processors with higher clock frequencies [5][7]. A property of the matrices involved in the computation of the filter has been used for an efficient and novel parallel computation of the algorithm. The LTI (Linear Time-Invariant) process considered to calculate the steady-state Kalman filter is a "discrete to Wiener process acceleration model" also known as a model with constant acceleration. It is easy to modify the equations proposed in this paper to increase or decrease the order obtaining a jerk (time-derivative of the acceleration) model or a velocity model for the target movement [10].

2 Prediction Filters

The above-mentioned ssKF filter is expressed as a predictive algorithm and is used to solve the delay problems introduced by the control scheme and used sensor (Td) in real-time systems [15], taking into account the sampling time (T(n) if it is considered as variable or T if it is considered constant) and the time needed to calculate the ssKF algorithm (Tcomp).

2.1 The $\alpha\beta\gamma$ Filter - Variable Sampling Time

The $\alpha\beta\gamma$-filter is based on a constant acceleration model and thus is better suited for the tracking of manoeuvering targets. In the x-direction, it is characterised by the following equations:

$$x(n,n) = x(n,n-1) + a(n)[x_m(n) - x(n,n-1)] \quad (1)$$

$$x(n,n-1) = x(n-1,n-1) + v_x(n,n-1)T(n) + \\ +0.5a_x(n-1,n-1)T^2(n) \quad (2)$$

$$v_x(n,n) = v_x(n,n-1) + \beta(n) \times [x_m(n) - x(n,n-1)]/T(n) \quad (3)$$

$$v_x(n,n-1) = v_x(n-1,n-1) + a_x(n-1,n-l)T(n) \quad (4)$$

$$a_x(n,n) = a_x(n-1,n-1) + y(n) \times [x_m(n) - x(n,n-1)]/T^2(n) \quad (5)$$

This filter is initialized by using the following procedure:

$$x(1,1) = x_m(1) \quad (6)$$

$$x(2,2) = x_m(2) \quad (7)$$

$$v_x(2,2) = (x_m(2) - x_m(1))/T(0) \quad (8)$$

$$x(3,3) = x_m(3) \quad (9)$$

$$v_x(3,3) = (x_m(3) - x_m(2))/T(0) \quad (10)$$

$$a_x(3,3) = (v_m(3,3) - v_m(2,2))/T(0) \quad (11)$$

where $T(0) = 40$ms for visual servoing applications, $T(0) = 1$ms for joint control in robotics, therefore, for a general case it can be named as: t_0. The additional variables are $v_x(n, n-l)$, the predicted velocity, and $a_x(n, n)$, the estimated acceleration at the nth data point. The smoothing constants $\alpha(n)$, $\beta(n)$, and $\gamma(n)$ are given by:

$$\alpha(n) = 3(3n^2 - 3n - 2)/n(n+1)(n+2) \tag{12}$$

$$\beta(n) = 18(2n - 1)/n(n+1)(n+2) \tag{13}$$

$$\gamma(n) = 60/n(n+1)(n+2) \tag{14}$$

This result can be derived using the same method as the one described in [9]. As n tends to infinity, $\alpha(n)$, $\beta(n)$, and $\gamma(n)$ tend to zero, and a target which changes its acceleration cannot be tracked. To overcome this problem, a constant, M, is chosen such that

$$\left.\begin{array}{l}\alpha(n) = \alpha(M)\\ \beta(n) = \beta(M)\\ \gamma(n) = \gamma(M)\end{array}\right\} \quad \text{for} \quad n > M$$

2.2 Variable Update Time

When T(n) is equal to a constant T, equations (1)-(5) reduce to the standard $\alpha\beta\gamma$-filter. The next step is therefore to derive a method of computing T(n) at each update. The algorithm for a variable update time presented by Cohen and explained afterwards by Munu [13] is based on an inverse relationship between the position residual error and the update time. The smoothed residual, $e_r(n)$, is described as:

$$e_s(n) = (1 - \alpha_R)e_s(n) + \alpha_R e_0(n) \tag{15}$$

where $e_0(n)$ is the residual at time point n normalized to the standard deviation of the measurement, and α_R is a smoothing constant. The update time, $T(n)$, is related to its previous value, $T(n-1)$, and the smoothed residual, $e_s(n)$ by

$$T(n) = \frac{T(n-1)}{\sqrt{|e_s(n)|}} \tag{16}$$

For $T(n)$ not to attain impractically small or large values, a threshold U is defined and a modified update time, $T(n)$, is chosen according to the following rules:

$$\begin{aligned}\tilde{T}(n) &= t_0 & U < T(n) < t_0\\ &= T(n) & U \le T(n) \le t_0\\ &= 6.25 \cdot t_0 & T(n) < t_0/16\end{aligned} \tag{17}$$

Cohen also suggested the use of a discrete set of values for the update time. As for the continuous case, this parameter is limited to maximum and minimum values of t_0 and $t_0/16$, respectively. If the position residual has a magnitude that is less than four times the noise standard deviation, a maximum value of t_0 must be chosen. If the magnitude of the position residual is greater than 256 times the noise standard deviation, a mean update time of $t_0/16$ must be chosen. For residual standard deviations between these extremes, the update time is chosen according to:

$$T(n) = \frac{4}{2^p} \quad if \quad 4^p < |e_s(n)| < 4^{p+1} \quad (p = 1, 2, 3) \tag{18}$$

In [6] Gardner and Mullen argue that $T(n)$ should be chosen according to

$$T(n) = \frac{T(n-1)}{\sqrt[3]{|e_s(n)|}} \tag{19}$$

i.e. they proposed an inverse cube root relationship as opposed to the square root relationship of Cohen. The following proportionality is used with a set of discrete values

$$T(n) \propto 1/\sqrt[3]{|e_s(n)|} \tag{20}$$

Equations (19) and (20) will be referred to as the $\alpha\beta\gamma$-cube root filter equations. The set of $\alpha\beta\gamma$-filter equations using the square root relationship of Cohen will thus be referred to as the $\alpha\beta\gamma$-square root filter equations. The process of selection of the update time from a set of discrete values was carried out in a similar way to the previous process using the update time algorithm of Cohen, i.e. a value is chosen by comparing the position residual with the measurement noise standard deviation.

2.3 The $\alpha\beta\gamma$ Filter - Constant Sampling Time

Assuming that the estimator can be calculated with constant T, the equations described in subsection 2.1 can be rewritten as follows: the steady-state Kalman filters considered can be obtained by expression (22) and (23). Expression (21) shows the innovation required to know if the filter is working properly.

$$Inn_k = y_k - \mathrm{C} \cdot \hat{z}_{k|k-1} \tag{21}$$

$$\hat{\mathbf{z}}_{k|k} = \hat{\mathbf{z}}_{k|k-1} + \mathrm{K} \cdot Inn_k \tag{22}$$

$$\hat{z}_{k+1|k} = \mathrm{A} \cdot \hat{z}_{k|k} \tag{23}$$

where $\hat{z}_{n|m}$ represents the estimate of z at time n given observations up to, and including time m. K is the steady-state filter gain, A is the system matrix and C is the output system array (the model considered is LTI).

Considering the particular case of a steady-state Kalman filter, the *discrete Wiener process acceleration model* (DWPA model) [1] (commonly referred as $\alpha\beta\gamma$), parameters needed are shown below:

$$\mathrm{K}_{abg} = \begin{bmatrix} \alpha_{abg} \\ \beta_{abg}/\mathrm{T} \\ \gamma_{abg}/(2\mathrm{T}^2) \end{bmatrix}; \quad \mathrm{A}_{abg} = \begin{bmatrix} 1 & \mathrm{T} & \mathrm{T}^2/2 \\ 0 & 1 & \mathrm{T} \\ 0 & 0 & 1 \end{bmatrix}; \quad \mathrm{C}_{abg} = \begin{bmatrix} 1 \\ 0 \\ 0 \end{bmatrix}$$

where α_{abg}, β_{abg} and γ_{abg} are obtained using expressions (27), (31) and (32) respectively (parameters obtained from [8]):

$$r_{abg} = \frac{\sigma_v \cdot T^2}{\sigma_w} \quad (24)$$

$$b_{abg} = \frac{r_{abg}}{2} - 3 \quad (25)$$

$$q_{abg} = \frac{2b_{abg}^3}{27} - \frac{b_{abg} \cdot c_{abg}}{3} - 1 \quad (26)$$

$$\alpha_{abg} = 1 - s_{abg}^2 \quad (27)$$

$$c_{abg} = \frac{r_{abg}}{2} + 3 \quad (28)$$

$$p_{abg} = c_{abg} - \frac{b_{abg}^2}{3} \quad (29)$$

$$s_{abg} = z_{abg} - \frac{p_{abg}}{3z_{abg}} - \frac{b_{abg}}{3} \quad (30)$$

$$\beta_{abg} = 2 \cdot (1 - s_{abg})^2 \quad (31)$$

$$\gamma_{abg} = 2 \cdot s_{abg} \cdot r_{abg} \quad (32)$$

$$z_{abg} = -\left(\frac{q_{abg} + \sqrt{q_{abg}^2 + \frac{4p_{abg}^3}{27}}}{2} \right)^{(1/3)} \quad (33)$$

The ssKF algorithm shown in the equations (21), (22) and (23) can be implemented in a sequential microprocessor using the blocks shown in Fig. 1. Evidently, the operations performed in a sequential processor have to be executed from left to right in Fig. 1, without any possibility neither to modify the order nor to carry out any other task at the same time. But the implementation shown in this figure, can be optimised if it is implemented in a parallel processor (e.g a FPGA) obtaining the schemes shown in Fig. 2 and Fig. 3 (depending on whether it is desired to propagate the prediction or not). The blocks placed to the same height represent the possibility of parallel calculation[1].

3 Execution Time Analysis

For high-performance applications, the time required to compute the ssKF can be relevant compared with the closed-loop sampling time. Therefore, an analisys of these two times is needed to design properly the implementation of the predictor. Considering the time needed to calculate the ssKF filter as Tcomp and the delay introduced by the control scheme and sensors used as Td, three different practical cases can be found:

(a) Tcomp $\ll$ Td
The time needed to calculate the ssKF (Tcomp) is very small compared with the sampling time (T) (this situation is called on-the-fly processing [3] [4]). That means, the computational cost of the filter is not a problem for the control system. For this case: A1 propagates T, Latency=T. See Fig. 2. This case can be implemented in a sequential processor therefore 1 can be applied too.

[1] Note that, Fig. 1 and Fig. 2 perform the same computation although they look different. In particular, Fig. 2 changes the order of the ssKF equations and makes one factorization (K2=IK·C) in order to take advance of the possible parallelization in the FPGA. Additionally, Fig. 2 also gives the innovation as an output, which implies no time cost because of parallelization.

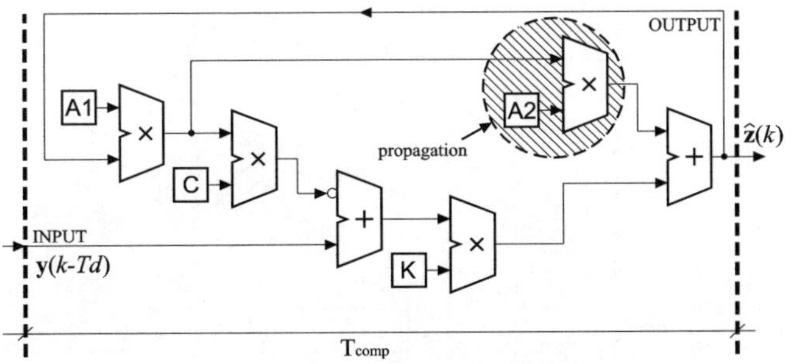

Fig. 1. Implementation of the ssKF in a sequential processor (including propagation). A1=$e^{\mathrm{A}\cdot\mathrm{T}}$ and A is the continuous state matrix or A_{abg} presented in subsection 2.3. A2 depends on the desired propagation.

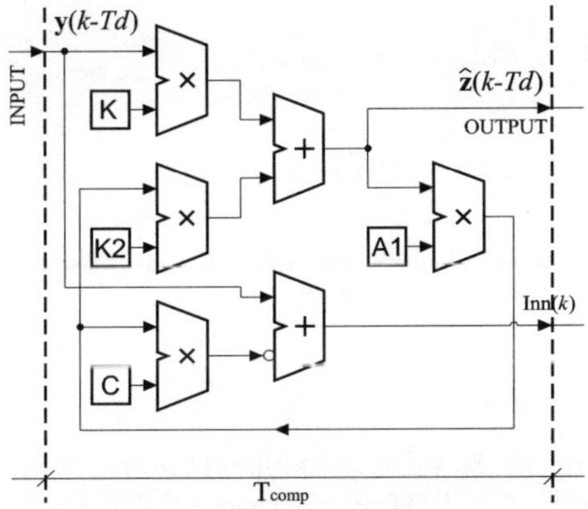

Fig. 2. Implementation of a ssKF optimizing the scheme shown in Fig. 1. Design for a parallel processor. Implementation of case (a) presented in subsection 3.

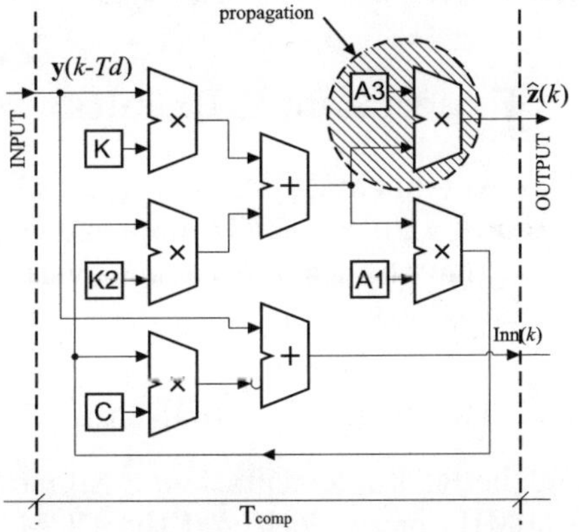

Fig. 3. Implementation of a ssKF used to implement the case (b). A1 propagates T≈Tcomp (A1=$e^{\mathrm{A}\cdot\mathrm{T}}$), A3 propagates 2T≈Tcomp+Td (A3=$e^{\mathrm{A}\cdot 2\mathrm{T}}$).

(b) Tcomp ≈ Td
The time needed to calculate the ssKF (Tcomp) is approximately the delay (Td). For this case, T=max(Td,Tcomp), Latency=2. A1 matrix propagates T(≈Tcomp≈Td), A2 propagates 2T(≈Tcomp+Td). See Fig. 2.

(c) Tcomp > Td
The sampling time of the control system is lower than the time needed to compute the ssKF algorithm. The solution is: segmentation. Figure 4 shows the ssKF segments: A4 propagate T1+T2+T3; A5 propagate T1+T2+T3+Td; Latency=T1+T2+T3+Td. The pipelined segments must be designed to be as similar as possible (attending to its execution time). The ideal case is: T1=T2=T3=Td=T.

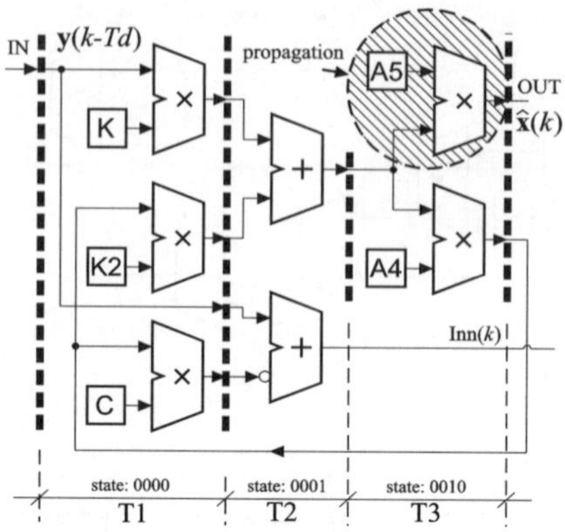

Fig. 4. Pipelined structure of the ssKF. The VHDL source code of this representation can be seen in Fig. 5. Dashed lines represent the registers used for pipelining.

4 Experimental Results

The experimental results of this article are focussed on validating the results obtained with the FPGA implementation using Matlab results as reference, verifying that the use of *fixed point* variables instead of *floating point* variables does not introduce a drift of the error with time.

4.1 Hardware for Real-Time Implementation

For the RT implementation, a Xilinx *Spartan3E* Rev.D (xc3s500e-4) board with a 50MHz main clock and the DCM (*Digital Clock Manager*) module is used, for example, to accelerate up to 250MHz the experiments obtaining a very low sampling time (T=4ns for steady-state Kalman filter).

4.2 Programming

The variables used are 32-bit fixed point (signed) with 16 bits for mantissa and 15 bits for integer (MSB bit to represent the sign). 32-bit fixed point numbers are the largest number implemented by Xilinx Synthesis Software and it can represent numbers up to $\pm 2^{15} = \pm 32768$ with precision of $1/2^{16} = 0,000015$.

4.3 ssKF Source Code

The Matlab code of the ssKF is shown in Fig. 5, whereas the VHDL code of the ssKF filter presented in Fig. 4 is shown in Fig. 5. This code can be easily modified to implement cases (a), (b) and (c).

Figure 5 shows the Matlab code for PC implementation of case (a) and Fig. 5 shows the VHDL code for the FPGA implementation of case (c). It is easy to obtain other cases modifying these codes.

```
                                (1) Code for MATLAB:
xa=[y(1);(y(2)-y(1))/T]; tic;
for i=1:samples
  inna=y(k-1)-Ca*xa;
  xa=xa+Ka*inna;
  xam=[xam Aa*xa];
end; time_comp_ssKF=toc/samples;

                                (2) Code for FPGA (VHDL):
process(clk2)
  variable aux:array2x1:=(to_sfixed(-0.00865129623056,
  nfix, -nfrac),to_sfixed(0.0117426027285, nfix,
  -nfrac));--(uRk(1);(uRk(2)-uRk(1))/T)
  begin
  if clk2'event and clk2='1' and SyncHW='1' and Send='0'
  and Conv='0' and iter<=73 then
    if (state="0000") then
      if (iter=1) then tmp1 <= prodArray(C,aux); else tmp1<=prodArray(C,tmp5); end if;
      tmp2 <= uFk(iter);
    elsif (state="0001") then tmp3 <= (tmp2-tmp1);
    elsif (state="0010") then tmp4 <= prodArray(kv,tmp3);
    elsif (state="0011") then
      if (iter=1) then tmp5 <= AddArray(aux,tmp4); else
      tmp5<=AddArray(tmp5,tmp4); end if;
    elsif (state="0100") then xk(iter) <= ProdMatrix(A1,tmp5); iter <= iter+1;
    end if;
  end if;
end process;
```

Fig. 5. (1) Fastest source code for Matlab of the ssKF filter for blocks shown in Fig. 1. Note that implementations of Fig. 2, 3 and 4 are parallel implementations. **(2)** Pipelined source code for VHDL of the ssKF filter for case (c). The non-pipelined code is quite similar, removing if/elsif/else/end if; sentences and changing signals to variables.

4.4 Experimental Data

Table 1 shows the time required to compute each filter for both platforms (PC-Intel Core 2 Duo-2.13GHz-Matlab & Spartan3E-xc3s500e-50MHz-VHDL). Notice that the area used in the Spartan-CPU is already 40%, showing that the xc3s500e might be a limited device for complex applications. However, the VHDL code could be "as it is" ported to a bigger FPGA (for example a Xilinx Virtex-5). The maximum frequency achieved by parallelizing tasks (but not pipelining) is 96.24MHz, and introducing pipelining this implementation works up to 250MHz. Latency is 3T for case (c) and the slowest segment is the first one (T1=3.99ns).

Table 1. Time (and resources) required to obtain the ssKF algorithm

Language:	MATLAB	VHDL		
Processor:	Intel Core 2 Duo	Spartan3E		
Implementation of:	Fig. 1	Fig. 1	Fig. 2&3	Fig. 4
Time:	3.55 μs	18.70 ns	10.39 ns	4 ns

4.5 Computational Error

An important issue of this implementation is that of the limited precision due to the fixed-point numbers used in the FPGA. The error stays within stable margins ($\pm 2.5 \cdot 10^{-5}$m), i.e. the error is not increased along with algorithm iterations.

5 Conclusion

In this paper, a theoretical study (see sub-sections 2.1 and 2.2) on the effects of a variable sampling time in ssKF filters is presented. These effects justify the use of a hardware implementation (that does not require operating system) for the real implementation of high-performance applications. Considering the sampling time as a constant, the equations of the ssKF filter can be established as it is shown in section 2.3. Later on, the three practical cases that can be found in a real implementation are analysed (subsection 3). This paper presents a substantial improvement in the implementation of the ssKF filter for processors with parallel and pipelining capabilities (in this case, a FPGA). Table 1 shows a significant difference (about 900 times) in the time required to calculate the filter. Notice that the clock frequency of the Intel Cores 2 Duo is 2,13GHz whereas the Spartan3E clock frequency is 50MHz. Moreover, it is demonstrated that the error signal (caused by using a fixed-point implementation of the filter) is stable with time.

References

1. Bar-Shalom, Y., Li, X., Kirubarajan, T.: Estimation with Applications to Tracking and Navigation. John Wiley & Sons, New York (2001)
2. Becker, K.: The alpha-beta-gamma tracking filter with a noisy jerk as the manouvering model. IEEE Transactions on Aerospace and Electronic Systems 2(29) (April 1993)
3. Chroust, S., Vincze, M.: Improvement of the prediction quality for visual servoing with a swithching kalman filter. International Journal of Robotics Research 22(10-11), 905–922 (2003)
4. Corke, P.: Visual Control of Robots: High performance Visual Servoing. Research Studies Press, Taunton (1996)
5. Cumplido, R., Jones, S., Goodall, R.M., Bateman, S.: A high-performance processor for embedded real-time control. IEEE Transactions on Control Systems Technology 13(3) (December 2005)
6. Farina, A., Studer, F.A.: Radar data processing. Introduction and tracking. Why, Research Studies Press (1985)
7. Goodall, R.: Perspectives on processing for real-time control. Annual Reviews in Control 25, 123–131 (2001)
8. Gray, J.E., Murray, W.: A derivation of an analytic expression for the tracking index for the alpha-beta-gamma filter. IEEE Transactions on Aerospace and Electronic Systems 29(3) (July 1993)
9. Hu, M.H., Mayer, M.P.: Clod form solution of a recursive tracking filter with u-priori velocity initialisation. IEEE Transactions on Aerospace Electrononic Systems AFS (21), 262–264 (1985)
10. Kalata, P.: The tracking index: A generalized parameter for alpha-beta and alpha-beta-gamma target trackers. IEEE Transactions on Aerospace and Electronic Systems AES-20(2), 174–182 (1984)
11. Kawase, T., Tsurunosono, H., Ehara, N., Sasase, I.: Alpha-beta tracking filter combined with maneuver-driven circular prediction. Electronics and Comunications in Japan 80(10) (1997)

12. Mechler, E., Rusell, J., Preston, M.: The basis for the optimum aided-tracking time constant. Journal of Franklin Institute 248, 327–334 (1949)
13. Munu, M., Harrison, I., Wilkin, D., Woolfson, M.S.: Comparison of adaptive target-tracking algorithms for phased-array radar. IEE Proceedings-F 139(5) (October 1992)
14. Pérez, C., García, N., Sabater, J.M., Azorín, J.M., Reinoso, O.: Object trajectory prediction. application to visual servoing. In: European Control Conference, Kos, Greece, pp. 2105–2111 (2007)
15. Vincze, M.: Real-time vision, tracking and control dynamics of visual servoing. In: IEEE International Conference on Robotics and Automation, San Francisco, CA, USA, pp. 644–649 (2000)

A Parallel Plugin-Based Framework for Multi-objective Optimization

Coromoto León, Gara Miranda, and Carlos Segura

Dpto. Estadística, I. O. y Computación
Universidad de La Laguna
La Laguna, 38271, Santa Cruz de Tenerife, Spain
cleon@ull.es, gmiranda@ull.es, csegura@ull.es

Summary. This work presents a parallel framework for the solution of multi-objective optimization problems. The framework implements some of the best known multi-objective evolutionary algorithms. The framework architecture makes usage of configuration files to provide a more extensive and simple customization environment than other similar tools. A wide variety of configuration options can be specified to adapt the software behaviour to many different parallel models, including a new adaptive model which dynamically grants more computational resources to the most promising algorithms. The plugin-based architecture of the framework minimizes the final user effort required to incorporate their own problems and evolutionary algorithms, and facilitates the tool maintenance. The flexibility of the approach has been tested by configuring a standard homogeneous island-based model and a self-adaptive model. The computational results obtained for problems with different granularity demonstrate the efficiency of the provided parallel implementation.

Keywords: Multi-objective optimization, evolutionary algorithms, parallel optimization, island-based models, plugin-based frameworks.

1 Introduction

Many real-world engineering problems are based on the optimization of more than one objective function. In this kind of *multi-objective optimization problems* (MOPs) a solution optimizing every objective might not exist. Since exact approaches are practically unaffordable, a wide variety of *multi-objective evolutionary algorithms* (MOEAs) have been designed [1] with the aim of obtaining an approximated solution set, as close as possible to the *Pareto front*.

Through the application of parallel schemes, the time invested in performing a multi-objective optimization can be decreased [2]. *Parallel multi-objective evolutionary algorithms* (pMOEAs) can obtain better or similar solutions than sequential approaches in less time. In the pMOEA island-based model the population is divided into a number of independent subpopulations or demes. Each subpopulation is associated to an island and a MOEA configuration is executed over each subpopulation. Each island (or processor) evolves in isolation for the majority of the pMOEA run, but occasionally some solutions can be migrated between

J.M. Corchado et al. (Eds.): DCAI 2008, ASC 50, pp. 142–151, 2009.
springerlink.com

neighbour islands. Many variants of island-based models have been proposed in the literature [3]. Although several specific EA-based frameworks have been proposed for the solution of MOPs [4, 5, 6], not many support parallel schemes [7] and even less provide an easy customization of the parallel models. Currently, Paradiseo [7] is one of the most widely used frameworks for implementing parallel evolutionary models. However, as the parallel models are specified in the code, when several models must be tested, or different number of islands are going to be used, the user must develop different codes for each of the models. Moreover, the island-based model is not easily customized in Paradiseo. This causes that final users have to make a big effort in order to incorporate new ideas to the models, thus hindering the use of pMOEAs to non-expert users.

This work presents a parallel plugin-based framework for the solution of MOPs implemented with the MPI tool. It is focused on offering an easy-customizable island-based model. As a novelty, it provides mechanisms to easily adapt the behaviour of the islands along the executions [8]. A selection of the literature best known MOEAs have been incorporated, and others can be integrated through the use of plugins. By the customization of the configuration files and the development of new plugins, users can adapt the framework to their requirements. The execution model can be specified through a configuration file, thus avoiding the necessity of implementing different codes for each model. The software has been designed in such a way that the user can easily specify the problem requirements and customize the configuration of the MOEAs that will participate in the problem solution.

The remaining content of the article is structured in the following way: The essence of the framework user interface is described in section 2. Section 2.1 shows the plugin interface designed to incorporate new problems in the framework. Section 3 briefly explains the internal operation of the developed framework. The computational study is presented in section 4. Finally, the conclusions and some lines of future work are given in section 5.

2 Framework User Interface

The user interface makes it possible to properly configure the framework for solving a particular problem in an easy and intuitive way. Mainly, the user tasks are two: specify the execution model to be used and, if necessary, develop the corresponding C++ plugins for defining new individuals, algorithms or functionalities, as specifying new migration methods, try to direct the island to different space regions, etc. The framework defines a set of interfaces for the different customizable operations. User plugins must subscribe such interfaces. Moreover, the framework contributes with a variety of libraries that provide a set of standard operations in evolutionary computation, e.g., mutation, crossover, selection and crowding operators.

After implementing the required plugins, the framework combine them allowing the usage of sequential and parallel solvers. For sequential solvers, the user only has to specify the problem name, the stopping criterium (e.g. number of individual evaluations), and the MOEA to be used with its associated parameters (population size, mutation and crossover rates, etc.). Every algorithm instance with its associated parameters is called *configuration*. For parallel solutions, a set of execution parameters and the configurations for each of the MOEAs taking part in the model are fixed through a *configuration file*. The execution parameters for the customization of the parallel models include: *Execution model*, *Stop criterion*, *Rate of initial individuals*, *Migration probability* and *Number of individuals to migrate* among others. *Execution model* describes how to map the algorithm configurations to the available islands along the execution. There are four different types of mapping: `cyclical`, initially takes the first configuration of every algorithm, then takes the second configuration and so on. Once all instances have been executed the process is repeated from the beginning. `No_change` chooses the initial configurations as in the cyclical case but maintains their execution on the islands until the end. `Elitist` selects the configuration with better statistics or success expectations. By default, the algorithm with better success expectations is the one with the best ratio of inserted individuals in the global solution per number of performed evaluations. `Probabilistic` randomly chooses a configuration whose probability is proportional to its quality statistics (measured as in the elitist case). The user is not restricted to use the same mapping along the complete execution, being able to specify a set of mappings, together with the indications of when to use each one. Each used configuration specifies a local stop criterium. When a local stop criterium is reached, the mapping algorithm is executed to decide which configuration must be executed on the idle island. *Stop criterion* indicates the end of the execution. It can be established on the basis of the execution time, the number of completed configuration executions, or the total number of evaluations. *Rate of initial individuals* is the ratio of individuals of the initial population that are taken from the global solution when the mapping algorithm produces a change of configuration on an island. This parameter allows to speed up the convergence of the solutions. *Migration probability* is the probability of performing a migration from one execution island to any other after every generation. *Number of individuals to migrate* is the maximum number of individuals to migrate at every migration.

Figure 1 shows two different configuration files to tackle one of the instances of the ZDT test suite [9]. The first file customizes the framework to execute a homogeneous island-based model. Only one algorithm configuration is specified: NSGA-II with mutation rate 0.033, crossover rate 0.9 and population size 25. Such configuration is executed on every slave island during the whole parallel execution. The mapping algorithm specifies that the algorithm executing on each island is not changed along the entire execution (80000 evaluations). In this case, as only one configuration has been specified, the selection of the mapping algorithm has no effect. The second file configures a self-adaptive parallel model. In this configuration, three different algorithms will be used during the execution. Only one configuration has been defined per algorithm. Each configuration has specified a local stop criterium of 330 evaluations. First, 1980 evaluations are equally distributed among the algorithms. If executing with 3 slave islands - one

```
Model: [EVALUATIONS,80000,NO_CHANGE]
Init_percent_of_individuals: 100
Migration_probability: 0.05
Number_of_individuals_to_migrate: 2
Max_global_front_size: 200
Max_final_solution_size: 200
Send_results_per_generation: no
Individual: ZDT1

Algorithm: NSGA2
Type_stopping_criterion: EVALUATIONS
Value_stopping_criterion: 10000
Max_local_front_size: 25
Solution_source: archive
0.033 0.9 25 25
```

(a) Homogeneous model

```
Model: [EVALUATIONS,1980,CIRCULAR]
       [EVALUATIONS,21040,PROBABILITY]
Init_percent_of_individuals: 100
Migration_probability: 0.05
Number_of_individuals_to_migrate: 4
Max_global_front_size: 150
Max_final_solution_size: 100
Send_results_per_generation: yes
Individual: ZDT1

Algorithm: NSGA2
Type_stopping_criterion: EVALUATIONS
Value_stopping_criterion: 330
Solution_source: internal
0.033 0.9 33
Algorithm: SPEA2
Type_stopping_criterion: EVALUATIONS
Value_stopping_criterion: 330
Solution_source: archive
0.033 0.9 33 33
Algorithm: IBEA
Type_stopping_criterion: EVALUATIONS
Value_stopping_criterion: 330
Solution_source: archive
0.033 0.9 33 0.002
```

(b) Self-adaptive model

Fig. 1. Configuration file examples for island-based parallel models

per algorithm - each configuration is executed twice. After that, achieved results will be used to fairly distribute the remaining evaluations among the algorithms with better success expectations. A probability mapping algorithm is specified for the rest of the execution.

2.1 Plugin Interface for the Problem Specification

One of the most typical requirements of final users is being able to incorporate their own problems (individuals) inside the framework in an intuitive way. The user must implement a plugin in order to solve new problems with the framework. The interface designed to specify the problems include the following methods:

- `bool init (const vector <string> &params)`: the user must specify the number of variables, their ranges, and the number of objectives in the problem, as well as the optimization directions of each objective. Any other initialization required by the problem, as loading an instance file, are performed in this method.
- `void evaluate (void)`: calculates the objective values.
- `void mutation (double pm)`: specifies the gene mutation. The user can make use of some of the mutation operators implemented inside the framework, or even implement new ones.
- `void crossover (Individual* ind)`: specifies the crossover operator. As with the mutation operator the user can make use of some of the crossover operators implemented inside the framework.

- `void restart (void)`: is an optional method. It is used to specify how to assign values to the initial individuals. For instance, a heuristic to assign such values could be implemented in this method. If the method is not implemented, a random assignation of the variables is performed.

3 Parallel Implementation

The parallel scheme consists of a *coordinator* process and as many *islands* or *slaves* processes as specified by the user. All the communications among the processes have been done using the message passing interface tool MPI. To improve the flexibility of the scheme, synchronous and asynchronous communications have been implemented. Figure 2 represents an approximated behaviour of the framework - the exact operation will depend on the configuration files and plugins specified by the user.

3.1 Coordinator

The coordinator - Figure 2 (a) - is in charge of starting the execution of an algorithm configuration on every idle slave, depending on the mapping method specified by the user. First, the coordinator reads and stores the framework setup which has been specified by the user through a configuration file. Then, the coordinator initiates all the slave processes and assigns to each of them an algorithm instance. The first configuration selections are done in such a way that every algorithm configuration is executed at least once. When all the configurations have been executed, the coordinator begins to apply the mapping criterion given in the framework customization for deciding, every time a slave gets idle, which is the next configuration to be executed. In order to increase the quality of the final solution, every time an algorithm execution is going to begin, the coordinator sends a subset of the current global solution to the slave. While the slaves are locally executing their algorithms, the coordinator keeps waiting for slave completions. After every algorithm completion, the slaves send to the coordinator the set of obtained solutions, so the coordinator is able to update the global solution. The size of this global solution is fixed by the user. This solution size is maintained by applying the NSGA-II crowding operator among slave completions.

3.2 Islands

Every slave - Figure 2 (b) - represents an *execution island*. The aim of an execution island is to search for problem solutions from an initial set of individuals applying the MOEA configuration specified by the coordinator. The MOEAs currently provided by the framework are: SPEA [10], SPEA2 [11], NSGA [12], NSGA-II [13], and two variants of IBEA [14]. Each provided algorithm has a set of parameters

```
initConfiguration (configFile);
initAllIslands ();
while (1) {
   idleIsland = recvLocalSolution();
   if (stop()) {
      for (i = 0; i < numIsland-1; i++)
         recvLocalSolution();
      sendMessageFinalize();
      printGlobalSolution();
   }
   config = selectionConfigAlgorithm();
   initIsland(config, idleIsland);
}
```

(a) Coordinator Pseudo-code

```
recvMigrationConfig();
while (1) {
   data = recvDataConfig();
   if (data == messageFinalize())
      break;
   initAlgorithmParams(data);
   recvInitialPopulation();
   while (!finishAlgorihtm()) {
      runGeneration();
      sendMigration();
      recvMigration();
   }
   sendLocalSolution ();
}
```

(b) Island Pseudo-code

Fig. 2. Pseudo-codes for the coordinator and the slave islands

that customizes its behaviour: mutation rate, crossover rate, population size, etc. The framework interface allows the user to specify the parameters to apply for each one of the algorithms through the configuration file. In this way, the user can tune the algorithms in order to improve their behaviour for a certain problem specification.

An initial set of individuals, taken from the global solution, are used to fill a part of the algorithm initial population. Thus, the slaves begin to evolve with populations closer to the problem solution region. The goal of this operation is to speed up the algorithm convergence. In cases where the mapping algorithm indicates that an island must continue executing the same algorithm configuration that it was already executing, the population of the last execution generation is taken as the new initial population, thus continuing the normal behaviour of the EA. Slaves execute different algorithm configurations while maintaining the partial solutions in their *local archive.* All the partial solutions in a local set are kept until the current instance execution is finished. At that moment, the local archive is unified into the *global solution* that is managed by the coordinator process. During the algorithm executions, the slave processes share part of their local solutions (*migration*). Considering that each island could explore different solution space regions, and that achieved solutions could have different qualities, migration allows to enrich the local solutions of the slaves and thus ensure, in the long term, a better scheme effectiveness [2].

4 Computational Results

As stated before, the software can be adapted to different parallel models. The achieved speedup depends on many factors: evolutionary algorithm, granularity of the problem, number of generations or evaluations to perform, crossover and mutation operators, migration scheme, and, in general, on any of the customizable parameters provided by the framework. It is known that island-based parallel executions of MOEA algorithms produce different results from the ones obtained by the sequential schemes [3]. That topic is out of the scope of this paper. The analysis presented here is performed only in terms of time. Specifically, the parallel homogeneous island model has been chosen to perform the scalability analysis. This model has been successfully applied in a wide variety of scenarios [15]. The MOEA selected for the homogeneous tests is NSGA-II. The speedup is calculated comparing the sequential NSGA-II algorithm and the parallel homogeneous island model executed through the proposed software. Both approximations are configured to perform the same number of individual evaluations.

Executions have been done over a dedicated Debian GNU/Linux cluster of 20 dual-core nodes with 1 Gb RAM and a Gigabit Ethernet interconnection network. Each processor is an Intel® Xeon™ 2.66 Ghz. The MPI implementation used was MPICH version 1.2.7 and the C++ compiler was gcc 4.1.3. Each type of execution was repeated thirty times and average values considered. Same settings as in [16] were used for all the experiments: simulated binary crossover with $\mu_c = 10$ and probability $p_c = 0.9$ and polynomial mutation with $\mu_m = 50$ and probability 1/30. An asynchronous migration scheme with unrestricted topology was specified. The migration probability was fixed to 0.05. The population size was fixed to 200 in the sequential executions whereas the subpopulation sizes in the parallel executions were set up by equally distributing the initial population among the available execution islands. Figure 1-(a) shows an example of a configuration file for a NSGA-II homogeneous execution using 8 slave islands and a total of 80000 evaluations.

Tests have been performed on problems with different granularity. Such granularity is associated to the evaluation of a single candidate solution, because this is the most costly step for real problems. Three type of problems have been considered: fine-grained problems (evaluation times of 10 μs), medium-grained problems (evaluations times of 100 μs), and coarse-grained problems (evaluations times of 1000 μs). When 20000 evaluations are specified, sequential execution times ranges from 0.77s to 21.7s for fine-grained and coarse-grained problems, respectively. There exists an almost linear relation between the time and the number of evaluations for all the problems.

Figure 3 shows the speedup - up to 32 processors - achieved for the considered problems when 20000, 80000, 320000 and 1280000 evaluations are performed. For 20000 evaluations, the coarse-grained problems attain an acceptable speedup. However, the behaviour is worse when using fine and medium-grained problems, limiting the number of useful processors to 8. The reason is that the sequential execution times are very low, and so, the overhead of the tool is too high to deal with such fine-grained problems when using more than 8 processors. When 80000 evaluations are specified, linear speedup is achieved for coarse-grained problems. Here, the scalability is also limited when dealing with fine and medium-grained problems. Now, using 16 processors makes sense for medium-grained problems, but for fine-grained problems using more than 8 processors does not have a positive effect on speedup. If the number of evaluations are increased up to 320000, the framework is able to speed up the computation for medium and coarse-grained problems, even when many processors are used. But again, fine-grained

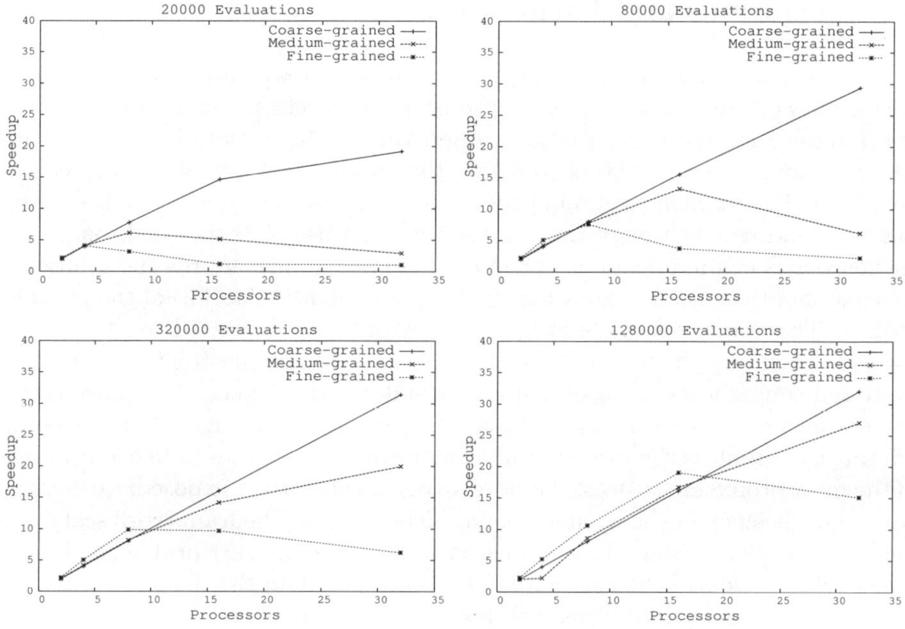

Fig. 3. Speedup for NSGA-II homogeneous parallel model

problems show a very limited speedup when using more than 8 processors. Finally, when 1280000 evaluations are executed, the speedups are linear for medium and coarse-grained problem but still limited for fine-grained problems.

Some conclusions can be drawn from the previous analysis. On one hand, there are some limitations when dealing with fine-grained problems. If the optimization desired can be performed with a low number of evaluations, the parallel model is useful when using less than 8 processors. If the number of evaluations to perform increases, the speedup improves, and then, it is useful to involve more processors in the computation. On the other hand, those limitations are not too severe, as they are restricted to problems that can be sequentially solved in few time. As the grain of the problem increases, the achieved speedup also increases, being reasonable to use up to 32 processors or even more. Some superlinear speedups appeared during the experiments. The main reason for such behaviour is that NSGA-II operators are quicker when applied to subpopulations, than when applied to an entire population. Most superlinear speedups appeared when dealing with fine-grained problems because for them the computational cost of each generation is mainly governed by the NSGA-II operators. On the other hand, when using coarse-grained problems the computational cost is governed by the evaluation of individuals, thus being more unusual the appearance of superlinear speedups.

5 Conclusions and Future Work

In this work a parallel framework for the solution of MOPs has been presented. The framework implements a set of the best-known MOEAs and provides a flexible interface for its customization, supporting configuration files and plugins. The user interface also allows to define the specific features of the problem to be solved. The implementation of the parallel interactions inside the framework has been addressed through the MPI tool, making use of both, synchronous and asynchronous communications. The presented framework affords the solution of MOPs without the need of knowing the internal operation details of the provided MOEAs. The main advantage of the framework when compared to other MOEA software resides in the flexibility for the specification and customization of standard and some more-advanced parallel island-based schemes. As a novelty, the framework provides methods for tuning the way different algorithms cooperate offering a dynamic assignation of the computational resources to take advantages of the most promising MOEAs. From the experimental study exposed, we demonstrate the efficiency of the framework parallel schemes. The framework scalability analysis has shown some limitations with very fine-grained problems. Typical real-world problems have associated grains greater than the highest grains here considered, so these limitations will not have a strong effect.

The main line of future work targets the extension of the set of available MOEAs. It is also necessary to better tune certain internal behaviours of the framework and to improve the current configuration framework interface in order to provide to the users an easier and more intuitive way of obtaining good results for their problems.

Acknowledgement. This work has been supported by the EC (FEDER) and the Spanish Ministry of Education and Science inside the 'Plan Nacional de I+D+i' with contract number (TIN2005-08818-C04-04). The work of Gara Miranda has been developed under grant FPU-AP2004-2290. This work was also supported by the HPC-EUROPA project (RII3-CT-2003-506079), with the support of the European Community - Research Infrastructure Action - under the FP6 "Structuring the European Research Area" Programme.

References

1. Coello, C.A.: An Updated Survey of Evolutionary Multiobjective Optimization Techniques: State of the Art and Future Trends. In: Angeline, P.J., Michalewicz, Z., Schoenauer, M., Yao, X., Zalzala, A. (eds.) Proceedings of the Congress on Evolutionary Computation, vol. 1, pp. 3–13. IEEE Press, Los Alamitos (1999)
2. Veldhuizen, D.A.V., Zydallis, J.B., Lamont, G.B.: Considerations in engineering parallel multiobjective evolutionary algorithms. IEEE Trans. Evolutionary Computation 7, 144–173 (2003)
3. Cantú-Paz, E.: A survey of parallel genetic algorithms. Technical report, IlliGAL 97003. University of Illinois, Urbana-Champaign (1997)

4. Bleuler, S., Laumanns, M., Thiele, L., Zitzler, E.: PISA — a platform and programming language independent interface for search algorithms. In: Evolutionary Multi-Criterion Optimization. LNCS, pp. 494–508. Springer, Heidelberg (2003)
5. Emmerich, M., Hosenberg, R.: TEA - A Toolbox for the Design of Parallel Evolutionary Algorithms in C++. Technical Report CI-106/01, SFB 531, University of Dortmund, Germany (2001)
6. Gagné, C., Parizeau, M.: Genericity in Evolutionary Computation Software Tools: Principles and Case Study. International Journal on Artificial Intelligence Tools 15, 173–194 (2006)
7. Liefooghe, A., Basseur, M., Jourdan, L., Talbi, E.G.: ParadisEO-MOEO: A Framework for Evolutionary Multi-objective Optimization. In: Obayashi, S., Deb, K., Poloni, C., Hiroyasu, T., Murata, T. (eds.) EMO 2007. LNCS, vol. 4403, pp. 386–400. Springer, Heidelberg (2007)
8. León, C., Miranda, G., Segura, C.: Parallel Hyperheuristic: A Self-Adaptive Island-Based Model for Multi-Objective Optimization. In: Genetic and Evolutionary Computation Conference. ACM, New York (to appear, 2008)
9. Zitzler, E., Deb, K., Thiele, L.: Comparison of multiobjective evolutionary algorithms: Empirical results. Evolutionary Computation 8, 173–195 (2000)
10. Zitzler, E., Thiele, L.: An Evolutionary Algorithm for Multiobjective Optimization: The Strength Pareto Approach. Technical Report 43, Computer Engineering and Networks Laboratory (TIK), Zurich, Switzerland (1998)
11. Zitzler, E., Laumanns, M., Thiele, L.: SPEA2: Improving the Strength Pareto Evolutionary Algorithm for Multiobjective Optimization. Evolutionary Methods for Design, Optimization and Control, 19–26 (2002)
12. Srinivas, N., Deb, K.: Multiobjective Optimization Using Nondominated Sorting in Genetic Algorithms. Evolutionary Computation 2, 221–248 (1994)
13. Deb, K., Agrawal, S., Pratab, A., Meyarivan, T.: A Fast Elitist Non-Dominated Sorting Genetic Algorithm for Multi-Objective Optimization: NSGA-II. In: Deb, K., Rudolph, G., Lutton, E., Merelo, J.J., Schoenauer, M., Schwefel, H.-P., Yao, X. (eds.) PPSN 2000. LNCS, vol. 1917, pp. 849–858. Springer, Heidelberg (2000)
14. Zitzler, E., Künzli, S.: Indicator-Based Selection in Multiobjective Search. In: Yao, X., Burke, E.K., Lozano, J.A., Smith, J., Merelo-Guervós, J.J., Bullinaria, J.A., Rowe, J.E., Tiňo, P., Kabán, A., Schwefel, H.-P. (eds.) PPSN 2004. LNCS, vol. 3242, pp. 832–842. Springer, Heidelberg (2004)
15. Coello, C.A., Lamont, G.B., Veldhuizen, D.A.V.: Evolutionary Algorithms for Solving Multi-Objective Problems. Genetic and Evolutionary Computation (2007)
16. Branke, J., Schmeck, H., Deb, K., Maheshwar, R.: Parallelizing multi-objective evolutionary algorithms: Cone separation. In: IEEE Congress on Evolutionary Computation, pp. 1952–1957. IEEE Press, Los Alamitos (2004)

Radix-R FFT and IFFT Factorizations for Parallel Implementation

Pere Marti-Puig[1], Ramon Reig Bolaño[1], and Vicenç Parisi Baradad[2]

[1] Department of Digital Information and Technologies
University of Vic (UVIC)
C/ de la Laura, 13, E-08500, Vic, Barcelona, Spain
[2] Department of Electronic Engineering
Politechnical Univerity of Catalonia (UPC)
Av. Victor Balaguer s/n, E-08800, Vilanova i la Geltrú, Barcelona, Spain
pere.marti@uvic.cat, ramon.reig@uvic.cat, Vicenc.Parisi@upc.edu

Abstract. Two radix-R regular interconnection pattern families of factorizations for both the FFT and the IFFT -also known as parallel or Pease factorizations- are reformulated and presented. Number R is any power of 2 and N, the size of the transform, any power of R. The first radix-2 parallel FFT algorithm -one of the two known radix-2 topologies- was proposed by Pease. Other authors extended the Pease parallel algorithm to different radix and other particular solutions were also reported. The presented families of factorizations for both the FFT and the IFFT are derived from the well-known Cooley-Tukey factorizations, first, for the radix-2 case, and then, for the general radix-R case. Here we present the complete set of parallel algorithms, that is, algorithms with equal interconnection pattern stage-to-stage topology. In this paper the parallel factorizations are derived by using a unified notation based on the use of the Kronecker product and the even-odd permutation matrix to form the rest of permutation matrices. The radix-R generalization is done in a very simple way. It is shown that, both FFT and IFFT share interconnection pattern solutions. This view tries to contribute to the knowledge of fast parallel algorithms for the case of FFT and IFFT but it can be easily applied to other discrete transforms.

Keywords: Fast Fourier Transform, Parallel algorithms, Fast algorithms.

1 Introduction

For the purpose of parallel processing, we require that a process be organized in a set of elementary operations that can be done simultaneously. There should be as few distinct types of elementary operations as possible. The parallel capability required should also be as simple and regular as possible. The maximum advantage of using regularity in FFT parallel implementations is achieved when an entire stage can be calculated completely in parallel. A fast transform algorithm can be seen as a sparse factorization of the transform matrix. We refer to each factor as a stage. The matrix dimensions of a stage are the same as the original transform matrix ones. Typically, in each row and each column of a stage there are only R values different to zero and the rest of its elements are exactly equal to zero. Number R is called the radix of the decomposition and usually is a power of two. From this observation, we can see that in a radix-R stage the basic operation consists in computing groups of R outputs from

J.M. Corchado et al. (Eds.): DCAI 2008, ASC 50, pp. 152–160, 2009.
springerlink.com

groups of R inputs. In radix-2 factorizations this basic operation is called a butterfly. Considering that N is the length of the transform, it is necessary to compute N/2 butterflies to accomplish a stage. In this work we present two general radix-R families for both FFT and IFFT in which R is a power of 2. If these algorithms have a regular interconnection pattern between stages then the inputs and outputs for each stage are addressed from or to the same positions, and the factors of the decomposition, also called stages, have the property of having their non zero elements in exactly the same positions. The first regular interconnection pattern for a discrete transform was presented by Pease in [1] for a radix-2 FFT and he refers to it as a parallel algorithm. In [2] Sloate presented the basis for a unified theory through which the various versions of the FFT algorithms can be formulated. In [2] the stages are defined by three basic operations: permutation, combination and multiplication. In the same work he presented a particular solution for a radix-4, N=1024, regular interconnection pattern factorization which can also be seen as a particular case of our solutions. In this work we follow the matrix representations for FFT provided by [5][6][7]. The presented approach is reminiscent of the factorization approach recently presented in [9]. Interesting tendencies in the field of fast discrete signal trans-forms can be found in [8]. In this paper, parallel factorizations of size N, being N a power of R and R a power of 2, are derived from the well-known Cooley-Tukey factorizations. Cooley-Tukey factorizations are also obtained from the basic recursion properties of FFT and IFFT. Factorizations with the same interconnection pattern as FFT are also obtained for the IFFT. This result shows that the same parallel hardware architectures, with the appropriate complex coefficients feeding the multipliers, can be used to calculate the FFT and the IFFT. The paper is organized as follows: in section 2 the used notation is presented. In the first part of section 3, the radix-2 Cooley-Tukey factorizations are obtained from the recursion properties of both FFT and IFFT and, in the second part, a method of extending the radix-2 factorizations to radix-R ones is shown. In section 4, taking the radix-R Cooley-Tukey factorizations from section 3, we proceed to derive the two general radix-R regular stage to stage algorithms by introducing permutation matrices between factors in a correct way. In this section it is also shown that the new factorizations exhibit a regular stage-to stage interconnection pattern. Finally some conclusions are presented.

2 Used Notation

Since we always deal with square matrices in what follows, an N×N square matrix is denoted by a bold capital letter with subscript N. The number N is a power of two. The elements of matrix $\mathbf{A}_N$ positioned at the row m and the column n are denoted by a_{mn}. Sometimes we will use the notation $\mathbf{A}_N=\{a_{mn}\}$. A column vector is represented by a bold small letter and, since its length can always be known from the context in this paper, its subscript indicates the position of the column in a matrix. The N×N identity matrix is denoted by $\mathbf{I}_N$ and it can be written by its column vectors e_i as $\mathbf{I}_N = [\mathbf{e}_1\ \mathbf{e}_2 \cdots \mathbf{e}_n]$. An even-odd permutation matrix $\mathbf{P}_N$ in terms of vectors $\mathbf{e}_i$ takes the form $\mathbf{P}_N = [\mathbf{e}_1\ \mathbf{e}_3 \cdots \mathbf{e}_{n-1}\ \mathbf{e}_2\ \mathbf{e}_4 \cdots \mathbf{e}_n]$. $\mathbf{P}_N$ is often used in this paper since permutation matrices involved in it can be written in terms of $\mathbf{P}_N$.

We will sometimes find it useful to divide a given matrix into sub matrices. Most of the times we will use the Kronecker product to show a particular matrix structure.

The symbol ⊗ stands for the right Kronecker product and, for arbitrary square matrices $\mathbf{U}_M$ and $\mathbf{V}_N$, the Kronecker product $\mathbf{U}_M \otimes \mathbf{V}_N$ is an MN×MN matrix that can be written using the elements u_{mn} of matrix $\mathbf{U}_M$ as:

$$\mathbf{U}_M \otimes \mathbf{V}_N = \begin{bmatrix} u_{11}\mathbf{V}_N & \cdots & u_{1M}\mathbf{V}_N \\ \cdots & & \cdots \\ u_{M1}\mathbf{V}_N & \cdots & u_{MM}\mathbf{V}_N \end{bmatrix}.$$

As mentioned above, all the permutation matrices can be written in powers of the even-odd permutation matrix $\mathbf{P}$ previously defined. Another possibility, not used here, is to write the same permutation matrices by using the commutation matrix defined in [3] [4]. Next, we recall some useful property involving the Kronecker product and the above defined even-odd permutation matrix $\mathbf{P}_N$.

$$(\mathbf{A}_M \otimes \mathbf{B}_N)(\mathbf{C}_M \otimes \mathbf{D}_N) = \mathbf{A}_M\mathbf{C}_M \otimes \mathbf{B}_N\mathbf{D}_N, \tag{1}$$

$$\mathbf{P}_{2^n}^n = \mathbf{I}_{2^n}, \tag{2}$$

$$\mathbf{P}_{2^n}^{n+n_1} = \mathbf{P}_{2^n}^{n_1}, \tag{3}$$

$$\mathbf{I}_{2^{n1}} \otimes \mathbf{I}_{2^{n2}} = \mathbf{I}_{2^{n1+n2}}. \tag{4}$$

The Kroneker product of any matrix $\mathbf{U}$ of size $2^{n1} \times 2^{n1}$ by any matrix $\mathbf{V}$ of size $2^{n2} \times 2^{n2}$ commutes with the powers of the permutation matrices $\mathbf{P}$ as follows [4]:

$$\mathbf{U}_{2^{n1}} \otimes \mathbf{V}_{2^{n2}} = \mathbf{P}_{2^{n1+n2}}^{n1}\left(\mathbf{V}_{2^{n2}} \otimes \mathbf{U}_{2^{n1}}\right)\mathbf{P}_{2^{n1+n2}}^{n2}. \tag{5}$$

Finally, the factorization of an arbitrary matrix $\mathbf{M}_N$ in terms of n factors (or stages) $\mathbf{E}_N(i)$ is written as follows:

$$\mathbf{M}_N = \prod_{i=1}^{n} \mathbf{E}_N(i) = \mathbf{E}_N(n) \cdots \mathbf{E}_N(1). \tag{6}$$

3 Cooley-Tukey Factorizations

3.1 Radix-2 Cooley-Tukey Factorizations

Suppose that N is a power of 2 and j denotes the square root of -1. The Fourier transform matrix $\mathbf{F}_N$ is defined as:

$$\mathbf{F}_N = \left\{ e^{-j\frac{2\pi}{N}(m-1)(n-1)} \right\} \qquad m,n = 1:N. \tag{7}$$

The Inverse Fourier transform matrix $\mathbf{F}_N^H$ -a scale factor is omitted-, is related with the hermitian of $\mathbf{F}_N$. Let us consider the following well-known recursion properties involving matrices $\mathbf{F}_N$ and $\mathbf{F}_{N/2}$:

$$\mathbf{F}_N = \mathbf{B}_N\left(\mathbf{I}_2 \otimes \mathbf{F}_{N/2}\right)\mathbf{P}_N, \tag{8}$$

$$\mathbf{F}_N = \mathbf{P}_N{}^T \left(\mathbf{I}_2 \otimes \mathbf{F}_{N/2}\right)\mathbf{B}_N{}^T, \tag{9}$$

$$\mathbf{F}^H{}_N = \mathbf{P}_N{}^T \left(\mathbf{I}_2 \otimes \mathbf{F}_{N/2}{}^H\right)\mathbf{B}_N{}^H, \tag{10}$$

$$\mathbf{F}^H{}_N = \mathbf{B}^*{}_N \left(\mathbf{I}_2 \otimes \mathbf{F}^H{}_{N/2}\right)\mathbf{P}_N. \tag{11}$$

Matrix **B** is defined using the identity matrix **I** and the diagonal matrix **A** as:

$$\mathbf{B}_{2^i} = \begin{bmatrix} \mathbf{I}_{2^{i-1}} & \mathbf{A}_{2^{i-1}} \\ \mathbf{I}_{2^{i-1}} & -\mathbf{A}_{2^{i-1}} \end{bmatrix}. \tag{12}$$

where the diagonal matrix **A** is:

$$\mathbf{A}_N = diag\left\{ e^{-j\frac{2\pi}{N}(i-1)} \right\} \qquad i = 1:N. \tag{13}$$

In expressions (9-11) upper index T, H and * denote the transpose, hermitian and complex conjugate respectively. The well-known radix-2 Cooley-Tukey factorizations can be obtained from (8-11) when these recursions are iterated. They can be written in the presented notation taking into account that the stop criterion of the recursive process is:

$$\mathbf{F}_2 = \mathbf{F}_2^H = \begin{bmatrix} 1 & 1 \\ 1 & -1 \end{bmatrix}. \tag{14}$$

By (8), we have:

$$\mathbf{F}_N = \prod_{i=1}^{n}\left(\mathbf{I}_{2^{n-i}} \otimes \mathbf{B}_{2^i}\right)\prod_{i=1}^{n}\left(\mathbf{I}_{2^{i-1}} \otimes \mathbf{P}_{2^{n-i+1}}\right) = \prod_{i=1}^{n}\left(\mathbf{I}_{2^{n-i}} \otimes \mathbf{B}_{2^i}\right)\mathbf{R}_N. \tag{15}$$

And by (9):

$$\mathbf{F}_N = \prod_{i=1}^{n}\left(\mathbf{I}_{2^{n-i}} \otimes \mathbf{P}_{2^i}^T\right)\prod_{i=1}^{n}\left(\mathbf{I}_{2^{i-1}} \otimes \mathbf{B}_{2^{n-i+1}}^T\right) = \mathbf{R}_N \prod_{i=1}^{n}\left(\mathbf{I}_{2^{i-1}} \otimes \mathbf{B}_{2^{n-i+1}}^T\right). \tag{16}$$

Note than the permutation matrix $\mathbf{R}_N$ known as the bit-reversal permutation matrix, appears in (15) and in (16) is written in two different ways. That is:

$$\mathbf{R}_N = \prod_{i=1}^{n}\left(\mathbf{I}_{2^{i-1}} \otimes \mathbf{P}_{2^{n-i+1}}\right) = \prod_{i=1}^{n}\left(\mathbf{I}_{2^{n-i}} \otimes \mathbf{P}_{2^i}{}^T\right) \tag{17}$$

Matrix $\mathbf{R}_N$ is well-known to be equal to its inverse. In fact we have $\mathbf{R}_N = \mathbf{R}_N^{-1} = \mathbf{R}_N^H = \mathbf{R}_N^T = \mathbf{R}_N^*$. Computing $\mathbf{R}_N$ only means a very simple hardware-made reordering. Note we have $n = \log_2 N$ factors or radix-2 stages. The factorizations obtained from (8-9) for the FFT can be rewritten by changing $\mathbf{R}_N$ in the left part of the next equations by $\mathbf{R}_N^{-1}$, in order to present them as a product of n radix-2 stages. This is:

$$\mathbf{F}_N \mathbf{R}_N^{-1} = \prod_{i=1}^{n}\left(\mathbf{I}_{2^{n-i}} \otimes \mathbf{B}_{2^i}\right). \tag{18}$$

$$\mathbf{R}^{-1}{}_N\mathbf{F}_N = \prod_{i=1}^{n}\left(\mathbf{I}_{2^{i-1}} \otimes \mathbf{B}^T{}_{2^{n-i+1}}\right) \tag{19}$$

$$\mathbf{R}^{-1}{}_N\mathbf{F}^H{}_N = \prod_{i=1}^{n}\left(\mathbf{I}_{2^{i-1}} \otimes \mathbf{B}^H{}_{2^{n-i+1}}\right) \tag{20}$$

$$\mathbf{F}^H{}_N\mathbf{R}^{-1}{}_N = \prod_{i=1}^{n}\left(\mathbf{I}_{2^{n-i}} \otimes \mathbf{B}^*{}_{2^i}\right) \tag{21}$$

3.2 Radix-R Cooley-Tukey Factorizations

When the transform matrix $\mathbf{F}_N$ has $R^E \times R^E$ dimensions, E being an integer and R the F power of 2 ($R=2^F$) it is possible to find radix-R factorizations easily by observing that radix-R stages can be written as F products of consecutive radix-2 factors. Such a result comes directly from the used notation. The radix-R equivalent factorizations to expressions (18-21) can be written as:

$$\mathbf{F}_N\mathbf{R}^{-1}{}_N = \prod_{i=1}^{E}\prod_{f=1}^{F}\left(\mathbf{I}_{2^{EF-(i-1)F-f}} \otimes \mathbf{B}_{2^{(i-1)F+f}}\right) \tag{22}$$

$$\mathbf{R}^{-1}{}_N\mathbf{F}_N = \prod_{i=1}^{E}\prod_{f=1}^{F}\left(\mathbf{I}_{2^{(i-1)F+f-1}} \otimes \mathbf{B}^T_{2^{E-iF-f+1}}\right) \tag{23}$$

$$\mathbf{R}^{-1}{}_N\mathbf{F}^H_N = \prod_{i=1}^{E}\prod_{f=1}^{F}\left(\mathbf{I}_{2^{(i-1)F+f-1}} \otimes \mathbf{B}^H_{2^{E-iF-f+1}}\right) \tag{24}$$

$$\mathbf{F}_N\mathbf{R}^{-1}{}_N = \prod_{i=1}^{E}\prod_{f=1}^{F}\left(\mathbf{I}_{2^{EF-(i-1)F-f}} \otimes \mathbf{B}^*_{2^{(i-1)F+f}}\right) \tag{25}$$

Expressions (22-25) can be operated in order to find more compact expressions but for our purpose it is not necessary. If we see (22) we realize that the family of solutions provided for the radix-R factors **E**(i), where i goes from 1 to E, take the form:

$$\mathbf{E}_N(i) = \prod_{f=1}^{F}\left(\mathbf{I}_{2^{EF-(i-1)F-f}} \otimes \mathbf{B}_{2^{(i-1)F+f}}\right) \tag{26}$$

Note also that the matrices $\mathbf{E}_N(i)$ are sparse matrices with R non-zero elements in each row and each column. In a similar way we can obtain the radix-R stages **E**(i) for (23) (24) and (25) .

4 General Radix-R Pease Factorizations

We can obtain the radix-R equal interconnection pattern factorizations starting from expressions (22-25) with the introduction of the appropriate permutation matrices between stages in order to change interconnection patterns without changing the result of the full factorization. As the derivation method is exactly the same for the families obtained from (22) and (25) and for the families obtained from (23) and (24) only the first two are shown. Let us begin with (21). If we introduce the powers of permutation matrices **P** in the following way we do not change the final result because the permutation matrices introduced between them are always the permutation matrix and its inverse. See below:

$$\mathbf{F}_N\mathbf{R}_N^{-1} = \prod_{i=1}^{E}\prod_{f=1}^{F}\mathbf{P}_N^{(i-1)F+f}\left(\mathbf{I}_{2^{EF-(i-1)F-f}} \otimes \mathbf{B}_{2^{(i-1)F+f}}\right)\mathbf{P}_N^{-(i-1)F-f+1}. \tag{27}$$

It is interesting to make an approximation to (27) for the radix-2 case when F is 1 and E is the log2 of N. In all cases, note that for i=1 and f=1 the first stages are obtained by post-multiplying with the identity matrix **I**. For i=E and f=F the last stages are also obtained by pre-multiplying by the identity matrix **I** as $EF=\log_R N \bullet \log_2 R = \log 2N = n$, so

$$\mathbf{P}_{2^n}^0 = \mathbf{P}_{2^n}^{EF} = \mathbf{P}_{2^n}^{-EF} = \mathbf{I}_{2^n}. \tag{28}$$

The factorizations in (27) have regular interconnection pattern stage-to-stage as shown above. They can be simplified using the Kroneker product property (5) that allows the following equality:

$$\mathbf{I}_{2^{EF-(i-1)F-f}} \otimes \mathbf{B}_{2^{(i-1)F+f}} = \mathbf{P}_N^{EF-(i-1)F-f}\left(\mathbf{B}_{2^{(i-1)F+f}} \otimes \mathbf{I}_{2^{EF-(i-1)F-f}}\right)\mathbf{P}_N^{(i-1)F+f}. \tag{29}$$

By combining (29) and (27), with properties (2-4) we have:

$$\mathbf{F}_N\mathbf{R}_N^{-1} = \prod_{i=1}^{E}\prod_{f=1}^{F}\left(\mathbf{B}_{2^{(i-1)F+f}} \otimes \mathbf{I}_{2^{EF-(i-1)F-f}}\right)\mathbf{P}_N. \tag{30}$$

And the result is:

$$\mathbf{E}(i) = \prod_{i=1}^{E}\prod_{f=1}^{F}\left(\mathbf{B}_{2^{(i-1)F+f}} \otimes \mathbf{I}_{2^{EF-(i-1)F-f}}\right)\mathbf{P}_N. \tag{31}$$

Let us continue with the family obtained from (22). Now to get solutions with regular interconnection pattern stage-to-stage we have to introduce the powers of matrix **P** as follows:

$$\mathbf{R}_N^{-1}\mathbf{F}_N = \prod_{i=1}^{E}\prod_{f=1}^{F}\mathbf{P}_N^{-(i-1)F-f}\left(\mathbf{I}_{2^{(i-1)F+f-1}} \otimes \mathbf{B}^T_{2^{EF-F(i-1)-f+1}}\right)\mathbf{P}_N^{(i-1)F+f-1}. \tag{32}$$

This operation doesn't change the result of $\mathbf{R}_N^{-1}\mathbf{F}_N$ because, as in the previous case, the two matrices introduced between radix-2 factors are always a permutation matrix and just its inverse. Using the property (5) we have:

$$\mathbf{I}_{2^{(i-1)F+f-1}} \otimes \mathbf{B}^T_{2^{EF-(i-1)F-f+1}} = \mathbf{P}_N^{(i-1)F+f-1}\left(\mathbf{B}^T_{2^{EF-(i-1)F-f+1}} \otimes \mathbf{I}_{2^{(i-1)F+f-1}}\right)\mathbf{P}_N^{EF-(i-1)F-f+1} \tag{33}$$

and as $\mathbf{P}_N^{-1}=\mathbf{P}_N^T$ and $\mathbf{P}_N^{EF}=\mathbf{I}_N$, expression (32) simplifies to:

$$\mathbf{R}_N^{-1}\mathbf{F}_N = \prod_{i=1}^{E}\prod_{f=1}^{F}\mathbf{P}_N^T\left(\mathbf{B}^T_{2^{EF-(i-1)F-f+1}} \otimes \mathbf{I}_{2^{(i-1)F+f-1}}\right). \tag{34}$$

That is the other result we are seeking.

$$\mathbf{E}'(i)=\prod_{f=1}^{F}\mathbf{P}_N^T\left(\mathbf{B}_{2^{EF-(i-1)F-f+1}}^T\otimes\mathbf{I}_{2^{(i-1)F+f-1}}\right) \tag{35}$$

The factorizations given by (30) and (34) have the property of having the same interconnection pattern stage-to-stage. A way to show that the stages in (31) and (35) have an identical interconnection pattern stage-to-stage is based on replacing matrix **B** by another simpler matrix **B+** having its non-zero elements in the same positions since the interconnection pattern is given by the position of the non-zero elements in each sparse matrix. To form **B+**, we replace in (12) the diagonal matrix **A** by the identity matrix **I** in the following way:

$$\mathbf{B}_{2^i}^+=\begin{bmatrix}\mathbf{I}_{2^{i-1}} & \mathbf{I}_{2^{i-1}}\\ \mathbf{I}_{2^{i-1}} & -\mathbf{I}_{2^{i-1}}\end{bmatrix}=\mathbf{F}_2\otimes\mathbf{I}_{2^{i-1}}. \tag{36}$$

As an example, if we replace **B** by $\mathbf{B}^+$ in the factors in (31) we will show, using (2-4), that the modified stages are independent of *i*. This is:

$$\left(\mathbf{B}_{2^i}^+\otimes\mathbf{I}_{2^{n-i-a}}\right)\mathbf{P}_{2^{n-a}}=\left(\mathbf{F}_2\otimes\mathbf{I}_{2^{i-1}}\otimes\mathbf{I}_{2^{n-i-a}}\right)\mathbf{P}_{2^{n-a}}=\left(\mathbf{F}_2\otimes\mathbf{I}_{2^{n-a-1}}\right)\mathbf{P}_{2^{n-a}}. \tag{37}$$

In a similar way, we can obtain the same kind of factorizations for the IFFT. Then for IFFT (except for a constant) we have the first family of solutions as:

$$\mathbf{R}_N^{-1}\mathbf{F}_N^H=\prod_{i=1}^{E}\prod_{f=1}^{F}\mathbf{P}_N^T\left(\mathbf{B}_{2^{EF-(i-1)F-f+1}}^H\otimes\mathbf{I}_{2^{(i-1)F+f-1}}\right) \tag{38}$$

$$\mathbf{E}''(i)=\prod_{f=1}^{F}\mathbf{P}_N^T\left(\mathbf{B}_{2^{EF-(i-1)F-f+1}}^H\otimes\mathbf{I}_{2^{(i-1)F+f-1}}\right) \tag{39}$$

and the second family:

$$\mathbf{F}_N^H\mathbf{R}_N^{-1}=\prod_{i=1}^{E}\prod_{f=1}^{F}\left(\mathbf{B}_{2^{(i-1)F+f}}^*\otimes\mathbf{I}_{2^{EF-(i-1)F-f}}\right)\mathbf{P}_N. \tag{40}$$

$$\mathbf{E}'''(i)=\prod_{f=1}^{F}\left(\mathbf{B}_{2^{(i-1)F+f}}^*\otimes\mathbf{I}_{2^{EF-(i-1)F-f}}\right)\mathbf{P}_N. \tag{41}$$

As from the indices *m,n* of the non-zero elements $e_{m,n}$ in each sparse matrix representing a stage, we can observe that the *n* input element is needed to calculate the *m* output element in the i-th stage.

Another way to see that the stages defined in (31), (35), (39) and (41) have the same interconnection pattern for large values of N is by representing each sparse matrix of a given factorization as an image in which the zero elements are represented by one colour and the non-zero elements with another colour.

In the case of equal stage-to-stage factorizations, once a factorization is given, all the images representing a factor are equal. In Fig. 1, the equal interconnection pattern stage-to-stage factorizations for an input vector **x** to an output vector **y** when N=16 and R=4 is represented. In Fig. 2 the same representation is done for N=16 and R=4.

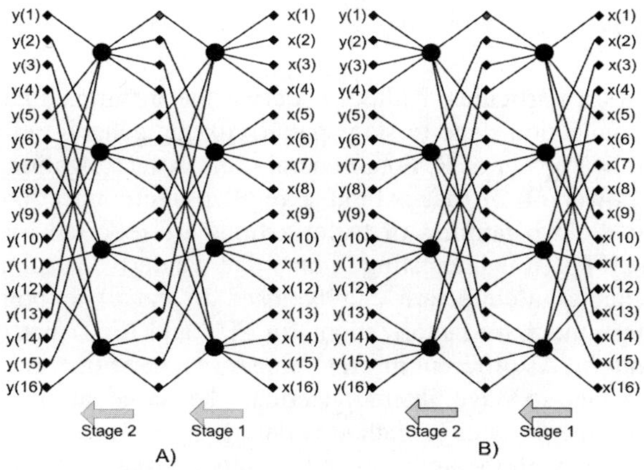

Fig. 1. Equal Interconnection stage pattern architectures for N=16 and R=4

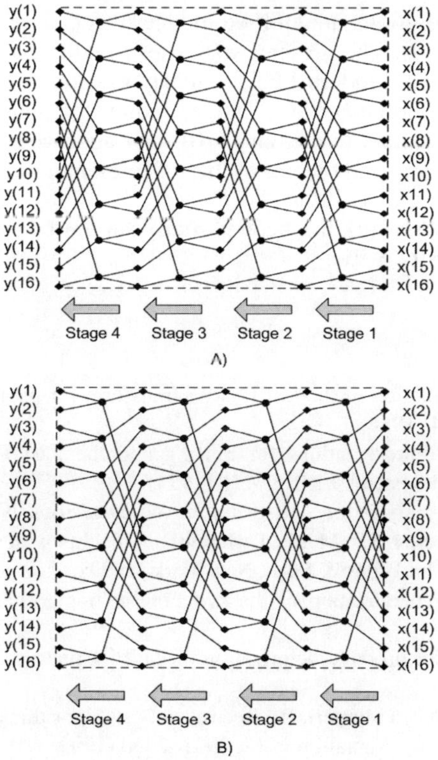

Fig. 2. Equal Interconnection stage pattern architectures for N=16 and R=2

It is interesting to note that factorizations given by (30) for FFT and by (40) for IFFT share the architecture A), and factorizations given by (34) for FFT and (38) for IFFT share the architecture B).

5 Conclusions

This article shows a particular method to derive the different FFT/IFFT topologies with equal interconnection stage-to-stage pattern for the general radix-R case, R being a power of 2 and N a power of R. Four general radix-R factorizations for any length N (where N is a power of 2) that exhibit a regular interconnection pattern between stages is presented. Two families of factorizations are reported for the FFT and another two for the IFFT. It is interesting to note that we derive two families of architectures but a particular architecture can be used to calculate both FFT and IFFT transforms because there are factorizations for FFT and IFFT that share the same interconnection pattern. As different discrete transforms have factorizations with a Cooley-Tukey type stage-to-stage interconnection, the same argument can easily be extended to them. The derivation method is based on the introduction of the even-odd permutation matrices with the appropriated powers between the stages. It will be also interesting to extend these kinds of factorizations, using the Kronecker product properties to the two dimensional case. The use of the even-odd permutation matrices to write all other permutation matrices involved in the derivation process and the presentation of the different solutions in a unified manner offers a new point of view of parallel FFT/IFFT algorithms. The present work tries to contribute to the understanding of fast parallel algorithms. Modern FFT architectures tend to optimize the number of operations and tend to implement much larger radices combined with other methods to manage efficiently the data in memory [10]. Our approach offers very regular and efficient memory managing and the possibility of implementing high radices.

Acknowledgements. This work has been partially supported by the cost center R008-R0904 from the University o Vic.

References

1. Pease, M.C.: An adaptation of the fast Fourier transform for parallel processing. J. Assoc. Comput. 15, 252–264 (1968)
2. Sloate, H.: Matrix Representations for Sorting and the Fast Fourier Transform. IEEE Trans.on Circuits and Systems 21, 109–116 (1974)
3. Schott, J.R.: Matrix Analysis for Statistics. John Wiley & Sons, New York (1996)
4. Magnus, J.R., Neudecker, H.: Matrix Differential Calculus with Applications in Statistics and Econometrics. John Wiley & Sons, New York (1999)
5. Glassman, J.A.: A generalization of the fast Fourier transform. IEEE Trans. Comp. 19, 105–116 (1970)
6. Drubin, M.: Kronecker product factorization of the FFT matrix. IEEE Trans. 20, 590–593 (1971)
7. Granata, J., Conner, M., Tolimieri, R.: Recursive Fast Algorithms and the Role of the Tensor Product. IEEE Trans. Signal Proc. 12, 2921–2930 (1992)
8. Egner, S., Püschel, M.: Automatic Generation of Fast Discrete Signal Transforms. IEEE Trans. on Signal Proc. 49, 1992–2002 (2001)
9. Marti-Puig, P.: A Family of Fast Walsh Hadamard Algorithmsw with Identical Sparse Matrix Factorization. IEEE Signal Proc. Let. 13, 672–675 (2006)
10. Johnson, S.G., Frigo, M.: A Modified Slitp-Radix FFT with Fewer Arithmetic Operations. IEEE Trans. Signal Processing 55, 111–119 (2007)

Improving Evolution of XSLT Stylesheets Using Heuristic Operators*

P. García-Sánchez, J.J. Merelo, J.L.J. Laredo, A.M. Mora, and P.A. Castillo

GeNeura Team. Departament of Computer Architecture and Computer Technology
University of Granada, Spain
pgarcia@geneura.ugr.es

Summary. This paper presents a new version of an evolutionary algorithm that creates XSLT programs from scratch, given a single example of their intended input and output. XSLT is a general purpose, document-oriented functional language, generally used to transform XML documents (or, in general, solve any problem that can be coded as an XML document). Previously, an efficient solution to the problem was proposed; in this paper, we improve on those results by testing different fitness functions, adding a new operator and changing the type of output document that can be obtained. Results show that the best results are obtained without considering the length of the XSLT program and including this new operator. The new type of output files used is also more realistic, and improves on old results.

Keywords: Stylesheets, XML, XSLT, Evolutionary Algorithms.

1 Introduction

Since the Information Technology industry has settled on different Extensible Markup Language (XML) dialects as information exchange format, there is a business need for programs that transform from one XML set of tags to another, extracting information or combining it in many possible ways; a typical example of this transformation could be the extraction of news headlines from a newspaper in Internet that uses XHTML. XSLT stylesheets (XML Stylesheet Language for Transformations) [1], also called *logicsheets*, are programs designed for this purpose: applied to an XML document, they produce another. There are other possible solutions: programs written in any language that work with text as input and output (using, for instance, regular expressions) or SAX filters [2], that process each tag in a XML document in a different way, and do not need to load into memory the whole XML document. However, they need external languages to work, while XSLT is a part of the XML set of standards (in fact, XSLT logicsheets are XML documents) which can be integrated within an XML framework; that is why XSLT is, if not the most common, at least a quite usual way of transforming XML documents. XSLT make use of XPath expressions [3] to select nodes from the source document.

* Supported by projects TIN2007-68083-C02-01, P06-TIC-02025 and OTRI-1515.

J.M. Corchado et al. (Eds.): DCAI 2008, ASC 50, pp. 161–170, 2009.
springerlink.com

The amount of work needed for logicsheet creation is a problem that scales quadratically with the quantity of initial and final formats. For n input and m output formats, $n \times m$ transformations will be needed. Considering that each conversion is a hand-written program and the initial and final formats can vary with certain frequency, any automation of the process means a considerable saving of effort on the part of the programmers.

So, the problem is to find the XSLT logicsheet that, from one input XML document, is able to obtain an output XML document which contains exclusively the information desired from the first one. This information may be sorted in any possible way (possibly in an order different to the input document). In this work, an Evolutionary Algorithm (EA) [4] to resolve this problem is presented. The logicsheet will be evolved using evolutionary operators that will take into account the structure of the program and its components.

In a previous work [5], we published the results of an initial version of this method for XSLT evolution, testing different document structures and operators. In this paper we will try to improve on those results, by using XML output documents with tree structure, instead of plain text-only documents. This means that output documents are composed of several nodes, which makes it easier to compare them with each other. So, the output XML will be a complete XML document with a (possibly sorted in a different way) list of nodes present in the original document.

The rest of the paper is structured as follows: the state of the art is presented in Section 2. Section 3 describes the solution presented in this work, with the novel elements introduced. Experiments with the automatic generation of XSLT stylesheets for different examples are described in Section 4, and finally the conclusions and possible lines of future work are presented in Section 5.

2 State-of-the-Art

To our knowledge, there are few works related to the application of genetic programming techniques to the automatic generation of XSLT logicsheets; one of them, by Scott Martens [6], presents a technique to find XSLT stylesheets that transform a XML file into HTML by using genetic programming. Martens works on simple XML documents and uses the UNIX diff function as the basis for its fitness function. He concludes that genetic programming is useful to obtain solutions to simple examples of the problem, but it needs unreasonable execution times for complex examples and might not be a suitable method to solve this kind of problems.

Schmidt and Waltermann [7] approached the problem taking into account that XSLT is a functional language, and using functional language program generation techniques on it, in what they call *inductive synthesis*. First they create a non-recursive program, and then, by identifying recurrent parts, convert it into a recursive program; this is a generalization of the technique used to generate programs in other programming languages such as LISP [8], and used thoroughly since the eighties [9].

A few other authors have approached the general problem of generating XML document transformations knowing the original and target structure of the documents, as represented by its DTD (Document Type Definition): Leinonen et al. [10, 11] have proposed semi-automatic generation of transformations for XML documents, but user input is needed to define the label association. There are also freeware programs that perform transformations on documents from a XSchema to another one. However, both XSchemata must be known in advance, and are not able to accomplish general transformations on well formed XML documents from examples.

In our previous work [5], we presented an evolutionary algorithm to obtain an XSLT program that extracts information represented in a output XML from an input XML. Several XSLT structures and operators were presented and studied. The main inconvenient of that work is the output XML file is a list of text elements instead of XML nodes, which would be much more useful to perform real XML transformations. Additionally, the existing operators apparently led to situations where evolutionary changes were quite difficult, so a new operator is proposed. In this paper, the XML output document includes a set of nodes (text and tags) extracted from the input document, which can be additionally processed to change the required output tags.

3 Methodology

The algorithm described here evolves XSLT stylesheets, which are generated (from an initial, random population) using a set of operators and evaluated using a fitness function that is related to the difference between generated XML and output XML associated to the example. The way the algorithm works is shown in Figure 1.

The solution has been programmed using JEO [12], an evolutionary algorithm library developed at University of Granada as part of the DREAM project [13], which is available from `http://www.dr-ea-m.org`.

Since the search space of possible stylesheets is exceedingly large, language grammar must be considered in order to restrict it and avoid syntactically wrong stylesheet generation. Due to this, transformations are applied to a predetermined stylesheet structure which was selected among three different ones in previous work [5]. An example of this structure is shown in Figure 2. This type of structure is more constrained than other types; and search is thus easier, since less stylesheets are generated. Despite the constraints, mutation and crossover are much more disruptive, generating a rougher landscape than before.

The operators may be classified in two different types: the first one consists in operators that modify XPath routes in the attributes of the XSLT instructions (`apply-template` and `copy-of`); and the other are the operators used to modify the XSLT tree structure. In order to ensure the existence of the elements (tags) added to the XPath expressions and XSLT instruction attributes, every time one of them is needed it is randomly selected from the input file. These operators, whose names should be self-descriptive, have been described in more detail in

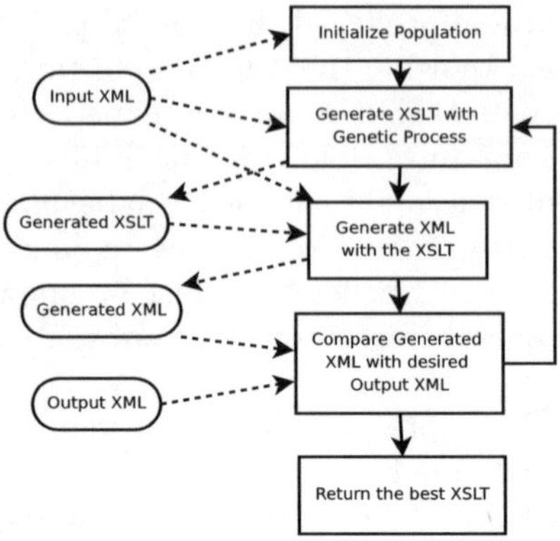

Fig. 1. This figure shows how the algorithm works. Each individual of the population is an XSLT stylesheet whose fitness is computed from matching the generated to the target XML.

```
<?xml version="1.0"?>
<xsl:stylesheet version="1.0"
  xmlns:xsl="http://www.w3.org/1999/XSL/Transform">
  <xsl:output indent="no" method="xml"/>
  <xsl:template match="/">
    <grammar>
      <xsl:apply-templates select="/grammar"/>
    </grammar>
  </xsl:template>
  <xsl:template match="/grammar">
    <xsl:copy-of select="div[3]"/>
    <xsl:copy-of select="div[1]/h3[5]"/>
    <xsl:copy-of select="h1"/>
  </xsl:template>
</xsl:stylesheet>
```

Fig. 2. Example of an XSLT generated by the algorithm

our previous work [5], so we refer the interested reader to that paper; a complete list, with the rates used, is shown in Table 1. When an operator of the second group of the table is selected to modify an individual, another operator is selected randomly from the first group to be applied together.

However, it was seen in our previous experiments these operators are not enough to perform an efficient search; sometimes the XSLT search converges into a bad solutions when we want to select ordered but alternated items from a node. So it is necessary to add a new operator to increase the diversification of this solution. The new operator proposed, XSLTreeMutatorSplitTemplate, expands a random `copy-of` node into a list of complete `copy-of` with all cardinalities (as shown in Figure 3). The result of applying the new XSLT and the previous is the same, but it is easier for genetic operators to modify the list of copy-of than the generic one (modifyng, adding or removing XPath and/or tags).

Since in this paper output is a fully formed XML document, fitness has been changed to be the XML difference between the desired and the obtained output,

```
<xsl:template match="book">
    <xsl:copy-of select="chapter"/>
</xsl:template>
```

```
<xsl:template match="book">
  <xsl:copy-of select="chapter[1]"/>
  <xsl:copy-of select="chapter[2]"/>
  <xsl:copy-of select="chapter[3]"/>
  <xsl:copy-of select="chapter[4]"/>
</xsl:template>
```

Fig. 3. A template before (left) and after (right) split mutator is executed. The number of chapters in the input XML were 4.

that is, the difference in nodes between the desired T and the actual document X. This difference breaks down in insertions (nodes in X but not in T) and deletions (nodes in T but not in X). We will leverage this vectorial structure of fitness so that evolution can profit from it: instead of using a single aggregative function, as we did in previous papers [5], fitness is now a vector that includes the number of node deletions and additions needed to obtain the target output from the obtained output, and the resulting XSLT stylesheet length. The XSLT stylesheet is correct only if the number of deletions and additions is 0; and minimizing length could help removing useless statements from it. So, fitness is minimized by comparing individuals as follows: An individual is considered better than another if the number of deletions is smaller; if the number of additions is smaller, being the number of deletions the same or if the length is smaller, being the number of deletions/additions the same.

Separating and prioritizing the number of deletions helps guide evolution, by trying to find first a stylesheet that includes all elements in the target document, then eliminating unneeded elements, while, at the same time, reducing length. However, this last element introduces selective pressure towards small stylesheets, which might hinder discovering the correct one, so we have also tested in this paper whether we should consider length or not as a part of the fitness.

4 Experiments and Results

To test the algorithm we have performed several experiments with 7 different XML input and output files. The algorithm has been executed 30 times for each input XML. Every experiment took 200.76 seconds in average to finish. The same input file was used for several experiments: a RSS feed from a weblog (`http://geneura.wordpress.com`) and an XHTML file. All input and output files and programs used in this experiment are available from our Subversion repository: `http://tinyurl.com/6nxv8c`.

The computer used to perform the experiments is a Centrino Core Duo at 1.83 GHz, 2 GB RAM, and the Java Runtime Environment 1.6.0.01. The population size was 128 individuals for all runs, generated using the input XML as information source. The termination criteria was set to 300 generations or until a solution was found, and selection was performed via a 5-Tournament; 30 experiments were

Table 1. Operator priorities (used for the roulette wheel that randomly selects the operator to apply) used in the experiments

Operator	Priority
XSLTTreeMutatorXPathSetSelf	0.1
XSLTTreeMutatorXPathRemoveBranch	0.17
XSLTTreeMutatorXPathAddFilter	0.18
XSLTTreeMutatorXPathMutateFilter	0.18
XSLTTreeMutatorXPathRemoveFilter	0.2
XSLTTreeMutatorXPathAddBranch	0.16
XSLTTreeMutatorAddTemplate	0.2
XSLTTreeMutatorMutateTemplate	0.10
XSLTTreeMutatorRemoveTemplate	0.12
XSLTTreeAddApply	0.1
XSLTTreeMutateApply1	0.1
XSLTTreeMutateApply2	0.14
XSLTTreeRemoveApply	0.1
XSLTreeMutatorSplitTemplate	0.05
Probability of crossover	0.25
Probability of mutation	0.5

run, with different random seeds, for each input document. The XML and XSLT processors were the default ones included in the JRE standard library. The operator rates used in the experiments, which were tuned heuristically, are shown in Table 1. The crossover and mutation probability have been set to 0.25 and 0.5, after several experimental runs, whose results are shown in table 2.

Due to the use of the new mutation operator we have performed the experiments using 3 different configurations. The first is the algorithm without the

Table 2. Average generations/standard deviation to find an optimal solution (in less of 300 generations), using different mutation/crossover rates

Mutation/Crossover	Average Generations	Solutions found
0.5/0.25	107.8 ± 128.74	21
0.5/0.5	159.8 ± 142.7	15
0.5/0.75	135.4 ± 137.2	18
0.25/0.25	131.67 ± 131.85	19
0.25/0.5	134.57 ± 129.99	19
0.25/0.75	168.67 ± 134.11	15
0.75/0.25	126.27 ± 135.23	19
0.75/0.5	159.77 ± 142.71	15
0.75/0.75	178.7 ± 141.13	13

new operator (XSLTreeMutatorSplitTemplate), the second one, using the operator and the third without considering the length of the stylesheet in the fitness function. This helps to keep the solutions which have the same number of deletions and insertions but larger size caused by the use of the operator (expanding the copy-of tags). The breakdown of results per input file is shown in Table 3.

Table 3. Number of times, out of 30 experiments, a solution is found within the predefined number of generations without the split mutator, using the split mutator with normal fitness, and using split mutator without considering of the length of the generated stylesheet.

Input file	Without Split	With Split	With Split w/o length
1	26	25	25
2	30	30	30
3	30	30	30
4	0	24	24
5	2	30	29
6	30	30	30
7	10	9	13

Examples 1, 2, 3 and 6 are complete and ordered lists of elements of one or several nodes, and the algorithm can easily create a logicsheet that extracts all the childrens of a specific node. Example 7 takes specific and repeated elements from distinct nodes, and different expressions are needed to extract each one of them, so the generated logicsheet is more complex. Finally, examples 4 and 5 focus into portions of ordered and unordered fragments of an XML section (i.e. the 3rd and 6th chapters of a book) so the population converges to solutions with all the elements (selecting all chapters of a book) due to the way the fitness works. Obtaining solutions for these examples is quite difficult without using the new operator (two times is found for example 5 and none for 4), but this is fixed when we use it; instead of selecting all chapters of a book (`book/chapter`), it selects all chapters using XPath location (`book/chapter[0], book/chapter[1]...`), so it is easier for the algorithm to change this into better solutions.

On the other hand, overruling the XSLT length comparison in fitness gives different solutions the same chances to evolve, so the algorithm maintains more diversity and finds solutions in less time than the cases comparing that length (see Table 4). However, the generated XSLT may contain useless statements, that could produce wrong XMLs in a a production environment.

When a solution is found, the number of generations and time used to find it also varies, as shown in Table 4. In general, the exploration/exploitation balance seems to be biased towards exploration. Being such a vast and rough search space makes that, after a few initial generations that create stylesheets with a small difference from the target, mutations are the main operator at work.

Table 4. Average generations/standard deviation to find an optimal solution (in less of 300 generations), without the split mutator, using the split mutator with normal fitness, and using split mutator without considering of the length of the generated stylesheet

Input file	W/o Split	With Split	With Split w/o length
1	62.86 ± 102.03	83.33 ± 112.81	66.33 ± 106.85
2	1.5 ± 1.25	1.6 ± 1.30	1.1 ± 1.29
3	4.13 ± 2.06	3.13 ± 1.94	3.83 ± 2.90
4	-	81.03 ± 112.25	71.68 ± 107.24
5	289.9 ±45.09	26.83 ± 5.35	36.03 ± 50.19
6	15.43 ± 5.74	20.43 ± 9.88	19.0 ± 11.12
7	232.17 ±105.48	245.46 ± 96.92	214.0 ± 112.90

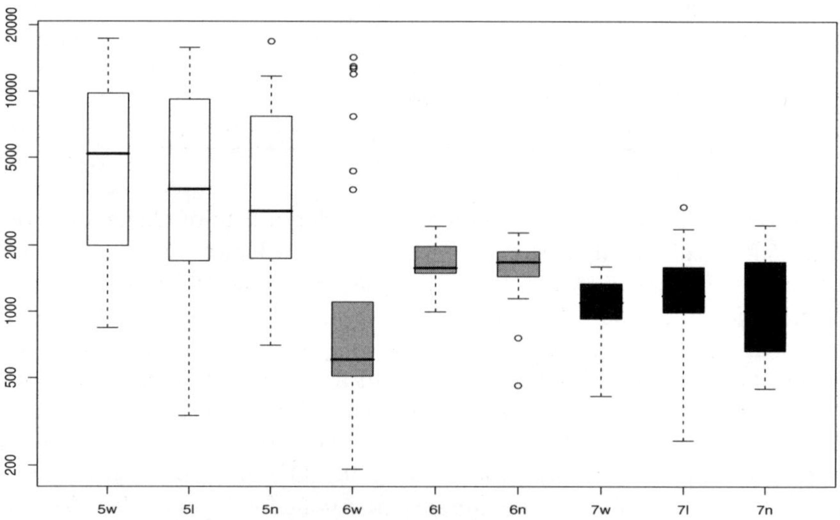

Fig. 4. Logarithmic boxplot of the number of evaluations needed to find the best individual in examples #5, #6 and #7 without split mutator (w), with split mutator considering length (l) and without this feature (nl)

5 Conclusions, Discussion, and Future Work

In this paper we present the results of an evolutionary algorithm designed to search the XSLT logicsheets that is able to make a particular transformation from an input XML document into a desired output one; one of the advantages of this application is that resulting logicsheets can be used directly in a production environment, without the interaction of a human operator. It tackles a real-world problem found in many organizations and it is open source software, available from `http://tinyurl.com/5lwjcn`.

The experiments have shown that the search space is particularly rough, with mutations in general leading to huge changes in fitness. The hierarchical fitness used is probably the cause of having a big loss of diversity at the beginning of the evolutionary search, leading to the need of a higher level of explorations later during the algorithm run. This problem will have to be approached via explicit diversity-preservation mechanisms, or by using a multiobjective evolutionary algorithm, instead of the one used now. A deeper understanding of how different operator rates affect the result will also help; for the time being, operator rate tuning has been very shallow, and geared towards obtaining the result. In addition, results shown in this paper can be used as a baseline for future versions of the algorithm, or other algorithms for the same problem. At any rate, unlike what was mentioned in the pioneering paper [6], solutions can be found effectively and efficiently using EAs.

However, there are some questions and issues that will have to be addressed in future papers:

- Using the DTD (associated to a XML file) as a source of information for conversions between XML documents and for restrictions of the possible variations.
- Adding different labels in the XSLT to allow the building of different kinds of documents such as HTML or WML.
- Testing evolution with other kind of tools, such as a chain of SAX filters.
- Obviously, testing different kinds and increasingly complex set of documents, and using several input and output documents at the same time, to test the generalization capability of the procedure.
- Tackle difficult problems from the point of view of a human operator. In general, the XSLT stylesheets found here could have been programmed by a knowledgeable person in around an hour, but in some cases, input/output mapping would not be so obvious at first sight. This will mean, in general, increase also the XSLT statements used in the stylesheet, and also in general, adding new types of operators.

References

1. Clark, J.: XSL transformations (XSLT), version 1.0, W3C recommendation November 16 (1999), `http://www.w3.org/TR/xslt.html`
2. Wikipedia: Simple API for XML — Wikipedia, the free encyclopedia (2007) (accessed, March 21, 2007)
3. Clark, J., S.,, DeRose, o.: XML Path Language (XPath) Version 1.0. W3C Recommendation 16 (1999)
4. Goldberg, D.E.: Genetic Algorithms in Search, Optimization, and Machine Learning. Addison-Wesley, Boston (1989)
5. Garcia-Sanchez, P., Laredo, J.L.J., Sevilla, J.P., Castillo, P., Merelo, J.J.: Improved evolutionary generation of XSLT stylesheets. ArXiV database 1926 (2008), `http://arxiv.org/abs/0803`
6. Martens, S.: Automatic creation of XML document conversion scripts by genetic programming. In: Genetic Algorithms and Genetic Programming at Stanford, p. 269 (2000)

7. Schmid, U., Waltermann, J.: Automatic synthesis of XSL-transformations from example documents. In: Hamza, M. (ed.) IASTED International Conference on Artificial Intelligence and Applications, pp. 252–257 (2004)
8. Biermann, A.W.: The inference of regular LISP programs from examples. IEEE Transactions on Systems, Man and Cybernetics 8(8), 585–600 (1978)
9. Biermann, A.W., Guiho, G. (eds.): Computer Program Synthesis Methodologies. Reidel, Dordrecht (1983)
10. Leinonen, P.: Automating XML document structure transformations. In: Proceedings of the 2003 ACM Symposium on Document Engineering, pp. 26–28 (2003)
11. Kuikka, E., Leinonen, P., Penttonen, M.: Towards automating of document structure transformations. In: Proceedings of the 2002 ACM Symposium on Document Engineering, pp. 103–110 (2002)
12. Arenas, M.G., Dolin, B., Merelo-Guervós, J.J., Castillo, P.A., de Viana, I.F., Schoenauer, M.: JEO: Java Evolving Objects. In: Proceedings of the Genetic and Evolutionary Computation Conference, p. 991 (2002)
13. Arenas, M.G., Collet, P., Eiben, A.E., Jelasity, M., Merelo, J.J., Paechter, B., Preuß, M., Schoenauer, M.: A framework for distributed evolutionary algorithms. In: Guervós, J.J.M., Adamidis, P.A., Beyer, H.-G., Fernández-Villacañas, J.-L., Schwefel, H.-P. (eds.) PPSN 2002. LNCS, vol. 2439, pp. 665–675. Springer, Heidelberg (2002)

Management Ubiquitous of Messages and Documents Organizational through Intelligent Agents

Rosa Cano[1], Juan G. Sánchez[2], and Cristian Pinzón[3]

[1] Departamento de Sistemas y Computación, Instituto Tecnológico de Colima
Av. Tecnológico s/n, 28976, Villa de Álvarez, Colima, México
rdegca@gmail.com
[2] Departamento de Informática y Automática, Universidad de Salamanca
Plaza de la Merced s/n, 37008, Salamanca, España
juangabriel@usal.es
[3] Universidad Tecnológica de Panamá
Av. Manuel Espinosa Batista, Panamá
cristian.pinzon@utp.ac.pa

Abstract. In the context of artificial intelligence, the multiagent systems are an alternative solution to address complex problems and distributed. PAINALLI is an multi-agent architecture for managing messages and documents organizational anywhere and anytime. To formalize, the management makes use of international standards such as ISO 15489 and MoReq specification. The ubiquity gained in the architecture is due to the use of intelligent agents deployed in fixed and mobile technologies: personal computers, smart phones and personal digital assistants.

Keywords: Multi-agent architecture, ubiquitous computing, messages and documents management, MoReq, ISO 15489.

1 Introduction and Motivation

This paper presents PAINALLI, a multi-agent architecture developed to provide automatic management of internal documents and communication processes in business. The aim of PAINALLI is to formalize and speed up communication among members of an organization, which is commonly based on exchange of messages and documents. Messages are written based on several templates which contain all required and satisfactory elements in order to achieve a better understanding of what is said; documents are attached to the messages after having been cataloged, manually or automatically, depending on their origin. If the document's origin is paper, cataloging is done manually, and if it is electronic, it will be cataloged automatically by the system. The cataloging process is supported by the ISO 15489 standard, this norm establishes requirements for the creation and maintenance of documents: authentic, reliable, honest and available, as well as the context or system that must be managed [1]. It helps to ensure quality in an organization, Formby means of a specific guide for managing documents used as a basis for quality systems. An e-mail system is just a communication device, not a document management system. PAINALLI merges an architecture for managing messages (email) and document management.

Sending and receiving documents and messages can be done from both mobile and desktop platform. Such feature is one of the PAINALLI's advantages over other related

J.M. Corchado et al. (Eds.): DCAI 2008, ASC 50, pp. 171–179, 2009.
springerlink.com

products, using personal computers, PDAs (Personal Digital Assistant) or smart phones. The use of a formal process to design the architecture, the use of templates as a base for message encoding, the feedback that ISO 15489 standard and MoReq, this specification describes a model of requirements for managing electronic documents and archives and affects especially in functional requirements for managing electronic documents and archives through a system of electronic document management file (SGDEA) [2] both provided for documents management and the support of desktop and mobile technologies make PAINALLI a robust system, reliable in its functional performance and employment.

The communication process is a key element in current organizations. In order to understand it we will take the elements quoted in the definition introduced by [3]: message, channel, sender, receiver, transmission, encoding and decoding, meaning, feedback and communication effects. The last five ones are the most relevant elements for this project, because they are the focus of our investigation work's hypothesis. A message encoding must be formalized so the receiver can decode it and interpret the exact meaning of what the sender is expressing. Any message's aim, being accompanied by a document or not, is at generating one or more actions as a result or effect from the communication. For this purpose, the sender must receive some feedback from the receiver; we will call this messages exchange as "conversation". At the present time, organizations demand tools not only for making the communication process easier. It is essential to add them some obligatory and sufficient features to provide them with intelligence, being able to support every element of the process.

A very useful and suitable technology to develop multi-platform systems consist on agents and multi-agent systems (MAS), which are composed by several agents interacting with each other, making able together to reach the desired functionality [4]. An agent is an entity that must have certain characteristics, like: autonomy, situation, reactivity, proactivity, social ability, learning capacity, mobility or organization [5]. BDI agents have got the mental states of Beliefs, Desires and Intentions [6]. It has likely been the most spread and studied model among agents reasoning models, and because of that it is the most advisable one for agent-based applications development [7]. A critical drawback for development an agent-based architecture is that, at the present time, we lack of clear standards or completely developed methodologies that set up the steps to realize correct analysis and design [8]. Based on the experience of the BISITE (Biomedicine, Intelligent Information Systems and Educational Technology) research group from the University of Salamanca, a suitable combination for multi-agent architectures development is to use the Gaia methodology and the AUML modeling language.

One of the advantages provided by the multi-agent systems is the ability of agents to run on mobile devices. Nowadays there is a growing need to find more effective ways to offer services in mobile devices [9] such as digital personal assistants and mobile phones, using communication technologies like GPRS (General Packet Radio Service), UMTS (Universal Mobile Telecommunications System), Bluetooth, etc. The versatility of mobile devices offers an opportunity that has to be taken into account in order to make the personnel an organization come closer. By means of those, a person will be able to, whenever or wherever he is access organizational information as if he was in his own office. Software systems that integrate the use of mobile technologies to their solutions, as PAINALLI system does, are more robust because provide users the possibility of accessing the system from any place and at any moment they may require.

2 Related Work on Organizational Communication

Nowadays, society demands more and better products and services, and as a result organizations need to count on business strategies in order to provide a way for their employees communicate and fulfill their clients and workmate's expectations more efficiently. The internal communication process in an organization is a key factor for its members being able to interact, because by means of such interactions, it is possible to achieve the organization's strategy goals. The vast amount of information generated in an organization, product or its domestic efforts of the interaction it has with other outside entities, usually is generated and stored on paper, the main drawbacks associated with this are the unknown the amount of information generated and the lack of standards for: development, classification, management and retrieval of documents. Through computer systems it is possible to solve these problems, which allows formalizing and systematizing processes associated with the generation, use, storage and retrieval of documents.

It is necessary that organizations count on a mailing system that guarantees an appropriate use of itself, the pure fact that electronic mail arrives sooner doesn't mean that the effort in mechanical details is going to last shorter [10]. This reasserts the proposal presented with PAINALLI, electronic mail per se is not the solution to communication problems; formalization is required in order to make the communication process efficient.

The purpose of establishing a CDMS (Corporate Documents Management System) is to provide a documents management model and a software tool that guarantees access to information and its availability in order to improve productivity, mind capital and knowledge [11]. Formalizing the process of creating a CDMS requires using some standard or methodology as a reference. PAINALLI was developed using ISO standard 15489:2001 and MoReq [2].

In the IV study about internal communication in Spain [12], elaborated by "Infopress", the "Instituto de Empresa" and the "Capital Humano" magazine, presented the following results: 68.7% of organizations are equipped with an internal communication plan. Descendant information is more effective than ascendant. The main capabilities that directives must improve are empathy with employees (81.1%) and the ability to expose appalling (79.3%). Directives say that company results would improve in 90.7% if there was better commitment with communication, and also that it would let decisions being taken faster in 85.6%, those decisions affecting directly to the results account. In view of such facts, we can say that a software system such as PAINALLI, which supports the internal communication system, would present a significant impact for the organization and its economic resources.

The 9th AIMC (Communication Media Investigation Association) survey on Internet users shows that [13]: 75.1% of users polled use a program to block pop-ups and 76.1% to eliminate or filter unwanted email (spam). 51.8% of them have a personal email account at work or college. The most employed email software is the Microsoft Outlook (40.6%), followed by webmail (25.1%). Subsequently to desktop computers and laptops, 94.7% and 51.5% respectively, mobile phones with 20.5% and PDAs with 9.3% are the most employed equipments to access the Internet. PAINALLI is a platform that avoids non wanted emails and pop-ups. It is specifically designed to manage messages among members of an organization. Besides, it is built with the required and adequate structure to formalize communication, thus giving users the possibility of accessing their messages and documents through two technologies: wired or wireless.

In economic terms, the use of electronic mail and documents management reduce costs significantly. Apart from being cheaper, they represent a way of saving economic resources in organizations. Traditional archiving costs can be reduced in three aspects: salary, administrative and chance loss [14]. On the other hand, Boronat classifies costs in: localization and recovering, distribution and storing [15]. A critical drawback organizations might face is the incidence of unwanted email or spam (non-requested messages), besides, they can bring viruses. In July, 2007, Panda Software presented a report which said that 88% of emails in companies were spam [16]. A study carried out by Dimension Data over 524 companies along the USA and other 12 countries of Europe, Middle-East, Africa and Asia says that 99.6% of the employees and executives use email to communicate at work [17], in a study alike stresses the fact that the communication type must be formalized so the communication process becomes efficient and secure. Taking into account the needs of modern organizations, we have provided some functionalities.

3 Overview

PAINALLI is a multi-agent architecture structured by interactions of intelligent agents that manage sending and receiving messages and documents. The platform's development has been carried out using Java programming language and JADE framework (Java Agents Development Environment). Combining these tools allowed us to build an easy-to-use multi-platform system based on a distributed technology of well-known efficiency. The user accesses the system through an intuitive interface helped by ex profeso designed templates. Users are able to write messages and attach electronic documents to them. This kind of access is available using personal computers or mobile devices such as smart phones or PDAs. The user may send a message with PAINALLI using one of the templates available, which let him Inform, Ask, Request or Arrange a Meeting with the receiver, who in turn may answer the message and attach a document if he needs to. The owner of the conversation (sequence of messages about one specific subject) is the one who sends the first message and, as a consequence, he is the only one who may delete the conversation from the server and consider it concluded. PAINALLI provides advanced automatic conversation management, as well as knowledge extraction techniques which allow obtaining a better understanding of communication processes inside the organization and a support system for decision making.

Management of registered, created or received information, which is stored as pure or test information, by an organization in the exercise of its activities [18] is the scope of application of documents management realized in PAINALLI platform. This is achieved by means of the agents in charge of visualizing, cataloging, storing and recovering the electronic documents [19] sent as attached files. If the document is printed, it may be converted to an electronic format to be considered as that. MoReq specification and ISO standard 15489 [2] are the tools used to make PAINALLI's documents management more robust.

Users of PAINALLI have a personal agenda which contains every commitment acquired from conversations with other/s member/s of the organization. There is one special agenda called "Institutional Agenda" where users can publish information of institutional interest. One way of formalizing commitments is by means of an acknowledgement of receipt, got by the sender, when the message has been opened by the receiver.

Commonly, organizational structure is based on functional areas or departments and in addition there may be committees or groups integrated by people from different departments or areas. It is possible to send messages and documents to a specific person, a group of people or to everyone in the organization. People in charge of an area or department have the possibility of knowing the productivity level of each one of their subordinates. All information is stored in a database. Critical information is kept encrypted.

3.1 Model Proposed

The formalism and system-based approach achieved during analysis, design, implementation, testing and deployment of a software product is obtained by employing techniques, tools, methodologies and languages. In the case of PAINALLI we take the Gaia methodology as reference, which mainly aims at providing a system analysis method and at designing its structure from a series of initial requirements, obtaining several models that allow us to define and analyze the basic structure of the system [20]. The Gaia methodology incorporates novel software engineering mechanisms, based on the artificial intelligence perspective. Gaia analyzes the problems focusing on social and organizational concepts, which allows obtaining models adapted to the user behaviors. The models obtained with Gaia are refined and adapted using AUML, getting a series of diagrams that provide a model of the system close to implementation [21].

The aim is to generate a multi-agent architecture which supports sending and receiving messages, by means of ex profeso elaborated templates, and sending documents attached to messages. Figure 1 shows the architecture which models the agents

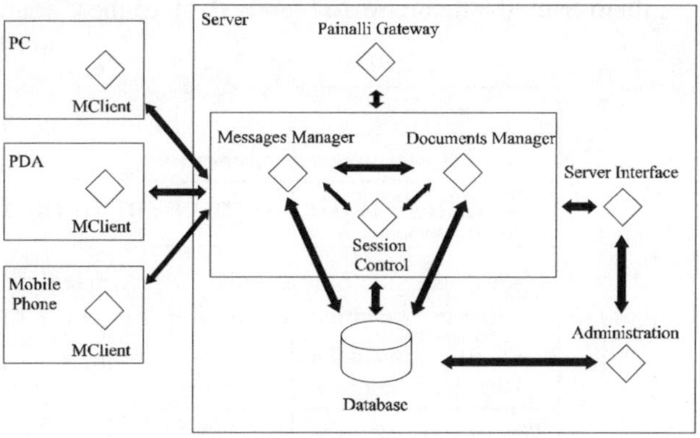

Fig. 1. System Architecture

who, through their performance and interaction, allow PAINALLI users to use the tools offered by the platform.

The Gaia agent model was developed based on roles observed in these kinds of problems, and the communicative acts among such roles. The agent types that make up the architecture are: PCClient, MClient, MessagesManager, DocumentsManager, AgileGateway, SessionControl, ServerInterface, and Administration. After obtaining the agents model and following the GAIA methodology, the acquaintance model was created as shown in figure 2. In this model it is possible to observe the communication flows among agents.

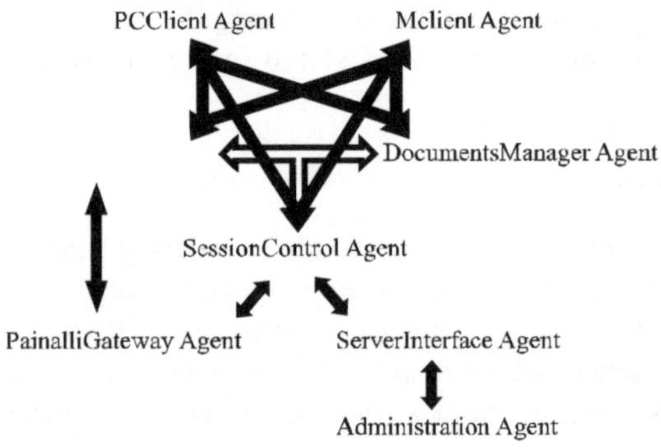

Fig. 2. Acquaintance Model

In order to analyze the system's functionality by means of agents collaboration, an interaction model was developed. Figure 3 shows an example of interaction diagram for opening a session, this interaction consists of two protocols. First, the Client role sends a login request to SessionControl, indicating its login and password. SessionControl checks the information in the database and, if the user exists and the password is correct, grants the Client access to the system. After that, the protocol "Get New Messages" is executed: the Client asks MessagesManager for its new messages, and the latter recovers them from the database and sends them to the Client.

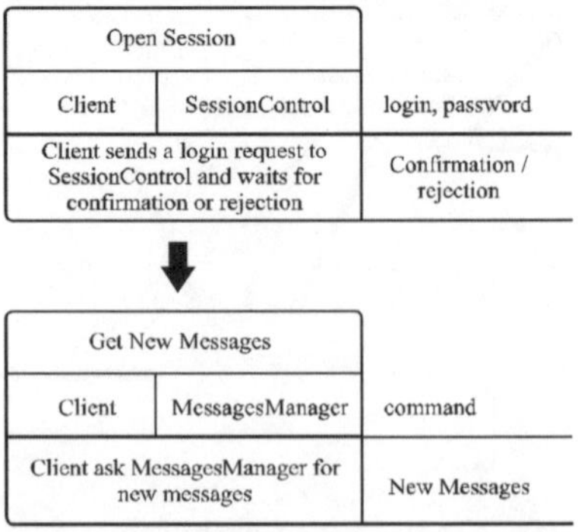

Fig. 3. Opening a session interaction diagram

3.2 AUML Diagram

Once the Gaia methodology finishes, it is necessary to detail the results obtained. The "Request" template contains the following elements: receiver(s), subject, message, format, date and time, and an optional attached file. The message is sent by agent A to the MessagesManager agent, which asks the SessionControl agent if agent A is logged in as a valid user. If so, it forwards the message to agent B, who is one of the receivers of the message. When agent B receives the message, an entry is created into the user's agenda that describes the request. When B has the information requested, he can hand it in to A, which includes another valid-user checked by the Messages-Manager and SessionControl agents, figure 4 shows the corresponding diagram.

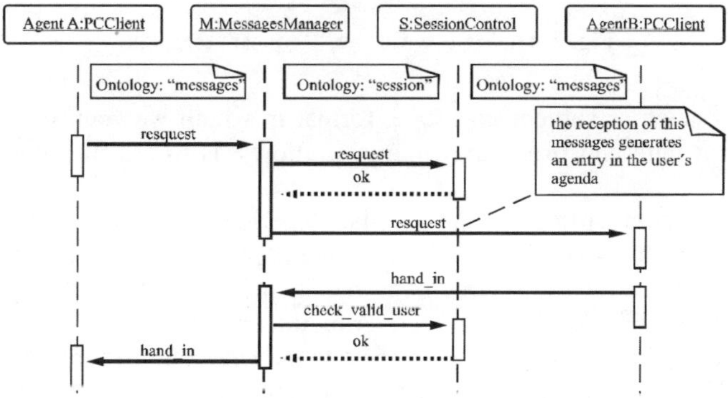

Fig. 4. "Request" template diagram

4 Results and Conclusions

Taking into account the consulted bibliography we can say that the organizations' communication process needs formalization and that a software system is a tool able to support and make it better through techniques and standards belonging to messages and documents management.

A message and document management system like PAINALLI can be developed as an intelligent agent's platform. MAS are a robust technology that supplies all required and sufficient versatility to carry out the messages and documents interchange by means of agents' interacting within wired or wireless networks.

Through the templates used in PAINALLI ("Arrange a meeting", "Ask", "Inform" and "Request"), users can communicate easier with their mates. In each message it is possible to attach a document, which is cataloged by the Documents Management agent.

In picture 5 we can see the dialog box for the "Request" template. On the left side of the dialog there is a "receivers panel", where we can pick the employees or groups which we want to receive the message. The upper side of this panel shows the company structured in groups and employees who belong to them, and the lower side shows all employees alphabetically sorted. The rest of the dialog lets us to enter the

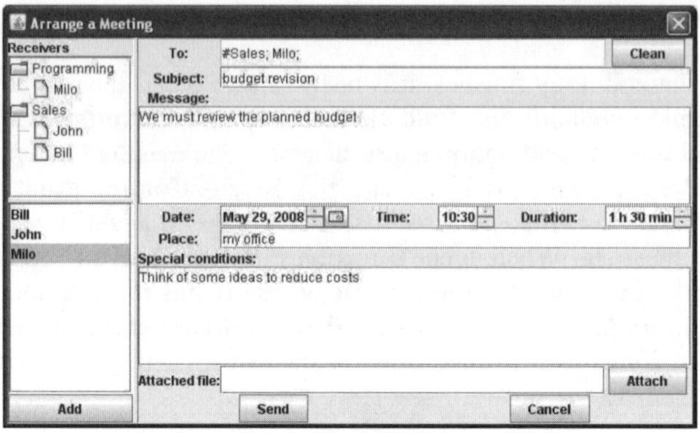

Fig. 5. Dialog box for the "Request" template

request details, such as subject, message, format in which we want to receive the information, limit date and time, and finally, it allows us to attach a document to the message.

We achieved competitive results on the experiments done with the prototype, where it was considered variables like: storage capacity, security, performance and utilization tests. With regard to document management, PAINALLI can carry out actions such as cataloging, storing and recovering documents.

PAINALLI has been evaluated by expert from Tulecom Group, Flagsolution and Matchmind. They have analyzed the proposal model and correspondence with other applications. Results are shown in table 1.

Table 1. Results table

	Gmail	**Yahoo**	**Lycos**	**Hotmail**	**PAINALLI**
Storage capacity	Excellent	Excellent	Very good	Acceptable	Excellent
Security	Good	Good	Acceptable	Good	Very good
Performance	Very good	Very good	Good	Good	Excellent
Utilization tests	Excellent	Very good	Good	Acceptable	Excellent
Total	**Excellent**	**Excellent**	**Good**	**Acceptable**	**Excellent**

In future works the prototype will be tested with different kinds of organizations (public and private) to contrast the results obtained with PAINALLI against other commercial and free-of-charge products of the market to know and demonstrate the efficiency of the platform.

Acknowledgements. This work has been supported by the MCYT project TIC2003-07369-C02-02. We also thank Tulecom Group, Flagsolutions and Matchmind for their support and collaboration.

References

1. Organización Internacional de Normalización. ISO 15489-1:2001. Información y Documentación. Gestión de Documentos. Parte 1: Generalidades. AENOR, España (2006)
2. Gómez Domínguez, D., Ruiz Rodríguez, A., Peis Redondo, E.: La gestión de documentos electrónicos: requerimientos funcionales. El Profesional de la Información 12, 88–98 (2003)
3. Krone, K., Janblin, F., Putman, L.: Communication Theory and Organizational Communication: Multiple Perspectives. In: Handbook of organizational communication. Sage Publications, Thousand Oaks (1987)
4. Bussman, S., Müller, H.J.: A communication architecture for cooperating agents. Computers and Artificial Intelligence 12, 37–53 (1993)
5. Wooldridge, M.J., Jennings, N.R.: Agent theories, architectures and languages: a survey. In: ECAI 1994 Workshop on Agent Theories Architectures and Languages (1994)
6. Haddadi, A., Sundermeyer, K.: Belief-desire-intention agent architectures, foundations of distributed artificial intelligence. In: Foundations of distributed artificial intelligence, pp. 169–185. Wiley-Interscience Publication, Chichester (1996)
7. Rao, A.S., Georgeff, M.P.: BDI agents: from theory to practice. In: Proceedings of the First International Conference on Multi-Agents Systems (1995)
8. Bajo, J., de Luis, A., Tapia, D.I., Corchado, J.M.: Sistemas multiagente inalámbricos basados en agentes CBR-BDI: de la teoría a la práctica. In: 5º Workshop internacional sobre aplicaciones prácticas de agentes y sistemas multiagente (2006)
9. Rigole, P., Holvoet, T., Berbers, Y.: Using Jini to integrate home automation in a distributed software-system. In: Plaice, J., Kropf, P.G., Schulthess, P., Slonim, J. (eds.) DCW 2002. LNCS, vol. 2468, pp. 291–303. Springer, Heidelberg (2002)
10. Flynn, N., Flynn, T.: Correo electrónico: cómo escribir mensajes eficaces. Editorial Gedisa, S. A. (2001)
11. De Inclán, M.: Actuaciones para la implantación de un sistema de gestión documental corporativa: experiencia del Banco de España. In: VIII Jornada de Gestión de la Información (2006)
12. Cabanas, C., Vilanova, N., Carazo, J.A.: IV Estudio sobre la comunicación interna en España. Technical report, Instituto de Empresa, Observatorio de Comunicación Interna e Identidad Corporativa, Grupo Inforpress y Capital Humano (2005)
13. AIMC, Navegantes en la red: 9a encuesta AIMC a usuarios de Internet. SERSA (2007)
14. Martínez Sereno, V.: Integración de sistemas de gestión electrónica documental en la empresa: evaluación y metodología de implantación. In: VI Jornadas Españolas de Documentación (1998)
15. Boronat, F., Cicuéndez, R., Lloret, J.: Sistema de gestión electrónica de documentos del servicio de reprografía de la EPSG. In: XX Simposium nacional URSI (2005)
16. Panda Security, `http://www.canal-ar.com.ar`
17. Clarín, `http://www.clarin.com`
18. Bustelo, C.: Gestión de documentos: enfoque en las organizaciones. Anuario ThinEPI, 141–145 (2007)
19. García Pérez, A.: La gestión de documentos electrónicos como respuesta a las nuevas condiciones del entorno de información. ACIMED 9, 190–200 (2001)
20. Wooldridge, M.J., Jennings, N.R., Kinny, D.: The Gaia methodology for agent-oriented analysis and design. Autonomous Agents and Multi-Agent Systems 3, 285–312 (2000)
21. Bauer, B., Huget, M.P.: FIPA modeling: agent class diagrams. Working Draft, foundation for Intelligent Physical Agents (2003)

A Multiagent Based Strategy for Detecting Attacks in Databases in a Distributed Mode

Cristian Pinzón[1], Yanira De Paz[2], and Javier Bajo[1]

[1] Departamento Informática y Automática, Universidad de Salamanca, Plaza de la Merced s/n 37008, Salamanca, Spain
[2] Universidad Europea de Madrid, Tajo s/n 28670, Villaviciosa de Odón, Spain
cristian_ivanp@usal.es, yanira@usal.es, jbajope@usal.es

Abstract. This paper presents a distributed hierarchical multiagent architecture for detecting SQL injection attacks against databases. It uses a novel strategy, which is supported by a Case-Based Reasoning mechanism, which provides to the classifier agents with a great capacity of learning and adaptation to face this type of attack. The architecture combines strategies of intrusion detection systems such as misuse detection and anomaly detection. It has been tested and the results are presented in this paper.

Keywords: Multi-agent, SQL injection, Security database, case-based reasoning, IDS.

1 Introduction

The exponential growth of the computer network and the increase in the interconnection between networks has extended the offer of new services within the cyberspace [1]. The information volume with a sensitive value for the organizations is stored on information structures denominated databases and this information generally is transmitted across computer network. Databases are the core of many information systems, reason for which databases are increasingly coming under large number of attacks. Every day are founded new vulnerabilities in security systems intended to protect databases. These vulnerabilities are used by hackers in order to carry out attacks on the stored data. A special intrusion type within of databases is the SQL injection attack, which occurs when the intended effect of a SQL sentence is changed by inserting SQL keywords or special symbols [2].

Nowadays, the majority of approaches had addressed the problem of SQL injection attack from a centralized perspective, such as the one described by [3] and [2]. However, the solutions are limited to solve only a part of the problem. Regarding this, other approaches had implemented strategies based on intrusion detection systems in order to block a SQL injection attack, such as [4] and [5]. These proposals have as main drawbacks the highest error rate and a limited capacity of learning and adapting when changes occur in the patterns of attacks.

Our proposal aims the SQL injection attacks in a distributed, dynamic and flexible mode. This proposal is founded in a hierarchical multiagent architecture using agents based on the BDI (Belief, Desire and Intention) model [6]. Agents are typically integrated into multiagent systems or agent societies, exchanging information and resolving problems in a distributed way [7]. Agents can be characterized through their capacities such as autonomy, reactivity, pro-activity, social abilities, reasoning, learning and

J.M. Corchado et al. (Eds.): DCAI 2008, ASC 50, pp. 180–188, 2009.
springerlink.com

mobility [6]. Our proposal incorporates classifier agents supported by a Case-based reasoning mechanism (CBR) [8] that includes a mixture of neural networks capable of making short term predictions [9]. Our multi-agent architecture is adequate to block the SQL injection attack, because it is designed for working in distributed and dynamic environments.

The rest of the paper is structured as follows: section 2 presents the problem that has prompted most of this research work. Section 3 focuses on the details of the multiagent architecture, the different levels of the architecture, the interaction possibilities and communication between the agents; section 4 explains in detail the classification model integrated within the classifier agent. Finally, section 5 describes how the classifier agent has been tested inside a multi-agent system and presents the results obtained.

2 SQL Injection Attacks Description

The impact of a SQL injection attack in a database has many consequences within of the organization and individuals. Personal, financial and legal information is compromised when this type attack is carried out. A SQL injection attack takes place when a hacker changes the semantic or syntactic logic of a SQL text string by inserting SQL keywords or special symbols on the original SQL command that will be executed at the database layer of an application [10], [2]. The results of this attack can produce unauthorized handling of data, retrieval of confidential information, and in the worst possible case, taking over control of application server. One particular inconvenient of the SQL injection attack is the biggest number of variants. Some strategies can be extremely complex due to the high number of variables that they can generate, thus making their detection very difficult.

Some approaches based on firewall and intrusion detection system (IDS) are a few effective due the strategy of detection, which requires an updated patterns database. Other approaches more specifics to face SQL injection attacks are founded in a technique of string analysis, some carrying out static analysis such as JSA (Java String Analyzer) [3]. Other more complex using dynamic and hybrid analysis is AMNESIA (Analysis and Monitoring for Neutralizing SQL Injection Attacks) [2]. These approaches generally have as main drawback that aim just one part of the problem, moreover the approaches based on models for detecting SQL injection attacks are very sensitive. With only slight variations of accuracy, they generate a large number of false positive and negatives.

Several approaches based on artificial techniques and hybrid systems propose a novel alternative. Web Application Vulnerability and Error Scanner (WAVES) [11] uses a black-box technique which includes a machine learning approach. Valeur [4] presents an IDS approach which uses a machine learning technique based on a dataset of legal transactions. These transactions are used during the training phase prior to monitoring and classifying malicious accesses. Rietta [5] put forward an IDS at the application layer using an anomaly detection model. Finally, Skaruz [12] proposes the use of a recurrent neural network (RNN). The detection problem is became a time serial prediction problem. Generally, this approaches present as main problem, generating a large number of false positive and false negative. In the case of the IDSs systems, they are unable to recognize unknown attacks because they depend on a signature database that requires a dynamic updating.

The multi-agent architecture aims the problem from a distributed context more dynamic and flexible to work according to the device used to execute a SQL injection attack. The incorporation of a CBR technique to the classifier agents allows them to offer a robust learning and adaptable solution that respond to an unlimited number of variants of SQL injection attacks. Finally, a combination of strategies based on intrusion detection system that covers misuse detection and anomaly detection grants to the architecture a highest level performance in the tasks of classification. It is important to highlight that the combination of these strategies reduce at a minimum the false positives and false negatives.

3 The Agent Based Architecture

Agents are characterized by their autonomy; which gives them the ability to work independently in real-time environments [13]. Agents are especially adequate to work in dynamic and distributed environments when they are integrated into multiagent systems [7]. Environments especially complex for protecting are the application that working with a relational database to provide information to different users. A

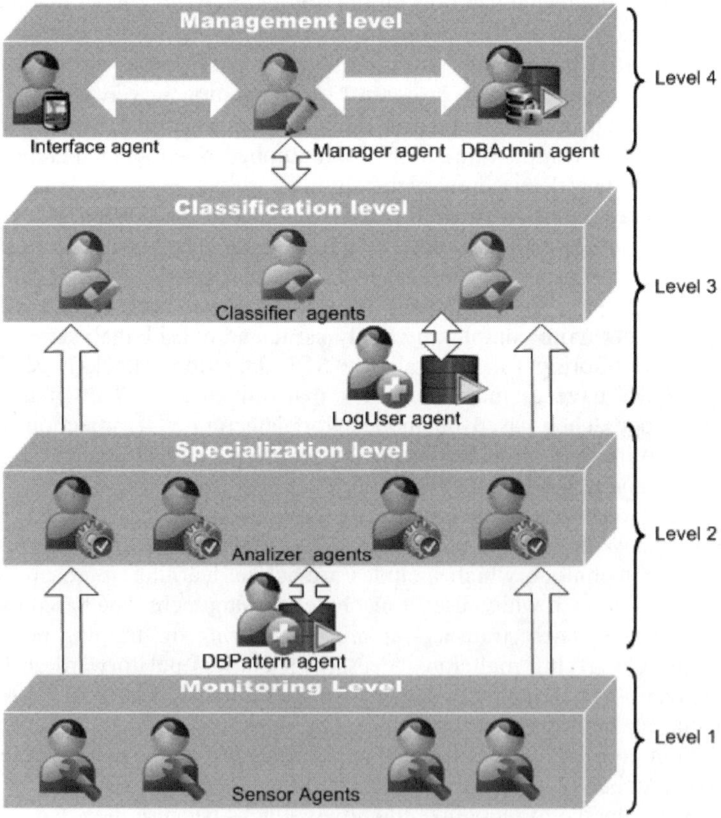

Fig. 1. Multi-agent architecture with BDI agents

distributed multiagent architecture presents a great capacity of error recovering, allowing using autonomous agents for decision making and adapting a new variant of a SQL injection attack.

Several architectures have been proposed to build deliberative agents, most of them based on the BDI (Belief, Desire, Intention) model [6]. In this model, the intern structure of the agents and their election capacities are based on mental aptitudes [14]. The main characteristic of the architecture proposes in this paper is the incorporation of CBR-BDI deliberative agents, which are capable protecting databases of the applications by means of SQL queries classification. CBR-BDI classifier agents use an intrusion detection techniques (anomaly detection) [15], identifying elements of SQL injection in the database queries.

Our proposal is a distributed hierarchical multiagent architecture integrated for 4 levels with distinct BDI agents. The hierarchical structure allows distributing tasks on levels of the architecture, defining specific responsibilities, even though the interaction and communication between the agents is continuous in order to request services and delivering results. The architecture presented in figure 1 shows the four levels and BDI agents organized according to their roles.

Next, the details of the functionality of each agent located at each level of the architecture are described:

- Sensor agents: These agents are located at the level 1 of the architecture. They are incorporated at each device with access to database. They have been assigned 3 functions: the capture of datagrams, the ordering of the TCP fragments for extracting the SQL query string, and the syntactic analysis. The tasks of the Sensor agents end when the results (the SQL string transformed by the analysis, the result of the analysis and the user data) are sent to the next level of the architecture.
- Analyzer agents: These agents are located at the level 2. They receive the transformed SQL string, the result of the analysis and user data. This information is assigned to a specific agent. The Analyzer agent receives the data sent by the Sensor agent and then it executes searching process and compares known SQL string patterns. These SQL string patterns are stored in a previously built database. The Analyzer agent works in coordination with the DbPattern agent for searching and saving SQL string patterns on the database. The creation of the Analyzer agents is dynamic and it depends on the workload at a given time. When an Analyzer agent is created it receives an instance number. The Analyzer agent finishes its task when it sends results to the next level. The results consist of the SQL string transformed, the result of the analysis, the user data and the result by misuse detection.
- DbPattern Agent: this agent is located at the level 2. It is the responsible to save the new SQL string patterns on the database and search for patterns when the Analyzer agent requests it.
- Classifier agents: These agents are in charge of carrying out the task of classification of SQL queries. These agents are located at the level 3, and implements a technique based on anomaly detection. A classifier agent incorporates a CBR mechanism, which allows the evaluation of the new SQL query through the search of similar cases stored in the memory cases, looking for a solution to the new problem. At this level there exist *n* instances of the classifier agents and their creation is

dynamic and depends of the workload at a given time. Classifier agents work together with the LogUser agent and Manager agent for carrying out their tasks.

- LogUser agent: this agent is located at the level 3, records the actions of the user and searches for the user profile (the historical profile and the user statistics) when it is requested by a Classifier agent.
- Manager agent: this agent works at the level 4. It has many tasks: monitors, evaluates and coordinates the classifiers agents. It ensures the capacity for errors recovering and the distribution of the workload within the architecture. An expert through of the Manager agent allows evaluating the results of the classification and the final decision that have been carried out. Moreover, this agent is in charge of managing the alert of attacks and the coordination of the required actions to take over an attack when it has been detected. A Classifier agent is promoted to be a Manager agent when the Manager agent fails. This agent is selected by means of a voting method between the classifier agents.
- DbAdmin agent: this agent is also located at the level 4. It is in charge of executing the query on the database and obtaining the results.
- Interface agent: this agent is located at the level 4. The Interface agent allows the interaction of the user of the security system with the architecture. The interface agent communicates the details of an attack to the security personnel when an attack is detected. This agent has the capacity to work on mobile devices such as PDAs, mobile telephones and laptops. This capacity allows a ubiquitous communication. Moreover, the interface agent provides an access mechanism for carrying out certain adjustments in the configuration of the architecture and the verification of the status of the Manager agent.

4 Classifier Model of SQL Injection Attacks

The CBR-BDI classifier agent [16] incorporates a case-based reasoning system that allows the prevention and detection of anomalies by means of a prediction model based on neural networks, configured for short-term predictions of intrusions by SQL injections. This mechanism uses a memory of cases which identifies past experiences with the corresponding indicators that characterize each of the attacks. This paper presents a novel classification system that combines the advantages of the CBR systems, such as learning and adaptation, with the predictive capabilities of a mixture of neural networks. These features make the system very appropriate for its use in dynamic environments. Working with this type of systems, the key concept is the case". A case is defined as a previous experience and is composed of three elements: a description of the problem; a solution; and the final state. A CBR cycle consists of four steps: retrieve, reuse, revise and retain.

The elements of the SQL query classification are described as follows:

(a) Problem Description that describes the initial information available for generating a plan. As can see in table 1, the problem description consists of a case identification, user session and SQL query elements. (b) Solution that describes the action

Table 1. Structure of the problem definition and solution for a case of SQL query classification

Problem Description fields		Solution fields	
IdCase	Integer	Idcase	Integer
Sesion	Session	Classification_Query	Integer
User	String		
IP_Adress	String		
Query_SQL	Query_SQL		
Affected_table	Integer		
Affected_field	Integer		
Command_type	Integer		
Word_GroupBy	Boolean		
Word_Having	Boolean		
Word_OrderBy	Boolean		
Numer_And	Integer		
Numer_Or	Integer		
Number_literals	Integer		
Number_LOL	Integer		
Length_SQL_String	Integer		
Cost_Time_CPU	Float		
Start_Time_Execution	Time		
End_Time_Execution	Time		
Query_Category	Integer		

carried out in order to solve the problem description. As can see in the table 1, it contains the case identification and the applied solution. (c) Final State that describes the state achieved after that the solution has been applied.

In the following section, the performance of the classifying system is described in detail. The proposed mechanism is responsible to classify SQL database queries made by users. When a user makes a new request, it is checked for pattern matching. This being the case, it is automatically identified as an attack. In order to identify the rest of the SQL attacks, the mechanism uses CBR, which must have a memory of cases dating back at least 4 weeks, and store the variables showed in the Table 1.

The first phase of the CBR cycle consists of recovering past experience from the memory of cases, specifically those with a problem description similar to the current SQL query. In order to do this, a cosine similarity-based algorithm is applied, allowing the recovery of those cases which are at least 90% similar to the current SQL query. The cases recovered are used to train the mixture of neural networks implemented in the recovery phase; the neural network with the sigmoidal function is trained with the recovered cases that were an attack or not, whereas the neural network with hyperbolic function is trained with all the recovered cases, including the suspects. A preliminary analysis of correlations is required to determine the number of neurons of the input layer of the neuronal networks. Additionally, it is to normalize the data (i.e., all data must be values in the interval [0.1]). The data used to train the mixture of networks must not be correlated. With the cases stored after eliminating correlated cases, the entries for training the mixture of networks are normalized. It is considered to be two neural networks. The result obtained using a mixture of the outputs of the networks provides a balanced response and avoids individual tendencies (always taking into account the weights that determine which of the two networks is more optimal).

With la mixture of the neural network we mean to detect attacks, so if one only network with a sigmoidal activation function was used, then the result provided by the network would tend to be attack or not attack, and no suspects would be detected. On the other hand, if only one network with a hyperbolic tangent activation was used, then a potential problem could exist in which the majority of the results would be identified as suspect although they were clearly attack or not attack. The mixture provides a more efficient configuration of the networks, since the global result is determined by merging two filters. This way, if the two networks classify the user request as an attack, so too will the mixture; and if both agree that it is not an attack, the mixture will as well. If there is no concurrence, the system uses the result of the network with the least error in the training process or classifies it as a suspect. In the reuse phase the two networks are trained by a back-propagation algorithm for the same set of training patterns (in particular, these neural networks are named Multilayer Perceptron), using a sigmoidal activation function (which will take values in [0.1], where 0 = Illegal and 1 = legal) for a Multilayer Perceptron and a hyperbolic tangent activation function for the other Multilayer Perceptron (which take values in [-1.1], where -1 = Suspect, 0 = illegal and 1 = legal). The response of both networks is combined, obtaining the mixture of networks denoted by y^2; where the superscript indicates the number of mixtured networks

$$y^2 = \frac{1}{\sum_{r=1}^{2} e^{-|1-r|}} \sum_{r=1}^{2} e^{-|1-r|} y^r \tag{1}$$

The number of neurons in the output layer for both Multilayer Perceptrons is 1, and is responsible to decide whether or not there is an attack. The error of the training phase for each of the neural networks can be quantified with formula (2), where P is the total number of training patterns.

$$Error = \frac{1}{P} \sum_{i=1}^{P} \left| \frac{Forecast_P - Target_P}{Target_P} \right| \tag{2}$$

The review stage is performed by an expert, and depending on his opinions, a decision is made as to whether the case is stored in the memory of cases and whether the list of well-known patterns has to be updated in the retain phase.

5 Results and Conclusions

The problem of SQL injection attacks on databases supposes a serious threat against information systems. This paper has presented a new classification system for detecting SQL injection attacks which combines advantages of multiagent systems, such as autonomy and distributed problem solving, with the adaptation and learning capabilities of CBR systems. Additionally, the system incorporates the prediction capabilities that characterize neural networks. An innovative model was presented that provides a significant reduction of the error rate during the classification of attacks. To check the validity of the proposed model, a series of test were elaborated which were executed on a memory of cases, specifically developed for these tests, and which generated attack consults. The results shown in Table 2 are promising: it is possible to observe different techniques to predict attacks at the database layer and the errors associated with misclassifications. All the techniques presented in Table 2 have been applied under similar conditions to the same set of cases, taking the same problem into account in order to obtain a new case common to all the methods. Note that the technique proposed in this article provides the best results, with an error in only 0.5% of the cases.

Table 2. Results obtained after testing different classification techniques

Forecasting Techniques	Successful (%)	Approximated Time (secs)
CBR-BDI Agent (mixture NN)	99.5	2
Back-Propagation Neural Networks	99.2	2
Bayesian Forecasting Method	98.2	11
Exponential Regression	97.8	9
Polynomial Regression	97.7	8
Linear Regression	97.6	5

As shown in Table 2, the Bayesian method is the most accurate statistical method since it is based on the likelihood of the events observed. But it has the disadvantage of determining the initial parameters of the algorithm, although it is the fastest of the statistical methods. Taking the errors obtained with the different methods into account, after the CBR-BDI Agent together with the mixture of neural networks and Bayesian methods we find the regression models. Because of the non linear behaviour of the hackers, linear regression offers the worst results, followed by the polynomial and exponential regression. This can be explained by looking at hacker behaviour: as the hackers break security measures, the time for their attacks to obtain information decreases exponentially. The empirical results show that the best methods are those that involve the use of neural networks. With a mixture of two neural networks, the predictions are notably improved.

Acknowledgments. This development has been partially supported by the Spanish Ministry of Science project TIN2006-14630-C03-03.

References

1. Kandula, S., Singh, S.: Argus: A distributed network-intrusion detection system. In: Proceedings of the 3rd International SANE 2002 Conference, Netherlands (2002)
2. Halfond, W., Orso, A.: AMNESIA: Analysis and Monitoring for Neutralizing SQL-injection Attacks. In: 20th IEEE/ACM international Conference on Automated Software Engineering, pp. 174–183. ACM, New York (2005)

3. Christensen, A.S., Moller, A., Schwartzbach, M.I.: Precise Analysis of String Expressions. In: 10th International Static Analysis Symposium, pp. 1–18. Springer, Heidelberg (2003)
4. Valeur, F., Mutz, D., Vigna, G.: A Learning-Based Approach to the Detection of SQL Attacks. In: Conference on Detection of Intrusions and Malware and Vulnerability Assessment, Vienna, pp. 123–140 (2005)
5. Rietta, F.: Application layer intrusion detection for SQL injection. In: 44th annual Southeast regional conference, pp. 531–536. ACM, New York (2006)
6. Woolridge, M., Wooldridge, M.J.: Introduction to Multiagent Systems. John Wiley & Sons, Inc., New York (2002)
7. Corchado, J.M., Bajo, J., Abraham, A.: GerAmi: Improving Healthcare Delivery in Geriatric Residences. Intelligent Systems 23, 19–25 (2008)
8. Corchado, J.M., Laza, R., Borrajo, L., De Luis, Y.A., Valiño, M.: Increasing the Autonomy of Deliberative Agents with a Case-Based Reasoning System. International Journal of Computational Intelligence and Applications 3, 101–118 (2003)
9. Ramasubramanian, P., Kannan, A.: Quickprop Neural Network Ensemble Forecasting a Database Intrusion Prediction System. In: 7th International Conference Artificial on Intelligence and Soft Computing, Neural Information Processing, vol. 5, pp. 847–852 (2004)
10. Anley, C.: Advanced SQL Injection. In: SQL Server Applications (2002) (accessed March 02 (2008), `http://nextgenss.com/papers/advancedsqlinjection.pdf`
11. Huang, Y., Huang, S., Lin, T., Tsai, C.: Web application security assessment by fault injection and behavior monitoring. In: 12th international conference on World Wide Web, pp. 148–159. ACM, New York (2003)
12. Skaruz, J., Seredynski, F.: Recurrent neural networks towards detection of SQL attacks. In: 21th International Parallel and Distributed Processing Symposium, pp. 1–8. IEEE, Los Alamitos (2007)
13. Carrascosa, C., Bajo, J., Julian, V., Corchado, J.M., Botti, V.: Hybrid multi-agent architecture as a real-time problem-solving model, Vol. Expert Systems With Applications 34, 2–17 (2008)
14. Corchado, J.M., Pavón, J., Corchado, E.S., Castillo, L.F.: Development of CBR-BDI Agents. In: Advances in Case-Based Reasoning. Springer, Heidelberg (2004)
15. Abraham, A., Jain, R., Thomas, J., Han, S.Y.: D-SCIDS: distributed soft computing intrusion detection system, Vol. Journal of Network and Computer Applications 30, 81–98 (2007)
16. Pinzon, C., De Paz, Y., Cano, R.: Classification Agent-Based Techniques for Detecting Intrusions in Databases. In: 3rd International Workshop on Hybrid Artificial Intelligence Systems (2008)

Towards the Coexistence of Different Multi-Agent System Modeling Languages with a Powertype-Based Metamodel

Iván García-Magariño

Software Engineering and Artificial Intelligence
Facultad de Informática
Universidad Complutense de Madrid, Spain
ivan_gmg@fdi.ucm.es

Summary. The diversity of Multi-agent System(MAS) methodologies and modeling languages(MLs) is increasing more and more. Although most of the MAS concepts are shared among the MAS MLs, these MAS concepts use different notations and have semantic particularities for each MAS ML. This paper presents a meta-modelling solution and tool support for the coexistence of the MAS ML diversity. This solution is based on a inter-ML metamodel, which provides a mechanism for the MAS specification interchange amongst different MAS modeling languages. This metamodel is based on the *powertype* pattern; in which, the *clabjects* represent subtypes and are instantiated at the model. The MAS designer can change the clabjects properties to indicate the particularities of each concept, depending on the ML the MAS designer is used to.

Keywords: multi-agent systems, software engineering, model-driven development, metamodel, powertype.

1 Introduction

The great diversity of *Multi-agent System*(MAS) methodologies and *Modeling Languages*(MLs) is described by Bernon[2]. There are several MAS methodologies, each one with its own ML, like Tropos, MAS-CommonKADS, INGENIAS, MASSIVE, GAIA, MaSE, AALAADIN, Agile PASSI, PASSI and ADELFE. On the other hand, AUML is an agent-oriented ML which is not associated with any particular agent-oriented methodology. Most of the concepts, such as the agents, roles, goals and interactions, are shared among the MAS MLs. However, the concrete syntax is quite different in each MAS ML, and there are slight differences in the meaning of the shared concepts. The aforementioned diversity hinders the understanding and communication among the MAS experts of different MAS methodologies. Further, the model interchange amongst MAS MLs is a challenging issue, because most of the tool support for MAS methodologies just serializes the specification to a particular MAS ML.

The goal of the presented work is to facilitate the comprehension and interchange of the MAS specifications amongst the different MAS MLs. This paper presents a metamodel that defines a generic MAS ML that includes the existent

J.M. Corchado et al. (Eds.): DCAI 2008, ASC 50, pp. 189–193, 2009.
springerlink.com

MAS MLs, based on the following facts. Most of the concepts are shared among the MAS methodologies, but the same concept has slightly different meanings in each MAS ML. In addition, the generic language allows the user to define subtypes of the concepts, depending on the MAS ML the designer is used to.

Next section briefly introduces the powertype pattern. Then, Section 3 presents the MAS Inter-ML metamodel as the solution for the coexistence of the aforementioned diversity. Finally, Section 4 indicates the conclusions and future work.

2 Brief Description of the Powertype Pattern

The *powertype pattern*[4, 6, 5] is presented and used several times in the metamodeling literature. The powertype pattern includes the two following metamodeling elements.

- A meta-element that describes the class facet. This meta-element is the *partitioned type.*
- A meta-element that describes its object facet. This meta-element is the *powertype.* It represents a subtype or a kind of a concept. The recommended notation[4] indicates that the name must have a meaningful suffix, such "kind".

The instances of the powertype represents *concept kinds.* It is crucial to understand that each concept kind is simultaneously instance of the powertype and subtype of the partitioned type. These concept kinds are usually called *clabjects* or *subtypes* at the powertype literature[4].

3 MAS Inter-ML Metamodel

As mentioned before, some concepts are shared among several MAS MLs. However, these concepts have some particularities in each MAS ML. For this reason, this paper proposes to define a metamodel which provides a mechanism to define several *kinds* of these shared concepts. In this manner, the MAS designer can establish the particularities of an used concept to let the other MAS designers know these particularities. With the presented solution, the understandability amongst different MAS MLs improves. In order to define these kinds of shared concepts, the powertype pattern[4] is used.

For example, Figure 1 shows the presented solution for the *Goal* concept. In this case, the *Goal* concept is partitioned for the following reason. There are several kinds of goals in MAS MLs. For instance, the *Service Goal* is a kind of goal which is never destroyed or satisfied. This kind of goal is used by INGENIAS ML, but there are other kinds of goals in other MAS MLs. The MAS designer should indicate the *Goal Kind* when defining a goal. In this case, *Goal* is the partitioned type and *Goal Kind* is the powertype. The *Service Goal* is the clabject. The *Service Goal* is simultaneously instance of Goal Kind and subtype of Goal (see powertype literature[4, 6, 5]). In this example, the *EvaluateDoc* entity is the goal. This goal pursues to evaluate the relevance of a document and

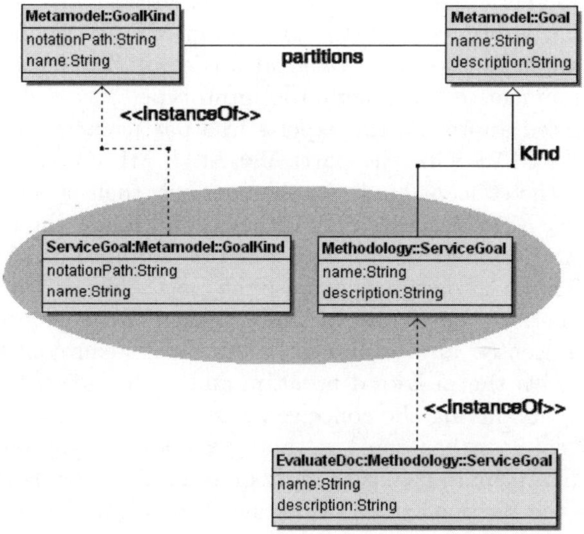

Fig. 1. An Example of Powertype pattern for MAS MLs. In this example, the *Goal* type is partitioned.

is taken from the MAS presented in [3]. The EvaluateDoc goal is instance of the Service Goal clabject.

Furthermore, several particularities of the MAS concepts can be defined with attributes of the powertype. Examples of these particularities are the concrete syntax and the execution particularities.

Firstly, the concrete syntax of the concept kinds can be defined with the *notationPath* attribute. This attribute indicates the visual notation of the concept. For example, INGENIAS uses a circle to denote the agent Goals. However, in other MAS MLs, other visual notations are used. The advantage is the following. Other MAS designers can understand the model design even if they do not know the concrete syntax of the corresponding MAS ML. The other designers just have to look the concept kind definition to understand the model design.

Secondly, some execution particularities can be indicated by the MAS designer with the powertype. For instance, the non-service goals are destroyed when they are satisfied. The Non-service goal kind is defined with an attribute called *condition-of-satisfaction*, whose value is "On Task Execution". This value means that the goal is destroyed on the execution of whatever task that satisfies the goal. In addition, the agent mental states contain several facts. These facts are created and removed in several manners depending on the MAS MLs. All the particularities of creation and deletion of these facts are specified with several Fact kinds.

This paper proposes the integration with existent CASE tools, and the creation of a new generic CASE tool for the presented inter-ML metamodel. In this line, a preliminary modernization of the INGENIAS Development Kit(IDK)[7] is done. Moreover, a generic CASE tool prototype is created with a Model-driven Development(MDD) approach from the MAS inter-ML metamodel. The *ECore* meta-modelling language and the *Graphical Modeling Framework*(GMF) were used for the development of the generic tool prototype.

With the presented approach, the experts in a particular MAS ML can continue specifying the MAS with the particular MAS ML. With a minor effort, they should define the concept kinds. In this manner, their specification can be understood by every MAS designer even if they do not know the particular ML. Moreover, the MAS designers do not have the obligation of using only concept kinds of one ML. The MAS designers can use the most appropriate concept kinds of several MAS MLs for a particular MAS, by means of the generic tool. In addition, with the presented approach, each MAS expert can continue designing the MAS designs with the preferred notation and, even, new notations can be used by defining or modifying the concept kinds.

The presented solution has some advantages above existent solutions. For instance, the AUML [1] main goal is to unify all the MAS MLs. However, AUML does not indicate the particularities of the agent concepts. Thus, all the agent design details with AUML cannot be easily understood because the MAS concept meanings, such as Goal or Fact meanings, which depend on the ML the designer is used to. By the same token, with the presented approach, concept kinds can be defined to indicate certain particularities of MAS concepts.

4 Conclusions and Future Work

On the whole, this paper presents a powertype-based metamodeling solution to overcome the MAS modelling language diversity problem. This solution is based on the presented inter-ML MAS metamodel, which defines the inter-ML MAS modelling language. Moreover, a generic tool prototype is developed for the presented modelling language. This tool is created with a Model-driven Development(MDD) approach.

In future works, the presented MAS inter-ML can be evaluated by contrasting opinions of several MAS experts. In addition, the modernization of the existent MAS tools is also left for future work. The goal is to modernize the MAS tools so that they can export the models to the presented inter-ML MAS ML. As an initial step, a preliminary modernization of the INGENIAS Development Kit(IDK) tool is already done to export the models for the presented inter-ML.

Acknowledgments

This work has been supported by the following projects: *Methods and tools for agent-based modeling* supported by Spanish Council for Science and Technology with grant TIN2005-08501-C03-01; *Methods and tools for agent-based modeling* supported by Spanish Council for Science and Technology with grant TIN2005-08501-C03-03 and Grant for Research Group 910494 by the Region of Madrid

(Comunidad de Madrid) and the Universidad Complutense Madrid; and *Methods and tools for agent-based modeling* project with grant TIN2005-08501-C03-01 and Grant for Research Group 92354, supported by the Universidad Complutense Madrid in 2008.

References

1. Bauer, B., Muller, J.P., Odell, J.: Agent UML: A Formalism for Specifying Multiagent Interaction. Agent-Oriented Software Engineering 1957, 91–103 (2001)
2. Bernon, C., Cossentino, M., Pavon, J.: An Overview of Current Trends in European AOSE Research. Informatica 29, 379–390 (2005)
3. García-Magariño, I., Agüera, J.R.P., Gómez-Sanz, J.J.: Reaching Consensus in a Multi-agent System. In: 6th International Workshop on Practical Applications on Agents and Multi-agent Systems, IWPAAMS 2007, Salamanca, Spain, November 12-13, 2007, pp. 349–358 (2007)
4. Gonzalez-Perez, C., Henderson-Sellers, B.: A powertype-based metamodelling framework. Software and Systems Modeling 5(1), 72–90 (2006)
5. Henderson-Sellers, B., Gonzalez-Perez, C.: Connecting Powertypes and Stereotypes. Journal of Object Technology 4(7) (2005)
6. Henderson-Sellers, B., Gonzalez-Perez, C.: The rationale of powertype-based metamodelling to underpin software development methodologies. In: Proceedings of the 2nd Asia-Pacific conference on Conceptual modelling, vol. 43, pp. 7–16 (2005)
7. GRASIA research group. INGENIAS Development Kit (July 2, 2008), `http://ingenias.sourceforge.net/`

Does Android Dream with Intelligent Agents?

Jorge Agüero, Miguel Rebollo, Carlos Carrascosa, and Vicente Julián

Departamento de sistemas informáticos y computación
Universidad Politécnica de Valencia.
Camino de Vera S/N 46022 Valencia, Spain
{jaguero,mrebollo,carrasco,vinglada}@dsic.upv.es

Abstract. In this paper, a new agent model "specially" designed for the recent Android[1] Google SDK is presented, where the Android mobile phone can be considered agent software. This agent model has an approach more practical than theoretical because it uses abstractions which allow to implement it on various systems. The appearance of Android as an open system based on Linux has created new expectations to implement agents. Now, agents may run in different hardware platforms, one approach useful in Ubiquitous Computing to achieve intelligent agent embedded in the environment, which can be considered the vision of the intelligent ambient. Finally, the proposed model abstractions that were used to design the Android agent has been employed to implement a simple example which shows the applicability of the proposal.

Keywords: Agent model, embedded agent, *Android Google*.

1 Introduction

The *Ubiquitous Computing* or *Pervasive Computation* [9] is a paradigm in which the technology is virtually invisible in our environment, because it has been inserted in the ambient with the objective of improving people's life quality, creating an *intelligent ambient* [5]. In the *Pervasive Computation*, awareness is becoming an habitual characteristic of our society with the appearance of electronics devices incorporated in all class of fixed or mobile objects (Embedded system), connected by means of networks to each other. It is a paradigm in which computing technology becomes virtually invisible as a result of being embedded computer artifact's into our everyday environment [6].

One approach to implement pervasive computing is to embed intelligent agents. An intelligent agent is a HW or (more usually) SW-based computer system that enjoys the following properties: autonomy, social ability, reactivity and pro-activeness [12]. Embedded-computers that contain these agents are normally referred to as *embedded-agents* [11]. Each embedded agent is an autonomous entity, and it is common for such embedded-agents to have network connections allowing them to communicate and cooperate with other embedded agents, as part of a multi-embedded agent system.

The challenge, however, is how to manage and to implement the intelligent mechanisms used for these embedded agents due to the limited processing power and memory capacity of embedded computational HW, the lack of tools for the development of

[1] Android is trademark of Open Handset Alliance, where Google is a member.

J.M. Corchado et al. (Eds.): DCAI 2008, ASC 50, pp. 194–204, 2009.
springerlink.com

embedded applications and the lack of standardisation. By these challenges and other found problems [8], a remarkable difference between the conceptual agent model and the implemented or expected agent has been detected. For example, it is known that Java is a language very used for the development of agents, but the difference between Java for personal computers (J2SE) and Java embedded devices (J2ME), produces big changes in the implemented agents that is solved often by adding new middleware layers, but reducing the agent functionality on many platforms [5].

But now, with the appearance of the SDK *Android* made by Google as platform for the development of embedded applications in mobile phones, it creates a new approach for implementing embedded intelligent agents, because it is an open source platform and the development of the applications is made with a new Java library (Java *Android* library) that is very close to Java for personal computers (J2SE) [1]. Furthermore, there exists the possibility that the *Android* Linux Kernel can be migrated to other platforms or electronic devices, allowing to such agents to be executed in a wide variety of devices.

To sum up, the basic idea is to present an agent model that can be designed using components or abstractions that can be deployed on any programming platform, such as the *Android* SDK that allows to implement such agent model. This will show the feasibility of implementing embedded agents using these abstractions, reducing the gap between the design of embedded agents and their implementation. The rest of the document is structured as follows. Section 2 describes the proposed *agent model*. The section 3 briefly explains the main components of the *Android* Platform. Section 4 details agent implementation in *Android*. Section 5 shows a simple example that demonstrates the viability to implement the model in the *Android* SDK. Finally, the conclusions of the present work are expounded in section 6.

2 Agent Platform Independent Model (APIM)

Major challenges in pervasive computing include invisibility or unawareness, proactivity, mobility, privacy, security and trust [5]. In such environments, HW and SW entities are expected to function autonomously, continually and correctly.

Traditionally, agents have been employed to work on behalf of users, devices and applications [11]. The agents can be effectively used to provide transparent and invisible interfaces between different entities in the environment. Agent interaction is an integral part of pervasive (intelligent) environments because agents acquire and apply effectively knowledge in their ambient.

At the moment, there is a large amount of agent models that provide a high-level description of their components and their functionality to define the agent model presented in this paper, some of the most used and complete agent model proposals have been studied. This study purpose was to extract their common features and adapt it to the current proposal. In this way, Tropos [3], Gaia [13], Opera [7], Ingenias [10] and AML [4] have been considered, because they are the most commonly used.

An agent model provides an abstract vision of its main components and their existing relationship. Figure 1 shows the agent model presented in this paper (*APIM)*.

The highest-level entity to be considered is the agent. At this level, organizations of higher order, group belonging rules or behaviour norms, are not taken into account.

Agent: An *Agent* has an identifier and a public *name*. The *Environment* is represented by means of its relationship with ambient or surrounding, allowing to define input and output ports to communicate with the outside. Agent's knowledge base is kept in its *belief* set and its *goal set*. The agent has two messages queues, *Input* and *Output*, to communicate with the outside, which temporally store incoming and outcoming messages respectively. Besides messages, the agent can be aware of event arrival, being stored in *EventQueue*. Lastly, the agent has a *State*, related with its life-cycle and with its visibility by other agents.

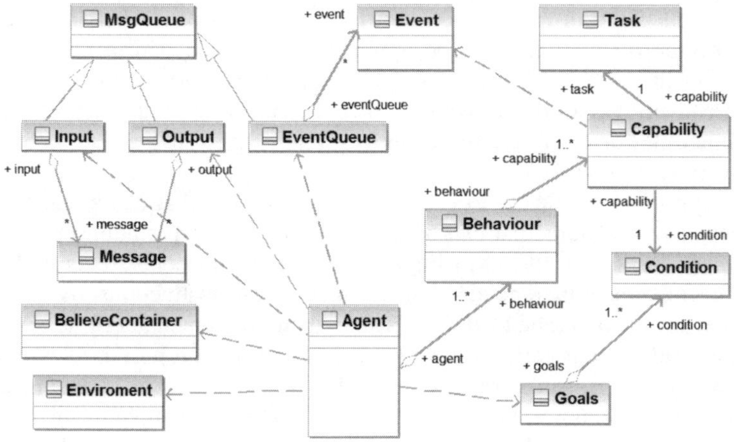

Fig. 1. *APIM* structure

With regards to the problem-solving methods, the agent has a set of core components. The *Capabilities* which represent the know-how of the agent and follow an event-condition-action scheme. To improve the efficiency of the agent, *Capabilities* are grouped into *Behaviours* that define the roles the agent can play. By doing that, can be kept active (ready) any *Capability* related with the current situation, avoiding overload the agent with unnecessary knowledge.

Behaviours: A set of *Behaviours* is defined in the agent to distinguish between different environments and attention focuses. Basically, *Behaviours* are used to reduce and to delimit the knowledge the agent has to use to solve a problem. So, those methods, data, events or messages that are not related with the current agent stage should not be considered. In this way, the agent's efficiency in problem-solving process is improved. A *Behaviour* has a *Name* to identify itself. It also has associated a *Goals set* that may be used either as activation or maintenance conditions (see Figure 2(a)). Lastly, a state indicating its current activation situation. More than one Behaviour may be active at the same time.

Capabilities: An *event* is any notification arriving to the agent informing that something that may be of interest for the agent has happened in the environment or inside the agent. It may have caused the activation of a new *Capability*.

The Tasks that the agent knows how to fulfill are modeled as Capabilities. Capabilities are stored inside Behaviours and they model the agent's answer to certain events.

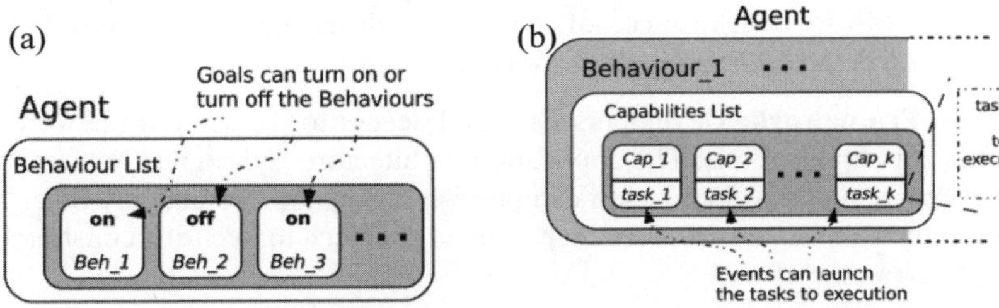

Fig. 2. (a) *Behaviours* in *APIM*, (b) *Capabilities* in APIM

A Capability is characterised by a Name that identifies it, its trigger Event, an activation Condition and the Task that has to be executed when the event arrives and the indicated condition is fulfilled (see Figure 2(b)). It is also indicated the State the Capability has. Only Capabilities belonging to current active Behaviours are executed.

All the Capabilities of an active Behaviour will be in a state marked as Active. When an event arrives, the Capability state changes to Relevant and its activation condition is evaluated. If this condition is fulfilled, the state passes to Applyable and the associated Task begins its execution. When this Task ends, the Capability returns to Active again and it remains waiting the arrival of new events. When a Behaviour becomes inactive, all its Capabilities stop their execution and change their state to inactive. It is assumed that all the Capabilities of an active Behaviour can be concurrently executed, so that the system has to provide the needed methods to avoid deadlocks and inconsistencies during their execution.

Task: The last component of the agent model is the Task. Tasks are the elements containing the code associated to the agent's Capabilities. One Task in execution belongs only to one Capability and it will remain in execution until its completion or until the Capability is interrupted because the Behaviour it pertains to is deactivated. It is not defined any method of recovering nor resuming of interrupted Tasks. On the other hand, the agent must have some mechanism of "safe stop" to avoid the agent to fall in inconsistent states.

3 Android Google: A New Platform for Mobile Devices

Android is a SW stack for mobile devices that includes an operating system, middleware and key applications. Developers can create applications for the platform using the Android SDK [1]. Applications are written using the Java programming language and they run on Dalvik[2] , a custom virtual machine designed for embedded use, which runs on top of a Linux kernel.

The main components of the *Android* operating system are:

Applications: *Android* will ship with a set of core applications including an email client, SMS program, calendar, maps, browser, contacts and others. All applications are written using the Java programming language. Every *Android* application runs in its

[2] Android Virtual Machine.

own process, with its own instance of the Dalvik virtual machine. Dalvik has been written so that a device can run multiple VMs efficiently.

Application Framework: Developers have full access to the same framework APIs used by the core applications. The application architecture is designed to simplify the reuse of components; any application can publish its capabilities and any other application may then make use of those capabilities (subject to security constraints enforced by the framework).

Libraries: *Android* includes a set of libraries used by various components of the *Android* system. For example, some of the core libraries support playback and recording of many popular audio and video formats, also the core Web browser engine and SQL engine (SQLite) for maintenance database.

***Android* Runtime:** *Android* includes a set of core libraries that provides most of the functionality for Java programming language. *Android* Runtime provides abstract components to create applications (see section **Error! Reference source not found.**)

Linux Kernel: *Android* relies on Linux version 2.6 as core system services such as security, memory management, process management, network stack and driver model. The kernel also acts as an abstraction layer between the HW and the rest of the SW stack.

There are four building blocks in an Android application: ***Activity***, ***Intent Receiver***, ***Service*** and ***Content Provider***. An application does not need to use all of them, but they can be combined in any order to create an application. Each application has a manifest file, called `AndroidManifest.xml`, which list all components used in the application. This is an XML file where you declare the components of your application and what their capabilities and requirements are:

Activity: Is the most common of the four Android building blocks. An activity is usually a single process with interface in an application. Each Activity is implemented as a single class that extends the Activity base class. The Activity displays a user interface composed of Views which responds to events.

Intent Receiver: It is an event handler, that is, it allows to define the reaction of the application to events (called Intents), such as when the phone rings, when the data network is available or when it's midnight. Intent Receivers do not display a UI (User Interface), although they may use notifications to alert the user if something interesting has happened. The application does not have to be running for its Intent Receivers to be called; the system will start the application, if necessary, when an Intent Receiver is triggered.

Service: A Service is a long-life code that runs without a UI. It is a process running in the background without interaction with the user for an indeterminate period of time. A good example of this is a media player application, the music playback itself should not be handled by an activity because the user will expect the music to keep playing even after navigating to a new screen. In this case, a Service will remain running in the background to keep the music going.

Content Provider: Applications can store their data in files, a database or any other mechanism. The Content Provider, however, is useful to share data with other

Android applications. The Content Provider is a class that implements a standard set of methods to let other applications store and retrieve the type of data that is handled by that Content Provider.

4 Implementing APIM in Android

The developing of *APIM* in *Android* was made using the above mentioned *Android* building block APIs (the API Version m5-rc14) [2]. There are four main components to model *APIM* agents: **Agent**, **Behaviour**, **Capability** and **Task**, which perform the functions described in section 3. Table 1 shows the *Android* blocks used for building components of the *APIM* model and other necessary components.

The design presented can be seen as an interface to implement the agent according to the users requirements and needs. This interface uses specific components provided by the API, as previously commented. Thereby this model inserts a new layer in the *Android* system architecture [1]. This new layer, called *Agent interface*, modifies the architecture as it is seen in the figure 3.

Table 1. The Android components used in the APIM model

APIM Components	Android Components	Overloaded methods
Agent	Service + Activity	onCreate(), onStart(), onDestroy()
Behaviour	IntentReceiver	registerReceiver(), onReceiveIntent()
Capability	IntentReceiver	registerReceiver(), onReceiveIntent()
Task	Service	onStart(), onDestroy()
Goals	Intents	IntentFilter()
Events	Intents	IntentFilter()
Beliefs	Contentprovider	–
ACL Communications	http	–

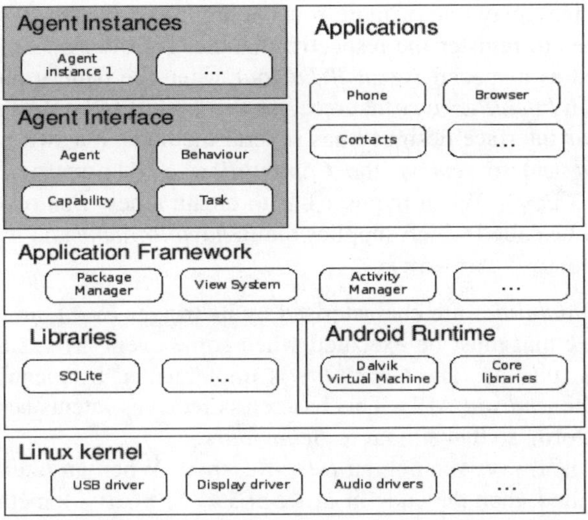

Fig. 3. Agent interface in *Android* System Architecture

Agent: the *Agent class* is designed to handle the arrival of events. Therefore an agent has to consider the changes in its environment (may be of interest for the agent) to determine its future actions activating and deactivating the appropriate *Behaviours* to respond to any internal or external situation. In this way, Agent class is implemented as one *Android Service* and one *Activity* as screen interface.

To implement the APIM model, some methods of Service class have to be overloaded. The `onCreate()` method allows initialise agent variables. After the `onStart()` method is executed that enabling the agent components. The agent is executed until the user decides to stop its execution. In this moment, the user call `selfstop()` or `stopService()` method, allowing effectively terminated the agent execution. Every agent component is stopped and destroyed (Tasks, Capabilities and Behaviours).

The agent interface designed has several methods that allow to implement the APIM, but there are two methods that are important to mention: the `init()` method, where the user may write the code necessary to initialise the agent, and the `run()` method, that activates roles that the agent has to play (active the Behaviours). The `init()` is executed within Service's `onStart()`, that is called when the agent starts for first time. The Agent class can also launch a UI (User Interface), one Activity, to interact with users and to show its internal state and progress.

Behaviour: The Behaviour class works as a container of the Agent Capabilities and it can group as many Capabilities as the user desires. All of them can be activated and deactivated when events arrive. Behaviours are implemented by mean of an IntentReceiver class of the Android APIs. This base class receives intents sent by events of the Android platform. An IntentReceiver have to be dynamically registered to treat intents, using `registerReceiver()` method.

The *IntentReceiver* will be running on the main agent thread. The Receiver will be called when any intent arrives and it matches with the intents filters, i.e. bind an intent to object that is the receiver of intent.

As the agent may play one or more roles at any moment, the *Behaviour class* can activate new roles to register the respective handler (of intents). For example, a role may be activated as a special Agent *Behaviour* when the battery phone is low. This can be done by an *IntentReceiver* that receives the intent LOW_BATTERY.

The *Behaviour* interface designed has several methods, but two main methods are provided to *add* and to *remove* the *Capabilities*: `add(capability)` and `remove(capability)`. When the user has to create a new *Behaviour*, the constructor method must be called which supplies the *Behaviour name* and its trigger *Intent* as `Behaviour(Name, Intent)`.

Capabilities: *Capabilities* are characterised by its trigger *Event*, an activation *Condition* and the *Task* that must be executed when some event arrives and the indicated condition that is fulfilled. The *Capability* is implemented by means of an *IntentReceiver* class of the *Android* APIs. This base class receives intents sent from events of the *Android* platform, so that similar to *Behaviours*.

A *Capability* is always running an *IntentReceiver*. When an intent arrives and the condition is fulfilled, then the code in `onReceiveIntent()` method is considered to be a foreground process and will be kept running by the system for manipulating the intent, in this moment then the *Task* is launched.

The *Capability* interface designed has one important method for matching a *Task* to its corresponding *Capability*, this is `addTaskRun(task)` method. When the user has to create a new *Capability* the constructor method must be called supplying the *Capability name* and its trigger *Intent* as `Capability(Name, Intent)`.

Tasks: Now, *Task* class is one special process to run as an *Android* Service. To implement the *Task*, some methods of *Service* class have to be overloaded. The `onCreate()` method allows initialise *Task* variables when it is launched the `onStart()` method allows to execute the user code, throughout a call to a `doing()` method that has to be overloaded by the programmer. Now, the main method of *Task* interface is `doing()`, in where the user writes the Java program to be executed.

Finally, the *intents* are used to model the **goals** that activate the *Behaviours* and **events** that allow execute the *Tasks* of a *Capability*. To manipulate and store the agent *beliefs*, the *Android ContentProvider* is used as a database. The *communication* between agents is implemented creating FIPA ACL messages.

5 Example

An example of two agents talking by means of a chat session is used to show the applicability of this proposal. So, a simplified Chat Session between two agents that send and receive ACL messages is proposal. This simple example is presented with academic goals, to explain and show how to use agent interface designed in *Android* platform only, this example does not wish to illustrate a complex agent's interaction.

The implementation of the agent was done in an *Android* emulator, because there is currently no real phones where applications can run at the moment. The first step of the design process is to identify the agents' roles. As agents simply send and receive information from each other, we consider modelling the agent with only one *Behaviour*, which is called *CHAT*. A simple chat session has one *Capability* where users send information whenever they want and another *Capability* that is waiting the arrival of any message. Therefore, two *Capabilities* are created: one to transmit a message and the other to receive it (see Figure 4(a)).

Each agent *Capability* has as mission to send or to receive messages. It is necessary to remember that a *Capability* receives intents. When the intent arrives and the condition is fulfilled, then the *Task* is launched. The *Capability* that sends messages is called *SendMsg*, and its *Task*, *task_Send*, transmits the information when users press the send button (see Figure 4(b)).

The *Capability* that receives messages is called *ReceiveMsg* and its *Task*, *task_Receive*, waits for the arrival of other agent messages. So, agents are ready to begin the process of communication and exchange of information in the Chat. Messages will be displayed on the phone screen. Now, to program the agent interface (for this preliminary implementation in *Android*) proceed as is explained below:

- Create one *Behaviour* with *name*= `CHAT`.
- Create one *Capability* for sending messages, with *name*= *SendMsg*, and the *condition* (intent) that wake it up. Then add the *Task* that permit send the ACL message: `task_Send`.

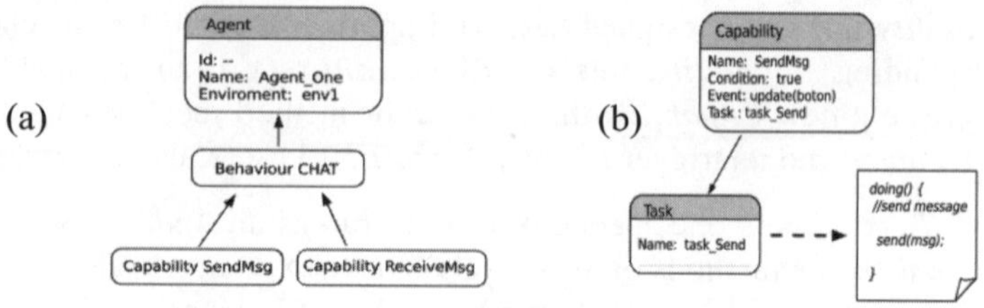

Fig. 4. (a) Agent Model for chat session, (b) Capability SendMsg

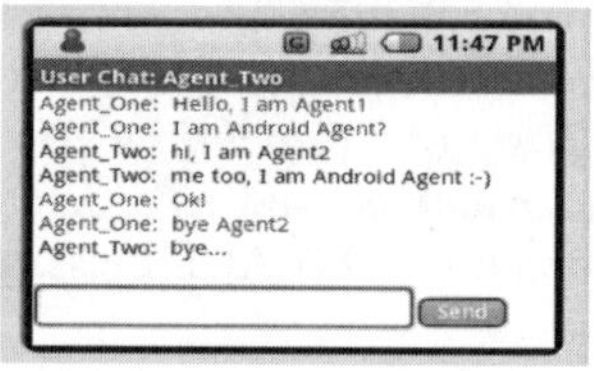

Fig. 5. Chat in the emulator screen

```
public class MyAgent extends Agent {
    public void init(){
     . . .
       //Create one Behaviour
       Behaviour myBehaviour= new Behaviour("CHAT");

       //Create two capabilities and its condition trigger
       Capability myCapabilityTX = new Capability("SendMsg");
       Capability myCapabilityRX = new Capability("ReceiveMsg");

       //Condition and intent trigger of send
       Condition mycondSend = new Condition() {
          @Override
          public boolean expression(Intent intent) {
             if (intent.getAction() == "Android.intent.action.MY_SENDMSG") {
                return true;
             }else { return false; }
      . . .
       //Set the condition different of null
       myCapabilityTX.setCondition(mycondSend);
       myCapabilityRX.setCondition(mycondReceive);

       //Create and add task that send and receive the chat msg
       Task myTaskTX = new task_Send;
       myCapabilityTX.addTaskRun(myTaskTX);
       Task myTaskRX = new task_Receive;
       myCapabilityRX.addTaskRun(myTaskRX);

       //Add the Capabilities to the Behaviour
       myBehaviour.add(myCapabilityTX);
       myBehaviour.add(myCapabilityRX);

       //Add Behaviour the agent and execute it
       addbehav(myBehaviour);
    }
}
```

Fig. 6. APIM Agents Chat

- Create other *Capability* for receiving messages, with *name= ReceiveMsg*, and the *condition* (intent) that wake it up. Then add the *Task* that permit receiver the ACL message: `task_Receive`.

- Add these two *Capabilities* to the *Behaviour*.
- Add the *Behaviour*, using `addbehav()` method. Now the agent is executed and the messages will be displayed on the emulator screen (see Figure 5).

```
public class task_Send extends Task {
    @Override
    public void doing() {
     . . .
        //create or get the id agent
        AID agentReceiver = new AID();
        agentReceiver.setName("AGENT TWO");
        agentReceiver.addAddresses("192.168.1.105");

        //Get the string (from screen chat) to send
        String content = getValueIntentDisplay("Display.varText");

        //Compose the ACL message to send another agent
        ACLMessage  msg = new ACLMessage(ACLMessage.INFORM);
        msg.setContent(content);
        msg.addReceiver(agentReceiver);
        send(msg);
     . . .
}
```

Fig. 7. Task for send Chat messages

Now, the program to implement the agent designed it is shows Figure 6 and to illustrate the Java code that the user write in the Task, shows the program to send Chat messages in Figure Fig. **7**, ie, the task: task_Send.

6 Conclusions

A general agent model to build intelligent agents in Android platform has been presented. This model can be easily adapted to almost any platform or architecture HW/SW. Moreover, the agent model has been implemented and tested in Android Google platform. The agent interface designed allows to implement embedded agents according to the users requirements.

Using of the *Android* platform demonstrated the utility and probed the feasibility of designing a platform independent agent. Using of the proposed model abstractions for *APIM* agent reduces the gap between the theoretical model and its implementation.

The embedded agent design meets with the functionalities that were thought for it. Besides, the *Android* platform promises to be a new platform to implement novel agent models. That is because Java API is very similar to the Personal Computer version, allowing to implement embedded agent-based approach with mechanisms even more advanced. This is a useful feature in Pervasive Computing. Additionally as *Android* platform is a Linux system, there is a high probability that the platform can be migrated to other different devices.

As future work, the services that the agent can deliver will be enriched and enhanced from this first version. The prototype has been developed using an emulator for *Android*. The evaluation of the performance of presented agent architecture will be done when the first mobile phone using *Android* system will be launched.

While this article was being written appeared a Jade[3] version for *Android* system. Though the authors have not deeply evaluated Jade in the *Android* architecture, it have to be underlined that this paper's agent model presents a conceptually different model to Jade's one, because this model is more integrated with *Android's* building block than Jade one.

References

1. Android, S.D.K.: An Open Handset Alliance Project, Web Site (January 2008), `http://code.google.com/android/`
2. Android SDK, Download the Android SDK, Web Site (January 2008), `http://code.google.com/android/download_list.html`
3. Castro, J., Kolp, M., Mylopoulos, J.: A Requirements-Driven Software Development Methodology. In: Conference on Advanced Information Systems Engineering (2001)
4. Cervenka, R., Trencansky, I.: The Agent Modelling Language–AML. Whitestein Series in Software Agent Technologies and Autonomic Computing (2007) ISBN: 978-3-7643-8395-4
5. Cook, D., Das, S.K.: How smart are our environments? An updated look at the state of the art. Pervasive and Mobile Computing 3(2) (2007)
6. Davidsson, P., Boman, M.: Distributed monitoring and control of office buildings by embedded agents. Information Sciences 171(4) (2005)
7. Dignum, V.: A model for organizational interaction: based on agents, founded in logic. Ph.D Dissertation, Utrecht University (2003)
8. Doctor, F., Hagras, H., Callaghan, V.: A type-2 fuzzy embedded agent to realise ambient intelligence in ubiquitous computing environments. Information Sciences 171(4) (2005)
9. European Research Consortium for Informatics and Mathematics (ERCIM NEWS), Special: Embedded Intelligence. Number 67 (October 2006)
10. Gomez Sanz, J.J.: Modelado de Sistemas Multi-Agente. Ph.D Thesis, Universidad Complutense de Madrid, Spain (2002)
11. Hagras, H., Callaghan, V., Colley, M.: Intelligent embedded agents. Information Sciences 171(4) (2005)
12. Wooldridge, M., Nicholas, R.: Jennings, Agent Theories, Architectures, and Languages: a Survey. In: Wooldridge, Jennings (eds.) Intelligent Agents. Springer, Berlin (1995)
13. Zambonelli, F., Jennings, N., Wooldridge, M.: Developing Multiagent Systems: The Gaia Methodology. ACM Transactions on Software Engineering and Methodology 12, 317–370 (2003)

[3] http://jade.tilab.com/

Genetic Algorithms for the Synthesis and Integrated Design of Processes Using Advanced Control Strategies

Silvana Revollar[1], Mario Francisco[2], Pastora Vega[2], and Rosalba Lamanna[1]

[1] Departamento de Procesos y Sistemas
Universidad Simón Bolívar
Sartenejas, 89000, Venezuela
[2] Departamento Informática y Automática
Universidad de Salamanca
Plaza de la Merced s/n, 37008, Salamanca, Spain
srevolla@usb.ve, mfs@usal.es, pvega@usal.es, lamanna@usb.ve

Abstract. This work presents a real-coded genetic algorithm to perform the synthesis and integrated design of an activated sludge process using and advanced Multivariable Model-based Predictive Controller (MPC). The process synthesis and design are carried out simultaneously with the MPC tuning to obtain the most economical plant which satisfies the controllability indices that measure the control performance (H∞ and l1 norms of different sensitivity functions of the system). The mathematical formulation results into a mixed-integer optimization problem with non-linear constraints. The quality of the solutions obtained evidence that real-coded genetic algorithms are a valid and practical alternative to deterministic optimization methods for such complex problems.

Keywords: Process synthesis, Integrated Design, Genetic Algorithms, Model Predictive Control.

1 Introduction

It has been broadly recognized, the importance of considering controllability issues at the early stages of plant design. As a consequence, a number of methodologies for the simultaneous process and control system design have been developed [7], [11], [4], [15]. These methodologies leads to optimization problems that involves economical and operational decisions about the plant dimensions and working point, process specifications, control system performance and, in the more thorough formulations, the structure of the plant and the control scheme.

Most of the approaches use conventional PID controllers, only few works as [12], [13], [4] have addressed to the application of advanced control techniques as Model-Based Predictive Controllers (MPC). The reason is that this advanced control schemes involve solving an optimization problem on-line, leading to a drastic increase in the complexity of the design framework [12], [13].

The mathematical formulation for the combined synthesis and integrated design results into a mixed-integer non linear optimization problem (MINLP), moreover, it translates into a mixed-integer dynamical optimization (MIDO) if dynamical simulations are required to evaluate the control performance indices. The solution of these problems involves the application of advanced algorithms to handle the continuous

J.M. Corchado et al. (Eds.): DCAI 2008, ASC 50, pp. 205–214, 2009.
springerlink.com

and discrete variables together with the evaluation of the controllability indices. Deterministic optimization techniques has been applied for solving this kind of MINLP [8], [7], [12], [13], but, decomposition algorithms are generally necessary[13], and the complexity of the optimization procedure makes their implementation considerably complicated, even in the presence of good results.

The genetic algorithms have been applied successfully as a non conventional alternative for solving process engineering MINLP [2], [16], and, specifically from the synthesis and integrated design of an activated sludge process [11]. They are powerful optimization techniques suitable for complex problems were discontinuities and non convexity are present. Conventional genetic algorithms are binary coded [14], however, the real coding allows to use large domains for the variables, and increases the capacity for the local tuning of the solutions [3]. Nevertheless, an actual drawback is the difficulty to handle the constraints because the stochastic search operators frequently produce infeasible solutions [14].

The aim of this work is to perform the integrated synthesis and design of a process using multivariable model-based predictive controllers (MPC) that will be tuned automatically, using frequency domain methods as controllability indexes as was proposed in [17]. It involves the solution of a mixed sensitivity problem with constraints, however, avoids the time-consuming dynamical simulations required by the typical optimization-based tuning techniques [4].

The activated sludge process of the Manresa's plant was selected to apply this methodology. The optimization model results into a complex MINLP including continuous, integer and binary variables, together with non linear constraints and controllability norms evaluated through the linearization of the process model. A real coded genetic algorithm was selected for its solution, due to its robustness and the reasonable effort required for its implementation.

The paper is organized containing, first, the description of the controllability metrics used for the MPC tuning, the formulation of the optimization problem in section 3. The description of the genetic algorithm proposed to solve the problem is presented in section 4, followed by the analysis of the results in section 5. Finally, conclusions and future work are included.

2 MPC Formulation and Controllability Metrics

The basic MPC formulation consists of the on-line calculation of the future control actions by solving the following optimization problem subject to constraints on inputs, predicted outputs and changes in manipulated variables

$$\min_{\Delta\hat{u}} V(k) = \sum_{i=Hw}^{Hp} \left\| \hat{y}(k+i \mid k) - r(k+i \mid k) \right\|_{Wy}^{2} + \sum_{i=0}^{Hc-1} \left\| \Delta\hat{u}(k+i \mid k) \right\|_{Wu}^{2} \tag{1}$$

where k denotes the current sampling point, $\hat{y}(k+i \mid k)$ is the predicted output vector at time $k+i$, depending of measurements up to time k, $r(k+i \mid k)$ is the reference trajectory, $\Delta\hat{u}$ are the changes in the manipulated variables, Hp is the upper prediction horizon, Hw is the lower prediction horizon, Hc is the control horizon, Wu and Wy are positive definite matrices representing the weights of control moves and the weights of the set-point tracking errors respectively. In this work, the matrices Wy and

Wu are considered diagonal but not time dependent, so the error vector is penalized at every point in the prediction horizon and the control moves $\Delta\hat{u}(k+i\,|\,k)$ are penalized at every point in the control horizon.

The optimization problem in (1) is a Quadratic Programming (QP) that gives a sequence of control moves $\Delta\hat{u}(k+i\,|\,k)$. The first component of this sequence is applied to the system at time *k*+1, and the optimization problem (1) is repeated at the next sampling time (receding horizon strategy). The MPC prediction model used here is a linear discrete state space model of the plant, the MPC Toolbox of MATLAB has been used for its implementation.

If the MPC controller is linear, constraints are not active, and the particular formulation of [9] is considered, the following transfer functions can be calculated from the closed loop system: Sensitivity function *S(s)* between the load disturbances (*d*) and the outputs (*y*); Control Sensitivity transfer function *M(s)* between the load disturbances (*d*) and the control signals (*u*):

$$S(s)=\frac{y(s)}{d(s)}=\frac{K_3G+G_d}{1+GK_1}\qquad M(s)=\frac{u(s)}{d(s)}=\frac{K_3-K_1G_d}{1+GK_1}\tag{2}$$

The first controllability index considered in this work is:

$$\|N\|_\infty=\max_w|N(jw)|\tag{3}$$

where *N* is a mixed sensitivity index that takes into account both disturbance rejection and control effort objectives. The function *N* is defined as:

$$N=\begin{pmatrix}Wp\cdot S\\ Wesf\cdot s\cdot M\end{pmatrix}\tag{4}$$

Wesf(s) is chosen to penalize control efforts adequately, and *Wp(s)* is chosen based on the spectra of disturbances to ensure proper disturbance rejection. For that, the following constraint (considering normalized disturbances) must be imposed:

$$\|Wp\cdot S\|_\infty<1\tag{5}$$

The maximum value of the manipulated variables (for the worst case of disturbances) can be constrained to be less than *umax*, by means of the l1 norm and the following condition:

$$\|M\|_1<u_{\max}\tag{6}$$

3 Formulation of the Optimization Problem

The activated sludge process is the second stage in a typical wastewater plant. It has been selected to study the integrated synthesis and design methodology. The complexity of the biological processes involved makes this case study very suitable to test this approach. The detailed model is presented in [10] for the real wastewater treatment plant (Manresa, Spain) used in this work.

The process takes place in a series of aeration tanks followed by settlers. In the aeration tanks or bioreactors, the activity of a mixture of microorganisms name biomass is used to reduce the substrate concentration in the water by converting the organic substrate into inorganic products, more biomass, and water. The dissolved oxygen required is provided by a set of aeration turbines. Water coming out from reactors goes to the settler, where the clean water is separated from the activated sludge which is recycled to bioreactors.

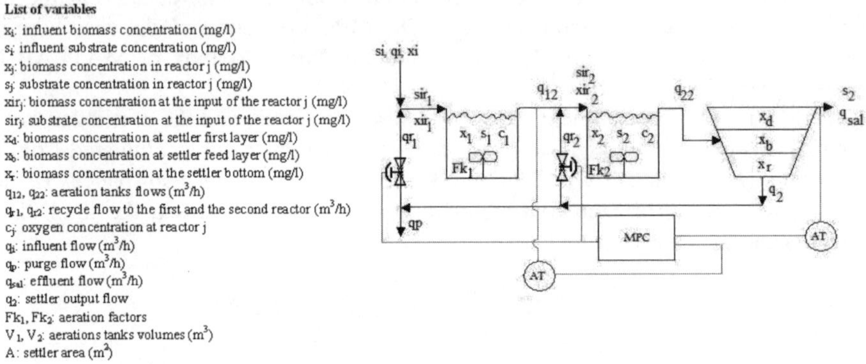

Fig. 1. Activated sludge process superstructure

The control of this process aims to keep the substrate at the output (s_1 or s_2) below a legal value despite the large variations on the incoming substrate concentration (*si*) using the recycling flows qr_1 and qr_2 as manipulated variables [10]. The frequency and magnitude of the disturbances at the *si* input make the control of the plant a difficult task. The set of disturbances used for evaluate the control performance while tuning the MPC has been determined by COST 624 program [1].

The superstructure of the process is presented in figure 1, the design alternatives consists in one or two aeration tanks and one secondary settler. The model of the process superstructure is a set of differential and algebraic equations which takes the appropriated values for each alternative according to the binary y_1. The problem is stated as a mixed integer non linear optimisation of the cost function where the objective is to obtain an economically optimal plant, with the best structure, dimensioning and stationary working point together with the MPC tuning parameters that allows the desired control performance.

The cost function represents the construction costs proportional to v and A, and the operational costs associated to the aeration factor and the recycle flow:

$$f = p_1 \cdot v_1^2 + p_1 \cdot v_2^2 + p_2 \cdot A^2 + p_3 \cdot Fk_1^2 + p_3 \cdot Fk_2^2 + p_4 \cdot q_2^2 \tag{7}$$

where v_1, v_2 are the reactor volumes and A is the cross-sectional area of the settler, Fk_1 and Fk_2 are the aeration factors and , is the overall recycle flow.

The mass balances in aeration reactors and decanter that must satisfy the steady state operation:

$$\left| v_1 \frac{dx_1}{dt} \right| = \left| \mu_{\max} Y \frac{s_1 x_1}{(K_s + s_1)} v_1 - K_d \frac{x_1^2}{s_1} v_1 - K_c x_1 v_1 + q_{12}(xir_1 - x_1) \right| \le \varepsilon \tag{8}$$

$$\left|v_1 \frac{ds_1}{dt}\right| = \left|-\mu_{\max} \frac{s_1 x_1}{(K_s + s_1)} v_1 + f_{kd} K_d \frac{x_1^2}{s_1} v_1 + f_{kd} K_c x_1 v_1 + q_{12}(sir_1 - s_1)\right| \le \varepsilon \tag{9}$$

$$\left|v_1 \frac{dc_1}{dt}\right| = \left|K_{la} Fk_1 (c_s - c_1) v_1 - K_{01} \mu_{\max} \frac{s_1 x_1}{(K_s + s_1)} v_1 - q_{12} c_1\right| \le \varepsilon \tag{10}$$

$$\left|v_2 \frac{dx_2}{dt}\right| = \left|\mu_{\max} Y \frac{s_2 x_2}{(K_s + s_2)} v_2 - K_d \frac{x_2^2}{s_2} v_2 - K_c x_2 v_2 + q_{22}(xir_2 - x_2)\right| \le \varepsilon \tag{11}$$

$$\left|v_2 \frac{ds_2}{dt}\right| = \left|-\mu_{\max} \frac{s_2 x_2}{(K_s + s_2)} v_2 + f_{kd} K_d \frac{x_2^2}{s_2} v_2 + f_{kd} K_c x_2 v_2 + q_{22}(sir_2 - s_2)\right| \le \varepsilon \tag{12}$$

$$\left|v_2 \frac{dc_2}{dt}\right| = \left|K_{la} Fk_2 (c_s - c_2) v_2 - K_{01} \mu_{\max} \frac{s_2 x_2}{(K_s + s_2)} v_2 - q_{22} c_2 + W_1\right| \le \varepsilon \tag{13}$$

$$\left|AL_d \frac{dx_d}{dt}\right| = \left|q_{sal} x_b - q_{sal} x_d - A \cdot nnr \cdot x_d \exp(aar \cdot x_d)\right| \le \varepsilon \tag{14}$$

$$\left|AL_b \frac{dx_b}{dt}\right| = \left|q_{22} x_2 - q_{sal} x_b + A \cdot nnr \cdot x_d \exp(aar \cdot x_d) - A \cdot nnr \cdot x_b \exp(aar \cdot x_b)\right| \le \varepsilon \tag{15}$$

$$\left|AL_r \frac{dx_r}{dt}\right| = \left|q_2 x_b - q_2 x_r + A \cdot nnr \cdot x_b \exp(aar \cdot x_d)\right| \le \varepsilon \tag{16}$$

where the W_1 term is used to cancel (13) when y_1=0.

The operation constraints for the activated sludge process are:

– Residence times:

$$2.5 \le \frac{v_1}{q_{12}} \le 8 \tag{17}$$

$$2 \le \frac{v_2 + (1 - y_1) \cdot W_2}{q_{22}} \le 6 \tag{18}$$

– Mass loads in the aeration tanks:

$$0.001 \le \frac{q_i s_i + q r_1 s_2}{v_1 x_1} \le 0.12 \tag{19}$$

$$0.001 \le \frac{q_{12} s_1 + q r_2 s_2 - (1 - y_1) W_3}{v_2 x_2} \le 0.12 \tag{20}$$

where the Wi terms adjust the relation to the actual number of bioreactors.

– Sludge age in the settler:

$$2 \le \frac{v_1 x_1 + v_2 x_2 + AL_r x_r}{q_p x_r 24} \le 10 \tag{21}$$

– Limits in hydraulic capacity:

$$\frac{q_{22}}{A} \le 1.5 \tag{22}$$

– Limits in the relationship between the input, recycled and purge flow rates:

$$0.07 \le \frac{q_p}{q_2} \le 0.3 \tag{23}$$

$$0.05 \le \frac{q_2}{q_i} \le 0.9 \tag{24}$$

Some logical conditions are imposed to guarantee the mathematical coherence of the model for each structural alternative. The chromosomes in the genetic algorithm are coded to ensure: for y_1=0 $\Rightarrow$ $v_2=0$, $x_1=x_2$, $s_1=s_2$, $c_1=c_2$, $Fk_2=0$, $qr_2=0$, for y_1=1 all the variables take values within their ranges.

The controllability constraints are the limits over the norms described by eq. (4), (5), (6), where the transfer functions are referred to s_2 as the output, si as the disturbance, and recycling flows qr_1, qr_2 as control variables. The parameter u_{max} is an upper bound for the magnitude of control variables.

These constraints: $\|N\|_\infty < 1$, $\|Wp \cdot S\|_\infty < 1$, $\|M\|_1 < u_{\max}$ ensure a satisfactory control performance of the plant with the tuned MPC.

The main difficulties when solving this optimization problem is the existence of continuous, integer and binary variables and the evaluation of controllability norms which implies the linearization of the process model for each possible solution. The genetic algorithms are particularly suitable, due to its robustness and the straightforward method to compute the objective function and constraints avoiding gradient evaluation, which makes it easy to implement computationally.

4 Genetic Algorithms

Genetic algorithms are stochastic optimization methods based on the principles of natural evolution [6]. The optimization process is carried out with a population of potential solutions (*chromosomes*) for the problem. A performance measure associated to the objective function (*fitness*) is assigned to each chromosome. The population evolves toward better regions in the search space by means of selection, crossover and mutation operators [6]. After several generations, the algorithm converges to the best solution of the problem. Conventional genetic algorithms are binary coded, however, the use of real parameters makes possible the representation of large domains, which is difficult to achieve in binary implementations, and, improves the effectiveness for strongly constrained problems [14], [3]. Another advantage is that slight changes in the variables reflects as slight changes in the objective function [3], which improves the local tuning of the solutions.

Here, a fixed length real coded chromosome is defined, which contains the continuous variables corresponding to the normalised process and controller weights (Wu=[Wqr1 Wqr2]), the integers for the normalized prediction and control horizon and a binary variable to set the structure of the plant:

$$[x_1,\ x_2,\ s_1,\ s_2,\ c_1,\ c_2,\ x_d\ x_b\ x_r\ qr_1\ qr_2\ qp\ Fk_1\ Fk_2\ v_1\ v_2\ A\ \ Wqr_1\ Wqr_2\ Hp\ Hc\ \ y_1]$$

The location of the variables in the chromosome is important for the objective function and constraints evaluation.

An appropriated technique to deal with constraints is necessary. A general approach borrowed from conventional optimization, is to incorporate the constraints into the objective function as a penalty term.

$$F(x) = f(x) + R\left(\sum_{k=1}^{p}\left[\max\{0, g_k(x)\}\right]^2\right) \tag{25}$$

where x is the chromosome, F is the fitness, f is the cost function, R is the penalty coefficient and p is the number of inequality constraints $gk(x)$. This strategy is used for the constraints over mass balances, operational specifications and controllability indices. The chromosome coding allows to shape the individuals according to the logical constraints mentioned in section 3.3 and the ranges for the variables.

The genetic algorithm starts generating randomly a population of a specific number of possible solutions, that contains the same quantity of individuals for both structural alternatives (y1=0 and y1=1). The roulette operator [6] and the arithmetic crossover [5], where the offspring *(z)* are obtained from a linear combination of the parents *x, y*, were selected, after the evaluation of different selection and crossover operators. Random mutation [6], which decreases proportionally to the generations' progress, is applied before manipulating the offspring. The population of the succeeding generations consists of 50% of the best individuals from the previous generation and 50% of the individuals generated by crossover.

The problem was solved using a population size of 100 individuals and 1000 maximum iterations. The mutation rate decreases with generations from 0.1 to 0.02 and the crossover probability was 80%.

5 Results

Two scenarios which differ in the demands on control performance were proposed for the integrated synthesis, design and control of the activated sludge. The case 1 refers to the problem with $\|M\|_1 < 1000$ and weight *Wp1* and the case 2 refers to the problem with $\|M\|_1 < 450$ and weight *Wp2*,

$$Wp_1(s) = \frac{8s + 19.2}{s + 0.0001} \quad Wp_2(s) = \frac{4.4s + 10.56}{s + 0.0001}$$

The genetic algorithm was run 10 times for each case, giving economically optimal feasible solutions for each run with an average computing time of 13307 seconds. It indicates that the proposed real-coded genetic algorithm is able to solve this complex

Table 1. Numerical results for integrated synthesis and design with MPC

	Case 1	Case 2		Case 1	Case 2
V1	5409.2	3442.7	Wu	[0.009 0.274]	[0.347 0.052]
V2	0	2819.2	Hp	7	7
A	1253.1	1147.0	Hc	2	4
S1	107.03	93.86	$\|N\|_\infty$	0.74	0.96
Residence times	3.73	2.53, 2.06	$\|M\|_1$	725.7	281.3
Mass loads	0.069	0.12, 0.08	$\|Wp \cdot S\|_\infty$	0.63	0.93
Hydraulic capacity	1.15	1.19	range(s1)	17.7	36.5
Sludge age	4.05	3.94	max(qr1)	563.9	254.5

problem in one step procedure avoiding the shortcomings of problem decomposition with a reasonable computational effort.

The comparison of the best results for the two scenarios is presented in table 1. In both cases, the transfer functions and weights are referred to disturbance *si*. It is observed that the solutions gives small economical plants as well as satisfies all the process and control constraints. It is important to notice the flexibility of the method for different conditions in the optimization problem, leading to different structural alternatives of the plant according to the limits imposed over the constraints.

The optimization case 1 produces a plant with better disturbance rejection because the weight Wp_1 is more restrictive for sensitivity function S. On the other hand, with a larger bound for $\|M\|_1$, the magnitude of control is more relaxed than in case 2 giving a wider range of action to the manipulated variable to reject disturbances. The values of $Wesf$ for qr_1 and qr_2 control sensitivity functions are fixed to:

$$Wesf_{qr1}(s)=\frac{0.0117s+0.14}{s+0.0004} \qquad Wesf_{qr2}(s)=\frac{0.0183s+0.22}{s+0.0004}$$

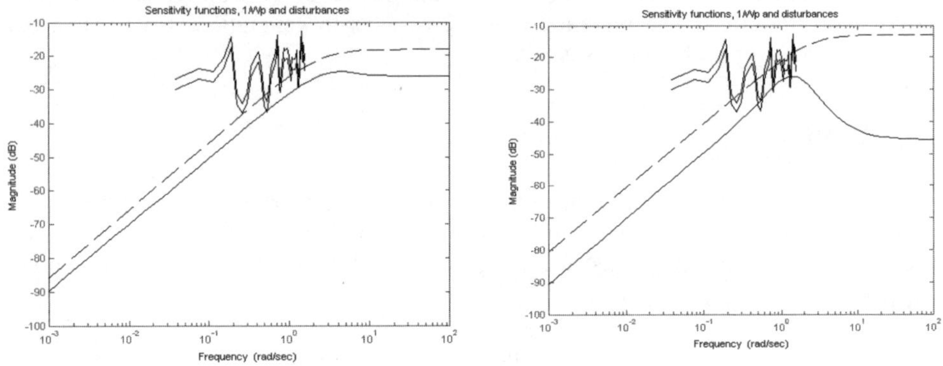

Fig. 2. Magnitude of sensitivity functions S (solid line) for cases 1 and 2 respectively, together with weights Wp-1 (dashed line) and the inverse spectrum of influent disturbances

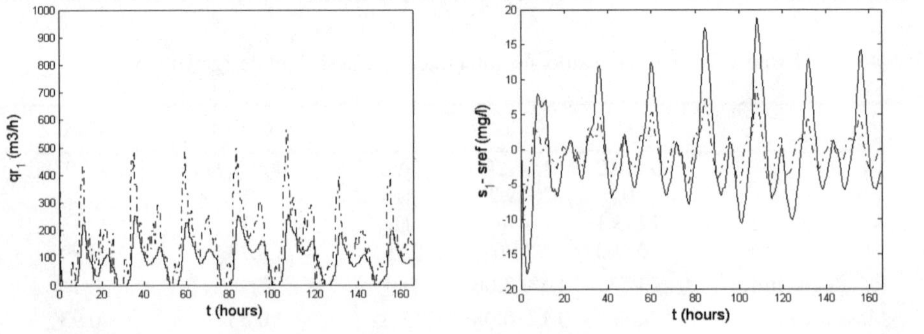

Fig. 3. Comparison of substrate responses and input flow for cases 1 (dashed line) and 2 (solid line) of synthesis

As for the plant configuration, in case 1 a plant with a single reactor has been obtained, but in case 2 the solution gives two reactors because of the stricter bound in $\|M\|_1$. In Figure 2 sensitivity functions S are presented for both cases. In the case 1 the inverse spectrum of disturbances is over *Wp-1*, and in case 2 this weight is a bit more relaxed representing worse disturbance rejection. In Figure 3 the dynamical responses of the optimal plants for both cases are presented, to illustrate the better disturbance rejection for case 1 as have been previously mentioned.

6 Conclusions and Future Work

In this work, the synthesis and integrated design of an activated sludge process with an advanced controller (MPC) was addressed. The problem was translated into a mixed-integer-non-linear optimization problem, with the evaluation of performance indices to ensure the most economical design with a suitable control performance.

The MINLP was solved using a real-coded genetic algorithm which leads to good quality feasible solutions with desired disturbance rejection, which is the main control objective. Different limits on control indices and parameters were tested, obtaining plants that have two or one bioreactors, with the corresponding differences in the investment and operation costs. On the other hand, the controllability norm based indices were actually considered as constraints in the formulation of the optimization problem, but they could also be considered as objectives in a multiobjective optimization problem including costs and controllability.

The optimization was performed in a single step procedure with a simple implementation of the GA, which was able to handle the different kind of variables and non linear constraints involved in the problem, showing to be an excellent alternative to deterministic methods. This is encouraging for the development of integrated design approaches with advanced control schemes which usually results in complex optimization problems difficult to solve with conventional techniques.

Acknowledgments

The authors acknowledge the support of Simón Bolívar University through project DI-CAI-002-05, of Spanish MCYT project DPI2006-15716-C02-01 and Samuel Solórzano Foundation Project of the University of Salamanca (Spain).

References

1. Copp, J.B.: The COST Simulation Benchmark: Description and Simulator Manual. Office for Official Publications of the European Community (2002)
2. Costa, L., Oliveira, P.: Evolutionary algorithms approach to the solution of mixed integer-non-linear programming problems. Comp. Chem. Eng. 25, 257 (2001)
3. Elliott, L., Ingham, D., Kyne, A., Mera, N., Porkashanian, M., Whittaker, S.: Reaction mechanism reduction and optimization for modeling aviation fuel oxidation using standard and hybrid genetic algorithms. Comp. Chem. Eng. 30, 889–900 (2006)

4. Francisco, M., Vega, P.: Diseño Integrado de procesos de depuración de aguas utilizando Control Predictivo Basado en Modelos. Rev. Iberoamericana de Automática e Informática Industrial 3(4), 88–98 (2006)
5. Gen, M., Chen, R.: Genetic algorithms and engineering optimisation. John Wiley and Sons, Chichester (2000)
6. Goldberg, D.F.: Genetic Algorithm in Search, Optimization and Machine Learning. Addison-Wesley, Reading (1989)
7. Kookos, I., Perkins, J.: An algorithm for simultaneous process design and control. Ind. Eng. Chem. Res. 40, 4079 (2001)
8. Luyben, M., Floudas, C.: Analyzing the interaction of design and control–1. A multiobjective framework and application to binary distillation synthesis. Comp. Chem. Eng. 18, 933 (1994)
9. Maciejowsky, J.M.: Predictive Control with Constraints. Prentice Hall, Englewood Cliffs (2002)
10. Moreno, R., De Prada, C., Lafuente, J., Poch, M., Montague, G.: Non-linear predictive control of dissolved oxygen in the activated sludge process. In: IFAC BIO, vol. 2, pp. 289–298. Pergamon Press, Oxford (1992)
11. Revollar, S., Lamanna, R., Vega, P.: Algorithmic synthesis and integrated design for activated sludge processes using genetic algorithms. ESCAPE, Barcelona (2005)
12. Sakizlis, S., Perkins, J., Pistikopoulos, E.: Parametric controllers in simultaneous process and control design optimization. Ind. Eng. Chem. Res. 42, 4545–4563 (2003)
13. Sakizlis, S., Perkins, J., Pistikopoulos, E.: Recent advances in optimization-based simultaneous process and control design. Comp. Chem. Eng. 28, 2069–2086 (2004)
14. Summanwar, V., Jayaraman, V., Kulkarni, H., Kusumakar, H., Gupta, K., Rajesh, J.: Solution of constrained optimization problems by multiobjective genetic algorithm. Comp. Chem. Eng. 26, 1481–1492 (2002)
15. Tlacuahuac-Flores, A., Biegler, L.: Integrated Control and Process Design During Optimal Polymer Grade Transitions Operations. Comp. Chem. Eng (in press, 2008)
16. Tsai, M., Chang, C.: Water usage and treatment network design using genetic algorithms. Ind. Eng. Chem. Res. 40, 4874 (2001)
17. Vega, P., Francisco, M., Sanz, E.: Norm based approach for automatic tuning of Model Predictive Controllers. In: Proceedings of ECCE-6, Copenhague (2007)

Genetic Algorithms for Simultaneous Equation Models

Jose J. López[1] and Domingo Giménez[2]

[1] Centro de Investigaciøn Operativa
Universidad Miguel Hernández, 03202, Elche, Spain
jlopez@umh.es

[2] Departamento de Informática y Sistemas
Universidad de Murcia, 30071 Murcia, Spain
domingo@dif.um.es

Summary. Traditionally, Simultaneous Equation Models (SEM) have been developed by people with a wealth of experience in the particular problem represented by the model. To obtain automatically a satisfactory SEM would help the experts in the development of such models. This paper presents an algorithm to obtain a satisfactory SEM from a set of variables. Because of the extensive number of SEM possible, exhaustive search methods are not well suited here. The algorithm combines genetic and a random search method. The behavior of the algorithm is studied and a parallel version in shared memory is developed and studied.

Keywords: Simultaneous Equations Models, High Performance Computing, Econometrics, Genetic Algorithms.

1 Introduction

Traditionally, Simultaneous Equation Models (SEM) have been developed by people with a wealth of experience in the particular problem represented by the model. The relationship between the variables is used to create the model, but this relationship depends on the criteria these experts choose. Examples of SEM can be found in econometrics [7], networks simulation [8], medicine [9], and even in the study of the air traffic in New York [11].

This paper studies how to obtain a SEM from a set of variables. The idea is to develop an algorithm which, given the endogenous and exogenous variables, finds the best SEM possible according to the criteria parameter for model comparison. The space of the possible solutions is very large since the number of equations of the best model is between one and the total number of endogenous variables. Because of that, exhaustive search methods are not well suited here.

The goodness of the model depends on the experts' criteria, but numerical methods can be used to represent it. Single equation regression models have made use of several goodness criteria, for example Akaike Information Criterion (AIC), corrected version of AIC (AICC), Schwarz BIC, Bozdogan ICOMP, etc. [1],[3],[12]. Some of them, particularly AIC, BIC and AICC, have been modified for multivariate regression models [2],[4], and have been adapted to SEM [6].

J.M. Corchado et al. (Eds.): DCAI 2008, ASC 50, pp. 215–224, 2009.
springerlink.com

Due to the large number of models to evaluate exhaustive search methods are discarded. Another possibility is to apply metaheuristic techniques [5]. This work, analyzes the solution of the problem via genetic algorithms. The solution is not necessarily the best, but the cost of finding this solution is much lower than the cost of finding the best one when using exhaustive search methods.

A basic version of a genetic algorithm is presented first. After that, a random search is used to improve this algorithm, so obtaining a hybrid metaheuristic. The idea is to use a random search method in some parts of the genetic algorithm to explore the space of solutions better. To reduce the execution time, a parallel algorithm in shared memory is developed and studied.

2 Simultaneous Equation Models

The scheme of a system with N equations, N endogenous variables (which influence and are influenced by the other variables) and K predetermined or exogenous variables (which influence but are not influenced by the system) [7] is:

$$\begin{array}{c} y_1 = \gamma_{1,1}x_1 + \ldots + \gamma_{1,K}x_K + \beta_{1,2}y_2 + \beta_{1,3}y_3 + \ldots + \beta_{1,N}y_N + u_1 \\ y_2 = \gamma_{2,1}x_1 + \ldots + \gamma_{2,K}x_K + \beta_{2,1}y_1 + \beta_{2,3}y_3 + \ldots + \beta_{2,N}y_N + u_2 \\ \ldots \\ y_N = \gamma_{N,1}x_1 + \ldots + \gamma_{N,K}x_K + \beta_{N,1}y_1 + \ldots + \beta_{N,N-1}y_{N-1} + u_N \end{array} \tag{1}$$

where $x_1, x_2, \ldots, x_K$ are predetermined variables, $y_1, y_2, \ldots, y_N$ are endogenous variables, and $u_1, u_2, \ldots, u_N$ are random variables. All these variables are vectors with dimension $d \times 1$ where d is the sample size.

The problem is to obtain $\gamma_{1,1}, \ldots, \gamma_{N,K}, \beta_{1,2}, \beta_{1,3}, \ldots, \beta_{N,N-1}$ from a representative sample of the model.

An equation is identified if the number of parameters in this equation is lower than or equal to $K+1$, i.e., $n_i - 1 \leq K - k_i$ where n_i is the number of endogenous variables in the equation and k_i is the number of predetermined variables in the equation. The parameters corresponding to an equation can be calculated only when the equation is identified (this condition is called order condition [7]).

Two different matrices can be defined, X $(d \times K)$ and Y $(d \times N)$, which contain the predetermined variables and the endogenous variables.

3 Two Step Least Square Algorithm

Two Step Least Square (2SLS) is the most common algorithm used to solve an equation in a SEM because it can be used when the equation is identified.

The expression of equation i is $y_i = \gamma_{i,1}x_1 + \ldots + \gamma_{i,K}x_K + \beta_{i,1}y_1 + \ldots + \beta_{i,i-1}y_{i-1} + \beta_{i,i+1}y_{i+1} + \ldots + \beta_{i,N}y_N + u_i$. Without considering the variables of the system which do not appear in the equation, it is: $y_i = \gamma_{i_1}x_{i_1} + \ldots + \gamma_{i_{k_i}}x_{i_{k_i}} + \beta_{i_1}y_{i_1} + \beta_{i_2}y_{i_2} + \ldots + \beta_{i_{n_i}}y_{i_{n_i}} + u_i$, where k_i and n_i referent the number of predetermined and endogenous variables in the equation.

Expressed in matrix form, equation i becomes $y_i = [X_i|Y_i]\delta_i + u_i$, where X_i and Y_i are matrices composed of the columns of X and Y which appear in the equation, and $\delta_i = (\gamma_i, \beta_i)^t$. 2SLS is used to solve this equation. First, the endogenous variables in the equation must be substituted by new variables called *proxy*. After that, Ordinary Least Square is used in the equation.

The *proxy* variable of y_j ($\hat{y}_j$) is calculated using all the predetermined variables in the system, with the expression $\hat{y}_j = X(X^tX)^{-1}X^ty_j$.

After changing the original variables for the *proxy* ones, the equation becomes $y_i = \gamma_{i_1}x_{i_1} + \ldots + \gamma_{i_{k_i}}x_{i_{k_i}} + \beta_{i_1}\hat{y}_{i_1} + \beta_{i_2}\hat{y}_{i_2} + \ldots + \beta_{i_{n_i}}\hat{y}_{i_{n_i}} + u_i$ (in matrix form $y_i = [X_i|\hat{Y}_i]\delta_i + u_i$), and then $\hat{\delta}_i = ([X_i|\hat{Y}_i]^t[X_i|\hat{Y}_i])^{-1}[X_i|\hat{Y}_i]^t y_i$.

Because in all the equations it is necessary to substitute the endogenous variables for the *proxy* ones, all of them can be calculated at the beginning by the expression $\hat{Y} = X(X^tX)^{-1}X^tY$.

4 The Problem: Find the Best SEM Given a Set of Values of Variables

Given two sets of endogenous and exogenous variables, i.e. a matrix Y ($d \times N$) of N endogenous variables (the columns of Y, $Y_1, ..., Y_N$, are the endogenous variables), and a matrix X ($d \times K$) of K exogenous variables (the columns of X, $X_1, ..., X_K$, are the exogenous variables), the problem is to obtain the best model which shows the relationship between all these variables. Some of these variables have simultaneous influences and will appear in different equations. One model is considered better than another if its criteria parameter is lower than that of the other. There are some different criteria parameters [6] and anyone of them can be used to select the model. In this work, AIC is used because it is one of the most used methods for comparing models. The expressions of AIC is:

$$AIC = d \ln |\hat{\Sigma}_e| + 2\sum_{i=1}^{N}(n_i + k_i - 1) + N(N+1) \tag{2}$$

where parameter $|\hat{\Sigma}_e|$ is the determinant of the covariance matrix of the errors e_i, $i = 1, ..., N$, where e_i is the difference between Y_i and the estimation of Y_i given in equation i.

5 Genetic Algorithms for Selecting the Best SEM

Genetic algorithms are used here to approach the best Simultaneous Equation Model. A population of SEM composed of chromosomes representing particular models is explored. Each chromosome represents a candidate to be the best model. If the chromosome can be evaluated (i.e., if the model can be solved) a value of the quality of the chromosome is calculated, and then it is compared with the rest of the population.

A chromosome is defined as a matrix with N rows and $N + K$ columns. In each row, an equation is represented using ones and zeros. If variable j appears in equation i, the value for the (i, j) position in the chromosome is one, and zero if not. The N first columns of a chromosome represent the endogenous variables and the other K columns represent the exogenous ones.

In equation i, the main endogenous variable will be the variable i. Thus, if equation i is in the system, the position (i, i) in the chromosome must be one.

For example, in a problem with $N = 2$ endogenous variables (Y_1 and Y_2) and $K = 3$ predetermined variables (X_1, X_2 and X_3), the model

$$\begin{aligned} y_1 &= \gamma_{1,1}x_1 + \gamma_{1,2}x_2 + \beta_{1,2}y_2 + u_1 \\ y_2 &= \gamma_{2,3}x_3 + \beta_{2,1}y_1 + u_2 \end{aligned} \tag{3}$$

is represented by the chromosome

$$\begin{matrix} 11110 \\ 11001 \end{matrix} \tag{4}$$

where the first two columns correspond to the endogenous variables Y_1, Y_2, and the other three columns to the exogenous ones X_1, X_2, X_3.

The endogenous variables are represented in the first N columns, and so the main endogenous of each equation falls in the main diagonal.

5.1 Defining a Valid Chromosome

The necessary conditions to be a valid chromosome are:

- C1: The model has to have at least one equation.
- C2: If the (i, i) element is zero, the column i will have only zeros. This means that when the equation i is not in the system, the position (i, i) in the chromosome is zero, and the i variable can not be in other equations, the i-column has only zeros.
- C3: Each equation in the model must have at least two variables, i.e., if the (i, i) element is one, there exists a j with $1 \leq j \leq N + K$ and $j \neq i$ that the (i, j) element is one.
- C4 (Rank condition [7]): Equation i is identified if it is possible to find a $(N-1) \times (N-1)$ matrix with full range where the columns are the unknown variables $\gamma_{1,1}, \ldots, \gamma_{N,K}, \beta_{1,2}, \beta_{1,3}, \ldots, \beta_{N,N-1}$ that do not appear in the equation. For example, if X_1 does not appear in the equation (i.e. $\gamma_{1,i} = 0$), one possible column of the matrix is $(\gamma_{1,1}, \ldots, \gamma_{1,i-1}, \gamma_{1,i+1} \ldots, \gamma_{1,N})^T$.

The order condition (section 2) is a necessary but not sufficient condition for a model. However, the rank condition is sufficient and necessary and subsumes the necessary order condition. For this reason, only rank condition is used.

When analyzing a chromosome floating point operations are not used. However the number of comparisons can be very high and would be taken into account. The cost of testing a chromosome depends on the number of conditions to be tested (since they are made one by one).

In the worst case, the cost of C1 is N, of C2 is $N(N-1)$ and of C3 is $N(N+K)$. The cost of C4 in each equation is that of finding a $(N-1) \times (N-1)$ matrix of range $N-1$ from the coefficients of the rest of the equations (using only variables which do not appear in this equation). The maximum number of possible matrices is $\binom{K+N-2}{N-1}$ (because the equation has at least two variables), and $(N-1)!$ comparisons are made in the worst case to know if the matrix has maximum range.

5.2 Initialization and EndConditions

Initially many valid individual solutions are randomly generated to form an initial population. The population size (called *PopSize*) is stated at the beginning. Each chromosome is generated according to the previous conditions. If some information is known, some models could be proposed for insertion in the initial population.

The process is repeated until the process reaches the maximum of iterations called $MaxIter$ or the best fitness is repeated over a number of successive iterations, called $MaxBest$. Both parameters are stated at the beginning of the algorithm.

5.3 Evaluating a Chromosome

Algorithm 1 shows the scheme of the fitness function of a chromosome. 2SLS algorithm is used to solved the complete system and the solution is used to estimate the main endogenous variable in each equation and to calculate the error variables. Matrices Y_c and X_c are formed by the columns of Y and X corresponding to the variables in chromosome c. Matrix $X_{c,i}$ is formed by the predetermined variables and the endogenous variables (which now take the values of the *proxys*) of the equation i of the system formed from c, excluding the main endogenous. Thus, matrix $X_{c,i}$ has size $d \times (k_{c,i} + n_{c,i} - 1)$ (where $k_{c,i}$ and $n_{c,i}$ are the number of exogenous and endogenous variables in equation i of chromosome c). In 2SLS, initially all the *proxys* are calculated and used when the equations are solved.

Algorithm 1. Chromosome evaluation algorithm

build the system using the chromosome c and the set of variables Y and X
$\hat{Y}_c = X_c(X_c^t X_c)^{-1} X_c^t Y_c$
for i=1...N_c **do**
  {solve each equation in the system}
  Compute $(X_{c,eq-i}^t X_{c,eq-i})^{-1} X_{c,eq-i}^t y_{c,i}$
end for
for i=1...N_c **do**
  {obtaining the error variables}
  Compute $e_{c,i}$=$y_{c,i}$- estimation of $y_{c,i}$
end for
Compute AIC {using (2)}

N_c is the number of equations in chromosome c. The cost of evaluating a chromosome is the sum of solving the system and calculating the AIC parameter:

$$
\begin{array}{c}
\frac{2}{3}K^3 + 2K^2(d+N) + 4NKd + \\
\sum_{i=1}^{N} \left(\frac{2}{3}(n_i + k_i - 1)^3 + 2(n_i + k_i - 1)^2(d+1) + 4d(n_i + k_i - 1) \right) + 2N^2d + 3N \\
\leq \frac{2}{3}K^3 + 2K^2(d+N) + 4NKd + N(\frac{2}{3}K^3 + 2K^2(d+1) + 4Kd) + 2N^2d + 3N \\
\approx O\left(K^2Nd + K^3N\right)
\end{array} \tag{5}
$$

Expression 5 considers the worst case (all the endogenous and exogenous variables of the problem are included in the chromosome).

5.4 Select the Best Ranking and Crossover

In each generation a proportion of the existing population is selected to breed a new generation. A comparison of the evaluations of all the chromosomes in the population is made in each generation, and only part of them (those which are in the best ranking) will survive. The number of chromosomes which survive in each population (called $SurvSize$) is preset.

For each two new solutions to be produced ("son" and "daughter"), a pair of "parent" ("father" and "mother") chromosomes is selected from the set of chromosomes. To combine the parents many different methods can be used. In this work three sorts of crossover are studied and the table 2 in section 7 show a comparison between these three methods.

Single Point Crossover

The traditional single point crossover where number e with $1 \leq e \leq N(N+K)$ of the total elements of the chromosome is chosen randomly.

Single Point Crossover Considering Equations

This crossover is similar to the *single point crossover* but it randomly chooses an equation instead of an element.

Crossover Inside an Equation

An equation e is chosen randomly and then two numbers $v1$ and $v2$ are generated randomly with $1 \leq v1 \leq N$ and $1 \leq v2 \leq K$. All the equations of the father go to the son except equation e, and the same is true for the mother and the daughter. Equation e is generated in the son using the first $v1$ endogenous variables of equation e of the father and the remaining endogenous ones of equation e of the mother and using the first $v2$ exogenous variables of equation e of the father and the remaining exogenous ones of the mother. Similarly, equation e is generated in the daughter using the rest of elements of equation e of the father and the mother.

5.5 Mutation

A small probability of mutation is considered in all the iterations. The mutation considered here is the bit inversion, i.e., a chromosome of the new subset generated in the crossover is chosen randomly, and a number, e, for the equation

Table 1. An example of the three methods of crossover. In each case, e, $v1$ and $v2$ have been chosen randomly and the number of endogenous variables and exogenous variables is $N = 3$ and $K = 5$.

parents		single point element $e = 10$		single point equation $e = 1$		inside an equation $e = 2$, $v1 = 2$, $v2 = 3$	
father	mother	son	daughter	son	daughter	son	daughter
11110110	10100100	11110110	10100110	11110110	10100100	11110110	10100100
11110101	01110100	11110100	01110101	01110100	11110101	11110100	01110101
01110110	11110110	11110110	01110110	11110110	01110110	01110110	11110110

(with $1 \leq e \leq N$) and a number, v, for the variable (with $1 \leq v \leq N + K$) are generated randomly. Then element v of the e equation of the chromosome is inverted. In each iteration a call to the mutation is made randomly (with a probability called P_{mut}, which is stated at the beginning of the algorithm).

6 Random Search

When a chromosome is mutated and then situated in a different part of the set of solutions, this chromosome normally does not have enough quality to survive long enough to create new chromosomes in this area, and perhaps best solution is next to it. To avoid this problem, a random search method is used in the mutation (algorithm 2) in the following way: a chromosome is created by mutation (line 4), after that the chromosome is evaluated and if its evaluation has enough quality (lines 5 and 6), i.e., if this value is lower than a parameter called SV ($Survivevalue$), the chromosome is included in the population. If not, a new chromosome is generated randomly (line 8). This process is repeated until a chromosome with enough quality is found or until a maximum of iteration (called $MaxRS$) is reached.

Algorithm 2. Scheme of the random search algorithm in the mutation function

1: Generate $1 \leq e \leq N$ and $1 \leq v \leq N + K$ randomly
2: $EndConditions = FALSE$
3: **while** $NotEndConditions$ **do**
4:   c1=Mutate(c) {invert the element (e, v) of the chromosome c}
5:   **if** $GoodChromosome$(c1) AND $Evaluation$(c1) < $Evaluation$(c) **then**
6:     $c = c1$
7:   **end if**
8:   Generate new $1 \leq v \leq N + K$ randomly
9:   **if** $Evaluation(c) < SV$ **then**
10:     $EndConditions = TRUE$
11:   **end if**
12: **end while**

7 Experimental Results

Some experiments have been carried out to tune some parameters of the algorithm. The set of variables used in the experiments have been created randomly (each variable is a vector of real numbers generated between 1 and 10). In the

Table 2. A comparison of the three crossover methods, for different sizes of the problem. In each case, t is the time to finish in seconds, $iter$ is the number of iterations that the algorithm has needed to finish, and $best\ fitness$ is the solution found. The population size is 100 and $MaxBest$=15.

problem size			crossover single point in elements			crossover single point in equations			crossover inside an equation		
N	K	d	t	$iter$	$best\ fitness$	t	$iter$	$best\ fitness$	t	$iter$	$best\ fitness$
10	15	50	3.03	48	2683.13	5.11	97	2732.90	0.67	20	2833.41
15	20	50	8.00	62	4548.68	6.73	53	4540.93	1.94	40	4709.50
30	40	100	58.33	50	21937.02	87.54	72	22120.11	9.47	17	22765.68
40	50	100	325.87	111	30956.78	294.19	102	31262.20	64.41	24	32975.04

Table 3. A comparison of the solution found by the genetic algorithm when varying the population size (*PopSize*), the number of endogenous variables (N), the number of exogenous (K). The sample size d=10, the crossover is "inside an equation" and $MaxBest$=$MaxIter$=150.

size		best fitness		
N	K	$PopSize = 100$	$PopSize = 500$	$optimum$
2	2	66.44	66.44	66.44
2	3	46.18	46.18	46.18
3	3	-177.03	-214.91	-216.68
3	4	-124.05	-213.16	-216.68
4	4	-99.73	-161.67	-218.59

Table 4. Execution time (in seconds) and speed-up of the algorithm in shared memory, when varying the population size (*PopSize*), the number of endogenous variables (N), the number of exogenous (K), the sample size (d) and the number of processors. The crossover is "inside an equation" and $MaxBest$=$MaxIter$=150.

$PopSize$	N	K	d	1 proc.	2 proc.	sp	4 proc.	sp	8 proc.	sp
100	10	15	50	4.22	2.51	1.68	1.62	2.60	1.04	4.06
100	30	40	100	40.74	26.21	1.55	16.24	2.51	12.31	3.31
100	50	65	150	217.79	152.19	1.43	102.27	2.13	63.81	3.41
100	70	90	200	709.05	417.62	1.70	277.15	2.56	185.88	3.81
500	10	15	50	21.31	11.55	1.85	7.50	2.84	4.70	4.53
500	30	40	100	201.29	115.71	1.74	62.30	3.23	47.47	4.24
500	50	65	150	1065.77	699.2	1.52	368.11	2.90	229.68	4.64
500	70	90	200	3580.94	1927.76	1.86	1076.45	3.33	699.21	5.12

experiments $SurvSize = \frac{PopSize}{2}$, $MaxIter = 150$, $MaxRS = N$, SV is equals to the evaluation of the chromosome situated in the position $\frac{PopSize}{4}$ in the ranking and the probability of the mutation is the 1 per cent.

Table 2 shows a comparison between the three crossover methods. The cost of the algorithm and also the number of iterations needed to finish are shown. The table shows that crossover inside an equation is the best option in cost but it obtains a solution which has a bit lower fitness function.

Table 3 shows a comparison of the best solution found by the algorithm when varying the population size (*PopSize*). When *PopSize* is increased the solution found is better (and more similar to the optimum solution found by an exhaustive search method) but the execution time is very large. Thus a parallel version of the algorithm is developed in shared memory.

Table 4 shows a comparison between the cost of the algorithm when varying the number of processors in shared memory. The *initialization* and the evaluation of the population (to determine if a chromosome is a valid one and the fitness function) are parallelized, because there are the parts with higher cost. To study the parallelism, the algorithm finishes when the maximum number of iterations ($MaxIter$) is reached.

8 Conclusions and Future Works

The problem of finding a satisfactory Simultaneous Equation Model from a set of data has not satisfactorily solved. For that an hybrid algorithm is proposed. Genetic and a random search method are combined to avoid to fall in local minimum and to speed-up the convergence. Different parameters are experimentally tuned. When the size of the model increases the execution time can become very large, and a parallel version of the algorithm could be preferable. A shared memory version, which allows us to efficiently use multicore processors in the solution of the problem, has been developed. For higher models the use of higher systems (networks of multicores) could be preferable, so we are working in the development of hybrid (message-passing plus shared memory) algorithms.

References

1. Akaike, H.: Information theory and an extension of the maximum likelihood principle. In: Petrov, B.N., Csaki, F. (eds.) Proc. 2nd Int. Symp. on Information Theory, Akademiai Kiado, Budapest, pp. 267–281 (1973)
2. Bedrick, E.J., Tsai, C.-L.: Model selection for multivariate regression in small samples. Biometrics 50, 226–231 (1994)
3. Bozdogan, H., Haughton, D.: Informational complexity criteria for regression models. Computational Statistics and Data Analysis 28, 51–76 (1998)
4. Fujikoshi, Y., Satoh, K.: Modified AIC and Cp in multivariate linear regression. Biometrika 84(3), 707–716 (1997)
5. Glover, F., Kochenberger, G.A.: Handbook of Metaheuristics. Kluwer, Dordrecht (2003)

6. Gorobets, A.: The Optimal Prediction Simultaneous Equations Selection. Economics Bulletin 36(3), 1–8 (2005)
7. Gujarati, D.: Basic Econometrics. McGraw Hill, New York (1995)
8. Harchol-Balter, M., Black, P.E.: Queueing Analysis of Oblivious Packet-Routing Networks. In: SODA: ACM-SIAM Symposium on Discrete Algorithms, pp. 583–592 (1994)
9. Henry, R., Lu, I., Beightol, L., Eckberg, D.: Interactions between CO2 Chemoreflexes and Arterial Baroreflexes. Am. Journal of Physiology 274 (Heart Circ. Physiol. 43), 2177–2187 (1998)
10. Mitchell, M.: An Introduction to Genetic Algorithm. MIT Press, Cambridge (1998)
11. Lu, I., Peixoto, J., Taam, W.: A Simultaneous Equation Model for Air Traffic in the New York Area. In: The Seventh Air Transport Research Society World Conference (2003)
12. Shi, P., Tsai, C.-L.: A note on the unification of the Akaike information criterion. J.R. Statist. Soc. B 60(3), 551–558 (1998)

Solving the Terminal Assignment Problem Using a Local Search Genetic Algorithm

Eugénia M. Bernardino[1], Anabela M. Bernardino[1], Juan M. Sánchez-Pérez[2], Juan A. Gómez-Pulido[2], and Miguel A. Vega-Rodríguez[2]

[1] Department of Computer Science, School of Technology and Management, Polytechnic Institute of Leiria, Apt. 3063, Morro do Lena, Alto do Vieiro, 2401-951, Leiria, Portugal
eugenia@estg.ipleiria.pt, anabelab@estg.ipleiria.pt

[2] Department of Technologies of Computers and Communications, Polytechnic School, University of Extremadura, Avenida de la Universidad s/n, 10071, Cáceres, Spain
sanperez@unex.es, jangomez@unex.es, mavega@unex.es

Abstract. Terminal assignment is an important issue in telecommunication networks optimization. The task here is to assign a given collection of terminals to a given collection of concentrators. The main objective is to minimize the link cost to form a network. This optimization task is an NP-complete problem. The intractability of this problem is a motivation for the pursuits of a local search genetic algorithm that produces approximate, rather than exact, solutions. In this paper, we explore one of the most successful emerging ideas combining local search with population-based search. Simulation results verify the effectiveness of the proposed method. The results show that our algorithm provides good solutions in a better running time.

Keywords: Terminal Assignment Problem, Genetic Algorithm, Local Search Algorithm.

1 Introduction

In recent years we have witnessed a tremendous growth of communication networks resulted in a large variety of combinatorial optimization problems in the design and in the management of communication networks. This is mainly due to the dramatic growth in the use of the Internet [1, 2]. One important problem in telecommunication networks is the terminal assignment (TA) problem. The objective is to minimize the link cost to form a network by connecting a given set of terminals to a given set of concentrators [3, 4]. This problem is formulated into a multi-objective optimization task with certain constraints (see Section 2). The TA problem is a NP-complete optimization problem. To deal with the difficulty, we use an approximate method to find a good solution. An approximate method is an algorithm that can find solutions to NP problems, and which runs quickly, but the algorithm no guarantee that the solution it will find is the best one. The existing, successful approximate methods fall into two classes: local search (LS) and population based search. There are many population-based optimization algorithms and various ways to handle the optimization issues. In this paper we explore one of the most successful emerging ideas combining local search with population-based search. We use a Local Search Genetic Algorithm (LSGA) for solving the TA problem. LSGA combines a Genetic Algorithm (GA) to

J.M. Corchado et al. (Eds.): DCAI 2008, ASC 50, pp. 225–234, 2009.
springerlink.com

explore several regions of the search space and simultaneously incorporates a good mechanism (LS algorithm) to intensify the search around some selected regions.

The paper is structured as follows. In Section 2 we describe the TA problem; in Section 3 we present the previous work; in Section 4 we describe the implemented algorithm; in Section 5 we present the studied examples; in Section 6 we discuss the results obtained and, finally, in Section 7 we report about the conclusions.

2 TA Problem

The TA problem involves determine what terminals will be serviced by each concentrator. The three constraints imposed in this article for solving the TA problem are: (1) the aggregate capacity requirement of the terminals connected to any one concentrator cannot exceed the capacity of that concentrator; (2) minimize the distances between concentrators and terminals assigned to them; (3) guarantee the balanced distribution of terminals among concentrators. To represent this problem we must consider the following aspects: (1) the terminal sites and concentrators sites have fixed and known locations; (2) the capacity requirement of each terminal is known and may vary from one terminal to another; (3) each concentrator is limited in the amount of traffic that it can accommodate; (4) the capacities of all concentrators and the cost of linking each terminal to a concentrator are also known.

Problem Instance

Terminals - a set `N` of n distinct terminals (`1, …, n`)
Weights - a vector `T`, with the capacity required for each terminal
Terminals Location - a vector `CT`, with the location (x,y) of each terminal
Concentrators - a set `M` of m distinct concentrators (`1, …, m`)
Capacities - a vector `C`, with the capacity required for each concentrator
Concentrators Location - a vector `CP`, with the concentrators location (x,y)

Figure 1 illustrates an assignment to a problem with N = 10 terminal sites and M = 3 concentrator sites. The figure shows the coordinates for the concentrators and terminal

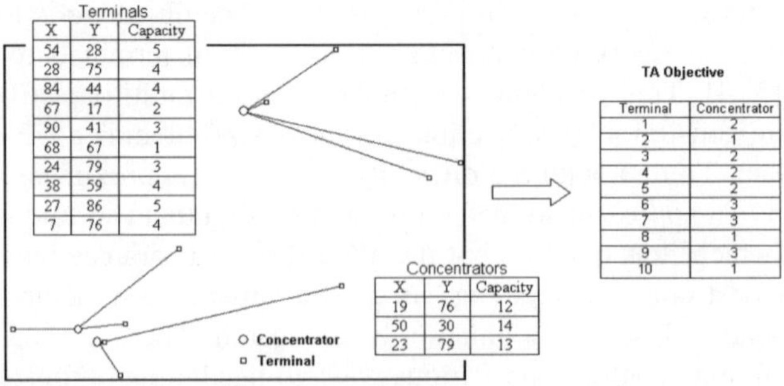

Fig. 1. TA Problem – example

sites and also their capacities. We consider a 100 x 100 Euclidean grid to represent the coordinates.

3 Previous Work

Some interesting approaches for the TA can be found in the literature. Atiqullah and Rao [5] proposed Simulate Annealing (SA) to find the optimal design of small-scale networks. Pierre et al. [6] proposed SA to find solutions for packet switched networks. Glover and Ryan [7] and Koh and Lee [8] adopted Tabu Search (TS) to find an appropriate design of communication networks. Abuali et al. [3] proposed a Greedy Algorithm and a Hybrid Greedy-Genetic Algorithm for solving the TA problem. Khuri and Chui [4] proposed a GA with a penalty function as an alternative method for solving the TA problem and compare the results with the Greedy Algorithm. Salcedo-Sanz and Yao [1] proposed two different Genetic Algorithms using Hopfield Neural Network and compare the results with the GA. Xu et al. [9] proposed a Tabu Search Algorithm and compare the results with the GA and the Greedy Algorithm. Salcedo-Sanz et al. [10] proposed to solve terminal assignment problems with Groups Encoding: the Wedding Banquet Problem. Yao et al. [2] proposed Hybrid Genetic Algorithms and compare the concentrator-based and terminal-based representations. Bernardino et al. [11] proposed a TS Algorithm and a Hybrid GA with a repair procedure and compare the results between them. Bernardino et al. [12] proposed a GA with multiple operators for crossover and mutation and compare the results with the traditional methods. In this paper, we compare our algorithm with the algorithms proposed by Bernardino et al. [11, 12].

4 LSGA

The LSGA is an evolutionary algorithm (EA) that applies a separate local search process to refine individuals. Our algorithm combines global and local search by

$$\begin{pmatrix} 0 & 1 & 0 \\ 0 & 0 & 1 \\ 0 & 1 & 0 \\ 0 & 1 & 0 \\ 0 & 1 & 0 \\ 0 & 0 & 1 \\ 0 & 0 & 1 \\ 1 & 0 & 0 \\ 0 & 0 & 1 \\ 1 & 0 & 0 \end{pmatrix}$$

Fig. 2. Binary representation - one position in the matrix indicates if a terminal `i` is assigned to a concentrator `j` (if `i` is assigned to `j` 1 else 0).

2	3	1	2	2	2	3	1	3	1

Fig. 3. Terminal Based Representation - each position in the vector corresponds to a terminal. The value carried by position `i` of the chromosome specifies the concentrator that terminal `i` is to be assigned to.

c1(t1,t2,t7) c2(t3,t9,t10) c3 (t4, t5, t6, t8)

Fig. 4. Concentrator Based Representation - is composed of a set of trees in one level, in each of which the concentrator is the root node and the terminals associated with the concentrator are the leaves. Each tree therefore indicates a concentrator together with its terminals.

using an EA to perform exploration while the local search method performs exploitation. Combining global and local search is a strategy used by many successful global optimization approaches, and this type of algorithms has in fact been recognized as a powerful algorithmic paradigm for evolutionary computing [13]. This method has proved to be of practical success in a variety of problem domains. This algorithm is also known as Memetic Algorithm, Hybrid EAs, Genetic Local Searchers, etc. [14].

Traditional GA often explores the candidate solution encoded in chromosomes and exploits those with better fitness iterally till the solution is reached. The local search by itself explores the solution space making specific moves in its neighbourhood. In our case, we combine those two aspects by using the chromosomes that are produced by genetic operators and optimize them by a local search algorithm.

The first step for the LSGA implementation involves choosing a representation for the problem. In the literature are used different representations. The most common are: (1) binary Representation (see Fig. 2); (2) terminal-based Representation (see Fig. 3); (3) concentrator-based Representation (see Fig. 4).

In this work, the solutions are represented using integer vectors. We use the terminal-based representation.

The main steps of the LSGA algorithm are given below:

```
Generate initial population
Evaluation
While TerminationCriterion()
       Selection
       Crossover
       Mutation
       For each solution in population
              Perform local search to get a new solution
       Evaluation
```

The LSGA starts to create a population with a group of individuals (candidate solutions) randomly or using a deterministic form. The deterministic form is based in the Greedy Algorithm proposed by Abuali et al. [3]. The Greedy Algorithm assigns terminals to the closest feasible concentrator. This kind of assignment can lead to infeasible solutions even if a feasible solution exists. This means that sometimes there are unassigned terminals that cannot be allocated to any concentrator.

Procedure Greedy:

```
while additional assignments of terminals to concentrators are possible
       For a randomly chosen terminal, say ti
              Determine the closest feasible concentrator ci
       Assign terminal ti to ci
```

The next step is to evaluate the fitness of each individual. The evaluation function (fitness function) gives the individuals a fitness value that reflects how optimal the solution is. The fitness function returns a fitness value based on: (1) the total number of terminals connected to each concentrator; (2) the distance between the concentrators and the terminals assigned to them and (3) the penalization if a solution is not feasible (the total capacity of one or more concentrators is overloaded). Some individuals are then selected based on their fitness, in our case, the lower the fitness, the higher the chance of being selected. These individuals then "reproduce" to create one or more

Table 1. GA Methods

Method	Definition
Selection	Tournament with Elitism (the elite size is 20% of the population
Selection	Tournament
Selection	Roulette
Mutation	Change order
Mutation	Change concentrator
Mutation	Change less distant concentrator
Mutation	Multiple
Crossover	One point
Crossover	Two points
Crossover	N points
Crossover	Uniform
Crossover	Reciprocal translocation
Crossover	Exchange Positions
Crossover	Exchange terminals of two concentrators
Crossover	Multiple

offspring, after which the offspring are mutated randomly. After we perform the local search to obtain new solutions. This continues until a certain number of generations defined by the user, have passed. We implement the same selection, crossover and mutation methods used in [12]. Table 1 summarizes the implemented methods.

The fitness function is based on the fitness function used in [11]:

$$fitness = 0{,}9 * \sum_{c=0}^{M-1} bal_c + 0{,}1 * \sum_{t=0}^{N-1} dist_{t,c(t)} + Penaliza$$

$$bal_c = \begin{cases} 10 & if\left(total_c = round\left(\frac{N}{M}\right)+1\right) \\ 20 * abs\left(round\left(\frac{N}{M}\right)+1-total_c\right) \end{cases}$$

$$Penaliza = \begin{cases} 0 & if\ (Feasible) \\ 500 \end{cases} \qquad total_c = \sum_{t=0}^{N-1} \begin{cases} 1 & if\ (c(t)=c) \\ 0 \end{cases}$$

$$dist_{t,c(t)} = \sqrt{(CP[c(t)].x - CT[t].x)^2 + (CP[c(t)].y - CT[t].y)^2}$$

c(t)= concentrator of terminal t t = terminal c = concentrator N = number of terminals
M = number of concentrators CP = concentrator location (x,y) CT = terminal location (x,y)

With the basic form of LSGA, it's possible to integrate different search techniques. We implement four different LS algorithms: (1) Local Search "Terminal-Based"; Local Search "Concentrator-Based"; Local Search "Random-Based" and Local Search "2 Concentrator-Based".

Local Search "Terminal-Based" - we generate a neighbour by swapping two terminals in the permutation. The neighbourhood size is N*(N-1)/2.

```
BEST-SOLUTION = INITIAL-SOLUTION
For x until number of terminals do
   For y until number of terminals do
      SOLUTION = interchange concentrator of x with concentrator of y
      If SOLUTION is best than BEST-SOLUTION
               BEST-SOLUTION = SOLUTION
```

Local Search "Concentrator-Based" - we generate a neighbour by assigning one terminal to one concentrator. The neighbourhood size is N*M.

```
BEST-SOLUTION = INITIAL-SOLUTION
For x until number of terminals do
   For y until number of concentrators do
      SOLUTION = interchange concentrator of x for y
      If SOLUTION is best than BEST-SOLUTION
               BEST-SOLUTION = SOLUTION
```

Local Search "2 Concentrator-Based" - we generate a neighbour by swapping two terminals between two concentrators C1 and C2 (randomly chosen). If isn't find a better solution then is created another set of neighbours. In this case, one neighbour results of assign one terminal of C1 to C2 or C2 to C1. The neighbourhood size is N(C1)*N(C2) or N(C1)*N(C2)+N(C1)+N(C2).

```
BEST-SOLUTION = INITIAL-SOLUTION
C1 = random (number of concentrators)
C2 = random (number of concentrators)
N = neighbours of INITIAL-SOLUTION (one neighbour result of inter-
change one terminal of C1 or C2 for one terminal of C2 or C1)
SOLUTION = FindBest (N)
If INITIAL-SOLUTION is best than SOLUTION
      N = neighbours of INITIAL-SOLUTION (one neighbour result of
      assign one terminal of C1 to C2 or C2 to C1
      SOLUTION = FindBest (N)
      If SOLUTION is best than BEST-SOLUTION
            BEST-SOLUTION = SOLUTION
Else
      BEST-SOLUTION = SOLUTION
```

Local Search "Random-Based" - we generate only one neighbour by assigning one terminal to one concentrator. The best solution between the neighbour and the initial solution is taken. The neighbourhood size is 1.

```
BEST-SOLUTION = INITIAL-SOLUTION
T = random (number of terminals)
C = random (number of concentrators)
SOLUTION = interchange concentrator of T for C
If SOLUTION is best than BEST-SOLUTION
      BEST-SOLUTION = SOLUTION
```

5 Studied Examples

In order to test the performance of our approach, we use a collection of TA instances of different sizes. We take 9 problems from literature [11, 12].

Table 2 presents the 9 problems that were used to test our algorithm.

Table 2. TA instances

Problem	N	M	Total T	Total C
1	10	3	35	39
2	20	6	55	81
3	30	10	89	124
4	40	13	147	169
5	50	16	161	207
6	50	16	173	208
7	70	21	220	271
8	100	30	329	517
9	100	30	362	518

6 Results

Table 3 presents the best-obtained results with our algorithm and three algorithms from the literature. The first column represents the problem number (Prob) and the remaining columns show the results obtained (Fitness, Time – Run Times) by the algorithms LSGA, GA with Multiple Operators (GAMO) [12], Hybrid Genetic Algorithm (HGA) [11] and Tabu Search Algorithm (TS) [11]. The algorithms have been executed using a processor Intel Core Duo T2300. The genetic algorithms were applied to populations of 200 individuals and the initial solutions were created using the Greedy Algorithm. The run times correspond to the average time that the algorithms need to obtain the best feasible solution. The values presented have been computed based on 100 different executions.

Table 3. TA instances

Problem	LSGA		GAMO		HGA		TS	
	Fitness	Time	Fitness	Time	Fitness	Time	Fitness	Time
1	65,63	<1s	65,63	<1s	65,63	<1s	65,63	<1s
2	134,65	<1s	134,65	<1s	134,65	<1s	134,65	<1s
3	270,26	<1s	270,26	1s	270,26	1s	270,26	<1s
4	286,89	<1s	286,89	1s	286,89	1s	286,89	<1s
5	335,09	<1s	335,09	1s	335,09	1s	335,09	<1s
6	371,12	1s	371,12	1s	371,12	1s	371,12	<1s
7	401,21	1s	401,21	2s	401,21	2s	401,49	1s
8	563,19	7s	563,19	8s	563,19	8s	563,34	1s
9	642,83	7s	642,83	8s	642,83	8s	642,86	2s

In the tests carried out with the LSGA, was verified that the selection methods tournament and tournament with elitism are the methods that obtain solutions with the least cost and in the least time. These methods were also the best methods used in GAMO and HGA. The best crossover method used is "One Point" (see Fig. 5, Fig. 6). The best mutation strategy was "Multiple" (see Fig. 7). Bernardino et al. [12] has proved that multiple mutation method is the best method. In each step, the multiple mutation method selects one operator based on the amount fitness improvements

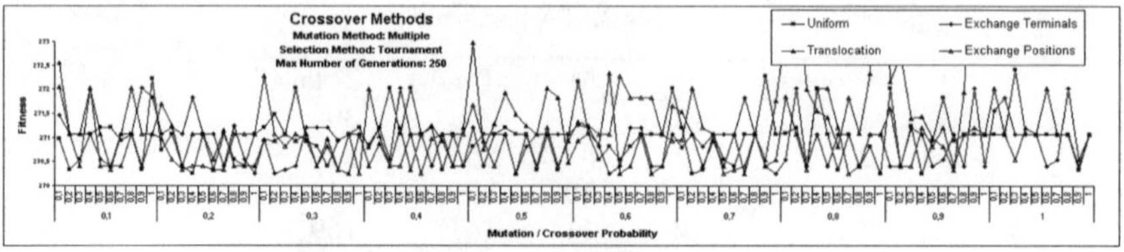

Fig. 5. Crossover Methods – GA – Problem 3

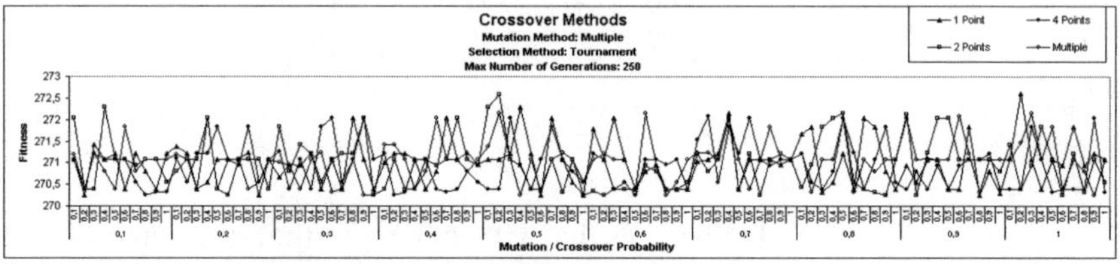

Fig. 6. Crossover Methods – GA – Problem 3

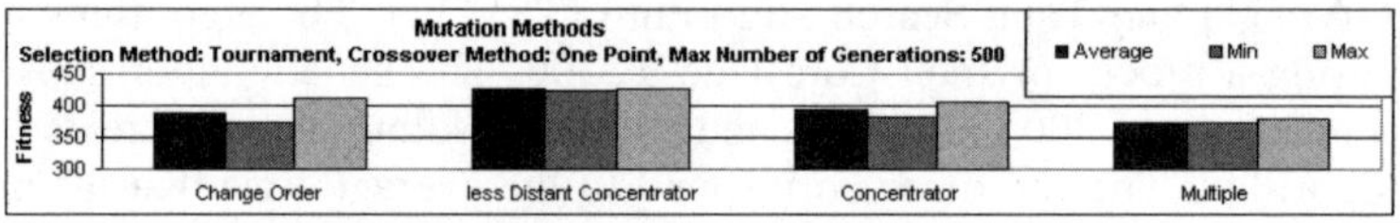

Fig. 7. Mutation Methods – GA – Problem 6

achieved over a number of previous mutations. A great advantage of the multiple mutation method is that it allows, through its execution, select mutation operators that are better adapted to the problem resolution.

In the LS "Terminal-Based" the neighbourhood size is very large for large-scale problems. These would waste a lot of computer time as you can see in Fig. 8. If we search only a part of the neighbourhood, the running time could be reduced dramatically. In the LS "Concentrator-Based" the neighbourhood size is also large in comparison with the LS "2 Concentrator-Based". In the LS "Random-Based" the neighbourhood size is only 1, which is smaller. The influence on the running time is also small. These last 2 methods prove to be more efficient because they can find a good solution in a better running time (Fig. 8, Fig.9).

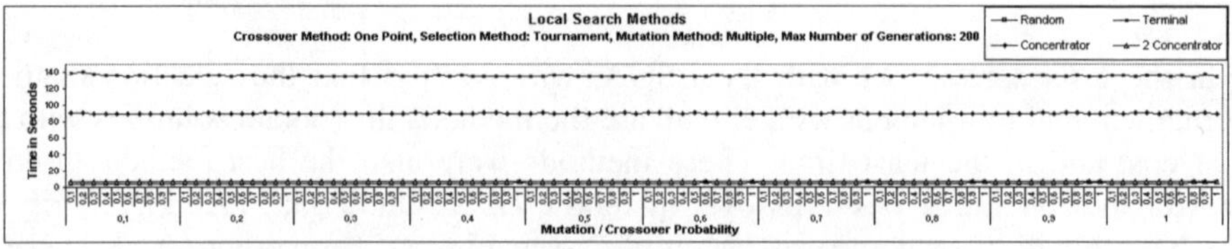

Fig. 8. Local Search Methods – Time Execution – Problem 2

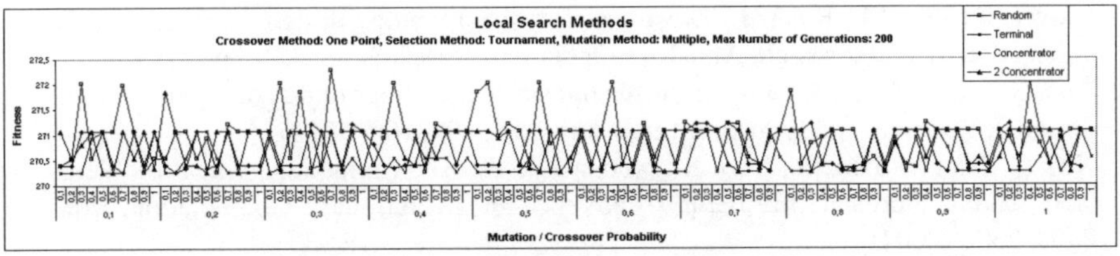

Fig. 9. Local Search Methods – Fitness – Problem 2

In comparison, the best algorithm is the LSGA because obtains better solutions in a better running time. Nevertheless, TS is the faster algorithm because can find a good solution in a better running time. The TS algorithm finds worst solutions for larger problems (7-9).

7 Conclusions

In this paper is presented a new Local Search Genetic Algorithm. The LSGA uses the candidate solutions that are produced by genetic operators and optimize them by a LS algorithm. This algorithm is used to solve a well-known problem, namely TA. The performance of' the proposed algorithm is compared with other algorithms using the same test cases (GAMO, HGA, TS). Relatively to the problem studied the LSGA presents better results. It can easily be seen that LSGA provides best solutions in a smaller execution time. The different LS algorithms implemented prove to be very efficient when combined with GA.

In any case, the implementation of new methods will permit to obtain better results. The implementation of parallel algorithms will speed up the optimization process.

References

1. Salcedo-Sanz, S., Yao, X.: A Hybrid Hopfield network-genetic algorithm approach for the terminal assignment problem. IEEE Transaction On Systems, Man and Cybernetics 34, 2343–2353 (2004)
2. Yao, X., Wang, F., Padmanabhan, K., Salcedo-Sanz, S.: Hybrid evolutionary approaches to terminal assignment in communications networks. In: Hart, W.E., Krasnogor, N., Smith, J.E. (eds.) Recent Advances in Memetic Algorithms and related search technologies, pp. 129–159 (2004)
3. Abuali, F., Schoenefeld, D., Wainwright, R.: Terminal assignment in a Communications Network Using Genetic Algorithms. In: 22nd Annual ACM Computer Science Conference, pp. 74–81 (2004)
4. Khuri, S., Chiu, T.: Heuristic Algorithms for the Terminal Assignment Problem. In: ACM Symposium on Applied Computing, pp. 247–251 (1997)
5. Atiqullah, M., Rao, S.: Reliability optimization of communication networks using simulated annealing. Microelectronics and Reliability 33, 1303–1319 (1993)
6. Pierre, S., Hyppolite, M.A., Bourjolly, J.M., Dioume, O.: Topological design of computer communication networks using simulated annealing. Engineering Applications of Artificial Intelligence 8, 61–69 (1995)

7. Glover, F., Lee, M., Ryan, J.: Least-cost network topology design for a new service: and application of a tabu search. Annals of Operations Research 33, 351–362 (1991)
8. Koh, S.J., Lee, C.Y.: A tabu search for the survivable fiber optic communication network design. Computers and Industrial Engineering 28, 689–700 (1995)
9. Xu, Y., Salcedo-Sanz, S., Yao, X.: Non-standard cost terminal assignment problems using tabu search approach. In: IEEE Conference in Evolutionary Computation, vol. 2, pp. 2302–2306 (2004)
10. Salcedo-Sanz, S., Portilla-Figueras, J.A., García-Vázquez, F., Jiménez-Fernández, S.: Solving terminal assignment problems with groups encoding: the wedding banquet problem. Engineering Applications of Artificial Intelligence 19, 569–578 (2006)
11. Bernardino, E.M., Bernardino, A.M., Sánchez-Pérez, J.M., Vega-Rodríguez, M.A., Gómez-Pulido, J.A.: Tabu Search vs Hybrid Genetic Algorithm to solve the terminal assignment problem. In: IADIS International Conference Applied Computing (2008)
12. Bernardino, E.M., Bernardino, A.M., Sánchez-Pérez, J.M., Vega-Rodríguez, M.A., Gómez-Pulido, J.A.: A Genetic Algorithm with Multiple Operators for Solving the Terminal Assignment Problem. New Challenges in Applied Intelligence Technologies 134 (2008)
13. MA HomePage, `http://www.ing.unlp.edu.ar/cetad/mos/memetic_home.html`
14. Moscato, P.: On Evolution, Search, Optimization, Genetic Algorithms and Martial Arts: Toward Memetic Algorithms. Caltech Concurrent Computation Program, C3P Report 826 (1989)

Solving the Ring Loading Problem Using Genetic Algorithms with Intelligent Multiple Operators

Anabela M. Bernardino[1], Eugénia M. Bernardino[1], Juan M. Sánchez-Pérez[2], Juan A. Gómez-Pulido[2], and Miguel A. Vega-Rodríguez[2]

[1] Department of Computer Science, School of Technology and Management, Polytechnic Institute of Leiria, Apt. 3063, Morro do Lena, Alto do Vieiro, 2401-951, Leiria, Portugal
anabelab@estg.ipleiria.pt, eugenia@estg.ipleiria.pt

[2] Department of Technologies of Computers and Communications, Polytechnic School, University of Extremadura, Avenida de la Universidad s/n, 10071, Cáceres, Spain
sanperez@unex.es, jangomez@unex.es, mavega@unex.es

Abstract. Planning optical communication networks suggests a number of new optimization problems, most of them in the field of combinatorial optimization. We address here the Ring Loading Problem. The objective of the problem is to find a routing scheme such that the maximum weighted load on the ring is minimized. In this paper we consider two variants: (i) demands can be split into two parts, and then each part is sent in a different direction; (ii) each demand must be entirely routed in either of the two directions, clockwise or counter-clockwise. In this paper, we propose a genetic algorithm employing multiple crossover and mutation operators. Two sets of available crossover and mutation operators are established initially. In each generation a crossover method is selected for recombination and a mutation method is selected for mutation based on the amount fitness improvements achieve over a number of previous operations (recombinations/mutations). We use tournament selection for this purpose. Simulation results with the different methods implemented are compared.

Keywords: Optimization, Genetic Algorithms, Ring Loading Problem.

1 Introduction

The ring is a popular topology for communication networks and has attracted much research attention. In a SONET ring, nodes (typically telephone central offices) are connected by a ring of fiber with each node sending, receiving and relaying messages by means of a device called an add drop multiplexer (ADM) that determines the actual bandwidth available along any edge of the SONET ring [1]. An important optimization problem arising in this context is the Ring Loading Problem (RLP). The goal of RLP is to minimize the maximum traffic load on an edge [1-9]. The RLP in which each demand is entirely routed in either of two directions (clockwise or counter-clockwise) is called the RLP without demand splitting (RLPWO). Otherwise, the problem is called RLP with demand splitting (RLPW), in which some portion of a demand is to be routed clockwise and the remainder to route counter-clockwise. Various approaches to solve the RLP are summarized by Schrijver et al. [1] and their algorithms compared in Myung and Kim [2] and Wang [3]. Cosares and Saniee [4] proved that the RLPWO is NP-complete. The intractability of this problem is a motivation for

J.M. Corchado et al. (Eds.): DCAI 2008, ASC 50, pp. 235–244, 2009.
springerlink.com

the pursuits of Genetic Algorithms (GAs) that produce approximate, rather than exact, solutions. In this article we report the results of the application of a GA using new methods including multiple intelligent operators.

The paper is structured as follows. In Section 2 we present the problem; in section 3 we describe the algorithm implemented while in Section 4 the studied examples; in Section 5 we discuss the computational results obtained and, finally, in Section 6 we report about the conclusions.

2 Problem Definition

Given a set of nodes connected along a bi-directional SONET ring, the objective is to determine a routing scheme which minimizes the bandwidth required to satisfy all the pair wise traffic demands.

For a given ring of n nodes in an arbitrary network, between each node pair `(ik, jk)` there is a demand `dk>=0` units. The RLP is formulated as follows:

$$d_{ijc} + d_{ijcc} = \text{total demand of a node pair } ij \qquad (1)$$

Fitness of each arc l =

$$\sum_{Cl(ij)\; i<=l;\, j>=l+1} d_{ijc} + \sum_{CCl(ij)\; i<=l;\, j<l+1 \;\; o \;\; i>=l+1} d_{ijcc} \qquad (2)$$

$$\forall l=1,\ldots,n; \quad \forall i=1,\ldots,npairs; \quad \forall j=1,\ldots,npairs; \quad i<=j \qquad (3)$$

Fitness Function = max {fitness of each arc (fit*l1*,…, fit*ln*)}

Constraint sets (1) in conjunction with constraints (3) state that each demand is routed in either clockwise (C) or counter clockwise (CC) direction. In RLPWO the demand just can flow in one direction. In RLPW the demand can be divided by the two directions. For an arc `l (l,l+1)` the load is the sum of `dk` for clockwise routed demands between nodes `i` and `j` such that `i<=l<=j-1` and counter clockwise routed demands between nodes `i` and `j` such that `i<=l>j-1` or `i>=l+1` (2). The objective is to minimize the maximum load on the arcs of a ring.

The initial population/individual can be created randomly or deterministically.

Pseudo code of the deterministic method:

```
FOR each pair
    Give a direction (C-demand value, CC-0)
pos = random (sizeIndividual)
FOR k=j=pos until npairs + pos
     IF (j > npairs)
        k=j-npairs
        change direction pair k
     IF fitnessNew_k > fitnessOld_k
        replace the previous value
```

Initially a deterministic strategy is followed and in a second phase is used the algorithm to optimize the solution.

3 Genetic Algorithms

The fact that the RLPWO is NP-Hard makes it a logical candidate for a heuristic search method like evolutionary algorithms (EA). GAs are EAs that are inspired by the natural process of reproduction [10]. Metaphors as chromosomes and population stand for solutions and solution set, respectively. Analogously, a single variable is often indicated as a gene. Mechanisms as recombination and mutation give rise to new offspring by manipulating the current population of solutions. Specially, mutation applies to a single solution (chromosome) while crossover creates new solutions from a pair of solutions selected in the current population. Following a standard Darwinian approach, selection extracts the most promising individuals in the current population. The stopping criterion for the algorithm is a pre-specified number of evaluations.

Table 1. Chromosome representation

Pair	Demand		
(1, 2)	15	15C	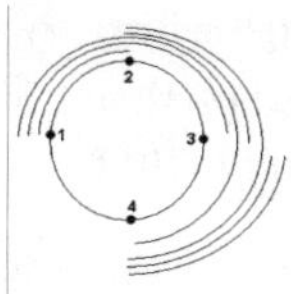
(1, 3)	3	3C	
(1, 4)	6	6C	
(2, 3)	15	15C	
(2, 4)	6	6C	
(3, 4)	14	14C	
Chromosome representation		15 | 3 | 6 | 15 | 6 | 14	
15C	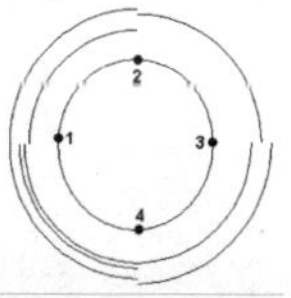	15C 0CC	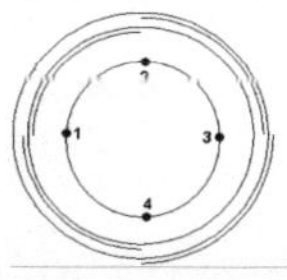
3CC		1C 2CC	
6CC		0C 6CC	
15C		15C 0CC	
6CC		3C 3CC	
14C		14C 0CC	
15 | 0 | 0 | 15 | 0 | 14		15 | 1 | 0 | 15 | 3 | 14	

The first step involves choosing a representation for the problem. In this work, the solutions are represented using integer vectors with the demand of clockwise direction (see Table 1).

We implement three well-known selection methods [11]: Roulette, Tournament and Tournament with Elitism (20% of the better individuals of the population are maintained in the following evaluation; the others are selected with the Tournament method).

We implement 8 crossover operators. One point, 2-points, 4-points, uniform and arithmetic (just for RLPW) are very well-known and widely used in practice. The crossover operator "Exchange positions" is based on 2-points operator, but in this case 2 points are obtained for each parent (see Fig. 1)

The "Differential" crossover operator is based on the demands subtraction of two genes. In RLPWO variant is necessary to do a demand conversion of the value. If the obtained value is less than half of the demand value then the exchanged value is 0 else is the total demand value of that gene (see Fig. 2).

15	0	6	15	6	0
0	3	0	15	0	14

⇩

15	3	0	15	6	0
0	3	0	15	6	0

Fig. 1. Exchange positions

15	0	6	15	6	0
0	3	0	15	0	14

⇩

0	3	0	0	0	14
15	0	6	0	6	0

Fig. 2. Differential

In the "Multiple" crossover operator, a set of available crossover operators is established initially. The crossover set contains the seven crossover operators applicable to the problem. In the initialization phase, each crossover operator has the same probability of being selected. From thereon and after every recombination is assigned a fitness value to the respective crossover operator based on its contribution to individuals fitness. This operator fitness value is used for recombination selection. The recombination operators are selected using a tournament selection. The program chooses three random operators. The operator with the higher fitness will win. After a predefined number of recombinations (NUM_PREV_OPS_RECOMB) the probabilities of each crossover operator are updated based on their contribution in the last recombinations.

Multiple Recombination:

```
FitIni1 = fitness(individual1)
FitIni2 = fitness(individual2)
if (first time)
      initialize numberOperationsR
      initialize fitnessPreviousOperationsR
      initialize fitnessActualR
      num_operations_totalR = 0
else
      If (num_operations_totalR = NUM_PREV_OPS_RECOMB)
           num_operations_totalR = 0
           For i=1 to NUM_OPERATORS do
              fitnessActualR[i] = fitnessPreviousOperationsR[i] /
                                    numberOperationsR[i]
           initialize numberOperationsR
           initialize fitnessPreviousOperationsR
num_operations_totalR = num_operations_totalR + 1
operator = random(NUM_RECOMB_OPERATORS)
for op = 1 to 3 do
      op = chooseRandomOperator()
      if (fitnessActualR[op] > fitnessActualR[operator])
            operator = op
switch(operator)
      case 1: recomb = Recombination1Point
      case 2: recomb = Recombination2Points
      case 3: recomb = Recombination4Points
      case 4: recomb = RecombinationUniform
      case 5: recomb = RecombinationArithmetic
      case 6: recomb = RecombinationExchangePositions
      case 7: recomb = RecombinationDifferential
run(recomb, individual1, individual2)
fitnessPreviousOperationsR[operator]=
      fitnessPreviousOperationsR[operator] + max(fitIni1,fitIni2) -
      max(fitness(individual1),fitness (individual2))
numberOperationsR[operator] = numberOperationsR[operator] + 1
```

We implement seven mutation operators. In "Change direction", one gene is selected randomly and its direction is exchanged. In "Change order" two genes are randomly selected and their directions are exchanged. In "Change direction-order" for 70% of the cases is applied "Change direction", to the rest "Change order". In "Change demand", one gene is randomly selected and its demand value is replaced with a random value of the total demand of the pair. In "Change demand split", one gene is randomly selected and its demand value is replaced with half of the demand value of the gene. In "Change bigger demand", is selected the gene with the biggest demand value and the demand value is replaced with a random value of the total demand of the pair (The last three mutation operators are only used in the variant RLPW).

In the "Multiple" mutation operator, a set of available mutation operators is established initially. The mutation set contains the three mutation operators applicable to the formulation RLPWO and six mutation operators to the formulation RLPW. In the initialization phase, each mutation operator has the same probability of being selected. From thereon and after every mutation is assigned a fitness value to the respective mutation operator based on its contribution to individual fitness. This operator fitness value is used for mutation selection. The mutation operators are selected using a tournament selection. The program chooses two random operators. The operator with the higher fitness will win. After a predefined number of mutations (`NUM_PREV_OPS_MUT`) the probabilities of each mutation operator are updated based on their contribution.

Multiple Mutation (RLPWO):

```
fitIni = fitness(individual)
if (first time)
      initialize numberOperations
      initialize fitnessPreviousOperations
      initialize fitnessActual
      num_operations_total = 0
else
      If (num_operations_total = NUM_PREV_OPS_MUT)
           num_operations_total = 0
           For i=1 to NUM_OPERATORS do
                fitnessActual[i] = fitnessPreviousOperations[i] /
                                        numberOperations[i]
           initialize numberOperations
           initialize fitnessPreviousOperations
num_operations_total = num_operations_total + 1
operator = random(NUM_MUT_OPERATORS)
for op = 1 to 2 do
      op = chooseRandomOperator()
      if (fitnessActual[op] > fitnessActual[operator])
            operator = op
switch(operator)
      case 1: mut = MutationChangeDirection
      case 2: mut = MutationChangeOrder
      case 2: mut = MutationChangeDirectionOrder
run(mut, individual)
fitnessPreviousOperations[operator]=
  fitnessPreviousOperations[operator] + fitIni - fitness(individual)
numberOperations[operator] = numberOperations[operator] + 1
```

4 Studied Examples

We evaluate the utility of the algorithms using the same examples produced by Bernardino et al. [12]. The studied examples arise by considering two different ring sizes with four different demand distributions. A ring in a telecommunications network will typically contain between 5 and 15 nodes. Thus, we consider the 10 node rings to be ordinary-sized rings and the 20 node rings to be extremely large rings.

The four demand cases considered are:

- Case 1: complete set of demands between 1 and 100 with uniform distribution;
- Case 2: half of the demands in Case 1 set to zero;
- Case 3: complete set of demands between 1 and 500 with uniform distribution;
- Case 4: complete set of demands between 1 and 500 with bimodal distribution.

The bimodal case is constructed by sampling from two separate intervals within the 1-500 range. The bimodal sample has 80% of its demands between 1 and 50 and the remaining 20% between 400 and 500.

It was generated 2 different problem instances for each case. This yields 8 instances for each ring size. For convenience, they are labeled `Cij`, where `1<i<4` represents the demand case and `1<j<2` represents the instance within a case.

5 Results

Table 2 presents the best results obtained with the GA. The first column represents the problem number (Problem), the second and the third columns show the number of nodes (Nodes) and the number of pairs (Pairs) and the fourth and fifth columns show the minimum fitness values obtained for the formulations RLPWO and RLPW. The GA has been executed using a processor Intel Core Duo T2300. The results of the GA

Table 2. Results

Problem	Nodes	Pairs	Algorithms	
			RLPWO	RLPW
C11	10	45	639	609
C12	10	45	721	721
C21	10	22	325	323
C22	10	22	449	447
C31	10	45	3350	3273
C32	10	45	3791	3694
C41	10	45	1933	1795
C42	10	45	1035	958
C11	20	190	2806	2806
C12	20	190	2849	2849
C21	20	99	1484	1484
C22	20	93	1447	1447
C31	20	190	12688	12688
C32	20	190	13327	13327
C41	20	190	6872	6872
C42	20	190	5993	5993

are generated using a population of 200 individuals. Since the best solutions produced are influenced by the generated initial population, the algorithms were tested using a randomly population and a deterministic population. The results reported are based on summary statistics from these experiments.

For the 10 node problems, are performed 5,000 evaluations. For 20 node problems it does 20,000 evaluations. These limits are selected based upon preliminary observation of several different values. Usually, it is the case that the best solution is observed well before reaching this limit. The timings reported in Table 3 present the minimum time it takes to reach the best solution (Times 10 and Times 20 – run time in seconds for the 10 and 20 node rings and Generations 10 and 20 – number of evaluations). The results have been computed based on 50 different executions using the best strategies and parameter values.

In the tests carried out with the GA, was verified that the selection method tournament with elitism is the method that obtains solutions with smaller cost and in

Table 3. Results – run times and number of evaluations of GA

Problem	Formulation	GA			
		Times 10	Times 20	Generations 10	Generations 20
C11	RLPWO	<0.5	<5	<10	<100
	RLPW	<0.5	<10	<30	<150
C12	RLPWO	<0.5	<5	<10	<50
	RLPW	<0.5	<10	<15	<100
C21	RLPWO	<0.5	<5	<5	<25
	RLPW	<0.5	<5	<5	<25
C22	RLPWO	<0.5	<5	<10	<25
	RLPW	<0.5	<5	<10	<25
C31	RLPWO	<0.5	<15	<10	<250
	RLPW	<0.5	<20	<20	<500
C32	RLPWO	<0.5	<15	<10	<300
	RLPW	<0.5	<20	<20	<600
C41	RLPWO	<0.5	<5	<5	<100
	RLPW	<0.5	<10	<20	<200
C42	RLPWO	<0.5	<5	<20	<50
	RLPW	<0.5	<10	<20	<100

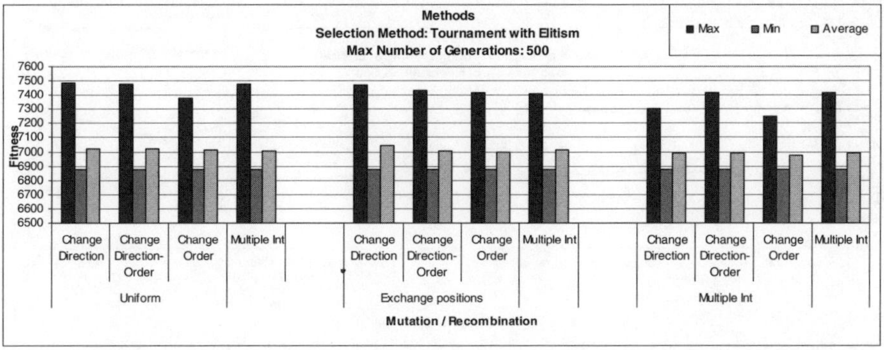

Fig. 3. Recombination and Mutation Methods of the problem C41 RLPW

smaller time. The best crossover methods used are the uniform and the multiple with probability in the interval [0.4-0.8]. The best mutation strategy was the multiple and the direction-order with probability in the interval [0.01-0.4].

Fig. 3 shows the results obtained for the problem C41 RLPW with the best cross-over/ mutation operators implemented. The multiple operators implemented proved to be very effective.

Fig. 4 shows all the combination probabilities with recombination uniform, mutation multiple and a number max of 500 evaluations. This combination was the best found.

We compare the efficiency of the multiple intelligent methods implemented with the multiple random methods used in [12]. Fig. 5 shows a comparison between the multiple crossover implemented in [12] and the multiple crossover method presented in this article. The multiple intelligent crossover method obtains better results. The multiple intelligent mutation method produces good results, but the difference from other good methods implemented is insignificant.

All the methods implemented proved to be very effective to solve RLPW and RLPWO problems. The commitment between convergence and diversity of the individuals is a constant in an EA and should be considered in the configuration of a

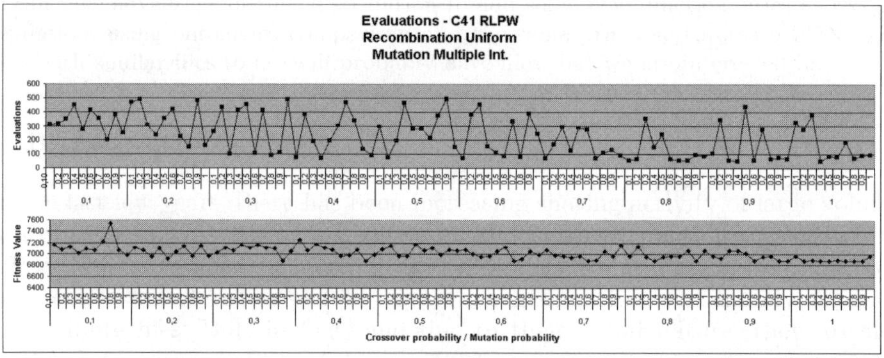

Fig. 4. Fitness value of the problem C41 RLPW

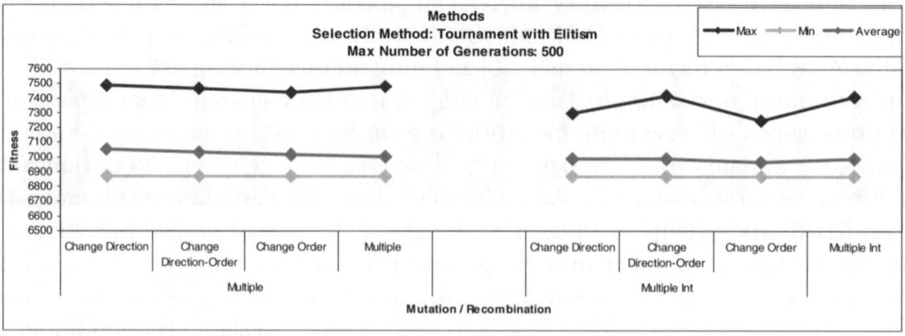

Fig. 5. Fitness value of the problem C41 RLPW

methodology of efficient optimization. The parameter values used to test the algorithms have been found using heuristics of test and error that have obtained promising results.

6 Conclusions

In this paper we present a GA with multiple intelligent operators for crossover and mutation. We present a multiple mutation method and a multiple crossover method. In each step, the multiple mutation method selects one operator based on the amount fitness improvements achieve over a number of previous mutations. The multiple crossover method works similar to multiple mutation method. We present a novel operator selection method similar to tournament selection. The implemented methods are used to solve a well-known problem, namely RLP.

The performance of' the different methods implement are compared. Relatively to the problem studied the multiple intelligent crossover method presents better results. It can easily be seen that this method provides best solutions. The multiple intelligent mutation method in comparison with other methods doesn't produce significant best results. A great advantage of the proposed method is that it allows, through its execution, select mutation operators that are better adapted to the problem resolution.

The different methods of crossover, mutation and selection implemented have driven to acceptable results. In any case, the implementation of new methods will permit to obtain better results. The implementation of parallel algorithms will speed up the optimization process.

References

1. Schrijver, A., Seymour, P., Winkler, P.: The ring loading problem. SIAM Journal of Discrete Mathematics 11, 1–14 (1998)
2. Myung, Y.S., Kim, H.G.: On the ring loading problem with demand splitting. Operations Research Letters 32(2), 167–173 (2004)
3. Wang, B.F.: Linear time algorithms for the ring loading problem with demand splitting. Journal of Algorithms 54(1), 45–57 (2005)
4. Cosares, S., Saniee, I.: An optimization problem related to balancing loads on SONET rings. Telecommunication Systems 3(2), 165–181 (1994)
5. Goldschmidt, O., Laugier, A., Olinick, E.V.: SONET/SDH ring assignment with capacity constraints. Discrete Applied Mathematics. Elsevier 129(1), 99–128 (2003)
6. Karunanithi, N., Carpenter, T.: A Ring Loading Application of Genetic Algorithms. In: Proceedings of the ACM Symposium on Applied Computing, pp. 227–231. ACM, New York (1994)
7. Lee, C.Y., Chang, S.G.: Balancing loads on SONET rings with integer demand splitting. Computers and Operations Research 24(3), 221–229 (1997)
8. Myung, Y.-S., Kim, H.-G., Tcha, D.-W.: Optimal load balancing on SONET bidirectional rings. Operations Research 45(1), 148–152 (1997)
9. van Hoesel, S.P.M.: Optimization in telecommunication networks. Statistica Neerlandica 59(2), 180–205 (2005)

10. Goldberg, D.: Genetic Algorithms in Search, Optimization and Machine Learning. Addison Wesley, Reading (1989)
11. Eiben, A., Smith, J.: Introduction to Evolutionary Computing. Springer, Berlin (2003)
12. Bernardino, A.M., Bernardino, E.M., Sánchez-Pérez, J.M., Vega-Rodríguez, M.A., Gómez-Pulido, J.A.: A Genetic Algorithm with multiple operators for solving the Ring Loading Problem. In: IADIS International Conference Applied Computing (2008)

Extending Korf's Ideas on the Pursuit Problem

Juan Reverte, Francisco Gallego, and Faraón Llorens

Department of Computer Science and Artificial Intelligence
University of Alicante
{jreverte,fgallego,faraon}@dccia.ua.es

Summary. The prey-predator pursuit problem is referenced many times in literature. It is a generic multi-agent problem whose solutions could by applied to many particular instances. Solutions proposed usually apply non-supervised learning algorithms to train prey and predators. Most of these solutions criticize the greedy algorithm originally proposed by Korf. However, we believe that the improvement obtained by these new proposals does not pay off with relation to their complexity.

The method used by Korf is a natural way to surround a prey without explicit communication between predators. The knowledge one predator has about others is limited just to what it can see. In Korf's model, agents are able to see the complete world at once. In this paper we propose to start from Korf's ideas and extend them to improve his model. First, we propose a simple extension of Korf's fitness function and we consider the problems related to a partial view of the world. Second, we propose a communication protocol to partially overcome them. The final results suggest that more work needs to be done, and we propose a way to follow-on.

Keywords: Prey-predator, multi-agent systems, communication.

1 Introduction

The Prey-predator pursuit problem emerges every time that a group of agents have to chase and surround another agent (or other agents) that tries to evade them [1]. The goal of the predator agents is to surround prey(s) without touching it, arriving to a final state position where predators impede any possible orthogonal movement to the prey, whilst the goal of the prey, as expected, is not to be captured.

This problem has been addressed many times in the literature from diferent points of view [2][3] [4][5][7].The first accepted solution was the one proposed by Korf [7]. Korf just combined 2 forces in the way that each predator is "attracted" by the prey and "repelled" from the closest predator. This solution keeps predators away from other predators while they get closer to the prey, thus chasing the prey like a circle stretching, in the best case. Korf's simple solution has got a lot of criticism and a great number of other solutions and alternatives have emerged. For instance, Haynes [3] used genetic programming to evolve coordinated behaviours of predators. Haynes compared diferences between communicating and non-communicating predators with respect to their

J.M. Corchado et al. (Eds.): DCAI 2008, ASC 50, pp. 245–249, 2009.
springerlink.com

success chasing the prey. He also co-evolved predators and the prey and found that a prey always going straight in diagonal is never caught by predators unless it is slower than them. Chainbi [2] used petri nets to coordinate predators while solving concurrency problems between them, while Jim and Giles [4] used a genetic algorithm and multi-agent communication using a blackboard. One of the most interesting alternatives was the one proposed by Katayama et al. [5]. They used an agent-oriented reinforcement learning algorithm, namely profit-sharing, with analytic hierarchy process (AHP) integrated. Their idea was to introduce primary knowledge to guide the agents when they start the learning process. It is a great idea to give some knowledge as a kind of "hints" to the agents when they start learning, but it seems not to be reasonable to continue giving these "hints" once agents have grown and developed their own knowledge. The solution proposed by Katayama et al. is to progresively take back the "hints", leaving the agents finally with their own knowledge. Analyzing the results shown by these alternatives to Korf [7] we conclude that they are not enough to state that Korf's ideas were not right. Korf idea was really simple: determine the next movement of each predator with a fitness function dependant on distance to the prey and distance to the nearest other predator. The fitness function simulated an "attractive force" to the prey and a "repulsive force" from the nearest predator. This idea represents the starting point of this paper. Section 2 explains the extensions we propose to Korf's ideas. Section 3 shows our implementation and results. Finally, section 4 sums up our conclusions and next under-development proposals.

2 Extending Korf's Ideas

Originally, Korf used a 100x100 discrete grid where all agents must occupy distinct positions. He stablished a rotatory turn system for the agents. At every turn one agent can move to an empty neighboring cell or remaining stationary. 90% of the times, prey moved to the neighboring cell that is furthest away from the nearest predator, remaining stationary the other 10%. The prey begins in the center of the board, and the initial position of the predators is randomly generated.

The first and simplest extension we propose to Korf's model is to change the fitness function to make each predator repel from all other predators it has in its viewing field. This should improve behaviour of predators when they are close, for instance, when they are on the point of catching the prey. Equation 1 shows the extended fitness function. This fitness function is calculated for each possible cell where the predator is able to move to, finally selecting the cell with maximum fitness. We assume X_p, Y_p as the location of the prey, X_i, Y_i as the location of predator i, and $d(x_1, y_1, x_2, y_2)$ as the manhattan distance between two cells.

$$f(x,y) = d(x,y,X_p,Y_p) - k \sum_{i=1}^{n} d(x,y,X_i,Y_i) \qquad (1)$$

In Korf's model, the most influent initial condition was the ability of agents to see the complete board. This is what makes his solution possible, but it is almost never happening in a real or virtual environment. It is common for agents to have a limited view of the environment as it would be the case in a real environment. In this more real situation, Korf's ideas are still interesting, but need to be extended and supported with communication between agents [4]. In our model, we consider that an agent located at x, y with a field of vision (FOV) of n cells means the agent is only able to preceive what happens in cells $\{(x', y')/x - n <= x' <= x + n, y - n <= y' <= y + n\}$. Take into account that an agent is only able to communicate with thouse inside its FOV.

This new situation does not prevent Korf's algorithm from catching the prey. However, the algorithm becomes slower. Namely, it takes more time to predators to find the prey. In strict sense, the probability of a predator indefinitely not finding the prey is not 0, and that is definitely a problem to overcome. Our propose is a simple yet effective protocol for communicating prey sights we have called Cascading Sight Notice (CSN, see figure 1). This protocol starts when a predator sees the prey. The predator sends a message to others in its FOV, telling them the relative location of the prey from its point of view. The hearing predators make an addition of the two vectors and then get aware of the location of the prey. These predators, once they have located the prey with respect to their own place, resend the message to other predators. The cycle continues until no predator is hearing or hearing predators already know where prey is.

3 Results

To validate our approach we compared it with Korf's method in different simulated environment conditions. For running the simulations we used Kok&Vlassis' Pursuit Domain [6]. This environment simulates a discrete, toroidal predator-prey environment with several configurable characteristics. Concretely, we used a 30x30 cells field, allowing agents to move diagonal, with the prey starting on the center and predators starting randomly placed. We lauched 4 predators to capture 1 prey. Finally, in case of collision, only predators colliding were

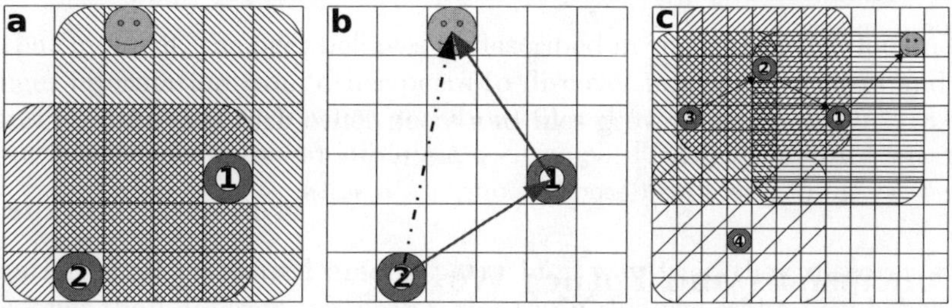

Fig. 1. a) Predator 1 sees prey, pred.2 sees pred.1. b) Pred.2 calculates prey location using pred.1 info. c) Preds.1, 2, 3 can figure out prey location, pred.4 cannot

penalized. Simulation was always ran for 250 consecutive episodes, and we got the average results.

With this configuration we ran 2 groups of 4 experiments each. Each group refers to a differently behaving prey whilst each experiment is about differently behaving predators. The 2 preys were: g1) randomly moving prey, g2) prey that moves to the cell most distant from menacing predators. The experiments ran were: e1) Korf's predators, e2) extended Korf's predators (using equ. 1), e3) Korf's predators using CSN, e4) extended Korf's predators using CSN.

Figure 2 shows results from the 8 experiments. We ran each varying the FOV from a minimum of 3 cells to a complete FOV of 15 cells (15 to each direction of a toroidal world covers the 30x30 field). Results show that our extended version of Korf's algorithm represents an improvement of an order of magnitude in most cases, with independence of the type of prey. This improvement is comparable to the ones achieved by most of the cited works in this paper. It is really curious to compare Korf's and Extended Korf's algorithms with their versions including CSN protocol described in section 2. Results show that CSN represents a small improvement that is only significant depending on type of prey and FOV. When FOV is complete, CSN is unnecesary; when FOV is at the bare minimum, CSN significance depends on the prey. This is due to the chasing movements predators do with the evading prey and the necessity of being inside FOV to communicate with others. The bigger the world, the more disperse predators are and the more difficult to communicate.

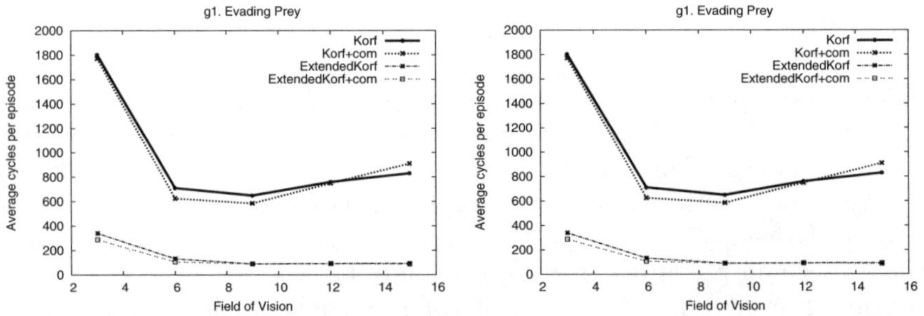

Fig. 2. Comparison between Korf's and Ext. Korf's models, with and without interagent communication

Notice that PDP treats collisions randomly replacing predators which collide and uses a toroidal world. These are two main differences between this work and Korf's that should be considered to compare results.

4 Conclusions and Futher Work

The Predator-prey pursuit problem is considered a good testbed for multi-agent collaboration. A great number of solutions have been presented. Recent solutions focus on machine learning and criticize Korf's greedy approach.

Although recent works have shown great results, we believed that Korf's idea still has a good potential to explore. Therefore, we proposed and developed a simple extension of Korf's algorithm. Moreover, as Korf did not take into account limited FOV of the agents, we also have developed a simple protocol (CSN) to overcome part of the problems caused by this limited FOV.

Our results show that with these two simple extensions, Korf's ideas get results comparable to most machine learning aproaches cited here. This supports our hipothesis that Korf's idea has a great potential. However, CSN is pretty simple and does not seem to be enogh. As future work we plan to develop a protocol to improve predators early finding and noticing the prey. This new protocol will also deal with concurrency problems and prevent collisions.

References

1. Benda, M., Jagannathan, V., Dodhiawalla, R.: On optimal cooperation of knowledge sources. Technical Report Tech. Rep. BCS-G2010-28, Boeing AI Center, Boeing Computer Services, Bellevue, WA (1986)
2. Chainbi, W., Hanachi, C., Sibertin-Blanc, C.: The Multi-agent Prey-Predator problem: A Petri net solution. In: Proceedings of the IMACS-IEEE-SMC conference on Computational Engineering in Systems Application (CESA 1996), Lille, France, pp. 692–697 (1996)
3. Haynes, T., Sen, S.: Evolving behavioral strategies in predators and prey. In: Sen, S. (ed.) IJCAI 1995 Workshop on Adaptation and Learning in Multiagent Systems, Montreal, Quebec, Canada, 20-25, 1995, pp. 32–37. Morgan Kaufmann, San Francisco (1995)
4. Jim, K.-C., Lee Giles, C.: Talking helps: Evolving communicating agents for the predator-prey pursuit problem. Artificial Life 6(3), 237–254 (2000)
5. Katayama, K., Koshiishi, T., Narihisa, H.: Reinforcement learning agents with primary knowledge designed by analytic hierarchy process. In: SAC 2005, pp. 14–21. ACM, New York (2005)
6. Kok, J.R., Vlassis, N.: The pursuit domain package. Technical Report Technical Report IAS-UVA-03-03, Informatics Institute, University of Amsterdam, The Netherlands (August 2003)
7. Korf, R.E.: A simple solution to pursuit games. In: Proceedings of the 11th International Workshop on Distributed Artificial Intelligence, Glen Arbor, MI, pp. 183–194 (February 1992)

Autonomous Artificial Intelligent Agents for Bayesian Robotics

Fidel Aznar, Mar Pujol, and Ramón Rizo

Departamento Ciencia de la Computación e Inteligencia Artificial
Universidad de Alicante.
Campus de Sant Vicent s/n, 03080, Alicante, Spain
fidel@dccia.ua.es, mar@dccia.ua.es, rizo@dccia.ua.es

Abstract. This paper addresses the problem of working with uncertain or incomplete information in multiagent systems. When we work with real life systems (like a robotic agent) we normally use models to "insert" our knowledge to the robot. Nevertheless, any model of a real phenomenon will always be incomplete due to the existence of unknown, hidden variables that will influence the phenomenon, causing the model and the phenomenon to have different behavioral patterns. In this paper we provide a new intelligent Bayesian agent architecture oriented towards Bayesian robotics that provides a framework for developing intelligent agent applications using Bayesian theory. This architecture allows the programmer to use a common framework for developing agents that needs to work with uncertain or incomplete information, using one data type, join probabilistic distributions, which allows to use a common exchanging uncertainty assessments in multi-agent systems.

Keywords: Bayesian Agents, Autonomous Intelligent Agents, Bayesian Programming.

1 Introduction

In recent years, probabilistic algorithms have become very popular in mobile robotics because they represent explicitly robot's uncertainty, using a robust mathematical background. A robotic agent could use a model of the environment representing the real universe where it will interact. Nevertheless, any model of a real phenomenon will always be incomplete due to the existence of unknown, hidden variables that will influence the phenomenon, causing the model and the phenomenon to have different behavioral patterns. Probabilistic inference helps to solve this problem taking into account incomplete and uncertain information. In addition, probabilistic approaches are usually more robust than their traditional counterpart [1].

Different efforts [1][2][3] have been taken to define programming frameworks for Bayesian programming, motivated by successful probabilistic methods for mobile control with the main goal of facilitate the development of such probabilistic software. Most probabilistic frameworks, however, focus only on structural or object programming.

Recent work with agents and multi-agent systems has encouraged robotic researches to take another step forward in the design of control architectures replacing modules with agents. Although exist some agent and multi-agent applications that use Bayesian inference [4][5][6][7][8], we have not found any architecture or framework

J.M. Corchado et al. (Eds.): DCAI 2008, ASC 50, pp. 250–258, 2009.
springerlink.com

for agent design, centered on Bayesian programming, that fully exploits its benefits. Bayesian programming agents will focus on discrete or continuous joint distributions as a basic information chunk for developing all the principles that an autonomous intelligent agent must follow.

In this paper a probabilistic agent framework, oriented to robot programming, will be provided. Firstly, we will review the most important design issues for developing good autonomous intelligent agents. Secondly, a definition and a specification for Bayesian robotic agents will be provided and also discussed with the previously commented design issues. Next, our framework will be briefly compared with recent agent architectures that use Bayesian inference. Moreover, a conceptual robotic example will be formalized using the proposed framework. Finally conclusions and future lines will be commented.

2 Agent Design Issues

Although an established theory regarding autonomous intelligent agents does not exist, and designing autonomous agents is an active research area, there are several principles that could be used as a guide for autonomous agent design [10]. These principles may help researchers to obtain better artificial intelligence systems. We will review some of the most important principles and use it in the next section to propose our framework.

- **Autonomy.** A perfect agent should work with no human supervision o instruction, being self-sufficient.
- **Emergence.** When programming an agent, the programmer has a specific idea about the task to be solved by this agent. However, to design an agent according to his view of the task may not always be the best solution because his view of the task depends on a human perspective. The agent may have a different embodiment and then a different perspective of task.

  One solution to these problems arises when the programmer design the system for emergence, when behaviors that are not programmed results from agent-environment interaction and from self-organization of the agent's control system.
- **Epigenesis.** This term define a process through which increasingly more complex cognitive structures emerge in a system as a result of interactions with the physical and social environment. The final behaviors are consequence of environment information and existing information. In this way the role of the environment is constructive rather than selective. Moreover, new tasks should be learned without requiring a redesign of control system, where human teachers may affect this learning process only as a part of the environment without interfering with its internal representation.
- **Parallel, loosely coupled processes.** For supporting emergence loosely coupled processes and little or not centralized resources are needed. When the control is decentralized and distributed, intelligent behaviors may emerge from the joint dynamics of the number of processes, each of which contributes to the overall function, as the agent interacts with the environment.

- **Sensorimotor coordination.** In classical AI the cycle perception-think-action explicitly separates in time perception from action, preventing the potential emergence of adaptive behaviors from their coordination. A continuous interaction between perception and action should exist, where perception guides action in interaction with internal state and action change internal state and the perspective of the environment as perceived by the agent. Moreover, as stated in [9], cognition and learning needs perceptual and effectors capabilities to be developed.
- **Goals.** An agent must be goal directed and must have a value system that would guide its behaviors. It must have predefined drivers that would induce the exploration of the environment. This exploration may lead to non-trivial sensorimotor patterns, as a consequence to self-organization.
- **Internal representation.** The action an agent perform should not be determined only externally, based on inputs, as in purely reactive systems, but neither be planned ahead, as in some classical AI systems. This representation should not be imposed by the user or the designer of the system, but should grown in the sensorimotor interaction of the agent with the environment.
- **Symbolic Communication.** Communication between agents requires interchanging symbols that must be understood by all the agents in communication. The meaning of these symbols should not be given by agent programmer but as in previous principle should grown in the sensorimotor interaction with the environment. Otherwise, the meaning cannot be accessible to the agent itself.

3 Bayesian Robotic Agents

In this section we will define an internal schema that provides support for Bayesian agents oriented towards robotic programming, briefly comparing our approximation with those proposed in [4][5][6][7][8].

Intelligent Bayesian Agents are a group of agents, designed for the formalization of complex robotic systems, which are based on probabilistic theory. A pure Bayesian agent is an agent that only uses Bayesian Inference for cognition and learning and its internal information is composed exclusively by join probabilistic distributions. Without lost of generality we will center this section with pure Bayesian agents, as is easy to extend a pure Bayesian agent with another kind of internal information or learning mechanism. However we have to underline that a wide range of robotic applications are solved only using Bayesian tools [2][3].

In figure 1, the internal schema of a Bayesian agent is presented. These agents use random variables and joint probabilistic distributions to model every step of perception, cognition and action.

Firstly, all variables must be specified and classified in one of the following groups: goals, behaviors, internal states, communications, perceptions and actions. Obviously more than one variable can be specified inside a group and one group could be empty. This classification shows the most important design issues presented in the previous sections. For example, communication variables specify the information symbols (Bayesian variables or distributions) to be emitted or received; state variables organize the internal representation of the agent...

Secondly, the joint probabilistic distribution of all variables should be specified. This helps to calculate the 'utility' of each variable. If the distribution is fully identified any question related to it could be asked. Variables decomposition shows the programmer knowledge about the problem but do not fix the behavior of the agent because usually some terms of the distributions must be learned. Bayesian learning allows the agent to increase more complex cognitive structures, enabling emergent behaviors and epigenesis. Two decompositions are proposed as a guide:

BAYESIAN AGENT

Identify agent information

$$AgentInformation = \begin{cases} Goals, Behaviors, States, Comunication \\ Perception, Action \end{cases}$$

Propose joint distribution decomposition of $AgentInformation$

$$P(G,B,S,C,P,A) = \begin{cases} 1)\ P(G \mid S) \times P(B \mid G,S) \times P(S \mid B,C,P,A) \times P(C \mid S) \times \\ \quad \times P(A \mid P,B) \times P(P) \\ 2)\ P(G) \times P(B \mid G) \times P(S \mid B,P,A,C) \times P(A \mid B,P) \times \\ \quad \times P(C \mid S) \times P(P) \\ \text{Previously based} \end{cases}$$

$\forall i, i \in AgentInformation$, Define the ' utility' of each variable

$$\begin{cases} \forall \text{question, question} \in \text{i} \\ \text{Solve the question using} : \underbrace{P(U \mid K) = \alpha \sum_{H} P(H,U,K)}_{\text{where } H \text{ are hidden, } U \text{ unknow variables and } K \text{ know variables. } \alpha \text{ is a normalization term}} \end{cases}$$

Use the distribution

Specify Bayesian inference method to be used

Fig. 1. Internal schema that define a Bayesian Agent

- The first decomposition reflects the interactions required for the basic design principles commented previously: in order to allow sensorimotor coordination the perception of the agent influences their actions in a reactive way. But actions are not only influenced by perceptions but also by behaviors, which raise the system from reactive level to hybrid level. Communications between agents require interchanging symbols that should be obtained from sensorimotor interaction, therefore communication variables influence the internal state of the robot. This state is obtained from the interaction of perception and actions, the agent communication and the behavior variables. In the same way, goals are driven by states, which allows agents to change dynamically its goals.
- The second one is a simplified version but more easy to use for Bayesian Inference (because this decomposition could be modeled and solved using Bayesian networks tools). In this case the agent cannot change its initial goals and also actions are not affected directly by a set of behaviors, only indirectly by agent states, which leads to a more simplified system that reduces the possibilities of emergence intelligence and epigenesis but still allows them.

Next, for each variable defined in the system, the 'utility' must be specified. We define the 'utility' as a set of questions that the agent must answer. These questions could be always specified using both, the full joint probabilistic distribution and general inference: $P(U \mid K) = \alpha \sum_{H} P(H,U,K)$. Once the question is specified all the terms must be defined, or approximated using discrete or continuous distributions, or even learned.

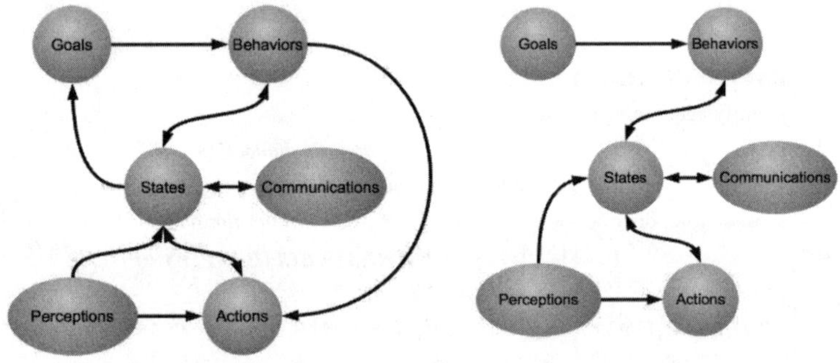

Fig. 2. Bayesian dependences between agent variables. Left figure represent these dependences taking into account the design issues previously commented. Right figure relax such assumptions for simplifying Bayesian inference providing a Bayesian Network as decomposition.

Finally, the programmer must specify the Bayesian inference method to be used in the agent and the way the information obtained from the distribution is used.

3.1 Comparing with Other Architectures

Now, we will present a brief review of the state of the art of Bayesian agents in scientific literature, comparing it with our proposed architecture. Three groups of applications will be presented as representative parts of this field.

The first group is related to another Bayesian agents frameworks. Santos, Fagundes and Vicari present in [8] an ontology for Bayesian agents interoperability. Moreover, a Bayesian agent architecture is presented. This paper shows a general ontology for working with Bayesian Networks. Compared with the proposed model this architecture does not provide any information about how defining Bayesian agents and what principles must be used for their design, it is only focused on how Bayesian agent will interact with its Bayesian Network ontology. This Bayesian architecture can be used with our model in the case of the result of the variables decomposition will be a Bayesian network, and then provide a tool for agent interoperability. We have not found any general agent architecture that uses Bayesian Theory guiding the user for agent design, problem solving and robotic programming.

The second one is related to purely Bayesian agents that are defined adhoc for a specific problem solving it without using an agent architecture. One of these applications is related to coordination and coalition formation in multiagent systems [4][5][6]. In these papers the Bayesian tool used for information decomposition and learning are based on MDP (Markov decision process). A good example of Bayesian

Inference is presented but no general framework for agent design is provided. However, these applications are pure Bayesian and thus could be modeled using our proposed architecture.

The third group is related to Bayesian agent applications, such as [7], where Bayesian tools are used for learning only as a part of the agent. The advantages of Bayesian Probability are only obtained in a limited way.

We want to emphasize the most important advantages of our proposed architecture, most of them derived from both, artificial agents and Bayesian theory: this architecture provides a common framework for comparing different Bayesian agent, it supports the most important design principles in intelligent agent systems, agents can use incomplete and uncertain information in a direct way when using the proposed model, a wide range of Bayesian inference tools could be used for learning or variable marginalization.

Finally, despite all these advantages, not all problems are suitable for being modeled using this framework, because Bayesian inference is, in general, a NP-hard problem. For this reason programmers must choose a distribution decomposition taking into account both, agent design and computational complexity.

4 A Proof of Concept Example

Imagine a robotic agent that must perform the following actions: when the agent detects a source of light it must move to it, moreover, it must avoid obstacles using its sonar sensors. In this section we will review a Bayesian agent, presented in figure 3, that performs these actions.

Once we identify agent variables and proposed a join probabilistic distribution decomposition for the problem we need to specify the utility of each variable. Related to goals, communications and perceptions there is no question that the agent need to know (for example, in this case the agent must use a fixed set of parallel goals, so we don't need to ask which goal will be used next).

The utility of behavior variable B is defined with the question $P(B \mid G,S)$. As the behaviors defined in this distribution are specified directly based on the agent goals modifying the distribution in a uniform way (we do not prefer any of them more than the other), we could omit these goals and only consider the internal state of the agent for obtaining the distribution of next behavior. For example this information could be useful in the case of communicating another agents the influence of phototaxy or obstacle avoidance in a specific place of the environment.

For obtaining which action must be send to the robot we could think to ask the question $P(A \mid P)$, which leads to a pure reactive system. We need to take into account the two behaviors proposed, so we include the behavior variable B using the marginalization rule. Finally we decompose $P(A,B \mid P)$, establishing that $P(A \mid P) = \sum_{B} (P(B \mid P) \times P(A \mid P,B))$. The first term of the distribution will obtain the probability of activating phototaxy or obstacle avoidance depending on the actual perception. The second, once the probability of each behavior is defined, use it with the actual perception to determine which action must be send to actuators.

AGENT Phototaxy and AvoidObstacles
Identify agent information

$$AgentInformation = \begin{cases} Goals: G = \{showPhototaxy, showColisionAvoidance\}; \\ Behaviors: B = \{phototaxy, colisionAvoidance\} \\ States: S = \{B_{t-1}\}; \\ Communication: \varnothing; \\ Perception: P = \{s_1, s_2, ..., s_7, photo_1, photo_2\}; \\ Action: A = \{vrot, vtras\} \end{cases}$$

Propose joint distribution decomposition of *AgentInformation*

$$\{P(G,B,S,C,P,A) = P(G) \times P(B \mid G,S) \times P(S \mid B,C,A,P) \times P(A \mid S,P) \times P(C \mid S) \times P(P)$$

$\forall i, i \in AgentInformation$, Define the 'utility' of each variable

$$\begin{cases} G,C,P = \{\varnothing\} \\ B = \{P(B \mid G,S) = P(B \mid B_{t-1})\}; \\ S = \{P(S \mid B,C,A,P) = P(S \mid B,P)\} \\ A = \{P(A \mid P) = \sum_B P(A,B \mid P) = \sum_B (P(B \mid P) \times P(A \mid P,B)) \end{cases}$$

Use

1) Ask actual behavior with: $P(B \mid B_{t-1})$
2) Determine the action to develop: $\sum_B (P(B \mid P) \times P(A \mid P,B))$
3) Redefine previous state with: $P(B_{t-1} \mid B,P)$

Statistic learn of $P(A \mid P,B)$

Fig. 3. Simple Bayesian Agent for obstacle avoidance and phototaxy behavior

Next, the use of each question must be defined. We first obtain the distribution of the actual behavior given the previous one. This information is not used in this simple example but, as commented before, could be used to communicate other agents the influence of both behaviors in a specific place. Then, we obtain the action to be sent to the robot given the influence of each behavior (determined by perception variable). Finally we review the previous state regards the actual state and perception. This is a smooth Bayesian filter that uses the perception and the actual state to recalculate the probability of being in the previous state.

All these distributions and decompositions were defined by the programmer (with their knowledge about the problem). Therefore, until know, we provide a fixed design that not accomplish epigenesis design principle. But even in this simple example we will see how we can achieve different behaviors using learning with the same agent structure.

We state previously that we need to solve $\sum_B (P(B \mid P) \times P(A \mid P,B))$. The first term is easy defined using a table were each perception defines the intensity of each behavior (if we are close to an obstacle *AvoidObstacles* will have much more probability than in empty space; in the same way if we see a brighter light we will go towards it with more intensity than if the light is smooth). But, how can we obtain which action to perform given the actual perception and behavior? How can we mix both behaviors? There is an easy solution to solve both problems: learning.

One way for learning this distribution is to control the robot using a joystick. When we control the robot performing phototaxy and collision avoidance behaviors we can collect for each perception which action is preferred. In this probabilistic framework we can learn a Gaussian distribution for each pair perception/action. As they are conditional independent, we can learn in separate ways the phototaxy behavior and the obstacle avoidance behavior to specify the distribution $P(A \mid P,B)$.

Even in this simple example, this model provides some advantages if we compare it with other Bayesian Agents systems: the programmer specifies the way the information is used but do not fix the system for doing a specific task. Although we learn phototaxy behavior, we could easy train the system to perform a different action such avoid light without changing a line of code (only we must train the robot moving it with the joystick and performing a light avoid behavior). Using this model emergence and epigenesis is supported by the system in a natural way. Obviously, a more complex multi agent system will provide a way to evaluate how using the proposed agents will support emergence, parallel processes and communication, but this is out of the scope of this paper.

5 Conclusions and Future Lines

In this paper an agent architecture that joins intelligent agents and Bayesian programming is presented. This architecture provides a common framework for design intelligent agents that use Bayesian inference. Agents defined with this model could work with uncertain and incomplete information in a well-established mathematic framework. In addition this architecture also support the most important design issues for intelligent autonomous agents.

Moreover we compared our agent architecture with other Bayesian agents and find no other model that provides a general framework for specifying Bayesian agents not only for robotic programming but also for general cognitive agents. This architecture provides join probabilistic distributions as common exchanging uncertainty assessments in multi-agent systems.

Furthermore, in this paper a robotic application has been provided and modeled as a proof of concept of our agent architecture.

Future lines will be related to analyze specific agent decompositions that take into account Bayesian inference method for reducing computational requirements. We are also modeling a robotic multiagent system using Bayesian agents where coordination and emergence are our investigation priorities.

References

1. Thrun, S.: Towards Programming Tools for Robots that Integrate Probabilistic Computation and Learning ICRA, pp. 306–312 (2000)
2. Park, S., Pfenning, F., Thrun, S.: A probabilistic language based upon sampling functions SIGPLAN, vol. 40, pp. 171–182. ACM, New York (2005)
3. Lebeltel, O., Bessière, P., Diard, J., Mazer, E.: Bayesian Robot Programming Auton. Robots, vol. 16, pp. 49–79. Kluwer Academic Publishers, Dordrecht (2004)
4. Chalkiadakis, G., Boutilier, C.: Coordination in multiagent reinforcement learning: a Bayesian approach. In: AAMAS 2003: Proceedings of the second international joint conference on Autonomous agents and multiagent systems, pp. 709–716. ACM, New York (2003)

5. Chalkiadakis, G., Boutilier, C.: Bayesian Reinforcement Learning for Coalition Formation under Uncertainty. In: AAMAS 2004: Proceedings of the Third International Joint Conference on Autonomous Agents and Multiagent Systems, pp. 1090–1097. IEEE Computer Society, Los Alamitos (2005)
6. Chalkiadakis, G., Markakis, E., Boutilier, C.: Coalition formation under uncertainty: bargaining equilibria and the Bayesian core stability concept. In: AAMAS 2007: Proceedings of the 6th international joint conference on Autonomous agents and multiagent systems, pp. 1–8. ACM, New York (2007)
7. Ueno, M., Okamoto, T.: Bayesian Agent in e-Learning ICALT, pp. 282–284 (2007)
8. Santos, E.R., Fagundes, M.S., Vicari, R.M.: An ontology-based approach to interoperability for Bayesian agents. In: AAMAS 2007: Proceedings of the 6th international joint conference on Autonomous agents and multiagent systems, pp. 1–3. ACM, New York (2007)
9. Steels, L., Kaplan, F., McIntyre, A., Looveren, J.V.: Crucial factors in the origins of word-meaning. In: Wray, A. (ed.) The Transition to Language. Oxford University Press, Oxford (2002)
10. Florian, R.V.: Autonomous artificial intelligent agents Center of Cognitive and Neural Studies (CONEURAL) (2003), `http://www.coneural.org/reports/Coneural-03-01.pdf`

A Motivation-Based Self-organization Approach

Candelaria Sansores[1] and Juan Pavón[2]

[1] Dep. Ciencias Básicas e Ingenierías, Universidad del Caribe, Cancún
SM. 78, MZ. 1, Lote 1, 77528, Cancún, Q.Roo, México
csansores@ucaribe.edu.mx
[2] Dep. Ingeniería del Software e Inteligencia Artificial, Universidad Complutense Madrid
Ciudad Universitaria, s/n, 28040, Madrid, Spain
jpavon@fdi.ucm.es

Abstract. In this paper we propose a self-organizing mechanism to model adaptive multi-agent systems. The mechanism is inspired in a psychological approach of motivation. Under this approach motivation is an internal state or condition that serves to activate or energize behavior and give it direction. Thus, we propose an intrinsic motivation or reason for engaging in a particular behavior called goal-achievement, in this way agents will have a drive to reach a clearly defined state. To increase the strength of this particular behavior we propose a reinforcement mechanism based on agents past experience as a feedback. The intention of this mechanism is to mimic the adaptation exhibited by entities of complex systems composed of multiple autonomous entities that make local decisions leading to a global organized behavior for the system. Such organization is reached autonomously in different ways, thus adapting to task and environment.

Keywords: Multi-agent systems (MAS), Self-organizing systems, Complex Adaptive Systems (CAS), Agent-based Modeling, INGENIAS Methodology.

1 Introduction

A self-organizing system consists of multiple autonomous components that make local decisions leading to a global coordinated behavior for the system without an external direction. Such systems can organize autonomously in different ways, thus adapting to task and environment. This adaptation is a dynamic process within the system due to its own interest to cope with environmental changes, maintaining in this way its internal organization autonomously. In consequence, these systems emphasize nonlinear interactivity and its internal organization cannot be determined merely as a sum of their elements [5].

Self-organization in multi-agent systems (MAS) has been proposed to design and implement artificial self-organizing social entities in order to provide insights of complex systems and to develop applications to solve complex problems [2]. Artificial self-organizing mechanisms come from different disciplines, for example, the stigmergy theory from the natural systems studies has inspired a coordination mechanism, and the AMAS theory [3] from the MAS discipline propose an approach based on cooperation, the former focuses on indirect communications through the environment as a mean to coordinate a collection of agents and the latter focuses on engineering microscopic aspects, i.e. the agents' cooperative behavior to achieve a global

J.M. Corchado et al. (Eds.): DCAI 2008, ASC 50, pp. 259–268, 2009.
springerlink.com

function that meets the systems' requirements. Both mechanisms deal with social aspects, as coordinated and cooperative interactions, and with the adaptation of the MAS to the environment.

In this paper we propose a self-organization approach that deals not only with social aspects but with cognitive too. The proposal is inspired in a psychological approach of *motivation* [4]. Under this approach motivation is an internal state or condition that serves to activate or energize behavior and give it direction. Thus, we propose an *intrinsic motivation* or reason for engaging in a particular behavior called *goal-achievement*, in this way agents will have a drive to reach a clearly defined state. For this, we propose an agent's architecture based on goal dynamics according to the theory of social action [1]. Under this theory goals are mental constructs which have a life cycle. We use this perspective to include a motivation state to these goals which value increases or decreases agents' behavior for pursuing those goals. Depending on the motivation strength (or intensity) agents will exhibit roles specialization which will guide the agents' adaptive behavior dynamically. The approach is complemented with a reinforcement mechanism that increases a given behavior. The proposal was designed and implemented for the INGENIAS MAS methodology [7] since its agent model supports quite well most of the social and cognitive aspects of this approach.

This paper is structured as follows. Section 2 introduces the self-organization approach. Section 3 presents how the model is instantiated in the INGENIAS MAS modeling language as well as its usage. Finally, Section 4 presents the conclusions and future work.

2 A Motivation Based Self-organization Approach

Self-organization proposed by this model is founded on the ability of the agents to change dynamically their behavior according to some reinforcement. This reinforcement comes from a positive or negative feedback as a property of self-organizing systems [5]. In [6] feedback is seen as a reward an agent receives from its actions. Alike [6], in our approach we do not use rewards; instead we propose a goal-based motivation and its reinforcer. The latter is based on agents' interactions feedback that increase or decrease agents' motivations strength. Also, as in [6] agents do learn an action policy but they do not try to maximize a reward, instead this policy guides their future activities. A good experience or previous positive interactions increment a determined agent behavior and negative ones or bad experiences decrement it.

Instead of an action policy (a selection of an action or task depending on the feedback) we propose a behavior policy. This means that agents are able to select dynamically a behavior represented as roles, that is, certain functionality an agent is responsible of, and which could imply more than one task. Thus, the consequence of motivation strength reinforcement is that agents adapt their capabilities through roles specialization.

2.1 The Agent Architecture Model

The previous mechanism for the adaptive behavior of individual agents is modeled as part of an agent's architecture. This is a deliberative architecture based on the goals an agent wants to achieve, the tasks it can perform to satisfy its goals and the roles it knows to play. Under this architecture, the decision to execute a given task or play a given role depends on the actual goal the agent is pursuing. If the agent drops this goal and activates another one, then it could happen to change its behavior dynamically. However, we provide another option here based on goal dynamics as in [1]. A goal as in [1] is a mental representation which has the potential to constrain the behavior of an agent towards its realization, whether or not this constraint is actually activated depends on the agent's beliefs. This is called belief-based goal processing and depending of the appropriate belief a goal may be activated, promoted, drop, suspended, etc. In this sense, we propose in this architecture the definition of a particular belief called motivation. Figure 1 illustrates all the elements of this self-organization approach.

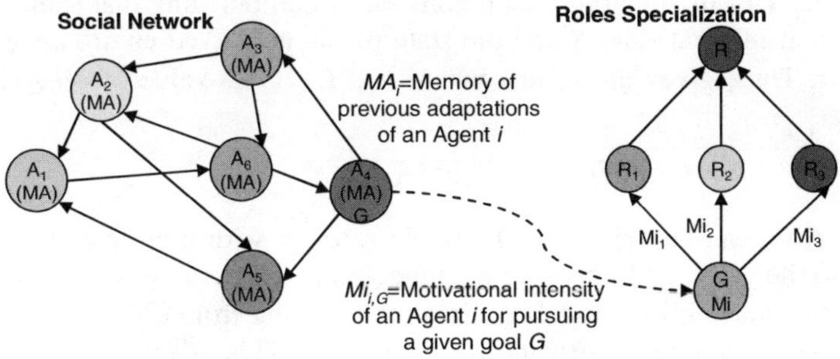

Fig. 1. Elements of the self-organizing model

The purpose of a motivation is that an agent pursues a given goal with a weaker or stronger intensity, and according to this intensity will be the specialized role the agent will play. In this way, the behavior of an agent can be adaptive even when pursuing a unique goal. Finally, the variation of strength of an agent's motivation will be achieved through the feedback of the agent's own past experience.

Basically, the mechanism is formed by: 1) a belief called motivation, endowing goals with a dynamic intensity for being pursued (*Mi* in Figure 1) 2) the behavior dynamic selection, achieved through roles specialization and according to a motivation for pursuing a given goal (on the bottom of Figure 1 we can observe how depending on *Mi* an agent plays a specialized role inherited from role *R*) and 3) the update of the motivation's intensity through a reinforcement mechanism that takes into consideration the agent's experience (modeled like a social network of previous adaptations in Figure 1), the state of the perceived environment and the current state of the agent.

2.2 The Reinforcement Mechanism

The reinforcement mechanism for dynamic adaptive behavior consists of the positive or negative feedback an agent receives from their own interactions with other agents. In the system, agents act and form an interaction network called social network. This network provides agents with a memory of adaptations or MA. The formation logic of this network will depend on the specific application domain. This network represents the experience of a particular agent or the positive or negative impact an agent can have over another agent during their interactions. Keeping this experience in its memory an agent will be able to reinforce or diminish its future behavior. The formation rules of the network are specified by the designer or domain expert who has knowledge about the system at design time, however, the evolution of the MA permits an adaptive behavior over time.

In Figure 1 we can observe that *MA* feedbacks the value of *Mi* associated with the goal an agent is pursuing. This affectation is performed dynamically during runtime. In accordance to *Mi* an agent will choose a strategy or role according to the mapping from different motivation values *Mi* to roles (we will see further on how this association is established). As a result, the behavior of the agent is adaptive depending on its motivation for pursuing a given goal. Consequently, the intensity of a motivation *Mi* of an agent *i* for pursuing a goal *G* in a given time *t* is defined as a function of: the agent's memory of its previous adaptations *MA* when pursuing that same goal *G*, the agent's current internal state *S* and the state of the perceived environment S_E in that same time *t*. Thus, given an agent *i* where $i \in 1 \ldots n$, the value of $Mi_{i,G,t}$ is given by function 1:

$$Mi_{i,G,t} = f_i\left(MA_{i,G,t}, S_{i,t}, S_{E,t}\right) \quad (1)$$

This function can be defined with simple rules or with more complex algorithms according to the problem domain we are modeling.

Finally, the adaptability model requires the mapping from different agent's behaviors to its corresponding motivations' intensities. This mapping is established at design time, even though an agent starting from its initial state can adopt those behaviors dynamically during its execution. This fact is thanks to the variation of the intensity of its motivations. The motivation-roles mapping is illustrated in Figure 2. In this figure roles are represented as nodes and the mappings from motivation to roles as the links of a graph. This graph is called *motivation-based behavior graph*. From this graph an agent can select a new behavior when it needs to adapt its current behavior. Although the mappings could be a simple motivation-role pair association, they are thought to be conditional entities. These entities could have parameters like agents' properties and those of their perceived environment. Since those properties also change during runtime, dynamic role selection is assured not only because motivation change but also as a function of the conditional entities' parameters.

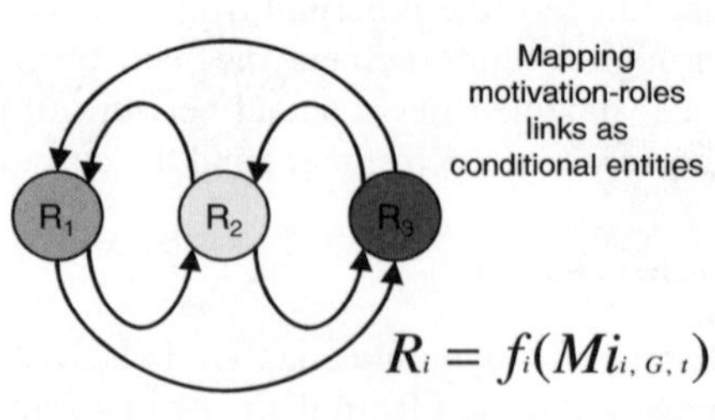

Fig. 2. Motivation-based behavior selection graph

Therefore, in Figure 2 we can observe that roles selection is represented as a function of *Mi*. Thus, given an agent *i* where $i \in 1...n$, the role *R* the agent will play is given by the function 2:

$$R_i = f_i(Mi_{i,G,t}) \tag{2}$$

This implies that the role selection mechanism can be implemented more adequately for each application domain defining this function to fit the adaptability requirements.

3 Implementing the Approach

Social systems are highly dynamic and complex. In particular, we are interested in observing the emergent behavior that results from the interactions of social individuals as a way to discover and analyze the construction and evolution of social patterns. The main reasons to choose INGENIAS as the implementation framework for this approach is that it supports well the specification of agent intentional behavior and organizational relationships, characteristics that are present in self-organizing social systems.

3.1 INGENIAS MAS Modeling Language

INGENIAS is a methodology for the development of multi-agent systems (MAS). Its development tools rely on its MAS modeling language, which is specified with a meta-modeling language, MOF (Meta-Object Facility), a standard by OMG. The language is structured in five packages that represent the viewpoints from which a MAS can be regarded: Organization, Agent, Goals-Tasks, Interactions, and Environment. The agent viewpoint describes the agent's behavior. It is determined by the agent mental state, a set of goals and beliefs as well as the roles it is able to play.

Also, an agent has a mental state processor, which allows the agent to decide which task to perform, and a mental state manager to create, modify and delete mental state entities. The goals-tasks viewpoint describes the relationship between goals and task, since agents are intentional entities; they act as they pursue some goals. Hence, it is possible to identify individual goals for agents, which could be refined into simpler goals up to a level where it is possible to identify specific tasks to satisfy them.

3.2 Extending the Language Meta-models

The INGENIAS MAS meta-models were extended to introduce new concepts envisaged for the adaptive agent architecture and for the reinforcement mechanism.

3.2.1 Goals-Tasks Meta-model

This meta-model was modified to include motivation *intensity* as a property of the goal entity. In this way, goals can be modeled defining a metaphoric intensity property for being pursued; this extension corresponds to the *Mi* belief of the proposed model. Figure 3 illustrates an extract of this meta-model. We can observe that there exits several types of relationships or associations between entities, for example a task

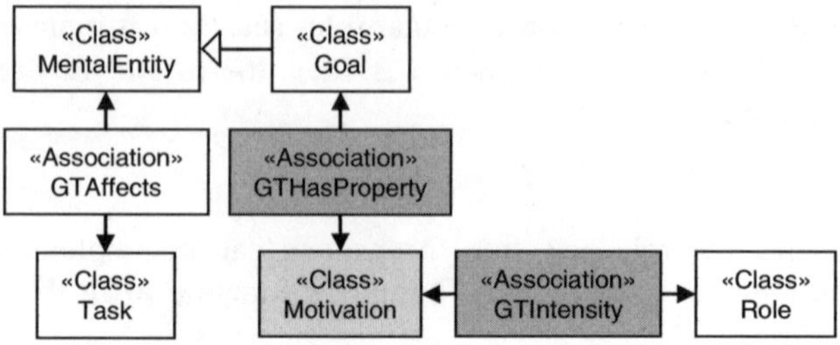

Fig. 3. Extended Goals-Tasks meta-model

can affect (*GTAffects*) the entities of an agent's mental state. Following this same criterion, we extended the meta-model as shown in Figure 3 with three new entities.

We included a class *Motivation* and two types of associations *GTHasProperty* and *GTIntensity*. *GTHasProperty* indicates that a goal has a property called *Motivation* and *GTIntensity* is an association between the class *Motivation* and the existing class *Role* to model the different possible mappings from motivation intensities to roles. Associations in INGENIAS can be decorated with a mental state pattern to indicate under which conditions they can exist. We took advantage of this characteristic to provide the possibility to include and edit association rules between motivations and roles to define the behavior mapping of the self-organizing model.

3.2.2 Organization Meta-model

This meta-model was extended to include a *social network* entity. A social network is a social dynamic structure constituted of nodes. These nodes are generally individuals or organizations. The network indicates the way the nodes are connected dynamically through relationships. This entity was included extending the *Group* concept of this meta-model. The fact that an agent belongs to this network group means that it is able to engage in relationships with other agents as well as other agents can create relationships with it.

The main purpose of this entity is to allow the designer to represent a network of relationships among the agents and define the formation rules, that is, whether an interaction is considered important for the feedback or not, if so, how to include it in the network. How a specific past experience or *MA* feedbacks positively or negatively an agent's motivation is dependent on the specific application and is described in terms

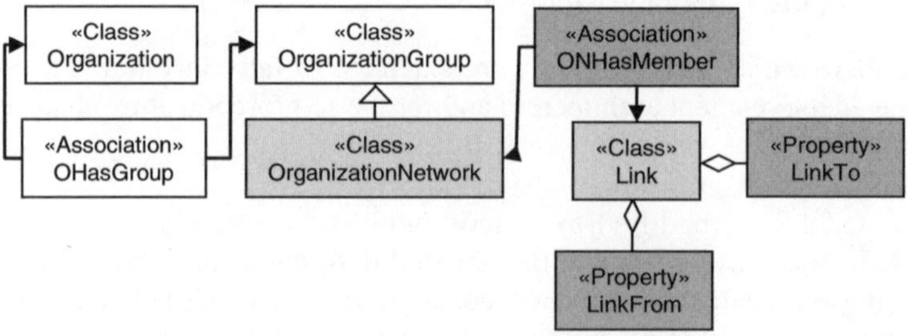

Fig. 4. Extended Organization meta-model

of the agents' interactions. Specifically, those subjectively considered bad or good experiences, and the evolution of their mental states. Figure 4 shows an extract of the organization meta-model together with the corresponding extensions.

In this diagram we can observe that an *organization* class is structured in *organizational group* classes through *OHASGroup* associations. Figure 4 also includes five new extensions to the meta-model. These extensions are: an *OrganizationNetwork* class, a *Link* class, the *LinkFrom* and *LinkTo* properties and the association *ONHasMember*. The *OrganizationNetwork* class inherits from *OrganizationGroup* class, so a network is practically a group where agents belonging to it are provided with a *Node* property, meaning they are members of the social network. Additionally, an *OrganizationNetwork* class can have *Link* members included through a new association called *ONHasMember*. *Link* entities also contain pointers to the nodes that they are to and from. These pointers are specified with the *LinkTo* and *LinkFrom* properties.

3.2.3 MAS Viewpoints

Finally, once the agents have learnt a behavior policy, they apply it to select dynamically a role to play. For this to happen, it is necessary to define the motivation-roles mapping, either as a simple motivation-role pair or as a function of *Mi*. It was necessary to define a new viewpoint called *Goals-Roles viewpoint*. In it we represent how an agent pursues a given goal and how the motivation's intensities are related with the roles an agent is able to play.

3.3 Applying the Self-organization Approach

The self-organization mechanism was implemented in an agent-based simulation model described in [8]. The model studies altruism among smart entities in a society. In this model, agents are endowed with social knowledge consisting of a memory of past interactions (e.g. food-sharing). Based on this memory, a small number of rules are defined for the modification of the values of agents' motivational intensity for pursuing goals. If an agent receives help, the force of the altruistic motivation or positive feedback increases, whereas it decreases if help is denied, and consequently resulting in a negative feedback. Figure 5 illustrates part of this model in an *organization viewpoint*.

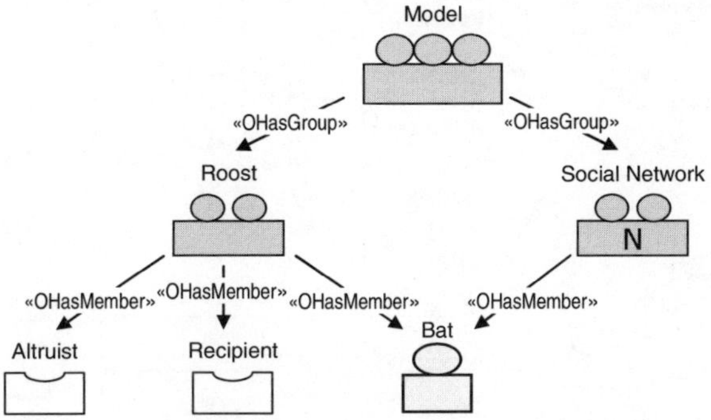

Fig. 5. The organization viewpoint of the self-organizing model

It simply structures the *model* in *roosts* and a *social network*. In INGENIAS the initial state of an agent (goals, beliefs, facts, etc.) as well as their responsibilities (tasks) and capabilities (roles) are described within the *agent viewpoint*. The evolution of the agents' mental state can be described too with mental state patterns using this viewpoint. This means that the mental states are modified by operations defined in these patterns. Figure 6 illustrates one of these patterns.

This pattern specifies the mental state that a *concrete agent* should have to trigger a positive feedback. From Figure 6 we can deduce that any agent of the system having these mental entities, that is, having the fact "*help_received*" when pursuing goal

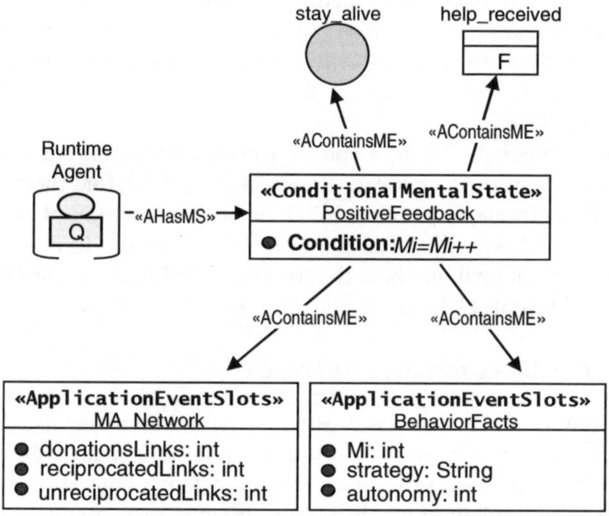

Fig. 6. A conditional mental state pattern

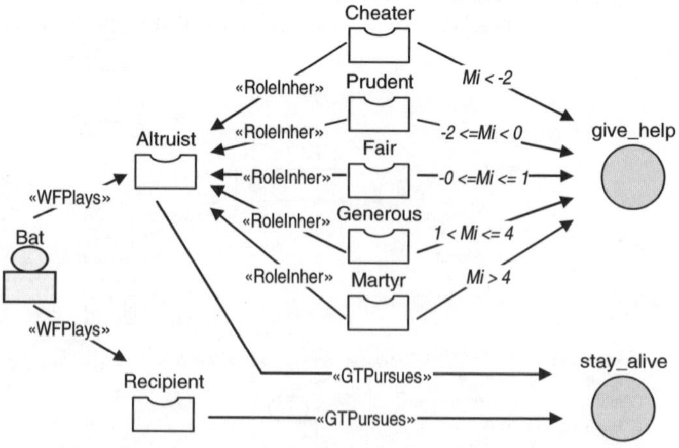

Fig. 7. The goals-roles viewpoint of the self-organizing model

"*stay_alive*", will receive a positive feedback. As a consequence, their motivations (*Mi*) for being altruists will increase. Figure 6 also illustrates that agents have mental entities that represent their memory of past experiences (*MA*).

Finally, Figure 7 depicts the motivation-roles mapping (in the form of simple rules) in the *goals-roles viewpoint*. This viewpoint is basically a diagram describing the motivation-based behavior. In this case, there is a major role of *altruist* which pursues goals *stay_alive* and *give_help*. Depending of the motivation intensity of the giving help goal (expressed with the *Mi* property), agents can play an inherited role of the altruist one: *cheater*, *prudent*, *fair*, *generous*, and *martyr*.

4 Conclusions

In this paper we reviewed the specification and implementation of a motivation-based self-organization approach using social and cognitive aspects to model autonomous adaptive agents. Even though the approach is based on the motivation an entity has to reach a goal, it demonstrated to be easily configurable to model a less-apparent reason such as altruism, for example, playing with the reinforcer and motivation intensity values.

The approach was implemented in the INGENIAS MAS modeling language to test its viability. The flexibility of this language allowed describing the approach easily as shown in this paper. However, not all the elements or concepts of the mechanism are part of the meta-models, leading to ambiguities at design time. Though the language proved to be especially expressive to allow a designer to specify the adaptive behavior of self-organizing MAS systems, the specification turns tedious when the types of agents and goals are numerous. Then, this approach is best oriented to model small societies.

As future work we envisage to explore new motivational theories to enrich the proposal and to include the whole approach as part of the INGENIAS meta-models.

Acknowledgments. This work has been developed with support of the program Grupos UCM-Comunidad de Madrid with grant CCG07-UCM/TIC-2765, the project TIN2005-08501-C03-01, funded by the Spanish Council for Science and Technology, and the Universidad del Caribe, Cancún Q. Roo., trough the program PROMEP, funded by the government of México.

References

1. Castelfranchi, C., Paglieri, F.: The Role of Beliefs in Goal Dynamics: Prolegomena to a Constructive Theory of Intentions. Synthese 155, 237–263 (2007)
2. Di, M., Serugendo, G., Gleizes, M.-P., Karageorgos, A.: Self-Organization and Emergence in MAS: An Overview. Informatica 30, 45–54 (2006)
3. Gleizes, M.-P., Camps, V., Glize, P.: A Theory of Emergent Computation Based on Cooperative Self-Organization for Adaptive Artificial Systems. In: 4th European Congress on Systemic (1999)
4. Geen, R.: Human motivation: A psychological approach. Wadsworth Publishing (1994)

5. Heyligen, F.: The science of self-organisation and adaptivity. In: The Encyclopedia of Life Support Systems, UNESCO Publishing-Eolss Publishers (2002)
6. Maes, P.: Modeling Adaptive Autonomous Agents. In: Langton, C.G., et al. (eds.) Artificial Life. MIT Press, Cambridge (1994)
7. Pavón, J., Gómez-Sanz, J., Fuentes, R.: The INGENIAS Methodology and Tools. In: Agent-Oriented Methodologies. Idea Group Publishing (2005)
8. Sansores, C., Pavón, J., Gómez-Sanz, J.: Visual Modeling for Complex Agent-Based Simulation Systems. In: Sichman, J.S., Antunes, L. (eds.) MABS 2005. LNCS (LNAI), vol. 3891, pp. 174–189. Springer, Heidelberg (2006)

Using Techniques Based on Natural Language in the Development Process of Multiagent Systems

Juan Carlos González Moreno and Luis Vázquez López

Escuela Superior de Ingeniería Informática
Departamento Informática
Universidad de Vigo
As Lagoas s/n, 32004, Ourense, Spain
jcmoreno@uvigo.es, eldelima@terra.es

Abstract. The last years have been very prolific in the study of methods, methodologies and even development process in the area of Agent Oriented Software Engineering. However, there are two aspects that are not developed sufficiently, the first is the acquisition and establishment of requirements and goals for Multiagent Systems from the descriptions of customers / users. The second is a successful implementation of reverse engineering to implementations undertaken in order to retrieve a detailed design of the application obtained. The concern of this work focuses on the first of these aspects. The proposal applies techniques of natural language processing, throughout the entire lifecycle of multi-system, to obtain a model and improve product required. One noteworthy point of this proposal is that the client is actively involved in the development process in a manner non-attendance. Lastly, the proposal allows maximize obtaining initial models for different projects, minimizing their involvement in this process.

Keywords: Natural Language, Initial Models, Multiagent Systems, Agent Oriented Software Engineering.

1 Introduction

As for the implementation of an information system, its realization is always an expensive task that involves a lot of effort. Thus, the presence of environments automated software provides a great advantage when it comes to deal with this work, because it represents a significant savings in both material and temporal.

INGENIAS [7] is a recent methodological proposal for the development of applications based on Multi Agents Systems.

In order to generate code automatically specification on starting out, you have to integrate all the features of the organization and not to be ambiguous.

To avoid the need for manual development models that develop a solution to the problem, is that the proposed system through a very active involvement by the client avoids a significant participation of the analyst.

It proposes an iterative process with the customer / user, which can describe this in writing and in natural language (in his own words) the problem we want it solved. The various iterations (do not need the involvement of the analyst) actively engage customers in the development process (can be made against the system by the customer from any location) and are accurate to better understand the domain of the application, the

J.M. Corchado et al. (Eds.): DCAI 2008, ASC 50, pp. 269–273, 2009.
springerlink.com

problematic and also to disembogues vague descriptions or impersonal. At each iteration the text introduced undergoes an analysis and subsequent transformation process leading to the award of problem a model based on INGENIAS.

The rest of the article has been structured as follows: The second section is devoted the proposed model. The following section presents our tool proposed. It continued section 4 makes a series of conclusions and future work, and lastly the classical sections of acknowledgements and references.

2 Our Proposal

The system raised multiagent is going to arise on the basis of the aims of the system and the first organization of the system in agents / roles. The aims are associated with cases of use described by means of scenes in those who develop individual tasks and interactions among the identified agents / identified roles.

As the description initial other that raises us the client / user to solve the problem is going to realize it in natural language not to force this one to have to be able to interpret the methodology of development, which in many cases would be so or more complex than the solution to the raised problem. Doing it in natural language, raises as principal disadvantage the possibility of that present ambiguity that is going to be solved by the involvement of an active way on the part of the client / user in the different phases of the process of product development, with this there are also going to avoid possible forgotten of characteristics that it must have the product or in coherences of what must have the end product developed with regard to the first realized description the organization model.

In the process of acquisition it differs between the aims of the System that indicate what is tried to obtain by the development, and the aims of an Agent who is what has, needs or this agent wants to reach to satisfy or to help to satisfy an aim of the system. As for the roles they are characterized as the set of actions that determine a distinctive behaviour of an agent. When multiagent refines [1] the graph of the system the roles they are constructed in cases of use that are in use for solving the aims indicated by the role. If the aims to resolving are a complex at the time the subsystems can be used as elements of refinement. On the other hand, the Roles realize Tasks that cover Activities and that satisfy the aims to resolving. In general the identification of all these elements of shaped comes from interviews, between the client and members of the equipment of development (analysts, engineers of requirements...) that remain reflected as descriptions in natural language in most of the cases.

In the process of identification of elements of shaped of the system multiagent from a description of text in Spanish the following characteristics are born in mind: The agents [5] are going to be extracted from the own names, the objects and the subjects of the passive verbs of the description.

The cases of use are going to correspond with the lexical verbs, due to the fact that the verb of the prayers that defines a case of use must be transitive. An idea similar to it, is the used in marked automatically of texts raised in [2].

In [4] plus information of automatic model generation for multi-agent systems.

3 Our Tool

Our shaped architectural it goes from the Model NLP4INGENIAS presented in [5] and the design of multiagent system architecture [3] whom might be summarized of the following way: Consisting of a morphologic analysis, a parsing and the alteration of the text in case of some mistake be detecting during the morphologic or syntactic analysis. Resting on an approach example STILUS. Due to this alteration sometimes the collaboration of the user is necessary. Finished this part of the process is obtained for each of the tried words: his type, his motto and his function inside the phrase that can store in format table.

Now already it is in conditions to do the process of extraction of the entities of the system consistent multiagent in:

The compound phrases are simplified using a few own rules of production. This process consists of being seeking for the verbs of request and to leave every verb with the corresponding complements as if it had been enunciated as a simple request. Another possibility that is being studied at this moment is the raised one in [6] by means of the marked one with the words in the own text.

The aims are identified following the criterion mentioned before: "the verbs are looked in gerund or participle and the infinitive verbs in phrases desiderative and conditional". It is possible to see more information since the aims of a description obtained in "Deriving Operational Software Specifications from System Goals".

It is looked the lexical verbs that determine the cases of use of the system. These verbs are those which are in indicative and that do not form a part of a verbal periphrasis.

Later there are identified the own names, the objects and passive subjects of the phrases that the agents / agents of the system determine. The roles associate to the subjects of adjectives or qualified names.

Transformation of the description of transparent form for the user to controlled language. [8]

Now he is already in conditions to indicate which is going to be the agents architecture that it is going to use for our system for the generation of the first phases of the process of the development of the cycle of life of the product requested by the client/user.

Applying the tool are extracted different entities to maintain the order of appearance. To verify the existence of entities or contradictory repeated scans synonyms and antonyms comparing the respective results.

So would the following:

Dealers heterogeneous form the Organization.

The agents form the homogenous group.

Adverbs tell us circumstances test which will be used to make test cases.

The common names not used as agents will be resources (If you are passive. Ni.: Drink) or Applications (Actions associated with active shares).

The names may have the following features: Role, Agent, and Appeal Application.

Verbs will lead to: Objectives, Case studies, Other (applications or tasks).

The DTD is the specification of this type:

<specification> <agent> ... + </agent> <objectives> .. +. </objectives>
<roles> ... + </roles> <usecase> ... + </usecase> <activities> ... + </activities>
<tasks> ... + </tasks> </specification>

Therefore the XML document used in the proposed model will be as follows. The model in XML integrating the offer in the IDK the INGENIAS.

Following the conclusion of the process ensures that, at least, for every paragraph of the description of the problem has been identified a goal and an instance of use associated with a role or an agent or an outside service that will be linked to an application / appeal.

All descriptions made by the customer / user to spend documenting goals and / or use cases identified, as well as the roles and agents identified.

4 Conclusions and Future Work

This paper presents a process of acquisition requirements of a problem posed by the user through a natural language description for the Spanish to get from this initial diagrams of the system to automatically develop and then carry them INGENIAS; its operation is based on the discovery of targets and agents / roles. The parts of a process of linguistic processing you get a lot of information. Because much of the information is redundant I know changing the text entered by the user to a controlled manner (in a manner transparent to the user).

A work in progress is the development of a knowledge base itself of the domains resolve to avoid, as far as possible, conducting much of the analysis in cases where domains are introduced and studied in specifications or similar as to its functionality.

The development of automatic software eliminates many of the traditional stages of the software development processes, there earns in time and efficiency.

Another useful tool to exploit is to use it to verify the design obtained. In this way we could check if they have created all the elements. Years ago that Visual Paradigm invites the analyst to see diagrams use cases with a semi-automatic system.

Acknowledgements. This work has been supported by the following project: Methods and tools for agent-based modelling supported by Spanish Council for Science and Technology with grant TIN2005-08501-C03-03.

References

1. Breaux, T., Vail, M., Antón, A.: Towards Regulatory Compliance: Extracting Rights and Obligations to Align Requirements with Regulations. In: Proc. of RE 2006, Washington, DC, USA, pp. 46–55. IEEE Comp. Society, Los Alamitos (2006)
2. Chitchyan, R., Rashid, A., Rayson, P.: Semantics-based Composition for Aspect-Oriented Requirements Engineering. In: International Conference on Aspect-Oriented Software Development (AOSD), Vancouver, Canada. ACM, New York (2007)
3. González-Moreno, J.C., Vázquez, L.: Design of Multiagent System Architecture. In: 3rd IEEE International Workshop on Engineering Semantic Agent Systems (ESAS 2008), COMPSAC 2008, Julio, Turku, Finland (2008)
4. González-Moreno, J.C., Vázquez, L.: Generación Automática del Modelo para Sistemas Multi-Agente" 3ª Conferencia Ibérica de Sistemas y Tecnologías de Información. In: CISTI 2008, Junio, Ourense, Spain (2008)

5. González-Moreno, J.C., Vázquez, L.: NLP4INGENIAS: Adquisición de Requisitos en INGENIAS a partir de descripciones en Lenguaje Natural. In: DESMA 2007 – CEDI 2007, Zaragoza, Spain, pp. 978–984 (2007) ISBN: 978-84-9732-613-1
6. Kiyavitskaya, N., Zeni, N., Breaux, T., Antón, A., Cordy, J., Mich, L., Mylopoulos, J.: Extracting rights and obligations from regulations: toward a tool-supported process. In: Proceedings of the twenty-second IEEE/ACM International Conference on Automated Software Engineering, Atlanta, USA, pp. 429–432 (2007) ISBN: 978-1-59593-882-4
7. Pavón, J., Gómez-Sanz, J., Fuentes, R.: The INGENIAS Methodology and Tools. In: Henderson-Sellers, B., Giorgini, P. (eds.) Agent-Oriented Methodologies, ch. IX, pp. 236–276. Idea Group Publishing (2005)
8. Zapata, C.M., Gelbukh, A., Arango, F.: UN-Lencep: Obtención Automática de Diagramas UML a partir de un Lenguaje Controlado. In: Avances en la Ciencia de la Computación. VII Encuentro de Computación ENC (2006) ISBN 968-5733-06-6

Semantic Overlay Networks for Social Recommendation in P2P

Alberto García-Sola and Juan A. Botía

Universidad de Murcia
agarciasola@um.es, juanbot@um.es

Summary. In P2P systems, nodes typically connect to a small set of random peers to query them, and they propagate those queries along their own connections. To improve that mechanism, semantic overlay networks influence those connections depending on the content of the peers, clustering peers in overlapped groups (Semantic Overlay Networks). We propose using ontologies for describing semantic information of both, the shared items and the user profile. Once the peers are grouped by semantic information, we can take advantage of that distribution to add some new functionalities as recommendation, using ontologies comparison and the same principle to create SONs: the users with similar files to us, will probably have files that we are interested in.

1 Introduction

On the last few years, there has been increasing sharing activity of large volume data objects such as video and software. Almost all this activity is located at P2P networks. It must be noticed that, despite current P2P systems include some social characteristics such as instant messaging, they were mainly designed to distribute files. But, in fact, and due to their social nature, they are not limited to file sharing. Shared data in P2P environments often has an implicit structured data model (i.e. an ontology) [6], due to its origin and relations to real world concepts. In this paper, the research focuses on taking some benefit from this two simple but yet important facts. And we work at the overlay structure level. More specifically, we work at the Semantic Overlay Network (SON) level. A discussion about SON can be found at [6]. The main idea behind SON consists on creation and management of a flexible network organization, improving query performance based on the semantic relations among peers. In order to do that, the notion of cluster of peers is used. Peers arrange into clusters according to the content they share. Clusters may overlap, because peers can contain different content and belong to several clusters. In this context, queries are distributed to relevant clusters only and flooded among relevant peers and cluster, and peers irrelevant to query do not receive any message, reducing unnecessary traffic while making queries quicker.

Working with SON implies taking into account a number of factors. The first one is **Classification hierarchies**. This concept refers to the classification that

J.M. Corchado et al. (Eds.): DCAI 2008, ASC 50, pp. 274–283, 2009.
springerlink.com

describes the SONs a peer belongs to. A peer may belong to many SONs, but each item must be indexed just in one SON. The second one is that of **Items and queries classification**. It consists on the mapping between a query or item and a SON. Another concept is **Peer assignment strategy**. It is used to make classification hierarchies assigning the SONs the peer belongs to. A misclassified item should not influence a misclassification of a peer but it can cause a classification in SONs the user is not interested in.

The following requirements are key to be satisfied by a SON. The first one is that SONs must have a **small population of peers**. The smaller the number of peers we need to search, the better the query performance. A classification should not be too much specific, since too few peers would be in the SON, neither too general, in order to avoid a SON hosting the majority of peers in the system. In a good classification peers must have connections with **small number of SONs**, the greater the number of SONs, the greater the cost for a peer to keep track of all of them. The second is that the systems compound by SONs must be **tolerant to classification errors**. There are many sources of errors in items classification, like user wrong classification, fakes and other. Two problems arise with wrong classified items. On the one hand we find the problem of profile creation, and, thus, peer classification in different SONs. A peer can still be correctly classified even if some of its items are misclassified. On the other hand, keyword queries must be classified into a SON, and, if that classification fails, no results will be obtained. This is avoided querying again a more general SON, as explained in section 3.5.

The rest of the paper is structured as follows. Section 2 justifies, through a brief state of the art in P2P systems, that there is a lot of work to do on semantic recommendation in P2P. Section 3 introduces our proposal for supporting it. Finally, section 4 outlines the most important conclusions we extracted and points out open issues we are involved in now.

2 Related Work

Queries routing in P2P networks is one of the important issues in this context. Traditional approaches to focus these problems are random walk, employing summarization, and grouping nodes with similar contents or sharing interests into groups. With respect to this last technique, Crespo and Garcia-Molina [5] introduce the notion of Routing Indices that give a promising "direction" toward relevant documents. The same authors were the first to talk about the concept of Semantic Overlay Networks a few months later in [6, 7]. We can find already there the notion of classification hierarchy but it is not designed in the sense of a modern ontology based on RDF or OWL-DL languages.

Tang and Xu use Semantic Overlay Networks (SONs) to develop pSearch [21], a decentralized non-flooding P2P information retrieval system. pSearch distributes document indices through the P2P network based on document semantics generated by Latent Semantic Indexing (LSI) [9], a technique based on natural language processing.

The work in [2] addresses the problem of building scalable SONs proposing GridVine [8], a semantic overlay infrastructure based on a decentralized access structure supporting the creation of local schemas with global semantic interoperability through pair-wise schema mapping and query reformulation. They use RDF/RDFS to encode meta-data and vocabulary definition. GridVine creates a semantic layer of higher abstraction over P-Grid [1], a self-organizing and fully decentralized access structure based on distributed hash tables (DHT). A similar approach is followed by INGA [14]. In all these works, semantic information is used for query routing and peer classification. We are interested on using semantic information shared by all peers for items recommendation, making possible a more social interaction. Our decision of using the same ontology for all peers is motivated by [12]. They use semantic similarity between a query and the expertise of other peers to select the appropriate peer to ask in queries, using a single ontology to represent every component in the system, remarked as a key concept to achieve good results.

We can find a number of works on semantic recommendation. One of the most relevant is AVATAR system [4]. It is a customized television recommendation system based on semantic information. They highlight the importance of choosing the right ontology. In [15] describe an ontology-based framework that captures information about the users' interaction with content units, in the form of the user feedback, and then use this information to recommend suitable content units to the author. Although none of these works are focused on semantic recommendation in P2P, neither in semantic overlay networks, they establish the basic lines of how we proceed in designing our system.

We propose a P2P recommender system based on semantic. In this sense, Tribler [16] is a social-based P2P system which exploits social phenomena by maintaining social networks and using these in content discovery, content recommendation, and downloading. They use the same concept of recommendation, but, in contrast with our proposal, Tribler do not use semantic information, recommendation is only based in the notion of Karma (i.e. a peer accept recommendation from other peer if the first peer has a good karma on the other peer). The idea is improving recommendation using a "less blind" recommendation based on semantic information. Moreover, interesting social concepts such as collaborative downloading are presented.

Our proposal combines some of the ideas described above using ontologies for semantic in contrast to most of current proposals. The main motivation of our work is the lack of proposals to take advantage of semantic overlay networks beyond improving search. Furthermore, no work has been done in semantic recommendation for P2P systems. So far semantic has been used in classification and assignment to SONs. We go a step forward.

3 Architecture

In this section we present our architecture proposal, depicted in *Figure 1* in the form of an abstract structural diagram.

Peer selection for query routing is a core task in peer-to-peer networks. Semantic overlay networks were born to decrease number of messages in search making use of semantic information to create semantic overlay groups. Then, finding relevant information in a SON can be done in the scope of a subset of peers with semantically similar content to the one we are searching for. Each query must be labeled with the corresponding SON (*Item Classifier in Figure 1*). In order to do this, we propose the use of two ontologies (*NIE Ontology* and *PIMO Ontology*). Please see details in section 3.1. The first one describes semantic information from shared items. The second one describes the user profiles, and hence, their preferences.

As shown in the figure, a peer may belong to more than one SON. Moreover, SONs may overlap. Deciding which SONs the peer belongs is done by the *SON Classifier*, previously described in the introduction. Profile information is filled using semantic information from the user items, and the created profile describes the SONs that the user will be in (see section 3.2 for details). Every peer in a SON shares its public profile with other peers in the SON, so every peer knows about other similar peers and how similar they are to itself (see section 3.3 for details). Then, sharing items list with their semantic information is enough in order to make accurate *recommendation* (see section 3.4). The peer which receives other items information from other peers can compare their semantic information with its own profile to determine how much does the item fit to itself, and from all received items a recommendation is built for the user. Finally, section 3.5 will illustrate how semantic searches are defined in the architecture.

3.1 Ontologies

The common definition of ontology is "a formal, explicit specification of a shared conceptualization" [11]. Ontologies are a powerful tool that still is not widely used in the P2P domain. We propose in this paper some uses of semantic using ontologies, such as *recommendation* or *enhanced queries*. A more consistent evaluation from files and/or user information is possible with ontologies. Due to using well known ontologies that everybody can access to, compatibility and interoperation with other applications is easier and assured. For instance, if a user uses an application which already offer semantic information such as a semantic desktop (e.g. Nepomuk [10]), semantic information will not be needed to be introduced or generated, user profile will be already in the system, and files will have all their semantic information.

Most SON proposals use semantic relations to describe items (e.g. [14]) instead of ontologies. Furthermore, that semantic information is only stored to describe items, but no one describes the user semantically. In contrast with them, we use ontologies, and not just for items, but also to describe user profiles.

Using the same two ontologies on all peers eases tasks which imply knowledge exchange. Information is described in the same way by all peers, letting us to do fast and accurate comparisons from different files and user profiles. However, chosen ontologies should be as expressive as possible and rich enough to do not change over time and let the user express any information regarding a file or his

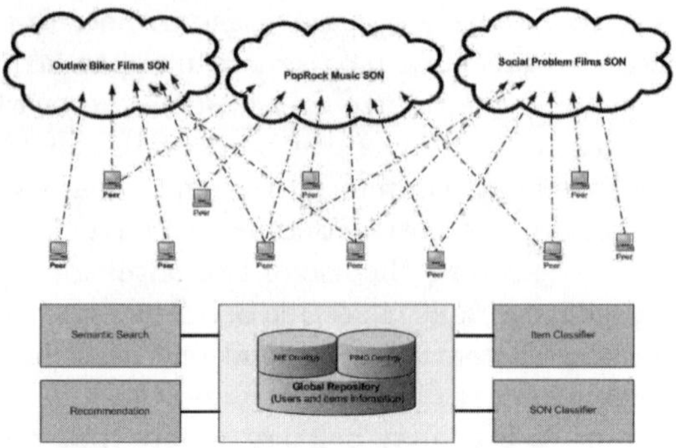

Fig. 1. System Architecture: Semantic Overlay Networks (above) and a single peer internal architecture (below)

own profile, otherwise, we would lose the advantages of using just the two same ontologies on all peers. These two ontologies are described below: In the first ontology (**Item Ontology**), we need to describe, somehow, meta-data about all files the user keep in its computer/share folder. **NIE Ontology** (Nepomuk Information Element Ontology) expresses the representation and content of a piece of data. NIE contains classes and properties for describing items, such as files (Word documents, images, PDFs), address book entries, emails, etc. NIE is based on existing formats for file meta-data such as EXIF for image meta-data, MPEG7 for multimedia annotations, ID3 for music files, iCal, and others. The second ontology (**User Ontology**) is based in **PIMO ontology** [19] from Nepomuk Project. This ontology let us describe the user and its preferences. The core application area of the PIMO is to allow individual persons to express their own mental models in a structured way. The problem of how meta-information is obtained, is the result of a combination of techniques. The first time an item is introduced in the P2P application to be shared, if no semantic information is supplied by the system, some information can be extracted automatically (such as ID3 information from multimedia items). This is done using Beagle++ [13].

Every item is unique and has a hash value which identifies the file, independently on where or when it was introduced in the system. Two files with the same data and hash value can have different semantic information associated to them, but the files are still the same. When a peer receives a file with some meta information different from the one it already has from a peer it trusts, that new information complements the existing one creating a more complete definition of the object: If received information is the same we already have, no action is done; If received information has any contradiction with the current information from the user (e.g. different song names) that information is ruled out and the peer decreases the trust of the other peer. This is done to avoid fakes and getting wrong information; If received information has no contradiction and some complementary concepts (e.g. they have the same artist, but one have not specified the song name and the other has), then, information is merged. This is done getting all information that is more concrete than current information or the peer has not defined. We replace our general information with the new one; If received information from a peer has less information than us (the inverse case from above), we send a message back to the peer letting it know our information. This foster propagation of semantic information.

Creation and update of users' ontology might is done automatically to ease user interaction and create a more consistent ontology with user files. Some work is done in automatically creation of user profile information based on user interaction with the system. However, sometimes the user must introduce manually that information, like done in other project as Nepomuk [10].

In real life, user preferences may change over time. This is more accentuated in P2P applications. The best way to take into account these changes is to make it automatically. To do this, we need a process to combine information from all files (NIE Ontology) to feed PIMO Ontology with the user profile (i.e. ontology populating).

Users will also have the possibility to introduce manually their preferences. This information given by the user is combined with the information retrieved from his files. Manually entered information will have preference over automatically gathered when that information is contradictory.

3.2 Bootstrapping

Finding other peers in a P2P system after first system execution is called bootstrapping. It depends on the P2P protocol. For instance, in the original BitTorrent [17] protocol, peers have to repeatedly connect to a tracker in order to discover other peers, restricting communication to peers within the swarms from files the peer is currently downloading from. Our proposal, independently of the P2P protocol, has some common steps when a peer joins the system:

1. Request for the network hierarchy. This depends on the protocol (e.g. flooding in Gnutella [18] fashion protocols, with a central directory, etc).
2. Based on its local items, the user profile is generated.
3. A peer classifier assigns the peer to specific SONs according to its profile.
4. The peer joins each selected SON by finding peers that belong to those SONs. This would depend again on the protocol (flooding, directory, etc).

From that moment, the peer behaves as described in the rest of the document. It exchanges information from other peers to discover new similar peers and their files, and its database with other peers and files grows continuously.

3.3 Sharing Meta-data

Information shall be shared among peers in a way that they discover similar peers sooner than different peers. In order to do this, we add a new behavior to the P2P protocol. Peers exchange messages with other peers from their peer list (stored in their locally stored global repository), sharing both shared items and known peers. Periodically, a peer requests another peer from their peer list the following information: user profile, information of other peers and information of meta-data about shared files.

User profile: This information lets the user "meet" other peers and compare itself with them to discover how similar they are. Knowledge about similar peers to us is the key concept for recommendation. This measure should not be a global similarity, but rather a specific one. Two peers are similar if they are similar at least in one aspect of their profiles (e.g. the just share the same preference in 80's Hard Rock). This information will not change very often in time.

Other peers' information: This information will help the system to discover new similar peers. Since we are asking a similar peer, probably he will answer us with similar peers to it and us. There are two alternatives. The first option is sending a list with the peers (and their profile) that are closer to our profile. Peers have this list already created, so almost no processing is needed. The other option is to sort the complete list of internal peers according to the requester, and just sending the closest to its profile. This will achieve better results, but will be more CPU demanding. Apart from similar peers, it will also send a list of fresh peers (i.e. alive peers recently discover), in order to motivate exploration. Information about peers includes their user profiles, URIs and Trust (the one from sender).

Shared files with its meta-data: This will help a peer know about other peers files, and hence, recommendation. Each file is tagged with the SON is assigned to. If too many files are shared, only those related to the SON we are asking for will be send.

The age of each peer is included in the message, as well as the last time it checked it was alive. To choose which peer a peer request this information, a peer always maintains in its repository an ordered list based on peers similarity, and take the first non previously chosen. Once a peer is chosen it is not used again for many hours, to aim exploration.

When a peer receives a petition, it enters into a FIFO queue, and periodically the peer responds a request as described above. Requests from peers that have recently requested are ignored unless we have changed our information (our internal database). Requests from peers with low trust are not answered unless no more peers remain in the queue. This avoids overloading the system.

3.4 Recommendation

A peer is subscribed to many SONs, and, from each one, it contains information from peers subscribed to that SON. Our item recommendation proposal is based on the simple, yet powerful principle taken from [20]: if a peer has a particular piece of content that one is interested in, it is very likely that it will have other items that one is interested in as well. This principle is called *interest-based locality*. This is the same idea SONs use to be created. But, once created, other peers from the SONs we are in will probably have similar items to ours, so recommendation can be quite straight forward.

Recommendation is done locally based on previously received information. In the locally stored global repository we have an internal list of peers for each SON, and, for each one, its files with their meta-data for that concrete SON, so, the only step we need to do is choose which files are closer to us. Potentially, every file from a user from a SON we are in may be interesting to us. We need to filter that information to the user, and only display the most interesting according to its profile.

Our propose is as follows, all items from all peers from a SON are chosen, and ordered by four metrics: How much does an item fits to our profile, in semantic terms; How much the profile of the owner of that item fits ours in semantic terms. In the case of more than one peer owning an item, the closer to us will be chosen. Since every peer use the same ontology this step is quite easy; How popular an item is inside that SON. A file that every peer in the SON owns will probably like the user, so should be recommended; Item rating. A low rating for an item means that even if some peers owns that file, they do not like it, so probably we will not like either. Recommendation is an option the user can enable/disable, and is the user the one who choose what to be recommended. If the user chooses automatic recommendation, the more files the user holds from a SON, a higher rate of recommendation that SON will have. The manual option let the user chooses the SON to be recommended from.

3.5 Semantic and Keyword Search

Searching in Semantic Overlay Networks is done inside a concrete SON. This reduces network overload and the time the query last. When the user launches a query, the *query classifier* classifies it into a SON. Then, the query is resolved into that SON (only forwarded to peers inside the SON). If insufficient results are obtained, the query if forwarded into a more abstract SON, where more peers are subscribed, and then, there are more chances to find it. In the worst case, the query is forwarded to the root SON (most general), where all peers can reply, avoiding no results when available.

Use of ontologies let the user go beyond keyword search taking advantage of semantic search. Search will depend on the used P2P protocol, since not all P2P protocols allow search (e.g. BitTorrent). For those which allow searching, protocol should be modified to allow semantic search, but this modification in minimal. The way the search is done would be the same, since the only change here is that, instead of query about a string, the query would be about a semantic string.

For instance, a user would be able to search items corresponding to *Led Zeppelin songs after 1984 recorded in live.* That query is sent in the same way a keyword query is sent, and, when received by other peers, instead of checking an exact text matching, it launches an ontology matching query against its local information. Whether the query matches results or not, they behave in the same way as the keyword search, the only difference is the way it checks if it has a valid result for the query.

4 Conclusions and Future Works

In this paper, we have presented a new proposal for a P2P system which eases data sharing which is based on an overlay network. An overlay network directly influences scalability in such kind of systems as peers are grouped into clusters. When we use some semantic based mechanism to decide how to group clusters and how to route queries, we have a semantic overlay networks. Little work has been done on using semantic overlay networks in the context of P2P. In the paper, it is included a review of what has been developed until now. It is clear that there are already a number of works which use semantic descriptions to arrange clusters and route queries properly. However, very little is done on the social dimension of P2P, i.e. very little is done on recommendation of shared items based on semantics. We work on this in the paper by offering a new P2P software architecture.

Starting from this paper, there are still many things to be done. An specific aspect that needs more work to be done is working on merging metrics used for recommendation in an adaptive manner [3]. How effective a recommendation is, will vary depending on the peer. The system must learn from past interactions with the user. Once an item is recommended, the user should choose how adequate that recommendation will be and this feedback might help the self-adaptive system to change it behaviour to take more accurate decisions in the future. More generic and long lasting goals follow. For example, we consider that future trends to socialize everything. With the advent of semantic desktops, which promote integration and global semantic sharing semantic recommendation will be transparent to the user. Our efforts are focused now on allowing global integration of semantics by designing and promoting the appropriate ontologies. Also, integration is an added value that we are considering. If we consider social desktops as Nepomuk, integration into the desktop of such a P2P system would be done in a natural and seamlessly way.

References

1. Aberer, K., Cudré-Mauroux, P., Datta, A., Despotovic, Z., Hauswirth, M., Punceva, M., Schmidt, R.: P-Grid: a self-organizing structured P2P system, pp. 29–33. ACM Press, New York (2003)
2. Aberer, K., Cudre-Mauroux, P., Hauswirth, M.: GridVine: Building Internet-Scale Semantic Overlay Networks, pp. 107–121. Springer, Heidelberg (2004)
3. Angeline, P.J.: Adaptive and self-adaptive evolutionary computations. Computational Intelligence, 152–161 (1995)
4. Cabrer, Y.B., Fernández, M.R., Solla, A.G.: AVATAR: Un sistema de recomendación personalizada de contenidos televisivos basado en información semántica.
5. Crespo, A., Garcia-Molina, H.: Routing indices for peer-to-peer systems, pp. 23–32 (2002)
6. Crespo, A., Garcia-Molina, H.: Semantic Overlay Networks (2002)
7. Crespo, A., Garcia-Molina, H.: Semantic Overlay Networks for P2P Systems. Springer, Heidelberg (2004)

8. Cudre-Mauroux, P., Agarwal, S., Aberer, K.: GridVine: An Infrastructure for Peer Information Management, vol. 11, pp. 36–44 (2007)
9. Deerwester, S., Dumais, S.T., Furnas, G.W., Landauer, T.K., Harshman, R.: Indexing by latent semantic analysis, vol. 41, pp. 391–407 (1990)
10. Groza, T., Handschuh, S., Moller, K., Grimnes, G., Sauermann, L., Minack, E., Jazayeri, M., Mesnage, C., Reif, G., Gudjonsdottir, R.: The NEPOMUK Project-On the way to the Social Semantic Desktop
11. Gruber, T.R.: Toward principles for the design of ontologies used for knowledge sharing, vol. 43, pp. 907–928. Academic Press, Inc., Duluth (1995)
12. Haase, P., Siebes, R., van Harmelen, F.: Peer Selection in Peer-to-Peer Networks with Semantic Topologies. Springer, Heidelberg (2004)
13. Iofciu, T., Kohlschutter, C., Nejdl, W.: Keywords and rdf fragments: Integrating metadata and full-text search in beagle++. In: Proc. of the Semantic Desktop Workshop held at the 4th International Semantic Web Conference (2005)
14. Loser, A., Staab, S.: Semantic Social Overlay Networks, vol. 25, p. 1. IEEE Institute of Electrical and Electronics (2007)
15. Nesic, S., Gasevic, D., Jazayeri, M.: An Ontology-Based Framework for Authoring Assisted by Recommendation, pp. 227–231 (2007)
16. Pouwelse, J.A., Garbacki: TRIBLER: a social-based peer-to-peer system: Research Articles, vol. 20, pp. 127–138. John Wiley and Sons Ltd., Chichester (2008)
17. Pouwelse, J.A., Garbacki, P., Epema, D.H.J., Sips, H.J.: The bittorrent p2p file-sharing system: Measurements and analysis. In: International Workshop on Peer-to-Peer Systems (IPTPS) (2005)
18. Ripeanu, M.: Peer-to-Peer Architecture Case Study: Gnutella Network. In: Proceedings of International Conference on Peer-to-peer Computing, vol. 101 (2001)
19. Sauermann, L., van Elst, L.: Pimo - a framework for representing personal information models. In: Pellegrini, T., Schaffert, S. (eds.) Proceedings of I-Semantics 2007, pp. 270–277. JUCS (2007)
20. Sripanidkulchai, K., Maggs, B.: Efficient content location using interest-based locality in peer to peer systems. In: INFOCOM 2003. Twenty-Second Annual Joint Conference of the IEEE Computer and Communications Societies, vol. 3. IEEE, Los Alamitos (2003)
21. Tang, C., Xu, Z.: Peer-to-peer information retrieval using self-organizing semantic overlay networks, pp. 175–186. ACM Press, New York (2003)

NAS Algorithm for Semantic Query Routing Systems in Complex Networks

Laura Cruz-Reyes[1], Claudia Guadalupe Gómez Santillán[1,2], Marco Antonio Aguirre Lam[1], Satu Elisa Schaeffer[3], Tania Turrubiates López[1,4], Rogelio Ortega Izaguirre[2], and Héctor J. Fraire-Huacuja[1]

[1] Instituto Tecnológico de Ciudad Madero (ITCM), 1ro de Mayo y Sor Juana Inés de la Cruz s/n, CP.89440, Tamaulipas, México
lcruzreyes@prodigy.net.mx, marco@marcoaguirre.com.mx
[2] Centro de Investigación en Ciencia Aplicada y Tecnología Aplicada (CICATA), Carretera Tampico-Puerto Industrial Altamira, Km.14.5. Altamira, Tamaulipas, México
cggs71@hotmail.com, rortegai@ipn.mx
[3] Facultad de Ingeniería Mecánica y Eléctrica (UANL), Avenida Universidad s/n. Cd. Universitaria, CP. 66450, San Nicolás de los Garza, N.L., México
elisa@yalma.fime.uanl.mx
[4] Instituto Tecnológico Superior de Álamo Temapache (ITSAT), Km. 65 Carretera Potrero del Llano-Tuxpan, C.P. 92750. Álamo, Veracruz, México
tania_251179@acm.org

Abstract. The modern distributed systems are acquiring a great importance in our daily lives. Each day more transactions are conducted through devices which perform queries that are dependent on the reliability, availability and security of distributed applications. In addition, those systems show great dynamism as a result of the extremely complex and unpredictable interactions between the distributed components, making it practically impossible to evaluate their behavior. In this paper, we evaluate the performance of the NAS algorithm (Neighboring-Ant Search), which is an algorithm for distributed textual query routing based on the Ant Colony System metaheuristic and SemAnt algorithm, improved with a local topological characterization metric and a classic local exploration method called lookahead, with the aim of improving the performance of the distributed search. Our results show that including local information like the topological metric and the exploration method in the Neighboring-Ant Search algorithm improves its performance 40%, in terms of the number of hops needed to locate a set of resources in a scale-free network.

Keywords: Internet, query routing system, ant colony system, scale-free, topology, peer-to-peer network.

1 Introduction

Over the past years, new communication models have emerged on the Internet that manage information in a distributed manner and offer significant advantages over centralized information-management systems. These systems are known as unstructured *peer-to-peer networks* (P2P) [1]. In a P2P network, a set of nodes form connections among themselves to offer their resources to the other peers within the network. The P2P system together with the underlying communication network (typically the Internet) forms a complex system that requires autonomous operation through mechanisms of intelligent search [2].

J.M. Corchado et al. (Eds.): DCAI 2008, ASC 50, pp. 284–292, 2009.
springerlink.com

Within the intelligent mechanisms currently successfully applied to several problems in distributed systems, there are algorithms based on *ants*. The metaheuristic of *ant colony system* (ACS) proposed by Dorigo in 1997 uses techniques to solve optimization problems based on graphs. In these algorithms are works under the category of ant colony routing, specifically designed for handling routing tables in telecommunications, currently there are few developed applications in the research line of semantic query routing [2].

In this paper, we evaluate the performance of the NAS algorithm (Neighboring-Ant Search), which is an algorithm for distributed text query routing based on the Ant Colony System metaheuristic and *SemAnt* algorithm, improved with a local topological characterization metric Degree Dispersion Coefficient (*DDC*) [3] and a classic local exploration method *lookahead* [4], with the aim of improving the performance of the distributed search. This algorithm allows intelligent navigation to locate textual information in a P2P system modeled as a complex network.

1.1 Ant Algorithms for Semantic Query Routing

Ant algorithms were inspired by the behavior of ants in search for food, because in performing the search, each ant drops a chemical called pheromone which provides an indirect communication among the ants.

The basis of the ant algorithms can be found in a metaheuristic called ACO (Ant Colony Optimization) [5]. They were proposed to solve problems modeled as graphs. The ACO algorithm and its variants (e.g. ACS) do not consider the dynamic of complex systems. The first modifications were designed for circuit-switched and packet-switched networks [6], where these algorithms need to know information about all nodes in the network to select the destination node, by contrast in the semantic query routing the goal is to find one or more nodes destinations for a query without having information from the complete network, only of the nodes requiring operate with local information on the network.

Another modification was for solving the problem of locate textual information on P2P networks, called *Semantic Query Routing* (SQR), which: given a network represented by a graph (G), a set of content (R) distributed in the nodes and a set of queries (Q) launched by the nodes, the goal is to maximize the number of query results and to minimize the number of hops necessary to satisfy these queries [2].

2 Related Work

Over the past few years, algorithms for semantic query routing on the Internet have received special attention [2, 6, 7, 8]. We summarize here some of the most relevant proposals for semantic query routing using ACS.

Michlmayr [2] proposes a distributed SQR-algorithm for P2P networks called *SemAnt*. She also includes an evaluation of the parameter configuration that affects the performance of the algorithm.

Yang et al. [7] propose an algorithm called *AntSearch* for non-structured P2P networks, where each pair node stores information on the level of success of past queries as well as on the pheromone levels of the immediate neighbors. The work of Yang et al. was motivated by the need to improve search performance in terms of the traffic generated in the network and the level of information recovery.

Babaoglu et al. [8] propose the *Gnutant* application based on the Anthill platform for the design, implementation and evaluation of P2P applications. The objective is to share documents with scalability like that of Freenet and the free search capabilities of Gnutella.

Di Caro and Dorigo [6] propose *AntNet* that is designed for packet-switched network. The ants collaborate in building routing tables that adapt to current traffic in the network, with the aim of optimizing the performance of the entire network. This algorithm needs information about all nodes that exist in the network in order to choose the destination nodes.

Many relevant aspects of these works were incorporated into the proposed NAS algorithm. The former work only learns from past experience, whereas NAS takes advantage of the environment where the search takes place in terms of the local exploration through lookahead and the local structural metric *Degree Dispersion Coefficient* (DDC) [3]. The DDC measures the differences between the degree of a vertex and the degrees of its neighbors. A node i is said to be a *neighbor* of a node j if they are connected in the network. The *degree* k_i of a node i is the number of neighbors it has. The DDC of node i is defined in equation (1), $\sigma(i)$ is the degree variation among i and its neighbors and $\mu(i)$ is their average degree.

$$DDC(i) = \frac{\sigma(i)}{\mu(i)}; \tag{1}$$

where $\sigma(i) = \sqrt{\frac{\sum_{j \in \Gamma(i)} [k_j - \mu_i]^2 + [k_i - \mu_i]^2}{k_i + 1}}$; and $\mu(i) = \frac{\sum_{j \in \Gamma(i)} [k_j] + k_i}{k_i + 1}$.

3 Neighboring-Ant Search (NAS)

3.1 Architecture

The architecture of NAS is shown in Figure 1. The environment is a static P2P complex network, having a set of linked nodes with local information about their neighbor nodes. The agent is an ant amending the environment with functions and parameters. The functions allow to characterize locally the network, with the goal of choosing a better approach towards the convergence of the solution.

In phase 1 of the architecture, the module *Evaluation of performance* evaluates through *lookahead* the current node and its neighbors. In case of achieving a good result, the functions of *global updating* (5) and *HOPS* (6) apply changes in the environment and provide knowledge of these results to the network. Otherwise the *state transition* of phase 2 is used to selects the next node to move (related to the *Generator of unexplored actions*). Both strategies carry out the characterization of the environment, making a local update in every movement.

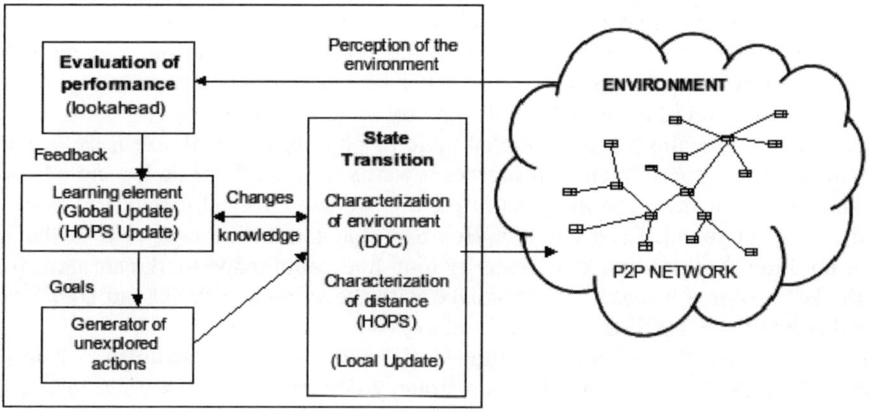

Fig. 1. NAS architecture

3.2 Experimental Setup

We generated a P2P complex network with the *scale-free network* model of Barabási et al. [9], where nodes are added one at a time with a fixed number d of connections each. The newly-arriving node chooses among the existing nodes to which to connect preferentially at random, giving preference to high-degree nodes. The resulting network has a small number of highly connected nodes while the majority of the nodes have degree close to d. With this model we generated networks with 1,024 nodes and all links between peers are bi-directional. The network topology and the content distribution in the network are considered static.

The application is a distributed search machine where each peer manages a local repository of resources and offers its resources to other peers. The "topics" of the resources were modeled as integer values from zero to 1,024, generated using the uniform-distribution generator of Repast [10]. Each node was assigned resources possibly repeated topics since the allocation of the distribution of data is Zipfian, which means that the frequency of the allocation of data is associated with a probability distribution scale-free [11].

Each node has a list of possible resources to search, which is limited by the range of values generated. The queries were generated to search a resource uniformly at random from 0 to 1,024. The time steps of the experiments were of 100 ms and the simulations were ran for 10,000 steps. During each time step, each node has a probability of 0.1 to launch a query.

3.3 Proposed Algorithm

The *Neighboring-Ant Search* (NAS) is a metaheuristic algorithm based on ACS algorithm [5], where a set of independent agents called ants cooperate indirectly and sporadically to achieve a common goal. Each ant represents a query and his operation is distributed, but indirect collaboration is achieved through globally maintained table of pheromones where the ants drop "trails" according to the quality of the content of the visited node with respect to the search query.

The algorithm has two objectives: guide queries towards nodes that have better connectivity using the *DDC* and minimize the hop count. Therefore, the more frequent is a query towards a resource, the better path is selected; that is to say, the degree of optimization of a query depends directly on its popularity.

Each node maintains a bidimentional table of pheromones τ of size n by R which contains the trail of R amount of resources towards its n neighbor nodes and a heuristic table *HOPS* of the same dimension to store the proportional distance in hops toward a concept found. Each entry in *HOPS* is updated with the inverse of the hop count until the evaluated node, divided by total hop count made by the ant until finding the last resource it seeks. τ is initialized with the value $\tau_0 = 0.009$ and *HOPS* with the value $HOPS_0 = 0.001$, that are selected to avoid division by zero.

The routing strategy of the algorithm is defined by the *state transition rule* (2), adopted from ACS [5]. An ant selects a strategy depending on a value q_0 that represents the importance between the *exploitation* and *exploration* strategies. It generates a random value q that represents the probability of selection of any of these strategies. These values will be compared, and if $q < q_0$, it will choose exploitation strategy; otherwise it will choose the exploration strategy. The initial value of $q_0 = 0.85$, meaning that more advantage is given to the exploitation strategy and the ant will choose as a first preference the better connected nodes.

Exploitation lets the ant choose from the set of neighbor nodes, a node that has a better contribution according to (2); that is, choose a node that has better connectivity and pheromone trail and a shorter distance towards a searched resource. The *DDC* and the heuristic value *HOPS* were included in this rule; in the initial stage, the DDC could guide the searches towards better connected nodes.

For the rule (2): q is a random number [0, 1], r is the current node where the ant k is located, u belongs to the set of neighbor nodes of r, V_k is the set of nodes visited by the ant k, t is the searched resource, β is the parameter that determines the relative importance between the *pheromone* and the *DDC* with *HOPS*, and S is a pseudo-random variable that is determinate by the *exploration strategy* (3). The initial value of $\beta = 2$ was taken from Dorigo and Gambardella [5].

$$s = \begin{cases} \arg\max_{u \in \Gamma(r) \wedge u \notin V_k} \left\{ \left[\tau_{r,u,t}\right] \cdot \left[DDC_u + HOPS_{r,u,t}\right]^{\beta} \right\}, & \text{if } q \le q_0 \text{ (exploitation)} \\ S, & \text{otherwise (biased exploration)} \end{cases} \tag{2}$$

The strategy of *exploration* (3) stimulates the ants to search for new paths. This strategy applies the technique of roulette wheel selection to select a node that has better connectivity and pheromone trail and a shorter distance towards a searched resource. The *DDC* and the heuristic value *HOPS* were included in this rule.

$$S = p_{r,u,t} = \frac{\left[\tau_{r,u,t}\right] \cdot \left[DDC_u + HOPS_{r,u,t}\right]^{\beta}}{\sum_{u \in \Gamma(r) \wedge u \notin V_k} \left[\tau_{r,u,t}\right] \cdot \left[DDC_u + HOPS_{r,u,t}\right]^{\beta}} . \tag{3}$$

Each time that an ant decides to move towards a node, it drops a trail in the pheromone table of each visited node, according to the local updating rule (4) in order to establish a tendency towards more popular nodes; where ρ is the factor of *local evaporation* of the pheromone, initialized with $\rho = 0.07$.

$$\tau_{r,s,t} \leftarrow (1-\rho)\cdot\tau_{r,s,t} + \rho\cdot\tau_0 \tag{4}$$

Each time that an ant locates a resource, the resource is retrieved and dispatched to the node that launched the query, through the global updating rule (5). The amount of dropped pheromone depends on the quality of the solution founded, through the importance between the resources found and the time-to-live (*TTL*) [2]: where α is the factor of *global evaporation* of the pheromone, initialized as $\alpha = 0.07$, w_d is the influence between the founded results and the time-to-live of the ant, $results_k$ are the found results by the ant *k*, *maxResults* is the maximum amount of results expected (fixed to 5 for all experiments), TTL_k is the partial time-to-live of the ant k until that moment, and *TTLmax* is the maximum amount of time-to-live for an ant in a query (fixed to 25 hops for all experiments).

$$\tau_{r,s,t} \leftarrow (1-\alpha)\cdot\tau_{r,s,t} + \alpha\cdot\left[w_d \frac{results_k}{maxResults} + (1-w_d)\frac{TTLmax}{2\cdot TTL_k}\right] \quad \forall r,s,t \in path\ k \tag{5}$$

The heuristic table *HOPS* in rule (6) is updated along with the *global updating rule* (5) using inverse of the hop count until the evaluated node $h_{r,s,t}$, divided by total hop count made by the ant until its last resource found h_k.

$$HOPS_{r,s,t} = \left(\frac{h_k}{h_{r,s,t}}\right)^{-1} \quad \forall r,s,t \in path\ k \tag{6}$$

Table 1. NAS algorithm pseudo code

```
for each query
   for each ant
      if Hits < maxResults and TTL_k > 0              // Phase 1
         if the neighbor from edge s_k has results    // Lookahead strategy
            append s_k to Path_k
            TTL_k = TTL_k -1
            globalUpdate
         else                                         // Phase 2
            s_k = apply the transition rule with the DDC
            if path does not exist or node was visited,
               remove the last node from Path_k
            else,
               append s_k to Path_k
               TTL_k = TTL_k -1
               localUpdate
            endif
         endif
      else
         Kill ant
      endif
   endfor
endfor
```

Another adaptation performed to the algorithm is the classic technique one-step lookahead [4], when an ant is located at a node *r*; this knows the set of neighbor nodes of level 1 of the node *r*. Through this technique the ant is responsible of requesting the expected resource to each of its non-visited neighboring nodes.

The NAS algorithm consists of two main phases. Phase 1 corresponds to the *evaluation of results* (lines 04-07 in de pseudo code of the Table 1). In this phase 8through the classic technique Lookahead, the ant *k*, located in a node *s*, queries to the neighboring nodes for the requested resource. If the resource is found, the result is retrieved. Through the *global updating* function given in (5) the path used by the ant is boosted. Similarly, *HOPS* (6) is updated. In the case that the evaluation phase had not found any result phase 2 is carried out.

Phase 2 of NAS algorithm corresponds to the *state transition* (lines 09-17 in the pseudo code on the Table 1), in which through *q*, a neighbor node *s* is selected by the *exploitation* (2) or *exploration* (3) function. In the case that there is no node towards which to move (that is to say, is in a leaf node or all neighbor nodes had been visited) a hop backward on the path is carried out, otherwise the ant adds the node *s* to its path, applying the *local updating* of the pheromone (4) and reducing TTL_k by one hop. The query process ends when the expected number of results has been satisfied or TTL_k is equal to cero. The ant is killed indicating the end of the query.

4 Results

In Figure 2(a), the number of hops used by the NAS algorithm with and without DDC to perform a set of queries is shown. The NAS algorithm without DDC begins with an average of 19 hops, and while the algorithm evolves, around the query 1,000 the hop count at decreases a range of 12 and 14. Incorporating DDC into NAS it begins with an average of 18 hops, and while the algorithm evolves, around the query 1,000 the hop count decreases at a range of 11 and 13. As a result of this decrease in the hop count we may conclude that the ant learns to select better paths to the requested resources.

In Figure 2(b), the average hit rate (*hit rate* divided by *maxResults*) by the NAS algorithm with and without DDC to perform a set of queries is shown. For both strategies start out around 85% to 90% of hits obtained by a set of queries, and while the algorithm evolves, around the query 10,000, we observe that the hit rate increases to a range of 88% to 95% by a set of queries

In Figure 2(c), the average efficiency of the NAS algorithm is shown as a function of the required hop count to satisfy a query (*average hops* divided by *average hit rate*). The algorithm without DDC in the first stage needs around of 5 average hops to find a resource. Around query 1,000, the average hop count decreases to a range of 2.7 and 3.4, and incorporating DDC, in the first stage needs around of 4.5 average hops to find a resource. Around query 1,000, the average hop count decreases to a range of 2.5 and 3.

These observations confirm the intuition that both the DDC and the lookahead, in the presence of a scale-free distribution allow a significant improvement to the search performance, which also implies that the NAS algorithm outperforms in such topologies the existing methods that do not incorporate local structural information.

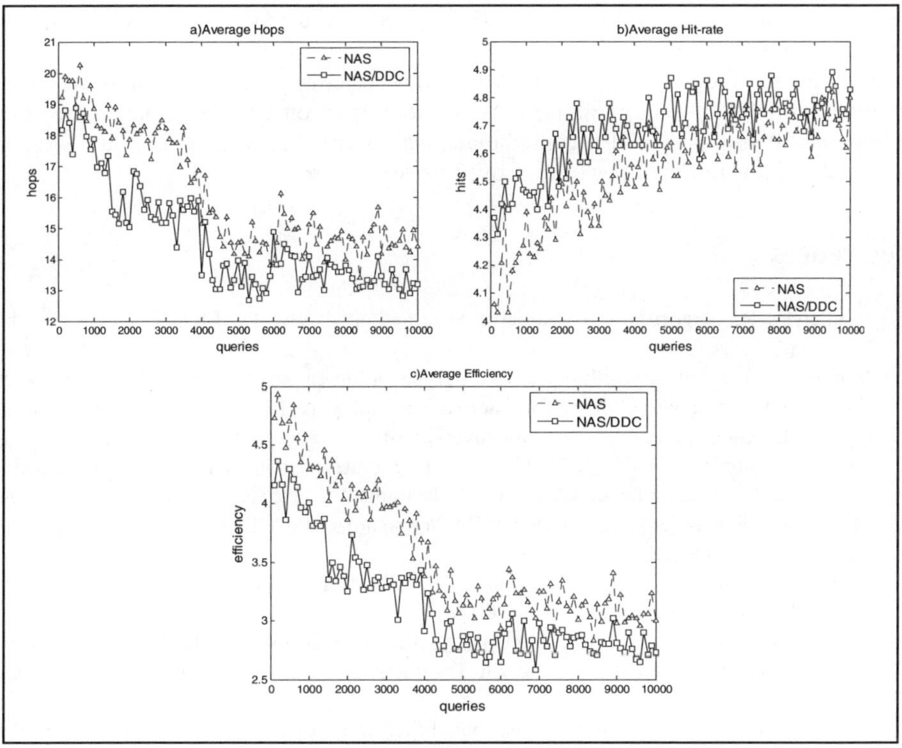

Fig. 2. Plots of the results on a scale-free network over 100 queries of (a) average hops, (b) average hit rate, and (c) average efficiency

Beside, the variability in the results is kept in a small range, and is due to the permanent exploration capability of NAS, which is needed to incorporate in the future others dynamic issues of complex networks (e.g. changes in the connections of the nodes and the information resources).

5 Conclusions and Future Work

We proposed an algorithm for semantic query routing based on existing ant colony algorithms, but incorporating measures of local network structure: lookahead and *DDC*. This combination results in better hit count with a lower hop count.

Using only one ant per query, we observe upon including local structural information in the algorithm that the hop count decreases 35% from the beginning to the reported stable stage, and the efficiency increases 35% in terms of the hop count to find a resource. This results yield an improvement of 8%. We also achieved an average good performance in hit rate: between 80% and 90%.

Nowadays there are a few applications developed for semantic query routing. An important factor for this study is the dynamism inherent in the complex networks, as changes in the contents of the repositories occur frequently, and peers can connect or disconnect at any moment time. Due to this, we need to extend the present experimental work to non-static networks.

We plan to study deeply the impact of the structural measure employed in the learning curve of ant-colony algorithms and the effect on the performance measures of hop and hit counts. We also contemplate using more than one ant per query to parallelize the algorithm in hopes of improved performance.

References

1. Wouhaybi, R.: Algorithms for Reliable Peer to Peer Networks. Ph.D Thesis, Columbia University, USA (2006)
2. Michlmayr, E.: Ant Algorithms for Self-Organization in Social Networks. Ph.D Thesis, Women's Postgraduate College for Internet Technologies, Institute of Software Technology and Interactive Systems, Vienna University of Technology, Austria (2007)
3. Ortega, R., Meza, E., Gómez, G., Cruz, L., Turrubiates, T.: Impact of Dynamic Growing on the Interner Degree Distribution. In: Thulasiraman, P., He, X., Xu, T.L., Denko, M.K., Thulasiram, R.K., Yang, L.T. (eds.) ISPA Workshops 2007. LNCS, vol. 4743, pp. 326–334. Springer, Heidelberg (2007)
4. Mihail, M., Saberi, A., Tetali, P.: Random Walks with Lookahead in Power Law Random graphs. Internet Mathematics 3, 1–3 (2007)
5. Dorigo, M., Gambardella, L.M.: Ant Colony System: A Cooperative Learning Approach to the Traveling Salesman Problem. IEEE Transactions on Evolutionary Computation 1(1), 53–66 (1997)
6. Di Caro, G., Dorigo, M.: AntNet: Distributed Stigmergy Control for Communications Networks. Journal of Artificial Intelligence Research 9, 317–365 (1998)
7. Yang, K., Wu, C., Ho, J.: AntSearch: An Ant Search Algorithm in Unstructured Peer-to-Peer Networks. IEICE Transactions on Communications 89(9), 2300–2308 (2006)
8. Babaöglu, O., Meling, H., Montresor, A.: Anthill: A Framework for the Development of Agent-Based Peer-to-Peer Systems. In: Proceedings of the 22nd International Conference on Distributed Computer Systems (ICDCS 2002). IEEE, Los Alamitos (2002)
9. Barabasi, A., Albert, R., Jeong, H.: Mean-Field theory for Scale-free Random Networks. Physica A 272, 173–189 (1999)
10. ROAD (Repast Organization for Architecture and Design). Repast Home Page. Chicago, IL, USA (2005), `http://repast.sourceforge.net`
11. Newman, M.: Power laws, Pareto distributions and Zipf's law. Contemporary Physics 46(5), 323–335 (2005)

CoDiP2P: A Peer-to-Peer Architecture for Sharing Computing Resources

D. Castellà, I. Barri, J. Rius, F. Giné, F. Solsona, and F. Guirado

Departamento de Informática e Ingeniería Industrial, Universitat de Lleida,
Jaume II 69, 25001 Lleida, Spain
{dcastella,ibarri,jrius,sisco,francesc}@diei.udl.cat

Summary. Peer-to-Peer (P2P) computing, the harnessing of idle compute cycles through the Internet, offers new research challenges in the domain of distributed computing. This paper presents CoDiP2P, a Computing Distributed architecture using the P2P paradigm. CoDiP2P allows computing resources from ordinary users to be shared in an open access by means of creating dynamic areas of computing resources in a completely distributed, scalable and fault tolerant way. This paper discusses its system architecture and evaluates its functionality by means of simulation.

Keywords: Peer-to-Peer computing, p2p simulation, distributed computing, job scheduling.

1 Introduction

During recent years, there has been an important increase in computing power requirements in different research fields (life and earth sciences, chemistry, etc.). Due to the expansion of the computer industry, this need has also been observed in business (i.e. financial models) and domestic scenarios (i.e. games). These new necessities have motivated the development of new distributed computing paradigms that use Internet as a single large virtual computer, reducing the elevated cost of the specialized environments, like supercomputers, needed to execute these applications with high computing requirements.

P2P computation [1, 2, 3] represents an emergent low-cost alternative to supercomputers and cluster systems, providing access to distributed computational resources in a scalable and fault-tolerant way. P2P architectures take advantage of the under utilization of personal computers, integrating them into a platform based on the sharing of computational resources between geographically distributed equals. While the P2P paradigm can not hope to serve as a totally general-purpose efficient parallel computer, it can still serve as an excellent platform with unlimited computational resources for solving a wide variety of computational problems: (1) Allowing processing-limited devices, such as wireless clients, to distribute their processing requirements to other machines in the network or (2) Executing embarrassingly parallel programs or (3) Distributing libraries with high computational requirements (i.e. a rendering library) through Internet.

J.M. Corchado et al. (Eds.): DCAI 2008, ASC 50, pp. 293–303, 2009.
springerlink.com

According to these challenges, in this article, we present CoDiP2P, a decentralized mechanism for distributing computation and resource management that takes into account the system heterogeneity by using the P2P paradigm. The design of CoDiP2P is focused on hiding system complexity from the programmers and users. Its hierarchical topology in managing and maintaining the system-growing capacity favors the good scalability of the system. The question of fault-tolerance has also been taken into account by providing self-organization to peers and avoiding centralized components (logic and physic). The system is able to manage its resources efficiently, independently of their heterogeneity, geographical dispersion and volatility.

The P2P performance was evaluated by means of the GridSim simulation environment [4]. Over this system, we evaluated the main functionalities of CoDiP2P (peer insertion, maintenance, job scheduling and peer output) in relation to the main P2P performance issues: scalability, sizes of the areas and network latency and bandwidth.

The remainder of this paper is outlined as follows. In Section 2, the related work is explained. In Section 3, the CoDiP2P architecture is discussed. The efficiency measurements of CoDiP2P are performed in Section 4. Finally, the main conclusions and future work are explained in Section 5.

2 Related Work

Internet computing [6] is classified in two paradigms: GRID and P2P computing. GRID computing involves organizationally-owned resources (mainly supercomputers and clusters), which are centrally managed by IT professionals and connected by high bandwidth networks. In contrast, P2P is targeting an open environment, one that is accessible to the average user and does not require membership of any organization. A P2P system is typically large, with thousand or even millions of users, and there is no centralized entity that controls the behavior of individual users. However, developing an architecture for P2P computing presents several research challenges.

The first P2P applications provided a means for file sharing via Internet using the excess storage capacity and network bandwidth available [3, 5]. In contrast, some current research projects, such as CompuP2P [7], JNGI [8] or our CoDiP2P, propose using the P2P paradigm for distributed computing. CompuP2P establishes a system of markets to regulate the requests for resources. Thus, it classifies the nodes between sellers and buyers of computational resources. The main problem is the high cost of maintaining updated market information at all time given that whenever the resources of a node are altered, the change has to be notified to the whole system. JNGI was created using the JXTA library [9] and is the most similar to our proposal. However, JNGI maintains a repository of the tasks to be launched in the system, which is only available to the manager of the system. Thus, the "all-equals" principle followed by any P2P system is infringed.

The CoDiP2P follows the P2P principles. Its architecture is decentralized and structured. It is decentralized because there is no any central server that manages the searches and any peer in the system can schedule and launch tasks. Likewise, it is structured because it creates a hierarchy of peers in the form of trees to maintain an efficient performance in the insertion and output of peers, task scheduling and managing the whole system.

3 CoDiP2P

This section presents the CoDiP2P architecture and its main functionalities. In order to explain these, some concepts must be previously introduced:

- **Area** A_i**:** Set of workers controlled by a manager.
- **Manager** M_i**:** One of the two roles that a peer can acquire. Its main goal is to manage peers in the same area and schedule tasks over the workers.
- **Worker** W_i**:** Responsible for executing tasks scheduled by its manager.
- **Area Size** $Size(A_i)$**:** Each area has a limited capacity of peers, which depends mainly on the network properties (BW and latency), the size of the exchanged messages and the system topology. It is assumed that all the system areas have the same size, denoted as *Size*.
- **Area Level** $Level(A_i)$**:** Given that CoDiP2P follows a tree hierarchy, the level is the growth unit of the system. The levels are made up of one or more groups of areas.
- **Major Node** N_{Major}**:** Due to the dynamism in a P2P system, a static access point is necessary for the new requesting peers entering into the system. This is a static node, called *Major Node*, which acts as incoming point for the new peers entering the system. Although it may seem that CoDiP2P loses part of the Peer-to-Peer philosophy due to its centralized nature, it is not true given that N_{major} is an entry point only for the initial process of creation of the system.
- **Replicated managers** RM_i**:** Due to the high probability that a manager M_i fails, each area A_i maintains a set of peers, name Replicated Managers, which have a special role for replacing M_i when it falls from the system. According to this, each RM_i maintains a copy of the same information kept by M_i. Thus, if M_i fails, then the oldest RM_i will replace it.

Fig. 1 shows the linked structure of peers in CoDiP2P based on a tree topology with area $Size$ of two peers, seven areas A_i throughout the system and leading to a deployment of three hierarchic levels. Note that this type of structure allows a manager of a lower area to be a worker in a higher area at the same time.

The main goals of the CodiP2P architecture are the following: **scalability** to ensure that the system supports the massive entry of peers; **distributed management** to harness the computing resources offered by the nodes making up the system; **fault tolerance**: CodiP2P must prevent the high possibility of a peer failure and replace it by ensuring the stability, robustness and performance of the global system; **self-organization**: each peer can be a manager

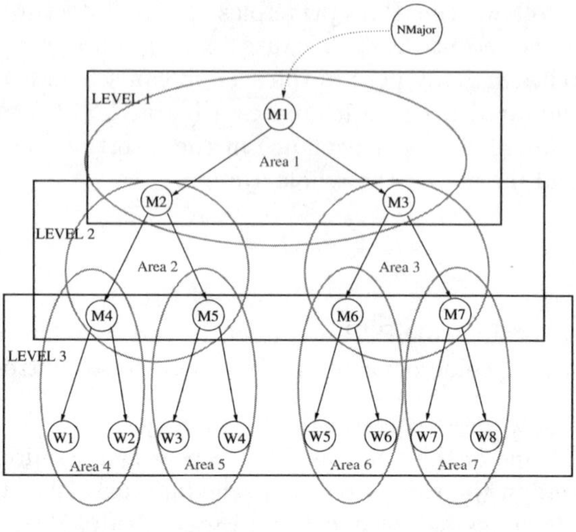

Fig. 1. Tree based structure of peers in CodiP2P

or a worker dynamically according to the needs of the system; **heterogeneous resource management**: the CodiP2P design is adapted to the heterogeneity of the resources to manage the scheduling and load balancing of tasks among the peers efficiently.

3.1 CoDiP2P Functionalities

This section explains the main operations of CoDiP2P: the *insertion* of new peers in the system, *the maintenance* mechanism to maintain the system updated, *the job launching and scheduling* strategies used by peers and the *peer output* mechanism used to balance the tree based structure when a peer leaves the system.

Peer Insertion

Whenever a new peer wants to enter the system, it first communicates with N_{major} and it returns the address of the root manager M_1 of the tree hierarchy. Next, M_1 begins to execute the following next exclusive conditional cases:

1. The first case checks if the current area A_i has a free site for the new peer. If it has, then the new peer proceeds to enter this area and becomes a worker.
2. The second case checks if there is a manager M_j in the area of the actual manager M_i which has a free site for locating the new peer. This case has the highest priority, given that to maintain the tree balanced, it is better to make good use of the existing areas than create a new area.
3. The third case checks if the actual manager M_i has any worker to change to a manager and create a new area A_j to locate the new peer.

Algorithm 1. Maintenance algorithm

1. **Each** T seconds
2. **for all** ($Peer_i$ connected to area A_i managed by M_i) **do**
3. **if** $Peer_i$ is a *Replicated Manager* RM_i **then**
4. M_i sends message ManagerAlive and Replicated Information to $Peer_i$
5. **else**
6. M_i sends message ManagerAlive to $Peer_i$
7. M_i receives statistical information from $Peer_i$
8. **end for**

4. The last case happens when there is no worker or manager with free sites in area A_i. So, the manager M_i has to check which is the manager M_j with fewer peers controlled directly and indirectly through all the branches of the tree. This heuristic ensures the balancing of the tree.

The cost of the insertion algorithm is determined by the number of petitions made by the new peer to the managers following the tree hierarchy. The worst case happens when the height of the tree is considerable and the new peer has to make a lot of petitions to managers located vertically in the structure. The cost is $\theta(h)$, where h is the number of levels of the tree. The cost in terms of number of peers yields a value of $\theta(\lg_{SIZE}(N)-1)$ where $SIZE$ is the area capacity and N is the number of peers in the system.

Maintenance of the System

The Alg. 1 shows how each period T, a set of messages is exchanged among each manager M_i and its workers to control the computational resources available in the area and the total underlying branch managed by M_i. Note that each M_i sends a kind of message depending on the role of the receiver peer. In addition, each peer, depending also on its role, sends its manager information about its local available computational resources (single peer case) or the available computational resources managed by such a peer (if the peer is also a manager of a lower area). The cost of the algorithm is $\theta(2N)$ where N is the number of peers in an area.

Job Launching and Scheduling

Each parallel job can be launched from any peer, denoted as $Peer_{job}$, independently of its role, by means of making a job request to its manager M_i. Next, M_i selects the workers in the system to execute the tasks making up the job (job scheduling). Both algorithms are shown in Alg. 2 and 3 respectively.

Both algorithms try to minimize the number of areas occupied to launch a specific job. Thus, the fragmentation of the system is decreased. Note that these algorithms assume that one worker is available when it is not executing a task and only one task is executed by a worker.

Algorithm 2. Launching algorithm

1. $Peer_{job}$ sends a job launching request message to his manager M_i
2. **if** *number of free workers* of $M_i \geq$ *number of tasks of job* **then**
3. M_i sends *Agree* message to $Peer_{job}$
4. M_i runs **Scheduling Algorithm**
5. **else if** *number of total workers* of $M_i \leq$ *number of tasks of job* **then**
6. M_i extracts from request job message the answer if it wants to wait for executing job in the actual area.
7. **if** $Answer = Wait$ **then**
8. M_i puts into its queue the *Job*
9. Wait until *Job* reaches its turn
10. **Goto** line 4
11. **else if** $Answer = NoWait$ **then**
12. M_i delegates with its upper manager to launch the job
13. **Goto** line 2
14. **else**
15. M_i delegates with its upper manager to launch the job
16. **Goto** line 2

The cost of launching a job is determined by the number of requests made between managers involved in the algorithm. Therefore, the worst case happens when $Peer_{job}$ is at the lower level of the tree and the algorithm has to cross all tree vertically because it requires a lot of available workers to execute the job. The cost in terms of peers yields a value of $\theta(\lg_{SIZE}(N) - 1)$, where N is the number of peers in the system. The cost of scheduling algorithm grows when the selected peers by launching algorithm are managers of a lower area and they also have to send all task requests to peers connected with it and then the scheduling algorithm will not finish until all the peers selected are just workers. Also, the cost of this algorithm yields a value of $\theta(\lg_{SIZE}(N) - 1)$.

Peer Output

When a peers leaves the system voluntarily or involuntarily, it is necessary to balance the system for a better performance of all the algorithms. So, the manager of the area to which the output peer belongs, has to determine if the peer is a worker or a manager of a lower area. If it is a worker, there is no problem in restructuring the system because, it is a final node. The restructuring operation, described in the Alg. 4, is applied by a replicated manager RM_i whenever the output peer is a manager.

The cost of Alg. 4 is determined by the number of peers who are affected by the fall of a peer. The worst scenario happens when the replicated manager RM_i selected to replace an output manager, is also the manager of a lower area. The cost of the algorithm yields a value of $\theta(\lg_{SIZE}(N) - 1)$, where N is the number of peers in the system.

Algorithm 3. Scheduling algorithm

```
1. Require: a manager M_i that executes the scheduling algorithm
2. if there is a manager M_j belonging to A_i whose number of available workers in
   the branch = number of tasks then
3.   M_j selects available workers belonging to A_j needed to execute Job
4.   if there is remaining task then
5.     M_j selects a set of lower managers needed to execute all remaining tasks.
6.     By each selected manager Goto line 2
7. else if there is a manager which has number of available workers in the branch
   > number of tasks then
8.   Selects a manager M_j with the lowest number of available workers
9.   By M_j Goto line 2
10. else
11.   while (remaining tasks > 0) do
12.     M_i selects the manager M_j with number of available workers most equal to
   remaining tasks
13.     By M_j Goto line 2
14.   end while
```

Algorithm 4. Peer Output algorithm

```
1. RM_i detects after Y seconds that there is no manager that controls A_i
2. RM_i sends a message to notify all peers belonging A_i that it will be manager.
3. if RM_i is also manager of a lower area A_j then
4.   RM_i selects the oldest replicated manager (RM_j) belonging A_j.
5.   RM_i notifies RM_j that it will be manager of A_j
6.   if RM_j is manager of lower area then
7.     By RM_j Goto line 2
8. RM_i becomes manager of A_i
```

4 Experimentation

In order to verify the performance of the CoDiP2P architecture, we used the GridSim simulator [4]. GridSim is a discrete-event toolkit simulation of application scheduling in grid environments. It supports modeling and simulation of heterogeneous grid resources (both time-space shared), users, applications, network, brokers and schedulers in a grid computing environment.

Although the GridSim simulator is based on grid environments, we used some features of this platform to simulate our CodiP2P architecture, such as the modelling of system-user entities (threads), like peers, and the association of each user entity with a grid resource of one machine or local task scheduling capabilities. Furthermore, all entities were connected by network links whose bandwidth and latency parameters can be specified. Nevertheless, we implemented some other specific features of a P2P environment, such as the peer fall. Given that when a peer fall it continues existing in the simulation environment, we had to implement a passive state where the peer does not accept any connection from another peer.

The next section shows the time cost of all main CoDiP2P functionalities in terms of number of peers in each area, index of a peer into the system, network parameters and size of parallel jobs.

4.1 Experimental Results

Firstly, we established the size of an area. According to the maintenance algorithm described in Alg. 1, the manager of each area periodically sends messages to its workers and viceversa to obtain information about them. So, the size of an area is restricted by the time spent on sending maintenance messages from a manager to its workers and the time to respond. If this time is higher than such a *Maintenance Period* then any new peer can not be lodged in this area.

Taking these arguments into account, the size of an area is limited by the bandwith and latency of the network. Thus, Fig. 2 displays the number of peers located in an area in terms of latency and bandwidth (left) and latency and maintenance period (right).

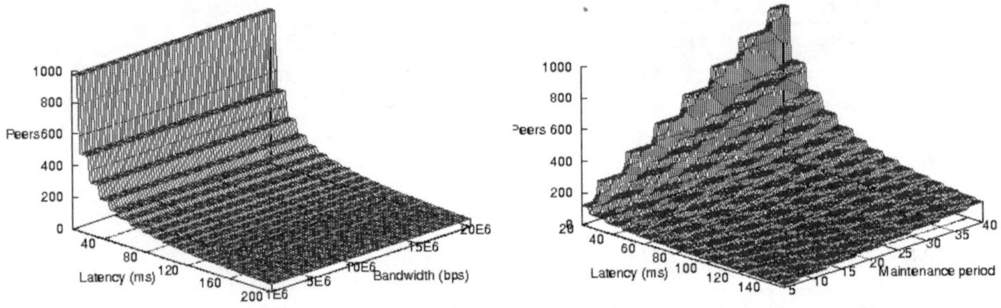

Fig. 2. Size of an area

The Fig. 2(left) shows that the latency is the most influential factor on the limit of an area because the messages are short and, as a consequence, the cost of modulating is lower than the cost of propagating the message. Likewise, Fig. 2(right) validates the good behavior of our simulation given that the size of the area is directly affected by the variation in the maintenance period. Therefore, the bad choice of this period can have negative consequences on areas, because the cost of maintaining all the system updated will be greater if the value of the size is too high, and on the other hand if the size is too low then the number of levels of the tree hierarchy will be increased.

Taking into account the average latency of Internet, the optimization of the cost of CoDiP2P functionalities and leaving a margin for not saturating the network , we found that the most optimal area size is between 20 and 30 peers and the maintenance period is around 20 seconds.

Next, we measured the cost of the insertion operation. Fig. 3(left) shows the insertion time of an incoming peer in terms of the index which a peer enters into the system and the size of the areas. The results were obtained under a network latency of 30ms and a bandwidth of 1Mbps. We can see that for small areas, the insertion time of each peer is higher because the hierarchy of the tree is increased and, as a consequence, the incoming peer has to be delegated from the top to the bottom of the tree. The diagonal jump that appears in the graph is due to the case that the incoming peer has reached the limit of the area and the manager has to create another area.

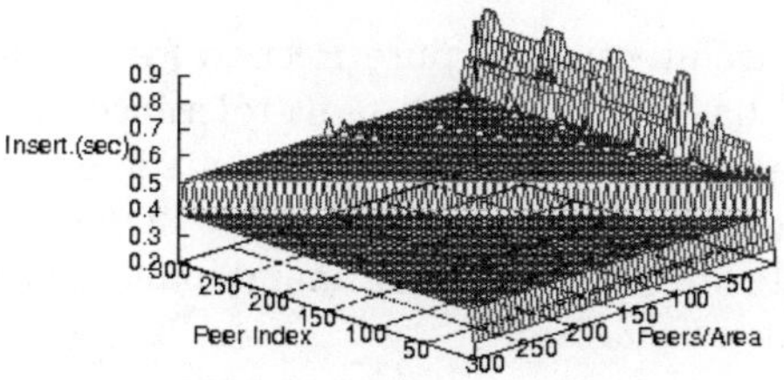

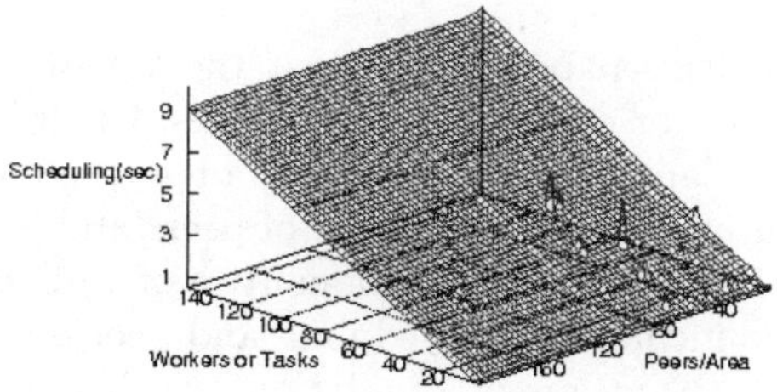

Fig. 3. (left) Cost time of peer insertion. (right) Cost time of job scheduling.

The next functionality to be evaluated was the scheduling time, which considered the cost of job launching as the cost of task scheduling over the peers. Fig.3(right) depicts the scheduling time varying the number of tasks making up a single job and the area size. We assume a system of 200 peers and a network latency of 30ms. The results show that the scheduling time increased according to the number of tasks of the job. In addition, we can see that the scheduling cost with smaller areas is somewhat higher than with larger ones because there are more managers involved in task scheduling. However, this difference was not very noticeable because the communication weight on sending the tasks to workers was much higher than the communication between managers to schedule the tasks.

Finally, the peer output mechanism was measured. This trial assumed a system composed of 100 peers and a network latency of 30ms. Fig.4 shows the

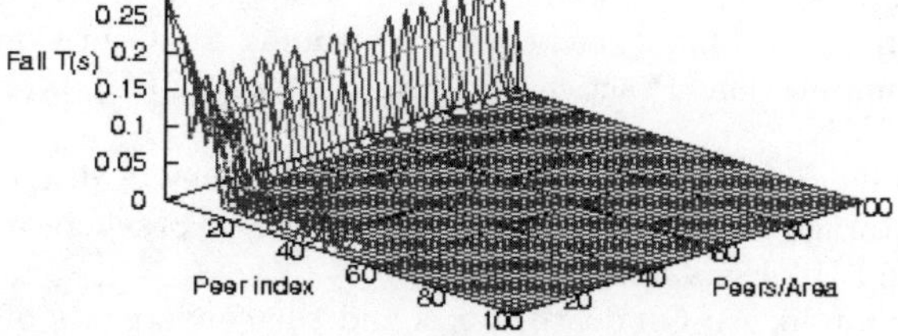

Fig. 4. Cost of leaving the system

restructuring time spent to balance the system when a single peer falls. Once all the system was created, the restructuring time was measured varying the index of the fallen peer from the first to the last one inserted into the system. We can see that the restructuring time was greater when the peer that failed was a manager (peers placed at the top levels of the tree). On the other hand, when a worker fails, the restructuring time is nil. So, smaller areas cause more peaks due to there being more managers in the system.

5 Conclusions and Future Work

In this paper, we present the CoDiP2P architecture, a pure Peer-to-Peer system for executing parallel jobs. CoDiP2P has a distributed, fault-tolerance and scalable architecture based on a tree hierarchy. The tree hierarchy is made up of areas of a fixed number of peers and each area is managed by a single manager. The four main functionalities of CoDiP2P are peer insertion, maintenance, job launching and scheduling and peer output mechanism.

By means of simulation, we demonstrate that the area size depends on the latency and it is not influenced by the network bandwidth. In addition, we show that the peer insertion, job launching and scheduling and output peer mechanisms decrease their time cost when the area size is increased. On the contrary, the communication weight related to the maintenance algorithm increased.

Nowadays, the CoDiP2P system is being developed using Java and the JXTA library. This real implementation will allow our simulator to be verified in relation to a real platform. Likewise, we are interested in designing a policy of Quality of Service (QoS) based on guaranteeing a specific satisfaction level to the users of the system. Another aspect to be improved is the system security. Implementing an authentication system and a secure communication channel for all main functionalities of CoDiP2P.

References

[1] Avaki (2006), `http://www.sybase.com`
[2] SETI@home (2006), `http://setiathome.ssl.berkeley.edu`
[3] The Gnutella Protocol Specication v0.6 (Document Revision 1.2) (2003), `http://rfc-gnutella.sourceforge.net`
[4] Buyya, R., Murshed, M.: GridSim: A Toolkit for the Modeling and Simulation of Distributed Resource Management and Scheduling for Grid Computing. Concurrency and Computation: Practice and Experience (CCPE) 14(13-15), 1175–1220 (2002)
[5] Clark, I., Sandberg, O., Wiley, B., Tong, T.W.: Freenet: A distributed anonymous information storage and retrieval system. In: Proc. Workshop on Design Issues in Anonymity and Unobservability (2000)
[6] Foster, I., Iamnitchi, A.: On death, taxes and the convergence of peer-to-peer and grid computing. In: Proc. of the 2nd. Int. Workshop on P2PSystems (2003)

[7] Gupta, R., Sekhri, V., Somani, A.: CompuP2P: An Architecture for Internet Computing Using Peer-to-Peer Networks. IEEE Transactions on Parallel and Distributed Systems 17(11) (November 2006)

[8] Ernst-Desmulier, J., Bourgeois, J., Spies, F., Verbeke, J.: Adding New Features In a Peer-to-Peer Distributed Computing Framework. In: Proc. of the 13th EuroMicro Conf. on Parallel, Distributed and Network-Based Processing (2005)

[9] JXTA Community Projects, `https://jxta.dev.java.net`

A Home-Automation Platform towards Ubiquitous Spaces Based on a Decentralized P2P Architecture

Sandra S. Rodríguez and Juan A. Holgado

Software Engineering Department. University of Granada
C/ Periodista Daniel Saucedo Aranda s/n, 18071, Granada, Spain
{sandra,jholgado}@ugr.es

Abstract. The vision of Home Environment is changing towards living interaction space populated of interconnected devices, services that encapsulate the functionality, and multiple interfaces through which the user can interact with these devices, in accordance with vision of ubiquitous computing. This paper presents a pervasive services platform for ubiquitous spaces based on distributed architecture P2P using JXTA technology, with an innovative approach, that favors the building of collaborative services from proactive entities, the peers. These ones are able to establish dynamic intercommunications synchronizing with others, form coalitions to cooperate with others for a common purpose, and are self-organized into groups.

Keywords: Home-Automation, Ubiquitous Computing, SOA, P2P, JXTA.

1 Introduction

Home-automation is an application domain where the problems of heterogeneity on devices, networks, hardware platforms and software frameworks have led to the proposal of multiple technologies and commercial products. Most of them are incompatible with each other and provide a partial management of some devices of home such as the surveillance, energy consumption or multimedia entertainment systems, making difficult an integrated solution to manage all of them. Other problems are the complexity, lack on security and high costs on installation, deployment and maintenance of these systems. To overcome these problems in our opinion we should change our vision towards pervasive spaces [1]. A pervasive space can be seen as an abstract logical environment where the devices provide compositions of functionality that users claim to consume. The development of software platforms that provide such abstraction should deal with the heterogeneity of devices, the mobility of devices or users, the seamless access to functionality and information resources, secure and reliable communications with fault tolerance, the interoperability between different networks, the evolving, extension and adaptations of functionality, and finally the interfaces between user and devices [2]. A large number of middleware technologies have emerged giving support to ubiquitous computing based on centralized and decentralized architecture such as Gaia, Aura, etc [2]. Distributed decentralized computing marked the next step toward pervasive computing by introducing seamless access to remote information resources and communication with fault tolerance, high availability, and security [3]. Currently, most middleware proposals are based on SOA (service oriented architecture) paradigm. SOA is an architectural style that establishes the

J.M. Corchado et al. (Eds.): DCAI 2008, ASC 50, pp. 304–308, 2009.
springerlink.com

organization model between the devices in terms of collaboration of loosely-coupled entities, the services. This work presents a new proposal for the design of a ubiquitous services platform to home-automation using Java technologies based on a decentralized P2P architecture. The new platform is based on JXTA [4], which is more scalable, interoperable, dynamic and extensible than other middleware platforms based on centralized client-server architectures. We are intent on leaving the classical model of residential gateway, leading to a large model, where services can be used beyond the living space.

The remainder of this paper is organized as follows. Section 2 presents the services platform for home-automation, its architecture and design decisions made during the platform development. Section 3 compares our approach with other related works. Finally, we present the conclusions of this work.

2 Dynamic Open Home-Automation Services Platform

A new distributed, ubiquitous, decentralized and dynamic platform is developed to facilitate the access, control and management of pervasive spaces from any computing device such as portable computer, portable device or embedded platform. This platform, named Dynamic Open Home Automation (DOHA), is based on JXTA architecture [4] and provides a full remote control of living environment in terms of services, according to SOA paradigm [5]. A service in the context of DOHA is an autonomous self-contained component which is able to perform specific activities or functions independently, that accepts one or more requests and returns one or more responses through a well-defined, standard interface. A service that provides functionality is a provider service, while the service that requests a service is a consumer service.

DOHA hides the physical distribution of devices as JXTA peers. This abstraction allows working with logical spaces based on services at high level. The cooperation between services at *services net* level involves communications between peers at *virtual net* level, and finally point-to-point communications between devices placed in different subnets at the lowest level, the *physical net*. DOHA is designed to ease interconnection of widely dispersed service nodes across the network, with loose coupling services and a dynamic model of operation.

2.1 DOHA Platform Design

We have taken some design decisions during the DOHA platform development to ensure robustness, scalability and security features, of great importance in ubiquitous computing systems. With these decisions we also guarantee the SOA principles of loose coupling, encapsulation, abstraction, reusability, composability, autonomy, optimization and discoverability.

Each service of DOHA is characterized by a multilayer architecture. The multilayer structure consists of several design layers that decouples the tasks performed by a service in components. This procedure facilitates the implementation and deployment of services, providing components to control the state of the service when a requested is accepted, the interactions of the service with the rest of services, etc. The *Interface Layer* guarantees the widespread access to services from any other element

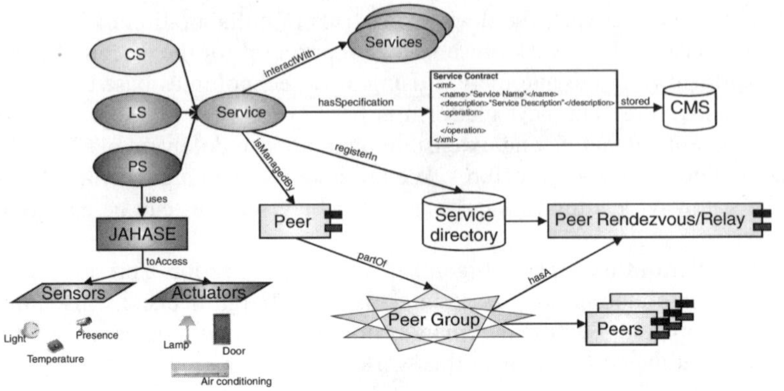

Fig. 1. Main elements of developed DOHA platform based in SOA principles

of the system. The *Application Layer* abstracts the functionality itself of a particular service. Finally, the *Interaction Layer* contains the logic needed to be able to communicate and collaborate with other services.

In Fig. 1 we can show the main ingredients to develop and deploy services in the DOHA platform. We have identified three types of system services depending on its role and responsibility: Customer Service (CS), Physical Service (PS) and Logic Service (LS). The applications may have different roles, and consequently can act with several service roles in the system. The *Customer Service* invokes the use of a specific service (provider service) or several services to satisfy the users requirements. Since the users will interact with them, they should provide a simple and natural way to modify the behavior of their living environment. The *Physical Service* is a provider service which interacts directly with hardware devices of the system (sensor or actuator). It can provide the state of the living environment, or change it, depending on the requests performed by other CS services. Finally, the *Logic Service* has specific functions to provide a specific task in either case, alone or in collaboration with other services. The services have a specification in XML, the *service contract*, to describe the services, its functionality and its methods. This specification is registered in a public directory of the service and it can be access via CMS (Content Manager Service) remotely. Although XML is used in the service contract, the developers are free to implement the data protocol more adequate to their purposes. The development of PS services requires the interaction with JAHASE as general platform to access hardware devices [7]. It hides the particularities of each hardware device such as sensors or actuators, and eases its management. Each service is managed by a peer that it is part of a peer group. The peer group is composed of many peers that it can interact with. The peer group has a Rendezvous/Relay peer allowing the discovery and communication between remote peers; thus, it eases the services interaction in a transparent fashion over different networks. We make the availability of services for multiple consumer services possible by the implementation of multithreaded peers and reliable communication channels between peers. This is possible with a controlled pool of channels to manage concurrently the requests made from other services. To more information about these aspects consult [8].

3 Discussion of Results and Related Works

In our study we have used the peer-to-peer architecture based on a SOA approach in order to obtain a system with the major characteristics associated with ubiquitous computing. In DOHA platform we have designed autonomous services capable of developing a collaborative behavior to carry out its functionality, with the publication and discovery mechanism associated with them. A set of adaptations have been made on JXTA middleware to optimize the system operation and obtain a behavior as independent and distributed as possible, such us the use of multithreaded pipes, detection and control of network failures, and the periodic search of advertisement to accelerate the discovery.

In the literature several studies in trends of construct ubiquitous systems from a peer-to-peer architecture can be found. The AMUN middleware (Autonomic Middleware for Ubiquitous Environment) is employed in the ubiquitous mobile agent system UbiMAS [9] and uses the JXTA as communication infrastructure based on an Event Dispatcher. In our approach, all services publish advertisements in the network when are available, and these events are filtered in each service, without the need of a centralized event dispatcher. Other similar approach is GAS-OS Architecture [10]. The GAS-OS Kernel Communication Module is responsible for communication between GAS-OS nodes and implements a P2P communication. In addition, DOHA complements the P2P features of JXTA with some design considerations such as multithreaded pipes, and detection and control of network failures. In CoCA a Collaboration Manager to share computing resources in a service platform is implemented [11]. The central figure of the Collaboration Manager might be a bottleneck in the system operations, which is resolved in DOHA using CMS among services. JXTA extensions are used in [12], proposing several modifications to the JXTA functionality that was proven successful in a mobile environment. In our case, the DOHA services platform has been tested in a home automation model with limited resource devices and we have obtained a satisfactory operation of the whole system [13].

4 Conclusions

With the development of DOHA we have been obtained a services platform, open, dynamic, reliable and extensible. The DOHA platform uses a peer-to-peer architecture instead of the traditional centralized client-server architecture that is generally used in the home-automation proposals. With the decision to use a P2P network based on JXTA technology, the DOHA platform has increased its modularity, scalability and independence characteristics.

The DOHA services platform eases the design, implementation and deployment of services that are independent in terms of functionality, based on the overall structure established. The platform is successfully applied to large-scale embedded devices [13], but when the memory resources are scarce, there is not space for JXTA middleware. A variation of JXTA for J2ME (CLDC-MIDP2) recently appeared is expected to be used.

References

1. Weiser, M.: Some Computer science issues in ubiquitous computing. Mobile Computing and Comunication 3, 12–21 (1993)
2. Saha, D., Mukherjee, A.: Pervasive Computing: A Paradigm for the 21st Century. IEEE Computer 36, 25–31 (2003)
3. Banavar, G., Beck, J., Gluzberg, E., Munson, J., Sussman, J., Zukowski, D.: Challenges: An Application Model for Pervasive Computing, pp. 266–274. Mobicom ACM Press, New York (2000)
4. Wilson, B.J.: JXTA. New Riders (2005)
5. Stojanovic, Z., Dahanayake, A.: Service-Oriented Software System Engineering: Challenges and Practices. Idea (2005)
6. Gong, L.: JXTA: A network Programming Environment. IEEE Internet Computing, 88–95 (2001)
7. Viúdez, J., Holgado, J.A.: Plataforma para el desarrollo de aplicaciones en entornos empotrados. In: I Symposium of Software Development, SDS 2007, pp. 147–162 (2007)
8. Rodríguez, S.S., Holgado, J.A.: A Services Platform for Home-Automation Based on Decentralized P2P Architectures. In: II Symposium of Software Development, SDS 2008, pp. 203–221 (2008)
9. Bagci, F., Schick, H., Petzold, J., Trumler, W., Ungerer, T.: Support of Reflective Mobile Agents in a Smart Office Environment. In: Beigl, M., Lukowicz, P. (eds.) ARCS 2005. LNCS, vol. 3432, pp. 79–92. Springer, Heidelberg (2005)
10. Drosos, N., Christopoulou, E., Kameas, A.: Middleware for Building Ubiquitous Computing Applications Using Distributed Objects. In: Bozanis, P., Houstis, E.N. (eds.) PCI 2005. LNCS, vol. 3746, pp. 256–266. Springer, Heidelberg (2005)
11. Ejigu, D., Scuturici, M., Brunie, L.: Hybrid Approach to Collaborative Context-Aware Service Platform for Pervasive Computing. Journal of Computers 3, 40–50 (2008)
12. Krco, S., Cleary, D., Parker, D.: Enabling ubiquitous sensor networking over mobile networks through peer-to-peer overlay networking. Science Direct 28, 1586–1601 (2005)
13. Rodríguez, S.S., Serrano, M.D., Holgado, J.A.: Control Domótico del Hogar sobre una Plataforma de Servicios Distribuida basada en JXTA. In: II Symposium of Software Development, SDS 2008, pp. 219–234 (2008)

Triplespaces as a Semantic Middleware for Telecommunication Services Development

David de Francisco[1], Marta de Francisco[2], Noelia Pérez[1], and Germán Toro[1]

[1] Telefónica Research and Development
{davidfr,npc,gtv}@tid.es
[2] Algor Consultoría y Sistemas
marta.defrancisco@algor.es

Summary. New generation of telecommunication services aims at achieving a customization and intelligent information management. Furthermore, most of these services are inherently collaborative. Developing this kind of services requires a middleware which can support intelligent information management in a decoupled and flexible way. This paper presents Triplespaces, a communication and coordination semantic middleware for the development of services based on persistent publication of information. Triplespaces manage information represented in RDF, supporting Web service standards. The authors motivate its benefits to develop new generation telecommunication services.

Keywords: semantics, middleware, co-ordination, triplespaces, telecommunications, marketplace, EAI.

1 Introduction

The increasingly impact of social Web based applications (Web 2.0), as well as the appearance of next generation mobile terminals, make telecommunication companies face a new business environment. This vision is supported by three observations: (1) users demand services which can offer a more useful information to them, (2) these services must intelligently manage the information they get from and provide to users, and (3) competence in telecommunication sector requires a fast time-to-market development of services.

Semantic Web is shaped as one of the main paradigms to follow in order to develop services taking knowledge management into account. The aim of the Semantic Web is to facilitate the knowledge reuse in different ways than those planned in advance by the author of this knowledge. This information reusing has the goal of creating user-oriented services, according to Tim Berners Lee's vision [1]. The first step in order to create these services has been taken by means of using microformats and ontologies [2], making use of W3C Web standards (RDF [3] and OWL [4] mainly). The standards mentioned above permit the information representation using different levels of expressiveness. The request of services which can effectively handle the knowledge produced by users entails the need to decouple the interactions, in order to ensure the scalability and a

J.M. Corchado et al. (Eds.): DCAI 2008, ASC 50, pp. 309–318, 2009.
springerlink.com

flexible implementation of such services. These requirements arise the need for an infrastructure for the communication and synchronization of services, called semantic middleware.

This paper has a double objective. On one hand, the authors motivate the need for a semantic middleware which can not only provide functional means of intelligently handling information, but also to simplify the development of services based on knowledge management. On the other hand, the Triplespaces platform is presented as a solution. With this aim in mind, a brief state-of-the-art is presented in Section 2, with a special emphasis on the paradigm that guides the development of Triplespaces. The physical and logical architecture of the built solution is presented in Section 3, which is being applied to the digital content marketplace use case briefly described in Section 4. In Section 5, the authors motivate the suitability of the proposed platform to its application in new business models in a telecommunication company. Conclusions and further work close the article in Section 6.

2 State-of-the-Art and the Triple Space Computing Paradigm

Space based computing, and more specifically tuple spaces [5], constitutes a valuable mechanism for the development of middleware platforms. The reasons are its high scalability and the use of coordination models (based on Linda language [6]). Coordination models synchronize the access to shared information in a transparent way. Linda defines three simple primitives: *out* (writing a tuple inside the space), *rd* (non-destructive reading of a tuple) and *in* (destructive reading of a tuple).

There are some works focused on the support of semantic information in tuple spaces, summarized in [7]. Semantic Web Spaces [8] involve an evolution of spaces with XML data support. The most outstanding, nevertheless, are based on the Triple Space Computing paradigm [9] [10]. This paradigm has brought about an Austrian project[1] with same name. CSpaces [11] was also born as an independent initiative based on this paradigm. More recently, the European project Triple Space Communication (TripCom , FP6)[2], is implementing an architecture with further functionality, which will be described later.

The Triple Space Computing paradigm is presented as a solution to communication problems between Web services derived from the point to point messaging followed by Web services [12]. Triple Space Computing proposes a solution based on the persistent publication of information exchanged by services. Triple Space Computing proposes an extension of Linda's coordination model [13]. This new coordination model uses RDF triples as elements for exchanging information. Using RDF implies an evolution from data exchange to knowledge exchange. The communication between services is decoupled in four different levels:

[1] See `http://tsc.deri.at/`

[2] This work is partially supported by EU funding under the TripCom project (FP6 - 027324). See `http://www.tripcom.org`

Temporal: services don't have to be active at the same time in order to communicate. The access to information can happen at different time from its publication, since it is persistent.

Spatial: services can be deployed in different computational environment. The communication infrastructure is distributed and independent of the sender and receiver implementation.

Reference: services can communicate anonymously, since the communication infrastructure acts as transparent mediator between them.

Schema: ontology-based knowledge representation defines a common vocabulary for applications. This grants the integration of heterogeneous data formats.

The next section describes the logical and physical architecture of the implementation designed in the TripCom project.

3 Logical and Physical Architecture

The physical architecture of Triplespaces is characterized for being distributed, decentralized and independent of the service implementation that uses it. These features facilitate the support of a wide number of scenarios and application domains. The architecture is composed by a distributed set of software nodes, called kernels [14]. These kernels interact among them using peer-to-peer communication [15]. Moreover, the architecture offers a simple and unique interface to the users of the platform. This interface is independent of the kernel to whom the user is connected, hiding all the complexity which derives from implementing the functionality described next.

A kernel presents the architecture showed in Figure 1. Starting from the lowest logical level component, we describe the functionality of each component briefly:

RDF Store: the information managed by the platform is represented in RDF triples. This component serves as physical storage and recovery environment of this type of information.

Triple Store Adapter: it is a wrapper of the physical storage environment which decouples the rest of architecture's components. It also includes functionalities to optimize the storage and recovery of information.

Security Manager: it manages security policies concerning the information management. These policies include the access control, authorization, authentication, writing permission, modifying permission and delete permission.

Mediation Manager: it performs the semantic mediation between the ontologies used by the services. These ontologies define the knowledge of the applications using the middleware.

Metadata Manager: it optimizes the distribution, writing and recovery of RDF triples by managing their metadata. An important contribution of the architecture is the distributed indexation of the information using Semantic Hash techniques (s-hash).

Transaction Manager: it has to manage the transactions on information operations, both local and distributed.

Distribution Manager: it efficiently distributes the information and the queries that require information from different sources.

Triple Space API: it is the interface and implementation of the communication model defined [13].

Management API: it is the interface and implementation to administer the kernels and security policies.

Web service API: they are the interfaces and implementations to invoke (Web service Invocation component), register (Web service Registry component) and discover (Web service Discovery component) Web services in the platform. The infrastructure defines mechanisms to deploy and invoke both Web services (WS specifications) and Semantic Web Services [16].

The described kernels offer two ways to be invoked: as a programming API (Triple Space API and Management API), and as Web service interface. This allows the integration of the platform with the most extended distributed technology nowadays: the Web services.

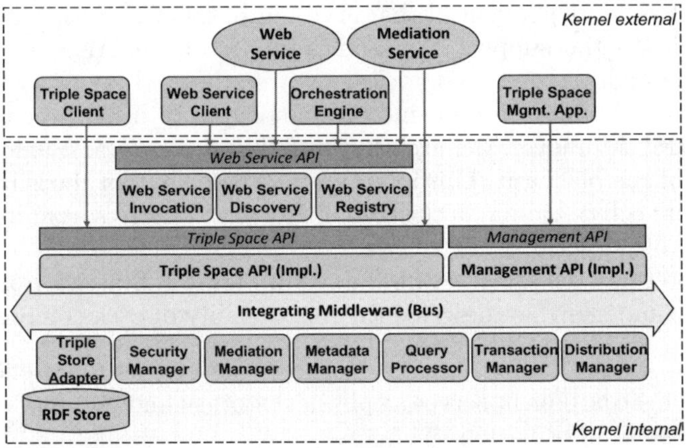

Fig. 1. Triplespace Logical Architecure [14]

The current implementation of the Triplespace kernel developed into the TripCom project is available for download as an open source project from SourceForge.net[3] uses Java as the programming language to allow its deployment in any operating system. The Triplespace kernel implementation uses Javaspaces[4] as the communication mechanism amongst the components that compose the kernel and, more specifically, the Blitz[5] implementation of the Javaspace specification. On the other hand, the distribution of the information amongst the

[3] See http://sourceforge.net/projects/tripcom/

[4] See http://java.sun.com/products/jini/2.0/doc/specs/html/js-spec.html

[5] See http://www.dancres.org/blitz/

kernels that compose the platform is based on P-Grid[6]. More detailed information can be found at the TripCom project's website.

The next section presents a digital content marketplace (an emerging telecommunication service) which is being implemented using Triplespaces as the underlying communication infrastructure. This prototype is both a proof of concept for the Triplespace architecture and a prototype for a future commercial implementation.

4 A Multimedia Content Marketplace Based on Triplespaces

The multimedia content marketplace use case is focused on the development of a marketplace to provide services based on multimedia contents. Several types of entities cooperate within this marketplace. The service suppliers are looking forward to obtaining digital contents to offer them to their customer through services. The content suppliers offer their contents through the developed marketplace. Users can subscribe themselves to services, providing feedback about the services and the contents that are receiving.

4.1 Information Hierarchy of the Use Case

In this use case, the communication between entities is a crucial aspect. The manager of the content marketplace needs to get in touch with these entities. This manager acts as a business mediator in which the involved entities trust. In [17], the benefits of applying Triplespaces as the underlying communication infrastructure in order to exchange semantic information in this use case are presented. A high level schema of the information hierarchy and information flow between actors can be seen in Figure 2. We stress out three aspects in this solution. First, the decoupled communication provided by the infrastructure allows a more flexible business model. Moreover, the use of subscription-notification mechanisms allows entities to obtain the necessary information in an asynchronous, precise (an actor can retrieve the relevant information for its businesses) and voluntary (an actor receive previously requested information) way. Finally, the use of knowledge in this scenario facilitates the automation in business negotiations. The content search and the negotiations among entities can be implemented through agents which make use of the shared knowledge.

4.2 Business Case and Communication Infrastructure Interaction

The functionalities provided by the architecture map business services from actors (Content Providers, Service Providers and Customers in our case) to communication functionalities offered by Triple Space Communication infrastructure via the Triple Space API. As an example of the translation process of business functionalities to communication functionalities offered by the infrastructure,

[6] See `http://www.p-grid.com`

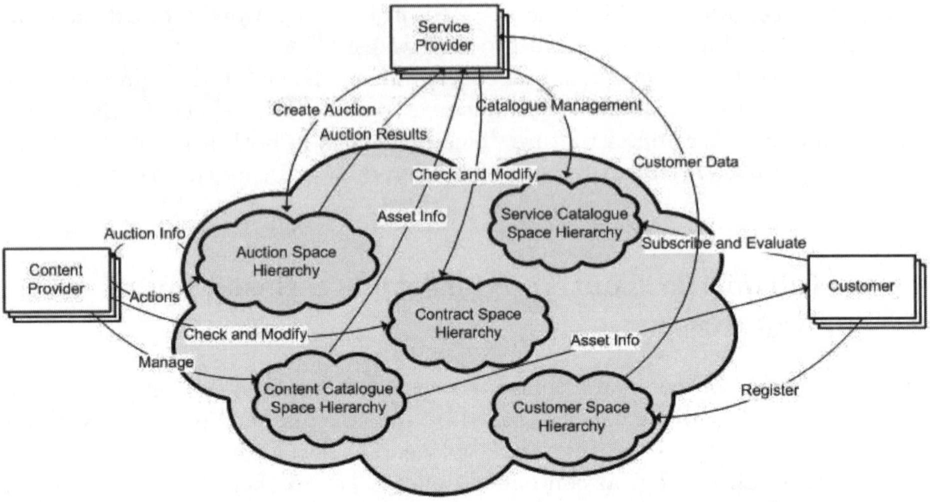

Fig. 2. Information Hierarchy and Flow of the Multimedia Marketplace

in this section we explain the auction management functionality. The translation is similar for the rest of functionalities provided by the architecture and is supported by these communication functionalities: information publication (out operation), blocking or not blocking information retrieval (in operation), space management (i.e, space creation and deletion, spaces policies management), subscription and notification (subscribe and notify operations).

The auction management scenario describes the behavior of an auction life cycle in which a Service Provider is looking for content in order to provide a content service. To accomplish it, the Service Provider starts an auction to get a provider for this content. Auction Participants will join the auction and perform binding (from an economical perspective) bids, which might be validated by the auction creator. Figure 3 depicts interactions between actors involved in an auction management life cycle from the infrastructure perspective in terms of Triple Space API operations.

The storyboard of these interactions is the following:

1. CPs are subscribed to the Auction Creation Space in order to get notified if any content being searched in any auction can be provided by themselves.
2. A SP publishes a new auction arrangement in the Auction Creation Space, in order to get a content to provide a service.
3. CPs subscribed to the space whose subscriptions are matched are notified of the new auction.
4. CPs interested in the auction can subscribe to it by publishing a request in the Auction Creation Space.
5. SP validates all subscriptions received from auction participants (i.e, checking internal black lists).

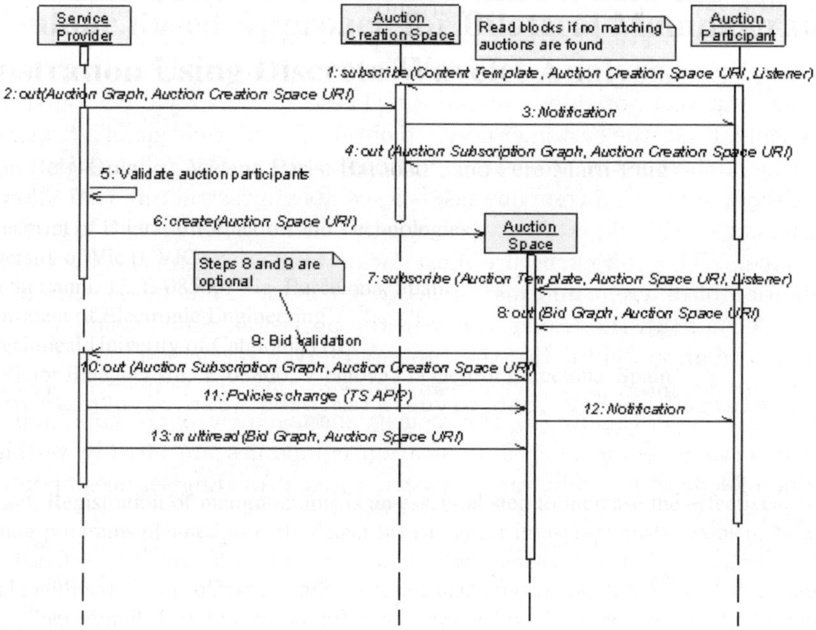

Fig. 3. Auction Management Interactions in Terms of Communication Primitives

6. SP creates an Auction Space allowing all validated CPs to write bids (no bid can be modified or deleted), as well as everyone to consult existing bids in the auction.
7. Subscription to auction management messages to get notified to things like winning bid change or auction end.
8. An Auction Participant writes a new bid.
9. The bid is validated by the SP following its own validation logic.
10. If the bid is rejected, the information of this bid is deleted from the Auction Space by the SP.
11. Once the auction timeout is reached (which is controlled by the Service Provider), policies of the Auction Space are changed, avoiding new bids to be published.
12. Auction Space notifies to Auction Participant the change of the usage policies.
13. SP reads all the bids published in the space in order to evaluate them.

5 Application of Triplespaces to Business Models

In the previous section a multimedia content marketplace has been presented. It is worth highlighting four characteristics that could be common to other business services. First, the information is generated by heterogeneous publishers. Second, the heterogeneous formats of the information published. Third, the need of synchronization in order to concurrently access to the information. Finally, complex communication patterns between publishers and subscribers. These features correspond to services focused on knowledge management, with a strong social and customer-oriented component. These services have two outstanding components. First, the customer needs to obtain the information in a transparent way. Second, the need of providing additional value to this information.

A telecommunication company needs a scalable infrastructure that allows the development of this type of services, which can be oriented to a critical mass of customers. This infrastructure must tackle four requirements. First (1), the combination of different information sources, solving the heterogeneity of these sources. Second (2), the transparency in the provision, combination and access to this information. Third (3), the support of Web services standards, so that services can be interoperable between different telecommunication companies and more reusable. Finally (4), the security management of the communication between services. Some telecommunication companies are already working in this kind of semantic middleware infrastructures, providing services upon them, such as Nokia [7].

The infrastructure presented in this article provides great flexibility and scalability in the implementation of this type of services. As we motivate next, it fulfills the above exposed requirements. The formal definition of the information through semantic technologies allows the mediation and combination of information (1). This mediation facilitates the interoperability among services inside and among telecommunication companies (3). The infrastructure offers mechanisms for publication, subscription-notification and blocking reading. These mechanisms allow to develop complex patterns of communication in a easy way for the programmer, such as a marketplace pattern [18] (2). Finally, the infrastructure provides mechanisms of security policy implementation (4).

6 Conclusions

In this paper the authors have motivated the need for a semantic middleware which can support the development of telecommunication new generation services based on knowledge management. These services are characterized by an intelligent information management and being user-orientated. As response to the need of a semantic middleware, Triplespaces, a scalable communication infrastructure between services based on tuple spaces has been presented. Triplespaces provide a decoupled communication and coordination infrastructure which can effectively handle semantic information. The information is represented in RDF, and it also supports Web services standards. Moreover, it provides synchronization in the information access and publish-subscribe mechanisms. All of these features provide transparency and flexibility for services developers.

The authors motivate the suitable applicability of this infrastructure to current telecommunication business models and more specifically, the potential benefits that Triplespaces can provide to a telecommunication company. In this

[7] See `http://research.nokia.com/tr/NRC-TR-2006-001.pdf`

sense, a multimedia content marketplace that is being developed within the TripCom project as a proof of concept has been presented. The following steps are focused on the evaluation of the scalability of Triplespaces through the simulation of marketplace competitors which can communicate over a distributed deployment of the application presented. Additionally, the implementation of transactionality in the knowledge management operations is being implemented currently.

References

1. Lee, B.T., Hendler, J., Lassila, O.: The semantic web. Scientific American (May 2001)
2. Gruber, T.R.: Towards principles for the design of ontologies used for knowledge sharing. In: Guarino, N., Poli, R. (eds.) Formal Ontology in Conceptual Analysis and Knowledge Representation, Deventer, The Netherlands, Kluwer Academic Publishers, Dordrecht (1993)
3. W3C: Rdf vocabulary description language 1.0: Rdf schema (February 2004), `http://www.w3.org/TR/rdf-schema/`
4. W3C: Owl web ontology language overview (February 2004), `http://www.w3.org/TR/owl-features/`
5. McLaughry, S.W., Wycko, P.: T spaces: The next wave. In: HICSS 1999: Proceedings of the Thirty-second Annual Hawaii International Conference on System Sciences, Washington, DC, USA, vol. 8, p. 8037. IEEE Computer Society, Los Alamitos (1999)
6. Gelernter, D.: Generative communication in linda. ACM Trans. Program. Lang. Syst. 7(1), 80–112 (1985)
7. Nixon, L., Simperl, E., et al.: Specification and implementation of a semantic linda model. tripcom internal deliverable (April 2007), `http://tripcom.org/docs/del/D3.1.pdf`
8. Tolksdorf, R., Bontas, E.P., Nixon, L.J.B.: Towards a tuplespace-based middleware for the semantic web. In: WI 2005: Proceedings of the 2005 IEEE/WIC/ACM International Conference on Web Intelligence, Washington, DC, USA, pp. 338–344. IEEE Computer Society, Los Alamitos (2005)
9. Fensel, D.: Triple-space computing: Semantic web services based on persistent publication of information. In: Aagesen, F.A., Anutariya, C., Wuwongse, V. (eds.) INTELLCOMM 2004. LNCS, vol. 3283, pp. 43–53. Springer, Heidelberg (2004)
10. Riemer, J., Martín-Recuerda, F., et al.: Triple space computing: Adding semantics to space-based computing. In: Mizoguchi, R., Shi, Z.-Z., Giunchiglia, F. (eds.) ASWC 2006. LNCS, vol. 4185, pp. 300–306. Springer, Heidelberg (2006)
11. Martin-Recuerda, F.J.: Towards cspaces: A new perspective for the semantic web. In: Proceedings of the 1st International IFIP/WG12.5 Working Conference on Industrial Applications of Semantic Web (IASW 2005), Jyvaskyla, Finland (August 2005)
12. Krummenacher, R., Hepp, M., et al.: Www or what is wrong with web services. In: ECOWS 2005: Proceedings of the Third European Conference on Web Services, Washington, DC, USA, p. 235. IEEE Computer Society, Los Alamitos (2005)
13. Simperl, E., Krummenacher, R., Nixon, L.: A coordination model for triplespace computing. In: 9th Intl. Conf. on Coordination Models and Languages (June 2007)

14. Martin, D., de Francisco, D., et al.: An architecture for a qos-aware application integration middleware. In: Proceedings of the 11th International Conference on Business Information Systems. Lecture Notes in Business Information Processing. Springer, Heidelberg (2008)
15. Minar, N., Hedlund, M., Shirky, C., O'Reilly, T., et al.: Peer to Peer: Harnessing the Power of Disruptive Technologies. O'Reilly and Associates Inc., USA (2001)
16. Cardoso, J., Sheth, A.P.: Semantic Web Services, Processes and Applications (Semantic Web and Beyond: Computing for Human Experience). Springer, New York (2006)
17. de Francisco, D., Pérez, N., et al.: Towards a digital content services design based on triple space. In: Abramowicz, W. (ed.) BIS 2007. LNCS, vol. 4439, pp. 163–179. Springer, Heidelberg (2007)
18. de Francisco, D., Elicegui, J.M., et al.: Using triple spaces to implement a marketplace pattern. In: Procceedings of the 1st Space Based Computing as Semantic Middleware for Enterprise Application Integration Workshop in 1st European Semantic Technology Conference, Viena, Austria, SpaceBasedComputing.org (May 2007)

Usage of Domain Ontologies for Web Search

Dulce Aguilar-Lopez, Ivan Lopez-Arevalo, and Victor Sosa

Laboratory of Information Technology, Cinvestav - Tamaulipas
Km. 6 Carretera Victoria-Monterrey
87276 Victoria, Tamaulipas, Mexico
{daguilar,ilopez,vjsosa}@tamps.cinvestav.mx

Summary. The Web is a wide repository of information available, but its heterogeneity, size and human oriented semantic supposes an obstacle in the search for desired information. Web search engines are a great help for accessing Web resources, nevertheless their classification algorithms are still limited since they only check the presence of a specific keyword or links, they do not analyse the semantic content of the resources. In recent years, several works are being related to convert the Web from an information space to a knowledge space by using common plans. One of the ways to achieve this purpose is the use of ontologies. The present paper proposes a methodology, where previously defined domain ontologies and the *WordNet* thesaurus are used to perform semantic searches obtaining suitable Web page results that really belong to the expected domain.

Keywords: Semantic search, ontologies.

1 Introduction

Nowadays most of the current search engines base their queries on keywords, linking, or index, they only verify if the keywords or links are inside the documents, and they do not pay attention on their semantic meaning. Thus they generate little prominent results for the theme that Web users are looking for, and finally, it represents a large quantity of lost time upon analyzing the results. A proposal to tackle this problem is to convert the information in knowledge by means of ontologies. With this, data can be used and understood for computers without need of human supervision. The definition commonly accepted of ontology is the proposal of Gruber [1], which say that is a matter of an explicit specification of a conceptualization. Specifically, an ontology is a common vocabulary for people and applications that work in a specific area of interest.

Within the most recent methodologies to perform semantic searches using ontologies are presented in Bocio *et al.* [2] who have developed a search module to find prominent Web pages for a domain of interest. They use domain ontologies and some parameters that the system user should introduce (name of the search engine, maximum number of resulting pages, language, and others). Another work is presented by Gao *et al.* [3], they use, as semantic structure,

J.M. Corchado et al. (Eds.): DCAI 2008, ASC 50, pp. 319–328, 2009.
springerlink.com

an ontology and the weight vectors crossing algorithm that they propose to analyze the initial information and store preliminary results based on keywords in an set of concepts. Its proposal builds a vector of weights according to the influence of this set inside the ontology. Ramachandran *et al.* [4] have developed a tool that uses the ontology of the LEAD project [5], which contains important concepts of the Atmospheric Science, besides definitions and relations among atmospheric phenomena, parameters, data, and services. The use of this ontology extends the capacities of search of their tool *Noesis* in a catalogue of metadata and Web resources upon using it for the keyword searches. Sánchez-Ruenes [6] implemented a tool to build an ontology from a chosen domain by means of obtaining related keywords from the Web. Then the concepts of the ontology are used to perform new searches and select the pages that belong to the domain.

The rest of the document is organized as follows: Section 2 describes the approach, describing the tools; Section 3 presents preliminary results obtained with the approach, Section 4 contains a summary of the future work, Section 5 discusses the conclusions and Section 6 presents the acknowledgments.

2 Ontological Approach

The main objective of this work is to obtain a Web page search method by using domain ontologies, through which the results will be obtained from a really chosen domain. In spite of the existence of the methodologies before mentioned, it does not exist a methodology that uses an overall solution including a mix of thesaurus and ontologies, which is our main approach to tackle the problem.

In our approach, it is considered as premise that well-defined ontologies are previously built, whose are used in semantic searches in the Web for english language. In addition, it includes the thesaurus $WordNet$ [7] and the search engines *Excite*[1], *Google*[2], $HotBot$[3], $Metacrawler$[4], and MSN[5].

The proposed methodology is supported by the architecture shown in Figure 1. It is composed by 5 main modules, whose are described as follow.

- **Input data.** The application requires the keyword to search as well as the domain wherein this search will be performed. The domains are predefined according to the ontology collection.
- **Web search.** In this module is performed the search of the keyword, using the search engines *Excite*, *Google*, $HotBot$, $Metacrawler$, and MSN. The appropiate search string for each search engine is built indicating the parameters that each one requires. The resulting Web pages of each search engine are stored temporarily.

[1] `http://search.excite.com/`
[2] `http://www.google.com`
[3] `http://www.hotbot.com/`
[4] `http://www.metacrawler.com/`
[5] `http://www.msn.com/`

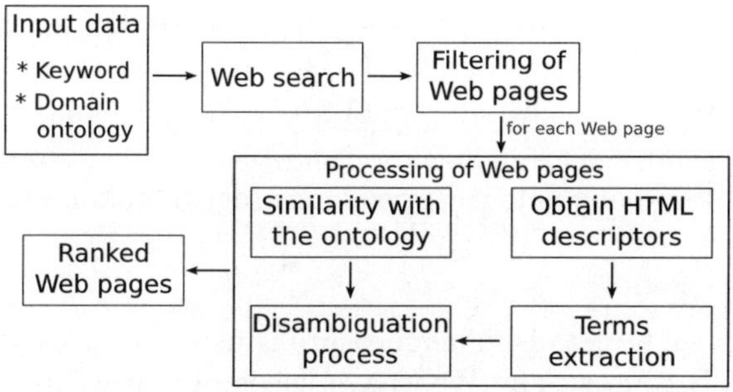

Fig. 1. Architecture modules

- **Filtering of Web pages.** In this module a unique list with the resulting Web pages of each search engine is built, eliminating duplicate and unavailable resources at moment of search (unreachable after 10 seconds).
- **Processing of Web pages.** For each Web page collected, the following is performed:
  - The HTML page is converted into a XML file descriptor containing the relevant text, eliminating the publicity that it could contain in order to have a cleaner content of the page.
  - The tool *WVTool* [10] is used to extract a list of prominent terms of the text obtained in the previous step. The *stop words* of the language are eliminated (articles, prepositions, etc.).
  - A disambiguation process over the term list of each page is carried out by means of the *WordNet* thesaurus. In this process are considered the senses (meanings) of the terms included inside the ontology and the terms of the Web page. They will be obtained using the glosses obtained from *WordNet* for each term, a gloss is a definition or example sentences for a term. The senses of the Web page will be compared with the senses of the ontology, obtaining the number of common senses between them. This process is a work in progress and it is planned to be improved.
  - The computation of the similarity between each Web page and the ontology is performed according to the Generalized Cosine-Similarity Measure [9]. In this measure, two collections A (ontology) and B (the Web page) are considered, whose are represented by the vectors $\overrightarrow{A} = \sum_i a_i \overrightarrow{l_i}$ and $\overrightarrow{B} = \sum_j b_j \overrightarrow{l_j}$ where $a_i = W(l_i) * Count_A(l_i)$ for $i = 1...n$ and $b_j = W(l_j) * Count_B(l_j)$ for $j = 1...n$; n is the total number of nodes of the collection, l means the leaves of the collections, and $W(l_i)$ is the weight of the node l_i. For two terms (l_i and l_j) taken of the collections before mentioned, the dot product is given by $\overrightarrow{l_i} \cdot \overrightarrow{l_j} = \frac{2*depth(LCA(l_i,l_j))}{depth(l_i)+depth(l_j)}$. The *depth* of a node is the number of edges on the path from the root to that node and LCA is the Lowest Common Ancestor, the node of greatest depth that is the ancestor for two nodes. For obtaining the similarity

between the collections is necessary to calculate the dot product of them by $\overrightarrow{A} \cdot \overrightarrow{B} = \sum_{i=1}^{n} \sum_{j=1}^{n} a_i b_j \overrightarrow{l_i} \cdot \overrightarrow{l_j}$. Then, the measure of the cosine similariy between two collections is given by the formula:

$$sim(A, B) = \frac{\overrightarrow{A} \cdot \overrightarrow{B}}{\sqrt{\overrightarrow{A} \cdot \overrightarrow{A}}\sqrt{\overrightarrow{B} \cdot \overrightarrow{B}}}$$

See [9] for more details. This measure is used to rank each page.

- **Ranked Web pages.** The Web page list is organized in descending order according to the score assigned in the previous module, eliminating those that have a score equal to 0, so that finally, they be shown to the user.

Following, a brief description of the tools used in the implementation is presented.

WordNet [7] is a tool that combines the advantages of an electronic dictionary and an online thesaurus used for meaning disambiguation, semantic labeling, and so on. It has been considered to use *WordNet* because it allows to obtain relationships among common terms. *WordNet* saves a significant effort to the application since it implements all the processes mentioned before. The Java API for accessing WordNet called JWNL (Java WordNet Library) has been used because it provides application-level access to WordNet data. The search engines *Excite*, *Google*, *HotBot*, *Metacrawler*, and *MSN* have been chosen to perform the search of the related terms in the domain ontologies. According to the surveys of Dujmović *et al.* [13] and Sánchez-Ruenes [6], such search engines have the best results based on performance, operating capacity, cover, answer time, quantity of queries per day or IP direction, and frequency of updating. The Word Vector Tool (*WVTool*) [10] is a flexible Java library for statistical language modeling. It is used to create word vector representations of text documents in the vector space model [11]. In the vector space model, a document is represented by a vector denoting the relevance of a given set of terms for such document. Terms are usually natural language words. In this work, *WVTool* is used to build a term vector from the Web page text. The resulting vector will be used to perform the disambiguation process.

3 Preliminary Results

The Figure 2 presents the proposed algorithm which performs the semantic search from Web resources.

Table 1 includes the topics of the 80 ontologies that have been compiled from the repositories Protege[6], Dumontier Lab[7], and SchemaWeb[8], in order to test the proposed methodology. Although the development of the application is still in

[6] http://protege.stanford.edu/
[7] http://dumontierlab.com/
[8] http://www.schemaweb.info/

Algorithm 3.1. SEMANTICSEARCH(*keyword*, *domain*)

```
for i ← 1 to totalEngines
  do {< links >←< links > +GETLINKSXENGINE(keyword)
< taxonomy >← GETTAXONOMYONTOLOGY(domain)
for i ← 1 to sizeof(< links >)
  do {< termsxLink >← GETTERMSXLINK()
< sensesOnt >← GETSENSESWITHWORDNET(taxonomy)
for i ← 1 to sizeof(< links >)
  do {< sensesxPage >← GETSENSESWITHWORDNET(link_i)
     {dis(Page_i) ← GETCOMMSENSES(< sensesOnt >, < sensesxPage >)
for i ← 1 to sizeof(< links >)
  do {similarityPage(i) ← GETSIMILARITY(< taxonomy >, < termsxLink >)
ELIMINATESIMILARITYZERO(< links >)
for i ← 1 to sizeof(< links >)
  do {CALCULATESCORE(links_i)
ORDERBYSCORE(< links >)
```

Fig. 2. Proposed algorithm for performing semantic searches

Table 1. Topics of the compiled ontologies from Protege, Dumontier Lab, and SchemaWeb

	Topics	
Amino acids	Employment	Physic
Animals Classification	Food	Pizza
Areas of research	Geography	Restaurant data
Beer	Material properties	Science Fields
Biological services	Medical terminology	Things for babies
Biosphere	Music	Types of plants
Breast cancer	Periodic table	Wines

progress, some tests have been succesfully performed. It is important to mention that the disambiguation process has not been totally implemented. The *Wine* ontology has been taken for the preliminary tests, a fragment of it is shown in Figure 3. It includes some classes, subclasses, instances, relations and restrictions of the ontology.

A set of 31 tests were performed with the methodology for the keyword "*red wine*" with the *Wine* ontology. The left-hand side of Figure 4 shows the statistics obtained from the runtimes of these tests, and the right-hand side shows a comparative graphic with the runtimes of the test. As it can be observed, the runtimes do not vary so much between each test. The test with the median runtime of the previous set is taken for showing the details of the test. The resulting

Fig. 3. Fragment of the Wine ontology

Best	259.758 s.
Worst	321.209 s.
Mean	287.478 s.
Median	285.549 s.
Variance	151.399 s.

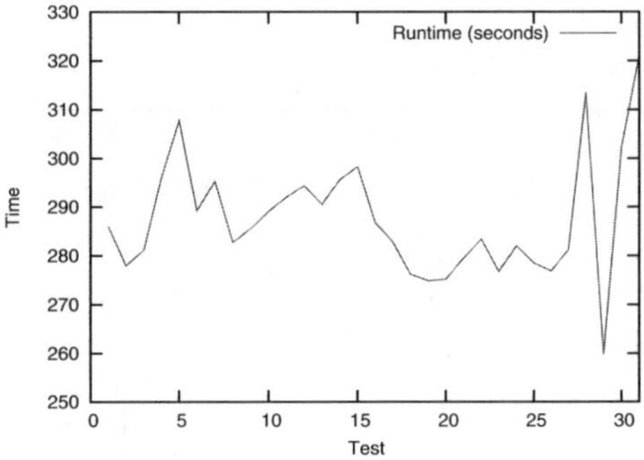

Fig. 4. Statics of runtimes of the 31 tests performed with the keyword *red wine* within the domain *Wines*, and a comparative of runtimes

Web pages[9] obtained from the search engines included in *Search Web* module are shown in Table 2.

As can be seen in Table 2, duplicate Web pages exist, also some links are unavailable. These results look like if the search engines were using similar strategies. That situation produces the same Web pages and the same broken links problem. This makes tedious and time consuming the process of select the best ones. After finishing the last two modules of the methodology, *Processing of Web Pages* module (with neither the disambiguation process) and *Ranked Web pages* module; the pages shown in Table 3 are obtained, whose are in descending order by their score, the similarity of each Web page was computed with the ontology according to the Generalized Cosine-Similarity Measure described in the previous section. Table 4 presents a comparative runtimes for the keyword

[9] The links were taken from each search engine, which are originally presented to the user. The number of links was reduced for presentation. Within the application, the 20 most promising links are considered.

Table 2. Preliminary results by Search Engine

URL	Search Engine
www.healthcastle.com/redwine-heart.shtml	Excite, Google, Metacrawler
wine.about.com/od/redwines/a/redwinebasics.htm	Excite, Google,
www.anconaswine.com/main.asp?request=PROMOGROUP&name=Our+Favorites	Excite
wine.about.com/od/redwines/A_Guide_to_Red_Wines.htm	Excite, Google, Metacrawler
en.wikipedia.org/wiki/Wine	Excite, Google, Metacrawler
www.ynhh.org/online/nutrition/advisor/red_wine.html	Excite, Google, Metacrawler, MSN
www.cancer.gov/newscenter/pressreleases/redwine	Excite, Google, Metacrawler
en.wikipedia.org/wiki/Red_Red_Wine	Google, Metacrawler
www.classicwines.com/Red-wine	Google, MSN
www.advance-health.com/redwine.html	HotBot
www.wine.com/wineshop/product_list.asp?N=7155+124	Metacrawler
en.wikipedia.org/wiki/Red_wine	Metacrawler
en.wikipedia.org/wiki/Red_wine_headache	Metacrawler

Table 3. The resultant Web page ordered by their score

Score	URL
0.37169	en.wikipedia.org/wiki/Wine
0.31664	www.advance-health.com/redwine.html
0.28362	wine.about.com/od/redwines/a/redwinebasics.htm
0.27767	www.healthcastle.com/redwine-heart.shtml
0.27648	www.ynhh.org/online/nutrition/advisor/red_wine.html
0.24994	www.classicwines.com/red-wine
0.21154	www.wine.com/wineshop/product_list.asp?N=7155+124
0.08112	en.wikipedia.org/wiki/Red_Red_Wine
0.07025	www.anconaswine.com/main.asp?request=PROMOGROUP&name=Our+Favorites

of the test, it considers the runtime for the search of the selected search engines and the proposed methodology.

If a user performs a search for a specific term in some of the used search engines it would take long time in analyzing the results seeking those that really are of interest for him/her. On the other hand, using our application, only will be able to visit the relevant pages since by means of the ontology, the results obtained from such search engines have been filtered. Despite the fact that this application can take longer time to respond compared to a popular search engine, the results

Table 4. Comparative runtimes for the keyword of the preliminary test, this includes the selected search engines and the proposed methodology

Search engine	Runtime
Excite	3.002 s.
Google	0.750 s.
HotBot	1.634 s.
Metacrawler	1.916 s.
MSN	2.876 s.
Proposed methodology	285.549 s.

obtained using our methodology will have better quality since the broken links and the pages outside of the domain are eliminated, which avoids the tedious task to visit them. Because of this, it is considered that the application offers better characteristics than a normal Web search engine in spite of the runtime disadvantage. This work is still in progress, that is why our results are not being compared with ones obtained by similar methodologies.

4 Future Work

The methodology shown in the section 3 (Figure 1), has not been totally implemented yet. The efforts are directed to the following activities.

- The disambiguation process will be implemented completely by means of *WordNet* using the proposed method in [12].
- This application will be implemented with JSP and Java Beans, so that any user can use it through the Web. It diseminates and facilitates the use and evaluation of the application.

5 Conclusions

The methodology described in this paper includes previously built domain ontologies, which allows a better understanding of the semantic of the words to be searched. It also considers additional tools like Web search engines, a text parser (*WVTool*), and a thesaurus (*WordNet*). This in order to obtain a useful tool for semantic search of Web pages really belonging to the domain of interest. As it can be observed in the tables of the Preliminary Results section, the proposed methodology eliminates results obtained from the selected search engines, those pages that are unavailable, which avoids the user the tedious task to visit them without success. Moreover, the list of pages is ordered according to the relation that keep with the domain ontology selected by the user, with it, some superfluous pages of the original list are also eliminated. The implementation of the methodology is still in progress. A restriction of the proposed methodology is that searches will be done only inside the domains restricted by the ontology collection, in comparison with the known search engines that allow to the user performs a search for any domain.

Acknowledgement

This research was partialy funded by project number 51623 from "Fondo Mixto Conacyt-Gobierno del Estado de Tamaulipas".

References

1. Gruber, T.R.: Toward principles for the design of ontologies used for knowledge sharing. International Journal Human-Computer Studies 43, 907 (1993)
2. Bocio, J., Isern, D., Moreno, A., Riaño, D.: Semantically grounded information search on the WWW. In: Recent Advances in Artificial Intelligence, Research and Development (Proceedings of Setè Congrès Català d'Intelligència Artificial (CCIA 2004)), Barcelona, Catalunya, pp. 349–356. IOS Press, Amsterdam (2004)
3. Gao, M., Liu, C., Chen, F.: An ontology search engine based on semantic analysis. In: ICITA 2005: Proceedings of the Third International Conference on Information Technology and Applications (ICITA 2005), Washington, DC, USA, vol. 2, pp. 256–259. IEEE Computer Society, Los Alamitos (2005)
4. Ramachandran, R., Movva, S., Graves, S., Tanner, S.: Ontology-based semantic search tool for atmospheric science. In: 22nd International Conference on Interactive Information Processing Systems(IIPS). 86th American Meteorological Society Annual Meeting, Atlanta, GA, USA (2006)
5. Droegemeier, K.K., Gannon, D.D.R., Plale, B., Alameda, J., Baltzer, T., Brewster, K., Clark, R., Domenico, B., Graves, S., Joseph, E., Morris, V., Murray, D., Ramachandran, R., Ramamurthy, M., Ramakrishnan, L., Rushing, J., Weber, D., Wilhelmson, R., Wilson, A., Xue, M., Yalda, S.: Service-oriented environments in research and education for dynamically interacting with mesoscale weather. IEEE Computing in Science & Engineering 7, 24–32 (2005)
6. Sánchez-Ruenes, D.: Domain Ontology learning from the Web. PhD thesis, Universidad Politécnica de Cataluña, Departamento de Lenguajes y Sistemas Informáticos (2007)
7. Morato, J., Marzal, M., Lloréns, J., Moreiro, J.: WordNet applications. In: Proceedings of the Second Global Wordnet Conference, pp. 270–278 (2004)
8. Mierswa, I., Wurst, M., Klinkenberg, R., Scholz, M., Euler, T.: Yale: rapid prototyping for complex data mining tasks. In: Proceedings of the 12th ACM SIGKDD International Conference on Knowledge Discovery and Data Mining (KDD 2006), pp. 935–940. ACM Press, New York (2006)
9. Ganesan, P., Garcia-Molina, H., Widom, J.: Exploiting hierarchical domain structure to compute similarity. ACM Trans. Inf. Syst. 21, 64–93 (2003)
10. Wurst, M.: The Word Vector Tool User Guide (Consulted, February 1, 2008) (2008), `http://nemoz.org/joomla/mining/wvtool/wvtool.pdf`
11. Salton, G., Wong, A., Yang, C.S.: A vector space model for automatic indexing. Commun. ACM 18, 613–620 (1975)

12. Patwardhan, S., Pedersen, T.: Using WordNet based context vectors to estimate the semantic relatedness of concepts. In: Proceedings of the EACL 2006 Workshop Making Sense of Sense - Bringing Computational Linguistics and Psycholinguistics Together, Trento, Italy, pp. 1–8 (2006)
13. Dujmovic, J., Bai, H.: Evaluation and comparison of search engines using the LSP method. In: Computer Measurement Groups International Conference, Reno, Nevada, vol. 2, pp. 711–722. Computer Measurement Group (2006)

An Ontology for African Traditional Medicine

Ghislain Atemezing and Juan Pavón

Dep. Ingeniería del Software e Inteligencia Artificial
Universidad Complutense Madrid, Ciudad Universitaria s/n, 28040 Madrid, Spain
atemezing@yahoo.com, jpavon@fdi.ucm.es

Abstract. This paper describes an ontology for African Traditional Medicine (ATM), which is the basis for a knowledge management system, controlled by a multi-agent system. The interest of this problem, from the point of view of artificial intelligence and software engineering lies on the issues that arise from integration of the requirements of the different stakeholders in such a system and the diverse nature of concepts to be considered in such an ontology. One of these issues is the need to allow the ontology to evolve as far as experts provide more knowledge and the mechanisms for validation of such knowledge.

Keywords: African Traditional Medicine, Ontology, Multi-agent systems, Knowledge Management.

1 Introduction

African Traditional Medicine (ATM) is the result of diverse experience, mixing customs and knowledge about Nature, which has been transmitted by oral tradition along the history. Today, the availability of computers and networks in more and more places around the African continent opens the possibility to consider the support of knowledge systems for new practitioners, who can take benefit of ATM knowledge. Building an ATM knowledge management system requires first a formalization of ATM concepts and their relationships. From the software engineering point of view, this task implies several challenges: the specification of an ATM ontology, and the development of tools for allowing experts in ATM to build a knowledge base, validate such knowledge, and recover it when needed. This paper addresses the problem of defining an ontology for ATM that should be easy to extend as required, and that facilitates structuring and integration of knowledge from different complementary areas, as described below.

An attempt to structuring ATM concepts in order to bring these data into a well-structured representation [1] was not within the mark up of other medicine ontologies. Hence, the definition of an ATM ontology has to consider many aspects derived from traditional medicine particularities. First, there is a need to complete ontologies from other domains, e.g. from Botany and Medicine, which are incomplete in the scope of ATM. For instance, it is necessary to take into account the role of elements from Nature in rituals and traditional domains. This has not been addressed yet in other works, probably because the lack of concrete applications dealing explicitly with these domains. The problem of integration and extension of these ontologies in relation with ATM requires the consideration of an open ontology, that should be able to evolve

J.M. Corchado et al. (Eds.): DCAI 2008, ASC 50, pp. 329–337, 2009.
springerlink.com

with contributions from experts in different fields (medicine, botanic, rituals, etc.) and being able to contrast experts knowledge in order to validate new concepts.

Due to the lack of a common and structured vocabulary specifically dedicated to ATM, as it is a particular and sometimes efficient way of many traditional healers in Africa to contribute to health issues of the native population, a well structured computational representation of ATM domain can therefore be used to manage knowledge and information gathered from the field practices. In addition, using the same concepts for the description of this domain in other similar ontologies would facilitate interoperability among them.

To address these needs, this paper presents an ontology that describes the ATM domain, which can be used by experts of the field and the scientists' community with interests in the development and the expansion of a different way of treatment and cure. The benefit is the protection by the means of new technologies of many centuries of oral transmission knowledge, which is in the way of disappearance. The use of the ATM ontology intends to promote an harmonization and integration of data from diverse sources.

The rest of the paper is organized as follows. Section 2 reviews some medicine ontologies that are taken as reference for the definition of ATM ontology. This is followed by the presentation of the main concepts of ATM in Section 3. Section 4 presents how the ontology is implemented and its elements. Section 5 briefly describes the multi-agent system that is used to manage the ontology. Finally, Section 6 discusses how this work could be extended for knowledge management in other domains, and identifies open issues.

2 Medicine Ontologies

Ontology is a technical term denoting an artifact that is *designed* for a purpose, which is to enable the modeling of knowledge about *some* domain, real or imagined [2]. Ontologies play an important role in facilitating information retrieval for structured controlled vocabularies and relationships between terms. Since the development of the Gene Ontology (GO) for the annotation of attributes of gene products, many ontologies have been developed and many of these are available from the Open Biological Ontologies (OBO) site [3]. They include several ontologies from different aspects such of health, anatomy, environment, taxonomy, biological process. Thus, there exist ontologies on biological processes, cell types, environments, human diseases, infections diseases, pathogen transmissions and plant structures, among others. Each of these ontologies covers a specific domain with the purpose of being more precise and specialised. Hence, it is normal that for a large domain like medicine, a unique ontology does not exist to cover all the aspects.

The approach for management an ontology for ATM starts from considering some existing ontologies, part of the modern medicine ontologies, such as plant structure, human disease and disease transmission, could be used in the ATM ontology. In concrete, the following ontologies have been taken into account because they are closed and can be considered also as part of the domain, as medicine issue en general:

- The Plant Ontology (PO) [4] contains many classes, but for ATM the interest is mainly with *<plant structure>* class, and not others such as *<in vitro*

cultured cell, plant cell and tissue>. Figure 1 shows the relationships between concepts within that specific class by the use if Directed Acyclic Graph (DAG) viewer.

- The Pathogen Transmission Ontology (TRANS) [5] describes the means during which the pathogen is transmitted directly or indirectly from its natural reservoir, a susceptible host or source to a new host. It considers two types of transmission: direct and indirect. The former has three subclasses (congenital, contact and droplet spread) and the later structured in airborne, vector borne and vehicle borne.
- The Human Disease Ontology (DOID) [6] classifies diseases in five groups: behavioral disease, biological process disease, disease of anatomical entity, disease on environmental origin and syndrome. It also has at the top level class <*temp holding*> with head, neck, dermatologic disease, etc.
- The Infectious Disease Ontology (IDO) [7] deals with the means during which the pathogen is transmitted directly or indirectly from its natural reservoir, a susceptible host or source to a new host, en definitely the process transmission.

3 Concepts of African Traditional Medicine

African Traditional Medicine (ATM) is a complex system of cure in which disease is considered as a social illness, which is necessary to eradicate from the root. There is intervention of several actors from several domains, which turns complex and diversified exchanges and treated knowledge. This knowledge is passed on with oral way and is not structured [8].

Several actors can be identified in ATM, with specific roles and functions:

- The h*ealer* is a well-known and respected person, psychologist, botanist, pharmacologist and doctor. He knows the names of plants, animals and rocks.
- The fetishist predicts important events (misfortune or happiness) and is consulted to find the cause of a disease, to protect against certain misfortunes.
- The *Soothsayer* predicts and is seeing as the intermediary with the divinity. He generally diagnoses but can advise a healer to a patient.
- The *Magician* throws lots and makes use of black arts. But he is part of the actors as well.

In the process of treatment, a healer tries to reconcile the patient in all its integrity, as well physical as psychic, by using symbols that are a part of the universe and the life of every day of the patient during its interventions. For example, one pinched of ground collected in the market will represent the social activity of the patient; bits of gravel collected in the crossroads, public life and a washbasin of water, the river at the edge of which extends the village of origin.

In ATM, it is also necessary to distinguish symptoms from disease. When a patient meets a traditional doctor, he suffers from the evil of which one can attribute a name in human disease of the modern medicine. But for the traditional doctor, this patient is seen as a person who possesses a *symptom*, a sign of a social illness. *Social illness* expresses tensions (hidden or revealed) that could exist in the circle of acquaintances of the patient. Certain anthropologists, such as [9], introduce the concept of traditional model to express all that is lived in the traditional vision. The body in this model consists of two entities: a visible part and an invisible part.

The global step in the traditional model is the following one:

- Interpretation of the cause of the bad physical appearance (sort of diagnosis): the traditional doctor considers that the disease that the patient suffers can result from several sources: death ancestors, who continue to live and who sometimes show their *dissatisfaction*, acting on the alive, witchcraft, incest (the fact of falling under the yoke of the forbidden), twins who possess supernatural powers, destiny which is individual, God who here is a natural cause. [10]
- Phase of divination to know how to treat the patient.
- Prescriptions according to the cause of the disease: remedies with natural base, ritual products and other sacrifices to be done. [11]
- Follow-up of the patient evolution in the process of cure which sometimes can take years.

From the above description, some relevant concepts can be pointed out, as part of the ontology. Concepts such as:

- FUNCTION for the actors of this medicine: the healer, the fetishist or the soothsayer.
- PROCESS for all the different types of proposed process of treatment.
- SYMPTOMS for the role of the symptoms.
- DISEASE as it is considered in this medicine.

4 The ATM Ontology

The first design decision was whether to integrate all the aspects of ATM or only those that make it different from modern medicine. Although it may be relatively easy to design an ontology based on concrete facts such as names, birth dates, etc., it is considerately more difficult to design an ontology based on knowledge that is incomplete or not yet well understood, or that it does not have yet a common-agreed vocabulary [12].

The ontology consists of concepts or terms (nodes) that are linked by three types of relationships (edges). That means the ontology appears as a directed acyclic graph. The parent and child terms are connected to each other by *is_a* and *part_of* relationships. The former is a relation in which the child term is a more restrictive concept than its parent (thus divination *is_a* traditional_practice). The latter is used to show the inclusion relationships between concepts, for example that a potion_type is *part_of* a potion.

The rules for building the ontology are the same as those defined by the GO consortium. That is, each concept in the ATM Ontology has an identifier with the syntax ATM:nnnnn, where nnnnn is a unique integer, and ATM identifies the ATM Ontology. In addition, if there are precisely equivalent terms in other databases, for example in the Plant Ontology, the unique identifiers from these databases are included in the ATM Ontology. The present version of the ontology has an average depth of about five nodes.

The six top-level nodes of the ATM Ontology are *disease_conception, traditional_act, traditional_believes, traditional_intervenor, traditional_practices* and *traditional_treatment*. The *disease_conception* includes abnormal disease (disease caused by external agents such as spirits, etc.) and natural disease. All the actors of the medicine are included into the *traditional_intervenor* class. Regarding to the form of diagnosis, divination and other religion practices, they are classified as children of *traditional_practices* class. Concerning the way treatments are done, from the medicinal plants to rituals, we have the *traditional_treatment* class to structure the previous concepts.

It should be pointed out that, like many such resources, this ontology is not complete: although it contains the main concepts of traditional medicine, it is intended to be extended and completed by the experts in an incremental way. For example the

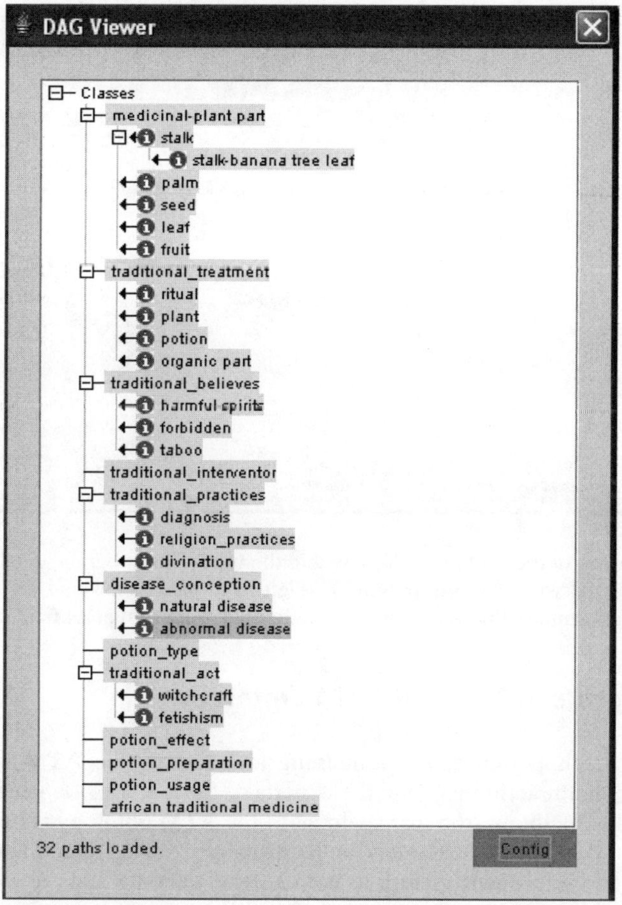

Fig. 1. ATM ontology view within a Directed Acyclic Graphs structure

categories identified as *traditional_believes* or *traditional_act* have to be much more populated. The ontology was constructed using the open source Java tool OBO-Edit, which is convenient for building ontologies that are consistent with the GO formalism. The resulting ontology [13] is available in the "OBO format" [14] and can be easily viewed.

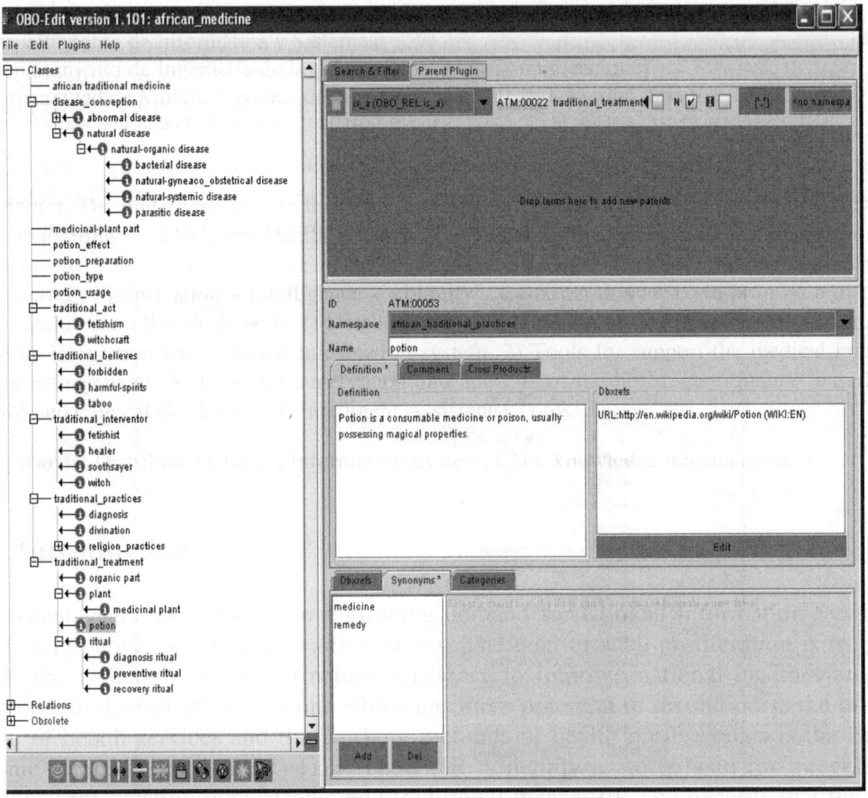

Fig. 2. A screenshot of the ATM Ontology within the OBO-Edit program, displaying all the information associated with the term potion. The left-hand panel shows all the top-level terms, together with the location of potion within the traditional_treatment classification.

5 Using Agents to Manage ATM Ontology

According to the importance of maintaining and sharing the ATM knowledge, for both the traditional practitioners and the experts of the domain, the need for having an environment that facilitates the management of the ATM ontology is necessary to better make use of the new techniques in Knowledge Management (KM) techniques. Based on that, a multi-agent system to help access, validate and recover ATM information, described in [15] has been designed and is currently being implemented. The system consists of five main types of agents: user, interface, ontology manager, profile, and broker agents.

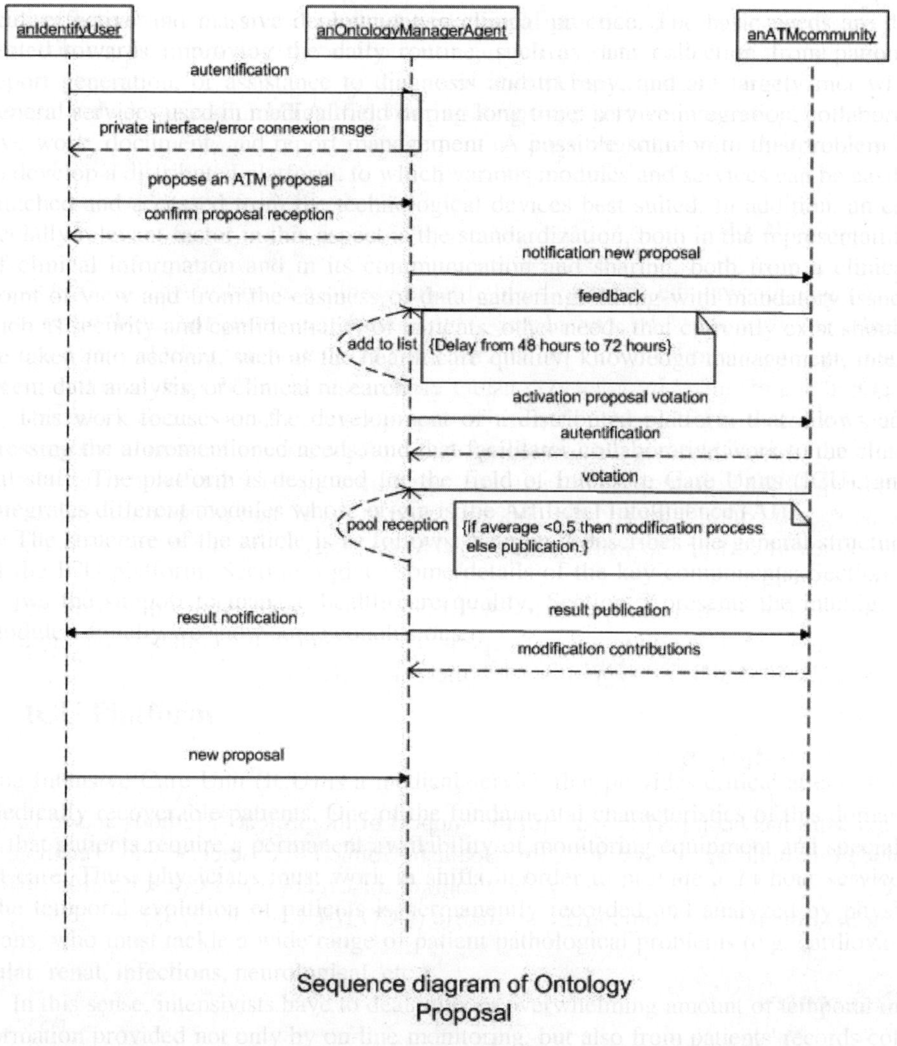

Fig. 3. Sequence diagram for management of a new proposal in the ATM Ontology

Each user of the system is managed by a *user agent*, which is in charge of the connexion, error reports and assignation of authentication to the registered users, and the notification of context information to the user from the system.

The *interface agent* facilitates the user's view of the system depending of the users preferences and communicates with the *user agent*. Also deals with the session connection.

The *ontology manager* manages ATM information proposal, modification and publication. That's says all the cycle of proposal to the publication of ATM information.

The *profile agent* keeps the users preferences of the system to give contextual information and suggestions to the system users.

The *broker agent* realizes the connection to the ontology, retrieve and modify knowledge. It is also in charge of doing the necessary conversion between different types or formats of manipulated knowledge.

Figure 2 gives a view of the proposal sequence of ATM information, showing the role of the Ontology Manager Agent in the process.

6 Conclusion

This paper presents the effort to build structured information for African Traditional Medicine. This starts by considering some ontologies from the field of modern medicine, structured following the Open Biological Ontologies (OBO), which are integrated and extended in order to consider elements from ATM, such as actors, treatment process and the general step in its way of curing illness. The identification of such elements is the basis for developing the ontology, reusing existent ontologies and applying OBO-Edit tool. This facilitates the follow-up of relations that exist with some specific ontologies and the concepts used in ATM, and points out some different choice of conception while building ATM ontology regarding to the particularity of the domain.

Some ideas used in this paper for ATM could be transposed to others domains or applications where a community of experts needs to build in a collaborative environment a knowledge based support system incrementally.

Acknowledgments

This work has been performed with the support of the project “Methods and tools for modeling multi-agent systems", by Spanish Council for Science and Technology (TIN2005-08501-C03-01), and of the Program for Creation and Consolidation of Research Groups UCM-Comunidad de Madrid (CCG07-UCM/TIC-2765).

References

[1] Fotso, L.P.: Table of Entities and Attributes of Data Bases in MEDITRA (Knowledge based on African Traditional Herbal Medicine). In: Rapport de Recherche no. 20, Université de Yaoundé I, Février (1999)

[2] The Gene Ontology Consortium. Creating the gene ontology resource: design and implementation. Genome Res. 2001 II, 1425–1433 (2001)

[3] OBO: Open Biological Ontologies, http://www.obofoundry.org/

[4] Plant Ontology, http://www.obofoundry.org/cgi-bin/detail.cgi?id= po_anatomy

[5] Pathogen Transmission Ontology, http://www.obofoundry.org

[6] Human Disease Ontology, http://www.obofoundry.org/cgi-bin/detail. cgi?id=disease_ontology

[7] Infectious disease Ontology, http://www.obofoundry.org/cgi-bin/detail.cgi?id=infectious_disease_ontology

[8] Atemezing, et al.: Modélisation multi agent massif d'un système d'aide à la décision en médecine traditionnelle. In: Annales de la Faculté des Arts, Lettres et Sciences Humaines, University of Ngaoundéré, pp. 199–216 ISSN: 1026-3325
[9] de Rosny, E.: L'Afrique des guérisons, Karthala, Paris (1992)
[10] Laplantine, F.: Anthropologie de la Maladie, Payot, Paris (1986)
[11] Tall, E.K.: Guérir à Cubatel: Interprétation de la maladie et pratiques thérapeutiques chez les Haalpulaaran dans la vallée du fleuve Sénégal. Ecole des Hautes Etudes en Sciences Sociales, Paris (1984)
[12] Bard, et al.: An ontology for cell types. Genome Biology 2005 6(2), Article R21 (2005), http://genomebiology.com/2005/6/2/R21
[13] http://www.bioontology.org/files/32145/african_medicine
[14] Gene Ontology: the OBO flat format guide, http://www.geneontology.org/GO.format.html#boflat
[15] Atemezing, G., Pavon, J.: Intelligent Environment for Medical Practices in African Traditional Medicine. In: 6th International Workshop on Practical Applications of Agents and Multiagent Systems (IWPAAMS 2007), Salamanca, Spain, pp. 101–108 (2007)

A Tool to Create Grammar Based Systems

Vivian F. López, Alberto Sánchez, Luis Alonso, and María N. Moreno

Dep. Informática y Automática
University of Salamanca
Plaza de la Merced s/n, 37008, Salamanca, Spain
vivian@usal.es, lalonso@usal.es, mmg@usal.es

Abstract. This paper describes an Integrated Development Environment (IDE) for the automatic generations of language-based tools to be used in the grammar-based systems, that adopts the functionality of the classic tools and other innovative solutions to ease the implementation of these systems in the new fields of grammar application. The IDE attenuates the complexity of the manual design of the grammar specification, improves the quality of the obtained product and sensibly diminishes the development time and cost. We tried to reduce the learning time for users not expert in the area of compiler generation.

Keywords: context-free grammars, syntax, formal language theory, parser, compiler generation, language-based tools, grammar-based systems.

1 Introduction

The increasing amount of Domain-Specific Languages (DSLs) [14] created in the different scopes of Science leads to the necessity of automatic design methods to generate analyzers and/or language translators that facilitate this task.

There is a perfectly defined theory that eases the development of this type of tools for programming languages like C, C++, Java, etc. The difficulties arise when a processor for a new language is needed, when the traditional methods are sometimes of little efficient since the time inverted in the design is very high with relation to the obtained results and the tools that are used do not support features like incremental language development or reusability.

Traditional applications allow to obtain a valid solution, but a revision becomes necessary, to ease and extend their functionality, by including new techniques as the handling of data in XML format, the using of graphical interface or the application of object-oriented concepts for grammar specifications.

The main goal of the paper is to create an Integrated Development Environment (IDE) for the automatic generation of language-based tools [5] that starts from the traditional solutions and facilitates the use of formal language theory in other disciplines: grammar-based systems (GBSs) [10]. These systems allow the development of different practical applications within the area of *grammarware*, where the concepts of the language analysis are applied to other disciplines, like software engineering, bioinformatics, evolutionary computation or neural networks.

J.M. Corchado et al. (Eds.): DCAI 2008, ASC 50, pp. 338–346, 2009.
springerlink.com

1.1 Related Work

In this section we briefly review relevant research for the work presented: YACC [15] a language used to specify grammars in parser generators. It generates a LALR(1) parser [7] written in C, from a grammar in Backus–Naur Form (BNF). The handling of the errors is made by inclusion of *token* "error" and the incorrect elements are replaced up to a synchronization point user defined.

CUP [3] analogous to Yacc, generates Java code. It can be used in combination with JFlex [3]. The production rules are oriented to procedures. SableCC [2], similar to the previous ones, in case of finding an error produces an exception with the location and the content of the error and finishes the execution.

Beaver (beaver.sourceforge.net/) similar to Yacc. When it locates an error, it sends a process based on techniques of backtracking that tries to recover the analysis by eliminating erroneous token or introducing the expected element.

A great amount of generators of top-down analyzers also exists, like ANTLR (www.antlr.org/), JavaCC [3], LISA [9]. As far as the error treatment is concerned, these tools normally use exception treatment mechanisms that the user can associate with the specification rules.

The remainder of this paper is organized as follows: Section 2 presents an overview of the design decisions described in the paper; Section 3 introduces the parser generator, including the techniques and algorithms involved; Section 4 illustrates the system description. Concluding remarks and future research work are given in Section 5.

2 Design Decisions

Based on this theoretical background, a system was developed that consists of several tools: editor, scanner generators, parser generators, compiler generators, and graphic tools (inspectors, visualizers, evaluators).

The traditional tools needs implicit information given by the user, to automatically generate a language-based tool. The aim of this approach is to remedy this situation by defining a language independent implementation. In this respect, we constructed the compiler generator GAS 1.0, that automatically generates a scanner and a parser for language specification. Taking as input the grammar specification (graphical or textual), the syntactic analysis tables are created, giving as result an decorated abstract syntax tree (DAST), or the syntactic error, as it is shown in Figure 1.

Based on [13] we try to automatically generate language-based tools from an extended language definition in a systematic manner by identifying generic and specific parts.

The system is designed by a technology of generic structure of components. The design of language-based tools is made as far as possible from the language details. It takes decisions based on the user provided data.

Since only the information about the language given by the designer is provided, it is not allowed to make assumptions about the language to analyze. This raises a serious disadvantage from the traditional point of view of the treatment of some processes within the analyzer, for instance the error treatment, where traditionally the intrinsic characteristics from the language are taken into account to carry out procedures that allow to solve such situation.

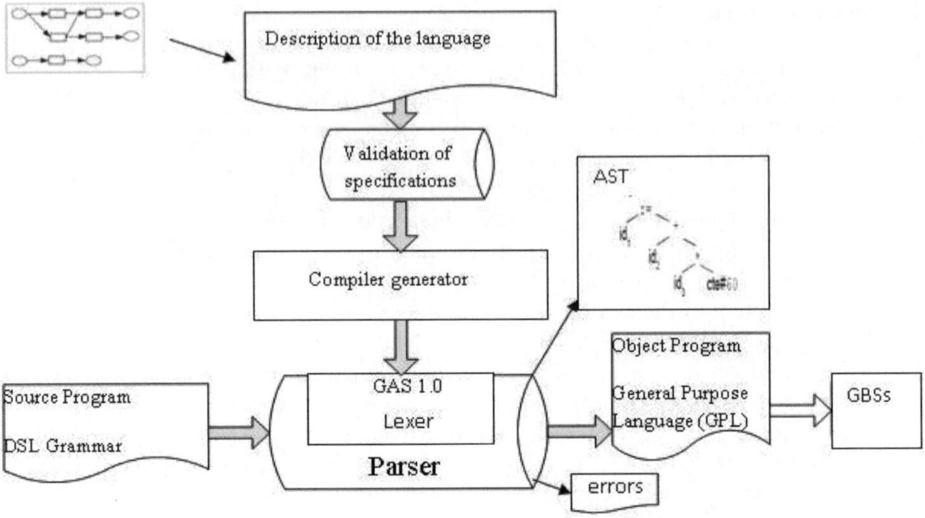

Fig. 1. The compiler generator GAS 1.0

2.1 Use of Patterns

Following the object-oriented paradigm the final set of classes is obtained by refining the preliminary set: classes that manage the grammar specification.

Considering the modularity in the developed system, patterns (***Layer)*** have been used, to establish a defined set of layers that logically structures the operation of the application with relation to the component modules. The packages are created as shown in Figure 2.

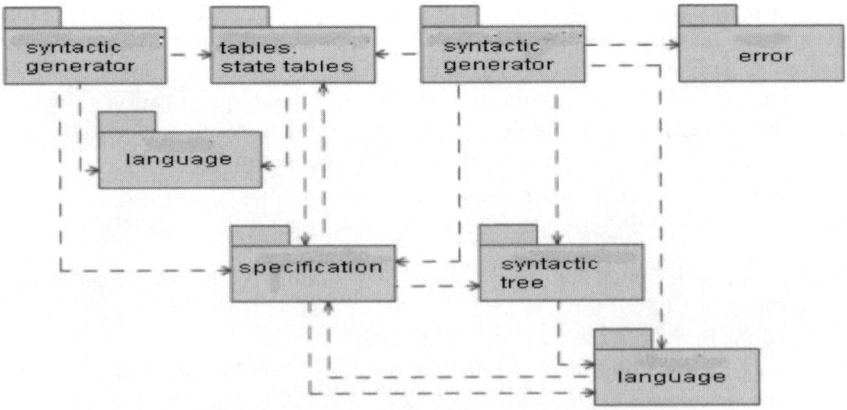

Fig. 2. Conceptual diagram of class

The aim of each package is defined based on the behavior of its elements, that allows to establish a stratified architecture [12], where each layer contains the previous one with a defined behavior that can be classified of the following form:

1) Classes for the input grammar management.
2) Classes to represent the language elements (grammar states and symbols).
3) Classes for operations associated to the translator (actions and transitions).

Figure 3 shows the conceptual diagram corresponding to the layers identified in the representation and use of the tables of actions and transitions generated for the parser.

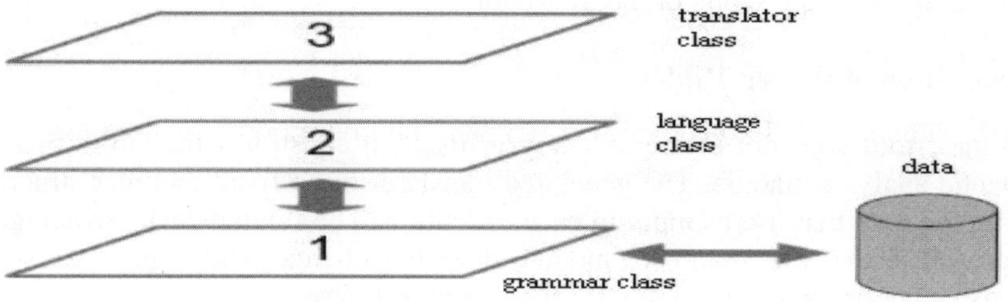

Fig. 3. Conceptual diagram of class

2.2 Abstraction

Following the Software Engineering principles, classes that can change were equipped with a layer of abstraction with the purpose of easing maintenance and reusability. In this way, if it were convenient to implement a new method not included in the present version (for example, the method of descending syntactic analysis), a class can be added to the module that implements the methods of the abstract class including the code of the algorithm.

2.3 Parametric Types

These are used in the implementation of the syntactic analysis tables, whose structure is common but they differ in the type of elements that can store. Thus, the skeleton of the structure has been uniquely defined, introducing the data related to the states and the grammar symbols: a stream of characters in the table of actions and an integer that represents an identifier of the state in the transition tables.

3 The Parser Generator

The system was conceived as an ascending LR(k) analyzer. This election is mainly justified because it allows analyzing a greater number of grammars than the descendent analyzers and makes the analysis without backtracking, what makes a greater efficiency of the operation of the constructed component.

3.1 Grammar Specification

In order to ensure correctness in the input elements previous verifications are made to their use, at two levels:

• **Run time Verifications:** made on grammar specification or input data. For example, the class *ReglaProduccion* does not allow to assign as the body of the production an empty set of elements and the *class Grammar* maintains the restriction that contained rules must be different, verifying it as new productions are added.
• **Static verifications:** the use of XML for the storage of the data and its validation by designed schemes allows checking that the external definitions fulfill the requirements for their use in the scope of the translator.

3.2 Generation of Parser Tables

The designed tool does not create an analyzer itself: it constructs the tables that use the syntactic analysis module. The generated translator has a fixed modular structure, modifying the data that use as input in each execution. The fundamental advantage of this approach is that the analyzing module does not change, and it provides a homogenous behavior modifying only the tables used as reference.

One of the main aspects to make the analysis procedure is the use of a state table directing the process. This approach provides a wider option set, since it introduces the possibility that the annotations to the table are instructions (for example, fragments of code of a General-Purpose Language) or conditions of error. The traditional mechanisms consider solely transitions or indications of acceptance and error.

For the construction of the table of syntactic analysis LALR, the algorithm described in [1] is extended, including the following elements:

Conflict resolution

When a value is detected to include the new combination *(state, grammar symbol)*, the type of conflict is verified and the option indicated by the user in the configuration is carried out. This is a fundamental difference of the tool with respect to the traditional applications, where a conflict of these type is treated as decided by the application creator and not by the user. In case the conflict cannot be solved, the algorithm finishes.

Allocation of values for error recovery

The algorithm goes through the set of states stored in the action table. If the option associated to the end of sentence character ($), is not empty or if it has a different value to "accept", it assigns this value to the remaining set of empty options for the state and the terminal symbols (it "fills" the row). This approach is based on the fact that the recovered value for a state and the end of sentence character, represents a valid action when the sequence of elements has been completely read, so that this value can be propagated to the rest of actions, turning the state a synchronization point within the language specified by the grammar.

Table 1. Syntactic analysis table for language $L(G_1)=\{c^n dc^n d, n \geq 0\}$

	Actions			Transitions	
State	c	d	$	<S>	<C>
0	d3	d4		1	2
1	-	-	a	-	-
2	d3	d4	-	-	5
3	d3	d4			6
4	r3	r3	r3	-	-
5	-	-	r1	-	-
6	r2	r2	r2	-	-

Table 1 is completed with the value *r1,* the empty cells in the row corresponding to state 5.

As it is shown in Table 2, when the analysis process was in state 5 and the lexical analyzer provided a symbol like c or d, the parser would reduce using the production 1 (S → CC), that is, it would emulate the behavior of a correct analysis where the elements that have been identified make a valid entry up to that point. This fact allows continuing until the remaining entrance is consumed, locating the errors that could have been introduced. It constitutes the fundamental difference with respect to the created table using the traditional method, where both cells would be left empty, producing the shutdown of the analyzer.

Table 2. Syntactic analysis table for LALR(1) generator

	Actions			Transitions	
State	c	d	$	<S>	<C>
0	d3	d4		1	2
1	-	-	a		
2	d3	d4		-	5
3	d3	d4		-	6
4	r3	r3	r3		
5	r1	r1	r1	-	-
6	r2	r2	r2	-	-

3.3 Construction of the Parse Tree

As the sequence of entry stream is crossed, the data structure (tree) is constructed containing the information of the tokens identified by the lexical analyzer. Thus, when executing an action "shift", it becomes the terminal element a node leaf. When the reduction of a rule A → *a* is made, being *a* a sequence of terminal and/or non terminal that conforms with the symbols of the constructed nodes, the symbol is introduced in the tree A as an intermediate node, father of nodes a. The necessity to show in the system intermediate results in the different stages of the analysis and the use of a decorated abstract syntax tree (DAST) as the intermediate representation, that will be used in later phases, forces the use of the tree to represent these results.

3.4 Error Recovery

We tried to free the user of the error treatment so that he/she can be centered of the grammar design. Instead of yielding to the user the responsibility of introducing a set of elements in the input specification, to establish what to do in this situation of an error, the locality principle is applied to take the information provided by the input specification, the values of configuration and the internal operation of the analyzer, taking as a reference the description from [4].

We suppose that the error is inside a correct sequence of elements that corresponds with the body of a grammar production rule, so that we should try to delimit this context from the elements introduced in the top of the stack and the following elements of the sequence to analyze.

The solution to the problem comes from the study of the error position context within the input sequence where the error is, combining it with the information stored in the tables of the parser. The following steps are followed:

1. Starting from the top of stack, elements are extracted until the first state that stores values in the table of transitions is located, this one being the new top of the stack.
2. The set of *actions [states, token]* of the input is obtained:
   a) If the set is empty, the following input token is consumed.
   b) If it is not empty, the optimal action is selected to continue the analysis.

3.5 Measurement of Complexity

In this section the possible application of an objective measurement of complexity to the structure of the grammar is introduced, that will allow to provide the user with a method to evaluate the design.

The definition of the measurement comes from the concepts exposed in [6] that will be used in the objective evaluation of the quality of the grammars:

- **Number of non terminals:** It allows to measure the size of an Context-Free Grammar, by applying *fine degree* metric whose use in the evaluation of the complexity of programs is centered in the number of procedures.
- **Ciclomatic complexity:** In [12] and [11] the complexity of McCabeis defined, or ciclomatic complexity of a flow graph.

In the tool these metrics have been implemented. The measurement of the number of nonterminal elements is trivial, since the set of the elements of the syntax is available at any moment. With regard to the ciclomatic complexity, in order to ease its understanding the graph associate is constructed.

4 System Description

The application has been developed in Java code. Version 1.5 is necessary to be able to use elements like the generic types. The system provides in addition:

- An project oriented scheme, that allows the grouping of related grammars for their management.
- A graphical interface for handling the specification and components of the grammar, so that the time used in its design is reduced.
- Auxiliary components for the representation of the specification and the measurement of the grammar complexity.
- Exhaustive information on the tables of syntactic analysis.
- A source code editor that allows to check the constructed specification, either from the content of a file or writing it of manual form.
- A set of components that facilitate step by step the consultation of the analysis operation, by simulation of the process from the input.
- Configuration mechanisms to customize, easing the handling and adapting it to novels or experts users.

5 Conclusion

The constructed system goes beyond the scope of the formal languages and can be used in the creation of the GBSs, easing the work of DSLs design. Unlike the traditional applications, this one provides a real support to the incremental development of the language that allows the reusability in the design of grammars, with the use of object oriented concepts.

The IDE attenuates the complexity of the design of the specification because it allows the graphic representation of different structures like tables, tree parser and grammars evaluations.

The design of language-based tools is made, as far as possible, free of the details of the language. It takes decisions based on the data provided by the user. In this respect a novel approach is adopted in the conflict resolution and error handling, facilitating greatly the design of the grammar:

- The mechanism of conflict resolution follows the configuration indicated by the user, whose fundamental advantage is that it extends the set of input grammars, because it is the tool who takes the decision related to the ambiguous structures that are the origin of the conflicts.
- The error treatment is made without the user including additional information. This allows that the user is centered in the design of the grammar, yielding the responsibility of the error handling to the application. Since the languages usually are structured in compilation units, it identifies the error within this unit, and it continues the analysis without affecting the rest of the process.

As a summary, it provides the base to create new components in application fields possibly different from the traditional ones. That is why as future lines of work we try to use in the GBSs, more precisely in data mining to classification with grammar genetic programming and the discovery of biological data.

References

1. Aho, A., Lam, M., Sheti, R., Ullman, J.: Compilers: principles, techniques and tools. Addison-Wesley, Reading (2007)
2. Gagnon, E.: SableCC, an object–oriented compiler framework. Master's thesis. McGill University, Montreal, Quebec (1998)
3. Gálvez, S., Mora, M.A.: Java a tope: Traductores y compiladores con LEX/YACC, JFLEX/CUP y JAVACC. Edición electrónica. Departamento de Lenguajes y Ciencias de la Computación, E. T. S. de Ingeniería Informática. Universidad de Málaga. Málaga (2005)
4. Grune, D., Bal, H., Jacobs, C., Langendoen, K.: Modern compiler design. Wiley, Sussex (2001)
5. Henriques, P., Pereira, M.V., Mernik, M., Lenič, M., Avdičaušević, E., Žumer, V.: Automatic generation of language-based tools. In: van den Brand, M., Laemmel, R. (eds.) Electronic Notes in Theoretical Computer Science, vol. 65. Elsevier Science Publishers, Amsterdam (2002)
6. López, V., Alonso, L., Moreno, M.: Aplicación de las métricas de calidad del software en la evaluación objetiva de gramáticas independientes de contexto inferidas. In: Moreno, M.N., García, F.J. (eds.) Actas del I Simposio Avances en Gestión de Proyectos y Calidad del Software, pp. 209–220. Salamanca (2004)
7. Louden, C.: Compiler construction. Thompson International (2004)
8. Luengo, M., Cueva, J., Ortín, F., Izquierdo, E.F.: Análisis sintáctico en procesadores de lenguaje. Cuaderno didáctico número 61. Departamento de Informática. Universidad de Oviedo. Servitec. Oviedo (2005)
9. Mernik, M., Lenič, M., Avdičaušević, E., Žumer, V.: Compiler/Interpreter Generator System LISA. In: Proceedings of the 33rd Hawaii International Conference on System Sciences (2000)
10. Mernik, M., Crepinsek, M., Kosar, T., Rebernak, D., Žumer, V.: Grammar-Based Systems: definition and examples. Univerty of Maribor (2004)
11. Piattini, M., Calvo-Manzano, J., Cervera, J., Fernández, L.: Análisis y diseño detallado de aplicaciones informáticas de gestión: una perspectiva de Ingeniería del Software. Ra-Ma, Madrid (2004)
12. Pressman, R.S.: Ingeniería del Software, un enfoque práctico. Quinta edición. McGraw-Hill, Madrid (2002)
13. Rangel, P., Varanda, M., Mernik, M.: Automatic generation of language-based tools using the LISA system. Elsevier Science, Amsterdam (2004)
14. Rebernak, D., Mernik, M., Wu, H., Gray, J.: Domain-Specific Aspect Languages for modularizing crosscutting concerns in grammars. In: GPCE Workshop on Domain-Specific Aspect Languages, Portland, OR (2006)
15. Levine, J.R., Mason, T., Brown, D.: Lex & Yacc. O' Reilly & Associates, U.S.A (1992)

A Contour Based Approach for Bilateral Mammogram Registration Using Discrete Wavelet Analysis

Ramon Reig-Bolaño[1], Vicenç Parisi Baradad[2], and Pere Marti-Puig[1]

[1] Department of Digital Information and Technologies
University of Vic (UVIC)
C/ de la Laura, 13, E-08500, Vic, Barcelona, Spain
[2] Department of Electronic Engineering
Politechnical Univerity of Catalonia (UPC)
Av. Victor Balaguer s/n, E-08800, Vilanova i la Geltrú, Barcelona, Spain
ramon.reig@uvic.cat, pere.marti@uvic.cat, Vicenc.Parisi@upc.edu

Abstract. Registration of mammograms is an essential step to increase the effectiveness of all screening programs planned to early detect breast cancer in asymptomatic women. New techniques based on data fusion of different spatial views, temporal analysis or sequences follow up, and multimodal data analysis, require accurate data registration: forcing spatial alignment of images taken from different views, at different times or from different natures. In this work we will focus on a new technique for automatic bilateral mammograms registration. However the same registration technique is also useful for temporal analysis of the same patient. Bilateral mammogram registration is a challenging task; the mammographic appearance of breast tissue may vary considerably, because of differences in breast compression and positioning, differences in imaging techniques, and changes in the breast itself; moreover, there are no clear landmarks in a mammogram, except for the nipple when it is visible. In our approach we detect the skin-line contour as a first step, we describe the contour as a chain-code, and we make a wavelet based correlation analysis, finally we get the matching of the contours. At the end we apply a global affine transformation.

Keywords: medical imaging, image registration, image alignment, wavelets transforms, multiresolution analysis, signal processing, image processing.

1 Introduction

The organization of a screening program to early breast cancer detection is such that asymptomatic women are invited for X-ray examination of both breasts on a regular basis. One of the most common projections is the mediolateral oblique (MLO), taken from an angled view, that shows part of the pectoral muscle. The images are interpreted by one or two radiologists. Breast cancer will manifest itself in the form of a mass or calcifications in the mammogram. When there are abnormal findings the woman is recalled for further examination [1]. The majority of research into computer-aided detection has focused on analyzing a single image to detect abnormalities. In clinical practice however, mammograms are often interpreted by comparative analysis: temporal and bilateral. Temporal comparative analysis involves comparing a pair of corresponding mammograms of the same breast, acquired at different times. It

J.M. Corchado et al. (Eds.): DCAI 2008, ASC 50, pp. 347–356, 2009.
springerlink.com

determines if any regions of the breast have undergone changes which may be indicative of a developing abnormality. Bilateral comparative analysis refers to comparing mammograms of the left and right breasts, representing the same view, acquired during the same screening session. Left and right mammograms from the same woman tend to exhibit a high degree of symmetry. A deviation from this symmetry can indicate the presence of an abnormality. Sometimes a Bilateral mammograms subtraction is carried to emphasize the abnormalities [2]. Inadequate positioning is the most frequent problem encountered when reading mammograms. To some extent suboptimal positioning can be compensated for by registration [3]. In this work we will present a new technique for automatic bilateral mammograms registration. However the same registration technique is also useful for temporal analysis. In our approach we detect the skin-line contour of the breast as a first step, we describe the contour as a chain-code, and we make a wavelet based correlation analysis, finally we get the matching of the contours. At the end we apply a global affine transformation.

2 Materials and Preprocessing Methods

We will work on MLO breast images from Mini-MIAS database [4]:

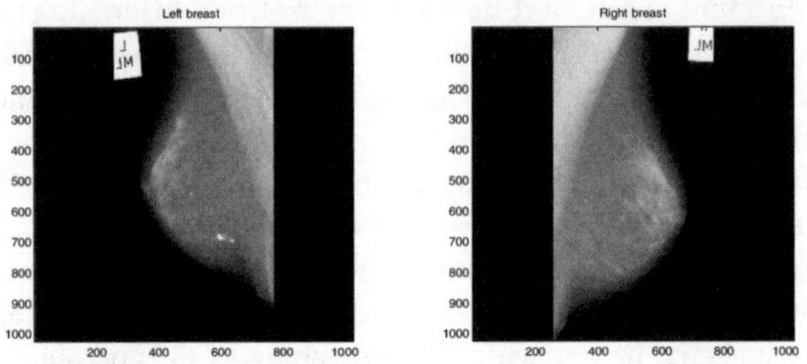

Fig. 1. MLO bilateral Mammograms of mini-MIAS Database (Mammographic Image Analysis Society) [4], (n- 75, n-76). Left breast (at left), Right breast (at right).

The bilateral registration is the most difficult process that has to be carried out for a proper bilateral subtraction. When this process is done manually, we need to identify pairs of related points at each breast, then obtain the global transform coefficients relating these pairs of points, and finally align the images. In our automatic registration approach of the right breast (working image) to the mirrored left breast (reference image), we follow the next steps.

2.1 Contours Extraction

A first set of pre-processing steps, such as: to mirror the left breast mammogram (reference image); then to take the right breast mammogram (working image) and finally to adjust their resolutions. A second set of steps consists on applying on both images the skin-line extraction process. Depending on the variations of mammograms this could be very critical to be carried out automatically [1], in this paper we presume we

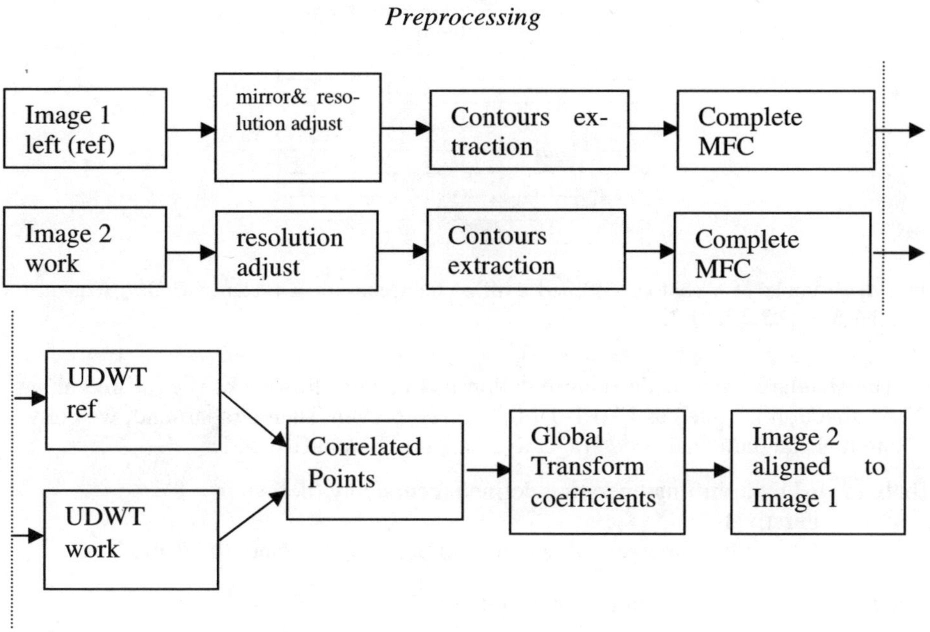

Fig. 2. Steps for bilateral automatic registration

have a good approximation to the contours of the skin-line (i.e. with [5] techniques, based on morphologic contours extraction of binary images).

2.2 Contours Codification with a Modified Chain Code

These skin-line contours are then codified with a modified chain code [6] or Freeman Code, leading to a first order descriptor of the image. Each contour is represented with a finite sequence, and the codes of the main contours are grouped with related information at each image. With a chain code any digital curve can be represented by an integer sequence:

$$[a_1, a_2, a_3 \dots a_N] \text{ with } \{a_i \in \{0,1,2,\dots 7\}\},$$

Each value depends on the relative position of the current edge pixel with respect to the next edge pixel. One unit corresponds to an angle of 45º. Thus a chain code value of 3, for example, indicates the next pixel is on the north-west (135º) direction.

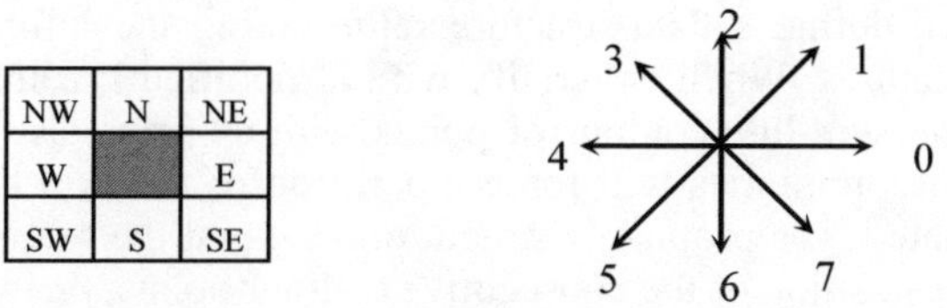

Fig. 3. A template for the Freeman contour codification (left); values assigned to each one of the eight possible directions from a pixel to the next pixel following the contour (right)

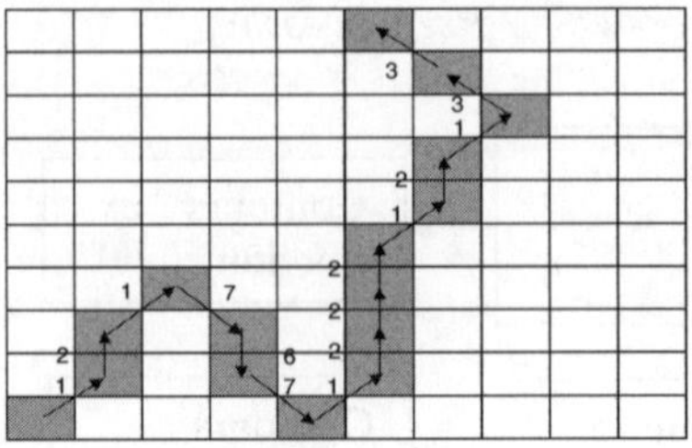

Fig. 4. Example of a contour codified with a chain code or a Freeman Code Sequence = {1,2,1,7,6,7,1,2,2,2,1,2,1,3,3}

The standard chain code representation has certain drawbacks (i.e., a line along -22.5° direction is coded as [7 0 7 0 7 0. . .]). To prevent such wraparound, we convert a length N standard chain code $[a_1\ a_2\ a_3\ ....\ a_N]$ into a modified code

$[b_1\ b_2\ .\ .\ .\ b_N]$ by a shifting operation defined recursively (define $b_0 = 8$)
. iteratively

b_i an integer value that minimizes $|b_i - b_{i-1}|$ and $(b_i - a_i)\mathrm{mod}8=0$

The line along -22.5' direction is then coded as [7 8 7 8 7 8. . .].

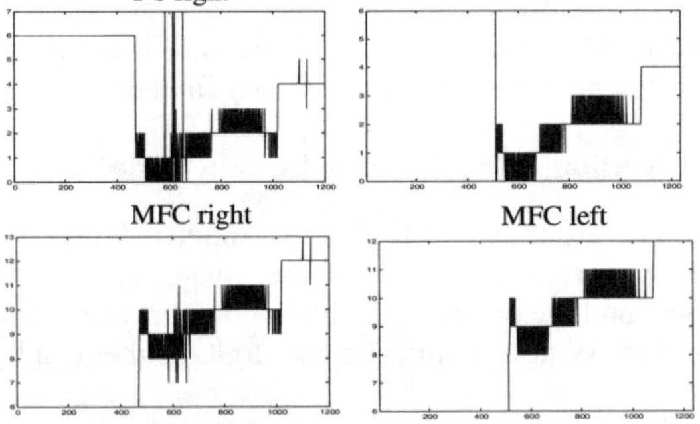

Fig. 5. Comparison of FC (Freeman Code) and Modified Freeman Code (MFC), on skin-line contours of MIAS database (n°75, n°76)

Skin-line data-structure at each breast

For each breast we define a data-structure representing the information of the extracted skin-line features. When we codify with a modified Freeman Code we could find ruptures on the skin-line contour or points with an imprecise one-pixel contour follow-up, these discontinuities will force a partition of the skin-line in several segments spatially related. We propose a structure to save al the relevant information of these segments: the position of the consecutive contour points, their FC and MFC, and also a pointer to other related segments.

Segment i						
x	y	FC	MFC	Node 1	Node 2	Node i

Fig. 6. Data structure for the segments of the skin-line

The data structure will save for each consecutive point of a segment:

- x, y point position on image
- FC: Freeman code or chain code
- MFC: Modified Freeman code.
- Node 1: zero if there is not a bifurcation of the segment. Different of zero means bifurcation to a specified segment number.
- Node 2: same with a second bifurcation
- Node i: we can have until 4 different bifurcations.

The end of a segment is when the following point is part of another segment (we codify this with the FC=-1), or when there are no more points (with FC=-2).

Skin-line complete MFC of each breast

We will apply an extensive search algorithm to unify all the segments in a single sequence for all the skin-line at each mammogram. We could prune the first and the last part of the sequence, corresponding to the limits of the mammogram information, and finally we obtain the complete contour codification; we have named this codification *the first order descriptor* of the breast skin-line.

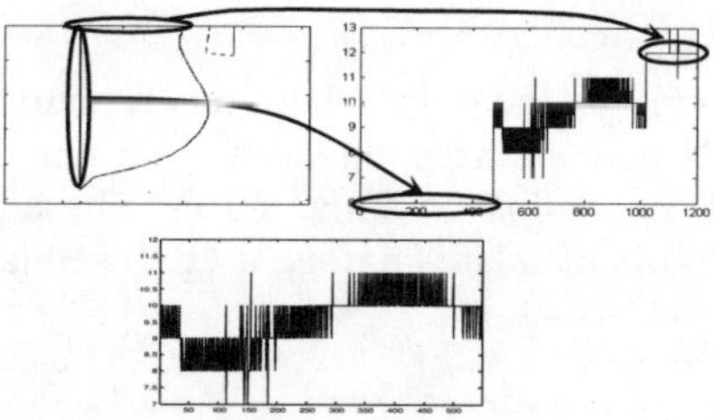

Fig. 7. At the top we have the right breast of MIAS. n.75, with his corresponding complete MFC, at the bottom we have the pruned sequence.

3 Undecimated Discrete Wavelet Transform (UDWT)

The second order descriptors are calculated with the Undecimated Discrete Wavelet Transforms [7] of the contours codes. The UDWT is implemented with the algorithm 'à trous', using a dyadic filter-bank on each approximate component (low-pass), and finally we obtain a multiresolution decomposition of each contour, from level 0 (high resolution) to level L (low resolution).

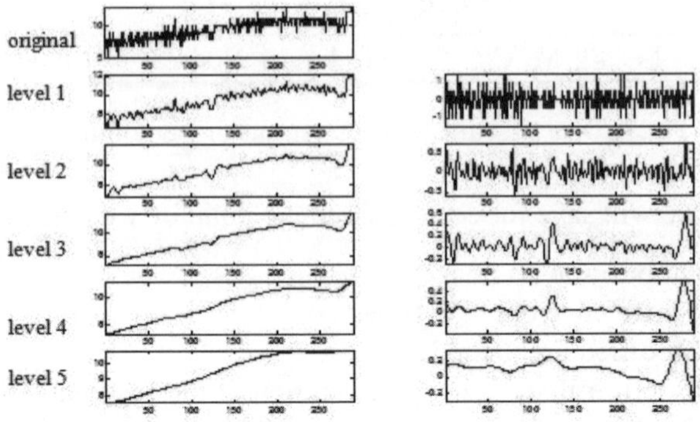

Fig. 8. Example of UDWT until level 5, on a codified skin-line contour. The approximate component at each scale is decomposed on a detail component (high-pass, at the right column), and an approximate component (low-pass, at the left).

4 Descriptors Alignment Using MCC

In order to find the matching points on the contours second order descriptors we calculate the normalized cross correlation (1) - the maximum of cross-correlation (MCC)- on each possible pair, using the *common level of lowest resolution approximation* of both contours.

$$\rho_{xy}(k) = \frac{r_{xy}(k)}{\sqrt{r_{xx}(0) \cdot r_{yy}(0)}} \qquad r_{xy}(k) = \frac{\sum_n \left((x(n) - m_x) \cdot (y(n-k) - m_y) \right)}{N} \tag{1}$$

Once it is calculated, we will get the point where this function is maximum. From this point we can derive the N points correspondence from one contour descriptor to the other, we name this pairs "*control point pairs*" *CPP* . These points are the equivalent of the manually obtained pairs of related points, used in semiautomatic methods.

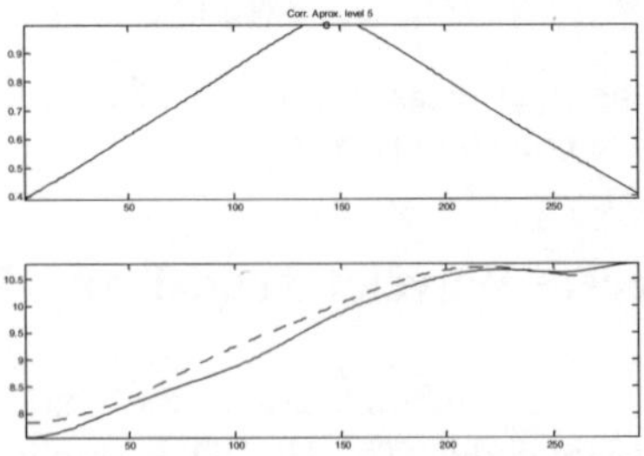

Fig. 9. The normalized cross correlation of the level 5 approximation of both contours. The top image is the correlation measure. And at the bottom image there are the level 5 approximations values of both contours.

5 Image Alignment Coefficients

The next step is finding the geometric transform coefficients relating the CPP (in this example we suppose an affine transformation) with direct equations like (2), relating a point on Image 2 (working image) and a corresponding point on Image 1 (reference image):

$$x_i = a_0 + a_1 \cdot X_i + a_2 \cdot Y_i \qquad y_i = b_0 + b_1 \cdot X_i + b_2 \cdot Y_i \tag{2}$$

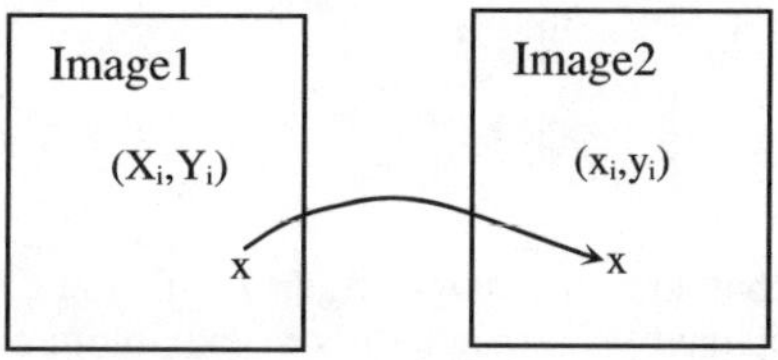

Fig. 10. A point to point correspondence of Image 1 (reference image) and Image 2 (working image)

We solve the overdeterminated system minimizing the MSE mean square error, we get an approximate set of coefficients.

$$\begin{bmatrix} 1 & X_1 & Y_1 \\ 1 & X_2 & Y_2 \\ . & . & . \\ 1 & X_n & Y_n \end{bmatrix} \begin{bmatrix} a_0 \\ a_1 \\ a_2 \end{bmatrix} = \begin{bmatrix} x_1 \\ x_2 \\ . \\ x_n \end{bmatrix} \qquad \begin{bmatrix} 1 & X_1 & Y_1 \\ 1 & X_2 & Y_2 \\ . & . & . \\ 1 & X_n & Y_n \end{bmatrix} \begin{bmatrix} b_0 \\ b_1 \\ b_2 \end{bmatrix} = \begin{bmatrix} y_1 \\ y_2 \\ . \\ y_n \end{bmatrix}$$

$$MXY \cdot a = x \qquad MXY \cdot b = y \tag{3}$$

$$a = (MXY^T \cdot MXY)^{-1} \cdot (MXY^T \cdot x) \qquad b = (MXY^T \cdot MXY)^{-1} \cdot (MXY^T \cdot y) \tag{4}$$

6 Image Alignment and Interpolation Results

With the working image mammogram (corresponding with the right breast), and the set of transformation coefficients, we can align the working image to the reference image, and leave it ready for the bilateral subtraction.

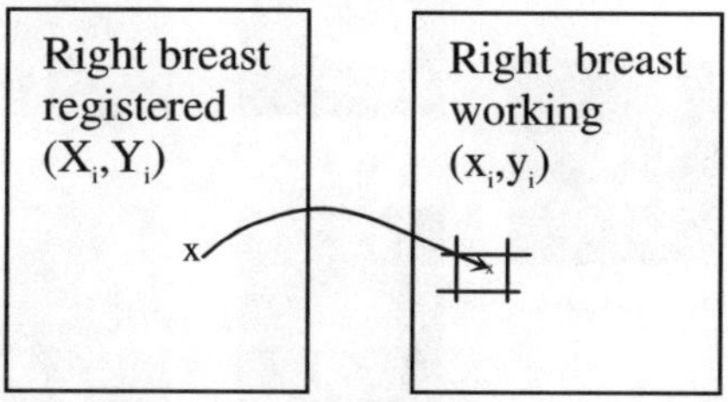

Fig. 11. Scheme for image registration and value interpolation

The inverse transform method gives the relation of each pixel of the registered image to a corresponding point at the working image space. With the transform coefficients (3), the point at the working image doesn't correspond to a point in the pixel

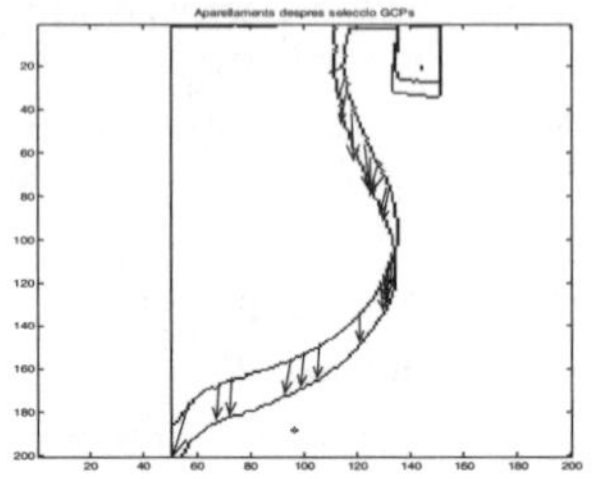

Fig. 12. We represent both contours on a single figure, with the CPP (Control Points Pairs). Each pair of related points obtained from neighbours of maximum correlation value is chained with an arrow.

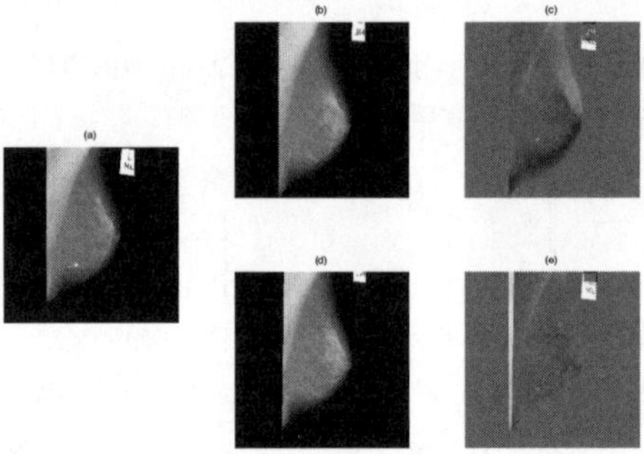

Fig. 13. At the first column we represent the reference image (a) – the mirrored left breast - , at the second column we represent the working image (b) – upper image, right breast- , and his registered version (d)- lower image -. At the last column we compare the direct bilateral subtraction (c)- the upper image -, (c)=(a)-(b) and the bilateral subtraction with the registered image (e)- bottom image -. (e)=(a)-(d).

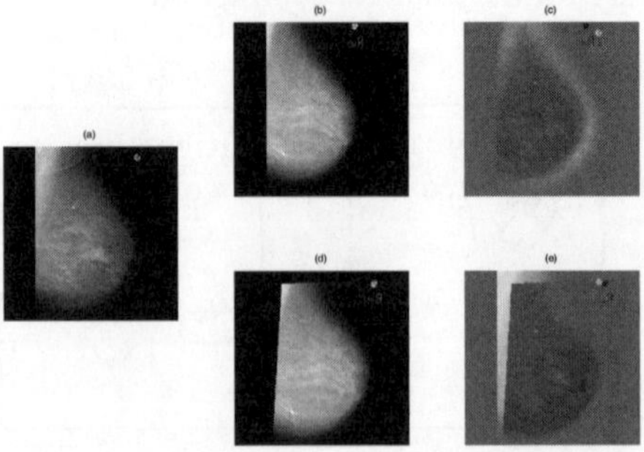

Fig. 14. Another example of bilateral subtraction from MIAS database (n69-70). With the registered image (e)=(a)-(d), or with the unregistered case (c)=(a)-(b).

grid; then, in order to generate the value at the registered image, we need to apply an interpolation with the surrounding pixels at the working image. We can see the results in Fig.13 and Fig. 14. In Fig. 12 we show the results of the registration just on the skin-line contours of the bilateral mammograms. In Fig. 13, and 14 we show the results of the bilateral registration of the complete mammograms using the same coefficients calculated for the contours registration. Finally at Fig. 13 (e) we show a Bilateral subtraction that should be used to emphasize the abnormalities for diagnose proposes.

7 Conclusion

The proposed method for bilateral mammogram registration offers some preliminary good results, both qualitatively and quantitatively. The method is robust and exportable to other kind of registration, for example on intra-modal data analysis like sequences follow up, or new techniques based on data fusion of different spatial views, temporal analysis or sequences follow up, and multimodal data analysis. The results are valid either with the nipple visible or not visible, some of the referenced methods [3] work only with images with the nipple or muscular textures present. Other methods are highly dependant on regular geometry of the breast, our method is completely robust to these factors. The mutual information methods are well proven for both: sequences registration or multimodality images; their major drawback is when there are tissue variations on the series. Our approach is also robust to these variations. To go a step beyond we need an objective comparative analysis with different methods, we are working with benchmark measures proposed by PEIPA, the Pilot European Image Processing Archive [8], in order to compare objectively with other methods. Our principal motivation to carry out this research is to validate our approach to image registration in another context. A similar method was successfully used for multimodal contour based registration of multimodal remote sensing images.

Acknowledgements. This work has been partially supported by Codesign Hardware-Software Group from the University of Vic: cost center R008-R0904.

References

1. van Engeland, S., Snoeren, P., Hendriks, J., Karssemeijer, N.A.: Comparison of Methods for Mammogram Registration. IEEE Trans. on medical imaging 22, 11 (2003)
2. Wirth, M.A., Narhan, J., Gray, D.: Nonrigid mammogram registration using mutual information. In: SPIE Medical Imaging: Image Processing, San Diego, USA, vol. 4684, pp. 562–573 (2002)
3. Guo, Y., Suri, J., Sivaramakrishoa, R.: Image Registration for Breast Imaging: A Review. In: Proceedings of the 27th Annual International Conference of the Engineering in Medicine and Biology Society IEEE-EMBS, vol. 3379, 3382 (2005)
4. Suckling, J., et al.: The Mammographic Image Analysis Society Digital Mammogram Database. Exerpta Medica. International Congress Series 1069, 375–378 (1995)

5. Serra, J.: Image Analysis and Mathematical Morphology. Academic Press, New York (1982)
6. Li, H., Manjunath, B.S., Mitra, S.K.: A contour-based approach to multisensor image registration. IEEE Transactions on image Processing 4(3), 320–334 (1995)
7. Dijamdji, J.P., Bijaoui, A.: Disparity Analysis: A Wavelet Transform Approach. IEEE Transactions on Geoscience and Remote Sensing 33(1) (1995)
8. PEIPA, the Pilot European Image Processing Archive, `http://peipa.essex.ac.uk/`

Application of Hidden Markov Models to Melanoma Diagnosis

Vicente J. Berenguer, Daniel Ruiz, and Antonio Soriano

Departamento de Tecnología Informática y Computación
Universidad de Alicante
03690 San Vicente del Raspeig, Alicante
vjberenguer@dtic.ua.es, druiz@dtic.ua.es, soriano@dtic.ua.es

Abstract. In this paper we present a clinical decision support system for melanoma diagnosis. Unlike other systems based on diagnosis obtained just from one image, in this work it is employed an image set, that represents the evolution of damaged tissues, taken in different instances of time (for example once a month). Therefore, the system analyses the image sequence extracting the affected area and using the gradient orientations histogram of each area to compose a description which allows achieving a decision about the input. Hidden Markov Models are proposed as classify method, obtaining classification rates of 77%.

Keywords: Decision Support System, melanoma automated recognition, image processing, Hidden Markov Models.

1 Introduction

Melanoma is the most deadly form of skin cancer. Although it involves just 4% of all tissue cancers, it is responsible of the major number of deaths, around 75%.

The main cause of melanoma is due to a long exposition to ultraviolet radiations, although skin type or other genetic factors can influence too. The most effective treatment is an immediate extirpation, but just when the melanoma had been detected in early phases. In other cases, if it is not diagnosed in time, the life expectancy is reduced up to less than one year. Therefore, it is fundamental to distinguish as soon as possible between benign lesions (as a simple spot or a mole) and melanomas. Dermatologists employ for their diagnosis several techniques which have been developed based on experience, among which it is emphasized to obtain the total dermatoscopy index based on the mnemonic ABCD [14], the rule of 7 points [2] and the method of Menzies [8]. All these techniques allow identifying symptom of a malignant lesion based on the observation of a series of characteristics. Furthermore, doctors indicate that evolution could be a crucial sign to decide if a lesion (i.e. a spot) is degenerating in a melanoma. Even so, in some cases could be a hard task the interpretation of these properties visually, and therefore, to make a right diagnose. Actually, classification rate from well experienced dermatologists is around 65% previously to the biopsy [7].

We propose a decision support system with two applications. On one hand, it allows patients to do periodical self-explorations with the aim of detecting, as soon as possible, a probable malignant lesion. On the other hand, to provide a second opinion with which the dermatologist could complement and/or be precise in his decision.

J.M. Corchado et al. (Eds.): DCAI 2008, ASC 50, pp. 357–365, 2009.
springerlink.com

The system bases its automatic diagnosis in three phases: detection, description and classification of the lesion. In the first phase a preprocessing of image sequences is done which allows us to identify the affected area properly. Afterwards, in the description phase, the histogram of gradient orientations from detected melanoma areas is calculated. Finally, this histogram is used as an observation for Hidden Markov Models, which are able to supply a diagnosis of the lesion, classifying it as a melanoma or a benign lesion.

2 State-of-the-Art

There are several researching groups which have developed works related to automatic diagnosis of melanoma.

In [16] a method for automatic recognizing is employed using a section of a microscopic image. An analysis of the shape and cell sizes is made using morphological math and geodesic algorithms. An automatic classification is performed with the values obtained in previous states using an interpretation technique without learning.

In [3] a system which employs images of microscopy epiluminiscence as input is presented. In the first place a segmentation of the affected area is done applying different types of algorithms (thresholding in the blue plane, searching 3D colour groups, etc.), and they use the best result from the fusion of the different methods. Afterwards they calculate a set of radiometric characteristics, and global and local parameters to describe the malignity of the lesion. They do a selection of more significant characteristics using statistical methods. Finally, a classification by the algorithm of K nearest neighbours is carried out.

In [13] the authors present a system for melanoma diagnosis taking into account border detection and quantification of asymmetry rate. Its outline detector employs a technique based on clustering with Fuzzy C-Means algorithm. Classification is based on rate of symmetry quantification uniquely with a six dimensions vector.

Hintz-madsen develops in his doctoral thesis a efficient system for diagnosing [5]. He uses an optimal thresholding technique which splits regions with lesion from tissue, and for classifying he uses a neural network in a probabilistic frame.

In [17] it is used segmentation by pixel adding with a previous processing based on fuzzy sets. An attribute series is extracted from the detected area and processed by a neuronal network to distinguishing between melanomas and benign lesions. Moreover, the authors remark optimization of description characteristic sets of the lesion.

In [12], authors introduce a system based on knowledge for carrying out an early diagnosis of melanoma. System detects the lesion using a thresholding technique based on the red component and the saturated component. Then, system extracts a colorimetric and geometric characteristic set, with which a diagnosis is perform using a voting system, taking into account the produced results by different instances of the K nearest neighbours algorithm.

Tim Lee presents in his doctoral thesis an algorithm for segmenting affected regions from images [7]. In this, after a multi-phase medium filter applied for removing deep noise, a thresholding is carried out. Finally, a system based on rules is applied over the result of this process, allowing the identification of the lesion.

In [6] a classification method for tissue lesions based on the use of artificial expert entities is proposed. This technique performs a preprocessing and a segmentation using colour statistical clustering and pixel adding algorithm. Next, a set of characteristics related with the lesion asymmetry, uniformity of outline and tissue, is determined and they will be employed for training a multi-agent classifier. The multi-agent is composed by a series of neural networks managed by a master entity which gives input vectors and generates a diagnosis adjusted to the output of its different components.

As we could appreciate, most of the researches related with automatic melanoma diagnosis are based on the analysis of a unique lesion capture. However, under the point of view of specialists, evolution could be a crucial sign in detecting task when a lesion is degenerating in a melanoma [1]. Principal indicators of a malignant evolution are: lesion increasement, shape changes, tissue changes, pigment expansion around the tissue lesion, bleeding lesion and apparition of pain, itching and smarting.

In [10] genetic algorithms are used to establish a correspondence between different images with same melanocytic lesion in different time instances. With this mechanism, algorithms detect possible variations in irregular lesion border, and predict a malignant evolution. Consequently, we propose a classifier based on Hidden Markov Models (HMMs), modeling image sequences of histograms of gradient orientations from identified lesion regions. Therefore each image in the sequence will match with an observation of the model, and each possible evolution type (benign or malign) will correspond of a HMM.

3 Proposed System

The main objective of the development is, taking into account the lesion evolution formed by a set of 4 of 5 captures, to determine if the lesion could be malignant or not. The system has three phases: detection, description and lesion diagnosis. The first phase presents two steps: in the first one a preprocessing of images for eliminating noise and no-desired objects is performed, and in the second one region of interest from the capture is extracted, producing a sequence of binary masks in its output. In the phase of description it will be obtained the gradient values from active image pixels in each binary mask, and the histogram of orientation of each affected region will be calculated. In the last place, these histograms will be discretized converting input sequence in an observable sequence for Hidden Markov Models, with which the system will be able to provide a diagnosis.

4 Detection of the Lesion

4.1 Preprocessing

Before carrying out the image processing to do a diagnosis, we must solve several important problems related to the image type which the system is dealing with. In the first place, captures of human tissue could present hairs, which may mislead the segmentation process. The immediate solution would be shaving the hair before imaging sessions, but it is a process which, besides increase costs and time for obtaining samples, it is uncomfortable for patient and it is impractical in many cases. In these cases we will apply the algorithm of hair deleting proposed in [7], called DullRazor. This method consists in three phases: hair identification, hair pixels replacing by around tissue pixels and a final image smoothing.

On the other hand, all digital images present impulse noise, which we will mitigate applying some filter. Here we apply a median filter. This will contribute, in addition, to remove small pores and to reduce possible reflections and shines which could appear in the dermatoscopic image, preserving region outlines. We have checked several sizes of filter (3, 5, 7, 9 and 11) and the best results can be obtained with a mask of size 7.

4.2 Segmentation

Segmentation is the main phase in an application of automatic image processing, due to that post-analysis depends on this for carrying out a correct diagnosis. The objective here is to split objects that constitute the scenario for obtaining the affected region of the image. Several applicable methods exist for this purpose, among them we emphasize pixel segmentation, region segmentation and border detection [4]. We will use thresholding, which is a pixel-based technique and not on the relation between image pixels. In particular, we will apply an adaptive threshold method, because the image set to segmentation is very heterogeneous, and to establish a global threshold for everyone will not be viable.

The developed technique tries to choose in each case, automatically, the threshold that will split the affected region from the rest of the image; for that reason we will apply the Otsu method [9]. This is a technique of adaptive thresholding with the aim of separating two classes, searching to maximize distance between both classes and to minimize areas formed by points associated to each class. The resultant image from preprocessing stage, converted to gray scales, will be the input for this algorithm; and the output will produce a binary image with segmented regions at high level (white) and the rest of the image at low level (black). An opening operation (an erosion followed by a dilatation) and afterwards, a closing operation (a dilatation followed by an erosion) with structured elements of size 3 and all elements with value 1, are applied with the objective of smoothing the outline of the region of interest from the resulted image. The algorithm of connected components labeling with 8-connectivity is applied to binary images after morphological transformations and the objective is to identify objects which have been extracted in the segmentation process. Once we have all objects labeled, we keep the ones with the biggest size, which will correspond with the affected area. Finally the extracted region could have holes due to the automatic thresholding, we will process the image with a hole filling algorithm. This calculates its negative, segment the image in binary form, identify the background and fill the rest of the image with value 1 (high level).

5 Description of the Affected Region

Here, as in the case of diagnosis based on a unique image, we need to characterize the extracted region in the segmentation phase in order to be understood by the system properly. A first approximation could be to establish the same characteristics that in image description stage. That is, a set of geometric describers, of colour, morphological, etc. which defines the extracted lesion of an image. Nevertheless, we think that to study the image evolution in this way does not make sense, since what we are interested in is the lesion changes (whether it is colour, shape, texture, etc.). In order to that, it will not be necessary to invest too much computational time for the extraction of features of the lesion. Otherwise, studying histograms of gradient orientations should be enough. In this kind of histograms we will have the number of times that each orientation is repeated in the image. This information will serve us to represent each lesion capture and determine possible changes respect to the rest of the images in the sequence.

The image gradient is calculated obtaining the histogram of orientations from the affected region:

$$\nabla I = \begin{bmatrix} G_x \\ G_y \end{bmatrix} = \begin{bmatrix} f(x+1, y) - f(x, y) \\ f(x, y+1) - f(x, y) \end{bmatrix} \tag{1}$$

The calculation can be broken down for each pixel (i, j) as follows:

$$\left|\nabla I(i, j)\right| = \sqrt{G_x(i, j)^2 + G_y(i, j)^2} \tag{2}$$

$$I(i, j)_\varphi = \frac{G_x(i, j)}{G_y(i, j)} \tag{3}$$

We should take into account that, in points without gradient, orientation $I(i, j)_\varphi$ will take arbitrary values, so when it is time to construct the histogram, we must employ just the points which their gradient value $|\nabla I(i, j)|$ exceed a threshold that we will establish depending on training samples. This histogram indicates the number of pixels which has a determined orientation, and each orientation can take value belonging to the range $-\pi$ to π. Due to memory and efficiency problems, the histogram will be discretized, so we represent an orientation set very similar with an average of the values. Therefore, the resultant histogram will be a normalized vector of real numbers between 0 and 1, which we will have to characterize with a integer number. So the lesion evolution will act as a sequence of discrete observation to employ as input for HMMs. We use the vector quantization algorithm of K-means [15]. This allow us to decide how many different classes we want (n) and group histograms (by likeness) in that n classes, each one defined by means of a prototype or average histogram. Hence, each time a new histogram is presented we can consider that it belongs to the class with the prototype nearer and, therefore, reduce the histogram to an integer number between 0 and n - 1.

6 Classification

The last stage of the system has the task to perform the appropriate inferences over the extracted information in previous phases in order to produce a diagnosis from the input. It means, the resultant sequence of normalized histograms from the description phase will be employed to decide which type of lesion is more approximated to the input, if a benign lesion or a melanoma. For that aim we will employ the Hidden Markov Models. These are a set of stochastic automata which permit characterize statistical properties of a sequence [11]. They consist in a finite number of states N and they can be specified by three parameters $\lambda=\{\Pi, A, B\}$; where A is a matrix of $N \times N$ elements a_{ij}, which indicate the probabilities of transition from the state i to j; $B=b_j(O_t)$ is the matrix of probabilities of observation, which represents the likelihood that the state j emits an observation O_t in the instance t; and $\Pi = \pi_i$ indicate the likelihood that the state i will be the initial state of the model.

Diagnosis through HMM is made up of two stages: a first part of training or learning, in which a HMM is built per each class of the domain and other part of classification, in which these models are employed to determine what class the input sequence is more approximated. A set of previously diagnosed lesion evolutions is used for the learning phase. These evolutions have been preprocessed to obtain sequences of histograms corresponding to affected areas. Once we have the histogram set, we will apply the K-means algorithm for calculating the prototypes of the n classes, and then to transform each capture sequence in a sequence of discrete observation. Afterwards, the sequences of each lesion type (benign or melanoma) will be taken and Markov Models will be make using the Baum-Welch algorithm. If we suppose that we have K observation sequences of a determined type, this algorithm allows us to determine the parameters of the HMM which reach the major likelihood to generate the observations. In order to help convergence of the algorithm we must initialize the HMM properly. As image evolutions are sequential, from one capture to other, changes can exist or not, but it will not be loop backwards. This is represented suitably with a left-right or Bakis model [11], it begins always from first state ($q_0 = 1$) and transitions between states are always, either to the same state or to the following ($a_{i,i} = 0.5$ y $a_{i,i+1} = 0.5$). As we do not know the observation probabilities we initialize them uniformly (equal likelihood for each class).

Once we have deduced each one of the HMM corresponding to each lesion type we can employ the system for diagnosing. The classifier takes as input an image sequence (lesion evolution), obtains the histograms of each capture and assigns a class to each histogram depending on the prototypes which were determined in the learning phase. The probability of the sequence taking into account each HMM will be calculated. Hence the type determined by the classifier for that input sequence will be the one modeled by the HMM with the maximum likelihood.

7 Experimentation and Results

In this work we have employed, as input to the system, images acquired by means of dermatoscopy technique or microscopy of epiluminiscence. The main reason is that, due to capture procedure, this technique reduces the reflections of the most superficial layer of the tissue and facilitates the description process of lesions.

A set of image sequences has been employed in the design of the classifier, with half of images representing benign lesions and the other half melanomas. The validation method "left-one-out" is used to test the system. We employ the measures of classification rate (CR), specificity (SP) and sensitivity (SE) for evaluating the performance. The first one represents the portion of samples over the total classified correctly. The specificity is a measure of the percentage which shows healthy cases detected correctly over total of benign lesions. And the sensitivity indicated the portion of melanoma samples detected as right over total of malign lesion samples.

An important parameter to be determined in the classifier design is the number of states of the Markov's models. Each one of the states of the model will represent a different part of the image sequence, it means, an image subsequence that does not differ notably. As we suppose that the system uses sequences with 4 and 5 samples, we present in the table 1 the classification results for models since 2 to 6 states.

Table 1. Results of the Hidden Markov Models according to the states number

States	CR (%)	SP (%)	SE (%)
2	58.82	62.50	55.55
3	85.12	91.67	77.38
4	83.45	88.22	75.87
5	83.33	87.53	76.34
6	61.53	75.43	63.44

In order to obtain this data we have considered that the orientation histograms from each image are discretized with 20 bins and the number of possible classes in each sequence is 4, since this value corresponds intuitively with the number of images that are different. In addition, in the training process we have run 100 iterations with the Baum-Welch algorithm with the aim of reaching the convergence of the model taking into account all training sequences.

We can observe that the optimum performance obtained is when we use Markov Models with 3 states, presenting a classification rate of 85.12%, and a specificity and sensitivity of 91.67 y 77.38% respectively. The classification measures obtained are upper than the ones proportionated by dermatologists, which are around 65%. Anyway, the most important measure we find in this work is the sensitivity, since in this case is more concerned a false negative (the system diagnoses a melanoma as a benign lesion) than a false positive.

8 Conclusions

We have presented in this text a decision support system for melanoma diagnosis taking a sequence of images from same lesion during a period of time as input. We base its logic in a first stage of preprocessing and segmentation by means of Otsu thresholding method. Afterwards, from the histograms of gradient orientation of affected areas we compose an observable sequence which will act as input to the Hidden

Markov Models. These will indicate likelihoods for each sequence, from which we will be able to make a diagnosis in two possible classes: benign lesion or melanoma. Although classification rate obtained, around 85%, is acceptable, the system provide low sensitivity measure, around 77%. One of the future objectives to study is to optimize the different stages of the system, with emphasis on the segmentation and diagnosis phase, with the aim to increase the sensitivity. Other critic element to take into account is to improve the sample database, since classification results that we obtain depend a lot on the quality of training sets.

Acknowledgements. This work has been granted by the Valencian Autonomous Government (GV07/126).

References

1. Abbasi, N.R., Shaw, H.M., Rigel, D.S., Fiedman, R.J., Osman, I., Kopf, A.W., Polsky, D.: Early Diagnosis of Cutaneous Melanoma. Revisiting the ABCD Criteria. Journal of the American Medical Association 292, 2771–2776 (2004)
2. Argenziano, G., Fabbrocini, P., Carli, P., De Giorgi, V., Sammarco, E., Delfino, M.: Epiluminescence microscopy for the diagnosis of doubtful melanocytic skin lesions. Comparison of the ABCD rule of dermatoscopy and a new 7-point checklist based on pattern analysis. Arch Dermatol 134, 1563–1570 (1998)
3. Ganster, H., Pinz, A., Rhrer, R., Wilding, E., Binder, M., Kittler, H.: Automated Melanoma Recognition. IEEE Trans. on Medical Imaging 20(3), 233–239 (2001)
4. González, R.C., Woods, R.E.: Digital Image Processing. Prentice Hall, Wilmington (2002)
5. Hintz-Madsen, M.: A probabilistic frame-work for classification of dermatoscopic images. Ph.D. dissertation, School of Computing Science, Simon Fraser University (1998)
6. Kreutz, M., Anschtz, M., Grnendick, T., Rick, A., Gehlen, St., Hoffmann, K.P.: Automated Diagnosis of Skin Cancer Using Digital Image Processing and Mixture-of-Experts. Bildverarbeitung fr die Medizin 357–361 (2001)
7. Lee, T.: Measuring border irregularity and shape of cutaneous melanocytic lesions. Ph.D. dissertation, School of Computing Science, Simon Fraser University (2001)
8. Menzies, S.W., Ingvar, C., Crotty, K.A., McCarthy, W.H.: Frequency and morphologic characteristics of invasive melanomas lacking specific surface microscopic features. Arch. Dermatol. 132, 1178–1182 (1996)
9. Petrou, M., Bosdogianni, P.: Image Processing. The Fundamentals. Wiley, Chichester (1999)
10. Popa, R., Aiordachioaie, D.: Genetic Recognition of Changes in Melanocytic Lesions. In: 8th International Symposium on Automatic Control and Computer Science (2004)
11. Rabiner, L.R.: A Tutorial on Hidden Markov Models and Selected Applications in Speech Recognition. Proceedings of the IEEE 77(2), 257–286 (1989)
12. Sboner, A., Blanzieri, E., Eccher, C., Bauer, P., Cristofolini, M., Zumiani, G., Forti, S.: A knowledge based system for early melanoma diagnosis support. Intelligent Data Analysis in Medicine and Pharmacology (2001)
13. Schmid-Saugeon, P., Guillod, J., Thiran, J.P.: Towards a computeraided diagnosis system for pigmented skin lesions. Computerized Medical Imaging and Graphics 27, 65–78 (2003)

14. Stolz, W., Rieman, A., Cognetta, A.B.: ABCD rule of dermoscopy: a new practical method for early recognition of malignant melanoma. European Journal of Dermatology 4, 521–527 (1994)
15. Theodoridis, S., Koutroumbas, K.: Pattern recognition. Academic Press, London (2003)
16. Thiran, J.P., Macq, B.: Morphological Feature Extraction for the Classification of Digital Images of Cancerous Tissues. IEEE Trans. on Biomedical Engineering 43(10), 1011–1020 (1996)
17. Zagrouba, E., Barhoumi, W.: An accelerated system for melanoma diagnosis based on subset feature selection. Journal of Computing and Information Technology 13(1), 69–82 (2005)

Intensive Care Unit Platform for Health Care Quality and Intelligent Systems Support*

M. Campos[1], A. Morales[2], J.M. Juárez[2], J. Sarlort[2], J. Palma[2], and R. Marín[2]

[1] Departamento de Informática y Sistemas
[2] Departamento de Ingeniería de la Información y las Comunicaciones,
Universidad de Murcia, Campus de Espinardo, CP: 30.100, Murcia, Spain
{manuelcampos,morales,jmjuarez,salort,jtpalma,roquemm}@um.es

Abstract. The underlying idea in this work consists on providing added values utilities that allow exploiting the Electronic Health Record (EHR) as something more than a simple information record. The key for providing added value to the clinical information systems is to exploit the synergy "Information + intelligence + ubiquity". Based on this idea, we propose a distributed architecture that deals with: 1) Database and an integration layer to exploit the data stored and its integration with external information system, 2) Tools for support the medical knowledge management, 3) Tools for supervision and analysis of the health care quality (based on EBM and Clinical Guidelines) 4) Intelligent Assistance Tools.

Keywords: Distributed Clinical Information System, CH4, knowledge management.

1 Introduction

In recent years, there has been a growing demand for Clinical Information Systems (CIS) by public and private health systems. The goal of such proliferation is to provide the technological infrastructure necessary to improve national health systems. Among all the initiatives, the one which has more potential to revolutionise the provision of health services and the working methods of health professionals is the Electronic Health Record (EHR) [12]. The EHR is mainly a mechanism for integrating medical information in electronic and traditional means [6]. The benefits that the use of such systems entail are countless [6,12]:

1. Improvement of the quality of services provided by the health system, keeping costs within reasonable margins.
2. Reduction of medical errors and time saving.
3. Reduction of the frontiers between inhospital and outpatient care, allowing the provision of services focused on the patient at a reasonable cost.
4. Promotion of the cooperation between health professionals and between them and their patients, since information sharing is easy.
5. Reduction of the redundancy of tests and patient mobility.

The architecture and objectives of the EHR systems are becoming increasingly complex in order to answer the real needs of physicians and patients and to achieve a

* This work has been supported by the Spanish MEC under the national project TIN2006-15460-C04-01, PET2006-406 and PET2007_0033.

J.M. Corchado et al. (Eds.): DCAI 2008, ASC 50, pp. 366–374, 2009.
springerlink.com

real, effective and massive deployment in clinical practice. The basic needs are directed towards improving the daily routine, such as data collection from patient, report generation, or assistance to diagnosis and therapy, and are largely met with general services used in medical field during long time: service integration, collaborative work, document, and report management. A possible solution to this problem is to develop a distributed platform, to which various modules and services can be easily attached and accessed from the technological devices best suited. In addition, an especially relevant factor in this aspect is the standardization, both in the representation of clinical information and in its communication and sharing, both from a clinical point of view and from the easiness of data gathering. Along with mandatory issues such as security and confidentiality of patients, other needs that currently exist should be taken into account, such as the health care quality, knowledge management, intelligent data analysis, or clinical research.

This work focuses on the development of a distributed platform that allows addressing the aforementioned needs, and that facilitates collaborative work to the clinical staff. The platform is designed for the field of Intensive Care Units (ICU), and integrates different modules whose origin is the Artificial Intelligence (AI).

The structure of the article is as follows: Section 2 describes the general structure of the ICU platform; Section 3 gives some details of the key components; Section 4 shows the support to manage health care quality; Section 5 presents the intelligent modules. Finally we show some conclusions.

2 ICU Platform

The Intensive Care Unit (ICU) is a medical service that provides critical attention to medically recoverable patients. One of the fundamental characteristics of this domain is that patients require a permanent availability of monitoring equipment and specialist care. Thus, physicians must work in shifts in order to provide a 24 hour service. The temporal evolution of patients is permanently recorded and analyzed by physicians, who must tackle a wide range of patient pathological problems (e.g. cardiovascular, renal, infections, neurological, etc.).

In this sense, intensivists have to deal with an overwhelming amount of temporal information provided not only by on-line monitoring, but also from patients' records collected from different hospital departments (e.g. laboratory results, radiology,... etc.).

It is a fact that computer-based systems are required in ICU to record and manage clinical data and also acquire and exploit medical knowledge. As far as we concerned, three computational dimensions should be managed in the ICU: (1) the management of the temporal dimension of clinical data, (2) the management of traditional information processes (EHR, integration to other information systems) and health-care information quality; and (3) the management of medical knowledge (acquisition, discovery and exploitation). However, a system that integrates the aforementioned aspects is not a simple issue since requires the experience in traditional medical software and AI techniques, especially considering the ubiquity and distribution of information and knowledge.

In this work, we propose a distributed architecture (see Figure 1) to deal with these issues based on the identification of the following components:

- Temporal database (TDB): we designed a database considering the temporal component of each event that could occur. These temporal components are represented by time points (something occurs in an instant) and time intervals (something that occurs during a period of time).
- Temporal Integration Layer (TIL): this layer allows to manipulate (insert, update, delete and query) the temporal information in a homogeneous way, and allows the intercommunication of the different modules by means of a messaging system.
- ICU information management system: a set of tools to manage the EHR information (physicians, nurses and assistants), covering the daily needs concerning patient demographic data for admission, evolution and discharge reports, nurse cares, therapy, microbiological test, etc.
- Hospital and Department Information Systems: Hospital Information System (HIS) and Department Information Systems also interact with the ICU. In this sense, the architecture provides interfaces and protocols to exchange information. In particular, by the use of international standards as HL7 (for messaging interchange) or ICD (to codify clinical terms).
- Health Care Quality Support: in order to improve health-care quality, some of the most sounded approaches in medical science are Evidence-Based Medicine (EBM) and its methodological development by the use of Clinical Practice Guidelines (CPG). This architecture must include pieces of software to support EBM and fulfil CPG.

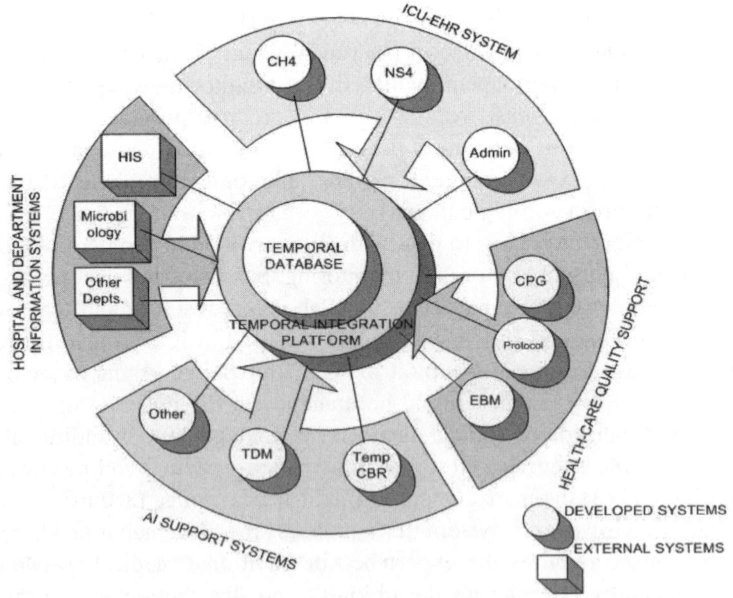

Fig. 1. ICU Platform architecture

– Artificial Intelligence Support: thanks to the availability of a large amount of medical information, this subsystem is designed to support clinical research and to assist physicians in their daily work.

In the rest of this paper, we describe in depth the different components of this architecture and present the different implementations in a real ICU.

3 The ICU-EHR System

The ICU-EHR System is divided in three subsystems: the Medical Information System (CH4), the Nursing Information System (NS4) and the Administrative Information System (Admin). Each one is designed to meet the needs of each type of staff in the ICU (physicians, nurses and administrative staff). Although they are different systems, they complement each other to give rise to a common EHR with a wealth of temporal clinical data.

CH4 (see Figure 2) is a tool designed for the physicians, who can register, search, and display information on patients from their admission to their discharge from ICU. Both administrative data and data concerning the patient's health (e.g. personal information, laboratory tests, treatments, diagnostics, scoring systems, etc.) can be managed.

A key element for the continuous supervision of the patient is the concept of problem. A problem is any deviation in the normal state of any variable of the patient, e.g. fever. These problems can be related with the rest of elements of the patient's clinical history, providing an easy way to explain the patient's evolution.

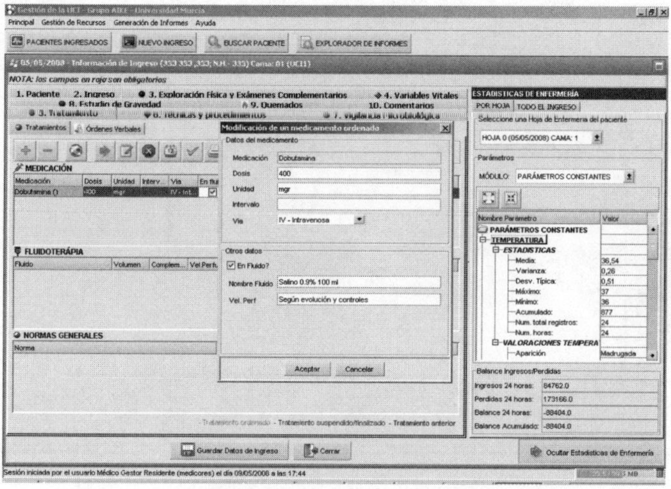

Fig. 2. CH4 physician application

With the objective of standardizing clinical tests and therapies, CH4 incorporates the *profiles*. Profiles collect a set of specific tests and treatments which are used to manage the most common cases.

Another remarkable feature of the system is its flexibility and extensibility for each specific ICU department, providing an easy way to add new parameters such as medical tests, treatments or new diagnostics. Therefore, it is completely adaptable to the usual way of working of different ICU services as well as different clinical protocols in use. These facilities reaches two objectives. The first one is to personalise the concepts and terminology to the concrete ICU where the platform is deployed. The second one is to incorporate standard terminologies (such as ICD). In this way, each ICU is able to organise the terminology in three steps: (1) select from standards the concepts used in the ICU domain, (2), create a mapping to the local (service) or regional terminology, and (3) organise those concepts in a hierarchy according to logical criteria.

The second tool, NS4, provides an interface similar to traditional nursing sheets (see Fig. 3), which are documents that record the patient's evolution, the care and treatments they receive. NS4 allows the nursing staff to manage easily the patients by means of report templates, facilities to search and browse patients, and a simple way to fill in data within the ICU box.

The whole ICU-EHR system is characterised by providing facilities in the management and automatic generation of medical reports in different formats. This characteristic is essential in the Administrative Information System, whose objective is the overall supervision of the ICU service. This supervision is carried out by means of reports reflecting the state of service (e.g. patients admitted and discharged, staff workload) or statistical summaries by the type of patients (e.g. mortality reports, infectious patients, patients with certain types of injuries).

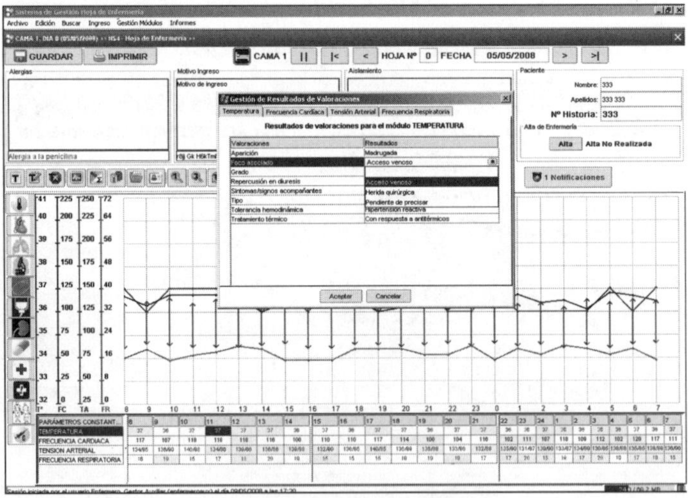

Fig. 3. NS4 nursery sheet

All these subsystems are intercommunicated across the temporal integration platform through mechanisms based on message queues, visual alerts, and notifications in the respective interfaces. In this way, the nursering staff is notified when a new event is triggered for a patient, e.g., when a new treatment is ordered.

4 The Hospital and Department Information Systems

The Temporal Integration Platform provides communication mechanisms for exchanging information between the ICU-EHR System and other systems where patient's information is stored. Regarding the ICU, the Hospital Information Systems (HIS), the laboratory information systems (clinical tests and microbiology), imaging diagnostic, and pharmacy are the main systems outside the ICU. Due to the heterogeneous nature of these systems it has been difficult to find a common framework for information exchange. For this reason, we have opted for combining standard and *ad-hoc* interconnection solutions based on the needs for each department. For example, the communication with the HIS to get affiliation data or with the laboratory in order to obtain the results of the clinical tests needs an exchange of HL7 messages. Moreover, the problem of using a common coding for laboratory testing and other elements of the clinical record has been solved by manual translation tables (or mappings). However, access to microbiology laboratory consist in a direct connection to a view of the database.

Obviously, interoperability is a key issue to be improved in this platform. A possible solution would be to use an integration engine based on standards for the exchange of clinical information (HL7) and medical terminology, encoding or classifications (SNOMED, UMLS, LOINC, ICD).

5 Health Care Quality Support

Nowadays, some of the most important problems in public health system are the increment of the cost, the maintenance and improvement of the health care quality. The basic principles for the health care quality consist of 1) reducing of the variability during resources utilization, 2) improving the efficiency of such resources, and 3) reducing the cost of the whole medical process.

In short, to get through the day with a high standard of quality, most professionals must have: (1) access to critics, debiases and influencers to help them identify relevant and irrelevant knowledge that can be applied to a patient; (2) reminder and alerting agents to support the memory, thus avoiding lapses and oversights of knowledge previously acquired; (3) sources from which they can acquire new knowledge to solve a problem not faced before, (4) a tool able to extract, store and share knowledge explicitly, e.g. corporate memory.

To cope with these problems, the clinical authorities foment the use of Evidence Based Medicine (EBM) [13]. EBM advocates the use of evidences gained from the scientific method to the clinical practice. The EBM has its greatest exponent in the elaboration of clinical guidelines that give an orientation for improving the health care quality and add better resource management.

A Clinical Practice Guideline (CPG) is a set of recommendations/rules developed in some systematic way in order to help professionals and patients during the decision-making process concerning an appropriate health-care pathway, by means of opportune diagnosis and therapy choices, on specific health problems or clinical conditions [5]. Thus, a CPG can be considered to be a powerful decision-making tool, which helps reducing the difference between the practical cases and the optimal procedure, and then facilitating the necessary mechanisms to improve the healthcare processes and to reduce their cost.

In our system we have opted for providing basic support to these tools through a practical implementation. Firstly, physicians and nurses should organise knowledge, mainly taxonomically, about all elements applicable to patients (treatments, tests, diagnoses, problems, cares, etc.). Secondly, users can add annotations (in the form of text or address file) in the definition of those elements, thus documenting evidence and motivations of the use of each element, for example, a medicament. Thirdly, therapy profiles allows to define a standardized combination of medicaments, fulfilling the recommendations provided by the GPC. Moreover, these profiles can also be adapted to the needs of each concrete patient. Finally, the use of profiles for tests and problem oriented clinical history provides a methodological procedure to collect all the evidence to support the treatment given to the patient.

6 Intelligent Modules

Thanks to the large amount of data recorded, the explicit knowledge can be exploited, and even increased through intelligent modules. The idea of exploiting the knowledge led to the creation of expert systems[14], but the only partial success was due to their discontinued development and use. Currently, physicians are more interested in support tools for research than in those classic systems experts for diagnosis assistance. Some paradigms such as Knowledge Discovery (KD) and Case-Based Reasoning (CBR) can be used for this purpose.

6.1 Knowledge Discovery

KD can be defined as the process of extracting not trivial, unknown, and potentially useful information implicit in the data [4]. This process incorporates techniques from database, statistics and machine learning with two main purposes: pattern discovery and prediction.

In the medical domain, the collection of the vast amount of data provided by the EHR, makes it possible to apply Data Mining (DM) techniques aimed at the discovery of new relationships between all the recorded data (pathologies, symptoms, treatments, ...). This is particularly true for the ICU, where continuous monitoring of patients generates an enormous amount of heterogeneous data: biological signals sampled periodically, data from clinical history, events and episodes that reflect states and trends.

In this domain, DM techniques have acquired great significance, with applications such as the patient clustering for identification of colorectal cancer risk groups [2], or temporal decision trees for predicting mortality in the ICU [15].

We are now integrating in the platform a module for KD that includes the techniques developed by our the research group for the discovery of sequential patterns [7] and complex temporal patterns [1].

6.2 Cased Based Reasoning

CBR tackles new problems by referring to analogous problems that have already been solved in the past. In other words, CBR is based on individual experience in the form of cases (episodic knowledge). Therefore, CBR seems to be an effective approach in medical domains since cases refer to patient episodes within the EHR. We highlight three major advantages for using CBR. First, the explicit experience is easy to be added to CBR systems, since they can insert, replace and eliminate cases at the explicit knowledge base. Second, the continuous integration of cases during the use of the CBR system allows an incremental acquisition of knowledge. Finally, another advantage of the CBR is its integration with CIS since the HER itself serves as the basis of cases. In the clinical domain, CBR systems have been effectively used for decision support in the treatment of patients on dialysis [11], or the diagnosis of cancer from mammograms [8].

We are currently working in the integration of a temporal CBR module developed by our research group [9].

7 Conclusions

In this article we have presented a platform that is part of a distributed clinical information system. This platform is specifically focused on a ICU and similar hospital services. The main benefits obtained are those of an EHR, such as easy information sharing and remote access to it, but besides, it also addresses other needs such as facilitating the daily routine.

Four different subsystems have been integrated in the platform around its kernel. The ICU-EHR subsystem provides conventional mechanisms to record patient data and to generate reports on them. There is a service integration subsystem, which maintains a flexible structure that allows using communication standards or *ad hoc* integration. In this modular approach, two subsystems that provide a competitive advantage have been included: the health care quality support subsystem and the intelligent modules subsystem.

The health care quality support system responds to current clinical practice needs, providing protocols to assist physicians, both in the nomenclature and in treatments and tests made to patients.

As for the intelligent modules, the approach consists of incorporating techniques to support the physician in the clinical practice and tools to enhance clinical research. KD techniques allow the extraction of implicit knowledge from data stored in the system, whereas CBR techniques allow a physician to locate, at any time, all medical records in the system that are similar to the patient's that is being treated, thus supporting decision-making with the largest available evidence. The results obtained so far show the promising future of these systems and their definite incorporation as an essential utility in the medical domain.

Acknowledgements. The authors would like to thank Francisco Palacios from the Intensive Care Unit service of the Hospital Universitario of Getafe, and the Biomedical Research Institute of the Hospital Universitario of Getafe.

References

1. Campos, M., Juárez, J.M., Palma, J.T., Marín, R.: Temporal data mining with temporal constraints. In: Bellazzi, R., Abu-Hanna, A., Hunter, J. (eds.) AIME 2007. LNCS (LNAI), vol. 4594, pp. 67–76. Springer, Heidelberg (2007)
2. Chen, J., He, H., Jin, H., McAullay, D., Williams, G., Kelman, C.: Identifying risk groups associated with colorectal cancer. In: Williams, G.J., Simoff, S.J. (eds.) Data Mining. LNCS (LNAI), vol. 3755, pp. 260–272. Springer, Heidelberg (2006)
3. DeClercq, P.A., Blom, J.A., Korsten, H.H.M., Pasman, A.: Approaches for creating computer interpretable guidelines that facilitate decision support. Artificial Intelligence in Medicine 31, 1–27 (2004)
4. Fayyad, U., Piatetsky-Shapiro, G., Smyth, P.: From Data Mining to Knowledge Discovery: An Overview. In: Advances in Knowledge Discovery and Data Mining. AAAI/MIT Press (1996)
5. Field, M.J., Lohr, K.N.(eds.): Guidelines for clinical practice: from development to use. National Academy Press, Washington (1992)
6. Gunter, T.D., Terry, N.P.: The emergence of national electronic health record architectures in the United States and Australia: models, costs, and questions. J. Med. Internet Res. 7(1), 3 (2005)
7. Guil, F., Juárez, J.M., Marín, R.: Mining Possibilistic Temporal Constraint Networks: A Case Study in Diagnostic Evolution at Intensive Care Units. In: Proc. of the Intelligent Data Analysis in Biomedicine and Pharmacology, IDAMAP 2006, Verona, Italy, pp. 7–12 (2006)
8. Hung, S.Y., Chen, C.Y.: Mammographic case base applied for supporting image diagnosis of breast lesion. Expert Systems with Applications 30(1), 93–108 (2006)
9. Juarez, J.M., Guil, F., Palma, J., Marin, R.: An uncertain temporal similarity proposal for temporal CBR. In: Roth-Berghofer, T.R., Göker, M.H., Güvenir, H.A. (eds.) ECCBR 2006. LNCS (LNAI), vol. 4106, pp. 210–219. Springer, Heidelberg (2006)
10. Knaup, P., Wiedemann, T., Bachert, A., Creutzig, U., Haux, R., Schilling, F.: Efficiency and safety of chemotherapy plans for children: Catipoa nation wide approach. Artificial Intelligence in Medicine (24), 229–242 (2002)
11. Montani, S., Terenziani, P., Bottrighil, A.: Exploiting decision theory for supporting therapy selection in computerized clinical guidelines. In: Proc. of the 10th Conference on Artificial Intelligence in Medicine, pp. 136–140 (2005)
12. Powell, J., Buchan, I.: Electronic Health Records Should Support Clinical Research. J. Med. Internet Res. 7(1), 4 (2005)
13. Sackett, D.L., Rosenberg, W.M., Gray, J.A., Haynes, R.B., Richardson, W.S.: Evidence based medicine: what it is and what it isn't. BMJ (British Medical Journal) 312(7023), 71–72 (1996)
14. Shortliffe, E.H.: MYCIN: rule-based computer program for advising physicians regarding antimicrobial therapy selection. Ph.D. thesis, Stanford University (1974)
15. Terenziani, P., Montani, S., Bottrighi, A., Molino, G., Torchio, M.: Clinical guidelines adaptation: managing authoring and versioning issues. In: Proc. of the 10th Conference on Artificial Intelligence in Medicine, pp. 151–155 (2005)
16. Toma, T., Abu-Hanna, A., Bosman, R.: Predicting mortality in the intensive care unit using episodes. In: Proc. of the 1st International Work-conference on the Interplay between Natural and Artificial Computation, pp. 447–458 (2005)

The Intelligent Butler: A Virtual Agent for Disabled and Elderly People Assistance

Gabriel Fiol-Roig, Diana Arellano, Francisco J. Perales, Pedro Bassa, and Mauro Zanlongo

Departamento de Matemáticas e Informática
Universitat de les Illes Balears
Carretera de Valldemossa, km 7.5, 07122 Palma de Mallorca, Spain
biel.fiol@uib.es, diana.arellano@uib.es, paco.perales@uib.es

Abstract. Social assistance constitutes an increasing problem in developed countries, which can be considered from two dimensions: the home and the hospital frameworks. Anyway, most of the tasks have to do with aiding people with limitations in complex environments as a hospital or a house. Intelligent agents constitute a powerful approach in designing computer systems making possible the interaction of the users (elderly and disabled people) with the elements of a domotics environment. Such a purpose can be achieved through the unification of artificial intelligence techniques, virtual reality, multimodal interfaces and digital nets with domotics services. This paper describes the design and implementation of the prototype for a virtual agent capable of attending disabled people in a home environment. The results are shown through a computer program that simulates the behaviour of the agent in developing some typical functions.

Keywords. Artificial intelligence, human-computer interaction, virtual home assistance, facial animation.

1 Introduction

Developing and integrating agent technologies in real world applications is a complex task, since not a few research and technological areas are involved. Some of these domains are distributed object technology, digital networks, 3D virtual words, virtual reality, multimodal interfaces, and artificial intelligence, among others. Such a complexity and diversity requires considering a first step where the behaviour of the agents is tested in a controlled virtual environment.

Our main purpose consisted in designing an intelligent agent, the butler, for aiding disabled and elderly people in a virtual domotics environment. The agent must be adaptive, with the ability to react in real time, to learn, remembering and evolving with the experience, and to decide in specific situations in an autonomous way. Under these considerations, the users are allowed to interact with the assistive environment through voice, gesture and touch. In this sense, a large amount of data must be processed and inferred by the butler, which demands an adequate knowledge representation and efficient inferential mechanisms. In [6] some theoretical considerations about a knowledge representation model supporting complex relations between perceptions and actions are described. Also, a multiagent system [11] with several agents cooperating among them is required.

J.M. Corchado et al. (Eds.): DCAI 2008, ASC 50, pp. 375–384, 2009.
springerlink.com

Related work can be found in [8] and [3], where a platform based on mobile agents combined with federated information management mechanism to create a flexible infrastructure for specialized care services, was developed. Attard et al. [2] illustrate a home system oriented ontology and an intelligent agent based framework for the rapid development of home control and automation. Velasco et al. [10] proposed an architecture for building a smart home environment using multiagent systems, demonstrating its effectiveness with an application example where multimedia contents follow the user movements throughout the house.

Section 2 describes the characteristics of the home environment; in section 3, the general aspects concerning the agent program are presented; finally, a practical implementation through a program prototype is discussed in section 4.

2 The Virtual Home Environment

Our task consisted of creating a virtual domotics house as the environment where the butler agent acts. The interaction with the house and the disabled user is through a set of domotics devices, in a reactive and continuous way, by receiving perceptions and taking the corresponding actions. Figure 1 shows a general scheme of such behaviour.

The environment progress is cyclic, with the agent receiving a set of perceptions – perception vector– coming from the domotics sensors of the house and the requests and/or the answers of the user. On the other hand, a daily planner provides the butler with the protocol related with the house and the disabled to be followed (user's lunch time and diet, medication, etc.). On the basis of the perception vector and the daily planner, the butler evaluates the house's and the user's states. As a result, a set of actions is planned –action vector–. The action vector contains three types of actions: actions on the environment (affect the disabled and the domotics house), actions on the daily planner (reprogram activities in it), and actions on the report of the disabled user (related to healthcare or diet considerations). Notice from figure 1 that actions on the environment may have direct consequences on the perception vector. At each time instant, the butler analyzes the perception vector and considers the perception with the highest priority, removing it from the perception. On the basis of the considered

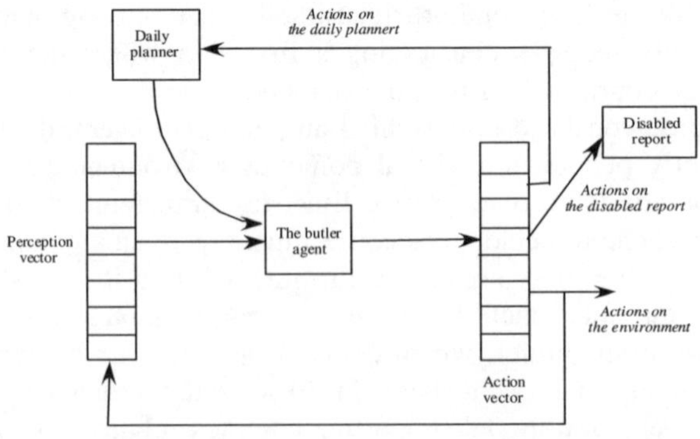

Fig. 1. Reactive behaviour of the butler

perception, the butler plans a set of actions to be performed; these actions will update the actual state of the environment.

2.1 The Environment Specification

The environment specification considers the set of components and devices capable of generating perceptions as well as generating and/or receiving a set of actions. These actions are connected with the corresponding perceptions, in such a way that they describe the natural progress of the environment.

The basic components of the environment are: the domotics house, the user (disabled or elderly person), the daily planner, the user's report, and the butler agent. The domotics house, the disabled person and the daily planner are in charge of generating perceptions. Actions are taken by the butler, whose effects are manifested on the domotics house, the disabled person, the daily planner and the report of the user. The functional character of the environment is structured in five modules [7]:

1. The healthcare module, which functions are related with the health and medical prescriptions for user's treatments.
2. The nursing module, in charge of planning the daily menu of the user.
3. The preferences and scheduling module, by which the user's activities and hobbies preferences are scheduled and reprogrammed.
4. The stock module, that informs about the goods available and those exhausted in the house, such as food or medication.
5. The control devices module, which manages the control devices of the house. This module is made up of two components:
   5.1. The house accommodation module, which covers the needs of accommodation of the user (light level, temperature, blinds, windows, etc.).
   5.2. The house security module, in charge of covering the security matters of the user and the house (warning of electrical failures, water or gas leaks, fires, unexpected visitors, etc.).

2.2 The Environment Progress

The way that the environment evolves is determined by the connection between perceptions and actions. The perception-action vector (PAV) is the basic mechanism connecting perceptions with the actions to be taken in certain moment.

The PAV can be described as a set of *n* mappings, P_i: $V_i \rightarrow \Pi(A)$, i=1..n, P_i being the i-th perception, V_i is the domain of values of P_i, A is the set of possible actions on the environment, and $\Pi(A)$ is the set of parts of A. That is, the PAV associates each P_i value with the corresponding subset of actions to be taken. The PAV can be modeled through an array structure, such as figure 2 illustrates. Each set of actions associated with a given value of a perception will be considered independent of the moment it occurred.

It can be observed from figure 2 that $V_1=\{V^{1,1}, V^{1,2}, \ldots, V^{1,s}\}$, and value $V^{1,1}$ is associated with set $\{a_i, \ldots, a_j\}$ of actions, value $V^{1,2}$ with set $\{a_k, \ldots, a_r\}$.

The PAV provides a synthesized overview of the world progress. However, a wider range of general properties extending the capacity of the environment progress

P_1	$V^{1,1}$	a_i
		...
		a_j
	$V^{1,2}$	a_k
		...
		a_r
	...	...
	$V^{1,s}$	a_p
		...
		a_v
...	...	...
P_n	$V^{n,1}$	a_t
	...	...
	$V^{n,w}$	a_z

Fig. 2. The perception-action vector

is also considered. Overcoming conflicting situations is a significant task. Conflicts appear when a goal can be reached through several possibilities, but only one of them can be considered. A particular case is when different subsets of actions are associated with the same perception –or subset of perceptions–.

Example: the Healthcare and the Control Devices Modules. In this prototype two modules have been considered: healthcare and control devices. Both modules contain

Perceptions				
Type	Source	Code	Description	Actions
Generals	House	1	Date	
		2	Time	
		3	Position where the user stays in the current moment	
Physiological Parameters -PP-	Daily planner	4	It is time to take the PP to the user –temperature, hearth rhythm, blood pressure,...-	A1
				A2
	User	5	The user wants its PP to be taken	A2
	House	6	The health center notifies by e-mail that the PP of the user must be taken	A1
				A2
		7	Blood pressure	A3
				A4
		8	Hearth rhythm	A3
				A4
		9	Body temperature	A3
				A4
		10	Weight	A3
				A4
Take medication	Daily planner	11	It is time to take the medication	A5
New doctor's prescription	User	12	User informs the butler of a new doctor's prescription	A6
				A7
				A3
User don't feel well	User	13	User has a headache	A8
				A3
		14	User has a cold	A8
				A3
		...	...	...
Artificial illumination	User	20	The user asks for a change in artificial illumination	A9
		21	User asks butler for Turn light on / turn light off	A10
		22	User asks butler for more / less light intensity	A11
...	...	...	...	...
Emergency situation	House	36	House sensors detect an emergency situation	A38
				A39

Fig. 3. Sketch of the PAV corresponding to the healthcare and control devices modules

a total of 36 perceptions coming from different components of the environment. Figure 3 shows a sketch of the environment progress in these situations, described in a PAV format. Perceptions in figure 3 are classified in several types, depending on their nature, and come from different sources of the environment. Each perception in figure 3 is identified by a code. Some of them are associated with more than one action, having a total of 39 different actions. Figure 4 illustrates the meaning of action symbols in figure 3.

2.3 Information Systems as Decision Mechanisms

The agent's decision mechanism requires a full knowledge representation model of the environment, describing actions in terms of perceptions.

In the case of environments with simple perception-action relationship, a decision model based on a priority mechanism is enough to decide which perceptions are first considered and which actions are first performed. Nevertheless, when complex perception-action relations are established, a more powerful decision model is needed.

An information system (IS) is an abstraction model that constitutes a powerful tool to describe concepts in terms of properties or. An IS should also provide a mechanism to perform the corresponding decision.

In the context of this work, an IS provides a functional approach to describe actions in terms of perceptions. In particular, we deal with Object Attribute Tables (OAT), which can be understood as models of information systems.

An OAT is a mapping from a set $NR=\{n\_e_1, n\_e_2,\ldots, n\_e_m\}$ of n-tuples, to a set $\Pi(C)$ of subsets of concepts. C is a set of concepts and $\Pi(C)$ the set of parts of C; that is, OAT: NR $\rightarrow \Pi(C)$. Tuples n_e_i are made up of *n* values corresponding to *n* attributes of a set $R=\{r_1, r_2,\ldots, r_n\}$. Thus, $n\_e_i= (v(r_1), v(r_2),\ldots, v(r_n))$, $v(r_j) \in V_j$, $j=1\ldots n$, $v(r_j)$ being the value of attribute $r_j \in R$, whose domain is V_j.

If set R of attributes corresponds to a set $P=\{p_1, p_2,\ldots, p_n\}$ of perceptions, and set C of concepts is identified by a set of $A=\{a_1, a_2,\ldots, a_w\}$ of actions, then we are talking about a mapping connecting perceptions with actions. Formally, $OAT:NP\rightarrow \Pi(A)$, NP being the set of n-tuples of values of the perception domains. Figure 5 shows a graphic illustration of the concept of OAT.

Symbol	Action meaning
A1	To notify the user about the imminent PP taking
A2	To take the PP to the user.
A3	To update the medical record
A4	If it is not normal, then contact the health center.
A5	Take the medication to the user.
A6	Buy the new prescribed medication
A7	To update the daily planner
A8	Give the user an aspirin
A9	Change the state of artificial illumination.
A10	Turn light on / turn light off.
A11	Increase /decrease light intensity
...	...
A38	Evacuate the user
A39	Call the emergency services

Fig. 4. Meaning of some action symbols in figure 3

NP \ P	p_1	p_2	…	p_j	…	p_n	A
n_e_1	t_{11}	t_{12}	…	t_{1j}	…	t_{1n}	$\{a_i, a_j,\ldots, a_k\}$
n_e_2	t_{21}	t_{22}	…	t_{2j}	…	t_{2n}	$\{a_t, a_v\}$
	………						
n_e_i	t_{i1}	t_{i2}	…	t_{ij}	…	t_{in}	$\{a_r, a_s,\ldots, a_p\}$
	………						
n_e_m	t_{m1}	t_{m2}	…	t_{mj}	…	t_{mn}	$\{a_x,\ldots, a_z\}$

Fig. 5. Object Attribute Table

Sets $\{a_p, a_q,\ldots, a_r\}$ of figure 5 represent subsets of actions, that is, $\{a_p, a_q,\ldots, a_r\} \subseteq A$, and t_{ij} is the value of perception *j* corresponding to the i-th tuple. The meaning of the i-th row of an OAT is as follows: «if t_{i1} is the value of perception p_1 and t_{i2} the value of p_2 and…and t_{in} the value of p_n, then perform the subset $\{a_r, a_s,\ldots, a_p\}$ of actions».

Object attribute tables have a higher descriptive power than perception-action vectors, since they allow expressing a subset of actions in terms of a set of perceptions. OAT also allows considering aspects closely related to complex environments, such as the existence of incomplete knowledge, the presence of a vast amount of perceptions, the capacity of handling quantitative and qualitative knowledge, and provides a way to extend the capabilities of agents towards tasks learning [4] [5]. Deciding what action must be performed in a given moment may require additional knowledge of the environment. This means that an extension of the model of the environment progress is required. Such extension is considered as an *internal state* of the environment [9]. The internal state may be useful for several purposes, for example, to adopt more precise actions or to avoid absurd decisions.

3 The Agent Program

The agent program is permanently monitoring the state of the disabled user, and it is based on a set of condition-action rules. At each step, the program takes the perception vector, the daily planner and the environment's state as the input data, and determines which OAT has to be considered. Based on the selected OAT, the agent decides the subset of actions to be performed.

Anyway, if two or more rules can be triggered in a given step –and so the corresponding object attribute tables must be considered–, only one of them will be selected. Such conflicting situations are overcome by associating a priority mechanism with rules. Rules with a higher priority are firstly evaluated. A given perception will only be taken into account when no perception with a higher priority is present in the perception vector. In this sense, perceptions to do with emergency situations have the highest priority, then those task referring to the health of the disabled are considered, and so on… Next, a simple sketch of a general agent algorithm is shown.

```
procedure BUTLER-AGENT(var:perception_vector; daily plan-
ner; environment's state; var action_vector);
```

```
begin
     if P36 then performing the emergency_situation OAT;
     else
          if P13 OR P14 OR… then performing the
                               user_don't_feels_well OAT;
          else
               …
end;
```

Symbols P_i of the above procedure refers to the i-th perception, according to figure 3.

A practical program prototype based on the discussed artificial intelligence approaches has been developed, showing the suitability and efficiency of the knowledge representation model adopted. Next section presents some aspects of the simulated environment, which the prototype is based on.

4 Simulation of a Home Domotics Environment

The main goal of this application is to simulate the interaction between a disabled person and his/her environment with the help of a virtual assistant.

The implemented modules of the system were the *control devices module* and the *healthcare module*. The *control devices module* takes into consideration various elements of the house: windows (*opening/closing*), curtains (*opening/closing*), lighting (*on/off*), and temperature (thermostat *on/off*). The *health care* module controls the state of health of the disabled person, monitors his physiological parameters and diagnosed diseases, and manages all the information related with medical prescriptions and medication times. The environment progress could be seen by providing the values for the parameters of each element of the control devices and healthcare modules –that is, the perceptions–, and creating simulated lists of tasks –the actions associated with the perceptions–. Thus, the behaviour of the intelligent agent could be evaluated.

Through a graphical interface, developed using C++ programming language and OpenGL, the user is able to change the status of each element. The interface of the application has been divided in 4 different windows. Figure 6 shows the graphical environment of the house with a disabled person in it.

Fig. 6. Corner view of the virtual house with the simulated disabled person

Figure 7(a) shows a window with the parameters of *time, environment, health parameters of the disabled person, the view of the camera and options*. At the moment, we are controlling the values of these parameters interactively, to see the results of the actions of the domotics elements and the butler as well. In the real system, the input to these parameters will come from the house itself and the disabled person. So, their simulation in this application is extremely important to refine the behaviour of the agent in extreme situations.

Figure 7(b) shows a window with the daily planner (*agenda*). The daily planner is implemented as a XML file, which contains the actions protocol to be followed by the butler. As the events of the daily planner occur they are shown in this window. Thus, we can evaluate the correspondence between the action of the butler and the action established in the daily planner.

Figure 7(c) shows the window corresponding to the *disabled record.* This file saves all the events that occur in the house as well as all the health related actions carried out by the butler. It is a very important component of the system, because it records all the events allowing us to detect any problem concerning the health of the disabled, so that it can be corrected in the real system.

Another important aspect of the simulation was the introduction of emotions in the butler behaviour. The considered emotions were the basic ones: *joy*, *sadness*, *anger*, *disgust*, *surprise*, and *fear*. The reason to do this was the need for an expressive and realistic interface in order to get interactions as natural as possible between the butler and the disabled person.

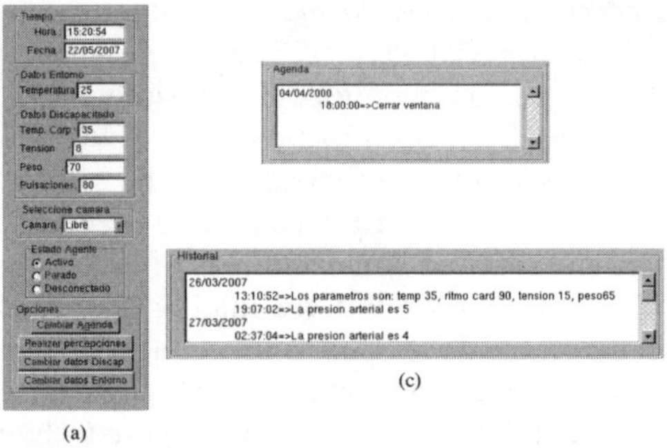

Fig. 7. (a) Interface controls for *time, environment data, disabled information, camera and options*; (b) Interface of the daily planner *–agenda–*; (c) Interface of the *disabled record*

Each emotion was associated with some events, which determine the conditions under which the emotion is triggered. Thus, an event like a noticeable elevation of the blood pressure of the patient elicits the emotion of *sadness* if we have a sensitive agent, or *surprise* if the agent did not expect it. The process to evaluate the recognition of emotions in facial expressions of synthetic faces is found in [1]

5 Conclusions

Some major issues affecting the design of an intelligent agent for home assistance have been discussed. First, the environment specification made up of five functional modules was described. Next, a first level specification of the way that the environment evolves was discussed. The underlying model of such specification is known as the perception-action vector (PAV). Then, the basic mechanisms supporting the decision-making process of the agent, known as information system (IS), were described.

Finally, a prototype simulating a home domotics environment has been presented. It provided a tool to visualize how the events of the world would be triggered and how the actions of the agent would be performed. Interactivity was achieved by a set of control interfaces, which allowed the user to change the parameters of the environment, of the health of the patient, and the daily planner. Thus, we could prove that situations are correctly handled by the underlying logic of the butler agent. The visualization of the emotional state of the butler was also a very important issue, because human-computer interaction needs a believable and realistic character, which is accomplished using emotional states.

In the final system, the environment itself will provide input parameters for the butler, and a complete virtual avatar capable of moving and talking will be responsible for the interaction between the disabled person and the agent.

Acknowledgments. This work has been partially supported by the Direccion General de Investigación del Ministerio de Educación, Ciencia y Tecnología, through the TIN2007-67993 and the TIN2007-63025 research projects. We are also grateful to Dr. M. Miró, Mr. G. Trias, Ms. C. Blanco, Mr. D. García, Mr. S. Lora, Mr. E. Sigg, Mr. J. Simó and Mr. A. Sobrino, for their collaboration.

References

1. Arellano, D., Lera, I., Varona, J., Perales, F.J.: Virtual Characters with Emotional States. In: 2nd PEACH Summer School (2008) (presented)
2. Attard, M., Montebello, M.: DoNet: a semantic domotic framework. In: Proceedings of the 15th international Conference on World Wide Web WWW 2006, pp. 997–998. ACM, New York (2006)
3. Camarinha-Matos, L.M., Castolo, O., Rosas, J.: A multi-agent based platform for virtual communities in elderly care. In: Proceedings of the IEEE Conference on Emerging Technologies and Factory Automation, ETFA 2003, vol. 2, pp. 421–428 (2003)
4. Fiol, G.: UIB-IK: A Computer System for Decision Trees Induction. In: Raś, Z.W., Skowron, A. (eds.) ISMIS 1999. LNCS, vol. 1609, pp. 601–611. Springer, Heidelberg (1999)
5. Fiol, G.: Inductive Learning from Incompletely Specified Examples. Frontiers in Artificial Intelligence and Applications 100, 286–295 (2003)
6. Fiol, G.: Intelligent Agent for Home Assistance. In: 11th IASTED International Conference on Artificial Intelligence and Soft Computing, pp. 102–107. ACTA Press, Calgary (2007)
7. Muñoz, C., Arellano, D., Perales, F.J., Fontaned, G.: Perceptual and Intelligent Domotic System for Disabled People. In: Proceedings of the 6th IASTED International Conference on Visualization, Imaging and Image Processing, pp. 70–75. ACTA Press, Calgary (2006)
8. Perales, F.J., Fiol, G., Varona, X., Miró, M., Fuster, P., Cerezo, E., Baldassarri, S., Remiro, V., Serón, F.J., Pina, A., Azkue, I.: El Proyecto INEVAI3D: Agentes Autónomos 3D, Escenarios Virtuales e Interfaces Inteligentes para Aplicaciones de Domótica y de Realidad Virtual. In: 1th CEDI Spanish Conference on Informatics, pp. 479–486. Thomson Editores Spain, Paraninfo (2005)

9. Russell, S.J., Norvig, P.: Artificial Intelligence: A Modern Approach, 2nd edn. Prentice-Hall, Englewood Cliffs (2004)
10. Velasco, J.R., Marsá-Maestre, I., Navarro, A., López, M.A., Vicente, A.J., de la Hoz, E., Paricio, A., Machuca, M.: Location-aware services and interfaces in smart homes using multiagent systems. In: The 2005 International Conference on Pervasive Systems and Computing (2005)
11. Wooldridge, M.: An Introduction to Multiagent Systems. John Wiley & Sons, Chichester (2002)

An Approach to Building a Distributed ID3 Classifier

Omar Jasso-Luna, Victor Sosa-Sosa, and Ivan Lopez-Arevalo

Center for Research and Advanced Studies
Laboratory of Information Technology
Cd. Victoria, Tam. Mexico
{jjasso,vjsosa,ilopez}@tamps.cinvestav.mx

Summary. Current applications from industry, science, and business are storing huge amount of data everyday. This data most of the time comes from distributed sources and are usually analysed for the organizations to discover knowledge and recognize patterns by means of Data Mining (DM) techniques. This analysis usually requires to put all information together in a big centralized datasets. Analysing this huge dataset could be very expensive in terms of time and memory consuming. For reducing this cost some Distributed Data Mining (DDM) architectures have been developed in recently years. This paper presents an approach to building a distributed ID3 classifier which takes only metadata from distributed datasets avoiding the total access to the original data. This approach reduces the computing time nedeed to build the classifier.

Keywords: Data Mining, Global Classifier.

1 Introduction

Nowadays complex business and industrial-scientific applications require to store a huge amount of data everyday. In most of the cases this big organizations tend to build centralized dataset or Data warehouse that collects all information from their geographically dispersed branches. This data is analysed to discover patterns or tendencies that represent knowledge, which is an important asset for any organization. Many organizations carry out the analysis of data employing DM techniques. There are some toolkits that implement these techniques. One of them is Weka[1], which is a widely used DM toolkit that contains a large collection of state-of-the-art machine learning algorithms written in Java. However, mining huge centralized datasets in a stand-alone approach requires powerful equipment working with high computer resources. Distributed Data Mining(DDM) is considered a feasible strategy to help with this issue. Some architectures are exposed in [3, 5, 4, 2]. Most of them propose to execute DM tasks over remote nodes.

They do not consider to build a classifier without having a global knowledge of the original data. This represents an important issue when some organizations are not able to share their information. As an example, we can find some

J.M. Corchado et al. (Eds.): DCAI 2008, ASC 50, pp. 385–394, 2009.
springerlink.com

health care institutions, which want to detect some potential patients who could get a type of cancer based on his/her history data. Some of these institutions do not have enough information to build a reliable classifier which could help with this process. Some other institutions have enough information for doing this work but they are not able to share it because of their privacy policies. This paper presents an architecture which allows this type of institutions to work together for building better classifiers using global knowledge without data privacy intrusion. This architecture is based on Web technologies and java components.

This paper is organized as follows: Section 2 describes some of related work. Section 3 shows the architecture of our approach. Section 4 describes briefly the Distributed ID3 Classifier process. Section 5 presents the implementation. Finally, Section 6 shows the preliminary results and Section 7 concludes the paper with ongoing and future work.

2 Related Work

DM can be defined as an infrastructure that uses a selection of different algorithms and statistical methods to find interesting and novel patterns within large datasets. It can be classified in three stages[4]. The first generation of tools provide users with a single DM algorithm operating on data stored in a local file. Examples include the use of classification algorithms such as C4.5[7], clustering algorithms such as K-means[8]. Such tools were provided primarily as standalone executables, obtaining input from the command line or via a configuration file. The second generation of tools combined a collection of different algorithms for DM under a common framework, and enabled users to provide input from various data sources. Some of these tools are described below:

- Weka[1] contains tools for classification, regression, clustering, association rules, visualization, and pre-processing.
- Illimine[10] is another DM tool developed in C language. It built-ins algorithms in data cubing, association mining, sequential pattern mining, graph pattern mining, and classification.
- Rattle[12] is a DM tool based on statistical language R[11].
- Rapid Miner[13, 9] features more than 400 operators for DM in java which can be used merging some of them.

Subsequently, third generation tools started to address the limitations that are imposed by the closed world model. Some examples of third generation tools are:

- Grid Weka[3], essentially a modification to the Weka toolkit that enables the use of multiple computational resources when performing data analysis.
- WekaG[5] is an adaptation of the Weka toolkit to a Grid environment. It is based on a client/server architecture. The server side defines a set of Grid

Services that implements the functionalities of the different algorithms and phases of the DM process.

- FAEHIM[4] consists of a set of DM Web services for DDM, a set of tools to interact with this services, and a workflow system used to assemble these services and tools.
- Weka4WS[2] extends Weka to support remote execution of the DM algorithms. In such way, DDM tasks can be executed on decentralized Grid nodes by exploiting data distribution and improving application performance.

Although most of these toolkits support DDM, they focus on executing DDM tasks over remote nodes to get a benefit from all their distributed computing resources. For carrying out this process, they need to upload complete data in the processing nodes without paying attention in data privacy. Our approach is based on an architecture, which build a Global Classifier through distributed datasets. This approach pays special attention in data privacy since our Central Computing Node are not able to access original data. One potential use of this approach could be to help health care institutions to share information, getting a better classifier which allows to detect dangerous illness in a patient at early stages.

3 Architecture for Building a Global Classifier

This architecture attempts to build a Global Classifier through a distributed datasets with similar structures. This approach is based on a set of java components executed on remotes machines, which interchange metadata with a Central Computing Node. This approach avoids the data privacy intrusion from third-party institutions by means of just interchanging metadata. The approach allows small and large institutions exchange data between them for building a better classifier that takes into account global knowledge. This architecture includes the following(see figure 1): a Central Computing Node (CCN) and a set of Local Nodes.

- Central Computing Node is the main component in this architecture and contains a Classifier Builder and a Global Classifiers Repository
  - A Classifier Builder is the responsible for receiving all metadata from the distributed Java Components, grouping them and built the Global Classifier
  - Global Classifiers Repository contains the built classifiers
- Local Node represents a server that contains the following components:
  - Java Components that access a private dataset and send only metadata to the CCN
  - The local dataset represents all the files like data bases or text files where the data is stored

A prototype of a Web application called TeamMiner implements this architecture. It allows users to be registered for collaborating in a Global Classifier

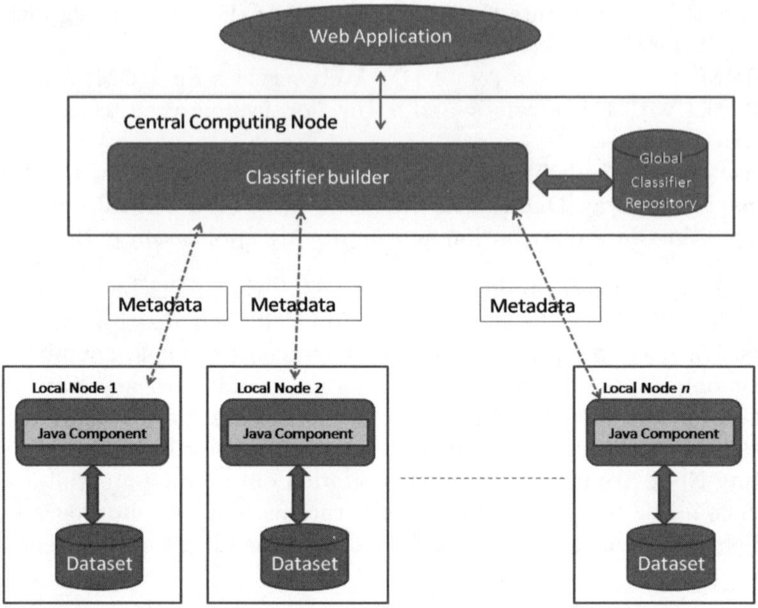

Fig. 1. Architecture

building process. This process consists in defining working groups where the interested institutions work together to build a global classifier. It is carried out by means of interchanging metadata between java nodes previously defined as a team member in our CCN. CCN gathers all metadata from java components without need of knowing the real data. This way to build a global classifier avoids original data intrusion and motivates organizations to participate in building a better global classifier.

3.1 Building a Global Classifier

The process of building a global classifier begins when a new local node registered at the CCN wants to obtain a global classifier. A registered local node asks CCN to obtain a global classifier using the local node dataset and other datasets with similar structures located in all of registered local nodes. This process follows the next steps(see figure 2):

1. A java component is registered to join to the group that study or analyze datasets with similar structures. These datasets must have the same number and type of attributes.
2. The java component requests to the CCN a Global Classifier. The CCN offers two options: to get the last global classifier stored in the Global Classifiers repository or to get a new one.

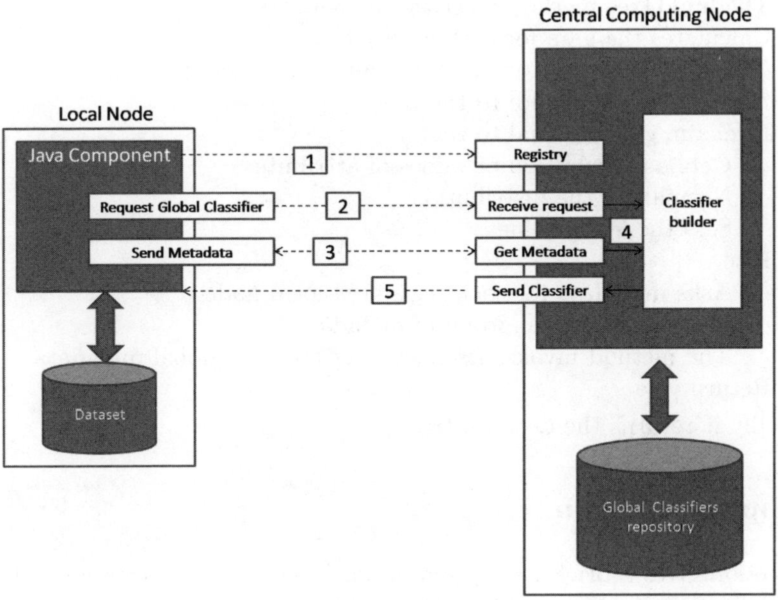

Fig. 2. Metadata interchange process

3. If the java component chooses to built a new classifier then the CCN begins a number of iterations with all the registered java components asking for metadata.
4. The metadata is sent to the CCN which interacts with all the local nodes asking for their metadata. This interactive process allows the CCN to built the Global Classifier.
5. Once the CCN has built a Global Classifier, it is saved and sent to the java component that request for it.

4 ID3 Global Classifier

The Web application TeamMiner which implements this architecture includes a classification algorithm to obtain decision trees. The ID3 algorithm[6] has been adapted for our architecture. It is implemented using RMI and Web Services technologies. This algorithm works as follow:

- The Classifier Builder obtains metadata from each registered local node(see figure 2)
- It creates a global group of metadata
- Invokes buildTree method sending global metadata as parameter(see Algorithm 5.2)

  - The buildTree method receives the metadata
  - Calculates the gain for each attribute
  - Chooses the attribute with best gain
  - Appends the attribute to the tree
  - If maxim gain is equal to zero
    · Gets a distribution of values of attributes
    · Normalizes the distribution
    · Sets the leaf's value
  - Else
    · Asks metadata for each registered local node
    · It creates a global group of metadata
    · The method invokes itself sending the new global metadata
  - Return tree
- Finally it returns the decision tree

5 Implementation

This section gives a brief description of our TeamMiner application, which is a Web application implemented using the architecture depicted in section 3. The first prototype of the TeamMiner application implements the ID3 algorithm described in section 4. The basic data structures used in this implementation are next described.

1. **Remote interface**
   Code below shows the remote interface of java components. It exposes two methods: *getAttributes*, it sends the attributes contained into the local sources, and *getMetadata*, it sends metadata wrapped into a Vector

```
public interface localInterface extends Remote{
public Vector getAttributes() throws RemoteException;
public Vector getMetadata() throws RemoteException;
public Vector getMetadata(int attribute, String value,
    int level, boolean leaf) throws RemoteException;
}
```

   This interface is implemented to interacting between the CCN and the local nodes
2. **Structure of metadata**
   Metadata are a summary of the real data obtained from structures of nested vectors. The Metadata Vector contains one vector for each attribute of the original source. Each attribute vector contains one vector for each value of attributes. Then, each value vector contains a vector for each value of the target class. These vectors contain the total number of related instances. An example of this structure is showed in figure 3.

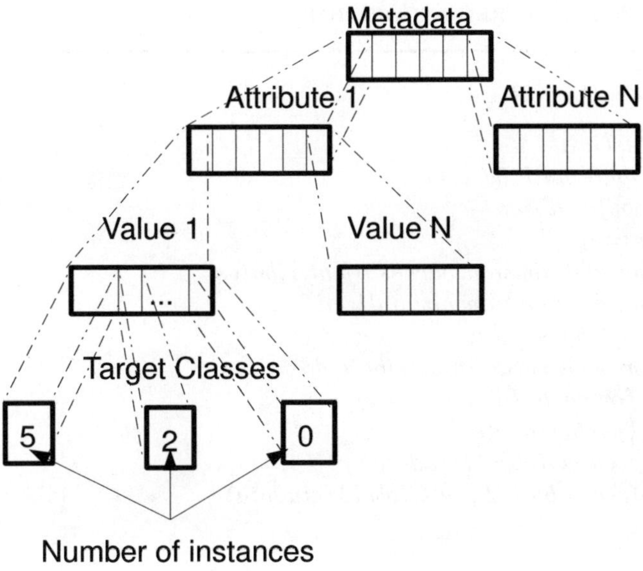

Fig. 3. Structure of metadata

3. **Central Node**
   CCN gathers metadata from all remote java components, grouping them into a global metadata group. Then, the Classifier Builder starts to build the classifier by means of getting the most significant attribute and requiring more metadata. Finally, when it gets the classifier, the CCN sends it to the requester java component. The mainly algorithms are shown in Algorithm 5.1 and Algorithm 5.2:

Algorithm 5.1. CLASSIFIER BUILDER ALGORITHM()

Classifier Builder
  for each Java Agent do
  {getMetadata
  end for
  createGlobalMetadata
  tree = buildTree(GlobalMetadata)
  return tree
end Classifier Builder

Algorithm 5.2. BUILDTREE ALGORITHM()

```
buildTree
    getAttributesGain
    choose attribute with best gain
    tree = attribute chosen
    if maxGain = 0
        then
            get distribution values of attributes
            get distribution normalized
            set leaf's value
        else
            for each value in attribute do
                for each Java Agent do
                {getMetadata
                createGlobalMetadata
                tree = buildTree(GlobalMetadata)
    end if
    return tree
end buildTree
```

6 Preliminary Results

We have tested our TeamMiner Web application by means of simulating the process of registering thre different hospitals as local nodes. Each hospital owns a different dataset. The dataset are real data obtained from private hospital. In the first evaluation the TeamMiner application built a classifier using only the individual dataset taken from each hospital. We mainly evaluated the hit rate[1] of the classifier. In the second evaluation the TeamMiner built a global classifier accesing the three registered local nodes. Results are shown in table 1.

Table 1. Preliminary results

Classifier type	Training set	Correctly Classified	Incorrectly Classified	Unclassified
Local	874	73.2	14.8	12
Local	1750	83.07	0	16.93
Local	4750	86.13	0.67	13.2
Global	7374	100	0	0

First column shows the classifier type. The size of training set(number of instances) is shown in the second column. Next columns 3 to 5 show the percent

[1] The percentage of instances that were correctly classified.

of correctly, incorrectly and unclassified instances respectively. All of them were tested using a test set of 750 instances. Last row shows the Global Classifier results.

We can see how the hit rate improves after building the global classifier.

7 Conclusions

Data classifiers are tools that allow users to predict events based on historical information. Some institutions, such as hospitals, can obtain benefits from data classifiers, specially for illness prevention. However, many of them do not have enough historical information to built a good data classifier. Institutions with similar interest, like health care, would like to exchange information for building data classifiers. ID3 Classifiers can be built using only summaries taken from the real data. These summaries d not reveal the original data. This approach is attactive for institutions which are not able to share the original information(like digital health records). Global classifiers help in tasks like illness prevention and developing of better clinical guides. Data classifiers performs better when they are built from global knowledge. This situation was the motivation for developing the TeamMiner Web application based on a distributed data mining architecture which keeps privacy of data and take advantage of global knowledge. As ongoing and future work, this prototype is going to extend other popular classification and clustering algorithms.

Acknowledgments

This research was partially funded by project number 51623 from "Fondo Mixto Conacyt-Gobierno del Estado de Tamaulipas"

References

1. Witten, H., Frank, E.: Data Mining: Practical machine learning tools and techniques. Morgan Kaufmann Publishers, San Francisco (2005)
2. Talia, D., Trunfio, P., Verta, O.: Weka4WS: A WSRF-Enabled Weka Toolkit for Distributed Data Mining on Grids. In: Jorge, A.M., Torgo, L., Brazdil, P.B., Camacho, R., Gama, J. (eds.) PKDD 2005. LNCS (LNAI), vol. 3721, pp. 309–320. Springer, Heidelberg (2005)
3. Khoussainov, R., Zuo, X., Kushmerick, N.: Grid-enabled Weka: A Toolkit for Machine Learning on the Grid. ERCIM 59, 47–48 (2004)
4. Shaikh Ali, A., Rana, O.F., Taylor, I.J.: Web Services Composition for Distributed Data Mining. In: International Conference Workshop on Parallel Processing, pp. 11–18. IEEE, Los Alamitos (2005)
5. Perez, M.S., Sanchez, A., Herrero, P., Robles, V.: Adapting the Weka Data Mining Toolkit to a Grid based environment. In: Szczepaniak, P.S., Kacprzyk, J., Niewiadomski, A. (eds.) AWIC 2005. LNCS (LNAI), vol. 3528, pp. 492–497. Springer, Heidelberg (2005)

6. Quinlan, J.R.: Induction of Decision Trees, Machine Learning, Hingham, MA, USA, vol. 1(1), pp. 81–106. Kluwer Academic Publishers, Dordrecht (1986)
7. Ross Quinlan, J.: C4.5: programs for machine learning. Morgan Kaufmann, San Francisco (1993)
8. McQueen, J.: Some methods for classification and analysis of multivariations. In: Proc. 5th Berkeley Symposium on Mathematical Statistics and Probability, pp. 281–2297 (1967)
9. Mierswa, I., Wurst, M., Klinkenberg, R., Scholz, M., Euler, T.: YALE: Rapid Prototyping for Complex Data Mining Tasks. In: 12th ACM SIGKDD International Conference on Knowledge Discovery and Data Mining (2006)
10. University of Illinois and Data Mining Research Group and DAIS Research Laboratory, IlliMine 1.1.0, `http://illimine.cs.uiuc.edu/`
11. Statistics Department of the University of Auckland, R Project 2.6.1, `http://www.r-project.org/`
12. Williams, G.: Rattle 2.2.74, `http://rattle.togaware.com/`
13. Artificial Intelligence Unit of University of Dortmund, Yale 4.0, `http://rapid-i.com/`

Techniques for Distributed Theory Synthesis in Multiagent Systems

Mª Cruz Gaya and J. Ignacio Giráldez

Universidad Europea de Madrid, C/Tajo s/n
28690 Villavicios de Odón, Madrid, Spain
mcruz@uem.es, ignacio.giraldez@uem.es

Abstract. Data sources are often dispersed geographically in real life applications. Finding a knowledge model may require to join all the data sources and to run a machine learning algorithm on the joint set. We present an alternative based on a Multi Agent System (MAS): an agent mines one data source in order to extract a local theory (knowledge model) and then merges it with the previous MAS theory using a knowledge fusion technique. This way, we obtain a global theory that summarizes the distributed knowledge without spending resources and time in joining data sources. New experiments have been executed including statistical significance analysis. The results show that, as a result of knowledge fusion, the accuracy of initial theories is significantly improved as well as the accuracy of the monolithic solution.

Keywords: Distributed Data Mining, ensemble techniques, MAS, Evolutionary computation.

1 Introduction

There are a lot of real problems where data are dispersed geographically, for example, the client data of different branches of a bank or the medical records from different hospitals. If we want to learn about a behavior pattern of the bank clients or about the symptoms of a disease, we need to take into account all the data sources. One option is to join all the data sources and run a learning algorithm over it. Another option is trying to obtain the knowledge model in a decentralized way. The first option is not always possible, or demands a lot of resources. The second one requires a distributed framework for mining the distributed data sources (in order to obtain local theories). And then, it requires merging the local theories into a single global theory.

Multi Agent Systems (MAS) have been used for distributed data mining successfully. For instance, in [1] MAS are applied to the credit card fraud problem, in [2] MAs are applied to carcinogenesis prediction and, recently, they have also been applied to artificial vision [3].

The merging process that obtains a single global theory from local theories can be considered a meta-learning process. Prodromidis et. all [4] define meta-learning as the technique that computes high level classifiers, called meta-classifiers, which integrate the base classifiers computed over independent and geographically disperse data sources, in some way. There are different well known techniques for meta-learning. [5] presents a study about different Ensembles of Classifiers (EoC) and the reasons for its base classifier accuracy improvement. These methods learn which base classifier to use for each sample. The training set of the meta-classifier is composed by the

J.M. Corchado et al. (Eds.): DCAI 2008, ASC 50, pp. 395–402, 2009.
springerlink.com

base classifiers predictions. The output is a new theory that can provide the final prediction given the base classifiers predictions for a new instance of the problem. In [1,2] the EoC are demonstrated to improve the theory accuracy. Recently, in PAKDD 07 [6] the EoC have been the classifiers better positioned.

Evolutionary computation has been used for combining classifiers. In [11], an evolutionary algorithm is used for combining several classifiers. The members of the population are trees that can decide what initial classifier to apply in each case. In [12], an evolutionary algorithm is used for selecting the base classifier set that will be combined.

In [13] the authors present a MAS framework with an evolutionary approach to merge theories, in order to obtain in a decentralized way, a global theory more accurate than the initial ones is presented. Based on the aforementioned work [13], new experiments are presented: new datasets extracted for the UCI Machine Learning Repository [8], a 10 fold cross-validation technique has been used in order to compute the accuracy and a statistical significance analysis has been done.

In section 2 the framework based on MAS (called Theory Synthesis Evolution: TheSynEv) is described; in section 3 we present experiments and the results obtained. The paper ends with the conclusions presented in section 4.

2 TheSynEv System

This section describes the TheSynEv System: a MAS that mines several geographically dispersed data sources in order to obtain, in a decentralized way, a global theory that summarizes all the knowledge using an evolutionary approach.

As shown in fig. 1, a software agent is executed at every remote data repository.

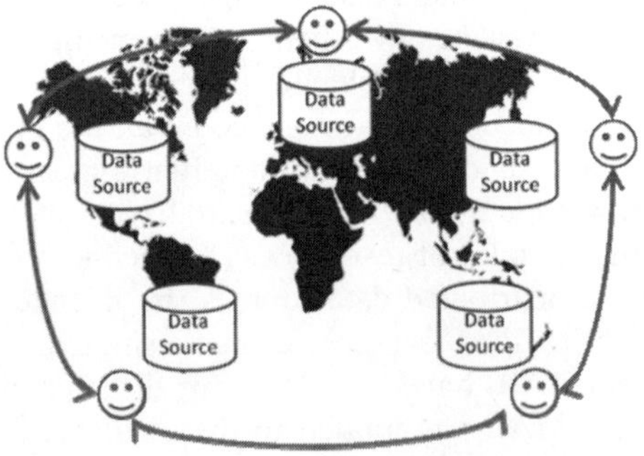

Fig. 1. The System layout. A software agent is executed at every remote data repository.

Each agent of the MAS can operate in two modes: learning or meta-learning. When an agent is in learning mode it learns from its data source and extracts the local theory that represents its knowledge model of the domain problem (fig. 2). Each agent does the same in a parallel way using a traditional learning algorithm (C4.5 [6]).

When the system operates in meta-learning mode it produces a global theory. Each agent asks for the global theory, waits for it, and executes a fusion process in order to add his local knowledge, resumed in his local theory, to the global system knowledge

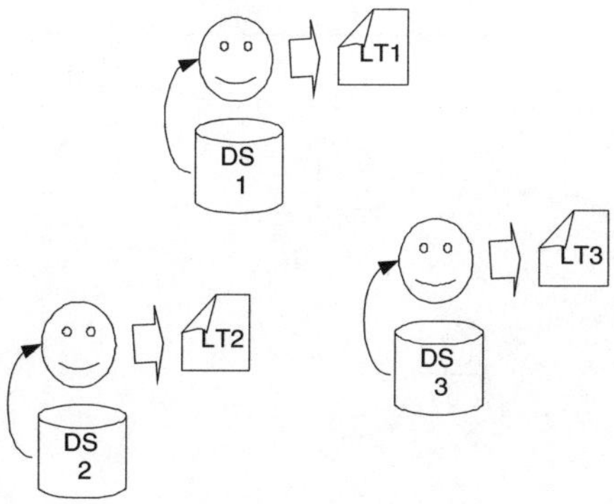

Fig. 2. System agents in learning mode. Each agent extracts his local theory (LT) using its data source (FD) as input.

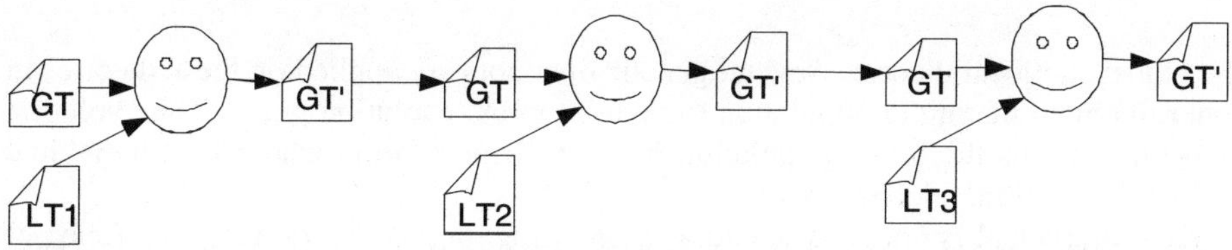

Fig. 3. System agents in meta-learning mode. Each agent merges the local theory (LTi) and the global theory (GT) and obtains the new system global theory (GT').

(global theory). This is done in a sequential mode: only one agent can merge its local knowledge with the global knowledge at a time. The fig.2. outlines TheSynEv in meta-learning mode.

Internally the system can be represented according to fig.3. In each site there is an agent capable of:

1. Learning a local theory from the local training data set. This local theory may be used to classify any problem instance. This local theory also represents the domain map estimated from the evidence of local data. A traditional algorithm C4.5 is used in order to obtain the local theory.
2. Adding its knowledge to the global system knowledge. The agent receives in a message the system global theory and modifies it adding the knowledge extracted from its local data source. The global theory is incrementally synthesized from the individual local theories provided by the remote agents.

The global theory improves the EoC operation in three aspects: (i) making a prediction without the participation of the base classifiers, (ii) offering an explanation in terms of the domain and (iii) solving knowledge contradictions.

On the other hand, the monolithic solution is not a good alternative because the contradiction between different rules and the length of the monolithic theory.

We use an evolutionary algorithm in order to merge theories. Each individual represents an theory. The genetic operators used are: mutation and crossover. The process is divided into two phases: growth and evolution. The first phase has as input

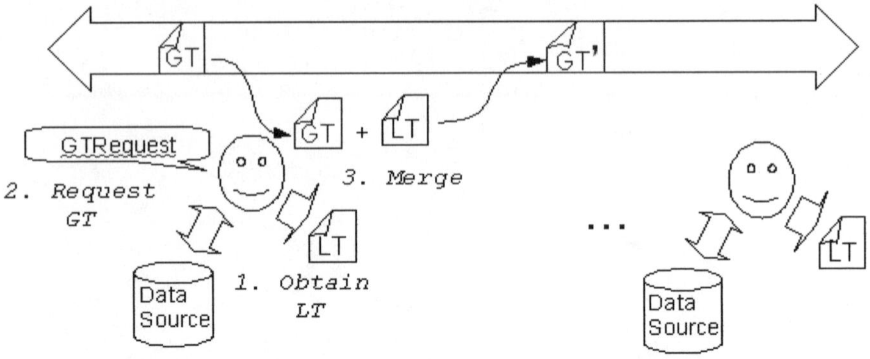

Fig. 4. Each agent extracts the local theory from its data source, asks for the global theory trough a GTRequest message, merges the local theory (LT) and the global theory (GT) modifying this. It sets the modified global Theory (GT') as the new system global theory.

the global and local theories, several genetic operators are applied on these theories in order to obtain the initial population to be used in the evolution phase. The evolution phase transforms the initial population into a final population whose best individual will be the resulting global theory.

Each individual is not represented as a bit string, but as several rule lists (one per class label). Each rule represents a region of the domain map.

For the evolution phase, an evolutionary algorithm [7] is used with the following parameters: i) the probability for applying a mutation operator, ii) the probability for applying a crossover operator, iii) the size of the population and iv) the number of generations to be generated. The output of the fusion process will be the individual of the last generation with highest fitness value, i.e., highest accuracy. The accuracy is measured over a test dataset.

3 Experiments

Our goal is to determine whether the accuracy of the resulting theory improves the accuracy of the initial theories (local and global) and the accuracy of the monolithic solution.

In [13] early experiments were presented with promising results. The authors now present the following innovations: 1) all experiments use 10 fold cross-validation, 2) additional new datasets are used 3) a statistical significance analysis is performed.

The experiments have been done over fifteen datasets of the UCI Machine Learning Repository [8]. The WEKA application [8] was used to preprocess the datasets by the following way:

1. the unsupervised attribute filter Discretize was used in order to discretize the range of numeric attributes in the dataset into nominal attributes using ten binds.
2. the unsupervised instance filter Randomize was used in order to randomly shuffle the instances.

3. attributes have been selected using information gain as attribute evaluator and ranker as search method. The ten best attributes have been selected using the attribute unsupervised filter Remove.
4. the dataset was partitioned into 10 subsets in order to compute de 10 fold cross-validation. One of these subsets was selected for being used as test set. The other ones were used as training set for global theory (45%) and local theory (45%).

Table 1. 10 fold cross-validation accuracies of 15 UCI datasets. The columns are: the name of the data set (Name) and the accuracies of the monolithic solution (Monolithic theory), the global theory (global theory), the local theory and the theory resulted of the TheSynEv System (Merged theory). All of them have been computed using 10-fold cross-validation. The accuracies that are improved in same way have been underlined.

Name	10 fold cross-validation averaged accuracy			
	Monolithic Theory	Global Theory	Local Theory	Merged Theory
Sonar	0,808809523809	0,837380952380	0,645714285714	0,990238095238
Weaning	0,804301075268	0,827634408602	0,777634408602	0,927311827956
Bands	0,911099365750	0,892970401691	0,876955602536	1
Ionosphere	0,925793650793	0,937380952380	0,866031746031	0,977142857142
Bupa	0,686974789915	0,750336134453	0,559159663865	0,956806722689
Haberman	0,734516129032	0,734516129032	0,734516129032	0,773763440860
Hepatitis	0,819583333333	0,7575	0,727083333333	0,954583333333
agaricus-lepiota	0,999453551912	0,998357653275	0,996171860925	0,999453551912
Ballons	0,814285714285	0,8125	0,635714285714	0,919642857142
Monks	0,839102564102	0,864743589743	0,716025641025	0,919871794871
Rds	0,930555555555	0,952777777777	0,918055555555	1
tic-tac-toe	1	1	1	1
Bridges	0,914545454545	0,914545454545	0,867272727272	0,952727272727
Crx	0,873913043478	0,871014492753	0,852173913043	0,881159420289
Text	0,907549498746	0,910557017543	0,906546992481	0,912808897243

The mutation probability and crossover probability values selected for the evolutionary algorithm were 0.4 and 0.1 respectively because the experiments in [13] suggest these are the adequate values. All of them use the same values for size population and number of generation (10 and 100) because the convergence of the evolutionary algorithm has been observed with these values. In order to compute the accuracy of the monolithic solution, the instances are joined in the two initial dataset portions for generating the monolithic theory.

Results are presented in table 1. For greater clarity the accuracies that are improved in same way have been underlined in table.1.

The experiments show that the accuracy of the merged theory is never worse than the accuracy of local, global and monolithic theories. In 13 out of 15 datasets the accuracy of the initial theories and of the monolithic solution are improved. The other datasets where this does not happen are agaricus-lepiota, where the accuracy improves that of the initial theories but equals the accuracy of the monolithic solution. In the tic-tac-toe dataset, the accuracy cannot be improved because it has the highest value. Thus, the improvements to the accuracies of the local theories, and to the accuracy of the monolithic solution, are demonstrated.

A t-test (student's t-test) [14] has been used in order to determine whether the improvement is statistically significant. The table 2 and the table 3 outlines the obtained results. Table 2 shows a 99,5% confidence level has been selected so when a "merged" value appeared in a cell of the table we can say that in the 99,5 % of the cases the merged theory is better than the other one. Table 3 shows a 95% confidence level.

The experiments show that:

- The accuracy of the merged theory never gets worse than global, local and monolithic theories. The values in tables are "merged" (the accuracy improvement of the merged theory is statistically significant) or "No difference" (there aren't any differences between them).
- Using a 99,5% confidence level in 10 out of 15 datasets the accuracy is improved significantly in some way.
- Using a 95% confidence level in 11 out of 15 datasets the accuracy is improved significantly in some way.

Table 2. Statistical significance with 99,5% confidence level. This table represents the statistical significance comparison between merged theory and monolithic, global an local theories. The first column is the name of the dataset (*name*), the following three (*Monolithic Th. vs. Merged Th., Global Th. vs. Merged Th., Local Th. vs. Merged Th.*) represents whether the Merged Theory improve in statistical significance way the other theory. The datasets improved are underlying.

	Statistical Significance with 99,5% confidence level		
Name	Monolithic Th. vs. Merged Th.	Global Th. vs. Merged Th.	Local Th. vs. Merged Th.
Sonar	merged	Merged	merged
Weaning	merged	Merged	merged
Bands	merged	Merged	merged
Ionosphere	merged	Merged	merged
Bupa	merged	Merged	merged
Haberman	No difference	No difference	No difference
Hepatitis	merged	No difference	No difference
agaricus-lepiota	No difference	No difference	No difference
Ballons	No difference	No difference	No difference
Monks	merged	No difference	merged
Rds	No difference	No difference	merged
tic-tac-toe	No difference	No difference	No difference
Bridges	No difference	No difference	Merged
Crx	No difference	No difference	Merged
text	No difference	No difference	No difference.

Table 3. Statistical significance with 95% confidence level. This table represents the statistical significance comparison between merged theory and monolithic, global an local theories. The first column is the name of the dataset (*name*), the following three (*Monolithic Th. vs. Merged Th., Global Th. vs. Merged Th., Local Th. vs. Merged Th.*) represents whether the Merged Theory improve in statistical significance way the other theory. The datasets improved are underlying.

	Statistical Significance with 95% confidence level		
Name	Monolithic Th. vs. Merged Th.	Global Th. vs. Merged Th.	Local Th. vs. Merged Th.
Sonar	Merged	Merged	merged
Weaning	Merged	Merged	merged
Bands	Merged	Merged	merged
Ionosphere	Merged	Merged	merged
Bupa	Merged	Merged	merged
Haberman	No difference	No difference	No difference
Hepatitis	Merged	merged	merged
agaricus-lepiota	No difference	No difference	No difference
Ballons	Merged	merged	merged
Monks	Merged	merged	merged
Rds	Merged	merged	merged
tic-tac-toe	No difference	No difference	No difference
Bridges	Merged	merged	merged
Crx	No difference	merged	merged
text	No difference	No difference	No difference.

4 Conclusion

This paper shows a method for merging theories using an evolutionary approach, with the goal of obtaining a merged theory with an accuracy greater than the accuracy of every local theories and also greater than the accuracy of the monolithic theory. The advantages over other meta-learning approaches are (i) the explanation in terms of the domain attributes, (ii) that base classifiers are not required to classify a new instance of the problem and (iii) the knowledge contradictions are solved.

The authors have improved the experimental design presented in [13] with the aim of obtaining stronger evidence to support the conclusions.

The accuracies obtained in 73% of the datasets are improved by those of the initial theories (local and global theories) and also the accuracy of the monolithic solution. In the cases in which the improvement isn't achieved, at least the accuracy of the monolithic solution is equaled. This improvement has been demonstrated statistically using the paired t-test.

We have found a method that learns a global theory taking account all the samples of the problem independently of the site in which they are, using knowledge learnt by remote agents. The experiments show that the accuracy of this global theory is in 27% of the cases, with a statistical significance confidence level of 95%, equal to the accuracy of the local theory and the monolithic solution theory (input theories) in the worst case and in a 73% of the cases is better than the input theories.

References

1. Giráldez, J.I.: Modelo de toma de decisiones y aprendizaje en sistemas multiagente, Tesis para el grado de doctor en Informática, Universidad Politécnica de Madrid (1999)
2. Stolfo, S., Prodromidis, A.L., Tselepis, S., Lee, W., Fan, W., Chan, P.: JAM: Java Agents for meta-learning over distributed databases. In: Third International Conference in Knowledge Discovery and Data Mining (KDD 1997), Newport Beach, California (1997)
3. Barandela, R., Sánchez, J.S., Valdovinos, R.: New applications of ensembles of classifiers. Pattern Analysis and Applications 6(3), 245–256 (2003)
4. Prodromidis, A.L., Stolfo, S.J., Chan, P.: Advances of Distributed Data Mining. Kargupta, H., Chan, P. (eds.). AAAI press, Menlo Park (2000)
5. Dietterich, T.G.: Ensemble Methods in Machine Learning. In: Kittler, J., Roli, F. (eds.) First International Workshop on Multiple Classifier Systems. LNCS, pp. 1–15. Springer, New York (2000)
6. Quinlan, J.R.: C4.5: Programs for Machine Learning. Morgan Kaufmann, San Francisco (1993)
7. Koza, J.R., et al.: Genetic Programming IV: Routine Human-Competitive Machine Intelligence. Kluwer Academic Publishers, Dordrecht (2003)
8. `http://www.ics.uci.edu/~mlearn/MLRepository.html`
9. Witten, I.H., Frank, E.: Data mining: practical machine learning tools and techniques with Java implementations. Morgan Kaufmann Publishers Inc., San Francisco (2000)
10. `http://www.ecmlpkdd2006.org/challenge.html`
11. Langdom, W.B.: Genetic Programing for combining clasiffiers. Presented at GECCO 2001, pp. 66–73. Morgan Kaufmann, San Francisco (2001)
12. Hung-Ren Ko, A., Sabourin, R., Britto, A.: Evolving ensemble of classifiers in random subspace. In: Proceedings of the 8th annual conference on Genetic and evolutionary computation GECCO 2006 (2006)
13. Gaya López, M.C., Giraldez Breton, J.I.: Experiments in Multi Agent Learning. In: Proceedings of the 3rd International Workshop on Hybrid Artificial Intelligence Systems HAIS 2008 (accepted for publication, 2008)
14. Mitchel, T.: Machine Learning. McGraw Hill, New York (1997)

Domain Transformation for Uniform Motion Identification in Air Traffic Trajectories*

José Luis Guerrero and Jesús García

Group of Applied Artificial Intelligence (GIAA)
Computer Science Department
Carlos III University of Madrid
Colmenarejo – Spain
{joseluis.guerrero,jesus.garcia}@uc3m.es

Abstract. In this paper, we will discuss the viability of a proposed algorithm to segment trajectories based on a study case of recorded opportunity traffic. This segmentation is the first step of the reconstruction process of the trajectory. Our algorithm will try to apply specific models for the three movement possibilities in our trajectories: uniform, turn and acceleration. We will cover specifically the parameters and viability (as a part of the general algorithm) of the uniform movement segmentation, centring our study in the appropriate descriptor attribute extracted from available samples expressed in its original domain. In particular, we detail a comparison between a general statistic such as a correlation coefficient against the residue of best linear unbiased estimator.

1 Introduction

Air Traffic Control (ATC) is a critical area related with safety and so requires from strict validation in real conditions. Validation and performance assessment of ATC centres is done with recorded datasets (opportunity traffic), used to reconstruct the necessary reference information [1]. The reconstruction process transforms multi-sensor plots to a common coordinates frame and organizes data in trajectories of individual aircraft. Then, for each trajectory, segments of different modes of flight (MOF) are identified, each one corresponding to time intervals in which the aircraft is flying in a different type of motion. These segments are a valuable description of real data, providing information to analyse the behaviour of targets objects (where uniform motion flight and manoeuvres are performed, magnitudes, durations, etc.).

This problem has been addressed from different perspectives such as multiple-model filters or machine learning techniques [2-4]. In this study we will discuss an approach to the segmentation of trajectories where the three possible movement modes (MM's from now on) are the following: uniform, turn and accelerated movements. The study case is the classification of MOF according to data coming from recorded opportunity aerial traffic, but the study may be applied to any study case where the variables we will manage through this paper (such as the covariance matrix to represent the noise in our values) are available.

* This work was supported in part by Projects MADRINET, TEC2005-07186-C03-02, SINPROB, TSI2005-07344-C02-02.

J.M. Corchado et al. (Eds.): DCAI 2008, ASC 50, pp. 403–409, 2009.
springerlink.com

The studied trajectories will have points with the following attributes: detection time, stereographic projections of its x and y components, covariance matrix, and real classification (this attribute only in simulated trajectories). With those input attributes, we will look for a domain transformation that will allow us to classify our samples into a particular MM with maximum accuracy, according to the model we are applying. In this paper we will centre our study in finding the best domain transformation to classify samples into a uniform MM.

Even though we won't include them in this study, we can't forget the non-uniform MM's. On the phase of the analysis covered on this paper, we will start from the unclassified points of the trajectory (coordinates x and y, in our case of study coming from an stereographic projection) and will end up with those points classified into uniform MM or declared as unknown (they will be analysed afterwards with the concrete models for the remaining MM's). The general algorithm applicable to any of these models, along with its parameters, will be explained in the second section of this paper. The third section will present two alternative domain transformations for the classification of the uniform MM, while the fourth will show the results using both transformations and the comparison between them. Finally we will present the conclusions which those results lead to, in the fifth section.

2 General Algorithm

The general algorithm will study sequentially the three models for the possible MM's we have defined before. Each trajectory (T_i) is defined as a collection of points, which are defined themselves by the following vector: $x_j^i = (x_j^i, y_j^i, t_j^i, R_j^i, C_j^i)$, $j \in \{1,\ldots,N^i\}$, where x_j^i, y_j^i are the stereographic projections of the point, t_j^i is the time of its detection, R_j^i is the covariance matrix and C_j^i is the real classification of the point, whose possible values are the three possible MM's. The application of each model will end up with the input unknown points divided into unknown and classified samples, and the unknown ones would we used as an input for the next MM model.

For each of the MM's models we will have to deliver a specific resolution method, making decisions about the following aspects:

1. ***Segmentation*** (segment study / independent study)
2. ***Segment extension*** (time / samples) and ***segment resolution*** (length of the segment, either in time or in number of points, depending on the previous decision)
3. ***Domain transformation*** used (a synthesized attribute to determine whether a point belongs to our MM model or not)
4. ***Criterion decision technique*** (based on a threshold over the transformed domain value)

The first step is segmentation, which will cover the decision of using an independent classification for each point or segments treated as indivisible units. The second step will have to determine the segments for our trajectory. A segment S_i may be defined as $S_k^i = \{x_j^i\}, j \in \{i_{k1}, \ldots, i_{kn}\}$. Segment extension will define how we choose the boundaries for our segments, by number of points (k_n) or by their time value $t_j^i \in (t_{k1}^i, t_{k1}^i + \Delta T)$. Segment resolution refers to the choice of the length of those segments, and how it affects our results. The third step will cover several approaches to possible domain transformations to be able to get a good classification. The last step is to get the value for the threshold used to classify each of those units. The criterion decision technique will be based on a constant or adaptative threshold value.

To be able to determine the quality of a model, we are going to introduce some terms in their classical meaning, which, to improve legibility, we will define here:

- **True positives (TP):** points belonging to our MM classified correctly.
- **False positives (FP):** points not belonging to our MM classified incorrectly (their should be unknown and are classified, while studying uniform MM, as uniform).
- **True negatives (TN):** points not belonging to MM, correctly classified unknown.
- **False negatives (FN):** points belonging to our MM classified incorrectly (while studying uniform MM, they should be uniform, but are classified as unknown).

  From these definitions we build the following rates:
- **True positives rate (TPR):** $TPR = \frac{TP}{TP+FN}$
- **False positives rate (FPR):** $FPR = \frac{FP}{FP+TN}$

We will use these definitions into three figures, which will allow us to compare different resolution methods: TPR and FPR versus a reference value (an increasing threshold) and TPR versus FPR (known as ROC curve).

The critical design criterion kept through this paper will be keeping a null FPR, trying to get a TPR as high as possible. This means that no MM model will classify another MM sample point into its own one, even though we will misclassify some of our own samples into the unknown category. We will use that criterion to choose the threshold in our classification.

3 Domain Transformations

In previous section, we have introduced the use of a descriptor attribute resulting of transforming the original data. On this section we define two alternative descriptors. One of them is the general correlation coefficient, which we will refer to as CC, and the other one will allow us to use besides the knowledge about the noise in the coordinate measures (in the form of covariance matrixes derived from ATC sensor models). This specific measure is called BLUE residue..

There is a wide bibliography on each of these estimators. The CC is a statistical descriptor which determines the lineal relationship between two variables (with values ranging from -1, which indicates a negative linear relationship, to 1, which indicates a positive one). Under uniform motion, variables x,y should have a maximum correlation magnitude and this value should drop for other types fo motion

$$CC = \frac{\sigma_{xy}}{\sigma_x \sigma_y} \qquad \sigma_x = \sqrt{\frac{\sum_{i=1}^{N}(x_i - \bar{x})^2}{n-1}} \qquad \sigma_{xy} = \frac{1}{N}\sum_{i=1}^{N}(x_i - \bar{x})(y_i - \bar{y})$$

In fact, we use the complement of correlation magnitude, 1-|CC(x,y)|.With the BLUE estimator, we will estimate an ideal uniform MM segment from the segment data by means of a weighted least squares [5], exploiting the noise information in the measures (matrix R_j^i) and add the difference of each one of the points in the segment to its estimated equivalent (which we will call the residue of that point). The measures belonging to uniform MM's will fit better to the estimated ones, so their residues will be lower. Also, to be able to clearly define the range of values in any trajectory (at least in uniform MM's) we will normalize the residue value (as we explained in the segment extension section). The model for uniform motion originating measurements is the following:

$$\vec{x}_m(k) = \begin{bmatrix} x_m(k) \\ y_m(k) \end{bmatrix} = \begin{bmatrix} 1 & t_k & 0 & 0 \\ 0 & 0 & 1 & t_k \end{bmatrix} \begin{bmatrix} x_0 \\ vx_0 \\ y_0 \\ vy_0 \end{bmatrix} + \begin{bmatrix} n_x(k) \\ n_y(k) \end{bmatrix} = H(t_k)\vec{\theta} + \vec{n}(k)$$

The first component $H(t_k)\vec{\theta}$) represents the ideal estimated uniform segment. The equation for the best estimator (BLUE) with minimum squared weighted residual is the following:

$$\langle\vec{\theta}\rangle = \begin{bmatrix} \langle x_0\rangle \\ \langle vx_0\rangle \\ \langle y_0\rangle \\ \langle vy_0\rangle \end{bmatrix} = \left(\sum_k H(t_k)^T R_k^{-1} H(t_k)\right)^{-1} \sum_k H(t_k)^T R_k^{-1} \overrightarrow{x_m}(k)$$

We can see we are introducing the noise information in that equation, in the form of its covariance matrix, R_k . Then, with estimator $\vec{\theta}$ we calculate the interpolated positions for the x and y components of points:

$$x_{int}(t) = \langle x_0\rangle + \langle vx_0\rangle t \qquad y_{int}(t) = \langle y_0\rangle + \langle vy_0\rangle t$$

Finally, with the previous values, we get the normalized residue of BLUE:

$$res = \frac{1}{kmax - kmin} \sum_{k=kmin}^{k=kmax} (x(k) - x_{int}(k) \quad y(k) - y_{int}(k))\, R_k^{-1} \begin{pmatrix} x(k) - x_{int}(k) \\ y(k) - y_{int}(k) \end{pmatrix}$$

It is very important to remember the design parameter for our algorithm: null FPR with a TPR as high as possible. So, a misclassification in the form of FP would not get the chance to be corrected, since those points are not reused with the following MM's models. Therefore, the criterion to select the most suitable descriptor is the capability to maximize TPR and keep FPR null.

4 Experiments

In the previous section we made the introduction to the two domain transformations we consider for our algorithm. Now we will show the results of the segmentation using a trajectory with a turn non uniform MM. The vale for the other algorithm parameters will be the following: sample delimitation and 61 samples resolution. In the following figures we will show the analyzed trajectory with its samples correctly classified, the descriptor values through the trajectory for both alternatives, along with their TPR and FPR figures, and finally a ROC curve comparison.

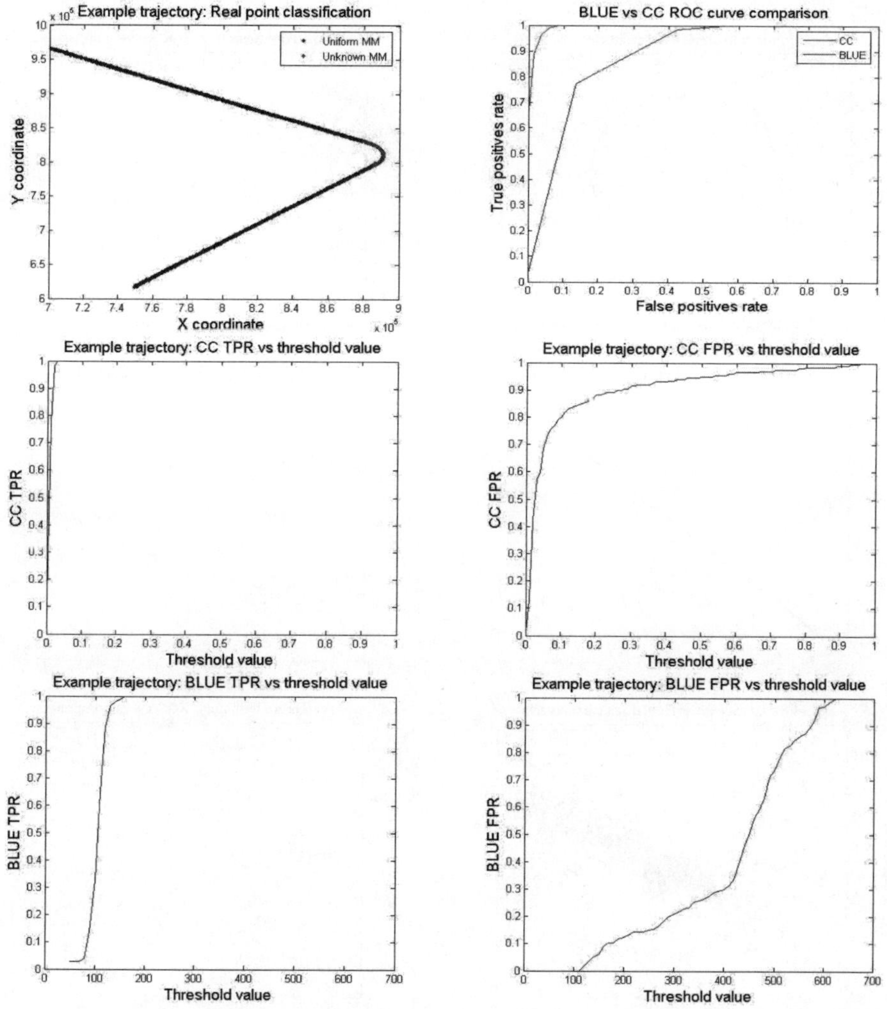

Fig. 1. Example of BLUE and CC evaluation

As we can see in the figures above, the BLUE estimator gives us a clear advantage in the classification over correlation coefficient. It is also important to realize that, using BLUE estimator, we have a range of values where FPR keeps a zero value while TPR increases, indicating that this measure value will serve our algorithm purpose. These results have been checked with a set of 54 simulated trajectories, including turns (as shown above), hippodromes, uniform and accelerated trajectories. We will show a ROC curve comparison for each of these possible trajectories, obtained by varying the classification threshold along a wide interval.

In all the examples we have used independent segmentation, with 31 point segments (uniform and hippodrome trajectories) or 111 point segments (accelerated trajectory).

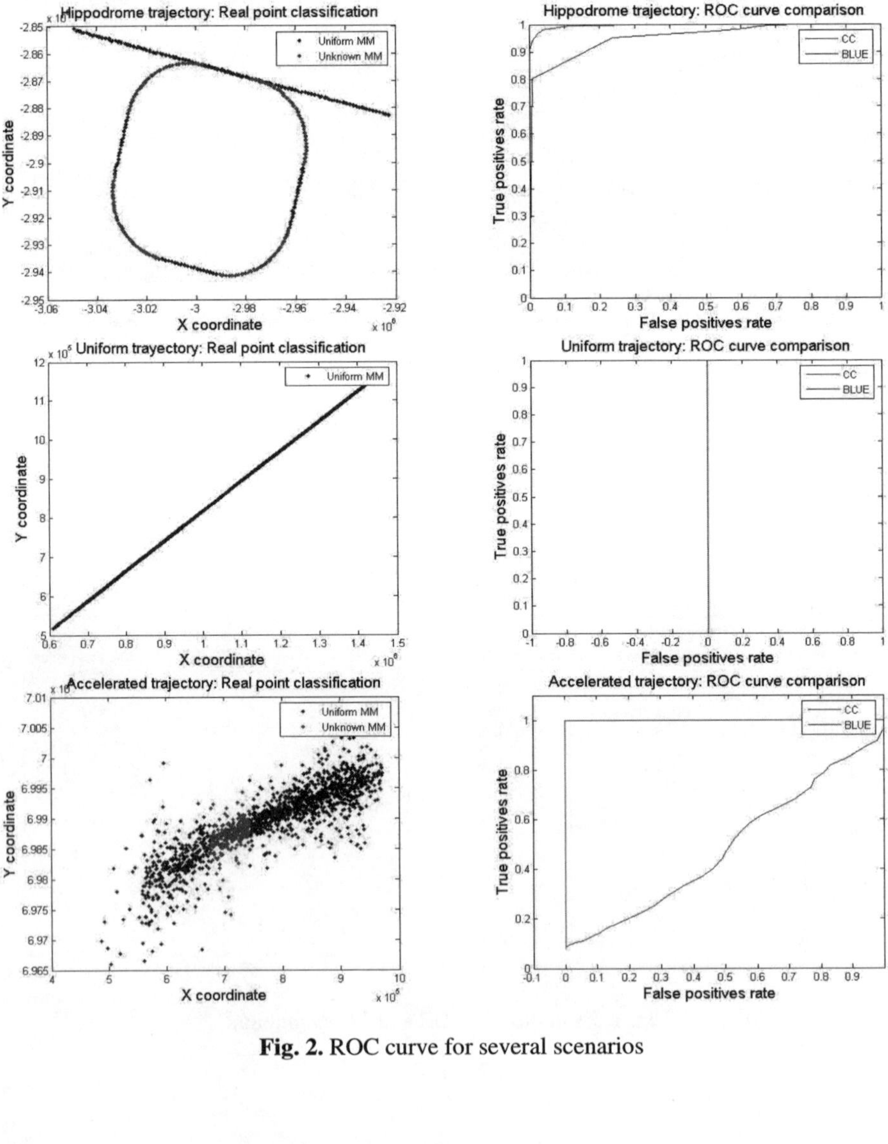

Fig. 2. ROC curve for several scenarios

As we can see in the figures above, the results displayed in the example turn trajectory are general to any of them, showing that using a BLUE residue for the domain transformation gets better results than CC (in uniform trajectories, having no possible FPR, their ROC curves are the same) and that those results follow, in any trajectory, our null FPR criterion with a reasonable TPR value.

5 Conclusions

In this paper we proposed and algorithm to be able to achieve trajectory segmentation into three possible MM's: uniform , turn and acceleration. We have specified an approach where we may use specific models for each of those MM's and apply them sequentially, based on a design criterion where each of these segmentations must have a null FPR. The first of these models is the uniform MM, whose parameters we have covered briefly, detailing a comparison for its descriptor attributes, showing that a BLUE residue is suitable for our design criterion, being able to get a TPR, with the right resolution values, over 99 %, while keeping a null FPR These results, having used the design criterion to get the threshold value, can't be applied to real trajectories, whose samples are not previously classified, so to be able to apply the algorithm to them we would need a resolution and threshold selection method.

References

1. Desmond-Kennedy, A., Gardner, B.: Tools for analysing the performance of ATC surveillance radars. Specifying and Measuring Performance of Modern Radar Systems (Ref. No. 1998/221), IEE Colloquium (March 6, 1998)
2. Garcia, J., Perez Concha, O., Molina, J.M., de Miguel, G.: Trajectory classification based on machine-learning techniques over tracking data. In: 9th International Conference on Information Fusion, Florence, Italy (July 2006)
3. Pérez, O., García, J., Molina, J.M.: Neuro-fuzzy Learning Applied to Improve the Trajectory Reconstruction Problem. In: International Conference on Computational Intelligence for Modelling, Control and Automation, CIMCA 2006, Sydney, Australia (December 2006)
4. Garcia, J., Molina, M., de Miguel, G., Besada, A.: Model-Based Trajectory Reconstruction using IMM Smoothing and Motion Pattern Identification. In: 10th International Conference on Information Fusion, Quebec, Canada (July 2007)
5. Kay, S.M.: Fundamentals of Statistical Signal Processing: Estimation Theory. Prentice-Hall, Englewood Cliffs (1993)

Techniques of Engineering Applied to a Non-structured Data Model

Cristóbal J. Carmona[1], María J. del Jesus[1], Pablo Guerrero[2], Reyes Peña-Santiago[2], and Víctor M. Rivas[1]

[1] Department of Computer Science and [2] Department of Animal Biology, Plant Biology and Ecology, University of Jaen, Campus Las Lagunillas s/n, 23071, Jaen
{ccarmona,mjjesus,pguerre,rpena,vrivas}@ujaen.es

Summary. The information developed pertaining to biodiversity studies tends to be scattered around many bibliographic references. A review of the Nordiidae family is being carried out, but the very data to be collected does not allow systematic access. The Nordiidae family, belonging to the animal taxon Nematodes, shows high diversity and an extraordinary ubiquity. Engineering techniques have been applied to turning this textual information into structured data, so that new knowledge can be discovered and data can be accessed through the net.

Keywords: Data Engineering, Automated Processing, Hypermedia System, Regular Expressions Extraction.

1 Introduction

One of the most important limitations in biodiversity studies is that the available information is dispersed throughout many bibliographic references, and frequently structured under different criteria. A revision of the family Nordiidae is currently been carried out by the Andalusian Group of Nematology, compiling the available information about the most representative genus and species. This updated information has been stored in a large number of unstructured documents. This work shows a software engineering procedure developed to translate this series of text documents into structured data, in order to improve their diffusion within the scientific community.

Nematodes (phylum Nematoda or Nemata) are an animal taxon showing high diversity [7] and extraordinary ubiquity, despite their simple morphological body plan [4]. The study of soil nematodes has received much attention during the last decades as they cause severe diseases in cultivated plants, meanwhile many free-living species are good (bio)indicators of soil quality (health). A set of text documents compiling available information of *Enchodelus* species has been prepared to describe and clarify their taxonomy. The information of each species consists of the following items:

1. **Nomenclature:** scientific name, and its binomen: authorship and date.
2. **Synonymy** (if it exists)**:** other synonymous scientific names and their corresponding bibliographic references.

J.M. Corchado et al. (Eds.): DCAI 2008, ASC 50, pp. 410–414, 2009.
springerlink.com

3. **Description:** morphological and morphometric data for both female and male, when available.
4. **Diagnosis:** a brief report of useful characteristics for species identification.
5. **Relationships:** a comparison of each species with its nearest relatives.
6. **Distribution:** data referring to the geographical distribution of the species.
7. **Type material:** number of type specimens and collections and/or places where they are deposited.
8. **Remarks:** additional comments on the species in question.

The complete procedure for turning the unstructured data into structured is detailed in the following steps: Section 2 describes the structure of the database storing the information. In Section 3 the different procedures used to generate structured data are listed. The final section shows conclusions and future prospects for this project.

2 Analysis of the Classification Structure

The revision of the family Nordiidae generated a series of documents to which database techniques have been applied. Thus, all the information is stored in a structured and easily accessible way.

Figure 2.a shows an example of a text document compiled from the original references. It includes taxonomic information (genus, sub-genus and species), authority, synonyms, and a list of bibliographic references. These references are grouped according to the name used by each author. After the nomenclature has

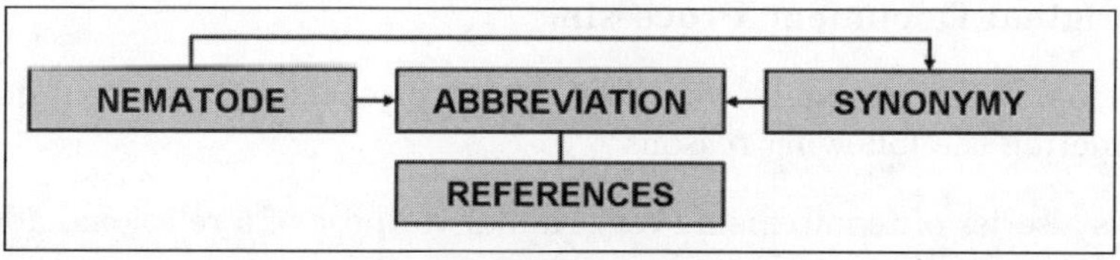

Fig. 1. Relational model for the system

Table 1. Tables generated from the revision with primary key indicated

NEMATODE(genus, species, subgenus, author_gen, year_gen, author_sp, year_sp, female, male, diagnosis, relationships, distribution, etymology, type_material, remarks, cod_icon)
SYNONYMY(gen_syn, sp_syn, gen_nem, sp_nem, subgenus, author_gen_syn, year_gen_syn, author_sp_syn, year_sp_syn)
ABBREVIATION(author, year, gen_nem, sp_nem, gen_syn, sp_syn, subgen_nem, subgen_syn,author_gen_syn, cite)
REFERENCES(id, author, year, nick, cite)

been established, the morphological features seen in section 1 are described (for instance, *Female*, *Male*, of *Diagnosis*)

The entity-relation diagram (shown in Figure 1) is used to obtain the normalized tables that will store all this information, in order to make possible its manipulation and visualization in different media. Normalized tables are shown in Table 1. Thus, NEMATODE is the main table and stores the valid name of the species together with all its features. The SYNONYMY table includes all synonymous names the nematode has received. The ABBREVIATION table stores all abbreviations for the references used in the original document. Finally, the REFERENCE table contains a full description of the references.

3 Data Processing Method

Introducing information into the database is a time-consuming task prone to mistake, due to the characteristics of the original text documents. Thus, data cannot be extracted in a straightforward way since fields are described by means of differences in style labels, and these labels are expressed and saved in an specific and proprietary wordprocessor format, making their processing quite difficult.

The automatic data extraction process is described in the following two sections: Firstly (see section 3.1) the justification for using a semi-structured document and the processing carried out in order to obtain this kind of document from the original. Secondly (see section 3.2) the procedure which obtains the items of the database from the semi-structured document.

3.1 Original Document Processing

A Relational Database can be considered as a type of semi-structured[1] document [8] considering the following reasons:

- It has a series of requirements very similar to those of a relational document.
- Values can be organized in the same way as in a database.
- Just as in a Relational Data Base, the attribute of a register may be another register in semi-structured documents.

Following the philosophy developed by Chris Bizer in *D2R-Map* [2] and *D2R-Server* [3] and Pérez de Laborda in *Relational.OWL* [5], where they transform a database into a semantic document (RDF), the data processing begins with the conversion of the original document into a semi-structured document. By means of word-processing tools, the un-structured document is transformed into an XHTML document, which is still complex to process since the branches generated rarely contain the same number of nodes.

[1] Semi-structured Data - http://ict.udlap.mx/people/carlos/is346/admon07.html

3.2 Semi-structured Document Processing

Since our original documents distinguished fields according to style labels, XHTML seemed to be the best solution for temporarily storing the information. In effect, XHTML allows the inclusion of style labels in the processed document. In this way, each different style identifies certain features, such as genus, species or author, which can be associated to the database fields. This association allows the identification of fields by means of detecting style patterns. In figure 2.b for instance, the *T1* style identifies the genus and species of the nematode, the *T2* style identifies the author of the species and so on. As values for fields are being obtained, the SQL insertion commands for a MySQL database are simultaneously generated.

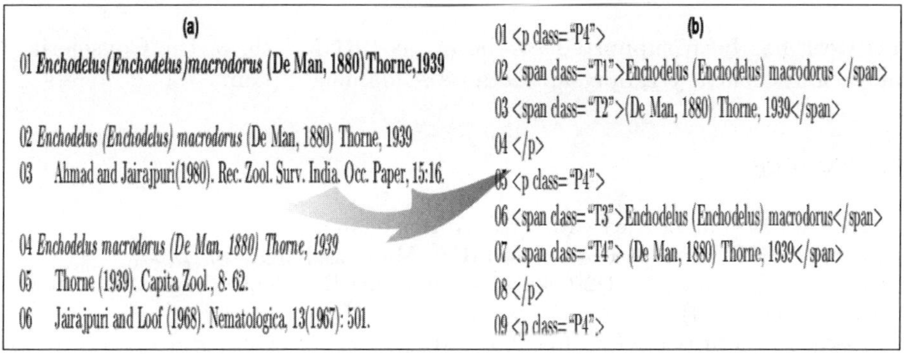
(a)
01 *Enchodelus(Enchodelus)macrodorus* (De Man, 1880)Thorne,1939
02 *Enchodelus (Enchodelus) macrodorus* (De Man, 1880) Thorne, 1939
03 Ahmad and Jairajpuri(1980). Rec. Zool. Surv. India. Occ. Paper, 15:16.
04 *Enchodelus macrodorus (De Man, 1880) Thorne, 1939*
05 Thorne (1939). Capita Zool., 8: 62.
06 Jairajpuri and Loof (1968). Nematologica, 13(1967): 501.

(b)
01 <p class="P4">
02 <span class="T1">Enchodelus (Enchodelus) macrodorus </span>
03 <span class="T2">(De Man, 1880) Thorne, 1939</span>
04 </p>
05 <p class="P4">
06 <span class="T3">Enchodelus (Enchodelus) macrodorus</span>
07 <span class="T4"> (De Man, 1880) Thorne, 1939</span>
08 </p>
09 <p class="P4">

Fig. 2. Original(a) and Semi-structured(b) document

The implementation is carried out in Java, since many free packages and components can be used in order to perform a pattern search. It is thus possible to complete the conversion from semi-structured document to items-insertion document in a short time. The algorithm developed for this project includes pattern recognition systems for finding links to other species or authors present in each field of the database. Due to implemented the highly variable bibliographic patterns of the original document, a complex process to extract regular expressions has been implemented. For instance: from "*author, year1, year2*", the process obtains two distinct links "*author year1*" and "*author year2*". Furthermore, the algorithm is designed to maintain the initial format when presented to the user. This processing is carried out using the following Java regular expression:

$$(\backslash\backslash([A-Z][a-z]+(-|\backslash\backslash s|,|[A-Z][a-z]+|and|\&)+\backslash\backslash d\{4\}(,\backslash\backslash s\backslash\backslash d\{4\})*\backslash\backslash)$$

4 Results and Conclusions

At the moment, access to the data pertaining to the Nordiidae family is achieved by means of a web page[2] that includes a sample revision of genus *Enchodelus*. The

[2] http://www.ujaen.es/investiga/nmundii

hypermedia system generated using engineering techniques allows us to access other species when referred to in the text by their valid names or synonyms. Cross references make it possible to navigate through the information of every species, visualizing the existing relationships among them.

Current work includes the development of a powerful system to perform complex searches. This will lead to an intelligent search system based also on frequent search patterns.

Accessibility to data will be also improved by means of *Ajax* [1][6], and also web services. The latter will help to develop useful tools able to join information from different sources, such as genetic and scientific publication databases.

Acknowledgments

This work has been supported by the project RNM-475, of the Consejería de Innovación, Ciencia y Empresa, Junta de Andalucía, Spain.

References

1. Babin, L.: Beginning Ajax with PHP. Apress (2007)
2. Bizer, C.: D2R Map - A Database to RDF Mapping Language (2003)
3. Bizer, C., Cyganiak, R.: D2R Server - Publishing Relational Databases on the Semantic Web (2004)
4. Brusca, R.C., Brusca, G.J.: Invertebrados, p. 1005. McGraw-Hill Interamericana, Madrid (2005)
5. de Laborda, C.P., Conrad, S.: Relational OWL. In: Hartmann, E. S., Stumptner, M. (eds.) Second Asia-Pacific Conference on Conceptual Modelling (APCCM 2005), vol. 43 (2005)
6. Hadlock, K.: Ajax for Web Application Developers. Kindle Books (2006)
7. Hawksworth, D.L., Kalin-Arroyo, M.T.: Magnitude and Distribution of Biodiversity, p. 1140. Global Biodiversity Assessment: 107-191 (1995)
8. Maier, D.: Theory of Relational Databases. Computer Science Pr. (1983)

A Connectionist Automatic Encoder and Translator for Natural Languages*

Gustavo A. Casañ[1] and Mª Asunción Castaño[2]

[1] Dept. Lengujes y Sistemas Informáticos
ncasan@lsi.uji.es
[2] Dept. Ingeniería y Ciencias de los Computadores
[1,2] Universitat Jaume I, Av./Sos Baynat s/n 12071 Castellón, Spain
castano@icc.uji.es

Abstract. One of the problems of the connectionist translator RECONTRA is the representation of the vocabularies of the languages implied in the task to be translated. In previous work, a simple connectionist model was used to provide automatic codifications for RECONTRA, but sometimes these codifications have shown not to be adequate for the translation task. In this paper we aim to extend the RECONTRA topology in order to integrate the creation of the codifications (for the languages to be translated) and the translation task in an unique connectionist architecture. To do that, a new hidden layer is added to the network, as it's known how a neural network develops its own internal representation of its input.

Keywords: Automatic Word Representation, Machine Translation, Neural Networks.

1 Introduction

Neural Networks are a possible approach to Machine Translation (MT), as the translation schemes presented in [1] and [2] have empirically shown. Encouraging results have been obtained for text-to-text limited domain applications using a simple Example-Based recurrent connectionist translator called RECONTRA (Recurrent Connectionist Translator) presented by Castaño in [3]. It directly carries out the translation between the input and the output languages and, at the same time, automatically learns the semantic and syntax implicit in both languages. In this approach distributed representations of both source and target vocabularies are required to approach large tasks [4]. Although Random Distributed Codifications (RDCs) can be used, the results are not very good. In [5], an automatic encoder, which used a simple neural model to create distributed codifications, was designed (see also [6] and [7]).

However, the codifications provided by these encoders, while better than RDCs, did not always lead to the expected good translation rates when tasks with vocabularies of medium or large size were approached. In this paper we try to integrate in an unique connectionist architecture the translator and the encoder. A similar integration was done previously for the creation of language models with neural networks [8] taking advantage of the known property of neural networks to create internal representations of their inputs. The simplest way to take advantage of this property with

* Partially supported by the Generalitat Valenciana Project number GV/2007/105.

J.M. Corchado et al. (Eds.): DCAI 2008, ASC 50, pp. 415–423, 2009.
springerlink.com

RECONTRA seemed to be adding a second Hidden Layer (HL) to the network. This layer would develop more adequate representations to the translation task than any other method.

The paper is organized as follows: first we describe the topologies of the extended RECONTRA translator and the encoder. The translation task to be approached and the experiments carried out with this task are presented later. And finally the conclusions of the experimental process are discussed.

2 The RECONTRA Translator

The basic neural topology of the RECONTRA translator is an Elman network [9], a Simple Recurrent Neural Network in which the activations of the preceding step in the hidden units of the networks are feedback as inputs in the hidden layer. The architecture of RECONTRA (which can be seen in Fig. 1) includes time delays in the input layer of the Elman network, in order to reinforce the information about past and future events.

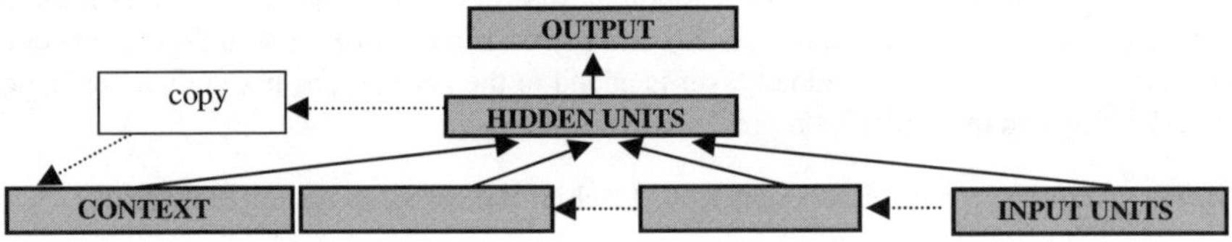

Fig. 1. The RECONTRA translator: an Elman network with input window

In this paper we propose to add a HL to the RECONTRA architecture, trying to force the network to separate the development of its own representation of the input vocabulary and the learning of the translation task. This strategy tries to take advantage of the known property of neural networks to select the more important characteristics and create internal representations of their inputs. This new layer would hopefully develop more adequate representations to the translation task than other methods. The resulting topology is shown in Figure 2. Some studies have shown how adding layers in a NN can apparently increase the results [10]. To show that this is not our case and that the representation potential is what is increasing the results, a few experiments with three layers are also presented. The extended RECONTRA is still an Elman network and its training process is the same as that used for non-extended RECONTRA:

The words of the sentence to be translated are sequentially presented to the input of the network, and the net has to provide the successive words of the corresponding translated sentence. The time delays of the input layer allow the net to simultaneously see the successive m+1+n words of the input sentence; for instance, the words i-m ... i-1 i i+1 ... i+n. In the next step, the words i-m+1 ... i i+1 i+2 ... i+n+1 are presented and so on.

Both the extended and basic (non-extended) RECONTRA translator are trained by using an "on-line" version of the Backward-Error Propagation (BEP) algorithm [11]; that is, the weights of the net are modified after each input is processed. The standard algorithm is modified to ignore the recurrent connections of the net in the process of adjusting the weights; these connections are considered additional external inputs to

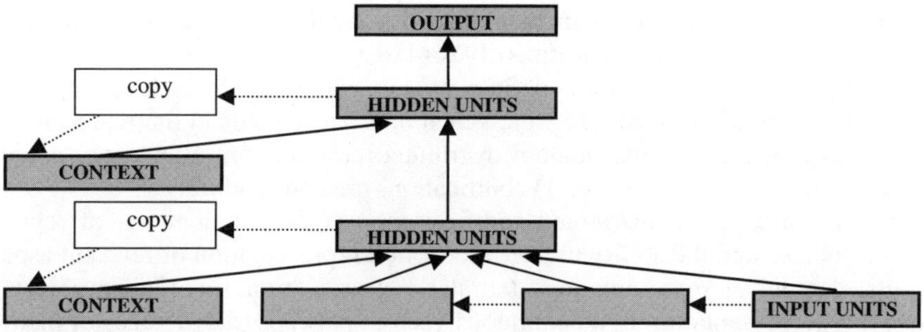

Fig. 2. The extended RECONTRA translator: an Elman network with input window and two hidden layers

the architecture. Consequently, the gradient of the error is truncated in the estimation of the weights; that is, it is not exactly computed. However, this learning method runs well in practice as it is shown in [3].

After inputs and target units are up-dated, the forward step is computed, the error is back-propagated through the net and the weights are modified. Later, the hidden unit activations are copied onto the corresponding context units. This time cycle is continuously repeated until the target values mark the end of the translated sentence. A sigmoid function (0,1) is assumed as the non-linear activation function and context activations are initialized to 0.5 at the beginning of every input-output pair.

The updating of the weights requires estimating appropriate values for the learning rate term and momentum term. The learning rate concerns the convergence speed towards a minimum in the error surface; and the momentum term is related to how the old weights change for the computation of the new change (it reduces some oscillation problems common with the BEP algorithm when the error surface has a very narrow minimum area). The choice of learning rate and momentum for the BEP algorithm is carried out inside the unitary bidimensional space which they define, by analyzing the residual mean squared error (MSE) of a network trained for 10 random presentations of the learning corpus. Training continues for the learning rate and momentum which led to the lowest MSE and stops after 150 training epochs.

The activations provided by the network are interpreted as the word associated to the nearest pre-established codification of the target vocabulary. Word error is computed by comparing the obtained and expected translations corresponding to every source sentence in the test sample using a conventional Edit-Distance procedure [12] and its expressed like the Word Accuracy Rate (WAR) in the tables.

3 Codification of the Vocabularies

There are two distinct types of representation used for symbolizing short-term data at the input/output of connectionist models: local and distributed representations.

In a local representation each concept (each word) is represented by a particular unit. This means that each input or output unit of the network represents one word of the vocabulary. To this end, one unit is on (his value is, for instance, 1) and the other units are off (value 0). The main problems with local representations concern inefficiency, as the networks grow enormously for large sets of objects.

A distributed representation is defined [13] as one in which each concept is represented over several units, and in which each unit participates in the representation of several concepts. The name random distributed codification (RDC) was adopted in this paper for binary (values 0 or 1) codifications created randomly.

Main advantages of distributed codifications are the efficient use of representational resources and the ability to have an explicit representation of relevant aspects of objects, analogical representations (similar representation for similar objects) and continuity (representation in a continuous vector space). Disadvantages of distributed codifications are the problem of creating adequate ones for each problem and how to represent certain associations (for instance, several words with the same meaning) and variable bindings (for instance, a word with several meanings). Also, translators in which distributed codifications are adopted usually require more training time.

3.1 Codification Generator for RECONTRA

A Multi-Layer Perceptron (MLP) with output window was presented in [5] as an encoder, to generate the codifications required for RECONTRA. It's based on the known capabilities of a network to develop its own internal representations of the input. The MLP has three layers of units and is trained to produce the same output as the input (a word of the vocabulary). After training, a word is presented to the MLP and the activations of the hidden units are extracted and assumed to be the internal representation developed by the MLP for this input word. Thus, the size of the hidden layer of the MLP determines the size of the codifications obtained.

In order to take into account the context in which a word appears, the previous and following words in a sentence are also shown at the output of the MLP. The resulting topology for the MLP encoder is shown in Figure 3. The importance of the input word over its context was increased by repeating the input words several times at the output window of the MLP encoder. According to this, a possible output window of size 8 for the *x-th* input word w_x could be $w_{x-2}\ w_{x-1}\ w_x\ w_x\ w_x\ w_x\ w_{x+1}\ w_{x+2}$, where w_{x-2}, w_{x-1} are the two previous words in its context and w_{x+1}, w_{x+2}, the two following words. To simplify the nomenclature, from now on, we will refer to such example of format for the output window as *x-2 x-1 x x x+1 x+2*.

The size of the codifications extracted from the MLPs has been automatically determined using a simple pruning method [15]. Moreover, the codifications extracted from these MLPs can be used to train other MLPs which lead to new codifications, in

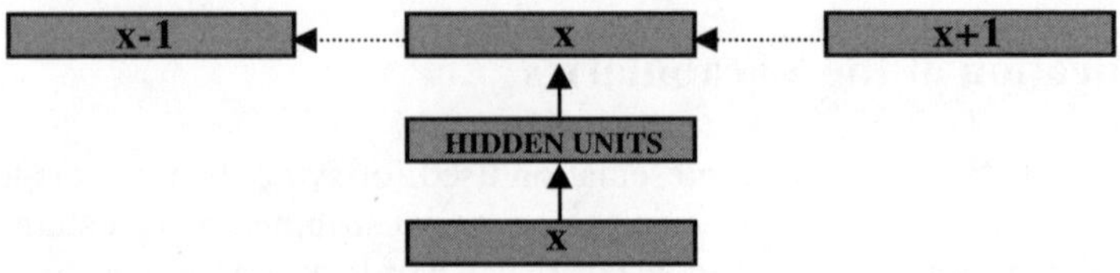

Fig. 3. Multilayer Perceptron with output delays (*x-1 x x+1*)

a sort of feedback. Finally, to mention that BEP has been used to train the encoders, but also a combination of BEP and a Scaled Conjugate Gradient method [14]. More details about the encoder, the types of output windows, and the pruning and the training methods used can be found in [5], [6] and [7].

4 The Traveler Task

The task chosen in this paper to test the extended RECONTRA translator was the Spanish-to-English text-to-text Traveler MT task (sometimes also called the *Tourist* task), which was designed within the first phase of the EuTrans project [16] and had been previously approached with basic RECONTRA. This task is restricted to a limited domain and approaches typical situations of a Traveler at the reception of a hotel in a country whose language he/she does not speak. The scenario has been limited to some human-to-human communication situations in the reception of a hotel: asking for rooms, wake-up calls, keys, the bill, a taxi, and moving the luggage; asking for information about rooms; asking about changing a room; notifying a customer about a reservation; signing the registration form; asking and complaining about the bill; informing reception about departure; and other common expressions. The task has 683 Spanish words and 513 English words. These are several examples of this task:

He de marcharme el día veintisiete de febrero a las siete y media de la tarde . # I should leave on February the twenty-seventh at half past seven in the afternoon .

¿ les importaría bajar el equipaje a recepción ? # would you mind sending the luggage down to reception ?

The corpora adopted in the translation task were sets of text-to-text pairs: a sentence in the Spanish language and the corresponding sentence in the English one. 10,000 pairs were used for training and 3,000 for test. The test-set perplexity of the Spanish sentences is 13.8 and 7.0 for the English ones. The medium size of the sentences in Spanish is 9.5 and 9.8 in the English ones. The best translation results previously obtained was 91.7% Word Accuracy Rate (WAR) with OSTIA, as reported in [17].

5 Experimental Results

Previous experimentation with the task [6] and non-extended RECONTRA suggested to adopt a translator with an input window of 9 words and 450 hidden units. This was the basic topology used in this paper. All the experiments were done using the Stuttgart Neural Network Simulator [18].

5.1 Codifications Employed

In order to compare the benefits of using extended RECONTRA over basic (non-extended) RECONTRA, in the experimentation of this paper we adopted codifications previously presented in different works by Casañ and Castaño ([5], [6] and [7]). In Table 1 we can see some of the characteristics: size of the codifications for the Spanish and English vocabularies (which are denoted by |Spanish|/|English|, respectively) and when codifications were extracted from MLPs, some of the characteristics of the networks: the use or not of pruning, the use of feedback and the output window format of the network. More details can be found in the original papers [5], [6] and [7]. The nomenclature adopted for the names of the codifications is as follows: a roman number for binary RDCs (the same as in [7]); in codifications extracted from MLPs trained with initial RDCs, *mlp* was added to the name of the original codification (for instance *V mlp*); if it was a codification extracted from a MLP in which the initial codifications were extracted from other trained MLP, we add mlp^2 to the name of the codification (*I* mlp^2); and when the MLPs have been pruned, a P is added (*IV mlp P*).

Table 1. Characteristics of the codifications: name adopted, sizes for Spanish and English (denoted by |Spanish|/|English|), the use of Feedback, Pruning and the output window format if it was extracted through MLPs

Codif. Name	Feedback	Pruning	|Spanish| / |English|	Output Format	WAR
IV mlp P	No	Yes	49/46	*x-1* **x x** *x+1*	79.7%
V	No	-	30/30	-	63.4%
V mlp	No	No	30/30	*x-1* **x x** *x+1*	76.5%
V mlp P	No	Yes	17/14	*x-1* **x x** *x+1*	52.5%
VI	No	-	30/30	-	67.5%
VI mlp	No	No	30/30	*x-1* **x x** *x+1*	75.3%
VII	No	-	30/30	-	67.4%
VII mlp	No	No	30/30	*x-1* **x x** *x+1*	76.3%
I mlp P	No	Yes	121/94	*x-1* **x x x x** *x+1*	78.2%
I mlp^2 P	Yes	Yes	121/94	*x-1* **x x x x** *x+1*	83.7%

5.2 Different Sizes of the Hidden Layers

First, we selected a pair of RDCs (one for the source language and another one for the target one): those denoted by *V* in Table 1. Then we trained the modified

Table 2. Translation rates using the codifications of Table 1 denoted by *V* and different sizes of the HLs of the extended RECONTRA model

| First HL |	| Second HL |	| Third HL |	WAR
30	450	-	52.3%
90	450	-	61.2%
270	450	-	71.3%
450	450	-	76.9%
600	450	-	77.7%
750	450	-	77.7%
270	270	-	68.7%
450	30	-	54.8%
450	60	-	62.1%
270	450	30	57.4%
270	450	270	69.2%
270	450	450	70.2%
270	270	450	68.5%

RECONTRA with these codifications using different sizes of the First hidden layer (HL) and the Second HL. The translation results obtained are shown in Table 2. The results achieved with the non-extended RECONTRA can be seen in Table 1. The translation rates achieved with the extended RECONTRA reveal a general improvement except when very small sizes of any of the two HL were used. Also, experiments with three HL confirmed that adding more layers did not improve the results.

5.3 Different Codifications

The codification which had provided previously the best results with the non-extended RECONTRA (which was denoted by $I\ mlp^2\ P$) was used for further experimentation, as well as some of those described in Table 1. For the codifications of sizes 30/30 only the results using 270 units in its first hidden layer are shown as this number of hidden units coincided with the number of input units. For the other codifications, different sizes of the first HL are tested, including using in some cases a HL layer with the same number of units as the input (|codification size|*number of input words: 441 and 1089). The translation rates achieved are shown in Table 3. For comparison purposes, the results with the corresponding non-extended RECONTRA (those in which the third column is "-") were also depicted. These results support that the extended RECONTRA is a better translator than the basic RECONTRA. They also point out that the benefits of the extended RECONTRA are larger when codifications of small sizes are adopted.

Table 3. Translation rates using different codifications of Table 1 with different RECONTRA models

Codification	|Spanish| / |English|	| First HL |	WAR
V mlp	30/30	-	76.5%
		270	80.9%
V mlp P	17/14	270	61.7%
		450	68.0%
VII	30/30	270	75.9%
VII mlp	30/30	270	79.1%
VI	30/30	270	76.1%
VI mlp	30/30	270	79.4%
		270	82.0%
IV mlp P	49/46	441	83.4%
		750	84.9%
I mlp P	121/94	270	81.4%
		1089	84.2%
I mlp^2 P	121/94	270	83.9%
		450	84.5%

5.4 Different Input Codifications

In the previous section we considered at the input and output of the RECONTRA models, codifications extracted through MLPs with the same characteristics. In this section we adopted for the output of the extended RECONTRA a codification extracted from a MLP trained with RDCs *V* (called *V mlp*) and different types of input codifications: Local, different RDCs (*V* and *VI*) and extracted from MLPs with (*V mlp P* and *IV mlp P*) and without pruning (*V mlp*). The translation results (which are shown in Table 4) are quite similar for the different input codifications adopted. This means that the extended RECONTRA is not quite dependant on the codifications of the input language, although the size still seems important.

Table 4. Translation rates using codification *V mlp* at the output of the RECONTRA models and different input codifications of Table 1

Input Codification	First HL	WAR
V mlp	270	80.9%
V mlp P	270	77.4%
V	270	81.7%
Local	270	85.4%
VI	270	81.7%
IV mlp P	270	82.6%

6 Conclusions

This paper approaches a text-to-text MT task with vocabularies of near 700 words using an extension of the RECONTRA translator. The new translator includes an additional hidden layer which encodes the input words. The translation results previously obtained using both RDCs and codifications extracted from MLPs are improved when the new hidden layer of RECONTRA is considered with output windows. But the resulting network is still dependant on the output codification adopted, which becomes the main factor in the results obtained.

Although the translation results are still not as good as the best ones obtained with other methods [17], this "new" network partially solves the problem of creating codifications for RECONTRA. The experimental results clearly show that two Hidden Layers provide better results than one (or three); the first one develops a representation of the input (the source vocabulary) more adequate to the task. The two main problems to be approached in the future are to resolve the dependence on the output codification (which the translator still doesn't learn); and the increased size of the resulting network and the instability problems it presents. Other possible direction is using different RECONTRA networks to create an only translation system (with ensembles or mixture of experts' techniques).

References

1. Koncar, N., Guthrie, G.: A Natural Language Translation Neural Network. In: Procs. of the Int. Conf. on New Methods in Language Processing, Manchester, UK, pp. 71–77 (1994)
2. Waibel, A., Jain, A.N., McNair, A.E., Saito, H., Hauptmann, A.G., Tebelskis, J.: JANUS: A Speech-to-Speech Translation System using Connectionist and Symbolic Processing Strategies. In: Procs. of the International Conference on Acustic, Speech and Signal Processing, pp. 793–796 (1991)

3. Castaño, M.A.: Redes Neuronales Recurrentes para Inferencia Gramatical y Traducción Automática. Ph.D. dissertation, Universidad Politécnica de Valencia, Spain (1998)
4. Casañ, G.A., Castaño, M.A.: Distributed Representation of Vocabularies in the RECONTRA Neural Translator. In: Procs. of the 6th European Conference on Speech Communication and Technology, Budapest, Hungary, vol. 6, pp. 2423–2426 (1999)
5. Casañ, G.A., Castaño, M.A.: Automatic Word Codification for the RECONTRA Connectionist Translator. In: Perales, F.J., Campilho, A.C., Pérez, N., Sanfeliu, A. (eds.) IbPRIA 2003. LNCS, vol. 2652, pp. 168–175. Springer, Heidelberg (2003)
6. Casañ, G.A., Castaño, M.A.: A New Approach to Codification for the RECONTRA Neural Translator. In: Procs. Ninth IASTED International Conference on Artifical Intelligence and Soft Computing, pp. 147–152 (2005)
7. Casañ, G.A., Castaño, M.A.: Tuning word codifications for the RECONTRA translator. In: Borajo, D., Castillo, L., Corchado, J.M. (eds.) Actas XII Conferencia de la Asociación Española para la Inteligencia Artificial, Salamanca, Spain. Universidad de Salamanca, vol. 2, pp. 351–354 (2007)
8. Bengio, Y., et al.: A Neural Probabilistic Language Model. Journal on Machine Learning Research 3, 1137–1151 (2003)
9. Elman, J.L.: Finding Structure in Time. Cognitive Science 2(4), 279–311 (1990)
10. Tamura, S., Tateishi, M.: Capabilities of a Four-Layered Feedforward Neural Network: Four Layers Versus Three. IEEE Trans. on Neural Networks 9(2), 251–255 (1997)
11. Rumelhart, D.E., Hinton, G., Williams, R.: Learning Sequential Structure in Simple Recurrent Networks. In: Rumelhart, D.E., McClelland, J.L., The PDP Research Group (eds.) Parallel distributed processing: Experiments in the microstructure of cognition, vol. 1. MIT Press, Cambridge (1981)
12. Marzal, A., Vidal, E.: Computation of Normalized Edit Distance and Applications. IEEE Transactions on Pattern Analysis and Machine Intelligence 15(9) (1993)
13. Hinton, G.E., McClelland, J.L., Rumelhart, D.E.: Distributed representations. In: Rumelhart, D.E., McClelland, J.L. (eds.) Parallel Distributed Processing: Explorations in the Microstructure of Cognition. Vol 1: Foundations. MIT Press, Cambridge (1986)
14. Möller, M.F.: A Scaled Conjugate Gradient Algorithm for Fast Supervised Learning. Neural Networks 6, 525–533 (1993)
15. Mozer, M.C., Smolensky, P.: Skeletonization: a Technique for Trimming the Fat from a Network via Relevance Assessment. In: Touretzky, D.S., Kaufmann, E.M. (eds.) Advances in Neural Information Processing, vol. 1, pp. 177–185 (1990)
16. Amengual, J.C., Castaño, M.A., Castellanos, A., Llorens, D., Marzal, A., Prat, A., Vilar, J.M., Benedí, J.M., Casacuberta, F., Pastor, M., Vidal, E.: The Eutrans-I Spoken Language System. Machine Translation, vol. 15, pp. 75–102. Kluwer Academic Publishers, Dordrecht (2000)
17. Prat, F., Casacuberta, F., Castro, M.J.: Machine Translation with Grammar Association: Combining Neural Networks and Finite-State Models. In: Procs. The Second Workshop on Natural Language Processing and Neural Networks, Tokio, Japan, pp. 53–61 (2001)
18. Zell, A., et al.: SNNS: Stuttgart Neural Network Simulator. User manual, Version 4.1. Technical Report no. 6195, Institute for Parallel and Distributed High Performance Systems, University of Stuttgart, Germany (1995)

Rewriting Logic Using Strategies for Neural Networks: An Implementation in Maude*

Gustavo Santos-García[1], Miguel Palomino[2], and Alberto Verdejo[2]

[1] Universidad de Salamanca
santos@usal.es
[2] Departamento de Sistemas Informáticos y Computación, UCM
miguelpt@sip.ucm.es, alberto@sip.ucm.es

Summary. A general neural network model for rewriting logic is proposed. This model, in the form of a feedforward multilayer net, is represented in rewriting logic along the lines of several models of parallelism and concurrency that have already been mapped into it. By combining both a right choice for the representation operations and the availability of strategies to guide the application of our rules, a new approach for the classical backpropagation learning algorithm is obtained. An example, the diagnosis of glaucoma by using campimetric fields and nerve fibres of the retina, is presented to illustrate the performance and applicability of the proposed model.

Keywords: Neural networks, rewriting logic, Maude, strategies, executability.

1 Introduction

Rewriting logic [8] is a logic of concurrent change that can naturally deal with states and with highly nondeterministic concurrent computations. It has good properties as a flexible and general semantic framework for giving semantics to a wide range of languages and models of concurrency. Indeed, rewriting logic was proposed as a unifying framework in which many models of concurrency could be represented, such as labeled transition systems, concurrent object-oriented programming, or CCS, to name a few [6, 9, 5].

Artificial neural networks [4] are another important model of parallel computation. In [7] it was argued that rewriting logic was also a convenient framework in which to embed neural nets, and a possible representation was sketched. However, and to the best of our knowledge, no concrete map has ever been constructed either following those ideas or any others. Our goal with this paper is to fill this gap. In our representation of neural networks we consider the evaluation of patterns by the network, as well as the training required to reach an optimal performance.

Since its conception, rewriting logic was proposed as the foundation of an efficient executable system called Maude [1]. Here we write our representation

* Research supported by Spanish project DESAFIOS TIN2006–15660–C02–01 and by Comunidad de Madrid program PROMESAS S–0505/TIC/0407.

J.M. Corchado et al. (Eds.): DCAI 2008, ASC 50, pp. 424–433, 2009.
springerlink.com

directly in Maude to be able to run our neural networks and apply them to a real case-study, the analysis of campimetric fields and nerve fibres of the retina for the diagnosis of glaucoma [3].

The paper is organized as follows. In Section 2 we review those aspects of Maude that will be used in our specification, mainly object-oriented modules and strategies. Section 3 introduces multilayer perceptrons and the backpropagation algorithm. Their specification in Maude, and an appropriate strategy for their evaluation and training, is presented in Section 4. The application of our implementation to the study of the diagnosis of glaucoma is considered in Section 5, and Section 6 concludes.

2 Maude

Maude [1] is a high performance language and system supporting both equational and rewriting logic computation for a wide range of applications. The key novelty of Maude is that besides efficiently supporting equational computation and algebraic specification it also supports rewriting logic computation. Mathematically, a rewrite rule has the form $l : t \longrightarrow t'$ *if Cond* with t, t' terms of the same type which may contain variables. Intuitively, a rule describes a local concurrent transition in a system: anywhere a substitution instance $\sigma(t)$ is found, a local transition of that state fragment to the new local state $\sigma(t')$ can take place.

Full Maude [1] is an extension of Maude with a *richer module algebra* of parameterized modules and module composition operations and with special syntax for object-oriented specifications. These object-oriented modules have been exploited in the specification of neural networks.

2.1 Object Oriented Modules

An *object* in a given state is represented as a term `< O : C | a1 : v1,..., an : vn >` where `O` is the object's name, belonging to a set `Oid` of object identifiers, `C` is its *class*, the `ai`'s are the names of the object's *attributes*, and the `vi`'s are their corresponding values. *Messages* are defined by the user for each application.

In a concurrent object-oriented system the concurrent state, which is called a *configuration*, has the structure of a multiset made up of objects and messages that evolves by concurrent rewriting (modulo the multiset structural axioms of associativity, commutativity, and identity) using rules that describe the effects of *communication events* between some objects and messages. The rewrite rules in the module specify in a declarative way the behavior associated with the messages. The general form of such rules is

$$\begin{array}{l} M_1 \ldots M_n \ \langle O_1 : F_1 \mid atts_1\rangle \ldots \langle O_m : F_m \mid atts_m\rangle \ \longrightarrow \\ \langle O_{i_1} : F'_{i_1} \mid atts'_{i_1}\rangle \ldots \langle O_{i_k} : F'_{i_k} \mid atts'_{i_k}\rangle \\ \quad \langle Q_1 : D_1 \mid atts''_1\rangle \ldots \langle Q_p : D_p \mid atts''_p\rangle \ M'_1 \ldots M'_q \quad \textit{if Cond} \end{array}$$

where $k, p, q \geq 0$, the M_s are message expressions, $i_1, \ldots, i_k$ are different numbers among the original $1, \ldots, m$, and *Cond* is a rule condition. The result of applying a rewrite rule is that the messages $M_1, \ldots, M_n$ disappear; the state and possibly the class of the objects $O_{i_1}, \ldots, O_{i_k}$ may change; all the other objects O_j vanish; new objects $Q_1, \ldots, Q_p$ are created; and new messages $M'_1, \ldots, M'_q$ are sent.

2.2 Maude's Strategy Language

Rewrite rules in rewriting logic need to be neither confluent nor terminating. This theoretical generality requires some control when the specifications become executable, because it must be ensured that the rewriting process does not go in undesired directions and eventually terminates. Maude's strategy language can be used to control how rules are applied to rewrite a term [2]. Strategies are defined in a separate module and are run from the prompt through special commands.

The simplest strategies are the constants `idle` and `fail`, which always succeeds and fails, respectively. The basic strategies consist of the application of a rule to a given term, and with the possibility of providing a substitution for the variables in the rule. In this case a rule is applied *anywhere* in the term where it matches satisfying its condition. When the rule being applied is a conditional rule with rewrites in the conditions, the strategy language allows to control how the rewrite conditions are solved by means of strategies. An operation `top` restricts the application of a rule just to the *top* of the term. Basic strategies are then combined so that strategies are applied to execution paths. Some strategy combinators are the typical regular expression constructions: concatenation (`;`), union (`|`), and iteration (`*` for 0 or more iterations, `+` for 1 or more, and `!` for a 'repeat until the end' iteration). Another strategy combinator is a typical 'if-then-else', but generalized so that the first argument is also a strategy. The language also provides a `matchrew` combinator that allows a term to be split in subterms, and specifies how these subterms have to be rewritten. Recursion is also possible by giving a name to a strategy expression and using this name in the strategy expression itself or in other related strategies.

For our implementation, the full expressive power of the strategy language will not be needed and all our strategies will be expressed as combinations of the application of certain rules (possibly instantiated), concatenation, and 'repeat until the end' iteration. For efficiency reasons, we have extended the previous strategy language with a new combinator `one`(S) which, when applied to a term t, returns one of the possible solutions of applying the strategy S to t.

3 Multilayer Perceptrons

A *neural network* is defined in mathematical terms as a graph with the following properties: (1) each node or *neuron* i is associated with a state variable x_i storing its current output; (2) each junction between two neurons i and k, called *synapse* or *link*, is associated with a real weight ω_{ik}; (3) a real *activation threshold* θ_i is associated with each neuron i; (4) a *transfer function* $f_i(y_k, \omega_{ik}, \theta_i)$ is defined for each neuron, and determines the activation degree of the neuron as a function of its threshold, the weights of the input junctions and the outputs y_k of the neurons connected to its input synapses.

Multilayer perceptrons are networks with one or more layers of nodes between the layer of input units and the layer of output nodes. These hidden layers contain neurons which obtain their input from the previous layer and output their results to the next layer, to both of which they are fully-connected. Nodes within each layer are not connected and have the same transfer function. In our case, the transfer function has the form $f(\sum_k \omega_{ik} y_k - \theta_i)$, where $f(x)$ is a sigmoidal function. It is defined by $f(x) = 1/(1 + e^{(\nu - x)})$, which corresponds to a continuous and derivable generalization of the step function.

3.1 The Backpropagation Algorithm

The accuracy of the multilayer perceptron depends basically on the correct weights between nodes. The backpropagation training algorithm is an algorithm for adjusting those weights, which uses a gradient descent method to minimize the mean quadratic error between the actual outputs of the perceptron and the desired outputs.

Let x_{ij}^k and y_{ij}^k be the input and output, respectively, for the i pattern of node j of layer k. Let ω_{ij}^k be the weight of the connection of neuron j of layer k with neuron i of the previous layer. By definition of the perceptron by layers, the following relationships are fulfilled $x_{ij}^k = \sum_l \omega_{lj}^k y_{il}^{k-1}$; $y_{ij}^k = f(x_{ij}^k)$.

The mean quadratic error function between the real output of the perceptron and the desired output, for a particular pattern i, is defined as $E_i = \frac{1}{2}\sum_{j,k}(y_{ij}^k - d_{ij}^k)^2$, where d_{ij}^k is the desired output for pattern i of node j of layer k. In order to minimize the error function we use the descending gradient function, considering the error function E_p and the weight sequence $\omega_{ij}^k(t)$, started randomly at time $t = 0$, and adapted to successive discrete time intervals. We then have $\omega_{ij}^k(t+1) = \omega_{ij}^k(t) - \eta\, \partial E_l / \partial \omega_{ij}^k(t)$, where η is the so-called *learning rate constant.*

We can conclude that $w_{ij}(t+1) = w_{ij}(t) + \eta \delta_j x'_i$, where x'_i is the output of neuron i, and δ_j is an error term for node j. For output neurons, it must be $\delta_j = y_j(1 - y_j)(d_j - y_j)$. For a hidden node j, $\delta_j = x'_j(1 - x'_j)\sum_k \delta_k w_{jk}$, where k ranges over all neurons in the layers above neuron j. Internal node thresholds are adapted in a similar manner.

4 Implementing the Multilayer Perceptron

In this section we focus on specifying a three-layer perceptron in Maude and designing a strategy for evaluation and training. In order to have a running net we need to specify the number of layers, the neurons in each of them, the weights of all links, and the input patterns which, in general, will be multiple. Whereas the object-oriented representation is very convenient for specifying their behavior, it is clear that introducing all these data directly in this form would be very cumbersome. Hence, we have decided to use matrices and vectors of values to specify thresholds and weights, and to define equations and rules to transform them into the object-oriented representation.

The core of our representation of perceptrons in Maude revolves around the definition of two classes to represent neurons and synapses as individual objects:

```
class Neuron | x : Float, t : Float, st : Nat .
class Link | w : Float, st : Nat .
```

Each neuron object carries its current activation value x, depending on its threshold t, and an attribute st that will be used to determine whether the neuron has already fired or not, that is, whether it is still waiting for input or has already output a value. Similarly, link objects store their numerical weight and contain an attribute st to flag whether some value has already passed through them or not. A net then is a "soup" (a multiset) of neurons and links.

Neurons and links are identified by a name. We define two operations that take natural numbers as arguments and return an object identifier: for neurons, the numbers correspond to the layer and the position within the layer; for links, the numbers correspond to the output layer and the respective positions within each layer of the neurons connected.

```
op neuron : Nat Nat -> Oid .    op link : Nat Nat Nat -> Oid .
```

The evaluation of the network is essentially performed by repeated application of the rules feedForward and sygmoid. Rule feedForward calculates the weighted sum of the inputs to the neuron, whereas sygmoid just applies the sigmoidal function syg (defined somewhere else in the code) to the net input. As can be seen in feedForward, the attribute st of links is assumed to be 0 prior to their firing and becomes 1 once the information has been sent from one neuron to the other. Hence, pending some kind of reset, links can only be used once. Similarly, the rule sygmoid sets the attribute st of a neuron to 1 once the sigmoidal function has been applied.

```
rl [feedForward] : < neuron( L, I) : Neuron | x : X1 , st : 1 >
   <link(s L,I,J) : Link | w:W, st:0> <neuron(s L,J) : Neuron | x:X2, st:0>
=> <neuron(L,I) : Neuron | x:X1, st:1> <link(s L,I,J) : Link | w:W, st:1>
   < neuron(s L, J) : Neuron | x : (X2 + (X1 * W)) , st : 0 > .
  rl [sygmoid] :   < neuron(L, I) : Neuron |  x : X , t : T , st : 0 >
  =>    < neuron(L, I) : Neuron |  x : syg(_-_(X, T), L) ,  st : 1 > .
```

Evaluation of a perceptron starts by obtaining an input pattern through the rule nextPattern, which is guided by the message netStatus. A message of the form netStatus(N0, 0, 0, N1) means that the s N0-th pattern should be considered, and then the following patterns until the N1-th.

```
msg netStatus : Nat Nat Nat Nat -> Msg .
crl [nextPattern] :   netStatus(N, N1, N2, N0)  =>
netStatus(s N, N1, N2, N0) inPatternConversion(s N, inputPattern(s N), 0)
outPatternConversion(s N, outputPattern(s N), 0)   if  N < N0 .
```

Before starting the feedforward process, the values of the neurons in the input layer and the corresponding weights are reset. After that, the rule `introducePattern` inserts the input pattern in the neurons of the input layer and removes them from the configuration.

```
rl [introducePattern] : < neuron(0, I) : Neuron | x : X , st : 0 >
inputPattern(N, I, X0) =>  < neuron(0, I) : Neuron | x : X0 , st : 1 > .
```

Once we are done with the evaluation of all patterns, we compute the error and mark the current object `net(N)` as completed.

```
rl [computeError] :    < net(N0) : Net | e : E , st : 0 >
< neuron(2, I) : Neuron | x : X0 , st : 1 >  outputPattern(N, I, X1, 0)
=> < net(N0) : Net | e : (E + ((_-_(X1, X0)) * (_-_(X1, X0)))), st : 0 >
< neuron(2, I) : Neuron | x : X0, st : 1 >  outputPattern(N, I, X1, 1) .
```

4.1 Backpropagation in Maude

For training the net we need neurons and links to hold additional information, namely the error terms δ_j and the adjusted weights $\omega_{ij}^k(t+1)$. Since evaluation is part of backpropagation, we define `NeuronTR` and `LinkTR` as *subclasses* of `Neuron` and `Link` with an additional attribute to store the extra information.

```
class LinkTR   | w+1 : Float .   subclass LinkTR < Link .
class NeuronTR | dt : Float .    subclass NeuronTR < Neuron .
```

Note that the rules for evaluating a net also apply to these new objects; the new attributes are simply ignored. The next step demands the evaluation of the error terms before adjusting the weights. The calculation of these δ_j depends on whether we are working with the output or hidden layers. For the output layer, the corresponding rule is straightforward:

```
rl [delta2] :  outputPattern(N, I, D, 1)
< neuron(2, I) : NeuronTR | x : X , dt : DT , st : 2 >
=>  < neuron(2, I) : NeuronTR | x : X ,
        dt : (X * ((_-_(1.0, X)) * (_-_(D, X))) ) , st : 3 > .
```

The case for the remaining layers is a bit more involved and is split in three phases: the rule `delta1A` initializes `dt` to zero, `delta1B` below takes care of calculating the sum of the weights multiplied by the corresponding error term, and `delta1C` computes the final product. Again, in all these rules the status attribute `st` is correspondingly updated.

```
rl [delta1B] :  < neuron(1, J) : Neuron | dt : DT1, st : 2 >
< link(2, J, K):Link | w:W, st:2 > < neuron(2, K):Neuron | dt:DT2, st:2 >
=>  < neuron(1, J) : Neuron | dt : (DT1 + (DT2 * W)), st : 2 >
< link(2, J, K):Link | w:W, st:3 > < neuron(2, K):Neuron | dt:DT2, st:2 > .
```

Once the error terms are available, the updated weights can be calculated: rule `link1` does it for the hidden layer and `link2` for the output layer. Finally the old weights are replaced by the adjusted ones with the rule `switchLink`. Here we show rule `link1`:

```
rl [link1] :  < neuron(0, I) : Neuron | x : X1, st : 1 >
   < link(1, I, J) : Link   | w : W, w+1 : W1, st : 1 >
   < neuron(1, J) : Neuron | dt : DT, st : 3 >
=> < neuron(0, I) : Neuron | x : X1, st : 1  >
   < link(1, I, J) : Link | w : W, w+1 : (W + (eta * (DT * X1))), st : 3 >
   < neuron(1, J) : Neuron | dt : DT, st : 3 > .
```

4.2 Running the Perceptron: Evaluation and Training

Our specification is nondeterministic and not all of its possible executions may correspond to valid behaviors of a perceptron. Hence, in order to be able to use the specification to simulate the evaluation of patterns we need to control the order of application of the different rules by means of strategies.

The main strategy `feedForwardStrat` takes a natural number as argument and applies to a `Configuration` (that is, a perceptron), chooses a layer `L'` and applies rule `feedForward`, at random positions and as long as it is enabled, to compute the weighted sum of values associated to each neuron at the layer. When all sums have been calculated, it applies the sigmoidal function to all of them by means of rule `sygmoid` which, again, is applied at random positions and as long as it is enabled.

```
strat feedForwardStrat  : Nat @ Configuration .
sd feedForwardStrat(L')  := one(feedForward[L<-L'])!; one(sygmoid[L<-s L'])!.
```

There are two auxiliary strategies. The strategy `inputPatternStrat` takes care of making the successive patterns available and of resetting the appropriate attributes of the neurons and links, whereas `computeOutput` is invoked to compute the error once a pattern has been evaluated.

```
strat inputPatternStrat  : @ Configuration .
sd inputPatternStrat  :=
 one(resetNeuron) ! ; one(resetLink) ! ; one(nextPattern)   .
strat computeOutput  : @ Configuration .
sd computeOutput  :=  one(computeError) ! ; setNet   .
```

Last, all these previous strategies are combined into the evaluation strategy, which inputs the next pattern, computes the values of the neurons in the hidden and the output layers, and returns the error:

```
strat evaluateANN  : @ Configuration  .
sd evaluateANN  :=  inputPatternStrat  ; feedForwardStrat(0)  ;
                     feedForwardStrat(1)  ;  computeOutput   .
```

Then, to force the evaluation of the first `M` patterns by the multilayer perceptron the following command would be executed:

```
(srew ann netStatus(0, 0, 0, M) using one(evaluateANN) ! .)
```

where the input patterns would have been suitable defined and ann would be a term of the form:

```
neuronGeneration(0, input0, threshold0, 0)
neuronGeneration(1, input1, threshold1, 0) linkGeneration(1, link1, 0, 0)
neuronGeneration(2, input2, threshold2, 0) linkGeneration(2, link2, 0, 0)
```

Similarly as for evaluation, we need to define an appropriate strategy for training the perceptron. Assuming we have already calculated the output associated to a pattern, we next must calculate the error terms, use them to obtain the adjusted weights, and transfer them to the right attribute. That can be easily done by applying the rules defined in the previous section in the right order.

```
strat backpropagateANN : @ Configuration .
sd backpropagateANN :=  one(delta2) ! ; one(link2) ! ;
    one(delta1A) ! ; one(delta1B) ! ; one(delta1C) ! ;  one(link1) !  ;
    one(switchLink) !  .
```

Finally, training a net consists in evaluating a pattern with the strategy `evaluateANN` and then adjusting the weights accordingly with `backpropagateANN`.

```
strat stratANN :  @ Configuration .
sd stratANN :=  evaluateANN ; backpropagateANN  .
```

5 Example: Diagnosis of Glaucoma

For the diagnosis of glaucoma, we proposed the use of a system that employs neural networks and integrates the analysis of the nerve fibers of the retina from the study with scanning laser polarimetry (NFAII;GDx), perimetry and clinical data [3]. In that work, the resulting multilayer perceptron was developed using MatLab.

We used the data from that project as a test bed for our specification of the backpropagation algorithm in Maude. Our results were equivalent and the success rate was of 100% but the execution time of our implementation lagged far behind, which motivated us to optimize our code. Since equations are executed much faster than rules by Maude and, in addition, do not give rise to branching but linear computations, easily handled by strategies, we simplified rules as much as possible. The technique used was the same in all cases and is illustrated here with the rule `feedForward`:

```
rl [feedForward] : C => feedForward(C) .
op feedForward : Configuration -> Configuration .
eq feedForward(C < neuron(  L, I) : Neuron | x : X1 , st : 1 >
        < link(s L, I, J) : Link  | w : W , st : 0 >
        < neuron(s L, J) : Neuron | x : X2 , st : 0 >)
=  feedForward(C < neuron(  L, I) : Neuron | >
        < link(s L, I, J) : Link  | w : W , st : 1 >
        < neuron(s L, J) : Neuron | x : (X2 + (X1 * W)) >) .
eq feedForward(C) = C [owise] .
```

The evaluation and training strategies had to be correspondingly modified since the combinator `!` was no longer needed. The resulting specification is obviously less natural, but more efficient.

6 Conclusions

We have presented a specification of multilayer perceptrons, in a two step fashion. First we have shown how to use rewrite rules guided by strategies to simulate the evaluation of patterns by a perceptron, and then we have enhanced the specification to make the training of the net possible. The evaluation process is straightforward, essentially amounting to the repeated application of two rules, `feedForward` and `sygmoid`, which further does credit to the suitability of rewriting logic as a framework for concurrency. The training algorithm requires more rules, but the strategy is also rather simple.

The simplicity of the resulting specification should be put in perspective. Our first attempts at specifying perceptrons made use of a vector representation like the one we have used here for inputting the data and similar to that proposed in [7]. Such representation was actually suitable for the evaluation of patterns but proved unmanageable when considering the training algorithm. The election of our concrete representation in which neurons and links are individual entities and which, at first sight, might not strike as the most appropriate, is of paramount importance.

In addition to the representation, availability of strategies turned out to be decisive. With the vector representation layers could be considered as a whole and there was no much room for nondeterminism, while the change to the object-oriented representation gave rise, as we have observed, to the possible interleaving of rules in an incorrect order. It then became essential the use of the strategy language to guide the rewriting process in the right direction.

As a result, our specification is the happy crossbreed of an object-oriented representation and the use of strategies: without the first the resulting specification would have been much more obscure, whereas without the availability of the strategy language, its interest would have been purely theoretical.

In addition to the novel application of rewriting logict to neural nets, the advantage provided by our approach lies on the allowing the subsequent use of the many tools developed in rewriting logic, such as the LTL model-checker or the inductive theorem prover [1, 5], to study the net.

The complete Maude code, the data used for the examples, and the results of the evaluation can be downloaded from `http://maude.sip.ucm.es/~miguelpt/`.

References

1. Clavel, M., Durán, F., Eker, S., Lincoln, P., Martí-Oliet, N., Meseguer, J., Talcott, C. (eds.): All About Maude - A High-Performance Logical Framework. LNCS, vol. 4350. Springer, Heidelberg (2007)
2. Eker, S., Martí-Oliet, N., Meseguer, J., Verdejo, A.: Deduction, strategies, and rewriting. In: Archer, M., de la Tour, T.B., Muñoz, C.A. (eds.) 6th International Workshop on Strategies in Automated Deduction, STRATEGIES 2006, Part of FLOC 2006, Seattle, Washington, August 16. Electronic Notes in Theoretical Computer Science, vol. 174(11), pp. 3–25. Elsevier, Amsterdam (2007)

3. Hernández Galilea, E., Santos-García, G., Franco Suárez-Bárcena, I.: Identification of glaucoma stages with artificial neural networks using retinal nerve fibre layer analysis and visual field parameters. In: Corchado, E., Corchado, J.M., Abraham, A. (eds.) Innovations in Hybrid Intelligent Systems, Advances in Soft Computing, pp. 418–424. Springer, Heidelberg (2007)
4. Lippman, R.P.: An introduction to computing with neural nets. IEEE ASSP Magazine, 4–22 (1987)
5. Martí-Oliet, N., Meseguer, J.: Rewriting logic as a logical and semantic framework. In: Gabbay, D. (ed.) Handbook of Philosophical Logic, 2nd edn., vol. 9, pp. 1–81. Kluwer Academic Press, Dordrecht (2002)
6. Martí-Oliet, N., Meseguer, J.: Rewriting logic: Roadmap and bibliography. Theoretical Computer Science 285(2), 121–154 (2002)
7. Meseguer, J.: Research directions in rewriting logic. In: Berger, U., Schwichtenberg, H. (eds.) Computational Logic: Marktoberdorf, Germany, July 29 – August 6, vol. 165, pp. 347–398. NATO Advanced Study Institute (1997)
8. Meseguer, J.: Conditional rewriting logic as a unified model of concurrency. Theoretical Computer Science 96(1), 73–155 (1992)
9. Talcott, C.L.: An actor rewriting theory. In: Meseguer, J. (ed.) Workshop on Rewriting Logic and its Applications, WRLA 1996. Electronic Notes in Theoretical Computer Science, vol. 4, pp. 360–383. Elsevier, Amsterdam (1996)

Integrated Approach of ANN and GA for Document Categorization

Karina Leyto-Delgado, Ivan Lopez-Arevalo, and Victor Sosa-Sosa

Laboratory of Information Technology, Cinvestav-Tamaulipas, Tamaulipas, México
{bleyto,ilopez,vjsosa}@tamps.cinvestav.mx

Summary. Nowadays the number of documents in any field, place or space is enormously growing. With these dimensions of data and documents stored or circulating on the web is desirable to have an appropriate way to categorize text documents according to their content. This need has motivated the implementation of several computational techniques whose main goal is the automatization of this process. This paper describes the integration of a SOM neural network and genetic algorithm for categorization of text documents.

Keywords: Document Categorization, Genetic Algorithm, SOM Neural Network.

1 Introduction

The large number of documents handled into organizations, institutions or on the web is growing. This situation is forcing to find new ways to organize this information. The document categorization is part of Information Retrieval, which according to Karypis and Zao [1] can be defined as:

Given a set S of N documents, we would like to partition $S_1, S_2 \cdots S_n$ *such that the documents assigned to each subset are more similar to each other than the documents assigned to different subsets.*

Several authors have tackled this problem by using different techniques [2] [3] [4]. The K-NN was used in the study by Yang [5] with the document corpus named Reuters 21578. The study covered 90 categories, with 7769 training documents and 3019 test documents. She used a vocabulary of 24240 words. Performance evaluations were carried out with the measures commonly used in the document categorization[1].

Venegas [7] showed the applicability of SVM to the categorization of academic documents. This study was done with a corpus of 222 documents. The vocabulary consisted of 2729 words. In the training phase 80% of the examples and a 20%

[1] *Recall(r), Precision (p),* F_1*, Micro-averaging, Macro-averaging* are measure the performance for the text categorization [6].

J.M. Corchado et al. (Eds.): DCAI 2008, ASC 50, pp. 434–442, 2009.
springerlink.com

for the testing phase were used. Four possible categories also were established a priori per each document. Similarly to the SVM experiment Venegas [7] applied a Bayes Naïve Classifier to the categorization of academic documents. This study was carried out with the same conditions that the previous technique, but he used a document frequency thresholding and information gain to reduce the matrix of terms.

Another used technique is neural networks, specifically those with competitive learning paradigm ART [9] and SOM [8]. Liu et.al.[10] used six document corpus for their categorization. These corpus have 160 documents with a different vocabulary for each corpus. They used a vector space model for each document. The number of output layer neurons was equal to the number of classes of input documents. The learning-rate was more important in this study. The study was measured by means of the F_1 measure. The best results were obtained with learning-rate varying between 0.6 and 0.7. The main goal was taking into account the semantic content of each document.

Genetic algorithms also have been used for document categorization. Liu et.al. [11] used a corpus of 230 texts. These included 3 possible categories. Each text included approximately 150 terms. The authors used the vector space model with inverse document frequency to determine the relative importance. Thus, they considered the chosen frequency term as candidates and then use genetic algorithm to re-evaluate their importance. The initial population was of 1000. The algorithm converged with 60 generations. This classifier achieved more than 80% of documents categorized.

This work presents two document categorization methods which combine a SOM neural network and a genetic algorithm. Only documents in english language are considered. The rest of the paper is organized as follows, in section 2 the proposed approach is described, where the technique for dimensionality reduction of vectors of terms is presented. It also describes the two approaches used for the static and dynamic methods of categorizations. Next, sections 3 and 4 present the obtained results and future work. Finally, some conclusions are presented.

2 Approach

This paper describes two interrelated methods for document categorization. The first one is a static method to categorize fixed corpus. The next one is a dynamic method to categorize documents that are not in the fixed corpus. This work uses the Reuters 21578 [12] document collection. This collection has been broadly used in previous works [2] [5] [10] . The hybrid approach considers account a comprehensive vocabulary for the collection, with a vector representation [13] of each document with inverse frequency for each term. Also the collection is reduced by means of a *stoplist* (no relevant words: propositions, articles, etc.). Subsequently, an overall reduction for all vectors is carried out to obtain a vocabulary for all documents. To reduce this vocabulary a subset of the corpus with words with greater frequencies is selected. After this, the vocabulary is used to represent each document in the static and dynamic methods.

The static method uses a genetic algorithm to categorize complete corpus of documents in any of the categories established a *priori.* A SOM neural network provides such categories to genetic algorithm. The SOM neural network obtains the number of categories of the 10% of the complete corpus [14] [15]. The dynamic method requires the performance of the genetic algorithm in the static method to refine the features in input and output neurons of the SOM. Then, the SOM neural network can categorize new documents (not in the original corpus). The performance of the dynamic method can only happen after the execution of the static method. The following sections describe the process.

2.1 Global Dimension Reduction (GDR)

The Global Dimension Reduction is performed to reduce the dimension of the vocabulary obtained from the initial corpus. This technique involves selecting a subset of documents representative of the corpus. This subset includes a 10% of document of the overall corpus. A third part of this subset will be represented by the largest documents of the corpus, the second third by documents with medium size and finally smaller documents. After this, from the subset of documents the terms of each document whose frequency is greater than the mean are selected. The repeated terms are eliminated and their frequencies are added. This creates a smaller vector for each document. This vector will be joined together to others to create a reduced vocabulary (GDR) (see Figure 1). Thus, the GDR is a bag of terms which gives a general idea of the most common terms in the overall corpus. The structure of the GDR is used both in the static and dynamic methods.

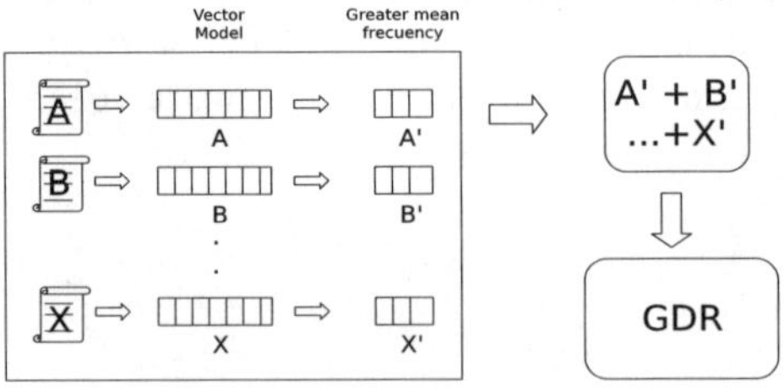

Fig. 1. Global Dimension Reduction

2.2 Neural Network Approach

This approach uses a SOM neural network. One of its main features is the mapping of input vector towards its output neurons with no supervision, given as output winner neurons (by means of competitive learning) that represent the categories of the input layer.

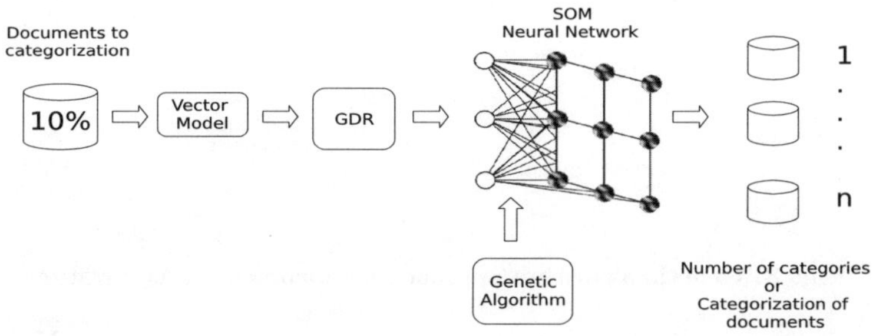

Fig. 2. Neural network approach for document categorization

As can be seen in Figure 2, a vector representation for each document is used. The input layer neurons of the SOM neural network have the same order of the bag of terms of GDR. Each document of the reduced corpus is represented by a vector of terms indicating the frequency of occurrence of every word in the vocabulary of such document. The input neurons receive the frequency of terms of each document according to the order of terms in the GDR. The number of output neurons is the same number of input neurons. This neural network internally uses a genetic algorithm to optimize the weights of its conexions. These weights will be used in further categorizations. This method is used to obtain the number of categories and in a subsequent manner, the categorization of new documents.

2.3 Genetic Algorithm Approach

A genetic algorithm has as one of its main features that it can explores a complete search space. In our approach, a genetic algorithm is used for the categorization of the complete corpus. This approach (see Figure 3) achieves document categorization base on a given set of documents (the complete corpus) and a number of defined categories (output of SOM).

$$cos(d_1, d_2) = \frac{d_1 \bullet d_2}{\parallel d_1 \parallel * \parallel d_2 \parallel} \tag{1}$$

$$similarityofgroup_j = \frac{1}{{n_j}^2}\left(\sum_{j=1,n=1}^{n_j} cos(d_i, d_j)\right) \tag{2}$$

$$averagesimilarity = \frac{\sum_{j=i}^{k} n_j * similarityofgroup_j}{N} \tag{3}$$

The genetic algorithm uses a group representation, the best individuals will be selected by means of the binary tournament. Finally, the operators crossover and mutation of the new population will be applied. The fitness is the average similarity of each group (Equation *2*), this measure uses the cosine similarity (Equation *1*) and the vector representation of pairs of documents in the group.

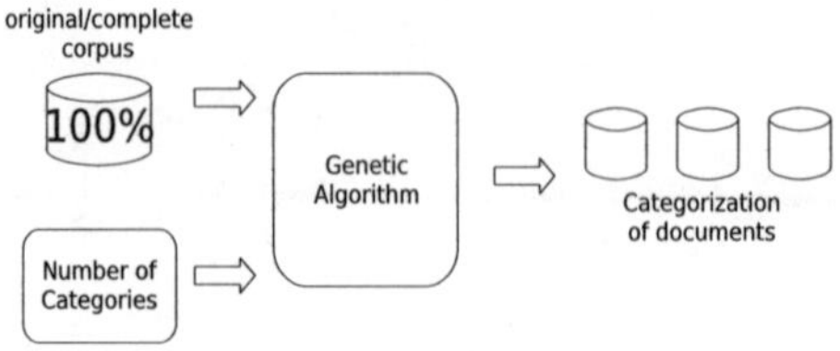

Fig. 3. Genetic algorithm approach to document categorization

The Equation *1* represents the separation between vector d_1 and d_2, where $d1$ and d_2 represent the vector model of each document to categorize. Then the average similarity of the overall group is obtained by means of Equation *3*.

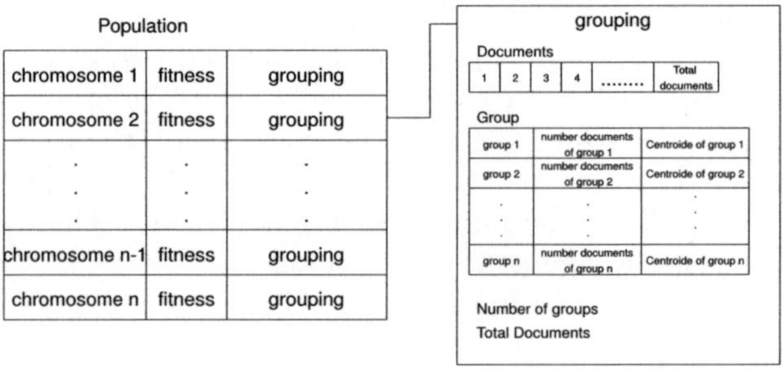

Fig. 4. Representation of one individual in the genetic algorithm

The chromosome structure can be viewed in Figure 4. The grouping column is the group representation of the groups of the chromosome. This structure has a vector of documents. This vector indicates the group of each document. The structure of the vector in group column indicates the number of group, the number of documents in the group, and the centroid of the group.

The SOM neural network and the genetic algorithm are combined within the static and dynamic methods proposed in this work. The characteristics of these two methods are described in the following section.

2.4 Static Method

The static method (Figure 5) categorizes a subset of documents that represents 10% of the complete corpus with the SOM neural network. The SOM obtains automatically the number of categories in which each classified document corresponds. This happens since the genetic algorithm requires this number to make the crossover and mutation operations. The idea is that initially each document corresponds to an aleatory assigned category. Such category is adjusted according to the evolution of chromosomes. Unlike to SOM, the genetic algorithm uses the complete corpus to obtain the category of each document in the overall corpus.

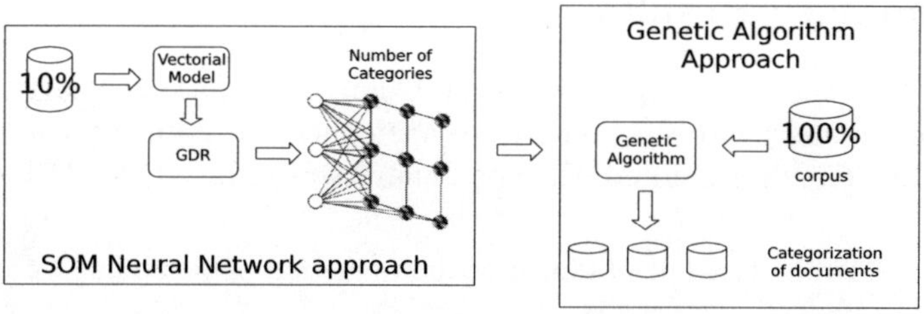

Fig. 5. Static method for document categorization

The particularity of the static method is that the original corpus can not increase, because this requires the complete execution of the overall method. The genetic algorithm can not be executed for a small number of documents. However, due to the SOM neural network nature, it can categorize new individual documents, which is carried out by the dynamic method.

2.5 Dynamic Method

As its principal task this method has to categorize documents that are not in the original corpus used in the static method. This method centralizes its role in the SOM neural network used in accordance with the genetic algorithm of the static method. Input neurons in the SOM depend on the features obtained from each of the groups resulting from the static method, the original GDR can be reduced. The terms of this GDR define the new order of the input neurons in the SOM. At this point, the result of the dynamic method is the categorization of new documents, not the number of categories. The performance of the genetic algorithm and the SOM is enhanced by means of several iterations for a same corpus. This is due to the aleatory component of weights values for the SOM and the individuals of the initial population in the genetic algorithm.

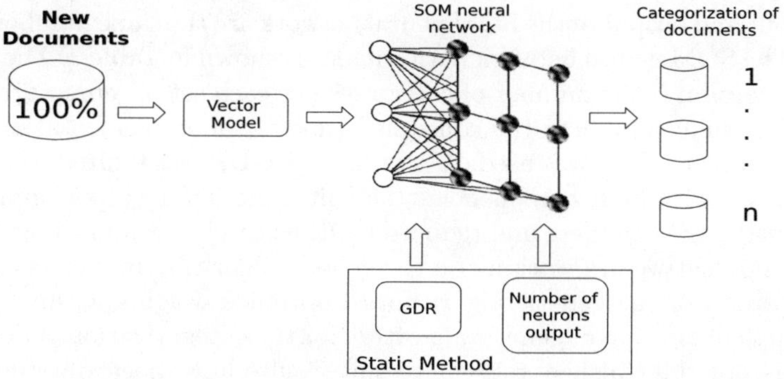

Fig. 6. Dynamic method for document categorization

3 Preliminary Results

At the moment, the SOM implementation (C language) has been carried out. The vector space model has been taken from the collection 20_newsgroups with the Bow tool [16]. The dimension of the vocabulary was reduced, Table 1 shows this.

Table 1. GDR obtained in the reduction of vocabulary

Complete corpus	10% of corpus	Complete vocabulary	GDR
200	20	1771	898
500	47	2584	1225
1000	100	3962	1713
5000	500	8545	3676

In Table 1, the complete corpus column is the total of documents in the corpus. The next column is a set of the representative documents. The complete vocabulary column contains the terms of the representative documents. The GDR column contains the total of common terms of the representative documents.

Table 2. Number of categeories obtained with the complete and limited GDR

	Complete GDR				Limited GDR	
10 % of Corpus	GDR	Obtain categories	Time (in second)	GDR	Obtain categories	Time (in second)
20	898	15	1.34	89	15	0.34
47	1225	28	13.75	122	25	1.45
100	1713	58	95.38	171	52	21.11
500	3676	308	1479.15	368	290	591.66

The input and output of the SOM neural network are the same number of terms of GDR. The SOM neural network performance is shown in Table 2. The obtained categories column is the number of categories per each set of representative documents. The time column is the consumed time of the implementation. At this point the complete GDR was used. After that, the GDR was limited to 10% of the words [14] [15] with higher frequencies. The differences in results is minimal (see Table 2), using the complete and reduced GDR as can be seen in Figure7.

At the moment we are working on the genetic algorithm implementation, the implementation of the genetic algorithm to optimize weights within the SOM, and the implementation of genetic algorithm for the categorization of documents. The last is not full finished yet. Successful results have been obtained with a population of 100 individuals and 250 generations. The genetic algorithm selects the best individuals in each generation with the fitness function (Equation *3*) for apply the crossover and mutation operators.

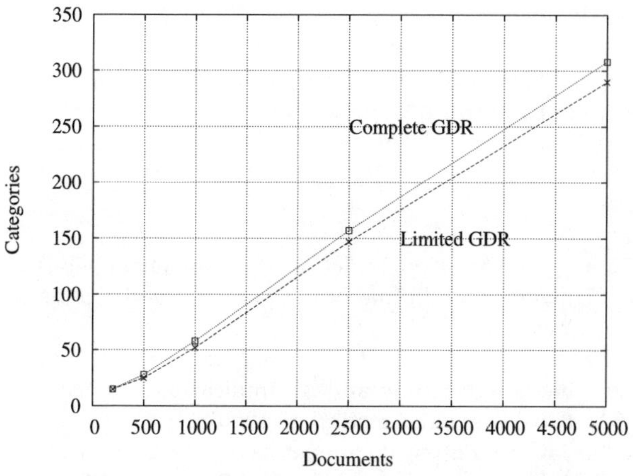

Fig. 7. Runtime performance using complete and limited GDR

4 Further Work

In a short term, we plan to do the following work:

1. The complete implementation of the genetic algorithm.
2. The integration of the SOM neural network and genetic algorithm approaches as an isolate module of software.
3. The categorization of documents will be carried out by means of the K-Means and Hierarchical algorithms for contrast.
4. The results of the K-Means and Hierarchical implementations will be analyzed and evaluated to identify, analyze, and evaluate differences respect to the approach here presented.
5. The integrated approach will be evaluated according to recall, precision, F1, micro-averaging, and macro-averaging measures.

5 Conclusions

The categorization of documents has been tackled by means of several techniques. Neural network and genetic algorithms are not the exception. The interest of this work is to integrate both techniques and to explore areas not covered in the application of these techniques in an isolate manner. The purpose of this work is to demonstrate the applicability of the neural network and genetic algorithm separately and then together to categorize electronic text documents. The implementation is not finished yet. Advantages of both approaches will be integrated to produce better results. The expected results will be a successful categorization of the documents. This categorization could be used in subsequent activities in *Information Retrieval.*

Acknowledgment

This research was partially funded by project number 51623 from "Fondo Mixto Conacyt-Gobierno del Estado de Tamaulipas".

References

1. George, K., Ying, Z.: Criterion Functions for Document Clustering: Experiment and Analysis. University of Minnesota, Department of Computer Science/Army HPC Research Center, pp. 1–40 (2002)
2. Hao, P.-Y., Tu, Y.-K., Chiang, H.: Hierarchically SVM classification based on support vector clustering method and its application to document categorization, vol. 33, pp. 627–635. Elsevier, Amsterdam (2007)
3. Kim, J.-H., Choi, K.-S.: Patent document categorization based on semantic structural information. Inf. Process. and Manag. 43(5), 1210–1215 (2007)
4. Yanjun, L., Soon, C., John, H.: Text document clustering based on frequent word meaning sequences. Data & Knowledge Engineering 64(1), 381–404 (2007)
5. Yiming, Y.: A study of thresholding strategies for text categorization. In: Proceedings of SIGIR 2001, pp. 137–145 (2001)
6. Aas, K., Eikvil, L.: Text categorization: A survey. Norwegian Computing Center, Report No. 947, P.B. 114 Blindern, N-0314 Oslo (1999)
7. René, V.: Las relaciones léxico-semánticas en artículos de investigación científica en español: Una aproximación desde el análisis semántico latente. PhD Thesis, Pontificia Universidad Católica de Valparaíso, Chile (2005)
8. Valluru, R., Hayagriza, R.: C++, neural network and fuzzy logic. MIS Press, New York (1995)
9. Sthepen, G., Gail, C.: A Massively Parallel Architecture for a Self-Organizing Neural Pattern Recognition Machine. Computer vision, graphics, and image processing 37, 35–115 (1987)
10. Yuanchao, L., Xiaolong, W., Chong, W.: ConSOM: A conceptional self-organizing map model for text clustering, vol. 71, pp. 857–862. Elsevier, Amsterdam (2007)
11. Chih-Hung, L., Cheng-Che, L., Wei-Po, L.: Document Categorization by Genetic Algorithms. In: 2000 IEEE International Conference on Systems, Man, and Cybernetics, Nashville (2000)
12. Hettich, S., Bay, S.: Reuters21578 Text Categorization Collection. University of California, Department of Information and Computer Science (1999), http://kdd.ics.uci.edu/databases/reuters21578/reuters21578.html
13. Baeza-Yates, R., Ribeiro-Neto, B.: Modern Information Retrieval. ACM Press Series. Addison Wesley, New York (1999)
14. Yiming, Y.: A comparative study on feature selection in text categorization. In: Proceedings of ICML 1997, 14th International Conference on Machine Learning, Nashville,TN (1997)
15. Fabrizio, S.: Machine Learning Automated Text Categorization. ACM Computing Surveys 34(1), 1–47 (2002)
16. Andrew, M.: Bow: A toolkit for statistical language modeling,text retrieval, classification and clustering (1996), http://www.cs.cmu.edu/~mccallum/bow

Analysis of Production Systems Using the VS-Diagram

Daniel Gómez, Jesús A. Trujillo, Enrique Baeyens, and Eduardo J. Moya

División Automatización y Control de procesos
Fundación CARTIF
Parque Tecnológico de Boecillo. Parcela 205. 47151 Valladolid, Spain
dangom@cartif.es, jestru@cartif.es, enrbae@cartif.es,
edumoy@cartif.es

Abstract. This paper presents a graphical tool, the VS (Virtual Supervisor) states space diagram based on the FPM (Finite Positions Machines) framework in order to analyze a manufacturing system and make it recover from a faulty situation (for example, a deadlock situation) to a safer operations sequence. The VS diagrams are intuitive and easy to obtain and they are shown here as another alternative that can be used to analyze a production system. They can be used as a complementary tool in conjunction with other alternatives based on well-known formalisms such as the Petri nets.

Keywords: PLC, Production Systems, Deadlock, Petri Nets, FPM, VS Diagram.

1 Introduction

Nowadays, in the high competitive global market, manufacturing systems must face rapid changes in production demand. In order to face those demands, current manufacturing systems operational conditions are required to be changed in a fast and cost-effective way, [5] and [7]. In order to introduce those changes, it must be taken into account that most of the production lines is controlled by Programmable Logic Controllers (PLC). Those PLCs are programmed using some of the IEC 61131-3 standard languages. One of the most used language of them all is the one based on the relay ladder logic schema, the ladder diagram. Programs based on ladder logic diagrams grow in number of lines magnitude, the more complex the controlled system is. This contributes to the fact that making a change in the operational specifications of the program can be a difficult task involving a lot of risk in the implications among the different system components. Even if the value of a binary or analog variable is found out, it doesn't involve the idea of knowing what is really doing nor the implications of what it represents to the rest of the system components, especially in a higher abstraction level. In fact, the more complex the system is, the more difficult can be to determine its operational conditions if the program logic is analyzed exclusively. And the use of the IEC 61131-3 standard languages is limited only to control the system but not to evaluate its qualitative, reliable and performance characteristics. Therefore, in order to change some design specifications of the manufacturing systems, it is necessary to apply formal methods that must be mathematically rigorous. In this way, any modification made in the system will be able to be analyzed, verified and validated.

J.M. Corchado et al. (Eds.): DCAI 2008, ASC 50, pp. 443–451, 2009.
springerlink.com

2 Preliminary Analysis: Initial Approach

In the modification of a manufacturing system, one of the interesting points to be analyzed is if it shows what is known in the literature as *deadlock*, and, therefore, estimate the *liveness* of the process. Depending on the used formal framework, different approaches to detect and analyze those problems have been proposed. A *deadlock* is a situation where a manufacturing system or part of it remains indefinitely blocked and cannot terminate its task [3]. It is usually caused by an inappropriate allocation of resources to concurrent executing processes, [7]. Thus, *liveness* means that for every reachable state the system can evolve to, every system activity can ultimately be carried out. Hence, to analyze *deadlock* phenomena, it is necessary to focus a Discrete Event Dynamic System (DEDS) model on the interaction between processes and resources. Therefore, a system is characterized by the interaction of a group of processes which are being executed concurrently and a group of resources that must be shared among the above mentioned processes which could lead the system to a *deadlock* situation.

3 Literature Found Solutions

Among the different formal frameworks that can be found in the current literature to solve the *deadlock* problem, one of the most extensively used is the Petri nets tool. For the definition and details on Petri nets, please see [6]. The approach to solve the problem of deadlocks with Petri nets has been treated in different ways in the literature. On the one hand, in order to identify if a block is present at a Petri net, the liveness of the net is checked. That is, the net is live if, from the initial marking and whatever marking has been reached from it, it is possible to fire ultimately any transition of the net by progressing through some further firing sequence. Consequently, liveness means that for every reachable state, the model can evolve in such a way that every transition can always fire in the future, [9]. Hence, a live Petri net guarantees deadlock-free operations. On the other hand, a dead marking or total deadlock is a marking such that no transition is enabled, and translating this idea to the automated manufacturing systems domain, a dead marking represents a state where no piece can move forward, [2]. Moreover, [7], a Petri net is said to be deadlock-free if at least one transition is enabled at every reachable marking. This definition does not guarantee that a circular wait condition involving a part of the system cannot occur. In general, deadlock freeness in Petri nets definition does not prevent a partial deadlock between a subset of resources and a subset of in-process parts.

The main strategies that are used to solve the problem of the detection and analysis of a deadlock in a manufacturing system are the following ones, [7]:

-*Deadlock prevention*: Prevention methods and algorithms which prevent and guarantee the necessary conditions for deadlock requirements not to be met.
-*Deadlock detection and recovery*: Algorithms that use a monitoring mechanism for detecting the deadlock occurrence and a resolution procedure to make the system deadlock-free by terminating those processes which have generated the situation or deallocating those resources that are necessary to end the situation.
-*Deadlock avoidance*: Algorithms that keep track of the current state of the system and calculate the possible future ones in order to prevent the system from evolving to a certain state that yields a deadlock situation.

Among the different methods and algorithms which use the Petri net framework to estimate the liveness of the analyzed process, it can be found those ones based on prevention techniques that require the calculation of the reachability tree. Recent algorithms are based on the detection of siphons, [1], in order to avoid a deadlock, [6]. Actually, what is done is to add (supervisory) control places which prevent the siphons to be emptied, situation which will lead the system to a deadlock. However, todate most of the attention has been paid to make the underlying Petri net models live without questioning whether or not all of the computed control places are necessary. It is often the case that the number of control places determined by these approaches is not minimal. Reducing it in order to reduce the complexity of the controlled system and to make the algorithms faster and more efficient from the computational point of view is an important issue that is being tackled nowadays, [4]. A third type of algorithms is based on the detection of circuits (circuit is a path from a node to the same node where every transition and node appears once, except for the last one) within the Petri net. They use a kind of colored Petri nets which are based on the resources allocation dynamics where every resource can be modeled with an only place [10]. Therefore, a Petri net, in order to prevent the system from evolving in such a way that it reaches a marking where the system is blocked, demands the use of external tools and algorithms. These have to estimate if the future reachable states are going to lead the system to such a situation. Hence, the Petri net formal framework requires a set of algorithms to analyze, detect and avoid the mentioned deadlock situation. That algorithms set belongs to what is known as the Supervisory Control Theory, [8].

4 Proposed Solution

This paper presents the use of the VS (Virtual Supervisor) states space diagram. The VS states space diagram is a visual and graphical tool used by the Finite Positions Machines (FPM) formal framework, [9]. FPM is a relatively new framework that can be used to design, analyze, verify and validate Discrete Event Dynamic Systems. For a formal definition of the FPM framework, please see [9]. VS states space diagrams represent the dynamic evolution of the process or system that is being analyzed through time and can be obtained from the PLC signals or SCADA (Supervisory Control and Data Acquisition) systems, almost in real-time. Formally, VS states space diagram comprises the set of states or actions that the system can carry out. The different states, places, positions, etc. (depending on which formal framework is used) can be seen on them and, therefore, used for the detection of the previously defined problem (deadlock) without using external algorithms and tools.

5 Obtaining the VS Diagram. A Practical Example

In order to show how a VS states space diagram can be obtained, this is presented using the following particular example: The system is a raw parts lifter-classifier which could be a part of a larger manufacturing system, see Figure 1. The first step is to measure the size of the raw parts that have been transported by a roller conveyor by means of a sensor called DTAM. In this case, there are only two sizes, large piece or small piece. The sensor emits a 0 signal if the piece is small and a logical 1 if the piece is large. Next, the piece is situated on a lifter device. The operations sequence starts when the piece has been detected by a proximity sensor called DPZA. At this point, the piece is raised by means of a cylinder A that has been activated thanks to its corresponding electrovalve. Then, the piece is classified: small pieces are placed on another conveyor by means of a cylinder B or they are placed on a third conveyor by means of a cylinder C if they are large. Cylinders A and B electrovalves are monostable and cylinder C is bistable. The process operations sequence is represented by means of the GRAFCET schema that can be seen on Figure 2. Obtaining from it the corresponding PLC IEC 61131-3 standard ladder diagram control program can be done in a very straightforward way. The value of the different signals that interact and

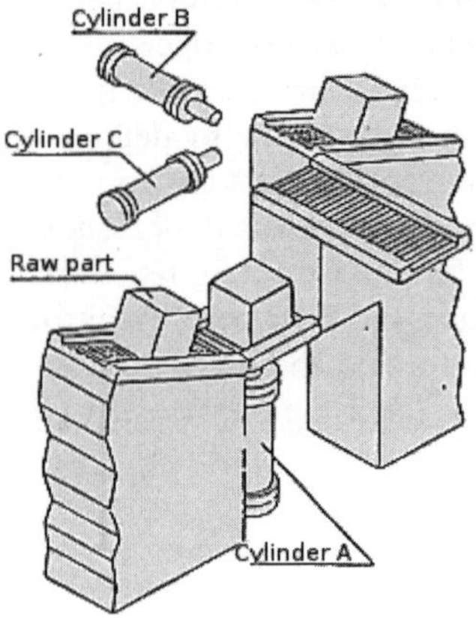

Fig. 1. Raw parts lifter-classifier

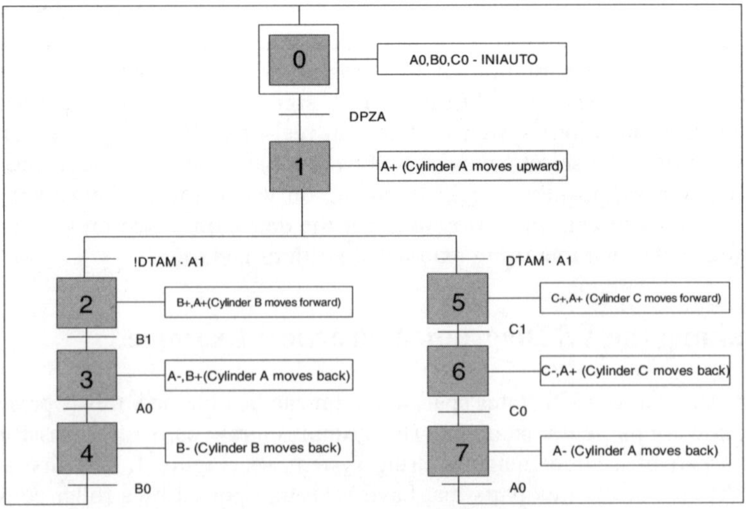

Fig. 2. Lifter-classifier GRAFCET

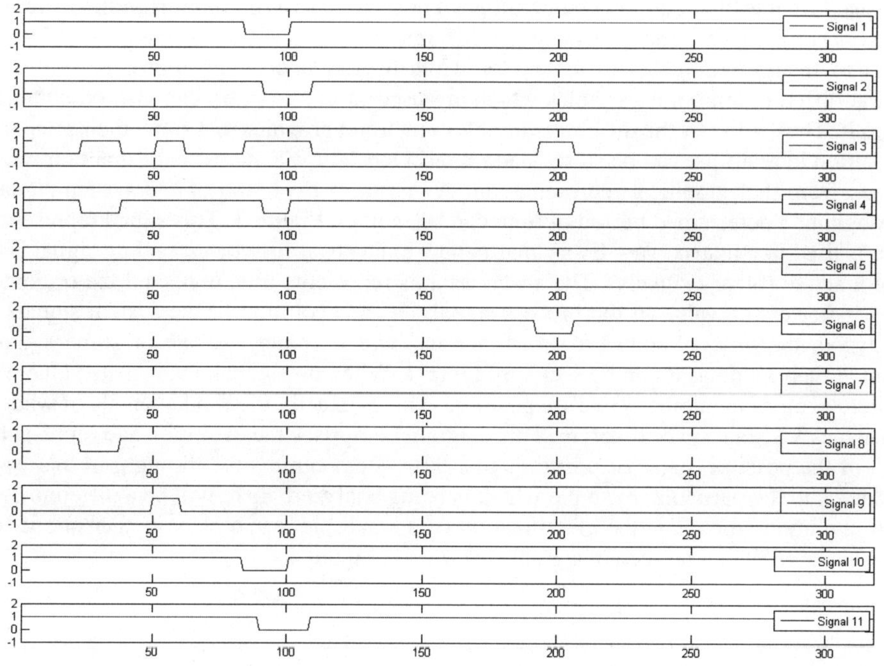

Fig. 3. PLC signals evolution throughout a period of time obtained from a SCADA system

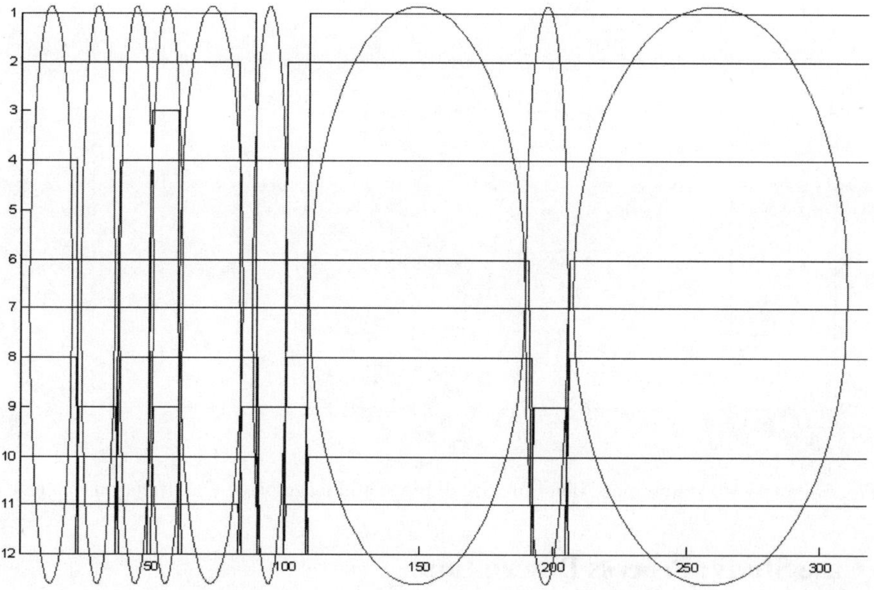

Fig. 4. Partial evolution of the system states sequence

are used by the PLC can also be obtained. Those signals will come from the different sensors (PLC inputs) and will go to the different actuators (PLC outputs).

Also, thanks to a SCADA (Supervisory Control And Data Acquisition), the evolution of the different temporal variables which are used in the program can also be obtained from it. Thus, a log of the different variables and trend graphics that show their temporal evolution in real time can be stored and represented. The VS states space diagram is obtained from that graphical representation. A graphical evolution of the system signals throughout a determined period of time can be seen on Figure 3. This signal representation is used in a matrix way. Every matrix row indicates if the corresponding signal is or is not active (binary signals). The abscissas axis represents time units and the ordinates axis indicates the value of the process signals. If the evolution of the system signals is analyzed, the process shows a set of 'actions' which change as one or some signals change their value as it can be seen on Figure 4. When one signal changes its value, the system evolves to another 'state' or 'position'. This is exactly the fundamentals of what is known as VS state space diagram. This diagram comprises the evolution and representation of the possible states or positions that the system can evolve through. In this way, and back to the particular example which is being analyzed, there will be a different system evolution when it is the classification of a small piece and also another one if the piece is large. It can be seen in a graphical way on Figure 5.

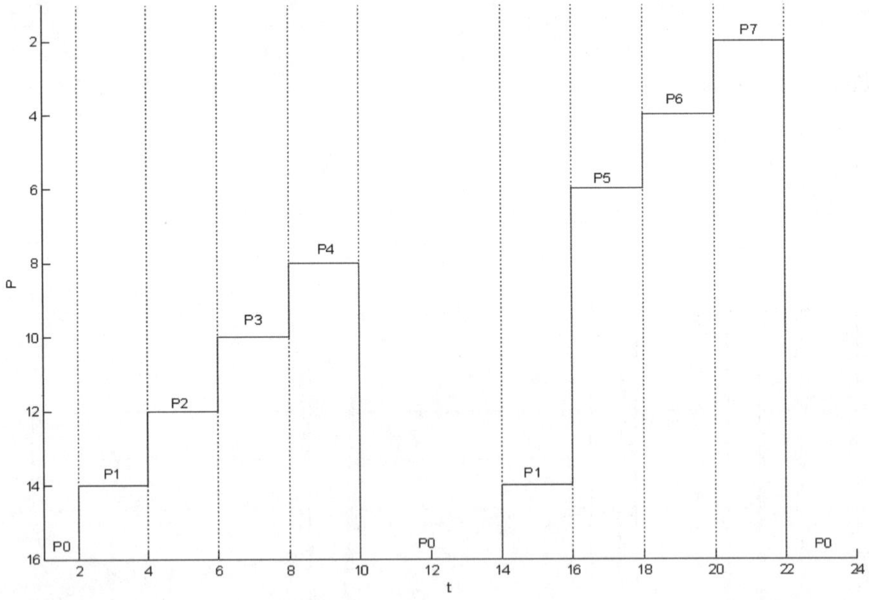

Fig. 5. Partial VS states space diagram. Small piece and large piece classification in a row.

6 Case Study: Process Description

The case study that is proposed in this paper in order to study and analyze a deadlock situation is a Flexible Manufacturing Cell which can be modeled using the Discrete

Event Systems theory and therefore, it will be able to be controlled and monitored by means of a Programmable Logic Controller (PLC).

To achieve this objective, the VS state space diagram will be used. The example is made up of a manufacturing cell composed of three robots (R1, R2, R3) where each robot can handle and move an only part or piece, four machine tools (M1, M2, M3, M4), three input buffers (I1, I2, I3) and three output buffers (O1, O2, O3)(see Figure 6).

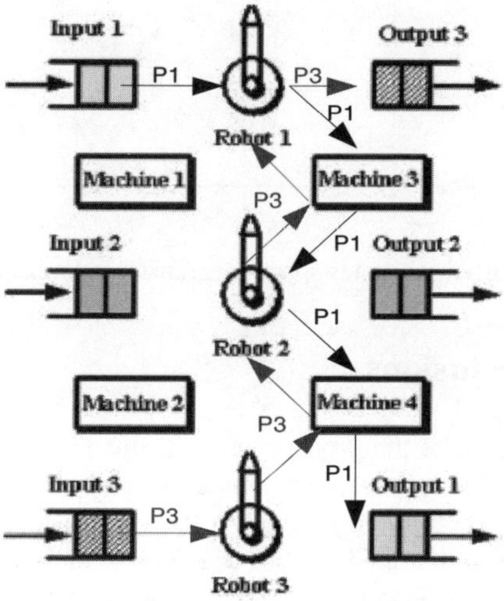

Fig. 6. Example of a manufacturing cell

7 Operational Conditions

The purpose of the cell is to process different types of raw parts or pieces in a flexible way. The processing sequences that three different types of pieces could follow can be the next ones: P1, raw part type 1, sequence: I1→R1→M3→R2→M4→R3→O1. P2, raw part type 2, sequence: I2→R2→M2→R2→O2. P3, raw part type 3, sequence: I3→R3→M4→R2→M3→R1→O3 (see Figure 6). All the machines and robots, in turn, are controlled by means of a supervisory controller made up of a PLC.

8 Deadlock Situation in the Process: Analysis Using the VS States Space Diagram

Objective: To obtain the corresponding final product from a raw part P1 and a raw part type P3. To process a P1 and a P3 piece, the system must follow the sequences described before. Separately, these will yield their corresponding VS states space diagrams. However, a deadlock situation may arise if both robots R1 and R2 have to use machine M3 at the same time. That is, if P1 piece reaches the step where it has to be mechanized by M3 (from R1) and, at the same time, a P3 piece has to use it as well (from R2). A *deadlock* has arisen. Its corresponding VS diagram can be seen on Figure 7.

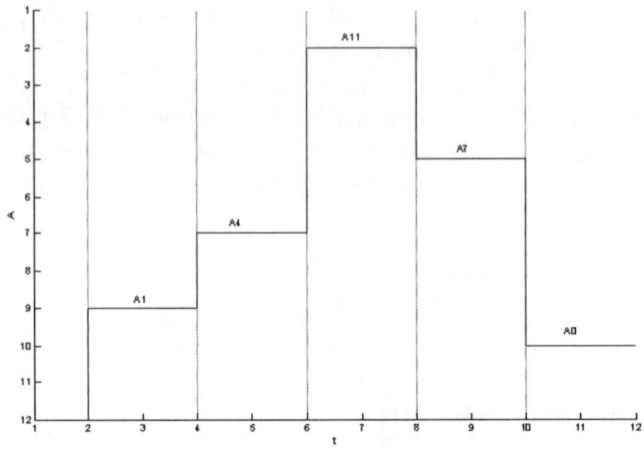

Fig. 7. Partial VS states space diagram. Deadlock situation.

9 Results and Conclusions

The VS states space diagram of the process when it runs normally and its main feature is liveness, that is, it is never led to a deadlock situation, can be considered as a pattern which can be monitored by a supervisory controller. At the very instant that the process goes off course and follows a pattern that leads it to a deadlock situation, the controller could take the proper measures while the process is still reversible in order to take it to a known state or position. From there, it will be able to resume to the normal operations sequence. Due to the fact that the VS states space diagram can also be obtained from the FPM system model, its use can guarantee a system validation in a rigorous and formal form. It can be seen that the deadlock situation is different from the beginning with respect to the actions sequences that the processes (which yield the final products) follow separately.

The VS states space diagrams are presented in this paper as a useful tool to detect a deadlock situation in a manufacturing process. This tool is used by the Finite Positions Machines (FPM) formal framework which guarantees the system validation. The VS states space diagram is a visual and graphical tool that can be used to analyze possible potential problematic states in the system evolution and sequence. Because of their simplicity, the use of the VS states space diagram is an intuitive alternative technique from the system analysis point of view.

References

1. Chu, F., Xie, X.-L.: Deadlock Analysis of Petri Nets Using Siphons and Mathematical Programming. IEEE Transactions On Robotics And Automation 13(6), 793–804 (1997)
2. Ezpeleta, J., Colom, J.M., Martínez, J.: A Petri Net Based Deadlock Prevention Policy for Flexible Manufacturing Systems. IEEE Transactions On Robotics And Automation 11(2), 173–184 (1995)
3. Lee, S., Tilbury, D.M.: Deadlock-Free Resource Allocation Control for a Reconfigurable Manufacturing System with Serial and Parallel Configuration. IEEE Transactions on Systems, Man and Cybernetics - Part C: Applications and Reviews 37(6), 1373–1381 (2007)

4. Li, Z., Zhou, M.: Control Of Elementary And Dependent Siphons In Petri Nets And Their Application. IEEE Transactions On Systems, Man, And Cybernetics-Part A: Systems And Humans 38(1), 133–148 (2008)
5. Li, Z., Zhou, M., Uzam, M.: Deadlock Control Policy For A Class Of Petri Nets Without Complete Siphon Enumeration. IET Control Theory And Applications 1(6), 1594–1605 (2007)
6. Murata, T.: Petri Nets: Properties, Analysis and Applications. Proceedings of the IEEE 77, 541–580 (1989)
7. Fanti, M.P., Zhou, M.: Deadlock Control Methods in automated Manufacturing Systems. IEEE Transactions On Systems, Man and Cybernetics - Part A: Systems and Humans 34(1), 5–22 (2004)
8. Ramadge, P.J., Wonham, W.M.: Supervisory Control Of A Class Of Discrete Event Processes. SIAM Journal Of Control And Optimization 25(1), 206–230 (1987)
9. Trujillo, J.: Finite Position Machine for Logic Control: Pattern Composition in Reconfigurable Manufacturing System. Thesis, Universidad de Valladolid (2004)
10. Wu, N., Zhou, M., Li, Z.: Resource-Oriented Petri Net for Deadlock Avoidance in Flexible Assembly Systems. IEEE Transactions On Systems, Man, And Cybernetics - Part A: Systems And Humans 38(1), 56–69 (2008)

A Systematic Methodology to Obtain a Fuzzy Model Using an Adaptive Neuro Fuzzy Inference System. Application for Generating a Model for Gas-Furnace Problem

Andrés Mejías, Sixto Romero, and Francisco J. Moreno

Escuela Politécnica Superior.
Universidad de Huelva
Ctra. Palos de la Frontera – Huelva 21819 La Rábida (Huelva), Spain
mjias@uhu.es, sixto@uhu.es, franmo@uhu.es

Abstract. In this paper we present a complete design methodology to obtain a fuzzy model with an Adaptive Neuro Fuzzy Inference System (ANFIS). This methodology consists of three phases: In phase I, the automatic selection of input variables and other parameters such as number and type of membership functions is made with a Genetic Algorithm with a special fitness function, obtaining a basic structure of the fuzzy model. The second phase is a massive training of the fuzzy model previously obtained. Finally, the third phase is a post-adjusting of the weights of the rules with a local search algorithm, based on an adjusted fitness function from the first phase. An application of the proposed design method for the gas-furnace time series, a well-known benchmark dataset used by many researchers in the area of neural networks and fuzzy systems is presented, and finally, we present a comparative with other Box-Jenkins models.

Keywords: Input variable selection, ANFIS, local search.

1 Introduction

The use of an ANFIS [6] system from a large input/output dataset presents serious problems to obtain a fuzzy model. One of the principal limitations is the computer system used. These forces us to choose the subset of input variables that will allow us obtain a model of reasonable complexity with our computer resources. In many cases, this is a difficult task.

If we use variables that are not necessary to generate the model, it is possible to significantly increase the complexity of the model. Moreover, learning and the precision of the model will be seriously affected. Previous works about input variable selection propose model-free [1] or model-based algorithms [3].

One of ANFIS's characteristics is that, in the first epochs of training, the e-rrors of the model generated, give an idea of the adequacy of the input variables chosen. Taking advantage of this capacity, we propose the first phase of obtaining input variables where different models (not all the possible models) are generated with a few epochs of training (a model-based method). Once the best one is chosen, it carries out a massive training in the second phase and with the obtained model, we adjust the weight of

J.M. Corchado et al. (Eds.): DCAI 2008, ASC 50, pp. 452–460, 2009.
springerlink.com

the rules by means of a local search algorithm, to improve as much as possible the number of rules and the errors presented by the model.

A brief description of ANFIS

ANFIS is a well-known universal aproximator when the number of rules is not restricted. It is composed of two approaches: neural networks and fuzzy logic. In this multi-layered neural network, the first layer carries out a fuzzification process. In the second layer, each node output is the product of the antecedent part of the fuzzy rules. The third layer normalizes the membership functions (MFs). The fourth layer executes the consequent part of the rules and the fifth layer computes the overall output as the addition of all incoming signals.

The concept of Genetic Algorithm

The GA is a method for solving optimization problems and simulates the pro-cess of biological evolution. The algorithm repeatedly modifies a population of individual solutions. Each individual is often a bit string and is referred to as a possible solution for a problem. At each step, the GA selects individuals at random from the current group of them (called a population) and uses them to produce the children for the next generation. The GA computes the function to be optimized with each individual, called fitness function, and tries to minimize this. Selection rules, Crossover rules and Mutation rules are used to select the indivi-duals, to exchange genetic material from two parents and to create new genetic material in the population respectively.

Direct Search

Direct search is a method for solving optimization problems that does not require any information about the gradient of the objective function. As opposed to more traditional optimization methods that use information about the gradient or higher derivatives to search for an optimal point, a direct search algorithm searches a set of points around the current point, looking for one where the value of the objective function is lower than the value at the current point. You can use direct search to solve problems for which the objective function is not di-fferentiable, stochastic, or even continuous.

We use a direct search algorithm called the generalized pattern search (GPS) algorithm, a pattern search algorithm that computes a sequence of points that get closer and closer to the optimal point. At each step, the algorithm searches a set of points, called a mesh, around the current point (the point computed at the previous step of the algorithm). The mesh is formed by adding the current point to a scalar multiple of a set of vectors called a pattern. If the pattern search algorithm finds a point in the mesh that improves the objective function at the current point, the new point becomes the current point at the next step of the algorithm.

The organization of this paper is the following: Section 2 shows a description of the methodology we propose to get fuzzy systems with ANFIS from a dataset with an extensive number of candidates input variables. Section 3 presents the application of our method to obtain a fuzzy model for the gas-furnace problem and a comparative with other fuzzy models obtained from the same dataset.

Finally, Conclusions and Future Works are summarized at the end of this paper.

2 The Proposed Method

The three phases of the proposed methodology to obtain a fuzzy model with ANFIS are:

Previous training (Phase I): The objective of this phase is select the input varia-bles among all the possible candidates that lead to a reasonable fuzzy system using a GA.

Habitually, ANFIS already gives, in the first epoch, an idea of the suitability of the input variables chosen. In most cases, the initial training and test errors already allow us to see if the selected variables are good (not necessarily the best), obser-ving the training and test errors [6]. In this phase, ANFIS is trained in each iteration during a very low number of epochs (typically from 1 to 5).

Extending the use of the GA, one can choose in addition a combination of a type of MFs and a number of these functions assigned to the input variables, lea-ving to the GA the choice of some necessary parameters of initialization, (normally we do not have an idea of which to choose if we try to model a system from an input/output dataset).

We use a fitness function, which can adapt to different computational scenarios and preferences. The principal variables included in its definition are:

- $\boldsymbol{e_t}$: *Training Error.*
- $\boldsymbol{e_c}$: *Checking error.* In our fitness function, the weight of errors in the equation depends on a coefficient α and we can change the influence of checking error and training error.
- $\boldsymbol{N_R}$: *Number of rules in FIS.* Rules needed to generate the ANFIS model in an iteration with input variables determined by the GA. A large number of input variables suppose a high number of rules and adjustable parameters in the model adjusted by ANFIS, and the curse of dimensionality problem arises. In our fitness function, a multiplicative factor γ allows more or less influence in the final value of the fitness function. If $\gamma = 0$, the number of rules has not influenced the function. If all input variables have the same number of MFs (N_{MFs}), then:

$$N_R = N_{MFs}^{\ N_V} \tag{1}$$

- $\boldsymbol{N_v}$: *Numbers of input variables selected by the GA.* Let us suppose the available number of input variables is V. A large number of variables is problematic for the adjustment process of the ANFIS system. In fact, we can have problems with the computational time or with the limits of the available computer. The number of variables N_v selected by the GA in each individual $P=(P_1, P_2 ... P_v)$ to form the training and checking vectors in our modeling scenario is:

$$N_V = \sum_{i=1}^{V} P_i \quad , P_i = \{0,1\} \tag{2}$$

We proposed that the number of input variables selected by the GA must be between two limits selected by the user, to assure that the computational time and the use of the available computer are reasonable. If the number of input va-riables selected by the GA is less or greater than upper and lower limits chosen by the user (Max_{Nv} and Min_{NV}) then it is not necessary to compute a model with this subset of input variables in ANFIS. The coefficient σ is included to increase the value of the fitness function. In this case, the algorithm rejects the individual and the ANFIS is not called in that iteration.

- ***Preferred Variables.*** It is possible that our experience points to a group of input variables that should be in the subset chosen by the GA to obtain a good model. In this case, the fitness function should favor this group of variables. A factor β_i is used to create predominant variables. If none of them exist, then:

$$\sum_{i=1}^{V} P_i \beta_i = 0 \tag{3}$$

A fitness function that includes all the commented parameters in the preceding list is shown in (4). Figure 1 shows a block diagram representing the structure of the system in phase I.

<u>Massive training and test (Phase II).</u> A massive training of the fuzzy model obtained previously is realized (with a higher number of epochs than in the previous phase) using the variables chosen in the previous step. This model will present a minor value of checking or test error.

<u>*Thin adjustment or post-training (Phase III)*</u>. ANFIS is a system that presents some limitations:

- Be first or zeroth order Sugeno-type systems.
- Have a single output, obtained using weighted average defuzzification.
- Have no rule sharing. Different rules cannot share the same output membership function.
- Have unity weight for each rule.

$$f_{fitness} = \begin{cases} \sigma\left[Max_{NV} - \sum_{i=1}^{V} P_i\right] & if \quad \sum_{i=1}^{V} P_i > Max_{NV} \\ \sigma\left[Min_{NV} - \sum_{i=1}^{V} P_i\right] & if \quad \sum_{i=1}^{V} P_i < Min_{NV} \\ \sum_{i=1}^{V} P_i \beta_i + \alpha e_c + (1-\alpha) e_t + \gamma N_R & if \quad Min_{NV} \le \sum_{i=1}^{V} P_i \le Max_{NV} \end{cases} \tag{4}$$

The last limitation is the one that we modify in the 3rd phase, by means of a GPS (Generalized Pattern Search) algorithm, which tries to modify the weight of each rule (in the interval [0,1]), to obtain a better adjustment of the final fuzzy model.

A difference with the 1st phase is that now we have a FIS already generated, and therefore our objective now is not to determine the structure of it, but only to adjust the weights of the rules. Thus, there are factors in the fitness function that do not make sense at this phase of the process.

Indeed, the number of rules (N_R), the lower and upper limits of input variables (Min_{NV} *and* Max_{NV}) or predominant variables are parameters that were needed only to

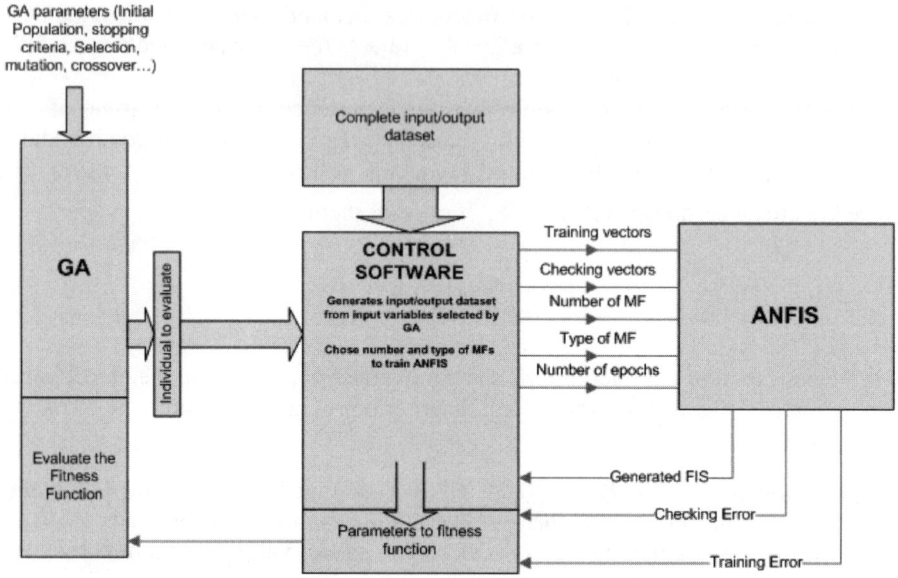

Fig. 1. Block Diagram of the system structure (phase I)

determine the FIS to use. Once the FIS is defined, we could improve the performance of the FIS by minimizing the objective function shown in (5).

$$f_{fitness} = \alpha \cdot e_c + (1-\alpha) \cdot e_t \tag{5}$$

3 Example: Generating a Model for Gas-Furnace Problem

The Box and Jenkins [2] model is widely used in specialized literature as a measure of the adequacy of the models that simulate the behaviour of the data in a gas-furnace. To compare our results with others extracted from various publications, we use our methodology to generate a model using the same dataset employed in [2].

To model Box-Jenkins's problem, previous values of the flow of methane $\{u(k); u(k-1); u(k-2);\ldots\}$ and concentration of CO_2 $\{y(k-1); y(k-2); y(k-3);\ldots\}$ are used traditionally to predict the present value of the concentration of CO_2, $y(t)$.

In this example, we will limit the number of input variables to 2 (a previous value of *u* and other one of *y*), and we will leave to the GA the selection of the type of MFs, to choosing among one of these types: gaussian, sigmoidal, bell or the product of two Gaussian MFs.

We will use 143 input vectors for training ANFIS and another 143 for verification or testing. The GA will also choose the number of MFs to use (voluntarily limited to 3 or 4). Depending on if the GA chooses 3 or 4 MFs in every input, we will have 9 or 16 rules respectively, according to the equation (2). The complete set of possible input variables will consist of 10 variables that form the vector *{u(k-6), u(k-5), u(k-4), u(4k-3), u(k-2), u(k-1), y(k-4), y(k-3), y(k-2), y(k-1)}*. The GA will choose two of them

(Max_{NV}=Min_{NV}=2 in (4)), and will handle individuals, each of them formed by a string of 13 bits. The meaning of this string is as fo-llows (from left to right): *Set of possible input variables* (bits 1 to 10); *MF type* ('00': Gaussian; '01': Product of two Gaussian; '10': Bell; '11': Sigmoidal functions); *Number of MFs in each input variable* ('0': 3 MFs; '1': 4 MFs).

Phase I: Figure 2 shows the results provided by the GA in the first phase. The interpretation of this chart is simple if we take into account the distribution of bits. Indeed, it can be seen that have been chosen bits 3 and 9, which indicates that the pair of entries chosen for the model are *u(k-4)* and *y(k-2)* respectively. Bits 11 and 12 are equal to zero, indicating that the MFs chosen are Gaussian functions. Bit 13 is active, which indicates that the number of MFs is 4. According to equation (1), this represents a total of 16 rules in the model generated by ANFIS. In this example, we have used a coefficient α = 0.3, to give more importance to the error test. A summary of the results of this first phase can be seen in Table 1.

Table 1. Results of the implementation of phase I

Selected inputs	Fitness value (α=0.3)	e_t	e_c	MF type	Number of MFs	Number of ru-les	Execution time
u(k-4), y(k-2)	0.7118	0.3305	0.8753	Gaussian	4	16	152.3 s.

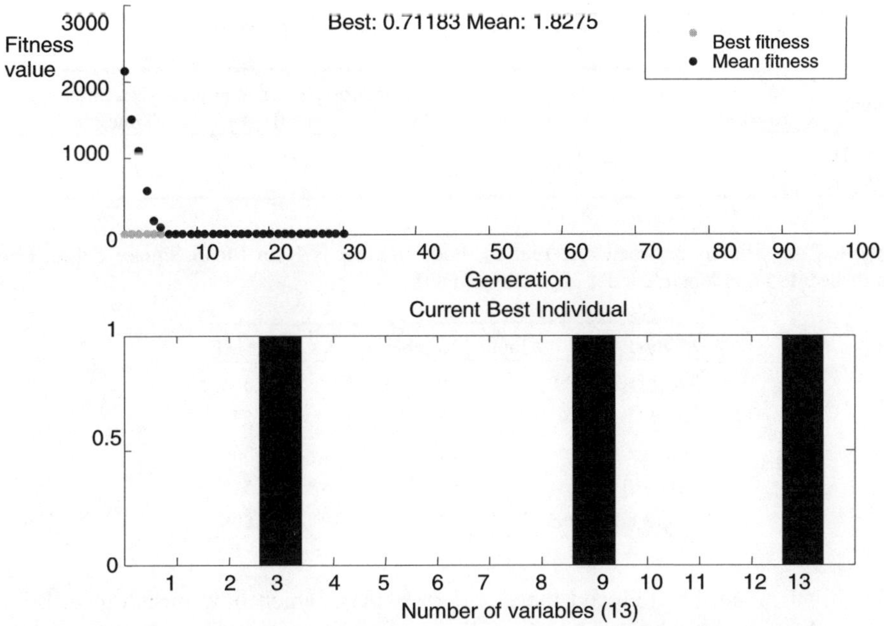

Fig. 2. Evolution of the fitness function value and combination selected by the GA

Phase II: In this second phase, and using the initial FIS obtained in Phase I, we train ANFIS over 100 epochs. The results are presented in Table 2.

Table 2. Results of the implementation of phase II

Inputs	Fitness value (α=0.3)	e_t	e_c	MF type	Number of MFs	Number of rules	Execution time
u(k-4), y(k-2)	0.6129	0.2766	0.757	Gaussian	4	16	0.7 s.

Phase III: In this final phase, we adjust the weights of the rules of the model obtained at the preceding stage (which by design is unity in all of them in ANFIS).

Table 3 summarizes the data obtained in the third phase. Comparing these results with those obtained in 2nd phase (see Table 2), we see that the final value of the fitness function has improved.

Table 4 shows a comparison between the results obtained after completing 3rd phase with other methods, all with two entrances and Sugeno type. We can observe that the comparative RMSE obtained by our method is very similar to those depicted, but that it is obtained with fewer rules.

Figure 3 shows the result of the GPS algorithm, where we see the evolution of the objective function and the weights adjusted to the rules. The exact values are presented in Table 5.

Table 3. Results of the implementation of phase III

Inputs	fitness (α=0.3)	e_t	e_t e_c	MF type	Number of MFs	Rules after phase III	Execution Time
u(k-4), y(k-2)	0.5822	0.3129	0.698	Gaussian	4	13	124 s.

Table 4. Comparisons between different methods (RMSE is Root Mean Square output error with respect to an independent set of checking data)

Method	Inputs	Number of rules	RMSE
Tong [10]	2	19	0.685
Xu [11]	2	25	0.573
Pedrycz [8]	2	81	0.566
Peng [9]	2	49	0.548
Proposed method	2	13	**0.698**

The final model has reduced the number of rules, which now turns out to be 13 (Rules 1,2 and 15 now have a weight of 0, which does not contribute to the output of the model generated). The eliminated rules are shown shaded in Table 5. Figure 4

Table 5. New weights of the rules once the 3rd phase is complete

Rule	Weight	Rule	Weight	Rule	Weight	Rule	Weight	Rule	Weight	Rule	Weight
1	0	4	0.937	7	0.617	10	0.750	13	0.984	16	0.602
2	0	5	0.781	8	1	11	0.617	14	0.641		
3	0.547	6	0.859	9	1	12	0.562	15	0		

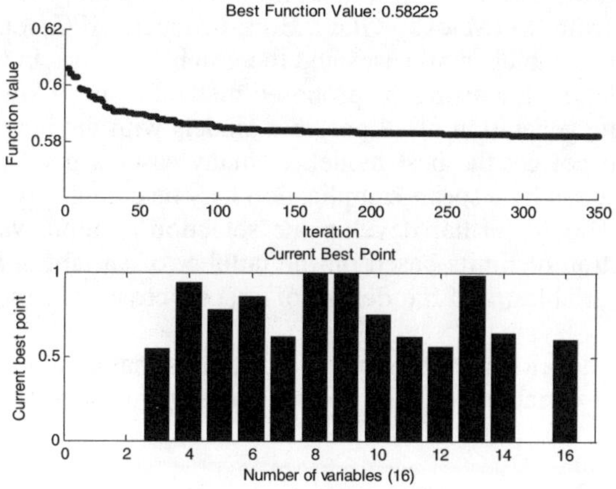

Fig. 3. Evolution of objective function and weights adjusted by GPS algorithm in phase III

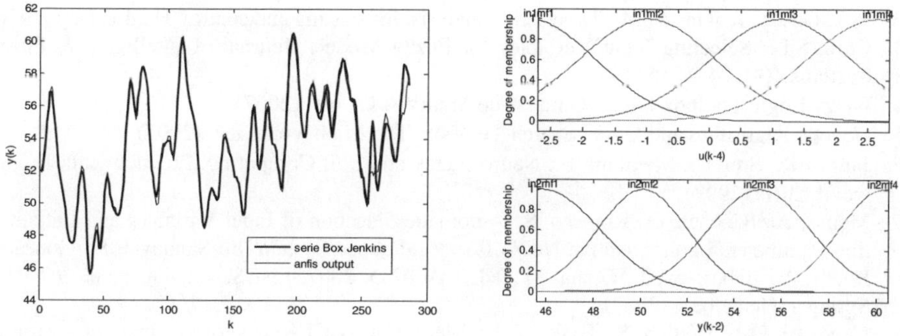

Fig. 4. Box-Jenkins series, output from ANFIS and input membership functions

shows Box-Jenkins series and the output provided by the final model adjusted by the proposed method. This figure also shows the input membership functions generated by ANFIS. The total time required to obtain the model (adding the three phases) was 277 seconds in an Intel Pentium M processor 735A (1.7 Ghz, 400 Mhz FSB, 2MB cache L2, 1 GB DDR2 RAM and Windows XP Prof.).

4 Conclusions and Future Works

We present a methodology for obtaining fuzzy models using an ANFIS system. This procedure consists of three distinct phases: in the first, a fitness function is designed to use with a GA which chooses the input variables as well as some design parameters (number and type of input MFs). The second is only a massive training of the model selected in the previous phase, and finally, in the third stage, the weights of the rules are adjusted using a GPS algorithm, which in some cases, as in the example shown, is able to reduce the number of rules generated initially by ANFIS. This method applied to different dataset from [8] (Mackey-Glass, Breast cancer, MPG, ionosphere) reduces the number of input variables in all cases and the number of rules in three of them.

The total computation time using the proposed method is much lower than we would need with a complete generation of all possible models with the input va-riables available. Even if we do not get the best model, in many cases a good model with errors within the limits required by a specific application may be completely valid.

Future research should further develop the selection of input variables, allowing the automatic selection of limits based on the number of variables, including statistical studies of the variables, and the design of special crossover, mutation and selection functions.

Furthermore, the development of genetic algorithms that assign a different number of MFs to each input variable would be desirable to generate a better model.

References

1. Back, A.D., Trappenberg, T.P.: Selecting Inputs For Modeling Using Normalized Higher Order Statistics and Independent Component Analysis. IEEE Trans. On Neural Networks 12(3), 612–617 (2001)
2. Box, G.E.P., Jenkins, G.M.: Time series analysis, forecasting and control, Holden Day (1970)
3. Chiu, S.L.: Selecting Input Variables for Fuzzy Models. Journal of Intelligent & Fuzzy Systems 4(4), 243–256 (1996)
4. Fuzzy Logic Toolbox User's Guide. The MathWorks, Inc. (2007)
5. Genetic Algorithm and Direct Search Toolbox. The MathWorks Inc. (2007)
6. Jang, J.R., Sun, C., Mizutani, E.: Neuro-Fuzzy and Soft Computing. Prentice-Hall, Englewood Cliffs (1997)
7. Mejías, A., Sánchez, O., Romero, S.: Automatic Selection of Input Variables and Initialization Parameters in an Adaptive Neuro Fuzzy Inference System. In: Sandoval, F., Gonzalez Prieto, A., Cabestany, J., Graña, M. (eds.) IWANN 2007. LNCS, vol. 4507, pp. 407–413. Springer, Heidelberg (2007)
8. Newman, D.J., Hettich, S., Blake, C.L., Merz, C.J.: UCI Repository of machine learning databases. Department of Information and Computer Science. University of California, Irvine (accessed June 12, 2007) (2007), `http://www.ics.uci.edu/~mlearn/MLRepository.html`
9. Pedrycz, W.: An identification algorithm in fuzzy relational systems. Fuzzy sets and systems 13, 153–167 (1984)
10. Peng, X., Wang, P.: On generating linguistic rules for fuzzy models. In: Yager, R.R., Saitta, L., Bouchon, B. (eds.) IPMU 1988. LNCS, vol. 313, pp. 185–192. Springer, Heidelberg (1988)
11. Tong, R.M.: The evaluation of fuzzy models derived from experimental data. Fuzzy sets and systems 4, 1–12 (1980)
12. Xu, C.W., Lu, Y.Z.: Fuzzy model identification and self-learning for dynamic systems. IEEE Trans. on System, Man and Cybernetics SMC-17, 683–689 (1987)

Evolving Machine Microprograms: Application to the CODE2 Microarchitecture*

P.A. Castillo, G. Fernández, J.J. Merelo, J.L. Bernier, A. Mora, J.L.J. Laredo, and P. García-Sánchez

Department of Computer Architecture and Technology
University of Granada, Spain
pedro@atc.ugr.es

Summary. The realization of a control unit can be done either using a complex circuitry or microprogramming. The latter may be considered as an alternative method of implementation of machine instructions that can reduce the complexity and increase the flexibility of the control unit. Microcode efficiency and speed are of vital importance for the computer to execute machine instructions fast. This is a difficult task and it requires expert knowledge. It would be interesting and helpful to have automated tools that, given a machine instruction description, could generate an efficient and correct microprogram. A good option is to use evolutionary computation techniques, which have been proved effective in the evolution of computer programs. We have developed a microarchitecture simulator of a real machine in order to evaluate every individual and to assign it a fitness value (to determine whether this candidate solution correctly implements the instruction machine). Proposed method is successful in generating correct solutions, not only for the machine code instruction set, but also for new machine instructions not included in such set. We show that our approach can generate microprogramms to execute (to schedule microinstructions) the machine level instructions for a real machine. Moreover this evolutive method could be applied to any microarchitecture just by changing the microinstruction set and pre-conditions of each machine instruction to guide evolution.

Keywords: computer architecture, microprogramming, microarchitecture, evolutionary computation techniques, optimization, automatic design.

1 Introduction and State-of-the-Art

The control unit (CU) in a microprocessor controls the flow of data through it, and coordinates the activities of the other units within it. In a way it controls what happens inside the processor, which in turn controls the rest of the computer. CU can be implemented using ad-hoc logic, which is difficult to design, or as a microprogram that is stored in a fast memory. This method of implementation is based on the observations that any complex operation is completely specified by a sequence of primitive operations (microinstructions) [1, 2, 3].

* Supported by projects TIN2007-68083-C02-01, P06-TIC-02025 and PIUGR 9/11/06.

J.M. Corchado et al. (Eds.): DCAI 2008, ASC 50, pp. 461–470, 2009.
springerlink.com

Microinstructions are selected by a microsequencer and the bits from those words directly control the different parts of the device, including the registers, arithmetic and logic units, instruction registers, buses and input/output. Building the CU using microcode, the hardware design becomes simpler. The use of microcode also makes the design of the CU much more flexible, so that existing instructions can be easily changed and new instructions introduced without the need for additional hardware [4, 5, 6]. According to Stallings, microcode is the dominant technique for the implementation of CISC processors [7].

However, microprogramming is a difficult task because writing good microcode requires an intimate knowledge of the hardware. The code is highly specific to each processor and few generic tools are available to assist the programmer. Compilation from higher level languages is generally not an option, since the generated code is usually not enough quality. At the microarchitectural level, both efficiency and correctness are vital [8, 9]. Any deficiencies at this level, either errors or latencies, affect the operation of the whole machine.

CU realization can be viewed as a set of microprograms, each of which implements a single machine code instruction. These microprograms are typically short, consisting of just a few microinstructions, and can be considered in isolation. It would be interesting and very helpful to have an automated tool that, given a machine instruction description, could generate an efficient and correct microprogram. A good option is to use evolutionary computation techniques, which have been proved effective in the evolution of computer programs.

Evolutive methods require evaluating each candidate solution in the population. In these problems the fitness evaluation requires not only the monitoring of the complete state space accessible at the microarchitectural level, but also the derivation of a fitness score that reflects the desirability of changes that take place within that state space. We have to assure that the execution of the code effects all of the state changes required at the microarchitectural level and at the same time it does not effect any undesired state changes.

Usually, microprogramming techniques have been based on heuristics, such as linear analysis, critical path, branch and bound, and list scheduling [10, 11]. Some authors have approached the microarchitecture design problem as an optimization of the control memory word dimension, in order to reduce memory size as well as execution time [12, 13]. Despite the optimizations on size, these approaches do not design microprograms for each machine code instruction (the harder task). To date, little research work has been done on the evolution of processor microcode. Some authors have applied evolutionary approaches to the generation of hardware designs, both for analogue and digital systems [14, 15, 16, 17, 18]. Paying attention to the program evolution problem, most genetic programming systems have focused on the evolution of programs in high-level languages [19, 20].

The bibliography yields some proposals to evolve machine code [11, 21, 22, 23, 24]. However, these systems tend to focus on very specific architectures, without the possibility of being modified or applied to other architectures. Jackson

[25, 26] proposes a genetic programming method to evolve an example microprogrammed system described in the widely used computer architecture text by Tanenbaum [27].

In this paper we propose an evolutive method that could be applied to any microarchitecture just by changing the requirements (pre-conditions) of each machine instruction to guide the method. We have analyzed in detail the microarchitecture of a real basic computer (Elemental Didactic Computer, CODE2) [28]. As an evolutionary problem, we have carried out the evolution of the microcode for every machine-level instruction, taking each one at a time. To test the proposed method, new machine code instructions have been defined, and we have searched for efficient microprograms that implement them.

The remainder of this paper is structured as follows: section 2 details the problem and the CODE2 machine architecture. In section 3 proposed method to evolve microcode is detailled. Section 4 describes the experiments and the results obtained, followed by a brief conclusion in Section 5.

2 Architecture of the Elemental Didactic Computer

At the assembly-language level code and data are transferred between main memory and the CPU. The CPU includes an arithmetic and logic unit (ALU) and several general purpose registers, program counter and stack pointer.

Below the assembler-level we find the microarchitecture-level. The CU (inside the processor) includes a memory of control where the microcode for all machine instructions is stored. Each control word stored in this memory includes some fields that along with the flag register control the sequencing logic to determine the next microinstruction to be executed. Bits in the control word are sent to the control bus as control signals.

CU manages the fetch-decode-execute instruction cycle. Some control word ask the memory interface to fetch the next machine instruction (pointed by the program counter register). When this arrives, the sequencing logic decodes it, and then passes control to the microcode responsible for realizing the machine instruction. When this is complete, the CU loops back to fetch the next machine instruction from main memory. CODE2 uses 29 signals to control the data logic elements and execute each instruction.

CODE2 implements a register-register architecture. At the assembly-language level this machine has 16 general purpose registers (r0 to rF) and 16 machine instructions. Some registers have a special role: rE register is usually used as stack pointer, and rD register as address register. The data path also includes the instruction register (IR) and the program counter (PC). The ALU is capable of the following operations: adition, substraction, logic NAND, logic right and left shift, and aritmetic right shift. The ALU also includes zero, sign, carry and overflow flags. Main memory size is 65536 16 bits words. Finally, CODE2 can handle 256 input ports and 256 output ports. There exists also a simulator of CODE2, available at http://atc.ugr.es/pedro/ev-micropr

3 Proposed Method

The proposed method we have implemented is based on an evolutionary algorithm [29]. In this approach, an individual encodes the sequence of microinstructions that every machine instruction implements. We have used a numeric representation (see Table 1 for the complete set). An individual encodes a list of numbers, each of one represents a microinstruction. As different machine instructions have variable lengths (number of microinstructions), individual sizes are randomly initialized.

Table 1. List of microinstructions used in our method. A numeric representation to encode the list of microinstructions in an individual is used.

0	AR ← H'00##IR(7:0) + RT	17	AR ← alu ← PC
1	AR ← H'00##IR(7:0)	18	DR ← alu ← PC
2	AR ← RF[RA]	19	DR ← IP(AR)
3	DR ← alu ← RF[RA]	20	esp ← 1
4	DR ← M(AR)	21	OP(AR) ← DR
5	M(AR) ← DR	22	PC ← alu ← RF[RA]
6	PC ← alu ← DR	23	RA ← H'D
7	PC ← RT+1	24	RA ← ra
8	RA ← H'E	25	RA ← rx
9	RA ← rs	26	RF[WA] ← H'00##RF[RA](7:0)
10	RF[WA] ← H'00##IR(7:0)	27	RF[WA] ← shl(RF[RA])
11	RF[WA] ← alu ← DR	28	RF[WA] ← shra(RF[RA])
12	RF[WA] ← shr(RF[RA])	29	RF[WA] ← RF[RA] - 1
13	RF[WA] ← IR(7:0)##H'00+RT	30	RF[WA] ← RF[RA] - RT
14	RF[WA] ← RF[RA] + RT	31	RF[WA] ← RT + 1
15	RF[WA] ← RF[RA] nand RT	32	WA ← H'E
16	RT ← RF[RA]	33	WA ← rx

Table 1 shows the microinstruction set used in our approach. In order to define microinstructions *DR* stands for "data register"; *AR* stands for "address register"; *WA* refers to the register to write to; *RA* refers to the register to read from; *M()* stands for "access to main memory"; *RF[]* refers to the register file (selects one of the 16 registers); *rs, rx, ra* refers to the registers specified in the machine instruction as source and destination; *alu, nand, shr, shl, shra* are the ALU operations; *RT* stands for "temporal register"; *IP()* refers to the input ports and *OP()* to the output ports; and *esp* corresponds to the "wait" signal.

As variation operators, our method uses mutation, crossover, gene insertion and gene deletion. Mutation randomly changes a gene (microinstruction) taking into account the mutation rate. Crossover carries out the multipoint crossover between two individuals, exchanging their microinstructions. Insertion operator inserts a new microinstruction into the individual, while the last operator (deletion), randomly removes a microinstruction.

The fitness evaluation is based on simulating microinstructions. For every machine instruction, we have defined the set of resources that should be used or avoided (operations allowed and forbidden to guide the evolutionary search). In order to evaluate all the individuals under the same conditions, the same set of register and memory values are used during the method run.

As an example we show the pre-conditions for a machine instruction:

```
ADDS R1,R3,RA
LM,          //memory positions that should be read
EM,          //memory positions that should be written
LR, A 3      //registers that should be read
ER, 1        //registers that should be written
EO           //out ports that should be written to
LI,          //in ports that should be read from
Esp, 0       //the "esp" microinstruction is not allowed
```

In this case, the machine instruction must read the R3 and RA registers, it must write into the R1 register, and it should not access the memory nor the IO ports. The last pre-condition prohibit the use of the microinstruction that stops the CPU.

The fitness function takes into account four values: 1) whether the implementation is correct, 2) how many forbidden operations are included, 3) how many allowed operations are included, and 4) individual length (the shorter, the better, if it is correct). In the case of non-correct solutions, the one that includes less forbidden operations is taken as the best.

For instance the following individual:

```
RA=ra             //id of the 1st register used
RT=RF[RA]         //using a temporary register
WA=rx             //id of the register used as destination
RA=rs             //id of the 2nd register used
RF[WA]=RF[RA]+RT  //the ALU carries out the operation
```

is coded as (24,16,33,9,14); the fitness value obtained after simulation was (1, 0, 3, 5). That means that this individual, the implementation using microinstructions, is correct (1) and the number of microinstructions is 5.

The Algorithm::Evolutionary library [30, 31] was used to implement the evolutionary algorithm. This library is available at http://opeal.sourceforge.net and `http://search.cpan.org/dist/Algorithm-Evolutionary` under GPL. We used this library due to the facility that it offers to evolve any object with a fitness function and use any genetic operator. One of the objectives of using Algorithm::Evolutionary is to make it easier to reproduce results, which is why the source should also be made available. The source code of the proposed method is available for download at `http://atc.ugr.es/pedro/ev-micropr`

4 Experiments and Results

The target of the experiment was to optimize the CODE2 machine instruction set. Each machine instruction is taken separately, the pre-conditions are set and

Table 2. Some machine instructions implementation using the CODE2 microinstruction set. We show the best solution obtained by the evolutionary algorithm (left) and the optimal implementation (right). In those cases where the proposed method found a correct but not optimal solution, the former could be obtained just by removing one or two microinstructions, as can be seen for instance in the NAND instruction.

Instruction	Obtained implementation	Optimal implementation
ADDS	RA=ra RT=RF[RA] WA=rx RA=rs RF[RA]=RF[RA]+RT	solution obtained is optimal
SUBS	RA=ra RT=RF[RA] WA=rx RA=rs RF[RA]=RF[RA]-RT	solution obtained is optimal
LLI	WA=rx RF[WA]=0x00FF##IR	solution obtained is optimal
LHI	WA=rx RA=rx RF[WA]=0x00FF##RF[RA]	solution obtained is optimal
SHL	WA=rx RA=rx RF[WA]=shl(RF[RA])	solution obtained is optimal
SHR	WA=rx RA=rx RF[WA]=shr(RF[RA])	solution obtained is optimal
SHRA	WA=rx RA=rx RF[WA]=shra(RF[RA])	solution obtained is optimal
NAND	RA = rs WA = rx RT = RF[RA] RA = ra RF[WA] = 0x00FF##RF[RA] RF[WA] = RF[RA] nand RT	RA = rs WA = rx RT = RF[RA] RA = ra RF[WA] = RF[RA] nand RT
OUT	RA = rs DR=RF[RA] AR=0x00FF##IR RA = rx DR=RF[RA] OP[AR]=DR	AR=0x00FF##IR RA = rx DR=RF[RA] OP[AR]=DR
IN	AR=0x00FF##IR WA=rx DR=IP[AR] RF[WA]=DR	solution obtained is optimal
BR	PC=RF[RA]	solution obtained is optimal
HALT	esp=1	solution obtained is optimal

the evolutionary method is run to design that machine instruction using a sequence of microinstructions (search for a correct implementation). The second experiment is based on defining and optimizing new instructions for CODE2 (the architecture was not originally designed to include these new machine instructions).

Experiments were set using 100 generations, and a population size of 100 individuals in order to avoid too long a run. A steady state [32] algorithm was used because it was empirically found to be faster at obtaining solutions than other selection algorithms. For each generation, the best n individuals of the population, those whose fitness is highest, are chosen to mate, using the genetic operators (mutation and crossover). The offspring replace the n worst individuals of the current generation.

The uniform crossover operator takes two individuals and swaps their genes (microinstructions). It uses an application rate of 0.4. Mutation operator was used with an application rate of 0.2. This operator changes the value of a gene (the microinstruction that gene codes). The insertion operator randomly selects a microinstruction to be inserted at random in the individual-chromosome. This operator was used with an application rate of 0.2. The deletion operator randomly selects a microinstruction to be removed from the individual-chromosome. This operator was used with an application rate of 0.2. These application rates were found empirically to obtain good results.

We conducted our experiments on an Intel Centrino with 1.66GHz and 1GB RAM. Each run, the evolutionary method takes about four minutes to obtain a solution for each machine instruction.

Obtained results show that in nearly all runs (8 out of 10 for every machine instruction), the evolutionary method is able to find a correct solution, that is, a

Table 3. AND and NOT machine instruction implementation using the CODE2 microinstruction set. We show the best solution obtained by the evolutionary algorithm (left) and the optimal implementation (right). Although the logical AND implementation is not correct, it is very easy to complete it to obtain the optimal solution (by adding only one microinstruction).

Instruction	Obtained implementation	Optimal implementation
NOT	RA = rs WA = rx RT = RF[RA] RF[WA] = RF[RA] nand RT	solution obtained is optimal
AND	RA = ra WA = rx RT=RF[RA] RA = rs RF[WA] = RF[RA] nand RT RA = rx RF[WA] = RF[RA] nand RT	RA = ra WA = rx RT=RF[RA] RA = rs RF[WA] = RF[RA] nand RT RA = rx RT=RF[RA] RF[WA] = RF[RA] nand RT

set of correct microinstructions. Table 2 shows implementations obtained using microinstructions for some machine instructions.

The second experiment is based on defining new machine instructions (not included in the original instruction set of CODE2). We have defined the logical NOT and the logical AND operations to carry out their optimization using the EA. The method was run for 10 times for each new instruction.

Table 3 shows the best solutions found for the logical NOT and the logical AND machine instructions. In the case of the NOT instruction, correct (and optimal) solutions were obtained. In the case of the logical AND instruction, although the solution obtained is not correct, just by including (by hand) an extra microinstruction we could obtain the optimal implementation.

5 Conclusions and Future Work

In this work, an evolutionary approach has been applied to the problem of designing the microarchitecture of a basic real computer. We have evolved microprograms that implement not only CODE2's machine code instruction set, but new machine instructions (see Table 3).

Obtained results show that EAs can optimize machine microcode; moreover it is usually possible to generate correct and efficient solutions.

Our approach needs some guidance introduced related to which resources should be used or avoided in the fittest solutions. This evolutionary guidance is derived automatically from the specifications of macroinstruction behavior, and does not require any "intelligent" input from human operators. Thus, the method could easily be applied to other microarchitectures by changing the preconditions and microinstruction set.

The evolutionary method automatically implements assembly-level instructions, needing neither expert knowledge nor good heuristics. Not only instructions included in the original machine code instruction set are optimized, but new instructions can be designed using our method.

As future lines of work, it would be interesting to add new complex machine instructions than these used in the second experiment, and designing their microprograms using the evolutionary method. In order to improve convergence speed and quality of solutions, it could be useful using either specific genetic operators or automatically defined functions [33]. We also intend to apply out method to more complex real architectures and to optimize a whole instruction set for a particular target.

References

1. Das, S.R., Nayak, A.R.: A survey on bit dimension optimization strategies of microprograms. In: MICRO 23: Proceedings of the 23rd annual workshop and symposium on Microprogramming and microarchitecture, pp. 281–291. IEEE Computer Society Press, Los Alamitos (1990)
2. Andrews, M.: Principles of firmware engineering. Computer Science Press, Rockville (1980)

3. Wilkes, M.V.: The growth of interest in microprogramming: a literature survey. ACM Computing Surveys 1, 139–145 (1969)
4. Hennessy, J.L., Jouppi, N.P.: Computer Technology and Architecture: An Evolving Interaction 24(9), 18–29 (1991)
5. Patterson, D.A., Hennessy, J.L.: Computer Organization and Design: The Hardware Software Interface, 2nd edn., San Francisco, California, EE, UU (1997)
6. Hennessy, J., Patterson, D.: Computer Architecture. A Quantitative Approach, 3rd edn., San Francisco, California, EE, UU (2003)
7. Stallings, W.: Computer Organization and Architecture: Designing for Performance, 6th edn. Prentice-Hall, Englewood Cliffs (2003)
8. Miller, S., Srivas, M.: Formal verification of the AAMP5 microprocessor: A case study in the industrial use of formal methods. In: Proc. Workshop Ind.-Strength Formal Specif. Tech (WIFT 1995), Boca Raton, FL, pp. 2–16 (1995)
9. Greve, D.: Symbolic simulation of the JEM1 microprocessor. In: Gopalakrishnan, G.C., Windley, P. (eds.) FMCAD 1998. LNCS, vol. 1522, pp. 321–333. Springer, Heidelberg (1998)
10. Landskov, D., Davidson, S., Shriver, B.D., Mallett, P.W.: Local Microcode Compaction Techniques. ACM Computing Surveys 12(3), 261–294 (1980)
11. Beaty, S., Whitley, D., Johnson, G.: Motivation and framework for using genetic algorithms for microcode compaction. In: Proceedings of the 23rd Annual Workshop and Symposium, MICRO 23. Microprogramming and Microarchitecture, Number IEEE Cat. No. 90TH0341-8, Orlando, FL, pp. 117–124. IEEE Computer Society Press, Los Alamitos (1990)
12. Kleir, R.L., Ramamoorthy, C.V.: Optimization strategies for microprograms. IEEE Trans. Computers C-20, 783–794 (1971)
13. Tabendeh, M., Ramamoorthy, C.V.: Execution time (and memory) optimization in microprograms. In: 7th Annual Workshop on Microprogramming Preprints Supplement, pp. 19–27 (1974)
14. Koza, J.R., Bennett-III, F.H., Andre, D., Keane, M.A., Dunlap, F.: Automated synthesis of analog electrical circuits by means of genetic programming. IEEE Trans. Evol. Comput. 1, 109–128 (1997)
15. Thompson, A., Layzell, P., Zebulum, R.S.: Explorations in design space: Unconventional electronics design through artificial evolution. IEEE Trans. Evol. Comput. 3, 167–196 (1999)
16. Miller, J.F., Job, D., Vassilev, V.K.: Principles in the evolutionary design of digital circuits Part I. Genetic Program. Evolvable Mach. 1, 7–35 (2000)
17. Torresen, J.: A scalable approach to evolvable hardware. Genetic Program. Evolvable Mach. 3, 259–282 (2002)
18. Schnier, T., Yao, X., Liu, P.: Digital filter design using multiple pareto fronts. Soft Comput. 8(5), 332–343 (2004)
19. Koza, J.R.: Genetic programming: On the programming of computers by means of natural selection. MIT Press, Cambridge (1992)
20. O'Neill, M., Ryan, C.: Evolving multiline compilable C programs. In: Langdon, W.B., Fogarty, T.C., Nordin, P., Poli, R. (eds.) EuroGP 1999. LNCS, vol. 1598, pp. 83–92. Springer, Heidelberg (1999)
21. Nordin, P.: A compiling genetic programming system that directly manipulates the machine code. In: Kinnear Jr., K.E. (ed.) Advances in Genetic Programming, pp. 311–331. MIT Press, Cambridge (1994)
22. Ahmad, I., Dhodhi, M.K., Saleh, K.A.: An evolutionary technique for local microcode compaction. Microprocess. Microsyst. 19(8), 467–474 (1995)

23. Nordin, P., Banzhaf, W., Francone, F.D.: Efficient evolution of machine code for CISC architectures using instruction blocks and homologous crossover. In: Spector, L., et al. (eds.) Advances in Genetic Programming, vol. 3, pp. 275–299. MIT Press, Cambridge (1999)
24. Kuhling, F., Wolff, K., Nordin, P.: A brute-force approach to automatic induction of machine code on CISC architectures. In: Foster, J.A., Lutton, E., Miller, J., Ryan, C., Tettamanzi, A.G.B. (eds.) EuroGP 2002. LNCS, vol. 2278, pp. 288–297. Springer, Heidelberg (2002)
25. Jackson, D.: Automatic synthesis of instruction decode logic by genetic programming. In: Keijzer, M., O'Reilly, U.-M., Lucas, S.M., Costa, E., Soule, T. (eds.) EuroGP 2004. LNCS, vol. 3003, pp. 318–327. Springer, Heidelberg (2004)
26. Jackson, D.: Evolution of processor microcode. IEEE Transactions On Evolutionary Computation 9(1), 44–54 (2005)
27. Tanenbaum, A.S.: Structured Computer Organization, 3rd edn. Prentice-Hall, Englewood Cliffs (1990)
28. Prieto, A., Pelayo, F., Lloris, A., Gomez-Mula, F.: Description and use of a simple didactic computer. EC Newsletter (Education in Computing Computers in Education) 2(1), 17–29 (1990)
29. Eiben, A., Smith, J.: Introduction to Evolutionary Computing. Springer, Heidelberg (2003)
30. Merelo, J.J.: Algorithm: Evolutionary (2006), `http://opeal.sourceforge.net`
31. Merelo, J.J.: Evolutionary computation in Perl. In: Münich Perl Mongers, YAPC: Europe: 2002, 2–22 (2002)
32. Whitley, D.: The GENITOR Algorithm and Selection Presure: Why rank-based allocation of reproductive trials is best. In: Schaffer, J.D. (ed.) Proceedings of The Third International Conference on Genetic Algorithms, pp. 116–121. Morgan Kauffmann Publishers, San Francisco (1989)
33. Koza, J.R.: Genetic Programming II: Automatic Discovery of Reusable Programs. MIT Press, Cambridge (1994)

A Model to Minimize the Hot Rolling Time of a Steel Slab Considering the Steel's Chemical Composition

Carlos A. Hernández Carreón[1], Héctor J. Fraire-Huacuja[2], Karla Espriella Fernandez[1], Guadalupe Castilla-Valdez[2], and Juana E. Mancilla Tolama[1]

[1] Sección de Estudios de Posgrado e Investigación,
Escuela Superior de Ingeniería mecánica y Eléctrica, Unidad Azcapotzalco, I.P.N.
Av. de las Granjas No. 682 Col. Santa Catarina
Azcapotzalco, México, D. F. C.P. 02550
cahc05@yahoo.com.mx, karla_rouge@yahoo.com, emtolama@hotmail.com
[2] Instituto Tecnológico de Ciudad Madero.
1° de Mayo y Sor Juana I. de la Cruz S/N.
89440-Cd. Madero, Tamaulipas, México
hfraire@prodigy.net.mx, gpe_cas@yahoo.com.mx

Abstract. This paper presents an optimization approach to deal with the problem of minimizing the hot rolling time of a steel slab. Unlike traditional approaches, this work also considers the chemical composition of the steel slab as a parameter, allowing the automatic setup for different steels of the hot rolling mill. To validate the approach discussed here, a six-stand rolling mill is modeled as an optimization constrained problem solving for six different steel types taken from real processes. The mathematical formulation and considerations for each presented case are fully described. The experimental evidence shows that the solution of the hot rolling scheduling problem requires a more efficient method than just a constrained nonlinear optimizer and that the proposed model properly simulates the hot rolling process.

Keywords: Evolutionary computation, genetic algorithms, rolling pass schedule.

1 Introduction

Hot rolling is one of the most important metalworking processes in comparison with any other forming process aimed to manufacture products of relatively large dimensions (i.e., sheets, strips, plates, foils, etc.), at high speeds [1]. The rolling mill reduces the slab thickness with two work rollers in a mill stand (Fig. 1). Due to the high operational costs of a rolling mill, it is not an acceptable approach to define the rolling schedule in an empirical way.

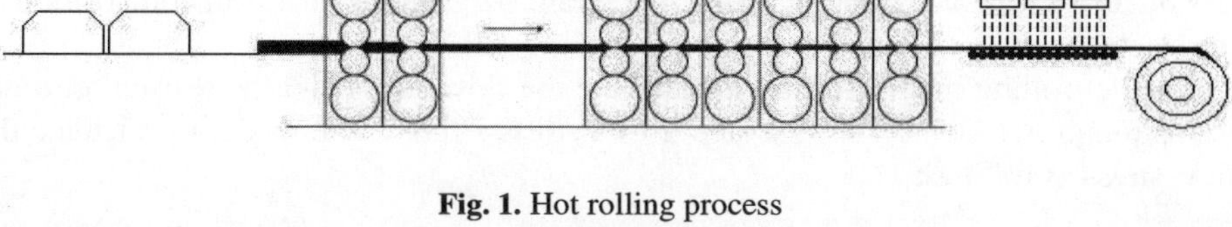

Fig. 1. Hot rolling process

J.M. Corchado et al. (Eds.): DCAI 2008, ASC 50, pp. 471–480, 2009.
springerlink.com

In essence, the hot rolling scheduling problem consist in determining the reductions for every rolling pass in order to obtain the final thickness, considering that the rolling power should be lower than the motor power.

2 Related Work

The hot rolling scheduling problem has been the subject of several research works. Nolle & Armstrong propose the optimization of a 7-stand hot rolling mill using simulated annealing (SA) and genetic algorithms (GAs) [2]. The goal was to optimize the surface quality of a steel slab. In this work, the SA shows a better performance than GAs. Oduguwa & Tiwari propose a general methodology to solve problems of sequential processes using GAs [3]. The proposal consists in a binary representation of the full set of parameters as a sub-set of strings using a multi-objective model, specifically applied to multi-pass hot rolling example. The goal was to maximize the system productivity by optimizing the rolling force. Chakraborti [4] applied GAs to the problem of minimizing the hot rolling time in a reversing mill stand by determining the optimum number of odd passes. In this work the efficiency of GAs to calculate a suitable hot rolling schedule, with respect to traditional methods, is demonstrated. Another contribution of Chakraborti [5] was the study of surface profiles of rolled slabs. In this case, two objective functions were applied to evaluate the wearing and deflection of the rolls as the main factors of the variation of the thickness during the rolling process. The GA produces good quality solutions with respect to the solutions corresponding to the industrial data [6]. Other approaches to determine hot rolling schedules have been applied, such as neural network [7, 8], fuzzy logic [9], and finite element methods [10]. Currently the more successfully approach to solve the hot rolling scheduling problem is the genetic algorithm approach.

In this work the chemical composition of the steel slab is incorporated, allowing the automatic setup of the rolling schedule for different types of steel to be rolled in the mill. To validate the approach, a six-stand rolling mill is modeled as an optimization constrained problem and a set of industrial cases is considered.

3 Hot Rolling Model

The process parameters to roll the steel are obtained using a rolling model. Fig. 2 shows the hot rolling schedule flowchart to calculate the parameters, for a rolling mill with n deformation passes of the roll stand i.

In this work we use the Hernández model to predict the flow stress of any kind of steel, independently of their chemical composition [11-14]. The flow stress parameters include temperature, strain, strain rate, grain size and the chemical composition of the steel.

The hot rolling model is used to calculate the flow stress and the resistance to deformation in two steps. The first step uses a stress-strain curve model to calculate the flow stress as follows:

Let

Z_i the Zener-Hollomon parameter in stand i,
A =(12.19 + 65.58·%C -49.05·%Nb) exp (7.076·Q) and
Q = 267,000-253552·%C +1010·%Mn +33,620.7·%Si + 35,651.28·%Mo +93,680.52·%Ti $^{0.5919}$+ 70,729.85·%Nb$^{0.5649}$+ 31,673.46·%V

where:

%C : percentage weight of carbon.
%Mn : percentage weight of Manganese.
%Si : percentage weight of Silicon.
%Mo : percentage weight of molybdenum.
%Nb : percentage weight of Niobium.
%Ti : percentage weight of Titanium.
%V : percentage weight of Vanadium.

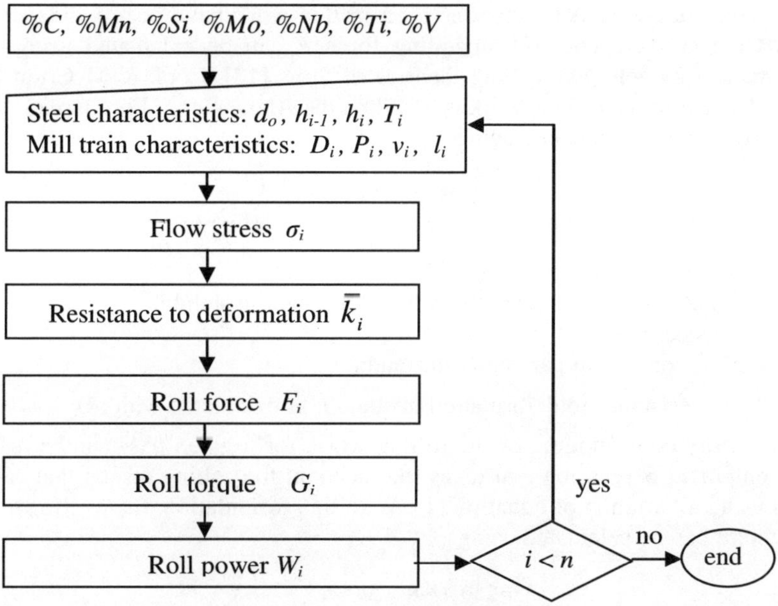

Fig. 2. Hot rolling schedule flowchart

Once the above parameters are determined, the flow stress is calculated using equation (1).

$$\sigma_i = B_i \cdot [1 - \exp(-C_i \varepsilon_i)]^{n_i} - B'_i \left\{ 1 - \exp\left[-K_i \left(\frac{\varepsilon_i - a\varepsilon_{p_i}}{\varepsilon_{p_i}} \right)^{m_i} \right] \right\} \qquad (1)$$

where:

a : empirical parameter with value 0.95.
ε_i: steel strain in stand i.
ε_p : peak strain for a given steel composition.

Z_i : Zener-Hollomon parameter in stand i .
B , B'_i, K_i, C_i, parameters dependent of Z_i/A.

$$A = (12.19 + 65.58{\cdot}\%C - 49.05{\cdot}\%Nb)\ \exp\ (7.076E{-}05{\cdot}Q) \tag{2}$$

$$Q = 267{,}000 - 253552{\cdot}\%C + 1010{\cdot}\%Mn + 33{,}620.7{\cdot}\%Si + 35{,}651.28{\cdot}\%Mo + 93{,}680.52{\cdot}\%Ti^{0.5919} + 70{,}729.85{\cdot}\%Nb^{0.5649} + 31{,}673.46{\cdot}\%V \tag{3}$$

In the second step, the resistance to deformation [15-16] is calculated as follows:

$$\bar{k}_i = \frac{1}{\alpha_i} \int_0^{\alpha} \sigma_i d\alpha_i \tag{4}$$

where:

α_i : angle between the steel and the roll in stand i.
σ_i : plane flow stress in stand i.

The constitutive model equations used in this work have been validated through industrial data [17]. The roll-separating force F can be calculated using different mathematical models, like rolling theories of Sims [18], Cook & McCrum [19] and Alexander & Ford [20]. This work is based on the Alexander & Ford model.

The roll force is calculated using:

$$F_i = \frac{X}{4} w \cdot L_i \cdot \left(\pi + z_{a_i}\right) \cdot \bar{k}_i \tag{5}$$

where:

X : parameter with value 1.07
L_i : contact arc between the roll and the steel in stand i.
w : steel width.
z_{ai} : geometrical parameter in stand i.
$\bar{k}_i$: resistance to deformation in stand I, as calculated with (4)

The energy consumption, or the rolling work, for a given pass can be determined by an empirical expression that takes into account the rolling torque that can be obtained using a variation of equation (1). With this formulation, the overloading of the main motor can be calculated using:

$$G_i = 250 \cdot w \cdot R'_i \cdot \Delta h_i \left(\pi + \frac{z_{a_i}^2}{z_{p_i}} \right) \cdot \bar{k}_i \tag{6}$$

where:

R'_i : roll radius with correction plane in stand i.
za_i : geometrical parameter in stand i.
zp_i : geometrical parameter in stand i.
Δh_i: difference between the final and the initial thickness in stand i.

Finally the rolling power can be calculated as follows:

$$W_i = 2 \cdot \pi \cdot G_i \frac{RPM_i}{60} \tag{7}$$

where:

G_i : roll torque in stand i.
RPM_i = revolutions turns per minute in stand i.

4 Instance Description

Industrial data was obtained from the Hylsa Monterrey Company and from the software HSMM of INTEG Process Group [21]. Table 1 shows the chemical compositions of the different type of steels considered in this work. The first column contains the steel identifier. Columns two to six indicate the contained percentage weight of carbon, manganese, silicon, niobium, titanium and vanadium.

Table 1. Used steels chemical composition

Steels Id.	% C	% Mn	% Si	% Nb	% Ti	% V
1	0.045	0.45	0.069	0.0056	0.002	0.080
2	0.038	0.300	0.009	0.005	0.002	0.002
3	0.082	0.480	0.045	0.036	0.002	0.002
4	0.071	0.758	0.014	0.023	0.013	0.003
5	0.0028	0.170	0.009	0.035	0.035	0.005
6	0.053	0.784	0.010	0.026	0	0

Considering six different types of steel, 17 cases using industrial data were defined, which can be consulted in [22] for further reference. Each case defines the parameters of a different rolling problem. A rolling problem consists in determining the intermediate reductions needed to roll the steel slab in order to obtain the desired final thickness in a 6-stand roll mill. Table 2 shows the parameters for each case: the data source, the number rolling stands (n), the instance name, the initial thickness (h_0), the final thickness (h_f), initial width (w) ,chemical composition (%C, %Mn, %Si, %Mo, %Nb, %Ti, %V), and for each rolling stand the roll diameter (D_i), the roll speed (v_i), the temperature (T_i), the grain size (d_{0i}). Also the source of the data is indicated (Hylsa Monterrey or software HSMM of INTEG Process Group), the motor power (P_i) and the inter-stand distance (l_i) are indicated.

Table 2. Hot rolling scheduling problem cases

Industrial Data: Hylsa		$n = 6$		Name: hyl001.txt		
$h_0 = 48$ mm		$h_f = 3.8$ mm		$w = 991$mm		
%C=0.053, %Mn=0.784, %Si=0.017, %Mo=0, %Ti = 0, %Nb = 0, %V = 0						
Roll Pass No.	1	2	3	4	5	6
D_i (mm)	752	764	758	492	456	474
v_i (m/s)	0.81	1.43	2.21	3.38	4.57	5.54
T_i (°C)	1010	987.64	964.25	942.7	927.32	908.27
D_{0i} (μm)	400	100	80	60	40	20
P_i (kW)	7000	7000	7000	7000	7000	7000
l_i (m)	3.5	3.5	3.5	3.5	3.5	

5 Formulation of the Optimization Problem

Once all the parameters of each individual case have been defined for the hot rolling scheduling problem, the goal is to determine the optimal intermediate thicknesses h_1, ..., h_{n-1} to minimize the total rolling time:

$$t = \sum_{i=1}^{n} t_i$$

To calculate the total rolling time t, the process time in each stand is added. These are calculated by adding the contact time between the roll and the steel, and the required time to transfer the steel slab from one stand to another. Thus, the formulation takes into account the roll radius R_i, peripheral roll speed v_i, inter-stand distance l_i, initial thickness h_i and final thickness h_{i-1}.

Therefore, the rolling time in a stand can be calculated as:

$$t_i = \frac{\sqrt{\Delta h_i \cdot R_i} + l_i}{v_i} \tag{8}$$

Where:

R_i : roll radius in stand i.
l_i : inter-stand distance
v_i : roll speed in stand i.
Δh_i: $h_{i-1} - h_i$

The problem includes the following two constraints:

1. In every rolling stand a reduction will be applied until the final thickness is obtained, defining that each intermediate thickness should be lower than the previous one, to avoid unnecessary rolling steps where no reduction of the thickness is produced. This constrain can be expressed as:

$$h_0 > h_1 > h_2 > h_3 > h_4 > h_5 > h_f$$

2. To achieve the thickness reduction in a rolling stand the obtained rolling power should be lower than the real motor power:

$$W_i < P_i \qquad \text{for } i = 1 \ldots. 6$$

6 Industrial Rolling Time Calculation

For each type of steel there is a rolling schedule given by the machine manufacturer. The rolling schedule indicates to the machine operator the reductions that must be applied to roll the steel slab. These reductions are a suboptimal solution of the problem solely based on empirical assumptions. This work considers the parameters for each individual case - as defined in Table 2, the rolling schedule proposed by the manufacturer and the intermediate process time (equation 8) to calculate the total industrial rolling time. Table 3 shows the reduction configurations for the steel specified in the Hylsa Monterrey data and the rolling time calculated with expression (8) for each stand.

Table 3. Total industrial rolling time

Rolling stand No.	D_i mm	v_i m/s	l_i m	Stand exit thick-ness, mm		Rolling time,s
				h_0	48	
1	752	0.81	3.5	h_1	26	4.433
2	764	1.43	3.5	h_2	14.3	2.494
3	758	2.21	3.5	h_3	9.31	1.603
4	492	3.38	3.5	h_4	6.03	1.043
5	456	4.57	3.5	h_5	4.55	0.769
6	474	5.54		h_f	3.8	0.0024
	Total industrial rolling time:					10.3444

The total industrial rolling time will be taken as the reference time to compare the rolling times obtained from the different solution methods.

7 Experimental Results

This section reports the experimental results of the model evaluation. The experiments were carried out with Microsoft Windows Server 2003 for Small Business, dual Xeon CPU 3.06 GHZ, 3.87 GB RAM and the compiler C++.

In the first experiment, the computational cost of the problem solution was studied. The five cases were solved considering different number of rolling passes, based on the Hylsa Monterrey data using a simplified version of the proposed model and a constrained nonlinear optimizer. Figure 3 shows the execution time to solve the hot rolling problem for 2 to 6 rolling passes. As it can be seen, for 6 rolling passes the execution time is increased to 400 CPU sec. Thus it can be conclude that the solution of the hot rolling scheduling problem requires a more efficient approach than the constrained nonlinear optimizer.

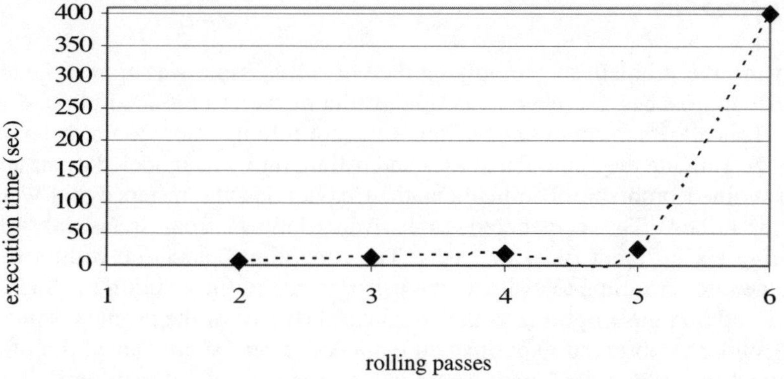

Fig. 3. Execution time with respect to the number of rolling passes

In the second experiment, the 17 defined cases were solved 30 times using a genetic algorithm. The genetic algorithm was configured using 40% of crossover, 50% of mutation, a population of 100 individuals and 100 generations. With the genetic algorithm the obtained results were: total solution average time of 42 CPU seconds and the average time to obtain the best solution of 23 CPU seconds.

Figure 4 shows the typical differences observed between the rolling schedules proposed by the manufacturer and the genetic rolling schedules generated for a given instance. In the graph, the rolling time and the thickness reductions for each one of the six stands are shown.

The global rolling time obtained solving the modeled problem with a genetic algorithm is 0.051% better than the industrial rolling time proposed by the manufacturer. These results show that the proposed model has the capacity of simulating the hot rolling process. Additionally the rolling schedule generated produces softer reductions than the rolling schedule proposed by the manufacturer. This characteristic of the solutions generated using the proposed model, reduces the equipment damage risks.

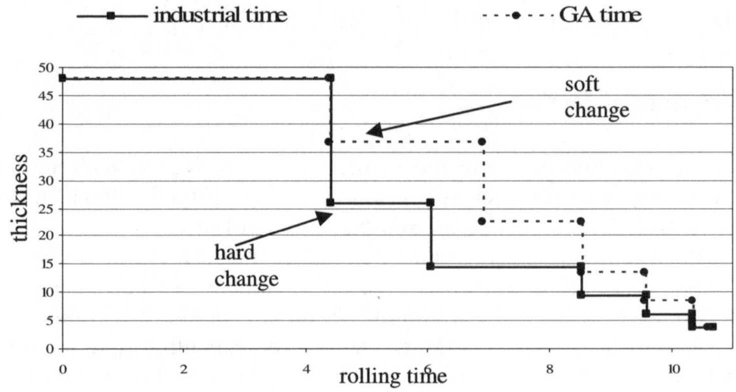

Fig. 4. Comparison of rolling schedule obtained with the GA vs. the industrial schedule

8 Conclusions

In this work the problem of minimizing the hot rolling time was approached. Unlike traditional approaches, the chemical composition of the steel is incorporated as a parameter. This allows the automatic setup of the hot rolling schedule for different types of steel. To validate the approach a six-stand rolling mill was modeled as an optimization constrained problem. The mathematical model and the instances set were completely described. The considered cases were defined from industrial schedules considering six different types of steels. The suboptimal industrial rolling time was used as the reference time to evaluate the performance of the solution methods. Using a constrained nonlinear optimizer, the computational cost of the problem solution was studied. With the observed experimental evidence it can be concluded that, the solution derived from the before mentioned optimizer is not yet optimal, requiring the investigation on alternative approaches, like the one presented here based on genetic algorithms. Finally, the results obtained from the genetic approach, indicate that the proposed model simulates the hot rolling process properly.

Acknowledgments

Thanks to Hylsa Monterrey or software HSMM of INTEG Process Group for their technical support. This research was founded by CONACYT (62233) and IPN (SIP 20082593).

References

1. Dieter, G.E., et al.: Chap 16-Rolling, Handbook of Workability and Process Design. ASM International, Materials Park (2003)
2. Nolle, L., et al.: Simulated Annealing and Genetic Algorithms Applied to finishing mill optimisation for hot rolling of wide steel strip. International Journal of Knowledge-Based Intelligent Engineering Systems 6(2), 104–111 (2002)
3. Oduguwa, V., Tiwari, A.: Sequential Process Optimisation Using Genetic Algorithms. LNCS, pp. 782–791. Springer, Berlin (2004)
4. Chakraborti, N., Kumar, A.: The Optimal Scheduling of a Reversing Strip mill: Studies using Multipopulation Genetic Algorithms and Differential evolution. Materials and Manufacturing processes 18(3), 433–445 (2003)
5. Chakraborti, N., et al.: Optimizing Surface Profiles during Hot Rolling: A Genetic Algorithms Based Multi-objective Optimization Practical Approaches to Multi-Objective Optimization. Comp. Mat. Sci. 37, 159–165 (2006)
6. Chakraborti, N.: Genetic algorithms in materials design and processing. Int. Mat. Rev. 49(3-4), 246–260 (2004)
7. Gorni, A.: The Application of Neural Networks in the Modeling of Plate Rolling Process, JOM-e (1992)
8. Nolle, L., et al.: Optimum Work Roll Profile Selection in the Hot Rolling of Wide Steel Strip using Computational Intelligence. LNCS, vol. 1625, p. 435 (1999)
9. Pataro, C., Herman, H.: Direct determination of sequences of passes for the strip rolling by means of fuzzy logic rules. In: Proc. 2d. Conf. on Intell. Proc. and Manuf. of Mat., IPMM, p. 549 (1999)
10. Shivpuri, R., Kini, S.: Application of Fuzzy Reasoning Techniques for Roll Pass Design Optimisation. In: Proc. 39th Mech. Work and Steel and Symp. on New Metal Form. Proc., p. 755 (1998)
11. Medina, S.F., Hernández, C.A.: General Expression of the Zener-Hollomon Parameter as a Function of the Chemical Composition of Low Alloy and the Microalloyed Steels. Acta Metall. et Mater 44(1), 137–148 (1996)
12. Medina, S.F., Hernández, C.A.: The Influence of Chemical Composition on Peak Strain of Deformed Austenite in Low Alloy and Microalloyed Steels. Acta Metall. et Mater 44(1), 149–154 (1996)
13. Hernández, C.A., et al.: Modelling Austenite Flow Curves in Low Alloy and Microalloyed Steels. Acta Metall. et Mater 44(1), 155–163 (1996)
14. Medina, S.F., Hernández, C.A.: Modelling of the Dynamic Recrystallization of Austenite in Low Alloy and Microalloyed Steels. Acta Metall. et Mater 44(1), 165–171 (1996)
15. Wusatowski, Z.: Fundamentals of Rolling. Pergamon Press, Oxford (1969)

16. Roberts, W.: Hot Rolling of Steel. CRC, Boca Raton (1983)
17. Hernández, C.A., Mancilla, J.E.: Evolución de los precipitados en un acero micro-aleado al Nb fabricado por laminación directa. Informe técnico sobre estudios metalúrgicos y por MET en planchones de acero 7095. HYLSA Monterrey. Com. privada (2003)
18. Sims, R.B.: The Calculation of Roll Forces and Torque in Hot Rolling Mill. Proc. AIME 168(6), 191–200 (1954)
19. Cook, P.M., McCrum, A.W.: The Calculation of Load and Torque in Hot Flat Rolling, Reporte B.I.S.R.A., Londres (1958)
20. Ford, H., Alexander, J.M.: Simplified Hot-Rolling Calculations. J. Inst. Met. 92, 397–404 (1963)
21. HSMM Rel. 3.0. INTEG process group, inc., AISI/DOE Technology (2006)
22. Espriella, F.K.: Optimización mediante algoritmos genéticos: aplicación a la laminación en caliente. Master Science Thesis. Instituto Politécnico Nacional (2008)

Less Expensive Formulation for a Realistic Routing-Scheduling-Loading Problem (RoSLoP)

Juan J. González-Barbosa[1], Laura Cruz-Reyes[1], José F. Delgado-Orta[1], Héctor J. Fraire-Huacuja[1], Guadalupe Castilla-Valdez[1], and Víctor J. Sosa Sosa[2]

[1] Instituto Tecnológico de Ciudad Madero
1°. de mayo s/n Col. Los Mangos, CP. 89440. Ciudad Madero Tamaulipas, México
[2] Centro de Investigación y de Estudios Avanzados (CINVESTAV), Cd. Victoria, Tamps, 87260, México
jjgonzalezbarbosa@gmail.com, lauracruzreyes@hotmail.com, francisco.delgado.orta@gmail.com, hfraire@prodigy.net, gpe_cas@yahoo.com.mx

Abstract. In this paper the Routing-Scheduling-Loading Problem (RoSLoP) is approached. This is a rich bin packing (BPP) and vehicle routing (VRP) problem formulated to satisfy the transportation requirements of a bottling company located in Mexico. The initial formulation of the problem uses 2^9 integer variables and 30 constraints making difficult to find the exact solution even for small instances. In this work it is proposed a transformation function that reduces the size of the problem formulation which allows obtaining the optimal solution of small instances using an exact algorithm. Experimental results of the performance evaluation of an approximated solution method, with regard to the optimal solution, are showed. It is important to emphasize, that this is the first time that this kind of evaluation is carried out for RoSLoP. In the experiments a set of 12 test instances were selected from the company database. The experimental evidence shows that the transformation function reduces 97% the number of customers orders. The percentage quality error for the traveled distance was 0% and for the vehicles used was 6.19%. Now these results can be used to evaluate the performance of any new approximation solution method of RoSLoP.

Keywords: Complexity, Routing-Scheduling-Loading Problem (RoSLoP), Vehicle Routing Problem (VRP), Bin Packing Problem (BPP).

1 Introduction

The distribution and delivery processes are inherent to many companies. Delivery of goods in appropriate time using the minimum quantity of resources reduces operation costs, yielding savings between 5% to 20 % in the total costs of the products [1]. Currently many researchers are working with real based transportation problems, formulating rich models and developing efficient algorithms to solve them. Many simple examples of reductions have been proposed, as the reduction of combinatorial optimization problems to the general zero-one integer linear programming problem [2]. Edmonds reduces graph theoretical problems ("covering all edges with a minimum number of vertex" and "finding a maximum independent set of vertex") to the general "set covering problem" [3]. Dantzig, Blattner and Rao propose to reduce the Traveling Salesman Problem (TSP) to the "shortest path problem" with negative edges

J.M. Corchado et al. (Eds.): DCAI 2008, ASC 50, pp. 481–489, 2009.
springerlink.com

lengths [4]. RoSLoP is a high-complexity problem associated with the transportation of bottled products in a company located in north eastern Mexico [5]. In this work a reduction method for RoSLoP, which permits diminish the dimensionality of the problem, is proposed. The general idea consists in the reduction of the BPP variants of RoSLoP into a one-dimension BPP (BPP1D), using a real number to represent an object in the domain of multiple variants.

The transformation method is used to improve the performance of the heuristic solution method proposed in [5], which solves the rich VRP-BPP with 11 VRP variants: Capacitated VRP (CVRP), VRP with Multiple use of the vehicles (VRPM) , Heterogeneus VRP (HVRP), VRP with Time Windows (VRPTW), Split Delivery VRP (SDVRP), site dependent VRP (sdVRP), road dependent VRP (rdVRP), VRP with Multiple Time Windows (VRPMTW), Constrained Capacity VRP (CCVRP), Depot Demand VRP (DDVRP) and Multiple Depot VRP (MDVRP); and three BPP variants: BPP with Constrained Capacity (BPPCC), BPP On-Line (BPPOn), and Multiple Destination BPP (MDBPP). Rich VRP-BPP and the included variants are defined in next sections.

2 Vehicle Routing Problem (VRP) Related Work

VRP is a classic *NP*_hard combinatorial optimization problem [6]. It consists in one or several depots, a fleet of m available vehicles, a set of n customers, and a graph $G(V,E)$. $V=\{v_0, v_1, v_2, \ldots, v_n\}$ is the set of vertex v_i, where v_0 represents the depot and v_i (for $i > 1$) represents the customers, each customer has a demand of products q_i required to the depot; and $E=\{(v_i, v_j) \mid v_i, v_j \in V, i \neq j\}$ is the set of edges. Each edge has an associated value c_{ij} that represents the transportation cost from v_i *to* v_j The problem consists of determining a set R of routes such that each route starts and ends at the depot, each vertex is visited only once by each route, and the length of each route must be less than or equal to L; minimizing the operation total cost.

The VRP variants add several constraints to the basic VRP such as capacity of the vehicles (Capacitated VRP, CVRP) [7], independent service schedules for the customers and the vehicles (VRP with Time Windows, VRPTW-VRPMTW) [8], multiple depots to satisfy the demands (Multiple Depot VRP, MDVRP) [9], customers to be satisfied by different vehicles (Split Delivery VRP, SDVRP) [10], a set of available vehicles to satisfy the orders (site dependent VRP, sdVRP) [11], customers that can ask and return goods to the depot (VRP with Pick up and Delivery, VRPPD) [12], dynamic facilities (Dynamic VRP, DVRP) [13], line-haul and back-haul orders (VRP with Backhauls, VRPB) [14], stochastic demands and schedules (Stochastic VRP, SVRP) [15], multiple use of the vehicles (VRP with Multiple use of vehicles, VRPM) [15], a heterogeneous fleet to orders delivery (Heterogeneous VRP, HVRP) [16], orders to be satisfied in several days (Periodic VRP, PVRP) [12], constrained capacities of the customers for docking and loading the vehicles (Constrained Capacity VRP, CCVRP) [5], transit restrictions on the roads (road dependent VRP, rdVRP) [5], depots ask for goods to another depots (Depot Demand VRP, DDVRP) [5], and vehicles that can end their travel in several facilities (Open VRP, OVRP).

A rich VRP variant is defined as an *Extended Vehicle Routing Problem*, which is applied to model real transportation problems [17]. This kind of problems are modeled using the classical formulation of Dantzig. However, it requires the addition of constraints that represent real transportation conditions, which makes most difficult the problem solution using exact algorithms, even for small instances.

3 Bin Packing Problem (BPP) Related Work

Bin Packing Problem is considered an intractable problem because the quantity of required resources to solve it grows up as an exponential or a high-grade polynomial function [18]. The BPP problem consists of an unlimited number of bins with capacity c and a number of items n, the size of each item are s_i. The objective is to determine the smallest number of bins m in which the items can be packed [19]. This is the definition of the one-dimension bin packing problem (BPP 1D).

In the literature has been described several variants of BPP: BPP with Capacity Constrained (BPPCC), in which the bins have a specific capacity and variable; in BPP Cardinality Constrained (BPPcC), the number of items that can be placed in a bin is limited; BPP On-line (BPPOn) establishes a known total number of elements to be accommodated at the beginning of the process; the BPP with Fragile Objects (BPPFO) where each item has a threshold of supported weight on itself to not suffer damage or deterioration; Multiple Destinations BPP (MDBPP), defines a set of items that can be unloaded in multiples destinations [20-24]. RoSLoP formulation solves the five defined BPP variants through the deterministic algorithm DiPro [5, 22]. The proposed method in this work focuses to build an exact solution considering only the BPPCC, MDBPP and BPPOn variants.

4 RoSLoP Definition

RoSLoP involves three tasks: routing, scheduling and loading. RoSLoP was formulated using two classical problems: VRP for routing and scheduling and BPP for loading. The specific elements of the case of study are the following:

- A set of facilities that includes customers C and depots D. Each facility has a finite capacity for the attention of the vehicles. The facilities have independent service schedules $[st_j, et_j]$, where st_j and et_j are the time when the facility j starts and ends its operation respectly. The travel time between a pair of facilities i and j is represented by t_{ij}. An integer variable x_{ijk} is used to assign an arc (i,j) to the route k. Depots have the possibility of request goods to other depots. A set of routes K_d that starts and ends at the depot d must be formed.
- A fleet of vehicles with heterogeneous capacity V_d, a service time $stime_v$ and an attention time tm_{vj}. The attention time depends of the capacity of the vehicle C_{vj} to visit a customer j and of the available people for docking and loading the containers $Pallets_v$ of the vehicle v where the goods will be transported. An integer variable y_{vk} is used to assign a vehicle v to a route k.
- A set of roads represented by the edges of a given graph. Each road has an assigned cost c_{ij}, a threshold of allowed weight $MAX_{Load_{vj}}$ for a determined vehicle v that travels towards a facility j, and a travel time t_{ij} from facility i to j.

- A set of orders Q to be deliver at the facilities of the customers, formed by units of products per each customer $I_j \in Q$. Each box (product package) has different attributes such as weight, high, product type, supported weight, packing type and beverage type.

Fig. 1 shows a graphical description of RoSLoP as VRP-BPP formulation.

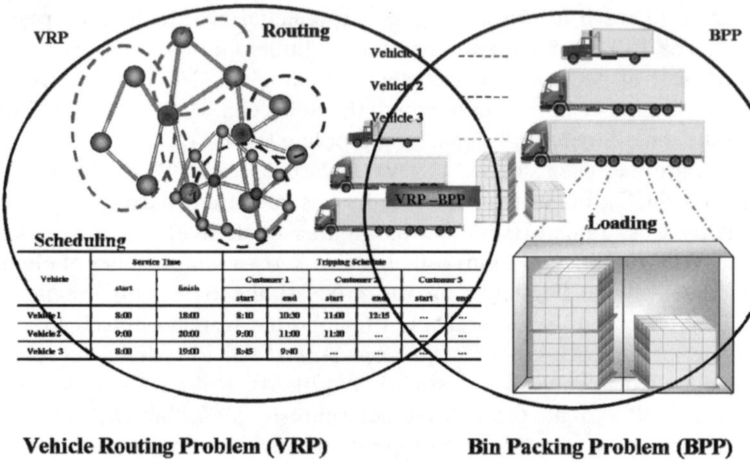

Fig. 1. Definition of RoSLoP

The objective of RoSLoP is to minimize the number of vehicles and the traveled distance, satisfying the orders of the customers.

The set Q is reduced to a linear representative dataset, which is created with elements in the domains of the variants rdVRP, CVRP, HVRP, BPPCC, BPPOn and MDBPP. Therefore, RoSLoP is solved as a rich VRP-BPP variant.

5 Reduction Method

The reduction method uses a transformation function that apply the learned experience from the case of study: "Bigger objects are the most weighted". It consists of two steps: *a*) the construction of loading units using the DiPro algorithm and *b*) the transformation of these units in a set of real numbers. The DiPro construction method creates the loading units, clustering the products into homogeneous platforms constituted by products of the same type, and heterogeneous platforms formed with different types of products with similar characteristics. Each platform is defined as a pair (w_i,h_i), where w_i and h_i are the width and high for each platform. Once that DiPro is invoked, each pair is transformed into a real number $item_i$, defined by the equation (1), which is used to solve the BPP variants related to RoSLop (BPPCC, BPPOn and MDBPP).

$$item_i = \frac{h_i w_i}{h_i + w_i} \qquad i \in I_j \tag{1}$$

The capacity C_{vj} of the vehicle v that visit the node j is transformed similarly. Each container of a trailer has two attributes: high $h_{pallet_{ij}}$ and weight $w_{pallet_{ij}}$ of the assigned load to deliver to the customer j. The width of the load is determined by a categorization of products to adjust the load into the containers. It is assumed that the weight of the assigned load to each container is uniform. Equations (2) and (3) are used to calculate the vehicle capacity, ensuring that the dimensions of the load objects and the vehicles are equivalent according to the VRP-BPP formulation.

$$w_{pallet_{ij}} = \frac{MAX_{Load_{vj}}}{Pallets_v} \qquad j \in C, v \in V_d \tag{2}$$

$$C_{vj} = \sum_{i=1}^{Pallets_v} \frac{h_{pallet_{ij}} w_{pallet_{ij}}}{h_{pallet_{ij}} + w_{pallet_{ij}}} \qquad j \in C, v \in V_d \tag{3}$$

These parameters are used to formulate RoSLoP as an integer problem.

6 Integer Formulation for RoSLoP

The objective of RoSLoP is to minimize the number of vehicles assigned to the routes and the total traveled distance, visiting all the customer facilities and satisfying the demands. Each route k is formed by a subset of facilities to be visited and has a length φ_k. Equation (4) defines the maximum covering set established by the use of variant HVRP. Equations (5) and (6) define the length and the travel time on a route k.

$$|K_d| = |V_d| \left\lceil \frac{\arg max(I_j)}{\arg min(C_{vj})} \right\rceil \qquad j \in C, K_d \in K, v \in V_d \tag{4}$$

$$\varphi_k = \sum_{j \in C \cup D} \sum_{i \in C \cup D} c_{ij} x_{ijk} \qquad k \in K, v \in V_d \tag{5}$$

$$t_k = \sum_{j \in C \cup D} \sum_{i \in C \cup D} t_{ij} x_{ijk} + \sum_{i \in C \cup D} \sum_{j \in C \cup D} \sum_{v \in V_d} tm_{vj} x_{ijk} y_{vk} \qquad k \in K_d \tag{6}$$

The equation (7) is the objective function of the problem. The problem consists in minimizing the number of assigned vehicles and the total length of the generated routes. Equations (8)-(10) solve the variants DDVRP and MDVRP through the solution of the associated Traveling Salesman Problem (TSP). Equation (8) restricts that each edge (i, j) on a route k to be traversed only once. Equation (9) ensures that a route k is continuous. Equation (10) optimizes the covering set problem related with the objective function.

$$\min z = \sum_{k \in K_d} \sum_{v \in V_d} \varphi_k y_{vk} \tag{7}$$

$$\sum_{i \in C \cup D} x_{ijk} = 1 \qquad k \in K_d, j \in C \cup D \tag{8}$$

$$\sum_{i\in C\cup D} x_{ijk} - \sum_{i\in C\cup D} x_{jik} = 0 \qquad k\in K_d, j\in C\cup D \tag{9}$$

$$\sum_{i\in C\cup D}\sum_{j\in C\cup D} x_{ijk} \geq 1 \qquad k\in K_d \tag{10}$$

Equations (11)-(13), compute the time used by a vehicle assigned to route k through the use of a_{jk} and l_{jk}, which represent the arrival and departure time respectively. Equation (14) ensures that the use of a vehicle does not exceed the attention time of the facility j. These equations permit solving the variants VRPTW, VRPMTW and VRPM. The variants CCVRP and SDVRP are solved using the equation (15), which ensures that two routes k and k' do not intersect themselves in facilty j.

$$t_k y_{vk} \leq stime_v \qquad k\in K_d; v\in V_d \tag{11}$$

$$l_{jk} \geq a_{jk} \sum_{i\in C\cup D} t_{ij}x_{ijk} + \sum_{v\in V_d} tm_{vj}y_{vk} \qquad j\in C\cup D, k\in K_d \tag{12}$$

$$a_{jk} \sum_{i\in C\cup D} x_{ijk} = \sum_{i\in C\cup D} t_{ij}x_{ijk} \qquad j\in C\cup D, k\in K_d \tag{13}$$

$$st_j \leq a_{jk} \leq et_j \qquad j\in C, k\in K_d \tag{14}$$

$$a_{jk} \leq l_{jk} \leq a_{jk'} \qquad k<k', \forall k, \forall k'\in K_d \tag{15}$$

Equations (16)-(18), combined with the linear transformation function, define the restrictions for variants CVRP, sdVRP, rdVRP, HVRP, MDBPP, BPPOn and BPPCC. Equation (16) establishes that a vehicle is assigned to a route k. Equation (17) ensures that vehicle capacities are not exceeded. Equation (18) establishes that all goods are delivered and all demands are satisfied.

$$\sum_{v\in V_d} y_{vk} \leq 1 \qquad k\in K_d \tag{16}$$

$$\sum_{i\in C\cup D}\sum_{r\in I_j} item_r x_{ijk} \leq C_{vj} y_{vk} \qquad j\in C\cup D, k\in K_d, v\in V_d \tag{17}$$

$$\left|I_j\right| - \sum_{i\in C\cup D}\sum_{r\in I_j} item_r x_{ijk} = 0 \qquad j\in C\cup D, I_j\in Q \tag{18}$$

The initial formulation of RoSLoP proposed in [25], uses 2^9 integer variables and 30 constraints. This new formulation requires 2^2 integer variables and 15 contraints. Both formulations can solve 11 VRP variants and three BPP variants.

7 Experiments and Results

In this section, the experimental results of the performance evaluation, with regard the optimal solution, of an approximation solution method of RoSLoP is showed. It is important to underline, that this kind of study is carried out for RoSLoP for the first time. Real instances were provided by the bottling company. They were solved using the metaheuristic algorithm, named Linear Heuristic-Based System for Assignment of Routes, Schedules and Loads (LHBS-ARSL), which uses the transformation function proposed in this work. To approximate the solutions of DiPro [22], the transformation function was applied to the heuristic algorithm developed in [5]. The metaheuristic algorithm was coded in C# and the results were compared with those obtained with the exact method reported in [26]. This method applies the proposed transformation function and uses the linear optimizer LINDO API v4.1 to obtain the optimal solution for each RoSLoP instance. A set of 12 instances were selected from the database of the company, which contains 312 instances classified by order date, 1257 orders and 356 products in its catalogues. Eight available vehicles were disposed for the experimentation in a graph with ten edges. The results are shown in Table 1.

Table 1 shows that the transformation function reduces the orders dataset size from 10457 to 238 items (97%). As we can see the LHBS-ARSL algorithm reaches a quality solution error for the traveled distance of 0% and for the vehicles used of 6.19%. Currently, these results are the reference to evaluate the performance of new approximation solution methods of RoSLoP.

Table 1. Experiments with real-world instances, provided by a bottling company

Instance	***C***	***Q***		**LHBS-ARSL**			**Exact Method**	
-	**-**	**boxes**	**units**	**Travel. Dist.**	**Used Vehic.**	**Time to best (CPU sec)**	**Travel. Dist.**	**Used Vehic.**
06/12/2005	4	6928	158	1444	4	12.03	1444	3
09/12/2005	5	7600	171	1580	5	15.64	1580	5
12/12/2005	7	11541	250	2500	6	19.58	2500	6
01/01/2006	6	9634	286	2560	7	15.31	2560	6
03/01/2006	4	5454	116	1340	4	8.96	1340	4
07/02/2006	6	12842	288	2660	7	20.03	2660	6
13/02/2006	5	9403	208	1980	5	15.93	1980	5
06/03/2006	6	9687	224	1960	5	16.49	1960	5
09/03/2006	6	12319	269	2570	6	18.70	2570	6
22/04/2006	8	16903	381	3358	7	18.56	3358	7
14/06/2006	7	10822	245	2350	6	16.09	2350	6
04/07/2006	7	12347	270	2640	6	16.92	2640	6
Average	**6**	**10457**	**238**	**2245**	**5.66**	**16.18**	**2245**	**5.33**

8 Conclusions and Future Work

In this paper the Routing-Scheduling-Loading Problem (RoSLoP) was approached. The initial formulation of the problem uses 2^9 integer variables and 30 constraints. In this work a transformation function that reduces the size of the problem formulation was proposed. The new formulation of RoSLoP now can be used to evaluate the performance of approximate algorithms with regard to the optimal solution. The reduction allows to obtain the optimal solution of small instances using an exact algorithm. In the experimental performance evaluation of the metaheuristic solution method used, a set of 12 test instances were selected from the database of the bottling company. The transformation function reduced the number of orders in 97%. The quality error observed for the traveled distance was 0% and for the vehicles number was 6.19%. Currently these results are the reference to evaluate the performance of new approximation solution methods of RoSLoP. A future work is the definition of a transformation method independent of the problem.

Acknowledgments. This research was supported by COTACYT, CONACYT and DGEST.

References

1. Toth, P., Vigo, D.: The vehicle routing problem. SIAM Monographs on Discrete Mathematics and App. Society for Industrial and Applied Mathematics (2000)
2. Dantzig, G.: On the significance of solving linear programming problems with some integer variables. Econometrica 28, 30–44 (1960)
3. Edmonds, J.: Covers and packings in family of sets. Bull. Amer. Math. Soc. 68, 494–499 (1962)
4. Dantzig, G., et al.: All shortest routes from a fixed origin in a graph. In: Theory of Graphs: International Symposium, Gordon and Breach, pp. 85–90 (1967)
5. Cruz, L., et al.: An Ant Colony System to solve Routing Problems applied to the delivery of bottled products. In: An, A. (ed.), pp. 68–77. Springer, Heidelberg (2008)
6. Dantzig, G.: Discrete-variable extremum problems. Operation Research 5, 266–277 (1957)
7. Shaw, P.: Using Constraint Programming and Local Search Methods to Solve Vehicle Routing Problems. In: Maher, M. (ed.) 40 Conf. on Principles and Practice of Constraint Programming, pp. 417–431. Springer, Heidelberg (1998)
8. Cordeau, F., et al.: The VRP with time windows. Technical Report Cahiers du GERAD G-99-13, Ecole des Hautes Etudes Commerciales de Montreal (1999)
9. Mingozzi, A.: An exact Algorithm for Period and Multi-Depot Vehicle Routing Problems. Department of Mathematics, University of Bologna, Bologna, Italy (2003)
10. Archetti, C.: The Vehicle Routing Problem with capacity 2 and 3, General Distances and Multiple Customer Visits. Operational Research in Land and Resources Manangement, 102 (2001)
11. Thangiah, S.: A Site Dependent Vehicle Routing Problem with Complex Road Constraints. Artificial Intelligence and Robotics Lab., Slippery Rock University, U.S.A (2003)
12. Dorronsoro, B.: The VRP Web. AUREN. Language and Computation Sciences of the University of Malaga (2005), `http://neo.lcc.uma.es/radi-aeb/WebVRP`
13. Bianchi, L.: Notes on Dynamic Vehicle Routing. Technical Report IDSIA - Institut Dalle Molle di Studi sull'Intelligenza Artificiale, Switzerland (2000)
14. Jacobs, B., Goetshalckx, M.: The Vehicle Routing Problem with Backhauls: Properties and Solution Algorithms. Techincal report MHRC-TR-88-13, Georgia Institute of Technology (1993)
15. Fleischmann, B.: The Vehicle routing problem with multiple use of vehicles. Working paper, Fachbereigh Wirtschafts wissens chaften, Universitt Hamburg (1990)

16. Taillard, E.: A Heuristic Column Generation Method for the Heterogeneous Fleet VRP. Institut Dalle Moli di Studi sull Inteligenza Artificiale, Switzerland. CRI-96-03 (1996)
17. Toth, P., Vigo, D.: An Overview of Vehicle Routing Problems. SIAM Monographs on Discrete Mathematics and Applications. The Vehicle Routing Problem (2000)
18. Garey, M.R., Johnson, D.S.: Computers and Intractability, a Guide to the Theory of NP-completeness. W. H. Freeman and Company, New York (1979)
19. Baase, S.: Computer Algorithms, Introduction to Design and Analysis. Editorial Addison-Wesley Publishing Company, Reading (1998)
20. Kang, J., Park, S.: Algorithms for Variable Sized Bin Packing Problem. Proc. Operational Research 147, 365–372 (2003)
21. Epstein, L.: Online Bin Packing with Cardinality Constraints. In: Proc. 13th European Symposium on Algorithms (2005)
22. Cruz, L., et al.: DiPro: An Algorithm for the Packing in Product Transportation Problems with Multiple Loading and Routing Variants. In: Gelbukh, A., Morales, A.F.K. (eds.), pp. 1078–1088. Springer, Heidelberg (2007)
23. Chan, W., et al.: Online Bin Packing of Fragile Objects with Application in Cellular Networks, tech. report, Hong Kong RGC Grant HKU5172/03E (2005)
24. Verweij, B.: Multiple Destination Bin Packing, tech. report, Algorithms and Complexity in Information Technology (1996)
25. Herrera, J.: Development of a methodology based on heuristics for the integral solution of routing, scheduling and loading problems on distribution and delivery processes of products. MS. thesis. Instituto Tecnológico de Ciudad Madero, México (2006)
26. Delgado, J., et al.: Construction of an optimal solution for a Real-World Routing-Scheduling-Loading Problem. Journal Research in Computing Science 35, 137–146 (2008)

Applying an Ant Colony Optimization Algorithm to an Artificial Vision Problem in a Robotic Vehicle

R. Arnay, L. Acosta, M. Sigut, and J. Toledo

Dep. Systems Engineering and Automatics, University of La Laguna. La Laguna, CP: 28204. España

Abstract. In this paper, a problem of artificial vision in a robotic autonomous vehicle consisting of the real time detection and tracking of non-structured roads is addressed by applying an Ant Colony Optimization (ACO) algorithm. The solution adopted tries to find some properties describing the probability that a pixel belongs to the boundaries of the road, and then formalize the road detection problem as an optimization one.

1 Introduction

Verdino is a low-cost self-guided electrical vehicle that is being developed for carrying out the passengers autonomous trasportation in a closed bioclimatic estate that is actually under construction in the ITER (Instituto Tecnológico y de Energías Renovables; Technological Institute of Renewable Energies) facilities located on the south of Tenerife, Canary Islands. This paper presents an artificial vision algorithm developed as part of the Verdino vehicle. Through the application of an ACO algorithm (Denebourg et al. 1983; Dorigo and Stützle 2004; Heck and Ghosh 2000) to solve a road detection problem in non-structured environments, the artificial ants will supply the lack of edges information under certain conditions in a traditional road segmentation approach. The ants will have a common memory that represents the acquired colony experience (Dorigo et al. 1996; Goss et al. 1990). The algorithm has been implemented under the OpenCv library, which provides complete support for image processing and allows the real time operation.

Some photographs showing the aspect of the non-structured road the vehicle Verdino will navigate are shown in Fig. 1. It can be observed the presence of strong shadows and dust that partially hides the borders. Due to the difficulties this kind of roads entails, traditional road detection will not work properly as there are no lane markes or other clues on which to base de detection.

Other authours have addressed the problem of navigating a robot vehicle on non-structured roads using different techniques. Some of them proposed a solution based on pattern recognition techniques (Crisman and Thorpe 1991), or faced this topic developing an algorithm based on the HSI colour space and 2D-spatial constraints (Sotelo et al. 2004). Others have proposed methods based on the extrapolation of the characteristics of a road pattern (supposed free of obstacles) to detect the road (Dahlkamp et al. 2006). Obviously, the pattern chosen determines the performance of this kind of methods, relying on some other tools (like laser based obstacle detectors) to perform this task. The approach proposed in this paper does not use supplementary

J.M. Corchado et al. (Eds.): DCAI 2008, ASC 50, pp. 490–497, 2009.
springerlink.com

Fig. 1. Real aspect of the roads in the ITER facilities

tools or rely on the uniformity of the road to perform an extrapolation. It tries to detect the road using information of the margins, so the proposed algorithm works under different initial segmentation conditions. Broggi and Cattani have also used the ACO methodology to detect roads (Broggi and Cattani 2006). In this paper, the initial segmentation process is different from the one they propose and the movement rule has been simplified, not implementing the backtracking process. Moreover, in the algorithm presented in this paper the offline pheromone contribution is influenced by a parameter proportional to the distance between the last pixel of the solution and the attraction point.

2 The ACO Approach

The use of an ACO metaheuristic to solve a road detection problem implies the formalization as an optimization problem (Dorigo and Stützle 2002). This means that some local properties of the image pixels have to be found, so a probability of these pixels to belong to the road can be established. The optimal solutions are the ones that overlap with the largest number of pixels belonging to the road boundaries. Two colonies of artificial ants, one for each side of the road (left and right), are to be executed (Heck and Ghosh 2000), so the agents, moving pixel by pixel, will try to find the optimal boundary curve on their side.

As a previous step to the application of the ACO algorithm some preprocessing procedures are executed to obtain the starting and final states, as well as the local properties of the road boundaries pixels. The most important step of the preprocessing consists of obtaining a 'distance image' from the original 320x240 RGB acquired image. This one contains information about edge and colour distance. Applying the Canny operator and some morphological operations, an edges map of the image is obtained. Making use of some road colour features taken from a HLS image obtained from the incoming RGB one and computing a distance to the average for every pixel marked as an edge in the previous step, it is possible to differentiate the contours as belonging to the road or not. The initial states of the two colonies of ants, that are updated frame by frame, are the pixels in the centre of the two starting areas marked with letters C and E in Fig. 2. They are placed in the periphery of the image, where a sufficient percentage of edges is present. The interest areas of each colony, marked with letters B and D in the same figure, are the pixels between the horizon and the initial state lines, while the horizon is marked with letter A.

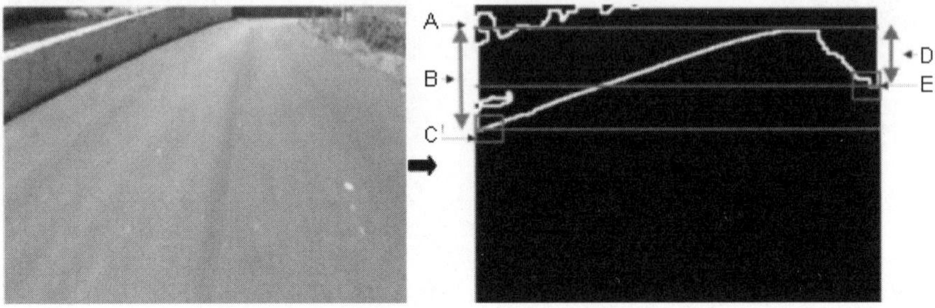

Fig. 2. Areas of interest obtained from the RGB incoming image

3 Application of the ACO Algorithm

3.1 The Point of Attraction

The set of feasible neighbours is computed by making use of the point of attraction. Since the algorithm final objective is the detection of the road in real time, the number of feasible neighbours of a given state must be small enough to make this possible. The question is how to choose the correct neighbours so its number do not degrade the overall performance but allows obtaining good solutions. The point of attraction is used to answer this question.

Given a pixel (i_a, j_a), an agent on $\sigma_h = < \sigma_{h-1}, (i_a, j_a) >$ can only move to the set of pixels with coordinates $\{(Centre(P, i_a, j_a) - range, j_a+1), \ldots, (Centre(P, i_a, j_a) + range, j_a+1)\}$. (i_a, j_a) represents the row-column coordinates of the current pixel, and *Centre*(P, i_a, j_a) is a function of the current pixel and the point of attraction $P = (i_p, j_p)$, that is the nearest integer to $r(j_a+1)$. *Range* determines the number of neighbours to explore, while $r(j)$ is the line connecting the point of attraction and the current agent position computed as indicated in Eq. (1), being $j_a \neq j_p$ always true.

$$r(j) = i_p + \frac{i_a - i_p}{j_a - j_p}(j - j_p) \tag{1}$$

The point of attraction, shown in Fig. 3, is computed frame by frame using information of road patterns previously obtained.

3.2 The Agents Motion Rules

The agents movement can be divided into two leves. The first one is the *random-proportional rule* for artificial ants (Dorigo and Stützle, 2004), in which the probability

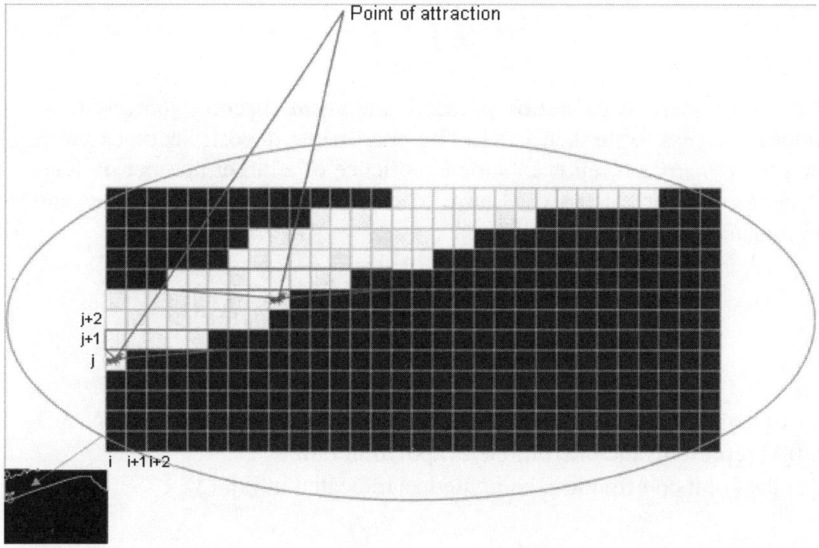

Fig. 3. Feasible neighbours (white pixels) given a point of attraction and an exploration range

that a given state is the next one depends on a parameter α with which it is possible to tune the balance between edge-exploitation and pheromone-exploitation in the agents behaviour, as it can be seen in Eq. (2).

$$p_{ij}^{k}(t)=\begin{cases}\dfrac{\left[\tau_{ij}\right]^{\alpha}\cdot\left[\eta_{ij}\right]^{1-\alpha}}{\sum_{l\in N_i^k}\left[\tau_{ij}\right]^{\alpha}\cdot\left[\eta_{ij}\right]^{1-\alpha}}, & if\ j\in N_i^k\\ 0, & otherwise\end{cases} \qquad (2)$$

Where:

- i is the current state,
- $p_{ij}^{k}(t)$ is the probability of the state $<\sigma_{i,j}>$ of being the next state,
- N_i^k are the neighbours of the agent k in the state i,
- α is a tuning parameter between edge-exploitation and pheromone-exploitation.

The second level of the rule occurs when the denominator of Eq. (2) is equal to zero, so the first one can not be applied. Then, the artificial ant makes a purely random movement.

3.3 Pheromone Updating

The N_a agents of the colony are divided into different subsets, N_{a1}, N_{a2},... N_{an}, each one characterized by different movement rules parameters and executed in sequence. Thus, the agents in subset N_{a1} have α=0, while the agents in subsets N_{a2},... N_{an} have a parameter alpha different from zero, as shown in Eq. (3), being n the number of subsets and α_p a predetermined value.

$$\alpha_i = \frac{i}{n} \cdot \alpha_p \tag{3}$$

As it can be seen, as execution proceeds the agents become more sensitive to the pheromone and less to the heuristics. The pheromone deposit becomes more reliable as time goes by, since it represents the experience of a larger number of agents. Once all the agents of a particular subset have reached the final pixels, the pheromone trails $\tau(t)$ are updated according to Eq. (4).

$$\tau_{ij}(t+1) = (1-\rho) \cdot \tau_{ij}(t) + \rho \cdot \sum_{k=1}^{N_{ax}} \Delta_{ij}^k(t) \tag{4}$$

Where:

- t is a time measure that represents the evolution of the pheromone deposit,
- $\rho \in (0,1]$ represents the pheromone evapotation ratio,
- Δ_{ij}^k is the k-ant contribution, computed as indicated in Eq. (5).

$$\Delta_{ij}^k(t) = \frac{Q}{L_K - L_{K\_best}} \tag{5}$$

In Eq. (5) Q a fixed parameter, L_k the cost of the k-agent, and L_{k_best} the cost of the current best solution. L_{k_best} is computed as shown in Eq. (6), being $edges_{ij}$ the brightness of the pixel (i,j) in the distance image, $|S_k|$ the number of components of the solution, and ς a parameter proportional to the distance between the final pixel of the solution and the point of attraction, making the solutions that end closer to the point of attraction less costly. As it can be seen in Eq. (6), the cost of a solution is inversely proportional to the brightness of the pixels that conformate it.

$$L_{K\_best} = \frac{\sum_{ij \in S_k} (255 - edges_{ij}) + \varsigma}{|S_k|} \tag{6}$$

3.4 Solution Extraction

When all the agents have obtained their particular solutions, two special ones, one for each colony, are executed. They start from the starting states and move to the neighbours with a larger pheromone deposit until they reach a final state. The paths obtained, which are the ones with the largest pheromone deposit, represent the road boundaries, as shown in Fig. 4.

3.5 Road Pattern Updating

The road pattern is as simple as two lines interpolated from the solution pixels sequence. It gives a simplified vision of the road and makes it easy to act the control mechanisms of the vehicle (traction, brake and steering systems). The pattern transforms frame by frame with an inertia that allows a stable detection, without sudden changes, in presence of noise in the capture, degraded edges detection or vibrations in

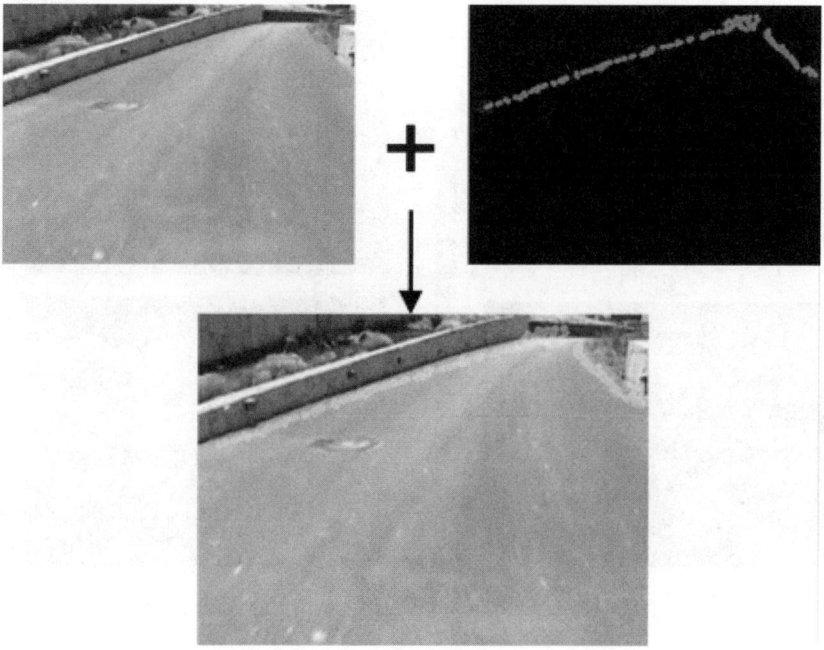

Fig. 4. Solution paths overlaped with the incoming RGB image

the vehicle. Both the point of atraction and the horizon line are computed from the intersection of the two lines of the pattern.

4 Experimental Results

If the road segmentation in the preprocessing step is good enough, just a few agents may be needed to achieve a good detection of the road boundaries. In this case, the use of a larger number of agents will not make the detection to improve. However, the initial detection almost never will be so good. In Fig. 5 it can be seen a case in which the initial segmentation of the road is poor, so when the number of agents increases the quality of the solution improves.

As it can be observed in Fig. 5, when the number of agents is too small there is no time for the pheromone bridges between edge zones to become *solid* enough for the majority fo the agents to follow them. If this number is increased, as the algorithm iterations go by, the better paths between edge zones (see the region stood out by the circle in Fig. 5) become more attractive to the successive generations of agents, so they finally become the preferred ones. Consequently, to increase the number of agents can improve the quality of the solution, but it will also make the processing time per fame to increase.

In Table 1 the average execution times per frame measured for different numbers of agents are shown. The hardware used is an Intel Corel Duo 2.0Ghz, 2GB RAM under Windows OS.

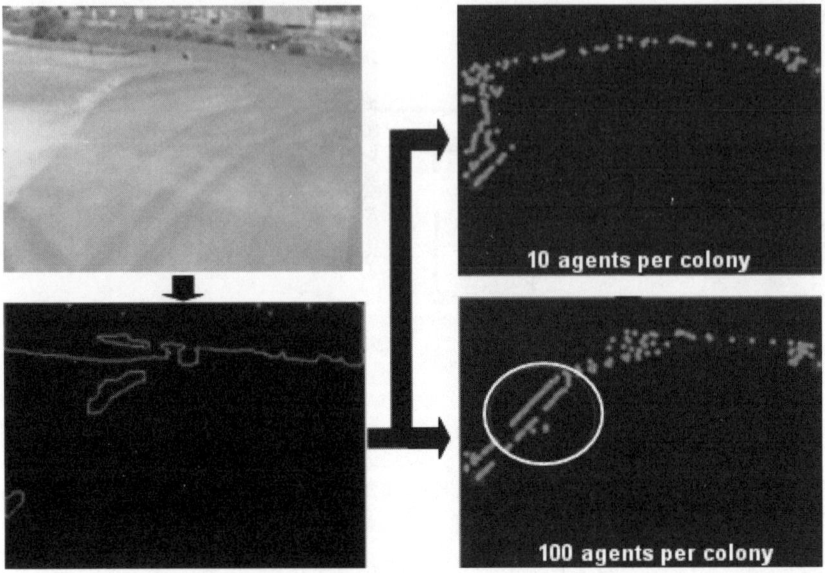

Fig. 5. Pheromone deposits obtained with 10 and 100 agents

Table 1. Average execution times per frame

Number of agents (per colony)	10	50	100	150	200
Time [ms]	16	31	47	78	94

In Fig. 6 three different situations are shown. As it can be seen, in all of them the ACO algorithm proposed by the authors detect the road borders.

Fig. 6. Three different sections of the road with the borders detected by the algorithm

5 Conclusions

In this paper the problem of detecting and tracking in real time a non-structured road has been addressed by using an ACO algorithm in which some new features, such as the polarization of the ants random movement by the attraction point, the adjustment mechanism of the parameter α during each frame processing, or the ability of changing the initial states frame by frame have been incorporated. These improvements complement the basic ACO algorithm and make it more suitable for the roads detection problem. If the quality of the images preprocessing is high a few agents per colony are enough to obtain good road patterns. Otherwise, more agents must be used and a compromise between a good road detection quality and short processing times must be reached with the purpose of ensuring the real time operation. In the case described in this paper, this compromise has been achieved since the correct detection and tracking of the non-structured roads in the ITER facilities is carried out in real time.

Acknowledgments. This work is supported by the Spanish Ministry of Education and Science, through the project SIBTRA (Low cost intelligent system for the transport and surveillance in non-structured ecological environments), with reference DPI2007/64137, and the ITER.

References

Broggi, A., Cattani, S.: An agent based evolutionary approach to path detection for offroad vehicle guidance. Pattern Recognition Letters 27, 1164–1173 (2006)

Crisman, J., Thorpe, C.: UNSCARF: A Color Vision System for the Detection of Unstructured Roads. In: Proc. of the IEEE International Conference on Robotics and Automation, California, pp. 2496–2501 (1991)

Dahlkamp, H., Kaehler, A., Stavens, D., Thrun, S., Bradski, G.: Self-supervised monocular road detection in desert terrain. In: Proc. of Robotics: Science and Systems, Philadelphia (2006)

Denebourg, J.L., Pasteels, J.M., Vergaeghe, J.C.: Probabilistic behaviour in ants: A strategy of errors? J. Theoretical Biology 105, 259–271 (1983)

Dorigo, M., Stützle, T.: The Ant Colony Optimization Metaheuristic: Algorithms, Applications and Advances. In: Glover, F., Kochenberger, G.A. (eds.) Handbook of Metaheuristics. Springer, New York (2002)

Dorigo, M., Stützle, T.: Ant Colony Optimization. MIT Press, Cambridge (2004)

Dorigo, M., Maniezo, V., Colorni, A.: The Ant System: Optimization by a Colony of Cooperating Agents. IEEE Trans. on Systems, Man and Cybernetics – Part B 26, 29–41 (1996)

Goss, S., Beckers, R., Denebourg, J.L., Aron, S., Pasteels, J.M.: How trail laying and trail following can solve foraging problems. In: Hughes, R. (ed.) Behavioral mechanisms of food selection. Springer, Heidelberg (1990)

Heck, P.S., Ghosh, S.: A study of synthetic creativity: behaviour modeling and simulation of an ant colony. IEEE Intelligent Systems 15, 58–66 (2000)

Sotelo, M.A., Rodríguez, F.J., Magdalena, L., Bergasa, L.M., Boquete, L.: A Color-Based Lane Tracking System for Autonomous Driving on Unmarked Roads. Autonomous Robots 16, 95–116 (2004)

Development of a Distributed Facial Recognition System Based on Graph-Matching

Rafael Espí, Francisco A. Pujol, Higinio Mora, and Jerónimo Mora

Specialized Processor Architectures Lab.
Dept. Tecnología Informática y Computación
Universidad de Alicante
P.O.Box 99, E-03080
Alicante, Spain
reb2@alu.ua.es, fpujol@dtic.ua.es, hmora@dtic.ua.es,
jeronimo@dtic.ua.es

Abstract. In this paper we present the methodology and the working plan of a PhD thesis about facial recognition. We propose a high level algorithm based on graph matching that will be adapted to parallel hardware architectures. This architecture must satisfy real-time constraints, and will be implemented through a multi-agent system and grid computing. Real time is incorporated at the low level by redefining the computer vision operators using imprecise computing techniques. The experiments have been performed with the BioID face database and the results show that our system has high recognition rates and low FRR and FAR rates.

Keywords: face recognition, EBGM, Gabor filters, distance measures.

1 Introduction

Facial recognition is the most usual method for a human being in the process of identifying other people, and it is an emerging AI research area where there are still many situations -such as occlusions or the variations of lighting conditions, pose, or facial expression, among others- that make it difficult to properly detect and recognize faces. In spite of the advances carried out in this field in recent years, this type of situations cannot be easily solved by a computer system yet.

The main goal of this PhD thesis is the development of a facial recognition system based on graph matching [1], Gabor filters [2] and linear subspaces [3], taking into account the data parallelism [4] and real time constraints inherent to these kind applications. We shall focus our research on detecting faces in a hardware architecture that implements an algorithm based on the Elastic Bunch Graph Matching (EBGM) method to match faces.

This paper is organized as follows: a revision of the main techniques of face recognition related with our proposal is outlined in Sec. 2. Then a description of the system we have implemented and the results of the experiments are shown in Sec. 3 and, finally, we shall conclude with some remarks to our work in Sec. 4.

J.M. Corchado et al. (Eds.): DCAI 2008, ASC 50, pp. 498–502, 2009.
springerlink.com

2 Background and a System Proposal

Two problems are tackled in the development of facial recognition systems: face detection and face authentication. In both problems, there are generally two types of methods [5]: methods based on feature extraction (featured-based) that use explicit data based on colour models and/or geometrical data, and global methods (image-based) where the information is focused on the recognition of patterns.

Global techniques work correctly when classifying frontal views of faces. However, they are not robust when the pose of the face changes because the global characteristics are sensitive to the movement and the rotation of the face. Hybrid methods apply combinations of local and global techniques, whose main goal is to compensate the changes of the pose allowing a flexible geometrical relationship between these components in the classification stage.

To overcome the pose problem, some kind of data alignment might be applied before classifying a face. The alignment of an input image of the face with a *reference* face requires the calculation of the correspondences between both images. In order to reduce the computational complexity of the problem, these correspondences are usually determined for a small number of points of the face, such as the centre of the eye or the mouth boundaries. According to these correspondences, the input image can be matched with an image of reference.

Methods based on EBGM [6], [7] use Gabor wavelets for feature extraction. These features are represented by a grid of points geometrically adjusted to the features extracted. The recognition is based on the wavelet coefficients, which are calculated for the nodes of a 2D elastic graph representing the grid containing the landmarks. This method combines a local and a global representation through the processing of a Gabor filter with several scales and several directions (jets), of a point set –called fiducial points– located in specific regions of the face. The location of the fiducial points is the most complex task of this method. These points depend on lighting conditions, the expression and the pose of the face.

In this PhD thesis, we propose a hybrid method based on the model suggested in [8] for facial recognition. In this model, combinations of local and global techniques are applied to build graphs based on feature vectors. The graphs are usually generated using the whole face without distinguishing the most discriminant regions. The information provided by non-discriminant regions or by the regions affected by partial or total occlusions has a big influence on the global result of the algorithm. For this reason, in this work a parametric model is suggested so that the more discriminant the regions are, the more relevant information they have. Initially, these regions are analyzed independently and then the global knowledge is built for the whole face joining the information of the local regions of the face.

This system must satisfy a series of real-time restrictions, where the time limits and the precision for obtaining the results will be defined for any specific hardware architecture. The operators to use in the system are defined so that the inherent data parallelism of image processing will be exploited. This data parallelism might be implemented using cluster architectures or grid computing [9], multi-agent technology [10], or a hybrid approach of both systems [11].

3 Description of the Face Recognition System and Experiments

In our work, we have chosen the well-known database BioID [12], which is widely used by the scientific community. Let us show now the main steps that are being currently implemented.As described before, we propose an algorithm based on the EBGM method. The recognition system consists of a database with the information of the faces of the users of the system and the algorithm of recognition, which will determine the identity of the person to be authenticated in the system.

The construction of the database consists in establishing a set of classes corresponding to the users, from information extracted from a set of input images. To obtain the information of each image, we take the following steps (see Fig.1):

1. Obtaining the enhanced face image in grey scale.
2. Detecting the features of the face applying an edge detector.
3. Obtaining a grid of landmarks from those features through the adjustment of a grid whose nodes are firstly distributed uniformly. The texture and geometrical information of each node of the grid is also encoded.

Fig. 1. Obtaining the grid points

The recognition algorithm determines the identity of the user identifying to which class of the database belongs. To determine the similarity with the captured face and each face stored in the database, a correspondence function establishes the degree of similarity between the face of the user that is going to be authenticated and each of the faces registered in the database. From the geometrical information of the points (coordinates, angles and distances), we propose two functions of correspondence: Match Cost Function (MCF) and Norm Vector Function (NVF). The MCF is calculated adding the match costs of each point of the input face F_1 with itsl corresponding point in the stored face F_2:

$$\text{MCF}(F_1, F_2) \equiv \text{MCF}(P, Q) = \sum_{i=1}^{n} C_{i\xi(i)} \,. \quad (1)$$

From both grids of nodes, we have two point sets $P = \{p_1, p_2, ..., p_n\}$ corresponding to the first face and $Q = \{q_1, q_2, ..., q_n\}$ corresponding to the second face. For each point i of both sets, a histogram 2D is computed. This histogram stores the distance $D = \{d_{i1}, d_{i2}, ..., d_{in}\}$ and the angle $\theta = \{\theta_{i1}, \theta_{i2}, ..., \theta_{in}\}$ of each point to the rest. The histogram bins are built uniformly in a polar logarithmic space. Each pair $(log(d_{ij}), \theta_{ij})$ will increase the corresponding histogram bin. Through these histograms, the correspondence between a point and a point is calculated:

$$C_{pq} = \sum_k \frac{[h_p(k) - h_q(k)]^2}{h_p(k) + h_q(k)} . \quad (2)$$

The point correspondence of P and Q is performed by minimizing C_{pq}:

$$\xi(i) : p_i \Rightarrow q_{\xi(i)} . \quad (3)$$

This function will give big values for different people and small values for the same person. The NVF is calculated adding the norm of the vector of differences among the matched points:

$$\text{NVC}(F_1, F_2) \equiv \text{NVC}(P, Q) = \sum_{i=1}^{n} \left\| \overrightarrow{p_i c_P} - \overrightarrow{q_{\xi(i)_i} c_Q} \right\| . \quad (4)$$

The global function GD is calculated adding the results of the previous functions. As the results generated by these functions are in a different dominium, these results must be normalized. Thus, matching photos belonging to the same person give results closest to 0, and matching photos of different people give high results, with a maximal value of 1.0.

$$\text{GD} = \frac{\lambda_1 \text{MCF}}{\max(\text{MCF})} + \frac{\lambda_2 \text{NVC}}{\max(\text{NVC})} . \quad (5)$$

For the experiments, we have used the Sobel edge detector with different thresholds. Then, the parameters in Eq.(5) are estimated, obtaining the best results for the combination λ_1=0.3, λ_2=0.5. The results of the experiments implemented so far can be observed in Table 1, where the most common error measures -False Acceptance Rate (FAR) and False Reject Rate (FRR)-, along with the correct recognition rate, are shown.

Table 1. Experiments results

Edge Threshold	63	64	65
Correct Recognition rate	94.10 %	95.55 %	95.14 %
FAR	2.43 %	1.83 %	2 %
FRR	3.47 %	4.62 %	2.86 %

4 Conclusions

In this work, an EBGM-based face recognition method has been defined. Experiments show that our method leads to very accurate face recognition results with low false acceptance and false reject rates.

As a future work, we are defining the multi-agent architecture for parallelizing the system and we are also developing a grid computing infrastructure to satisfy real-time constraints.

References

1. Wiskottz, L., et al.: Face Recognition by Elastic Bunch Graph Matching. In: Intelligent Biometric Techniques in Fingerprint and Face Recognition (1999)
2. Lyons, M., et al.: Coding facial expressions with gabor wavelets. In: Proceedings of Third IEEE International Conference on Automatic Face and Gesture Recognition, pp. 200–205 (1998)
3. Shakhnarovich, G., Moghaddam, B.: Face recognition in subspaces. Handbook of Face Recognition, pp. 141–168 (2004)
4. Kharbutli, M.M., et al.: Parallelization of Digital Image Processing Algorithms (2002)
5. Li, S.Z., Jain, A.K.: Handbook of Face Recognition. Springer, Heidelberg (2005)
6. Bolme, D.S.: Elastic Bunch Graph Matching (2003)
7. Gonzalez-Jimenez, D., Alba-Castro, J.L.: Shape Contexts and Gabor Features for Face Description and Authentication. In: Proc. IEEE International Conference on Image Processing ICIP 2005 (2005)
8. Espí, R., et al.: Reconocimiento facial basado en emparejamiento de grafos. In: Proc. II Simposio De Inteligencia Computacional (IEEE Computational Intelligence Society), Zaragoza, Spain (2007)
9. Berman, F., et al.: Grid Computing: Making the Global Infrastructure a Reality. Wiley, Chichester (2003)
10. Weiss, G.: Multiagent Systems: A Modern Approach to Distributed Artificial Intelligence. MIT Press, Cambridge (1999)
11. Shen, W., et al.: Adaptive Negotiation for Agent-Based Grid Computing. In: Falcone, R., Barber, S., Korba, L., Singh, M.P. (eds.) AAMAS 2002. LNCS (LNAI), vol. 2631, pp. 32–36. Springer, Heidelberg (2003)
12. BioID-Technology_Research, The BioID Face Database, http://www.bioid.com

Commitment Management in Real-Time Multi-Agent Systems*

Marti Navarro, Vicent Botti, and Vicente Julian

Departamento de sistemas informáticos y Computación
Universidad Politécnica de Valencia
Camino de Vera S/N 46022 Valencia, Spain
mnavarro@dsic.upv.es, vbotti@dsic.upv.es, vinglada@dsic.upv.es

Abstract. Today, real-time systems should be designed to be flexible and adaptable. For this reason, the multi-agent system paradigm seems to be appropriate for real-time system development. In the case of distributed real-time systems, complexity is added when a node must delegate a task with temporal restrictions to another node. This delegation can be modeled as a commitment act in a multi-agent system. This paper presents a commitment management framework that is specifically designed for real-time agents. This framework allows the agent to decide (in a bounded time) if it has the necessary resources and time to do the required task, and thus accept the commitment. This paper also presents an execution example that shows how this framework works.

Keywords: Real-time systems, Multi-agent systems, Commitment.

1 Introduction

Nowadays, the incorporation of a multi-agent system paradigm into real-time systems provides flexibility, adaptability and distribution. These characteristics help to achieve new challenges in this area when compared with typically rigid, traditional real-time applications. If we focus on distributed real-time systems, we can distinguish several independent nodes that work to achieve common goals in a temporally restricted environment. The development of new applications in the area of real-time distributed systems has added great complexity to systems of this type, especially in those requirements that refer to dynamic adaptation to changing conditions. In general, these systems have a physical distribution of the problem and their nodes must share restricted resources. Therefore, it is necessary for these nodes to work in a coordinated way and commit themselves to the fulfilment of specific tasks.

The agent paradigm is considered to be a very powerful and promising option that can be used in the development of complex systems, and thus in the development of real-time systems emerging the concept of real-time multi-agent systems [9]. In the case of this kind of systems, one important feature is related with the delegation of temporal bounded tasks among nodes or agents. If an agent delegates a task to another

* This work was supported by CONSOLIDER-INGENIO 2010 under grant CSD2007-00022 and from the Spanish government and FEDER funds under CICYT TIN2005-03395 and TIN2006-14630-C0301 projects.

J.M. Corchado et al. (Eds.): DCAI 2008, ASC 50, pp. 503–511, 2009.
springerlink.com

one with a determined execution time or deadline, the agent who commits itself to develop this task must not fail to fulfill this commitment. In hard, real-time systems a deadline violation may cause serious or catastrophic effects in the system. In soft real-time systems, it may downgrade the performance of the system or the commitment may not be useful.

Today, in the multi-agent system literature, there are works that formalize commitments among agents, in a semantical form, defining a commitment as the responsibility acquired by an agent to accomplish an action under certain conditions with respect to another agent [20]. Traditionally, the agent communication semantics is based on so-called agent's mental state [4] [12], but there are a number of drawbacks to using only mental concepts for specifying ACL semantics [15] [8]. For this reason, the communication based on social attitudes introduces the commitment concept. There are two types of commitments [14]: Psychological commitments, where the agent is committed to its beliefs and intentions, and social commitments, in which the agent is committed to other agents. There are many approaches that semantically formalize social commitments [16] [1] [5]. Other research use the theory of commitments for the definition of protocols [10] [6], or to define interactions within organizations [3] [18].

Nevertheless, current approaches have certain deficiencies in certain environments. Specifically, the applicability of multi-agent systems to real-time environments requires specific functionalities that are not currently available in existing commitment-based approaches. As commented above, in environments of this type, appropriate time management is necessary in communicative processes. This management includes strict control of available resources (mainly the time that is available without affecting critical responsibilities) when a commitment is accepted or rejected.

This work presents a commitment-based framework for real-time agents. The framework provides the necessary tools to allow the agent to decide, in a temporal bounded way, if it has the resources and the suitable time to commit itself to the execution of a specific task. The framework has been implemented and tested over the jART platform [11]. This platform is specifically designed to execute real-time multi-agent systems. The implemented systems are executed over a Real-Time Operating System. Systems developed through this platform assure the fulfillment of the temporal constraints that may exist.

The rest of the paper is structured as follows. Section 2 introduces the idea of real-time agents and a proposed model. Section 3 focuses on the commitment-based manager module. Section 4 shows an example and the analysis of the results obtained from this example. Finally, section 5 presents some conclusions and future works.

2 Designing a Real-Time Agent

According to Botti et al. [2], a Real-Time Agent (RTA) is an agent with temporal constraints in some of its responsibilities. Usually, a RTA is an agent composed of a series of tasks, some of which have temporal constraints. In these agents, it is also necessary to take into account the temporal correctness, which is expressed by means of a set of temporal restrictions that are imposed by the environment. The RTA must, therefore, ensure the fulfillment of such temporal restrictions. By extension, a Real-Time Multi-Agent System (RTMAS) is a multi-agent system with at least one RTA [9]. Systems of this type require the inclusion of temporal representation in the communication process, management of a unique global time, and the use of real-time communication [17].

It is well-known that a typical real-time system is made up of a set of tasks characterized by a deadline, a period, a worst-case execution time and an assigned priority. These temporal restrictions in the system functionality affect the features of an agent that needs to be modeled as a real-time system. For example, time unbounded problem-solving methods are a serious problem because their execution cannot be temporal restricted.

If agents must operate in a real-time environment, agent construction complexity is increased enormously. Evidently, different environments require different software structures. Therefore, in an agent context, an appropriate structure must be adopted in order to use agent features in real-time environments. From a user point of view, our real-time agent is mainly composed of a set of roles offering different services. Moreover, from an architectural perspective, the agent can be divided into different modules that provide the execution framework used to construct the final executable version of the agent.

From the **user model** point of view, our model is a service-oriented agent model that is principally made up of the following elements:

- *Set of roles*, this models the answer of the agent to different situations. A role is basically composed of tasks. The main reason for splitting the whole problem-solving method into smaller entities is to provide an abstraction that organizes the problem-solving knowledge in a modular and gradual way. When a role is active, all the tasks associated with this role are active, too.
- *Set of tasks*, a task is the execution object of an agent. Task execution allows the goals pursued by the agent to be achieved as a consequence of service activation. Each task has a set of attributes that defines the task behavior. These attributes are:
    - *Action*: Define the action that is performed when the task is executed.
    - *Temporal Constraints*: The temporal behavior of the tasks. The main characteristics are defined as:
        - *WCET (Worst-case execution time)*: indicates the estimated time needed to complete the task execution.
        - *Start*: instant of time when the task must be launched for execution.
    - *Resources*: Necessary resources to complete the task.
    - *Quality*: This parameter specifies the benefit or detriment to get the agent once this task is executed.
- *Set of services*, according to the role or roles that the agent plays, it can offer a set of services to the rest of agents in the system. Each service is basically defined as the following tuple:
    - *Interface*: Information given to agents of how they must invoke the service.
    - *Description*: Description of the offered service.
    - *Goal*: The goal that the service execution must satisfy.

- *Set of beliefs* comprising a world model (with all the domain knowledge that is relevant to the agent) and the internal state.

The **system model** provides software architecture for the real-time agent model. The main modules are the following:

- **Belief Manager Module:** This module manages all the mental states of the real-time agent. This information is stored in a time-stamped, frame-based blackboard. Temporal extensions have been incorporated, which allows time to be managed naturally.
- **In/Out Module:** This module makes it possible to interact with the environment. This module is used by physical robots to manage the actuators and reads its sensors. Due to the features of the environment, the perception and action processes are time-bounded.
- **Communication Module:** This module is in charge of coding/ decoding and sending/receiving the messages. The message is coded at this level. This codification converts the message into a useful load for the transport message. Additional and necessary information is added in order to support the platform communication model. This information is basically the service identification. This process is similar to the process described in the transport services of FIPA.
- **Commitment Manager Module:** This module decides if the real-time agent must commit to performing a service, and it controls the commitments acquired by the real-time agent for the correct conclusion of these commitments.
- **Control module:** This module is responsible for the real-time execution of the active tasks that belong to the real-time agent. The temporal requirements of the tasks are different; thus, the control module must employ different execution criteria for each one. Basically, this module is a set of extensions over a Real-Time Operating System that incorporates features for managing these real-time agents.

One of the main problems in real-time, multi-agent systems can appear when an agent needs to collaborate with other agents to perform a task. As commented above, if an agent delegates a task to another one with a determined execution time or deadline, the agent who commits itself to performing this task must not fail to fulfill this commitment. In order to manage this aspect, a real-time agent must include a commitment manager module. This commitment manager module (which is detailed in next section) examines the service requests that arrive to a real-time agent and determines whether or not it can complete the service. If the request is accepted, the agent that offers the service will be committed to perform the requested service in time.

3 The Commitment Manager Module

The commitment manager is a module incorporated in the real-time agent with the purpose of improving agent behaviour when the agent offers services in a real-time environment. This module performs two main functions: (i) To analyze whether or not the agent can satisfy the service requests of other agents. Once the service request is analyzed and accepted, the agent is committed to performing this service for the multi-agent system where the agent is located. (ii) To manage the different commitments that the agent has acquired in order to avoid possible commitment violations. As a last resort, the manager can cancel a commitment if agent integrity is in danger due to any unexpected error. The commitment manager is composed of two independent modules. These modules are the following:

- **The Resources Manager:** With this module, the agent can check if it has the necessary resources to execute the related tasks to achieve the goal associated to the service. If the agent does not have the necessary resources when the service request arrives, then the Resources Manager can determine when the agent would have these resources in the future. This analysis calculates when the agent should start the task execution with all the available resources.
- **The Temporal Constraints Manager:** Before the agent is committed to performing a service, it is necessary to verify if the agent can finalize the service before the deadline assigned to complete the service. The Commitment Manager module uses the Temporal Constraints Manager module to verify this. This module uses dynamic real-time scheduler techniques to determine if it is feasible to execute the task assuring its temporal constraints. If the agent cannot fulfill its temporal constraints, the temporal constraints manager can calculate the time instant when the task will probably finish. Using this temporal information, the agent can decide if it should commit with the agent that makes the request, or if it should enter into a negotiation process to change the deadline of the service. If the change of the deadline is not possible and the agent does not have time to execute the service then the request is refused and the agent do not commit to realise the service.

The Commitment Manager module works as follows:

1. When a service request arrives, the agent that offers this service uses the Commitment Manager module to extract the tasks that can achieve the goal fired by the invoked service.
2. For each task (or set of tasks) that can achieve the goal, the Commitment Manager must do the following:
   - To check if the agent has the necessary resources to perform the task, using the Resource Manager module to do this.
   - To analyze whether the task execution can be finished before the specified deadline. To do this, the agent uses the Temporal Constraint module.
3. Once the Commitment manager has verified that the agent has all the necessary resources related to the invoked service and, therefore, that service can be done in time, the agent is committed to performing it.
4. The commitment manager is in charge of ensuring that the acquired commitments are fulfilled. In the case that a commitment cannot be fulfilled, the agent should renegotiate the commitment or cancel it, keeping the potential negative effects to a minimum.

In Figure 1 is shown an Activities diagram with the steps that the manager commitments follow to accept or reject a commitment. The Commitment Manager analyzes the service request and makes a decision to accept or reject based on the current work load of the agent. When the agent that offer the service, called server agent, receives the request message, it analyzes whether it is possible to achieve the service requested by the client agent. The server agent makes a feasibility analysis to determine if it has the necessary resources and time to perform the service. If the feasibility analysis indicates that the agent can perform the service, the agent will be committed to the client agent to execute it. In contrast, if the server agent is executing other tasks when the service request arrives when the server agent makes the feasibility analysis, it verifies that it does not have enough time to execute the service before the deadline indicated by the client agent in its request. In this case, the server agent must reject the service request and the agent does not commit to realise it.

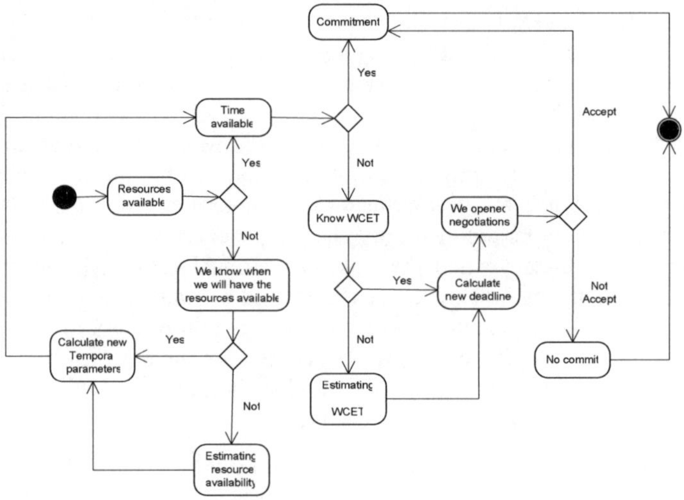

Fig. 1. Activities diagram

4 Application Example

A prototype of a mail robot example has been developed for our proposal. The problem to solve consists of the automated management of the internal and external mail (physical mail, non-electronic) in a department plant. The system created by this automation must be able to request the shipment of a letter or package from an office on one floor to another office on the same floor, as well as the reception of external mail at a collection point for later distribution. Once this service has been requested, a *Pioneer 2* mobile robot must gather the shipment and direct it to the destination. It is important to note that each mail or package distribution must be finalized before a maximum time, specified in the shipment request. The robot is modeled as a RT-Agent, called Robot Agent, which is in charge of controlling a physical robot and managing the mail list that this robot should deliver offering a mail delivery service that employs the commitment manager described above.

4.1 Tests and Results

To develop and execute the proposed real-time, multi-agent system, we have use the jART platform (which is especially designed for systems of this type) and RT-Java [14] as the programming language. Once the example has been implemented over the jART platform, several simulation experiments were conducted to evaluate different parameters in order to verify the use of the proposed commitment framework. A simulation prototype was implemented using a *Pioneer 2* mobile robot simulation software (specifically, the Webots simulator [19]). The simulation experiments were conducted to evaluate different aspects and to try to show the benefits of commitment framework integration in a real-time agent. The different experiments were tested on a robot agent without commitment management and on one that included the commitment management proposed above.

It is important to note that, in the case of the agent without commitment management, the received mail orders are stored in a request queue. This queue only stores up to five pending requests. When the robot agent receives a request, if it has space in the queue, the robot agent accepts the commitment associated to the request; otherwise the request is rejected. On the other hand, the other agent will decide its mail orders in accordance with its commitment management capability.

In each case, the mail or package distribution must be finalized before a maximum time, and the robot control behavior must guarantee robot integrity, which implies hard, real-time constraints. In the tests carried out, all requests have the same priority, therefore, an adequate metric to verify the improvement offered by the use of the commitments manager is to use the number of requests serviced by the Robot agent. Therefore, the greater the number of requests satisfied by the Robot agent will be the best result of the system. If requests have different priorities this metric is not correct. In this case, is more important to fulfill tasks with high priority than to fulfill a greatest number of low-priority tasks.

The first set of experiments investigates the commitment acceptance of the system according to package or mail arrival frequency. The simulation prototype was tested by increasing this frequency incrementally and by testing the number of un-accepted commitments. The tests consisted of groups of 10 simulations of five minutes duration. The Robot Agent received between 5 and 30 requests during these five minutes. Each experiment was repeated one hundred times and the results show the average obtained value. The obtained results are shown in Figure 2. This figure shows, how at a low request frequency, the Robot agent accepts all requests and is committed to fulfill them, independently of whether or not the Robot Agent incorporates the commitment manager. Nevertheless, when the request frequency is increased the agent with the commitment management capability works better, even with a high frequency rate. Obviously, these results must be contrasted with the percentage of successfully completed commitments. With respect to the agent without the commitment management capability, it can be observed how in the case of low average rates, the behavior is a little better at the beginning. This is because this agent accepts all requests while its request queue is not full, without taking into account whether or not it will be able to successfully carry out these requests. The Robot agent that has the commitment manager rejects requests sooner that the other one because it only accepts commitments if it is able to successfully complete the work associated with that commitment.

The second set of experiments investigates the success rate of accepted commitments according to package or mail arrival frequency (Figure 3). This figure shows that the commitment manager is very efficient maintaining the success rate close to 100% in contrast with the robot agent without the commitment manager. In this case, the success rate is lower when the number of requests is increased. Even when the saturation of requests in the system is very high, the agent with the commitment manager maintains a success rate of around 90%. The reason for this loss is the time consumed by the commitment manager module to analyze the requests and to control the execution of the accepted commitments. At the beginning, this execution time is bounded, but when the request arrival frequency is very high, this time is overrun, which affects the execution of the rest of the agent tasks. This problem must be improved in next implementations of the commitment manager module.

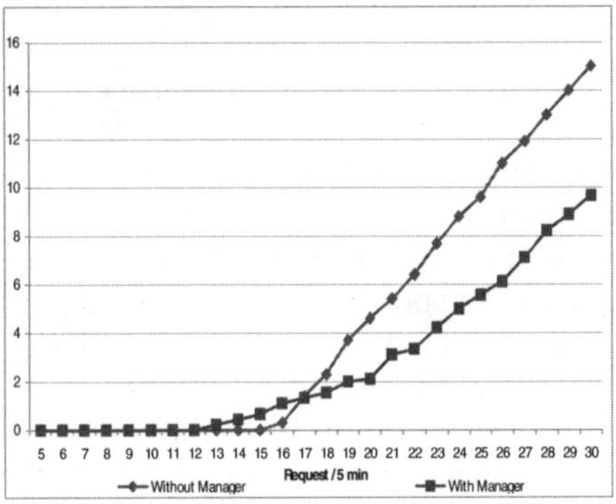

Fig. 2. Analysis of the number of unaccepted commitment

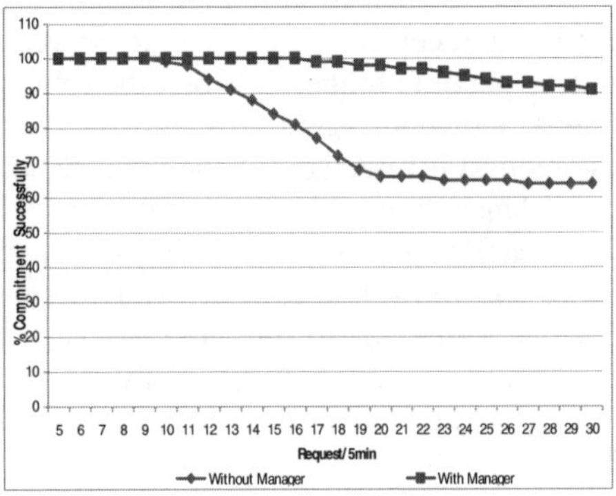

Fig. 3. Percentage of successfully ended Commitments

5 Conclusions

Typical techniques employed in multi-agent systems must be adapted for their correct use in real-time environments. In real-time, multi-agent systems the incorporation of strict temporal restrictions in agent interactions constitutes a fundamental aspect to be able to use systems of this type in environments with temporal constraints. In this paper, commitment management in real-time systems has been studied. As a consequence of this study, a new framework has been proposed and tested.

The proposed commitment-based framework permits a suitable management of the commitments that real-time agents acquire from other agents. It takes into account all the necessary resources (mainly time) to be able to carry out these commitments. With the commitment manager presented in this paper, it is possible to guarantee that agents work correctly according to their commitments in soft real-time systems.

The results obtained in the proposed example reinforce the proposal but further testing and more real examples to apply this framework are needed in order to validate the proposal. Future work will include the creation of a formal commitment-based semantics where real-time concepts will be included.

References

1. Bentahar, J., Moulin, B., Meyer, J.-J.C., Chaib-draa, B.: A modal semantics for an argumentation-based pragmatics for agent communication. In: Rahwan, I., Moraïtis, P., Reed, C. (eds.) ArgMAS 2004. LNCS (LNAI), vol. 3366. Springer, Heidelberg (2005)
2. Botti, V.J., Carrascosa, C., Julian, V.J., Soler, J.: Modelling agents in hard real-time. In: Garijo, F.J., Boman, M. (eds.) MAAMAW 1999. LNCS(LNAI), vol. 1647, pp. 63–76. Springer, Heidelberg (1999)
3. Carabelea, C., Boissier, O.: Coordinating agents in organizations using social commitments. Electr. Notes Theor. Comput. Sci. 150(3), 73–91 (2006)
4. Der Hoek, W.V., Wooldridge, W.: Towards a logic of rational agency. Logic Journal of the IGPL 11(2), 133–157 (2003)
5. Dunin-Keplicz, B., Verbrugge, R.: Collective commitments. In: ICMAS 1996, pp. 56–63 (1996)
6. Flores, R.A., Kremer, R.C.: A pragmatic approach to build conversation protocols using social commitments. In: AAMAS 2004, pp. 1242–1243 (2004)
7. Fornara, N., Colombetti, M.: Operational specification of a commitment-based agent communication language. In: AAMAS 2002, pp. 535–542 (2002)
8. Gaudou, B., Herzig, A., Longin, D., Nickles, M.: A new semantics for the FIPA agent comunication language based on social attitudes. In: ECAI 2006 (2006)
9. Julián, V.J., Botti, V.J.: Developing real-time multi-agent systems. ICAE 11(2), 150–165 (2004)
10. Mallya, A.U., Singh, M.P.: Semantic approach for designing commitment protocols. In: AC, pp. 33–49 (2004)
11. Navarro, M., Julián, V.J., Heras, S., Soler, J., Botti, V.J.: Multi agent systems over RT java for a mobile robot control. In: Corchado, E., Yin, H., Botti, V., Fyfe, C. (eds.) IDEAL 2006. LNCS, vol. 4224, pp. 1267–1274. Springer, Heidelberg (2006)
12. Rao, A.S., Georgeff, M.P.: Bdi agents: from theory to practice. In: ICMAS 2005, pp. 312–319 (1995)
13. RTJ: The real-time for java expert group, http://www.rtj.org/rtj.pdf
14. Singh, M.P.: Social and psychological commitments in multi-agent systems. In: AAAI Fall Symposium on Knowledge and Action at Social and Organizational Level (1991)
15. Singh, M.P.: Agent communication language: rethinking the principles. IEEE Computer 31(12), 40–47 (1998)
16. Singh, M.P.: An ontology for commitments in multiagent systems: toward a unification of normative concepts. Artificial Intelligence and Law 7, 97 (1999)
17. Soler, J., Julián, V.J., Fornes, A.G., Botti, V.J.: Real-time extensions in multi-agent communication. In: Conejo, R., Urretavizcaya, M., Pérez-de-la-Cruz, J.-L. (eds.) CA-EPIA/TTIA 2003. LNCS (LNAI), vol. 3040, pp. 468–477. Springer, Heidelberg (2004)
18. Udupi, Y.B., Singh, M.P.: Contract enactment in virtual organizations: A commitment-based approach. In: 21st AAAI National Conference on Artificial Intelligence, pp. 722–727 (2006)
19. WEBOTS: Cyberbotics page, http://www.cyberbotics.com/
20. Winikoff, M.: Designing commitment-based agent. In: IAT 2006, pp. 18–22 (2006)

Intelligent Streaming Server for Non Accessible Contents Stored on Web Servers to Disabled People: Signwriting Case

Rubén González Crespo[1], Gloria García Fernández[1], Oscar Sanjuán Martínez[2], Enrique Torres Franco[1], and Luis Joyanes Aguilar[1]

[1] Departamento lenguajes y sistemas informáticos, UPSAM, Madrid, Spain
{Ruben.gonzalez,gloria.garcia,enrique.torres, luis.joyanes}@upsam.net
[2] Departamento de Informática, UNIOVI, Oviedo, Spain
osanjuan@uniovi.es

Abstract. One of the main challenges a multimedia web server must face is to serve to its users the requested content in a fast and efficient way. In order to do so, the web server is required to meet several specifics needs from each client. Nowadays, accessing to multimedia content is becoming more and more popular among users with vision or hearing problems. That sort of users needs an adapted software/hardware client in order to reproduce multimedia contents, which in fact is a restriction for the transmission. Furthermore, this special client is often a wireless system which introduces even more hurdles into the problem. However, multimedia web servers don't take into account the special requirements of that type of clients, hence they process their request as if they were like any other. For the purpose of solving the needs of the disabled people accessing to multimedia content, intelligence could be added to the streaming server. In that way it will be capable of distinguish the clients' characteristics and the properties of the communication in order to provide the users with the most suitable content.

Keywords: multimedia server, accessibility, web server, intelligent system, wireless link.

1 Introduction

Video streaming services are nowadays widely available. One of their main uses is Entertainment, i.e. music clips, videos or news can be downloaded to our mobile phones or PDA's. But we must be aware that this kind of services are wider than just amusement. In fact, video streaming can be used for video surveillance tasks, rescue parties, or even e-learning.

Today we can access to streaming services using a wide range of devices such as notebooks, PDA's or mobile phones. The type of users that use these services are also very different among them, however there is one group that has the access more difficult than the others. We are talking about disabled people, for instance people that suffer from blindness or deafness. These users are oftenly unable to access to multimedia content, due to their need for an special device or software able to convert the multimedia content into accesible content. In other situations they are also unable to reach the data because the server can't meet the tailored service they need.

J.M. Corchado et al. (Eds.): DCAI 2008, ASC 50, pp. 512–520, 2009.
springerlink.com

In order to make 'traditional streaming servers' become adaptive to the requirements of every type of client, we are going to propose an architecture that helps the server in the streaming process. In that way every client obtains the most suitable content, since their needs are taken into account. We will focus on clients with birth deafness, which means that they are unable to talk. We will propose a mechanism that translates the audio tracks to sign writing, which is easier to understand than traditional writing.

1.1 Problems Faced by Disabled People Trying to Access to Multimedia Content

Almost every common device designed for non disabled people to access to multimedia content implements a large number of mechanisms for accessibility. The core idea under this is to achieve that all sort of content (general and multimedia) could be accessed by every type of systems and people. However the content can't be always adapted to the needs of people named as "different". This is the case of people who suffer from prelocutive deafness, which implies that they can't talk at all. In this situation, a based sign writing mechanism would be needed since deaf people who have never talk are unable to associate characters to phonetics signs, but can do it with graphic elements.

Our solution is going to provide the server with a seamless capability of decision. This feature will make the server to be able to distinguish whether it is required to serve multimedia content based in sign writing, or the standard one (assuming that the client will be having and standard Internet browser).

1.2 Problems from the Point of View of the Video Streaming

Aside of the problems that come with the type of content that is served (simple multimedia or sing writing multimedia content based), it is very important to point out that nowadays the access to this type of services is widely done through wireless devices (such as mobile phones). In that way we should take into account other restrictions that are introduced by the wireless link, i.e. bandwidth available or the interferences in the communication. So, if we want the video streaming to be the most suitable possible for our clients, an analysis of all the aspects of the communication (signal and content properties for instance) must be done.

In addition, inside a streaming session temporal constraints are very important. If the client that wants to access the service does it via a non-standard device, the needs for processing will be harder, which will make the transmission even slower.

2 Problem Formulation

2.1 Video Streaming

The main obstacles to beat for a multimedia server are :

- Network capacity issues. When it comes to a multimedia service, it is hard to estimate the total traffic volume a streaming transmission can amount. This happens because of the stream size (both in bandwidth and duration), that may change depending on its content. So, it is critical to provide the clients with a dynamic adaptation mechanism.

- Obstacles due to unguaranteed quality of service [7]. Generally speaking, wireless devices must be able to work in networks unable to guarantee a fixed quality of service (with the exception of GSM/UMTS circuit domain). This implies that both the bandwidth and the delay the user is given are supposed to change based on the network load. The adaptation to these parameters is critical. However, if bandwidth decreases during a long period of time, may sound reasonable to jump to another server, with the purpose of getting a higher quality of service.

When talking about a video stream addressed to deaf people, we must take into account that besides the 'normal' information stream we must send and additional stream, with has inside what is called 'sign writing' information.

2.2 Problems of the Deaf People Accessing to the Multimedia Contents

Audiovisual contents are composed of video and audio. Deaf people can visualize the visual content without any problem but they can't, obviously, hear the audio. As a result, most of the times they are unable to understand what they are watching.

An approach to solve their problem is to get from our multimedia server the audio stream (in a subtitles format) and the video stream separately. However, if we take this way we must face another problem, because as we will see in the following paragraph not all the deaf people are capable of understand the meaning of the textual content from a video.

Among deaf people all the sentiments are basically transmitted through the vision, so it is a visual and space based culture. Sign language is the communication mechanism that deaf people use, which is based in symbols. Every single symbol has a signification which anybody knows and that can be used to describe, sort, and catalogue experiences, concepts, or objects. [3].

This language is used to communicate but, what happens when written communication is needed? One solution might be writing the sounds on a speaking language, but, what about the people that have never hear a sound or a word? Sign writing, through literacy tools, wants to teach deaf people to write in their own sign language. So, why a sign language is needed? Is there any problem for deaf people to read in their mother languages?

There are many reasons to explain why is almost impossible for them to read text. The main one is that when we write a word we represent the phonetical sound of the word with characters. In that way for people who have never listen a sound, learning to read is just a memorization problem. It would be like trying to memorize a phone number for each word. So, sign writing is addressed to deaf people that already know sign language but don't understand the language enough in order to read large or complex texts.

This type of writing is named 'sign writing', and was developed by Valerie Sutton for the American Sign Language [4]. Basically, sign writing is just a simple graphic representation of the real symbol. It comprises representations for the hand movements, facial expressions, etc., using simple graphics that can be easily written. The graphical representation of signs is very alike to the real symbols used in sign language. As an example, the word 'where' in sign writing will be as follows:

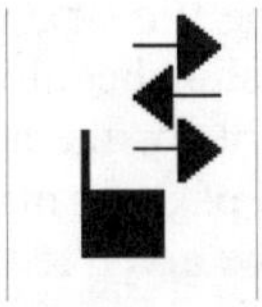

Fig. 1. 'Where' word representation

As described in the previous paragraph, it is very difficult for deaf people to understand common subtitles which came in plain text. However, it will possible for them to understand a movie based in sign writing.

Nowadays we have a possibility already implemented, a virtual signer. It is a person who would translate sign writing into sign language. However it will be an extra charge to the streaming session since if it is inside the video, the server should have two different versions of the same content. Besides, if someone visualizes the video with the signer it would be very difficult to follow the original film and the signer.

3 Problem Solution

The architecture proposal introduces two differences with the traditional one. The first is to include inside the streaming server the ability of analyze the communication, so as to the retransmission could be performed into the most suitable way. In order to achieve this objective, we are going to provide the server with a decision mechanism and with the suitable information which will allow the server to adapt the transmission to the needs of its clients.

The second modification is to include a middleware into the server that will allow the access to the multimedia content to deaf people through sign writing. The sounds will be translated into sign writing language through this module.

3.1 Multimedia Conversion Contents Based in SWMLSVG

Once the server has detected that the information into the streaming session is for a deaf person, the subtitles translation will be started. These subtitles would be stored into the server, as sign writing.

All the elements that are part of the sign writing are vectorial graphics, which and are easy to play with several applications widely used nowadays into the multimedia environment. We will use the markup language SWMLSVG (Sign Writing Markup Language Standard Vectorial Graphics), to bring together the sign writing and the solution expected.

SWMLSVG is the evolution of SWML language. In 2004 the first SSS (which stands Sing-Symbol-Sequence [5]) storage was developed for vectorial graphics SVG, which replaced all the storing sources developed by that time for sign writing (those were based in PNG or GIF graphics, which suffered from several constraints).

SSS marks the order in which sign writing symbols are arranged. These arrays of symbols are accessible through a lexicographical way. It could be said that SSS is the alphabetical order for this system, since it is the basement to build dictionaries and make translations. SSS-2004-svg is the file type used as storage space for the pictures in SVG format, which comprises the alphabet aforementioned.

Lets assume that the textual content for the audio or video tracks come in several detached tracks, for instance as a web page format. The overall goal of our application will be to perform the translation of the given text into sign writing. In order to do that we will use C++ as the programming language, and through a plug-in we will take the URL for the text and the database SSS-2004-svg, where the signs are stored. Searching through this database the application will be able to obtain the signs to be used, which will be loaded into the system memory.

At the same time we will use a dictionary, sbml.xml, which will be also loaded into the system memory. Here we will find all the words that are predefined into the sign writing, and also the rules to build them.

Once the two main structures are into the memory, the middleware take the text from the web page and obtain the words searching for each one of them as is explained next:

If the word that we are searching is in the system memory, it implies that the sign for the word has been already build. In that case the sign is taken and printed into the client screen.

If the word is not in the memory, it is because it hasn't been used yet. The application must build it from the scratch, by accessing to the dictionary and searching for the word to obtain the necessary information to build it. This information consists of a set of symbols and their coordinates.

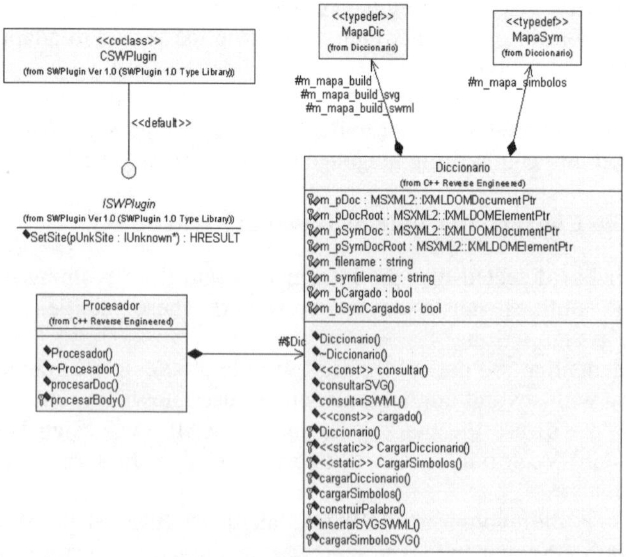

Fig. 2. Middleware classes

As soon as we have the information for the signs, the software will access to the file where the symbols are stored in order to build the word.

The symbols are represented using a vectorial format (.svg) and they have predefined a series of attributes, such as: Category, Group, Symbol, Variation, Filling and Rotation.

Once we have obtained the series of necessary symbols to compound the word, they will be placed in the corresponding coordinates as stated in the vectorial information retrieved from the database. This task must be completed by the plug-in as well as printing the result into the screen.

Design decisions: We have developed a plug-in, a middleware, using activeX technologies. The main class is called 'CSPlugin' and it uses the classes 'Procesador' and 'Diccionario'.

Processor (Procesador): The processor is responsible for building the translated track sent to the client. It receives the original text, parse it by searching for detaches words, and replaces them with SVG inline graphics. In order to do that the dictionary must be consulted since we must know the SVG inline sequence for each word.

The main problem to be faced with this solution is the necessity for the 'Adobe SVG Viewer' plug-in to be loaded in order to use the SVG graphics inside the HTML page.

Dictionary (Diccionario): The dictionary takes care of the words database and the SVG symbols loaded in the memory. First the words are translated into a symbols sequence, and then the SVG graphic is build for each word. The dictionary is read for the processor for each word that it is needed to translate.

Dictionary methods: Constructor, Destructor, Return the build (00-00-00..,x,y,00–00...) for a word, Return the SVG with the symbols for each word, Return the SWML for a word, The dictionary is already loaded into memory

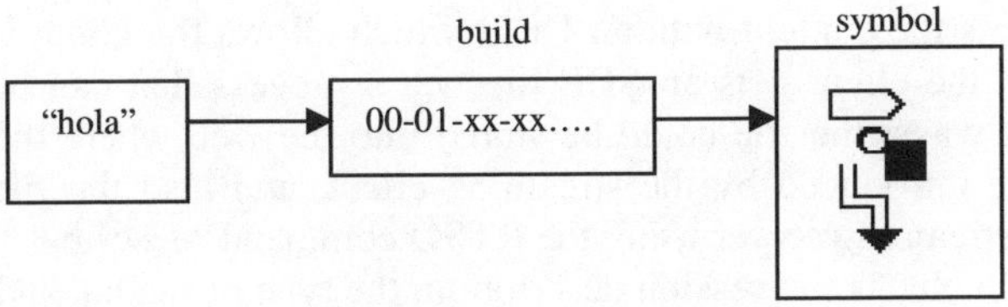

Fig. 3. Translation example

3.2 Modifications in Order to Optimize the Transmissions

Having an improved streaming server architecture [8] in order to obtain a more efficient transmission into the wireless environment, we are going to propose a change in the streaming server kernel. The idea is to make its features to be close to these a QuickTime has (i.e. an open programming frame).

In order to adapt the communications to the disabled client, we are going to introduce into the middleware a new module. This component will give the server the capability to take dynamic analysis-based and transmission parameter-aware decisions, in order to serve the most appropriate content to the client that is requesting access to the service.

This module will monitor the main parameters inside a streaming session sent to a disabled person client, for instance: client type that is accessing the service, the server could decide whether or not it is necessary to send sign writing information; available bandwidth, the server will adapt the transmission rate to the available bandwidth [10]; content required; bitrate needed.

While the transmission is going on, and especially if we are inside a wireless environment, it is not unusual for the media conditions to change because of different causes (i.e. interferences). This module will collect in real time information about the transmission, so all the data could be used later to change the transmission policies.

The event flow inside a streaming session into a wireless environment will be as follows:

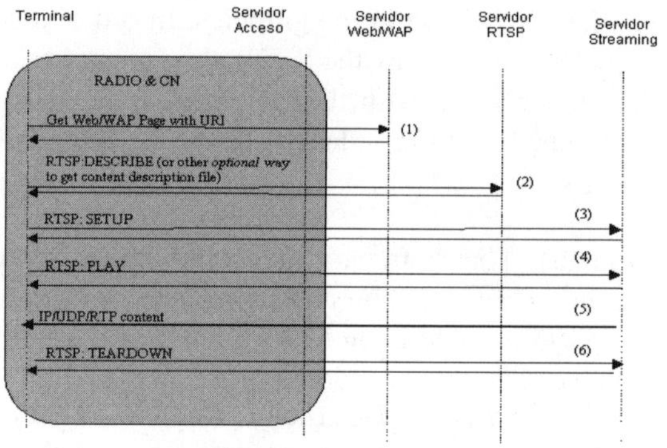

Fig. 4. Streaming session events

The user access the content with an URL, which allows the client to connect with the server. Then the client gets an SDP file, via a process that can be performed in several different ways: the file could be stored into the web where the client access, the file could be introduced by the streaming client, and also the file could be obtained from the streaming server using the RTSO command 'describe'.

The SDP file includes the session description, the type of media, and the bitrate for the retransmission. We will modify this file in order to store additional information, such as the name of the required subtitles stream.

The session establishment is a process in which the browser of the client requests the streaming server to begin a streaming session. The client points out the information requirements to establish the service sending a 'setup' message from the RTSP protocol. The server responses with the UDP and/or TCP port that would be used for the streaming. To start the session the client sends a RTSP 'play' message, which unleashes the streaming transmission.

The streaming conclusion could be sent by any peer of the session (the server, the client, or both). The termination is performed through a 'teardown' RTSP message.

In the previous diagram, the videostreaming client will send periodically information about its inner state. The data will comprise not only the video streaming status, but also the sign writing streaming reception and representation status.

The information reaching the server will be an XML document with the following structure:

```
<SignoWritingInfo>
        <SWReceptionBR>  </ SWReceptionBR >
        <TotalBW>   </ TotalBW >
        <SWBW>   </ SWBW >
        <SWPacketLoss>  </ SWPacketLoss >
        <timeStampSE>  </timeStampSE>
</ SignoWritingInfo >
```

- SWReceptionBR: Sign writing reception rate.
- TotalBW: Total Bandwidth dedicated to the video streaming.
- SWBW (sign writing bandwidth): Bandwidth dedicated to the sign writing streaming.
- SWPacketLoss: Sign writing packets loss.
- TimeStampSe: Last packet received from the sign writing.

4 Conclusions

The architecture proposed overcomes the video streaming problems for clients with hearing disabilities, by adding to the server a middleware. This middleware module is able to translate the subtitles tracks into sign writing, so that the prelocutive deaf people can understand them easily. In addition, the architecture proposes a mechanism so that the client can send additional information to the server in order to achieve a better transmission.

There is a drawback within the solution: if the client that is accessing the content does it through a web browser, the inline symbols can be represented only if the server is in quirks mode [6]. This feature can dislocate the new pages format, so as a future investigation line it would be interesting to propose a way to solve it.

In addition, the solution must run both in the server and also in the client side. A future approach would release the processing tasks from the client, so that small capacity clients could be used to access the information.

In addition, the solution runs entirely into the server, which must deal with all the processing tasks. This situation could reduce the computing power available for other issues, so it would be interesting to deliver some of the processing charge to the client side.

References

1. Lamport, L.: LaTeX: A Document Preparation System. Addison-Wesley Publishing Company, Reading (1986)
2. Ree Source Person. Title of Research Paper; name of journal (name of publisher of the journal), Vol. No., Issue No., Page numbers (eg.728—736), Month, and Year of publication (eg (October 2006)

3. Torres, S.: Curso de Bimodal, Sistemas Aumentativos de Comunicación, Universidad de Málaga (Consulta November 11, 2007) (2007) http://campusvirtual.uma.es/sac/_contenidos/
4. Parkhurst, S., Parkhurst, D.: Un sistema completo para escribir y leer las Lenguas de Signos, PROEL, Madrid (2000)
5. Rocha, A., Pereira, G.: Suppoting Deaf Sign Languages in Written Form on the Web. Escuela Informática de la Universidad Católica de Pelotas (2004)
6. Korpela, J.: What happens in Quirks Mode. IT and Communication, Finlandia (2007)
7. Manner, J., Burness, L., Hepworth, E., López, A., Mitjana, E.: Provision of QoS in Heterogeneous Wireless IP Access Networks. PIMR (2002)
8. Yufeng, S.: Cross-Layer techniques for adaptive video streaming over wireless networks. EURASIP journal on applied signal processing (2005)
9. QuickTime (2008), http://www.quicktime.com
10. Jinyang, L., Blake, C.: Capacity of Ad Hoc Wireless networks. In: Mobicom (2001)

A Modular Architecture for Navigation Applications Based on Differential GPS

S. Borromeo, M.C. Rodriguez-Sanchez, and J.A. Hernandez-Tamames

Electronic Technology Departament
Universidad Rey Juan Carlos
c/ Tulipán s/n, 28933, Móstoles, Spain
susana.borromeo@urjc.es, cristina.rodriguez.sanchez@urjc.es, juan.tamames@urjc.es

Abstract. Positioning systems play an important role in ubiquitous computing. There are several technologies to implement localization services, being GPS the most popular. This paper describes a modular architecture developed for sending and receiving GPS data for differential positioning. The hardware set-up for sender and receiver devices consists of an embedded PC, GPS, GSM/GPRS and WIFI modules. We proposed and evaluated three strategies that allow employing the different communication protocols for linking sender and receiver devices. The goals of our system are the modularity and portability, close to the real time, so we can use it in other applications that integrate GPS receptor or GSM/GPRS devices such as fleets control, tourism or health-care.

Keywords: Context-Aware Computing, Differential GPS, Ubiquitous Computing, Modular Software.

1 Introduction

According to Sahdbolt [1] ambient intelligent involves the convergence of several computing areas such as ubiquitous, pervasive computing, intelligent systems and context awareness. Moreover, advances in digital electronics over the last decade have made computers faster, cheaper and smaller. Furthermore, the revolution in communication technology has led to the development and rapid market growth of embedded devices, equipped with network interfaces. The Ubiquitous Computing is the answer (Mark Weiser 2002) defined as "enhances use by making many computers available through the physical environment, while making them effectively invisible to the user"', thus, location aware computing is only an aspect of ubiquitous computing.

There are many developments in localization technologies. Some of them use Bluetooth, GSM or RFID. "The active badge system" 3 presents a system designed by Olivetti Laboratories on 1990, which uses IR signal and does inquiries to a central server. "Cricket"4, developed by MIT, uses beacons to calculate geographical position of the user and they studied different methods for obtaining positions using beacons. "Nibble"5, it is a system without special hardware. It detects the position as a function of signal intensity. Architectural approaches for localization were proposed in CBR 6. Ultrasonic systems for computing positions by triangulation were presented in 78 obtaining absolute distances between sensors and receivers.

J.M. Corchado et al. (Eds.): DCAI 2008, ASC 50, pp. 521–525, 2009.
springerlink.com

Recently, there has been growing interest in GPS (Global Position Satellite) 9. GPS provides outdoor continuous and global covertures. GPS is also able to disseminate very accurate time references for universal synchronization. In other hand, further improvements in GPS signals (due to power, integrity monitoring and antennas), is a great opportunity to reinforce the enormous potential that GPS has shown up to date, making possible to reach, even, further location-aware applications at lower prices. Differential GPS is one of the most unknown applications, however it provide the high accuracy in positioning in relative measurements. The present work describes and implements a modular architecture to send and receive GPS data based electronics on-board. We investigate the best choice to implement a differential GPS platform oriented relative vehicle positioning with GPS electronics on-board. It means that our system enables a GPS receiver to determine its location, speed, direction, and time in a differential way.

In section 2, we describe the technical features of our system in terms of design, connectivity and data format. In section 3, we explain the development applications; in section 4, describes the results; finally, in section 5, we show some of the conclusions and further works.

2 Technical Features

Most of GPS receivers can relay position data to a PC or other device using the NMEA protocol. Our application supports, not only NMEA data format, but also GPS raw data, in order to improve the accuracy of a GPS application, thus, raw data contains additional information of satellite position

Another critical challenge for differential GPS, in terms of real time, is the connectivity strategy between on-board devices for transmitting raw data between them. Some choices can be Bluetooth, GSM/GPRS, WIFI (or any combination).

The hardware set-up of our system consists of the following elements:

- Sender Device. It consist of: three units: a GSM/GPRS module based on TC65 by the Siemens10, a GPS LAET-4T by UBLOX 11 and a computer or embedded PC.
- Receiver Device. It includes a GSM/GPRS module based on TC65 by the Siemens and a computer or embedded PC 12 13.

We propose 3 strategies (see Fig. 1.) to communicate the sender and the receiver devices:

1. Link based on GSM
The sender device sends GPS raw data (NMEA data are available too), by SMS to receiver device. We have implemented a firewall filtering SMS in order to accept only the "friend" messages. Data are stored in a SIM target (file or virtual buffer).
2. Link based on GPRS socket
The GPS information is sent by a GPRS socket. The firewall filters the received information using the IP and the enabled port socket. The software agent stores the data in a file or in a virtual buffer. The socket is available permanently.
3. Link based on WiFi TCP/IP
The GPS information is sent by TCP/IP socket (Wifi). The firewall filters the received information using the IP sender and enabled port socket. As option 2 the software agent in receptor stores data in a file or virtual buffer.

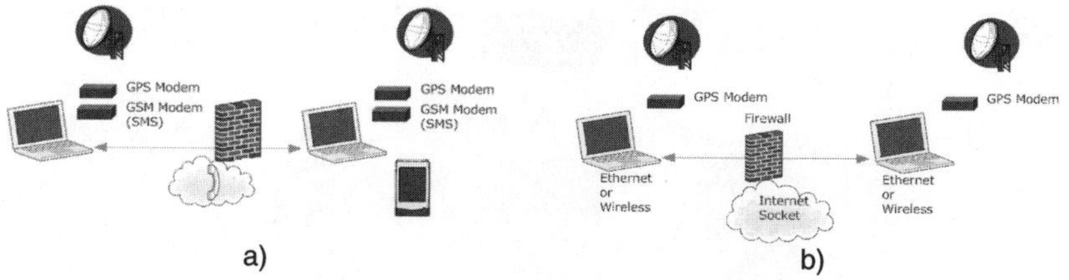

Fig. 1. a) GSM/GPRS link, b) WIFI or Ethernet link

3 Developed Applications

We have designed modular software consists of: libraries, modules or API's (Application Program Interface) and configuration files or drivers.

The applications are independent of the Operating System. They have been tested with GNU/Linux with minimal kernel version 2.4, CFLinux 14 and Windows XP. They have been developed using standard C/C++ language, so, we guaranteed a portable source code.

We have been developed different modules to integrate the different tasks of our system. According to the functionality we can divide the module five categories:

- M1: Management of hardware ports: open serial or USB port to read/write
- M2: Socket control: open the GPRS or Ethernet socket as receiver or sender.
- M3: Send GPS data.
- M4: Receive GPS data.
- M5: Data format of GPS. It can be NMEA or RAW-DATA.
- M6: Stop applications with security.

The configuration files allow set the different parameters of GPS module (output format and device attributes) and socket configurations (IP and port). Beside we have developed different drivers to connect GSM/GPRS and GPS modules to PC.

4 Results

The developed system has been tested in GNU/Linux with minimal kernel version 2.4, Windows XP and embedded system CFLinux.

We implemented two executable files: for sending coordinates (by SMS, GPRS socket or GPRS Ethernet/WIFI) and for receiving the coordinates. (see Fig. 2, Fig. 3) :

- Sender: Application to send GPS information. It can be NMEA coordinates or RAW-data (as mentioned above). Furthermore, can be sent by SMS, GPRS-Socket or Ethernet-Socket (WIFI).
- Receiver: Applications to receive information of incoming communications by SMS, GPRS-Socket or Ethernet-Socket (WIFI).

We test our system in different contexts using three strategies described in section 2 using two computers and with embedded PC too.

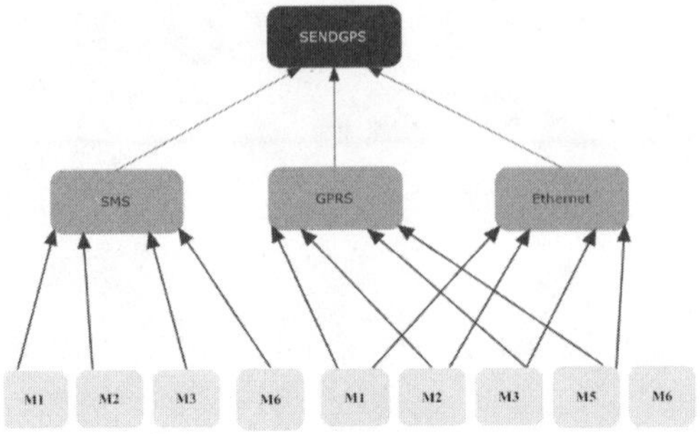

Fig. 2. Sender applications

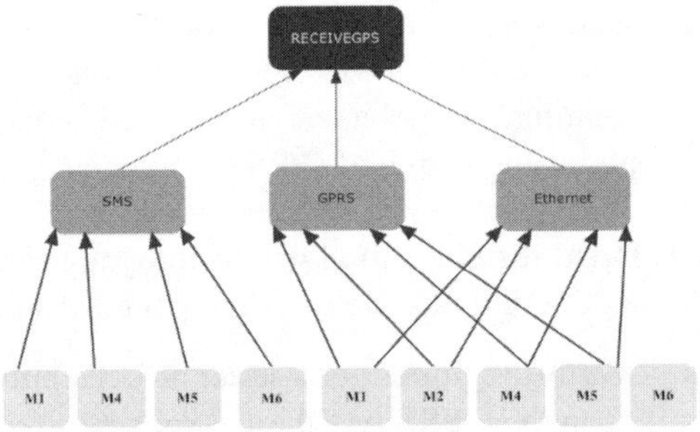

Fig. 3. Receiver Applications

Table 1. Tests

Test number	Node 1	Node 2	Option	Operating System	Measure delay Reception (seconds)
1	Computer	Computer	1	Linux/Windows	5
2	Computer	Computer	2	Linux/Windows	4
3	Computer	Computer	3	Linux/Windows	0,1
4	Laptop-Wifi	Laptop-Wifi	3	Linux/Windows	0,2
5	Embedded PC	Embedded PC	1	CFLinux	6
6	Embedded PC	Embedded PC	2	CFLinux	7

5 Conclusions

We propose three different strategies for communications in a differential GPS application (GSM, GPRS or TCP/IP socket by WIFI). Our approach is based on modular architecture, independent of operating system and it has been implemented in standard C/C++ language. As a result, we obtain a portable and reusable source code. The results (see Table 1) show that "option 3" is the best choice for real time. The system has capability for working close to real time and the Ethernet modules are compatible with Wireless IEEE802.11.

In further works, we will employ the different developed modules in other applications that integrate GPS receptor and GSM/GPRS devices. In fact, we are adapting these modules in several applications: help for people with disabilities 15, tourism and proximity marketing.

References

1. Shadbolt, N.: Ambient Intelligence. IEEE Computer Society, Los Alamitos (published)
2. Weiser, M.: The computer for the 21th century. IEEE pervasive computing 1 (2002)
3. Korkea-aho, M.: Context-aware applications survey, http://www.hut.fi/mkorkeaa/doc/context-aware.html
4. Dey, A.K., Abowd, G.D.: Cyberminder: A context-aware system for supporting reminders. In: Proceedings of Second Symposium on Handheld and Ubiquitous Computing, pp. 172–186. Springer, Heidelberg (2000)
5. Castro, P.: The nibble location system, http://mmsl.cs.ucla.edu/castop/nibble.html
6. Xu, L.D.: Case-based reasoning for AIDS initial assessment. Knowl. -Based Syst. 8(1), 32–38 (1995)
7. Nonaka, H., Date, T.: Ultrasonic position measurement and its applications to human interface. IEEE Transaction on Instrumentation and Measurement
8. Wehn, H.W., Belanger, P.R.: Ultrasound-based robot position estimation. IEEE Transaction on Robotics and Automation
9. Hightower, J., Borriello, G.: Location systems for ubiquitous computing. IEEE Computer, 57 66 (2001)
10. http://www.siemens.es
11. LEA-4T ANTARIS® 4 Programmable GPS Module with Precision Timing, http://www.u-blox.com/products/lea_4t.html
12. PC 104, http://www.pc104.org
13. Embedded System Design, http://www.embedded.com
14. CFLinux, http://www.cflinux.hu/
15. Borromeo, S., Rodriguez-Sánchez, M.C., Hernández-Tamames, J.A., Malpica, N.: Wireless Contextual Information Electronic System for People with Disabilities. In: BIODEVICES 2008, International Conference on Biomedical Electronics and Devices, Madeira, Portugal (2008)

Requirements for Supervised Fusion Adaption at Level 1 of JDL Data Fusion Model*,**

L.A. Lisboa Cardoso[1], Jesús García[2], and José M. Molina[2]

[1] Instituto de Pesquisas da Marinha
Rio de Janeiro, Brazil
cardoso@ipqm.mar.mil.br
[2] Computer Science Department-GIAA
Universidad Carlos III de Madrid
Colmenarejo, Spain
jgherrer@inf.uc3m.es@inf.uc3m.es, molina@ia.uc3m.es

Abstract. The JDL Data Fusion Model is revisited aiming explicit human intelligence retention, using supervision at every possible level of the original model. This subtle variation of the model would extensively accept direct human guidance, behaving as a tool for interactive visual perception enhancement. A software design for interaction at level 1 of the fusion model is proposed for implementation and evaluation. Typical applications expected are related with environmental analysis and assessment, particularly, in the tasks of data association and track fusion in multi-target multi-sensor scenarios.

1 Introduction

The JDL Data Fusion Model [1] has being a standard reference in target tracking and tactical situation assessment, typical defence applications. In fact, the model can be thought of as a general framework for emulating human reasoning, concerning environment perception, thus providing an automated decision making tool. In the model, the complex and apparently parallel mental processes exhibited by human intelligence are hierarchically organized as a sequence of growing abstraction levels, each taking input from the previous, lower levels [2]. In this report we present the possibility and the requirements for deriving a JDL-based model for intelligence capture through human-computer interaction (HCI). A system implementing this model would then accept direct human guidance at any level, exhibiting supervised learning properties and treating the human actor as an external Level 4 agent.

In the proposed model (see diagram in figure 1), both machine and man take sensorial information from the environment and process it until a decision or semantic synthesis is produced. The machine works hierarchically, executing from low level signal processing and feature extraction algorithms, at the bottom layer, to high level programming strategies, at the top layer. On the man side, perception and reasoning are similarly developed, to include perception induced by stimulus generated at intermediate machine processing levels. In some cases, the reality may lay out of human

* Funded by the Brazilian Navy and FINEP Contract Nº 0.1.05.0733.00.
** Funded by projects CICYT TSI2005-07344, CICYT TEC2005-07-186 and CAM MADRINET S-0505/TIC/0255.

J.M. Corchado et al. (Eds.): DCAI 2008, ASC 50, pp. 526–535, 2009.
springerlink.com

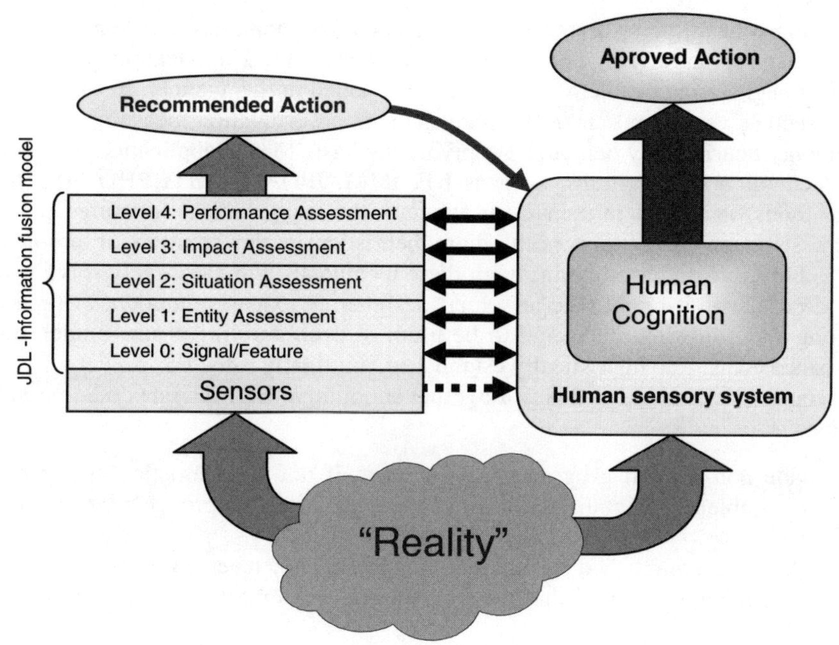

Fig. 1. JDL- model in support to intelligence augmentation

sensorial domain, with man relying only on machine translated stimulus. The machine, reciprocally, receives feedback at its different processing levels, adapting its algorithms and methods accordingly.

A successful system implemented with this paradigm would than deliver what has been called of "intelligence amplification" or "augmented intelligence" [3], in the task of environment analysis. Because of the high degree of generality of this conception and the presumable effort in implementing and testing it, this study is initially restricted to JDL level-1 interaction. At this processing layer, activities such as target tracking and identification are then to be performed through HCI in a symbiotic man-machine operation.

2 Interaction at JDL Level 1

Although having distinctively mathematical descriptions, the tracking of punctual targets and image objects in 2 or 3-dimensional spaces can be viewed as similar problems, under the human brain perspective, as both can be reduced to the problem of tracking visual information. A rough duality in between these two domains could be established by either associating object properties to target attributes, or the other way around, by abstracting the existence of a high-level mind processes that detect, identify and map objects of interest, in an given image sequence, into a suitable coordinate system.

Algorithms have being extensively developed for both cases, often using an explicit image-to-target conversion (detection) before tracking, leading to a large number of engineering problems successfully modelled in this manner. In these applications, such as vessel and air traffic management, an evaluation measure of success is commonly heuristically depicted, supplying the basis for the application of optimization methods and algorithms, such as KF, IMM, JPDA, PMHT, PHD [4],[5]. In these methods, as well as in their many variants, the mathematical structures and parameters selection tends to conceal more heuristics in the solution of problems. Whereas browsing and prototyping with these techniques will effectively solve many practical problems, it might take an expert designer an extensive amount of time instantiating and adjusting the model to be used, or even deriving a new model, until performance goals, also heuristically established, are finally satisfied.

The conjecture motivating this work is that an intuitive HCI schema could possibly be derived to:

- (i) provide non-explicit programming for most, if not all, heuristics in a general tracking problem, facilitating the system design in the approach of new posed problems.

If the design accomplished in this way would still not meet, by itself, a required quality level, it might happen that the system, under operator guidance through the HCI, could:

- (ii) provide means for continuous human interference with the model, so that it could be used as a visual processing augmenter.

In this latter, remedy approach, a low expertise human subject would still be able to adopt, adjust and run existing tracking models he does not precisely understand, interacting with the real world processes, continuous and intuitively, and boosting his own visual processing capacity, much in the same way hydraulic systems are designed to boost the power of human muscles.. Conversely, the machine, under his operation, would deliver smarter tracking services, relying on human perception skills when advantageous, possibly outperforming fixed design and fixed programming devices or systems.

Because of the wide range of implications of the above conjecture, it is still a hard task to investigate the concept in general. In this study, we preliminarily restricted the problem to the 2-dimensional multi-target multi-sensor tracking problem, abandoning, for the moment, any other generalized format of the conjecture. In this context, tracking limitations on the operator side appear when:

- (A-1) The kinematics of targets is incompatible to the human time perception or visual response envelope;
- (A-2) The number of targets becomes too large for them to be simultaneously handled;
- (A-3) The targets are too widely apart or too closely spaced to be visually tracked;
- (A-4) Targets get confused with excess background noise or other nearby targets;
- (A-5) Targets plots exhibit low confidence levels, even becoming provisionally unavailable, due to sensor limitations (e.g., blind zones);
- (A-6) The extension of time in which tracking activity should be sustained exceeds human attention capability.

On the other hand, typical heuristics and algorithm shortfalls are related with the difficult to:

- (B-1) Providing track initialisation;
- (B-2) Maintaining meaningful data association under unfavourable conditions, such as described for (A-4) or (A-5) above;
- (B-3) Maintaining trajectory estimation for complex, highly manoeuvring targets.

In the past, before computational processing and information fusion theory advances, tracking targets and summarizing target activity in track tables were functions accomplished by human operators, with reasonable success. With today's ever-growing machine capabilities in terms of precision, time response, computational load and complexity, human participation has been constantly minimized, sometimes totally eliminated from newer algorithms and systems, even though human reasoning can still be taken as a safe reference in the ultimate evaluation of target tracking systems, provided load and real-time is not critical. Thus, recombining the two elements in the proper manner may expectedly lead to better results than any single elements would exhibit.

Here we start establishing the requirements in order to experimentally investigate criteria and techniques for selection and parameterization of information fusion and tracking methods. Above all, these techniques should be based on intuitive human judgment, using supervised learning schemas or data domain mapping, from real world to a more human-tractable space-time representation.

3 Scenario Definition

Formally, the problem addressed is the detection and tracking of multiple targets in a L-dimensional space W, using measurements simultaneously obtained from M sensors detecting the targets, without all targets being necessarily detected by each sensor. The total number of targets N at detectable range is unknown and possibly variable in time. The detection process establishes targets existence, their respective estimated positions ("plots") and confidence levels. In some cases the detection also estimates other target features, such as radar cross section or Doppler deviation (a direct measurement of radial velocity). In order to give the operator a fair chance of contributing in the processing of tracks, all meaningful information should be efficiently translated into the HCI definition.

In the sensor model to be considered, detection errors are expected, either false positive, due to the unwanted detection of background noise ("clutter") or false negative, when missing a target within detection range. Also, position measurement errors are introduced by each sensor modelled, no independence in statistical behaviour being assumed. That is, not all sensors are able to detect all targets at all times: detection depends on relative position of targets to the respective sensor, among other factors, and it is generally represented by probabilistic modes.

A target j is intrinsically represented by a sequence {(position(k), time(k))}, denoting its actual, usually unknown trajectory. Target plots detected at each sensor are given by the sequence {(position(k), time(k), quality of detection(k))}, where the quality of detection refers both to the variance in kinematical estimated data and the confidence in the presence of such target. Target velocity and position are computed at individual sensors, commonly filtering the sequences of originally detected position estimates s (k),ij ("plots"), for instance, by using the Kalman Filter [6].

$$V^{(k)} = \left\{ v_j^{(k)} \mid v_j^{(k)} \text{ position of target "j" at time } t_k \; 1 \le j \le N^{(k)} \;\; k \ge 0 \right\} \quad (1)$$

$$S_i^{(k)} = \left\{ s_{ij}^{(k)} \mid s_{ij}^{(k)} \text{ plot "j" of sensor "i" at time } t_k \; 1 \le j \le n_i^{(k)} \;\; k \ge 0 \right\} \quad (2)$$

While some of the sensor measures $s_{ij}^{(k)}$ should lay in neighbourhood of the correspondingly target appearance at $v_u^{(k)}$, others should locate false positive targets. The data association problem, critical to the tracking-fusion processes, is the establishment, at every time t_k, of a correspondence in between the indexes u, $1 \le u \le N^{(k)}$, of the true position of targets or their tracking estimators, and j, $1 \le j \le n_i^{(k)}$, the indexes associated to sensor measurements.

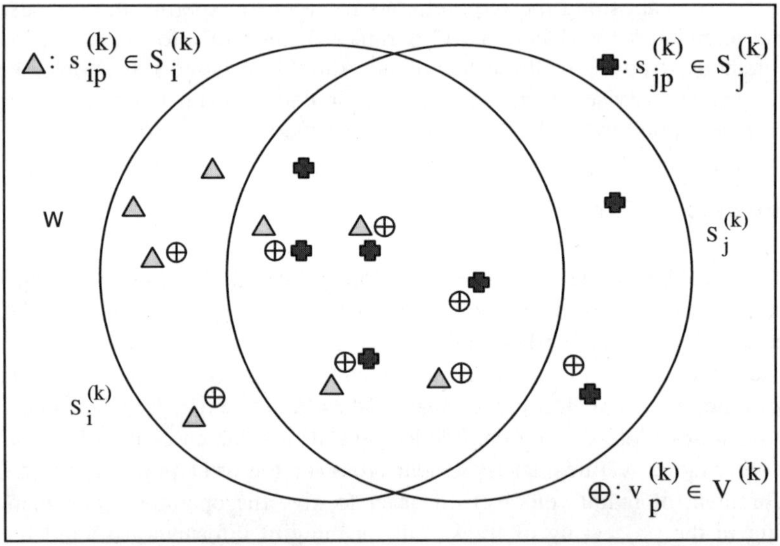

Fig. 2. The 2-dimensional data association in track-fusion problem: plots at time tk

This can be individually computed for each sensor i, when we end up on having different, but possibly similar, track estimations for every target j. For each target it can be created as many tracks as the number of sensors available. The track-fusion concept is related to further association steps in order to compute consolidated track estimations $r_{ij}^{(k)}$, using data from all sensors.

The fusion algorithm computes the coordinates of virtual targets (stars), each based on the sequence of associated plots $s_{ip}^{(k)}$ and $s_{jp}^{(k)}$. If the true reference position $v_p^{(k)}$ was

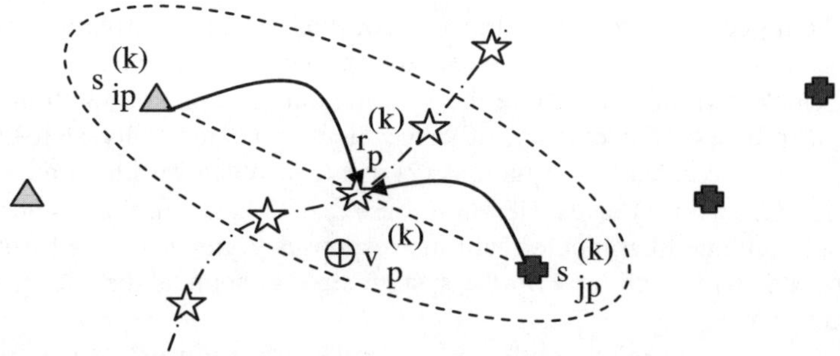

Fig. 3. Fusion by association of plots from different sensors to the same track

known, it could be used in the supervised training process. As $v_p^{(k)}$ is usually unknown, human interaction may provide means to create a reference set, based on expected geometric sequence behaviour.

4 Human Hypothesis and Interface Requirements

Since estimated position and speed may be geometrically represented for each target being tracked, human operators may easily introduce feedback in predictions. It shall be possible for an operator to graphically enter, by means of a pointing device, his preferred coordinates, speed and course for any target being tracked, on an intuitive basis, adjusting or even overriding system estimation. Due to the limited human response time, supervised training steps should be carried either on dilated time constants or on a frame-by-frame step mode. The training environment should then be designed to handle time buffering / deceleration when necessary, by the use of a user controlled time acceleration constant ξ. In terms of equations (1) and (2), the HCI and integrated simulation control shall then implement the mapping $\Xi: W \times T \longrightarrow W \times T^{(\xi)}$, such that:

$$\Xi\,(s_{ij}^{(k)},\ t_k) \;=\; (s_{ij}^{(k)},\ t_k / \xi),\quad \xi > 0,\quad k \geq 0 \tag{3}$$

Recent past values of sensor plots are relevant for intuitive reasoning and prediction, and that condition should be as well be represented in the graphics interface used for the training the system.

During the training phase, the HCI should enable the operator to:

1. Visualise the H most recent plots $\{s_{ij}^{(k)} \mid k_0 - H < k \leq k_0\}$ for all sensors i (the "history", as seen at simulation time t_{k_0} perspctive), in a graphically distinguishable time order. Evanescence has been a standard process for that in radar displays, one of the main applications in target tracking.
2. Whereas available, visualize available additional target information given by each sensor, other than position, for all recent plots.

3. Visualise trajectories of tracks, drawn according output of active tracking and fusion algorithms, either explicitly selected or previously learned.
4. Adjust clock rhythm ξ to handle the information presentation rate during the supervision process; if necessary, advance clock time manually step by step, by incrementing the index k in equations (2) and (3),. When working under a time acceleration (case $\xi > 1$) or deceleration (case $\xi < 1$) factor in this manner, the system shall continue to compute real time responses to input received from sensors. The operator will catch up with the system state as soon as the supervising intervention is over.
5. Analyse and modify relative influence of any sensor i towards the computation of each track j, given by $r_{ij}^{(k)}$, $1 \leq j \leq J^{(k)}$, where $J^{(k)}$ the number of associated tracks (ideally, but not necessarily, $J^{(k)} = N^{(k)}$). Interact until better or most intuitive justification model is obtained, without reduction of tracking performance. This could be possibly be done by graphically inspecting and changing computational weights and plot associations.
6. Modify tracks automatically generated by current algorithm and rules, according to own feeling, teaching the systems the expected fusion behaviour, if different from current output.

Compared to computer automated actions, a priori expectations regarding human performance, at this first level fusion task, are:

1. The operator will shortfall in numerical precision, while tracking well behaved, linear, non-manoeuvring trajectory targets.
2. The operator will not withstand performance in long runs. He might rather be able to favourably interact with trajectories estimation at critical times, for instance, during manoeuvres, plot misses or heavy clutter conditions.
3. The operator might be able to facilitate track initialisation tasks, this being one of major deficiencies in current well established data association and track-fusion algorithms and a common presupposition [7].
4. The operator might be able to improve measurement-to-track data association in confusing target trajectories, under the presence of confusing or false plots, either by explicitly providing data association points $q^{(k)}$ or by rejecting associations computed by algorithms, thus introducing ad hoc pruning criterion.

5 System Architecture and Implementation Issues

In order to address the issues above, quantify human potential, and clearly identify symbiotic opportunities for HCI, regarding level 1 of the JDL model, a software tool is proposed. With its interface strongly based in pointing devices, it has a minimum parametric configuration capability, to force graphical interaction. Whereas flexible pointing capabilities [8], such as multi-touch table-top devices, recently proposed for Multi-Actor Man Machine Interface (MAMMI) [9], are highly desirable, a prototype built over a standard PC and an optical mouse, was demonstrated as capable of reasonable real time data entry for trajectories definition. Figure 4 shows a preliminary version of the control panel of the system, to easy the task of creating, simulating and interacting with tracking scenarios. The specification depicts a system that behaves like a computer-aided drawing program with an embedded, resourceful, real time sensor simulation, and a tracking-fusion computation agent that can run known data association, tracking and fusion algorithms. Real data, if available, can be retrieved from secondary storage, making the simulations more realistic. The system design expected capabilities are such as:

1. Evaluation of human performance in lower level visual coordination tasks;
2. Identification of envelope of operation for man and machine;
3. Comparative evaluation, under human subjective judgment, of different tracking- and fusion algorithms;
4. Computer-aided design of tracking systems; and
5. to be, verified, supervised learning.

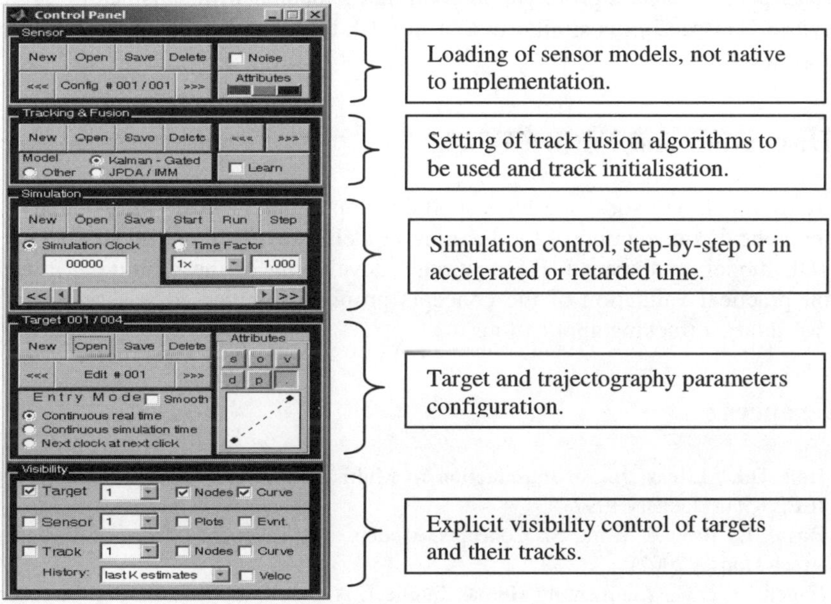

Fig. 4. Main command pad window for the proposed track-fusion interactive environment, as an auxiliary method to pointing device input

Whilst the configuration of sensors and the preparation of the scenario are conducted through the control panel, most of the simulation interaction is performed with the sole use of the pointing device over the plotting area, as shown in figure 5.

Frequently used in mathematical computation, the MathWorks MatLab software was adopted for the realization of a first prototype. In spite of its limited graphical

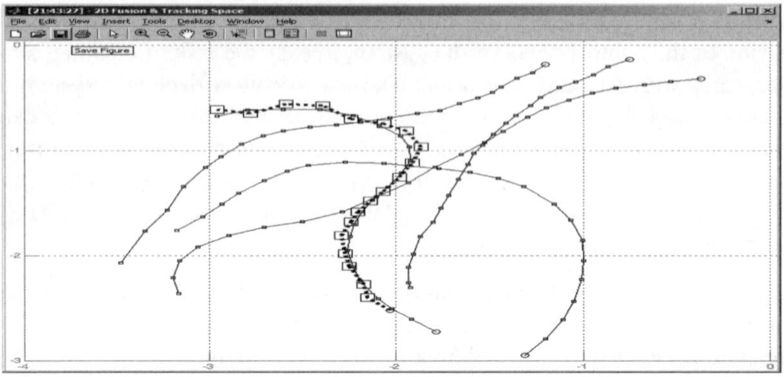

Fig. 5. Using the pointing device, trajectories are created with different semantic values, as in instantiating simulations and providing supervision for resident algorithms

interface performance, a prototype is being implemented with satisfactory qualitative behaviour. This design is similar to that in [10], however extends its HCI and model capabilities.

6 Discussion and Future Works

Future to this initial work, activities shall develop simultaneously in two areas: First, the theoretical construction of the learning models, adequate for cooperative use with the JDL model; and second, the continued developing of the prototype, in order to permit practical validation of the concepts proposed, aiming necessary advances in real word target tracking applications.

References

1. Hall, D.L., Llinas, J.: An Introduction to Multisensor Data Fusion. Proceedings of the IEEE 85(1) (January 1997)
2. Bossé, E., Roy, J., Wark, S.: Concepts, Models, and Tools for Informatio Fusion, ch. 4. Atech House (2007)
3. Engelbart, D.C.: Augmenting Human Intellect: A Conceptual Framework, Summary Report AFOSR-3233, AFOSR, Menlo Park, CA (October 1962)
4. Mahler, R.P.S.: Statistical Multisource-Multitarget Information Fusion. Artech House (2007)
5. Bar-Shalom, Y., Blair, W.D. (eds.): Multitarget-multisensor tracking, vol. 3, Appl. and Advances. Artech House (2000)
6. Welch, G., Bishop, G.: An Introduction to the Kalman Filter, TR 95-041, Univesity of North Carolina at Chapel Hill (July 24, 2006), `http://www.cs.unc.edu/~welch/kalman/kalmanIntro.html`
7. Puranik, S., Tugnait, J.K.: Tracking of Multiple Maneurvering Targets using Multiscan JPDA and IMM Filtering. IEEE Transactions on Aerospace and Electronic Systems 43(1) (January 2007)

8. Greenstein, J.S.: Pointing Devices. In: Helander, M., Landauer, T.K., Prabhu, P. (eds.) Handbook of Human-Computer Interaction, ch. 55, 2nd edn. Elsevier Science, Amsterdam (1997)
9. Vales, S., Chatty, S., Lemort, A., Conversy, S.: MAMMI Phase1 – Collaborative workspaces for En Route Air Traffic Controllers. In: 5th EUROCONTROL Innovative Research Workshops & Exhibition, Bretigny sur Orge, France (December 2006)
10. Alexiev, K.: A MatLab Tool for Development and Testing of Track Initiation and Multiple Target Tracking Algorithms. Information & Security 9, 166–174 (2002)

Applying Spatio-temporal Databases to Interaction Agents

Dolores Cuadra, Francisco Javier Calle, Jessica Rivero, and David del Valle

Departamento Informática, Universidad Carlos III de Madrid,
Avda. Universidad 30, 28911 Leganés, Madrid
{dcuadra,fcalle,jrivero,dvalle}@inf.uc3m.es

Abstract. In this paper, Natural Interaction will be approached for developing human-like interactive agents, and in particular interaction circumstances should be observed. With this aim, the Situation Model will be developed. For developing this model the support of an appropriate technology providing spatio-temporal description of objects and their evolution through time (both location and form evolutions) is essential. The technology meeting those requirements is the Spatio-Temporal Database. This work frames this sort of model in a Natural Interaction System, introduces the description of the model, and its design over a Spatio-Temporal Database Management System.

Keywords: Natural Interaction, Situation Model, Spatio-Temporal Databases, Spatio-Temporal Models, Knowledge Management based interaction.

1 Introduction

Natural Interaction (NI) [2] has been presented during last years as the research area standing for systems able to interact as human beings do. However, such paradigm is far from current interaction systems state of the art, so the actual challenge is to approach to a certain degree one or more of those behaviours to progressively attain a 'more natural' interaction. Users lacking of technological training will be a significant part of the users benefited by this interaction paradigm, but not the only, any other interaction disability could be overcome, or at least reduced its effect on interaction.

The path to reproduce humans' behavior is to acquire and formalize the knowledge and the reasoning mechanisms. First step is to divide it into smaller knowledge components which later integration will cover the whole needs. A classic approach involved a functional division into *interface*, *interaction*, and *application*, separating interactive abilities and external performance. A slight evolution pointed to differentiate the three major areas of the problem [4]: *interpretation*, *behavior*, and *generation*.

A particular knowledge need is that regarding the circumstantial aspects of the interaction. Through this paper a Situation Model (SM) proposal for representing the circumstantial aspect of the interaction will be introduced.

The paper is organized as follows: section 2 offers a brief description of the NI Cognitive Architecture as a framework for the SM. The section 3 presents different approaches for defining and representing circumstances in an interactive system and some issues on the design and development of a SM will be presented. Finally, the work concludes with some conclusions and current and future lines.

J.M. Corchado et al. (Eds.): DCAI 2008, ASC 50, pp. 536–540, 2009.
springerlink.com

2 Interaction System Framework and Related Work about the Situation Model

As aforesaid, human knowledge supporting interaction process is very complex and diverse. Because of this, it is necessary to divide it up into several specialized knowledge components (Fig. 1). The interface components (lower box in Fig. 1) are in charge of acquiring and interpreting the semantic content of user expressions describes through Speech Acts [12]. These acts will be let into the Dialogue Model Component (DMC), which has to update the dialogue state for every aspect (intentional, structural, contextual, etc.). Once updated, and during system's turn, the DMC will generate its own intervention. Often it will find that a specific interaction state requires external intervention before generating any intervention. Consequently, it will invoke the execution of the task and wait for the response.

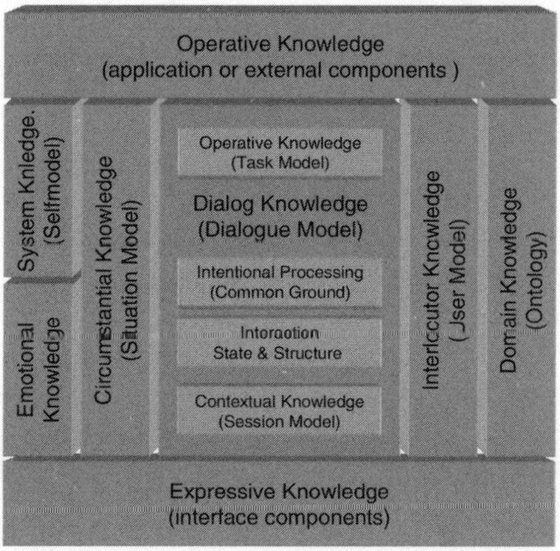

Fig. 1. Cognitive Architecture of the Natural Interaction System

During its processing, the DMC might need the contribution of reasoning mechanisms based on another type of human knowledge, for example, it might profit from the induction of some interlocutor features (User Model), the knowledge about the any circumstantial aspects (Situation Model), or the emotions affecting the dialogue (Emotional Model). There also should be a model to fix concepts referred by both participants through terms within their expressions, that is, Ontology. Finally, a model on behalf of system's own goals, named Self Model.

This cognitive architecture is in detail presented [5] and the agents platform is described in [7]. In summary, the characteristics of each knowledge model determine if it should be implemented through several agents (when it is appropriate to arrange several collaborating agents to assure the knowledge management) or not. The *Interactor platform* provides services, with inputs and outputs, and adding the session identifier (for storing the interaction state within each model) and all the needed parameters to represent the conditions of the service: the expiry-date, need-before, criticism and quality to make possible the IN System proposed.

Regarding the Situation Model, the circumstances of the interaction are very diverse, and could be differentiated quite some aspects following diverse criteria. According to Gee [7], there should be observed five main aspects for the situation: semiotic (signs in use), operative (task underlying the interaction), material (spatio-temporal), political (role of each interlocutor), and socio-cultural link. Most classic interfaces can do without a Situation Model because their requirements *fix* the circumstances so that they are well known. Many systems have some circumstance aspect modelled, fixing the rest. Hence, we could find several set of signs for choosing when interacting with a system, or several 'user profiles' with a different range of feasible circumstances arranged for each pair of roles. Systems capable of multiple task performances could also have the situation modelled as the state of execution for a required task or set of tasks [10]. Some other systems perform time or spatial-dependant tasks feed by a positioning component, such as the 'Navigator' in Cyberguide system [1], or have interaction influenced by multimodal input [6]. However, interaction systems rarely show different interactive behaviour depending on reasoning mechanisms based on situation knowledge and past, current, and predictions on future positions of interlocutor.

3 Situation Model Component

The situation material knowledge is designed following the object-relational databases methodology based on [3]. The *STOR* model provides it consistency, portability and easy integration. Moreover, the spatio-temporal multigranularity handling allows the objects representation with different space and time measures. According to the UML class diagram (Fig. 2), a user is looking for places, objects or persons of interest within a space referenced in a spatial system as common material aspects of every domain.

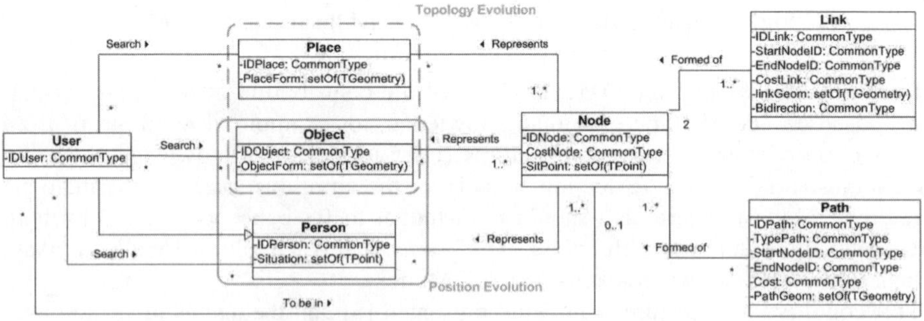

Fig. 2. The Material Knowledge representation

The UML class diagram has two outstanding parts. The first one (right part of fig. 2) is represented by one *net graph*. The *net graph* is composed by *Nodes* and *Links* between them which reflect the interaction space. A *Link* describes the practicable way from a particular node to another one. The *Path* class only observes ways which have been already covered. The second one (left part) consists of elements distributed on the net graph. *Person* and *Object* classes belong to moving objects. These elements cover the *graph net* and are characterized by one position and valid timestamp. A common constraint to every class is that the position is always valid with respect to the space defined.

The material aspect representation through spatio-temporal relational-objects allows defining Event-Condition-Action (ECA) rules [11] which will be triggered depending on the circumstance. The execution model of ECA rules follows three phases in sequence; event detection, condition test, and action execution. Once a rule is activated and its condition is evaluated to true, the predefined actions are executed automatically. For example, one feasible action to be executed could be to introduce a new subdialogue into the DMC (to be processed as appropriate).

According to the type of events, two active rules are proposed: triggers with temporal events and triggers with events caused by manipulation sentence in the databases (Insert, Update, and Delete).

Let P_{ij} be the User *i* in the location *j* within the *net graph.* The formal definition for each type of rules is:

(a) *RULE 1:* {*Event*: When t_k; *Condition*: Where P_{ij} is?;
Action: Send_Notif_Dialog Agent()};where t_k is the temporal event and the action is the message for Dialog Agent which will decide the interaction state.
(b) *RULE 2:* {*Event*: When DB operation; *Condition*: Where and when P_{ij} is?;
Action: Send_Notification_Dialog Agent()}; where the condition is a spatio-temporal expression. For instance, the user *i* is not in the j location (P_{ij}) at 11 a.m. Then, the SMC induces the DMC to initiate a subdialogue to warn the user.

According to the diagram presented and the spatio-temporal active rules, the material aspect implementation was developed on a DBMS for achieving autonomous behaviour.

4 Conclusions and Future Research

A Situation Model approach for NI has been introduced. Several improvements to the NI System have been introduced by the spatio-temporal databases. These are focused on three main directions: firstly, SMC is profited by the rest of the knowledge components across the cognitive architecture for filtering their own knowledge base; on second place, event triggering enables to program consequence execution when some circumstances are reached; finally, knowledge related to the situation (past and current circumstances or predictions on future circumstances), or planning actions sequences to achieve a particular circumstance from another.

Future works are focused on the inclusion of more circumstance aspects. Particularly, the ‘operative’ aspect, that is, to identify the task(s) underlying the interaction, is very significant and currently analysed to be taken into account. Thus, any circumstance-related process will consider the operative aspect, either for descriptions, knowledge filtering, triggering, constraining, path finding, etc.

Acknowledgments. This work has been partially supported by the Reg. Govt. of Madrid (MAVIR, S-0505/TIC-0267), and Spanish Min. of Ed. and Science (CIT-410000-2007-12).

References

[1] Abowd, G., Atkeson, C., Hong, J., Long, S., Kooper, R., Pinkerton, M.: Cyberguide: A mobile context-aware tour guide. ACM Wireless Networks 3, 421–433 (1997)
[2] Bernsen, N.O.: What is Natural Interactivity? In: Dybkjær, L. (ed.) Procs. of the 2nd Int. Conference on Language Resources and Evaluation (LREC 2000), pp. 34–37 (2000)
[3] Bertino, E., Cuadra, D., Martínez, P.: An Object-Relational Approach to the Representation of Multi-Granular Spatio-Temporal Data. In: Pastor, Ó., Falcão e Cunha, J. (eds.) CAiSE 2005. LNCS, vol. 3520. Springer, Heidelberg (2005)
[4] Blaylock, N., Allen, J., Ferguson, G.: Managing Communicative Intentions with Collaborative Problem Solving. In: Kuppevelt, J.v., Smith, R.W. (eds.) Current and New Directions in Disc. and Dialogue, pp. 63–84. Kluwer Academic Pubs., Dordrecht (2003)
[5] Calle, F.J., Martínez, P., Del Valle, D., Cuadra, D.: Towards the Achievement of Natural Interaction. Eng. the User Interface: from Research to Practice (2008)
[6] Cohen, P.R., Johnston, M., McGee, D.R., Oviatt, S.L., Pittman, J., Smith, I., Chen, L., Clow, J.: QuickSet: Multimodal interaction for distributed applications. In: Procs. of 5th Int. Multimedia Conf (Multimedia 1997), pp. 31–40. ACM Press, Seattle (1997)
[7] Cuadra, D., Rivero, J., Calle, F.J., Del Valle, D.: Enhancing Natural Interaction with Circumstantial Knowledge. Int. Trans. on Systems Science and Applications (2008); ISBN: 1751-1461
[8] Gee, J.P.: Introduction to Discourse Analysis. Routledge (1999)
[9] Meyer, S., Rakotonirainy, A.: A survey of research on context-aware homes. In: Proceedings of the Australasian information security workshop conference on ACSW frontiers (2003)
[10] Paternò, F.: Model-Based Design and Eval. of Interactive Apps. Springer, Heidelberg (2000)
[11] Paton, N., Díaz, O.: Active DB Systems. ACM Comp. Surveys 31(1) (1999)
[12] Searle, J.R.: Speech Acts. Cambridge University Press, Cambridge (1969)

Modeling of Customer Behavior in a Mass-Customized Market

Zbigniew J. Pasek[1], Pawel Pawlewski[2], and Jesus Trujillo[3]

[1] Department of Industrial and Manufacturing Systems Engineering, University of Windsor 401 Sunset Ave., Windsor, Ontario, Canada
zjpasek@uwindsor.ca
[2] Institute of Engineering Management, Poznan University of Techology, 11 Strzelecka St., Poznan, Poland
pawel.pawlewski@put.poznan.pl
[3] Department of Automation and Systems, University of Valladolid, Valladolid, Spain
jestru@esa.uva.es

Abstract. The use of market simulation models becomes increasingly widespread as it helps improve enterprise planning and product delivery. In case of mass customization approach, which aims to simultaneously target scores of individual customers offering them product variations tailored specifically to fit their individual needs, understanding customer behavior under growing market stratification conditions is critical. This paper describes a market simulator in which individual customers are represented as software agents, behavior of which is based on a set of predefined rules describing their behavior under conditions of abundant choice. Parameters in those rules can be either fixed or follow some probabilistic distributions. Customers/agents operate in a market offering them a line of products, which have a set of customizable features, presumably addressing the needs of customer population. The paper describes in detail the decision process of an agent, market fragmentation design, and presents example results which can be used to determine desired variation offering of a product.

Keywords: Product variety, decision making, customer satisfaction.

1 Introduction

A fundamental issue for companies defining portfolio of their products aimed at any particular market is to determine how many variations of a product or service to offer. General trend over the past 30 years indicates a continuously growing number of offerings available to the customers [2]. From an individual firm's perspective competing in a saturated market requires differentiation, and that directly leads to increased variety in the market offerings. Simplistically, more available variety satisfies more customers.

There are, however, costs associated with introducing more variety, such as, for example: manufacturing costs, handling costs, and the costs of space in the store that sells the product. The economic decision about the optimal number of product variations offered should be determined as a tradeoff between these costs and the customer satisfaction. While increased product variety may lead to increased sales, too many product variations will drive up firm's expenses due to increased complexities of

J.M. Corchado et al. (Eds.): DCAI 2008, ASC 50, pp. 541–548, 2009.
springerlink.com

inventory control and handling, manufacturing processes, and after-sale product support. In general, the number of products N in a family can be roughly determined by the following simplified formula [6], from

$$N = V^{C} \tag{1}$$

where C is product complexity (expressed by the available number of product features), and V is a number of product features that create the variations (it exponentially increases the total number of available products). It can be shown that if the total market size has an upper boundary, and the cost increases linearly with the number of product variations, then a point in which profits reach maximum can be calculated. Past that point, even though revenues can grow further, additional costs associated with handling the variation exceed the sale derivative, and therefore it is not economical to offer larger variety. This reflects the effects of the Law of Diminishing Returns.

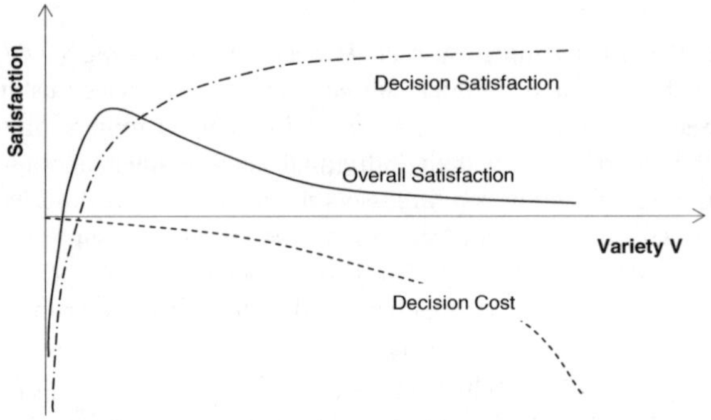

Fig. 1. Individual choice satisfaction

Economic arguments can precisely determine amount of product(s) variety that need to be offered to the market by an enterprise, given that the information about the market behavior is fairly accurate. However, recent research on individual consumer decision making [4, 10] indicates that decision-making behavior of individuals under conditions where choice is abundant is sometimes surprisingly counterintuitive. More choices, instead of better fulfilling customer needs, and hence increasing their satisfaction, lead to unanticipated effects, such as choice of no selections at all! It seems that the effort of making a choice decision not only carries its own costs (e.g., collection and analysis of necessary information necessary to make an informed decision), but also is under influence of various psychological effects which depend on the emotional profile of the individual (e.g., maximizers vs. satisficers). Thus from the psychological decision making point of view, there also exist an "optimal" amount of variety that a customer can emotionally and computationally handle (see. Fig. 1) [10].

This paper attempts to quantify how the combined effects of individual decision making under abundant choice conditions impact the model defining optimal variety on the firm's level in an effort to integrate the information flow between product design and marketing. These phenomena on an individual consumer level further modify the Law of Diminishing Returns effects, and quantify their impact.

2 Related Work

Individual consumer choice processes and their models are a source of continued interest, in particular in the areas of marketing and product line (portfolio) development. Most of the existing models are based on the choice axiom [8], which in principle is a multinomial logit model, a special case of a random utility model.

A growing interest over the past decade in the mass customization approach [9, 11, 3] underscores the importance of the individual consumer choices, in terms of both their structure and parameters. However, the industrial mantra of "more is (choice) better," derived from the times when the mass production paradigm reigned, does not seem to hold true anymore as many recent studies of consumer psychology indicate [4, 10]. The drive for availability of more choices is tempered by increasing complexity of potential choice decisions, connected unavoidable compromises, and psychological side effects. All of these cause the consumers to either delay (sometimes indefinitely) their decisions, or make conservative selections.

Since market surveys and comprehensive studies are usually expensive to conduct (not to mention that may also be unable to deliver correct answers to the burning questions), a natural solution is to turn towards modeling and simulation of consumer behavior. While early modeling approaches relied on presenting aggregate consumer behaviors, recent software developments have significantly accelerated transition to modeling individual consumers using agent-based concepts. The commercial market offers now a wide range of software tools enabling development of such models (Arena, AnyLogic, Simulink, etc.) [1, 5, 12].

3 Modeling Approach

The presented model is based on an imaginary market which operates under the conditions of a single manufacturer monopoly. The market population can be divided into a multitude of specific sub-segments (niches), ranging from one (uniform market) to as many as there are individual customers. Therefore, on one hand, there are a number of customers in the market under consideration, who may look for product(s) satisfying their needs, on the other – there is a producer who offers a line (or portfolio) of products aiming to satisfy these needs.

3.1 Customer Behavior Model

The decision process of each agent follows the state chart in Figure 2. After activation each agent enters the *Freeze* (idle) state, and remains there until it is told to proceed (a Boolean value of *Freeze* is then set to *FALSE*). The agent then moves to the first branch (circle marked *B1*), and subsequently decides which product to purchase, if any. Next it follows to the second branch, where, if it has purchased a product it goes to the Owned Product state and if it hasn't it travels to the No Product state. If no product is purchased, the agent will run through this process again to see if any new products have become available. If a product is purchased, the agent will use the product until it expires (determined in the user interface), then it will repeat the steps of looking for a new product to buy. It continues until the Boolean value of Freeze is set back to TRUE and all activity stops.

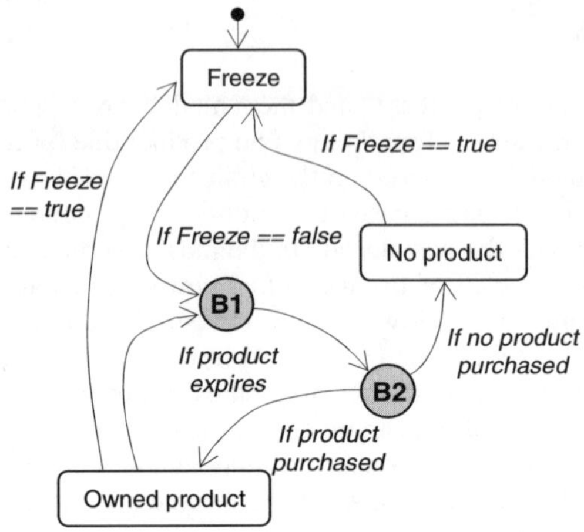

Fig. 2. Decision flow in an agent

The model was set up this way primarily to enable potential interruptions in the middle of an experimental run (e.g., for adjustment of parameters). Without this enhancement the structure of the model is fairly simple: if an agent doesn't own a product, it repeatedly tries to buy the one that best matches its criteria; if no match is found, the agent tries over and over again. If a product owned by an agent expires (either wears out of reaches the end of its useful life), the agent initiates an attempt to buy another product as a replacement.

3.2 Market Model

The model was designed to mimic the market of almost any simple product [7, 8]. In such a market several groups of consumers and up to 10 products can be created and their characteristics defined. The primary limitations in market model complexity setup are user-interface driven, but nominally the number of products can be infinite. The consumers/agents, given their predefined parameters, will choose a product that best fit their needs.

Market character is defined by the *Groups of People*, determining how many different consumer groups (or types of people with distinct characteristics) are to be considered in the model. This is used for populations with non-uniform distributions of features of its members, such as, for example, a population consisting of two subgroups: of very tall and of very short people.

Each group can independently have any number of agents, a uniform or normal distribution over their preferred feature values (such as width, weight, length, color, etc). The values that affect their decision making can also be setup independently for each group.

The tolerance parameter controls an individual consumer's willingness to buy products that are close to, but not exactly what they want. This is used to distinguish between consumers that would only be willing to buy an item that was exactly what they want or someone who will buy only the best-suiting item from available selection, even if it's far from what they really want. The budget determines the maximum amount that an agent will consider to spend. The compromise value determines how much consumers value a cheaper item with worse fit over a more expensive item that is closer to or matching their original preference. The complexity value is used to determine how much consumers can not make up their minds when more options are available to them.

Up to 10 different products with 7 different features each can be set up in the simulation. Each product has its own cost, durability (life-span) range, and functional features. The durability determines how long an item will last from the moment of purchase, and can follow either uniform or normal distribution and is expressed in relative terms (to other products).

3.3 The Agent Decision Process

The driving force behind the simulation is the decision-making-process of the agents. Each agent first considers the available budget. If the agent's budget is greater (or equal) than the price of the product, then the product is taken into consideration. A special case can also be setup with zero budget – price considerations are then excluded from the decision process. All considered products are compared using the agent's tolerance and perceived choice decision complexity. A product is considered (for purchase) using the following equation:

$$\prod_k (1 - \frac{|P_k - P_{dk}|}{R_C}) \geq 1 - TC^n \qquad (2)$$

where k is equal to the number of features a particular product may have, P_k is value of the considered product k, P_{dk} is value of the desired product k, R is the range of product values, T is customer's tolerance for product deviation from desired, C is a measure of decision complexity for the consumer (value between 0 and 1), and n is a number of products.

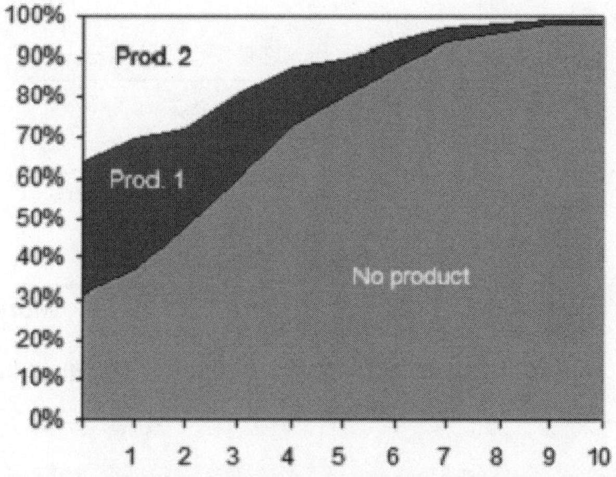

Fig. 3. Impact of consumer's decision making complexity on product market share

In this particular form of the decision function the tolerance has value between 0 and 1, making it independent from product values. It also considers all the features of a product together. If a product has many features deviating slightly from those desired by an agent, it may be equivalently desirable to the agent as a product with one feature significantly different. The complexity factor reflects consumer's expectation to find better fit through weighing of multiple product options simultaneously. It should have a value close to 1, so that when raised to a significant power it would not have an extremely significant effect.

It is worth noting that increase in the number of products also increases the complexity of agent's decision making, which in turn reduces number of products bought in a market. This can be clearly seen in Figure 3 showing these effects for a market with two products and varying levels of consumer–perceived complexity.

4 Exploration of Consumer Behavior

The simulator offers an opportunity to explore potential market behavior over a wide range of conditions, reflecting both the variety of consumer behaviors and their response to variety of product supply in the market. It also helps to understand any potential coupling effects present between the two.

The impact of the customer tolerance of the mismatch between the product that is desired and the one that is actually available can be readily explained using Figure 4. Tolerance T is expressed as a fraction of the available product feature range (e.g., $T = 0.1$ means deviation of ±10% of the product feature range is acceptable to an individual). The tolerance parameter is given in terms of lower and upper limits. Within the lower limit bounds (e.g., ± lower limit) it is assumed that all consumers would buy the not-quite-fitting product. Within the remaining range it is assumed that only 50% of consumers would make a buying decision.

The impact of the complexity parameter can be explored similarly. One example of its impact is shown in Fig. 3, and implies that even though more choices are available to consumers, the effort of making an informed decision is greater (e.g., it is harder to

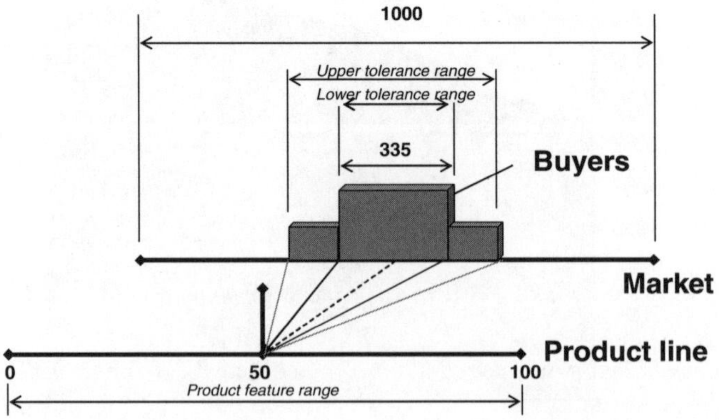

Fig. 4. Impact of customer tolerance to desired product mismatch on the number of buyers

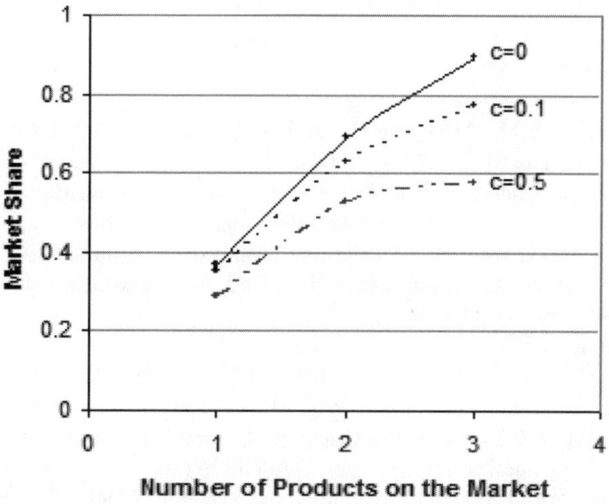

Fig. 5. Market share as a function of available product variants and their complexity

decide) and many may be expected to make no decision at all, thus reducing the number of potential sales.

5 Summary and Future Work

The paper describes a market simulation model intended as an aid in exploring the issues related to consumer decision making under the mass-customization conditions. These conditions are primarily characterized by increased pressure on the customer decision making by abundance of potential choices. A significant negative coupling effect can be observed, which reduces impact of the available product options on the potential market share of the manufacturers.

Implementation of the market model for mass-customization promises better understanding of the coupling effects between product line offered on the market, types of consumers, and their decision preferences. The next step in the process is to explore the impact these factors have on the economic performance of the enterprise. In particular, how careful design of the product line can influence optimization of profits and how enterprises can develop strategies of product introductions into the market under such conditions.

References

1. Borshchev, A., Filippov, A.: From System Dynamics and Discrete Event to Practical Agent Based Modeling: Reasons, Techniques, Tools (accessed January 29, 2006) (2006), `http://www.xjtek.com`
2. Cox, W.M., Alm, R.: The Right Stuff. America's Move to Mass Customization, Fed. Res. Bank of Dallas (accessed May 29, 2007) (1998), `http://www.dallasfed.org/fed/annual/1999p/ar98.pdf`

3. Gilmore, J.H., Pine, J.B.: Authenticity. Harvard Business School Press, Boston (2007)
4. Iyengar, S., Jiang, W.: The Psychological Costs of Ever Increasing Choice: A Fallback to the Sure Bet (assessed May 29, 2005) (2005), `http://www.columbia.edu/~ss957/`
5. Keenen, P.T., Paich, M.: Modeling General Motors and the North American Automobile Market (accessed January 26, 2006) (2005), `http://www.xjtek.com`
6. Koren, Y.: Global Manufacturing Revolution: Product-Process-Business Integration and Reconfigurable Systems (unpublished textbook manuscript, 2007)
7. Louviere, J.L., Woodworth, G.: Design and Analysis of Simulated Consumer Choice or Allocation Experiments: An Approach Based on Aggregate Data. Journal of Marketing Research 20(4), 350–367 (1983)
8. Luce, D.: Individual Choice Behavior. Wiley, Chichester (1959)
9. Pine, J.: Mass Customization. Harvard Business School Press, Boston (1999)
10. Schwartz, B.: The Paradox of Choice. Why More is Less, Ecco (2004)
11. Tseng, M.M., Piller, F.T. (eds.): The Customer Centric Enterprise: Advances in Mass Customization and Personalization. Springer, Heidelberg (2003)
12. Wooldridge, M.: An Introduction to MultiAgent Systems. Wiley, Chichester (2005)

Web Storage Service (WSS)

Hector Hernandez-Garcia, Victor Sosa-Sosa, and Ivan Lopez-Arevalo

Laboratory of Information Technology
Cinvestav Tamaulipas
Km. 6 Carretera Victoria - Monterrey
87276, Victoria, Tamaulipas, Mexico
{hhernandez,vjsosa,ilopez}@tamps.cinvestav.mx

Summary. This paper presents a Web Storage Service (WSS). This service is based on Web technologies and allows its users to manage a big and increasing amount of storage space. WSS creates a scalable virtual disk that continuing increases its storage space by means of integrating available heterogeneous file servers connected through Internet in a transparent way for users. The objective of WSS is to offer to its users a scalable and easy to use storage server which hides the concerns about file server heterogeneity and location.

Keywords: File Server, Web File Storage System.

1 Introduction

The use and storage of a large quantity of files in an efficient way has a fundamental role in activities of research, diagnostic and all kinds of processes that involve working in collaborating groups. Many institutions and individuals that work in projects in a collaborative way have the need to share and access to information obtained from their participating partners. When the quantity and the size of the files generated in these groups are very large, the idea to integrate them all in an only one storage server oversizes the capacity of current storage devices. For example, in the hospitals where the doctors obtain images of patients from x-rays, tomographies, etc, these images have an average size between 500 MB and 1 GB, considering situations where hospitals keeping hundreds of patients requiring more than one image, the storage requirement becomes an important constraint.

In spite of the advances in the storage systems, there is the dilemma to decide, what to do when these devices do not include enough space to store all the information generated? and how a user that has not knowledge about file servers can incorporate greater capacity of available storage using heterogeneous resources connected by Internet?. Protocols like FTP [1] and HTTP [2] allow creating services which let their users to exchange files between heterogeneous file servers. However, these services restrict their clients to take space from only one server at once. This means that the user storage capacity is limited by the storage capacity of the server in turn. Internet, as a great scale network, offers

J.M. Corchado et al. (Eds.): DCAI 2008, ASC 50, pp. 549–557, 2009.
springerlink.com

the opportunity to incorporate thousands of storage resources that would be able to contribute in the formation of a massive storage system. Unfortunately, one of the main factors for using those resources by inexpert users is that the connection between those storage systems is not transparent and requires some technical knowledge.

With the evolution of Web technologies, there are opportunities to build a Web storage service in which the users, accessing through a Web application, can obtain storage space that can scale in a transparent way. In this way, registered users in the WSS will see how their storage space automatically increases when the WSS attaches a new file server. A strategy for doing the WSS interoperable is to use what is known like Web services [3]. These services are components of software that are described and accessed in a standard way through the Web using technologies as XML [4], WSDL [5] and SOAP [6].

2 Related Work

Nowadays, there are several options that allow naive users to use heterogeneous file servers. Net2ftp [7] is a FTP client software, which allows its users to manage files through a Web application from any platform. However, its maximum storage capacity depends on only one FTP server. It does not implement distributed storage because the systems where data are stored are independent FTP servers. Therefore, if a user desires more space, he needs to request it in another server. DrFTPD [8] is a Web application which consists of a FTP client called *master* that allows users to store information in different FTP servers called *slaves*. In this way, user can get access to a distributed storage system. This system does not offer a transparent way to increase the available space. GMail Drive [9] is a shell namespace extension that creates a virtual file system around a Google Mail account allowing you to use GMail as a storage medium. The disadvantage is that in those servers the files reliability is not guaranteed by Google and users are not able to execute the client application in any platform. Other options like Java FTP [10] and JFTP [11] offer a Java Application Program Interface (API) that allows software designers to develop applications for a remote data storage. They have the advantage of being multiplatform because their Java-based environment, nevertheless their main disadvantage is that users need to have programming skills to use them. Another important disadvantage is that these APIs do not offer scalable distributed storage.

There are some related research projects like Jade [12], which is a logical system that integrates a collection of heterogeneous file systems. Jade hides to applications the different protocols used in several file systems. Jade allows users to define a private space name, where file systems can be mounted on a directory and a logical space name will be used instead. The main disadvantage of Jade is that its use is not addressed to users without technical knowledge.

Other distributed environments are focused on developers, such as Globus [13], Storage Resource Broker (SRB) [14], Dates box Cutter [15], DPSS [16], and BLUNT [17]. These solutions are not suitable for the integration of current data storage servers (NFS servers, CIFS, HTTP, WebDAV, etc.), forcing users to learn high level APIs, install new servers and modify or completely change their applications. Projects like Perez et al. [18] introduce a platform that allows users to access remote files through a low level middleware; the problem is that this platform is thought for a high technical developers.

The need of having a storage system that can manage a large quantity of files, and increase its space in a transparent way, was the reason to design and implement WSS. It allows naive users to manage files in a scalable storage service through an easy to use Web application.

3 Architecture

The modules that form the general architecture of WSS are: *users*, *WSS-manager* and *storage servers*, Figure 1 depicts the WSS architecture.

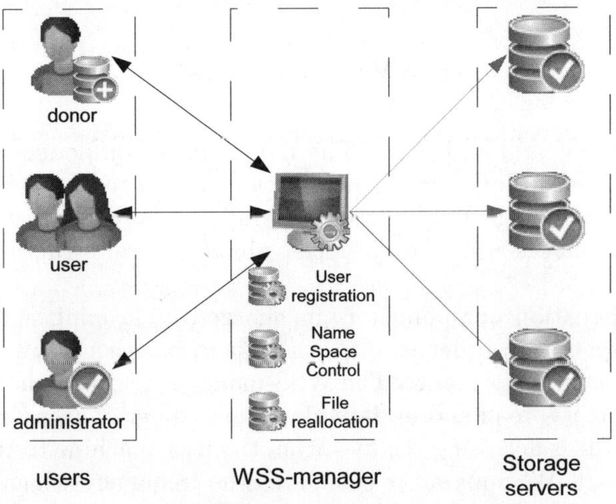

Fig. 1. WSS general architecture

The *users* module handles 3 types of users: administrator, storage donor and end user. The administrator is unique and is who can see the global system in order to supervise every operation. The storage donor is the user who donates storage space from their servers to the WSS, and the end user is the beneficiary of the available storage space.

The *WSS-manager* is responsible for having and verifying the registration of all type of users, the storage servers and the donated space, as well as the registration of space location, modification and access of all files stored in the servers. The WSS-manager is composed of three main components: *user registration*, *name space control* and *file reallocation*. The user registration component

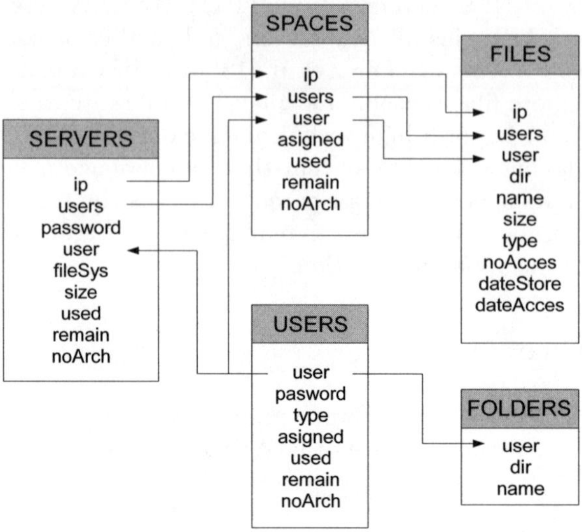

Fig. 2. WSS database design

controls both donor and end users. The name space component keeps a registration of the physical location of the files and is the responsible for mapping physical addresses to logical addresses, offering location transparency to users. This component makes use of a database which general structure is shown in the Figure 2.

The file reallocation component is in charged of keeping a registration of all files in the system in order to reallocate them in other storage servers that possibly offer better performance. The WSS-manager registers the date and time in which users create, read and update files, the number of accesses to these files and the time it takes accessing the files from the user machine to the destination server. In this way, WSS-manager calculates the frequency which the user uses his files. The equation 1 describes this frequency. *NA* represents the total number of accesses, *FA* is the date in which the file was created and *FU* is the date of the last access to the file. This equation helps WSS-manager to decide which files are more popular for users and possibly reallocate them to servers where the round trip time between the user machine and servers are better. Figure 3 shows this process. A file (f2) is moved from the server A to the server B without altering the logical name space seen by users.

$$frec = \frac{NA}{FU - FA} \tag{1}$$

The *storage servers* module is in charged of executing all file commands sent by users through the WSS-manager. Figure 4 depicts this process. The WSS-manager, after allocating storage space to an end user, creates a home folder with the name of the user; the WSS-manager does not allow replicated user names.

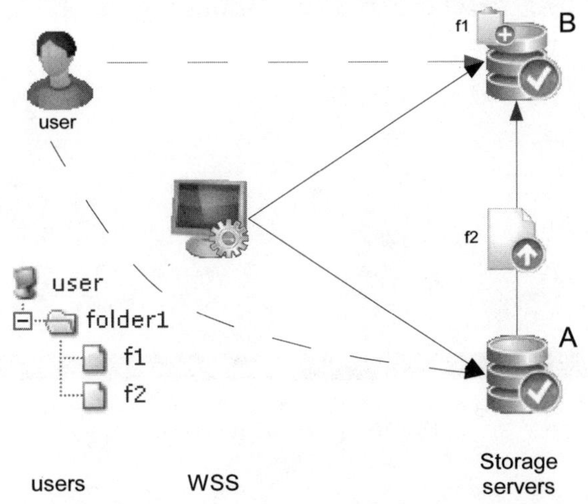

Fig. 3. File reallocation plan

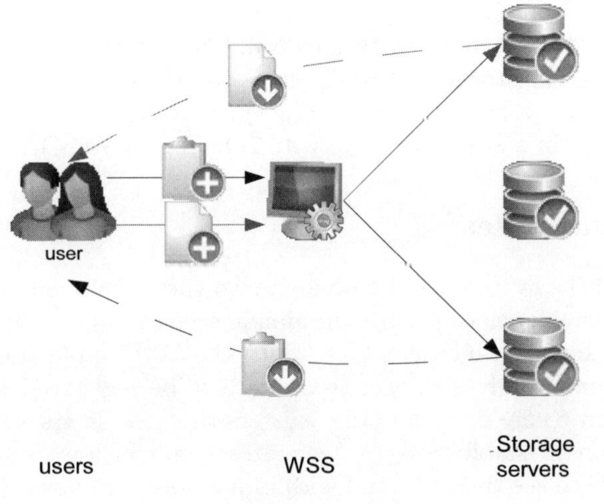

Fig. 4. WSS operation

The user storage space allocated in a server is identified with the ip address of the server where the space is taking from. The end user begins creating files in this space. However, in the future, the user files can be reallocated in different servers. End users do not have any knowledge about file reallocation because they still see the same logical path. When a file reallocation occurs, users can feel a better performance the next time they access to a reallocated file.

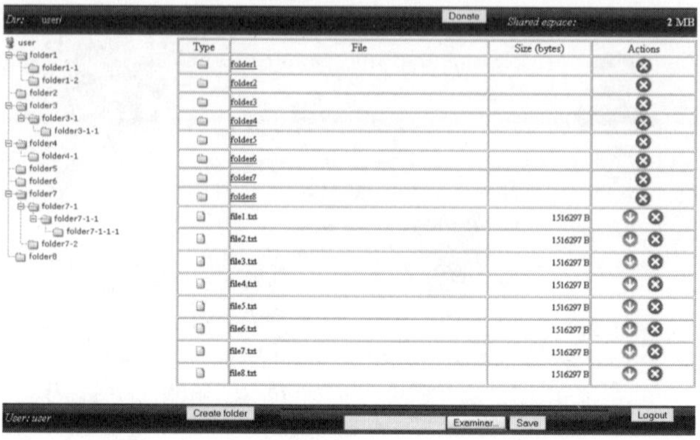

Fig. 5. WSS main screen

The way to communicate the users with the WSS-manager is through a Web application that allows users to manage their files in the donated storage servers. The communication protocol between the WSS-manager and the Web application is HTTP. The interaction between the Web application and the WSS-manager is done through asynchronous requests using AJAX [19], in order to offer a more efficient communication. The communication between the WSS-manager and the storage servers is done, so far, through the protocol FTP because most of our working servers today support this protocol. WSS will be also able to access, in a transparent way, to HTTP and WebDAV servers.

4 Operation Modes

The operation of the system can be divided into three main actions: interacting with the end user, interacting with the donor and the file reallocation action. The first action begins when a user get into the Web application, selects the registration option and gives the necessary data to be registered as an end user. The system, upon receiving this data, registers this user in its user registration database and than he is able to go to the main screen which is shown in Figure 5. Users will be able to see their files in logical units doing transparent the physical location of the different servers behind WSS. In the Web application main screen users can execute the typical operations with their files and folders: *create*, *store*, *access* and *delete*.

The second action occurs when a user is logged as donor. In this action, the user accesses to a registration option where he gives his data and grants all the information to get access in the storage server to be donated. WSS-manager receives this data and verifies the access to the server; if it is successful, the system includes the server and the user as storage donor. A storage donor user can take advantage of the same benefits given to end users. However, a donor user is allowed to donate more space taken from a server registered by him or to register a new storage server.

The file reallocation action is carried out by means of a special module of the WSS-manager. This module is executed in periods of time of low use and its function is to obtain the file access frequencies based on the *Equation 1*. Frequency information helps WSS to reallocate each file in the server which access time could be improved. This process will be successful when the selected server has enough space for a file to be reallocated; otherwise another possible server will be selected that offers the next best access time. This process can be seen in the Figure 3.

5 Preliminary Results

WSS, so far, is an asynchronous access Web tool that allows naive users to storage and manages large quantity of files. The main advantage of this tool is that offers a distributed and scalable storage system through a transparent and easy to use Web application. The current quantity of files stored in WSS oversizes the capacity of available storage in an individual server connected in our network. The file servers do not require of special procedures or additional configuration to be incorporated to the WSS system. Currently, it includes a set of FTP servers; nevertheless the system will be able to enclose HTTP and WebDAV servers available in our network. These servers will not need important changes in their configuration to be included in the WSS system. The file reallocation module using the simple strategy showed in the *Equation 1*, obtains a speed up of a 20% in access time compared with a random storage of the files. The cost of the file reallocation process does not impact to the main files management process, since it occurs in periods of time of low load in the system. We have prepared a test scenario consisting on 5 distributed servers running different operating systems: 2 Linux, 1 Solaris, and 2 Windows server. Each server had, in average, 4 GB of RAM memory and 150 GB of hard disk. There were different types of connections between servers ranging from 1 to 1.5 Mbps. We used 31 files to be transfered between clients and servers. The size of each file was about 117 MB. We evaluated three strategies to allocate files into servers (virtual disk). The first strategy consisted in using the Round-Robin method in which each server was selected, by the WSS-manager, one by one in a circular manner. In the second strategy, the WSS-manager used a Random allocation method to select a server. In the last strategy, the WSS-manager selected the servers taking into account their available space. In this case, a server having the greater available space was selected. In this test we only evaluated the file transfer time in the download operation. The test was divided into four steps. The first step consisted in allocating the files into the servers (uploading) using the 3 strategies. The second step consisted in generating a random file access frequency to the allocated files. The third step was in charged of reallocating the files in servers that offered better transfer time in step 1. The reallocation process considered the access frequency obtained from equation 1. In the last step we, again, measured the download operation in order to compare with the time obtained before the reallocation process. The average transfer time (ATT) was measured in seconds.

The results obtained using the different strategies are showed in Table 1. As we can see, the strategy which shows the better average file transfer time before the reallocation process is the Round-Robin. After the reallocation process all of the strategies got better results. However, the Round-Robin strategy still was the best. We wanted to find a simple strategy to allocate files which can be implemented in our WSS-manager. The results showed us that the Round-Robin strategy is a good option to start.

Table 1. Allocation strategies results

ATT (s) \ Strategy	Round-Robin	Random	GAS
Before reallocation	110.70	166.48	157.62
After reallocation	88.00	131.54	107.53

6 Future Work

In this paper a part of a greater project has been presented. It describes the WSS system. The main objective was to create a scalable Web storage system keeping in mind naive users. Naive user (e.g. physicians) means users who do not have technical knowledge to manipulate or configure file servers. As a work in progress this project will also include:

- An API that allows developers of applications to manipulate a distributed storage services in a transparent way through Web services technologies.
- A security module to encrypt user files in the servers.
- A file compression module used before transferring files between clients and servers. This module will try to improve the file transfer time.

7 Conclusions

Projects like WSS show us that thanks to the Web technologies protocols and languages that we can find nowadays, it is possible to develop a distributed storage system of easy use that allows increasing the storage space in a transparent way. Users can take advantage of their storage servers available most of time in their networks. These servers represent an asset for many institutions. The WSS allows no technical users to manipulate a big quantity of files and obtain more space without requiring special configurations in current working servers. An important advantage of the WSS is it offers an intelligent way to store files in the places where users more use them, offering a better response time.

Acknowledgment

This research was partially funded by project number 51623 from "Fondo Mixto Conacyt-Gobierno del Estado de Tamaulipas".

References

1. Postel, J., Reynolds, J.: File transfer protocol (ftp) rfc: 959. IETF (1985), `http://tools.ietf.org/html/rfc959`
2. Fielding, R., Gettys, J., Mogul, J., Frystyk, H., Masinter, L., Leach, P., Berners-Lee, T.: Hypertext transfer protocol, rfc: 2616. IETF (1999), `http://www.ietf.org/rfc/rfc2616.txt`
3. Salgado, J.F.A., Bueno, P.E.: Web service. In: I Jornadas de Ingeniería Web 2001 (2001), `http://www.informandote.com/jornadasIngWEB/articulos/jiw05.pdf`
4. Eastlake, D., Reagle, J., Solo, D. (extensible markup language) xml-signature syntax and processing, rfc: 3275. IETF (2002), `http://www.ietf.org/rfc/rfc3275.txt`
5. Chinnici, R., Moreau, J.J., Ryman, A., Weerawarana, S.: Web services description language (wsdl) version 2.0. W3C (2007), `http://www.w3.org/TR/wsdl20/wsdl20.pdf`
6. Box, D., Ehnebuske, D., Kakivaya, G., Layman, A., Mendelsohn, N., Nielsen, H.F., Thatte, S., Winer, D.: Web services description language (wsdl) version 2.0. W3C (2000), `http://www.w3.org/TR/2000/NOTE-SOAP-20000508/`
7. net2ftp: net2ftp - a web based ftp client (accessed on november 9, 2007) (2007), `http://www.net2ftp.com/index.php`
8. DrFTPD: Drftpd (accessed on November 9- December, 2007) (2007), `http://drftpd.org`
9. Drive, G. (2008), `http://www.viksoe.dk/code/gmail.htm`
10. Technologies, E.D.: Java ftp (accessed on november 8, 2007) (2007), `http://www.net2ftp.com/index.php`
11. JFTP: Jftp (accessed on november 9, 2007) (2007), `http://j-ftp.sourceforge.net/`
12. Rao, H.C., Peterson, L.L.: Accessing files in an internet: The jade file system. IEEE Transactions on Software Engineering 19, 613–624 (1993)
13. Vazhkudai, S., Tuecke, S., Foster, I.: Replica selection in the globus data grid. In: International Workshop on Data Models and Databases on Clusters and the Grid (DataGrid 2001). IEEE Computer Society Press, Los Alamitos (2001)
14. Baru, C., Moore, R., Rajasekar, A., Wan, M.: The sdsc storage resource broker. In: Proceedings of the International Conference in High Energy and Nuclear Phisys, Teatro Antonianum Padova Italia (February 2002)
15. Beynon, M., Ferreira, R., Kurc, T.M., Sussman, A., Saltz, J.H.: Datacutter: Middleware for filtering very large scientific datasets on archival storage systems. In: IEEE Symposium on Mass Storage Systems, March 2000, pp. 119–134 (2000)
16. Tierney, B.L., Lee, J., Crowley, B., Holding, M., Hylton, J., Drake Jr., F.L.: A network-aware distributed storage cache for data-intensive environments. In: Proceedings of the Eighth IEEE International Symposium on High Performance Distributed Computing, Redondo Beach, CA, pp. 185–193. IEEE Computer Society Press, Los Alamitos (1999)
17. Martínez, M., Roussopoulos, N.: Mocha: A selfextensible database middleware system for distributed data sources. In: Proceedings of the ACM SIGMOD International Conferenceon Management of Data, Dallas Texas (May 2000)
18. Pérez, J., García, F., Carretero, J., Calderón, A., Fernández, J.: A parallel i/o middleware to integrate heterogeneous storage resources on grids. In: Grid Computing. First European Across Grids Conference, Comput. Sci. Dept., Univ. Carlos III de Madrid, Spain, pp. 124–131. Springer, Heidelberg (2004)
19. Ajax: Openajax alliance - standardizing ajax development (accessed on February 11) (2008), `http://www.openajax.org/index.php`

Position Aware Synchronous Mobile Services Using A-GPS and Satellite Maps Provisioned by Means of High Demand Web Servers

Sergio Ríos Aguilar, Luis Joyanes Aguilar, and Enrique Torres Franco

Computer Science Faculty
Pontifical University of Salamanca
Madrid, Spain
sergio.rios@upsam.net, luis.joyanes@upsam.net,
enrique.torres@upsam.net

Abstract. This paper proposes a state-of-the-art framework for synchronous mobile location-aware content personalization, using A-GPS terminal-based/network assisted mobile positioning techniques and UAProf data processing at the origin server. Within this framework, in the dynamic generation of the response to the mobile device, the origin server accesses a Location Server in order to determine the geographic position of the mobile subscriber or gets it by means of SUPL data provided by an A-GPS mobile handset and, with a proper management of the CPI conveyed with the incoming HTTP request (for selecting or customizing the content being delivered to the client), provide him with location-sensitive content and services. The static satellite-view maps are served by means of high demand web servers.

Keywords: Location Based Services, Terminal Based Positioning, Network Assisted Positioning, A-GPS, SUPL, Web Server Storage.

1 Introduction

One of today's key technologies related to advanced mobile services development is the physical positioning of mobile terminals requesting services, because it really opens new frontiers in contents personalization. And moreover, the use of terminal based and network assisted solutions in order to achieve higher precission and lower adquisition times is the current direction of the state of the art research being conducted on this field.

In this paper, at first we will be presenting the fundamentals of Location Services and current location technologies,and we will go on with a description of the proposed platform designed to achieve full WAP/xHTMLMP-compliant contents personalization , together with integration services for spatial analysis, in terminal based and network assisted mobile services.

This implementation currently uses JSR-179 compliant Java code within the terminal, in order to obtain the local position measurement from the embedded positioning device and the A-GPS data from a SUPL-enabled network. [14]

This information is the feeded to the JavaEE core of the platform, which build XML-based spatial queries and forwards them to the available GNSS servers,

J.M. Corchado et al. (Eds.): DCAI 2008, ASC 50, pp. 558–564, 2009.
springerlink.com

processes their XML responses (GML, PoIX) through XSL transformations to the final WML or xHTML/MP contents delivered to the requesting client terminal.[1],[2]

2 Positioning Technologies

Currently available technologies for physical determination of the position of a mobile terminal fall into two broad areas: terminal based technologies and network based ones.

In the former case, the positioning intelligence resides in the mobile terminal or in its SIM/USIM card. Within these technologies we have those based/dependant on GNSS systems (Galileo, GPS, Glonass), those which use the mobile network operators (MNOs) infrastructure (i.e E-OTD, Enhanced Observed Time Difference) and finally, those hybrid solutions which constitute the main focus in this paper: terminal based and network assisted positioning, currently represented by A-GPS.[9],[11]

In the latter case, network bases solutions don't require the integration of the positioning intelligence within the mobile handset. So, this kind of positioning services can be provided to all existing handsets with no distinction, as there are no change sin hardware required. The tradeoff is the relative lack of precission, comparing to the forementioned terminal based solutions. Representative technologies in this area are CGI+TA (Celle Global Identity + Time Advance) and UL-TDoA (Uplink Time Difference of Arrival) [5],[7].

2.1 A-GPS and Enhanced A-GPS

The Assisted GPS technology appeared recently and represents a key turning point. The technology enables a powerful hybridization between a worldwide location means –GPS– and a mass-market communications means – GSM/UMTS.

Moreover, Assisted GPS comes in handy mixing the best of the two worlds, since it compensates for the major faults of GPS and GSM/UMTS: a purely network- based technology does not provide sufficient accuracy (80 meters at best), and pure GPS solutions suffer from long delays before position delivery (typically several minutes). The principle of Assisted GPS consists of coupling satellite positioning and communication networks, sending assistance data to the receiver integrated in the handset to improve its performance [7][10].

Compared with standard GPS, Assisted GPS offers (1) very short latency to get a position, by sending satellite data much more rapidly, than GPS itself; (2) Hence, very low power consumption; and (3) Increased sensitivity, therefore increased vailability of the location service, particularly in dense urban area and indoor environments.

Enhanced A-GPS is an improvement of the classic A-GPS positioning with new satellite technologies: EGNOS (European Geostationary Navigation Overlay Service), and Galileo. The Enhanced A-GPS provides EGNOS-based assistance data to GPS-enabled mobile phones, via GSM/GPRS or UMTS networks. This is obtained by incorporating an EGNOS/GPS reference receiver in the Enhanced A-GPS server, which receives the EGNOS correction messages.[11]

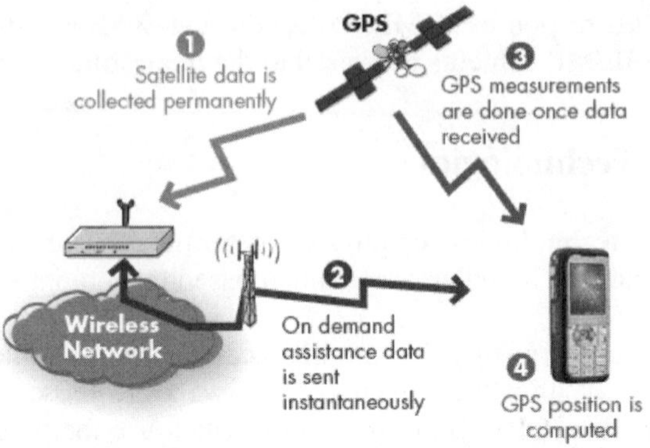

Fig. 1. A-GPS Operation Flow

EGNOS is a system consisting of three geo-stationary satellites and 34 ground stations in Europe, the Americas and Asia. EGNOS improves GPS & GLONASS position accuracy and availability by 'adding' measurements from its three geo-stationary satellites locally visible in Europe to the Galileo, GPS & GLONASS, and by providing ionosphere, orbitography and clock corrections.

2.2 A-GPS Servers User/Control Plane

There are two possible configurations for the A-GPS server, either user plane or control plane.

The user plane configuration is standardized by the Open Mobile Alliance (OMA) forum under the name of Secured User Plane Location (SUPL). SUPL is defining a set of functional entities and protocols for the GMLC, the SMLC/A-GPS server and the terminal. As an example, SUPL defines a direct connection between the GMLC and the SMLC/A-GPS server. The SMLC is embedded in the A-GPS server.[12],[13].

In the control plane configuration, standardized by 3GPP, the GMLC triggers the SMLC through the core network (when there is an Ls interface between the SMLC

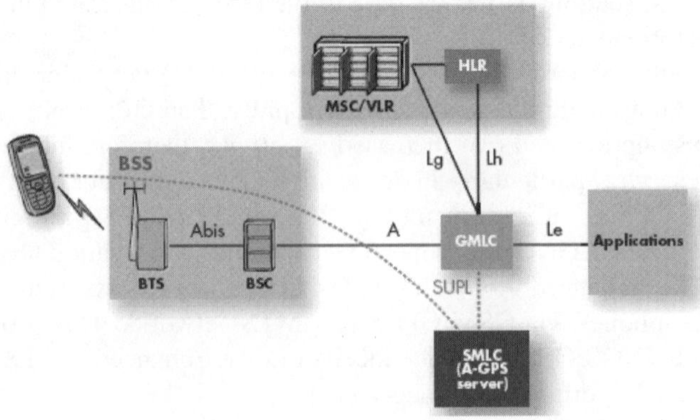

Fig. 2. OMA SUPL Operation in A-GPS

and the MSC), or through the radio network (with an Lb interface between the SMLC and the BSC).[13]

3 Accesing Terminal-Based Positioning Info

The Location API for JavaME specification defines an optional package, `javax.microedition.location`, that enables developers to write wireless location-based applications and services for resource-limited devices like mobile phones, and can be implemented with any common location method. The compact and generic J2ME location APIs provide mobile applications with information about the device's present physical location and orientation (compass direction), and support the creation and use of databases of known landmarks, stored in the device.

JSR 179 requires the Connected Device Configuration (CDC) or version 1.1 of the Connected Limited Device Configuration (CLDC). CLDC 1.0 isn't adequate because it doesn't support floating-point numbers, which the API uses to represent coordinates and other measurements. [14]

The Location API doesn't depend on any particular profile -- it can be used with MIDP or the Personal Profile. The hardware platform determines which location methods are supported. If it doesn't support at least one location provider, LBS won't be possible. Applications can request providers with particular characteristics, such as a minimum degree of accuracy. Some location methods may be free; others may entail service fees. The application should warn the user before any charges are incurred (i.e. when using A-GPS in user plane by means of SUPL), and cope with user's ano nymity concerns [16][17].

It is up to the application to determine the criteria for selecting the location method. Criteria fields include: accuracy, response time, need for altitude, and speed.

4 Location Aware Mobile Services (LAMS) Platform

The LAMS Platform has been developed as a proof-of-concept of the proposed service architecture, acting as an universal LCS (LoCation Services) client.(See Fig.3)

Once the platform gets the positioning info obtained using JSR-179, transcodifies this information to XML, using an specific XML application (LAMSX). With this information, the platform sends a spatial query to a GIS server in order to obtain reverse geocoding data which carries the desired Points of Interest (POIs) also in native XML format.[4],[5],[6]

At this point, the platforms enters a dynamic content generation phase, personalizing the overall response to the make, model and capabilities of the requesting handset, using UAProf as a reference for the involved transformations.[3]

If the provided service needs to include static contents, such as satellite imagery, the platform can redirect the necessary requests to a high demand web server platform, just to cope with the performance issues via agent-based distribution.

Finally, the outcome of this content generation phase is the end WML or xHTML/MP page targeted to the handset.

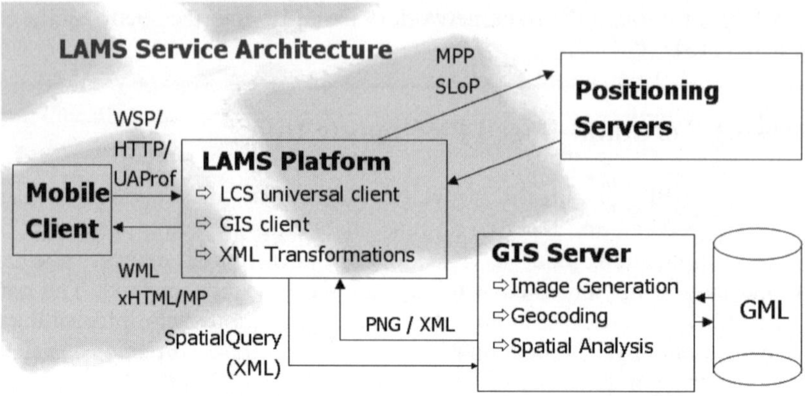

Fig. 3. LAMS Platform Service Architecture

4.1 Getting Device Capabilities and User Preferences: UAProf

In order to achieve a truly complete WAP/ xHTML MP contents personalization, we need to know the particular characteristics of the mobile terminal and the active user preferences.

To help in this purpose, the World-Wide Web Consortium (W3C) has defined mechanisms for describing and transmitting information about the capabilities of WAP mobile terminals and the display preferences of WAP users.[2] The User Agent Profile (UAProf) specification extends WAP 1.1 to enable the end-to-end flow of a User Agent Profile (UAProf), also referred to as Capability and Preference Information (CPI), between the WAP client, the intermediate network points, and the origin server. [3]

This specification uses the CC/PP model to define a robust, extensible framework for describing and transmitting CPI about the client, user, and network that will be processing the content contained in a WSP response. The Composite Capabilities/Preferences Profile (CC/PP) especification, in turn, defines a high-level structured framework for describing this information using the Resource Description Framework (RDF).

A new protocol, the CC/PP Exchange Protocol over HTTP enables CC/PP profiles to be transported over the HTTP 1.1 protocol with the HTTP Extension Framework.

The CPI may include hardware characteristics (screen size, image capabilities, manufacturer and model, etc.), software characteristics (operating system vendor and version, audio and video encoders, etc.), application/user preferences (browser manufacturer and version, markup languages and versions supported, scripting languages supported, etc.), WAP characteristics (WML script libraries, WAP version, WML deck size, etc.), JavaME capabilites and constraints, and network characteristics (bearer characteristics such as latency and reliability, etc.).

4.2 Getting Local Positioning Info

Only if the terminal based, network assisted positioning mode is active (i.e. using A-GPS), a JSR-179 enabled local application feeds these data to the LAMS platform, which in turn accesses the GIS application servers, and carries on with the foremen-tioned transcodification process.

Clearly stated, in this case there's no need to access the MNO's positioning servers, as the position data is now well known to both handset and platform. This has a direct impact in overall response times, shortened due to the faster operation of A-GPS and its improved TTFF.

Furthermore, the support of terminal based, network assisted positioning (by means of A-GPS and JSR-279), and network based positioning (by means of MNO´s positioning servers) provide an efficient way of integrating a positioning technology-agnostic module into the architecture established by Wu [15].

5 Summary

In this paper has been presented the practical use of Location Services, applied to the generation of position-aware contents to a WAP/ xHTML MP mobile terminal. This knowledge can be used along with the also described UAProf framework to achieve full personalization of the contents delivered to the end client. In this case, we are able to dynamically generate location-dependent, device-dependent and user preferences-aware contents.

Also, this paper demonstrates the feasibility of a time-constraint service provision when using terminal based technologies with the most up to date enablers, such as JSR 179, OMA SUPL and A-GPS at application level.

References

1. Wireless Application Protocol Architecture Specification, `http://www.openmobilealliance.org`
2. WAP WAE: Wireless Application Environment, `http://www.openmobilealliance.org`
3. Reynolds, F., Hjelm, J., et al.: Composite Capability/Preference Profiles (CC/PP): A user side for content negotiation. World Wide Web Consortiumm, `http://www.w3.org/TR/NOTE-CCPP`
4. Ríos Aguilar, S.: Generación dinámica de contenidos WAP para terminales móviles, Libro de Ponencias del Congreso Mundo Internet 2000 (February 2000)
5. Ríos Aguilar, S.: Position-aware WAP Contents Personalization. In: Wrox Wireless Developer Conference, Amsterdam (July 2000)
6. Ríos Aguilar, S.: Interfacing to a Mobile Positioning Center. Wireless Developer Network (September 2004), `http://www.wirelessdevnet.com`
7. Jochen, S., Agnes, V.: Location Based Services. Elsevier-Morgan Kaufman, Amsterdam (2004)
8. Jonathan, S.: Wireless Location Uses in the User Plane and Control Plane (May 2007), `http://www.lbszone.com/content/view/148/45`
9. Stojmenovic, I.: Handbook of Wireless Networks and Mobile Computing. John Wiley & Sons, New York (2004)
10. Katz, J.E., Aakhus, M.: Perpetual Contact. Mobile Communications, Private Talk, Public Performance. Cambridge University Press, Cambridge (2005)

11. AGILE project (May 2007), http://www.galileo-in-lbs.com/index.php?id=398
12. OMA-TS-MLP, Mobile Location Protocol, Open Mobile Alliance, http://www.openmobilealliance.org/
13. OMA-RD-SUPL-V2_0, Secure User Plane Requirements v2.0, Open Mobile Alliance, http://www.openmobilealliance.org/
14. JSR-179 Location API for JavaME. Sun Microsystems Java Community Process, http://jcp.org/aboutJava/communityprocess/final/jsr179/index.html
15. Shiow-Yang, W., Kun-Ta, W.: Effective Location Based Services with Dynamic Data Management in Mobile Environments. Wireless Networks 12, 369–381 (2006)
16. Axel, K., Georg, T.: Efficient Proximity and Separation Detection among Mobile Targets for Supporting Location-based Community Services. Mobile Computing and Communications Review 10(3) (2008)
17. Toby, X., Ying, C.: Location Anonymity in Continuous Location-based Services. In: Proceedings of the 15th International Symposium on Advances in Geographic Information Systems. ACM GIS 2007 (2007)

XML Based Integration of Web, Mobile and Desktop Components in a Service Oriented Architecture

Antonio Lillo Sanz, María N. Moreno García, and Vivian F. López Batista

Departamento Informática y Automática
Universidad de Salamanca
Plaza de la Merced s/n, 37008, Salamanca, Spain
mmg@usal.es

Abstract. Component autonomy and easy composition are two of the main purposes of Service oriented Architectures. Recently, some multilayer frameworks supporting service abstraction and tier-integration facilities have been developed. They are especially useful for developing ubiquitous software systems where the presentation layers for different visualization devices are decoupled from the business logic layer, but services provided by this one can be easily accessed. In this work, we present the practical experience in the deployment of new frameworks such as JavaServer Faces, Spring and Hibernate in a multilayer architecture for an application endowed with three types of user interfaces: Web, for accessing with a classic browser, mobile Web, for accessing through different mobile devices, and a desktop interface for the administration functionality, supplied as remote services from the server.

Keywords: SOA architecture, frameworks, web systems, mobile systems.

1 Introduction

Consolidation of object oriented technologies has impelled the development of component based software and its systematic reutilization through multiple applications. However, this systematization is not enough to build completely autonomous components that can be easily integrated in the applications. Concepts, such as modularity, extensibility and reusability have taken a real meaning when service oriented architectures (SOA) have appeared. SOA decouples reusable functions, which receive the name of services, by providing high level abstractions for building modular applications for large-scale, open environments. Applications are implemented and configured in a way that improves productivity and quality [3]. SOA is a flexible architectural framework that reduces costs, increases revenues, and enables rapid application delivery and integration across organizations [1].

Service oriented architectures have been popularized thanks to the great acceptance of Web services, but the underlying concepts of SOA can be applied for a solution without using Web services. A service oriented application has three main pieces: a registry of services, providers that publish their services on the registry and consumers that find services and invoke them [3]. These parts can be implemented in a multilayer architecture where lower layers provide services to higher layers [7]. Applications are then structured as autonomous software components that provide and/or consume services and can be developed in a parallel and independent way. Services

J.M. Corchado et al. (Eds.): DCAI 2008, ASC 50, pp. 565–573, 2009.
springerlink.com

are programming abstractions in which software developers can create different modules through interfaces with clear semantics. Semantic representations of services let easy customization of the modules and simplify software development. Information description standards such as XML are useful tools for building the semantic complements used for publishing and finding service interfaces.

In order to address the creation of SOA based applications, some multilayer frameworks supporting service abstraction and easy tier integration have been recently developed. They simplify the implementation and integration of software components, but also hide the complexity of the platforms they are developed for. One of the most popular platforms is the Java 2 Enterprise edition (J2EE), for which several known open source frameworks, such as *Struts*, *Spring*, *iBatis* or *Hibernate*, have been developed. In this work, we have combined two of them, *Spring* and *Hibernate*, with the *JavaServer Faces* (JSF) framework in a multilayer architecture that supports service abstraction. JSF is a Sun standard for the presentation layer of Web applications; however, it can be used in combination with Swing or SWT desktop applications without need of changing any line of code of the other layers, since XML configuration files are used for tier integration.

The rest of the paper is organized as follows. Section 2 contains a revision of main J2EE frameworks. In Section 3 the system architecture is described. Key aspects of the implementation of the system are showed in section 4. A comparative analysis of some popular frameworks is given in section 5 and finally, in section 6, we expose the conclusions.

2 Frameworks

A framework is a set of cooperating classes that make up a reusable design, which incorporates the control logic of an application piece. Contrary to the design patterns, frameworks can be implemented in a programming language, therefore, they also allow code reutilization.

The following advantages are derived of the use of frameworks [4]:

- Code for working directly with low level infrastructure is implemented by the framework.
- Structure and consistency that provide to the applications under development facilitates the work to developers joining the projects.
- Successful open source frameworks are better tested than in-house code.
- Well documented frameworks can promote best practice.

In the Web application area, frameworks provide additional benefits relative to quality attributes that Web applications must show [6], such as usability, scalability, security, performance, etc. Web application frameworks also give support to compatibility issues by providing automatic mechanism for transforming contents in a format suitable for several client platforms [2]. On the other hand, there is the trend for user interfaces of moving away from the traditional Web paradigm using web pages to simulate traditional desktop applications [2]. JavaServer Faces framework is the most representative of that tendency, but there are other frameworks such as the proposed by Puder [8], Echo and Jakarta Tapestry. Internationalization is another Web application issue supported by several frameworks and platforms. It consists of the language translation but also a cultural dimension must be included [5].

In the last years some open source frameworks have been developed, which can help developers to achieve the objectives mentioned previously. We describe next the J2EE frameworks that are placed among the most popular ones.

J2EE platform supplies java database connectivity (JDBC) mechanisms for doing java code portable between databases and other services designed to accommodate middleware vendors. It also offers standard services such as transaction and thread management between others. However, developers have to write very complex code to use them in their applications. In order to simplify these tasks, J2EE itself defines the Enterprise JavaBeens (EJB) framework, a component model created to let developers work with standard J2EE services. In addition, other open source Web application frameworks have been developed for the J2EE platform with the purpose of giving generic infrastructure solutions to generic problems.

2.1 Struts

The framework Struts is the proposal of Apache software Foundation to solve the deficiencies of JavaServer Pages (JSP) in the development of Web applications. JSP templates contained business logic and HTML code. Struts is a framework that implements the model-view-controller (MVC) architecture pattern which allows the separation between data, presentation and business logic. The Struts framework is formed by a java class hierarchy implementing functionality that can be extended. The view is constituted by the JSP pages and classes extended from *ActionForm*, which store the data introduced by the users in the forms. The model is formed by the classes containing the data obtained from the database. The controller is a specialized servlet provided by Struts that receives the user request, analyzes it, follows the configuration programmed in a XML file, and delegates the necessary tasks in the *Action*. The *Action* can instantiate or use the business objects for doing the task. According to the result, the controller will derive the generation of the interface to one or more JSPs, which can consult the model objects for doing the task.

2.2 JavaServer Faces (JSF)

A new alternative to Struts for the Web presentation layer is the Sun standard JSF. It is based on components and is implemented in the own J2EE 5, although there are other commercial and open source implementations. They provide basic components and libraries of additional components that extend basic functionality.

JSF is formed by the three components of the MVC pattern: controller (Faces Servlet), view (JavaServer Pages) and model (Backing Beans).

The JSF architecture defines a clean separation between application logic and presentation and at the same time it facilitates the easy integration between these two layers. This enables parallel and independent development of applications components that can be connected by the simple JSF programming model. JSF includes a set of APIs for representing user interface (UI) components and managing their state, handling events and input validation, defining page navigation, and supporting internationalization and accessibility. In addition, a JSP interface can be expressed inside a

JSP page by means of a JSF custom tag library. The UI component classes included with JSF technology encapsulate the component functionality, not the client-specific presentation, thus enable JavaServer Faces UI components to be deployed in different client devices.

2.3 Spring

The Spring framework combines *inversion of control* (IoC) and aspect-oriented-programming (AOP) with a service abstraction to provide a programming model in which application code are largely decoupled from the J2EE environment. IoC is a model in which the framework instantiates application objects and configure them for use. *Dependency injection* is a Java IoC that does not depend on framework APIs and thus can be applied to objects not known by the framework. Configuration is via JavaBean properties (*setter injection*) or constructor arguments (*constructor injection*). *Dependency injection* fails to apply declarative transaction management to select methods for security checking, custom caching, auditing, and so on. Spring provides a proxy-based AOP solution that complements *dependency injection* [4].

3 System Architecture

We have combined Spring, Hibernate and JavaServer Faces in a multilayer service oriented architecture in order to build a system formed by decoupled layers that can be independently developed and easily extended and modified. Every layer is supported by a framework: JSF for the presentation layer, Spring for the business logic layer and Hibernate for the integration layer. Figure 1 shows the architecture components.

In the presentation layer JSF implements the MVC pattern. The controller, Faces Servlet, detaches the presentation from control code since the information about the actions in the pages is provided in a XML file. Backing beans (model) demand services to the lower layer and provide obtained data to the JSP pages (view). The inte

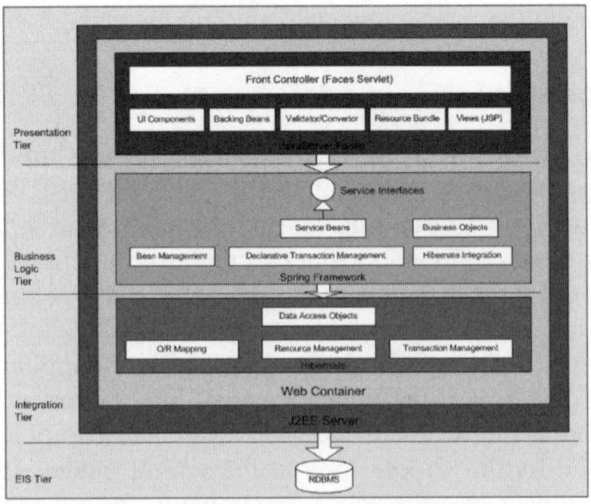

Fig. 1. System architecture

gration between model and view is carried out again through a XML file. Spring use the Service Locator pattern to supply service interfaces from the business logic layer to the presentation layer when backing beans request them. Once more a XML file links the interfaces with their implementation. Spring can publish the services by a configuration XML file and remote clients can demand these services. This SOA architecture is managed by Spring allowing transparency in publication and transaction management.

In the integration layer, Hibernate provides an ORM through XML files and, jointly with Spring, offers an API to simplify the treatment of persistent objects.

3.1 Service Locator Pattern

The presentation layer in the system architecture is composed by many backing beans that take charge of making petitions to the business logic layer services depending on the request received from the user interface. Since there are many services that can be requested, the number of dependences may be very large. This problem can be solved by means of the Service Locator Pattern whose purpose is to abstract the logic to localize the adequate services. This pattern is carried out by an interface and a bean managed by JSF, which finds the services from the Spring application context. In this way the backing beans in the presentation layer have one only dependency with the Service Locator Pattern interface. The bean of the pattern has the information about available services and Spring solve the dependencies by using the dependency injection. As a result, application code is independent of the concrete implementation of the service interface.

3.2 Server Application

We have developed an application endowed with three types of user interfaces: Web, for accessing with a classic browser, mobile Web, for accessing through different mobile devices, and a desktop interface for the administration functionality, supplied as remote services from the server. The system contains six subsystems: Clients, Vehicles, Damages, Automation, Reviews and Maintenance. Figure 2 shows the distribution of the classes into the three layers for the subsystem “Clients”. In the “Clients” subsystem the classes “ClientBean” and “UserBean” are the Backing beans of the Web interface in charge of invoking services through the Service Locator pattern, while JavaServerFaces resolves the dependency and assigns an implementation to the interface through a XML configuration file. For the mobile Web interfaces the beans are “ClientMobileBean” and “UsuarioMobileBean”.

The Service Locator pattern is implemented by the class “ServiceLocatorBean”. It has the interfaces of the client services and Spring resolves the dependencies between every interface and its implementation by means of a XML configuration file. The classes “ClientService”, “ClientServiceImpl”, “UserService” and “UserServiceImpl” are the interfaces and implementations of the services, which are published by Spring for their remote access, being this the SOA part of the architecture. The services use the business objects “Client”, “User” and “UserType” by delegating the data access in

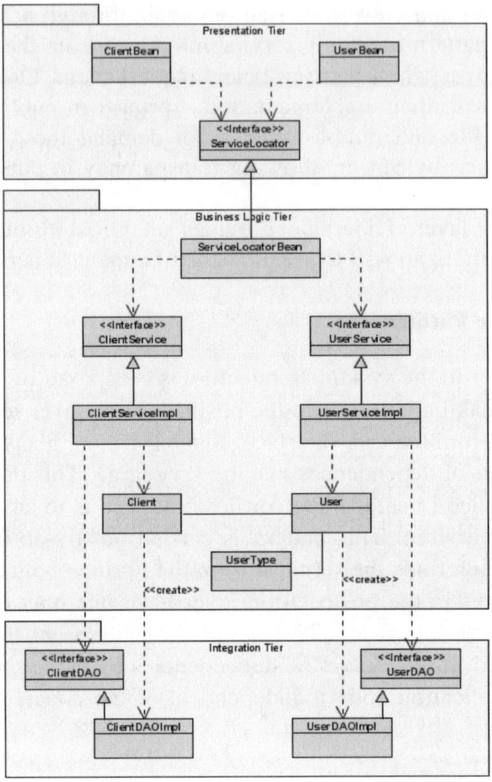

Fig. 2. "Client subsystem"

the DAO classes ("ClientDAOImpl" and "UserDAOImpl"). They use the DAO interfaces and Spring again resolves the dependencies. The framework Hibernate carries out the object-relational mapping by using a XML configuration file.

3.3 Client Application

The different interfaces provide different accesses. In the case of Web and mobile Web accesses the client is the Internet browser. In the case of the remote access to the administration functionality we have a desktop client application with Swing that invokes the remote services on the server, which contains all the application logic. The Spring framework provides the transparent access to the services through the Bean Factory. It provides the requested classes (beans) but it solves the existing dependencies between service implementations and service interfaces since the client application only know the service interfaces.

The desktop application uses the remote services provided by the server but also implements another services itself. Weighty process, such as document and graphics generation, are local processes that liberate the server of processing work and the dispatch of the elaborated information.

4 System Implementation

The development environment used for building the system was Eclipse 3.2. extended with different plugins to cover all the project aspects under the same environment (J2EE, JSF, Hibernate…). The Apache project MyFaces was selected as implementation of JSF.

The system is an application for a vehicle repair Spanish shop that includes functions for managing clients, vehicle damages, invoices, budgets and so on. The application maintains the state of repairs in real time, therefore in this way, the clients are informed at any time about the situation of their vehicle. Clients can access to the system by using different visualization devices and different languages (figure 3). In addition, the employees can use the administration functionalities from a remote client endowed with a desktop application with a Swing interface, only in Spanish language. Therefore this is a heterogeneous application that makes up the ideal case study to test the chosen frameworks and the proposed architecture.

The architecture, presented before, allows reusing without changes the services from the business logic layer for the three types of user interfaces (Web, mobile Web and desktop), since XML configuration files do the integration functions.

Web mobile interfaces were built following the W3C standard best practices [9] in order to optimize the visualization and allow users to interact with the system in a friendly way.

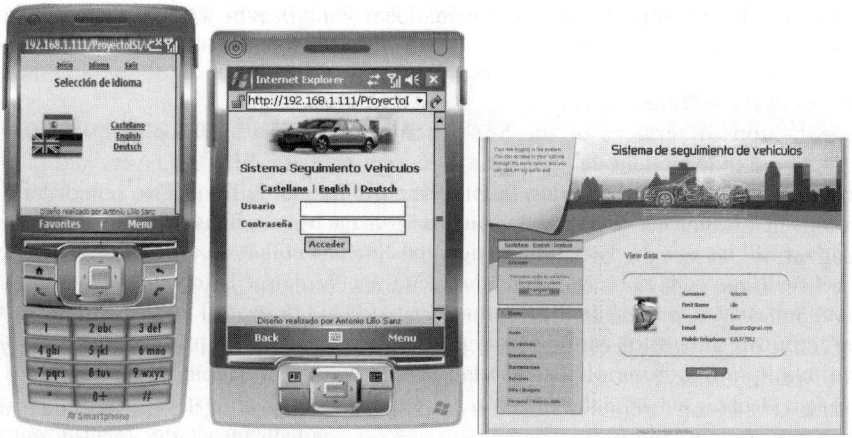

Fig. 3. System interfaces

As we could see in the previous section, several XML configuration files are necessary to connect different layers and components in the layers. We consider that the file *remoting-servlet.xml*, used for service publication, is the most representative in the SOA architecture. A fragment of this XML file is showed in the figure 4. In the file, a name for accessing the service is established, the Spring class that manages the remote services is also specified and, finally, service interface and its implementation are linked. Service implementations have dependencies with the DAO classes, which are specified in the *applicationContext.xml* file where an alias (id) is associated with the service implementation. Published services can be found through the Service Locator Pattern and, thus, used by remote clients. This service abstraction allowed us to develop a desktop application containing only user interfaces components and to leave the whole business logic residing in the server.

```
<?xml version="1.0" encoding="UTF-8"?>
<!DOCTYPE beans PUBLIC "-//SPRING//DTD BEAN//EN"
"http://www.springframework.org/dtd/spring-beans.dtd">

<beans>
      <bean name="/UsuarioServicio"
class="org.springframework.remoting.httpinvoker.HttpInvokerServiceExpo
rter">
        <property name="service" ref="usuarioServicio"/>
        <property name="serviceInterface"
value="com.proyectoisi.logicanegociopersistencia.servicios.UsuarioServ
icio"/>
      </bean>
```

Fig. 4. Fragment of the remoting-servlet.xml file

5 Conclusions

In this work, emergent technologies, such as the JSF, Spring and Hibernate frameworks have been integrated in a multilayer architecture and applied in the development of an application with different user interfaces. The objective has been to acquire practical experience that let us to obtain conclusions about the proclaimed advantages of these technologies.

Abstraction of services of the SOA architecture allowed collaborative work, by means of the independent development of components in different layers and the later integration through configuration XML files. At the same time, this aspect let us to develop an ubiquitous software system where the presentation layers for different visualization devices are decoupled from the business logic layer, whose services are reused for three types of user interfaces without changing any line of code. Besides the Web and Web mobile interface, a desktop application for a remote client was developed. It only contains presentation elements because the service publication facility of Spring allows all the control logic resides in the server. We can conclude that the frameworks and architecture proved lead to a reduction of the development time, to an increment of the software quality and maintainability of the system and to a better code reutilization.

References

1. Arsanjani, A., Zhang, L.J., Ellis, M., Allam, A., Channabasavaiah, K.: S3: A Service Oriented Reference Architecture. IT Professional-IEEE CS 9(3), 10–17 (2007)
2. Gitzel, R., Korthaus, A., Schader, M.: Using Esablished Web Engineering Knowledge in Model-Driven Approaches. Science of Computer Programming 66, 105–124 (2007)

3. Huhns, M.N., Singh, M.P.: Service-Oriented Computing (2005) Key Concepts and Principles. IEEE Internet Computing x, 75–81 (2005)
4. Johnson, R.: J2EE Development Frameworks. IEEE Computer 38, 107–110 (2005)
5. Marcus, A., Gould, E.W.: Cultural Dimensions and Global Web User Interfaces Design: What? So what? Now what? In: Proc. Of 6th Conference on Human Factors & the Web (2000)
6. Offutt, J.: Quality Attributes of Web Applications. IEEE Software, 25–32 (March-April 2002)
7. Petersson, K., Persson, T.: Software Architecture as a combination of Patterns, CrossTalk, October 25-28 (2003)
8. Puder, A.: Extending Desktop Application to the Web. In: Proc. Of 3rd International Symposium on Information and Communication Technologíes, Las vegas, June 8-13 (2004)
9. Rabin, J., McCathieNevile, C.: Mobile Best Practices 1.0, W3C (2006), (Last visited: March 24, 2008), http://www.w3.org/TR/mobile-bp/

Scalable Streaming of JPEG 2000 Live Video Using RTP over UDP

A. Luis and Miguel A. Patricio

Applied Artificial Intelligence Group (GIAA)
Universidad Carlos III de Madrid
Avda. Universidad Carlos III 22, 28270 – Colmenarejo, Spain

Abstract. This paper describes a scalable architecture for streaming JPEG2000 live video streams using Real Transport Protocol (RTP) over User Datagram Protocol (UDP). UDP makes the transmission faster and more efficient, due that it avoids the overhead of checking whether one packet has arrived or not. Moreover UDP is compatible with packet broadcast (sending to all on local network) and multicasting (send to all subscribers). However, with UDP the datagram´s may arrive out of order, appear duplicated, or go missing without notice. With the aim of solve this problem we propose the use of RTP with a new definition of one RTP payload header (RTP Payload Format for JPEG 2000 Video Streams) that it is being defined by the IETF. This payload and its use with RTP will become a new RFC standard. Our solution, thanks in part to JPEG2000 features, is scalable. The JPEG2000 live video server developed supports multiple clients and each one can display live video at a variety of resolutions and quality levels. Thus client devices with different capabilities, variety of screen resolutions, can all achieve a scalable viewing of the same content.

1 Introduction

Before local area networks proliferation, and more specifically the Internet, video surveillance domain was limited to local environments that centralize the display of all the cameras. Nowadays it is possible digitize the video signal and send it across networks and display it almost anywhere in the world. This technology is called live streaming and has emerged from networks communication evolution.

Although currently there are numerous applications and implementations of this technology, in this document we will study a specific type of live video streaming using the unusual codec JPEG 2000 [1] for image compression/decompression in order to support certain hardware used in our laboratories that provides this type of compression in real time. Moreover we need a live streaming with the minimal lag possible because this type of applications is used for video surveillance. We searched other similar investigations about live streaming and we based part of our work in some information found at [11] and [12].

Then, our main goal is provide a live video stream using JPEG 2000 codec through local networks or Internet. One server will be the responsible to provide access control and quality of service of the stream to clients. This architecture can be used as a remote real time high definition video surveillance system thanks to the use of JPEG 2000. The use of networks to send digital video signal let this system transmit video to practically anywhere in the world.

J.M. Corchado et al. (Eds.): DCAI 2008, ASC 50, pp. 574–581, 2009.
springerlink.com

It is very important in this kind of systems that use live video get the minimal time possible used to acquire, compress, transmit, and decode. If not, the live video in the client side will appear with some lag not suitable for security systems such as video surveillance. With our design we have obtained a very low lag.

Some codecs/applications tested before our server development shown that this lag was excessively high, in worst cases about 15 seconds. Codecs tested were MPEG4 [2], ASF [3], and OGG [4]. With this results and the use of specific image boards (Matrox Morphis boards [5]) that let us compress images with JPEG 2000 codec by hardware in real time, we have decided finally use this codec for video compression.

In the network side, these types of architectures are based on local area networks and Internet. With regard to transport protocols employed it depends basically on the type of streaming to be done, for applications such as Youtube, which streams a file, TCP is used because there is no concept of live video, there is better that a client can view the complete video without breaks. In the architecture we have developed it is necessary a real time transmission in order to obtain the minimal lag possible from live video, so the protocol used is UDP.

Since the use of UDP "as is" does not provide any mechanism for determining the correct sequence of packets received or whether it has lost some datagram, we need another transport protocol over UDP that can control these concepts in order that the client could receive the video frames consecutively for their correct display . The best protocol found for such transmissions is the Real Transport Protocol (RTP) [6], defined for the transmission of audio and video. The first version of this protocol was published in 1996 by the IETF; currently it's defined in RFC 3550 [6].

RTP needs to use an RTP payload in their messages in order to identify the type of content that is being transmitted through the channel, as JPEG 2000 is a relatively new codec and it isn't defined in first as a video codec, this type of payload isn't contemplated in RFC 3550. At the writing of this document it is being defined by the IETF a special payload for transmit JPEG 2000 frames over RTP [7] that uses some features that provide JPEG 2000. Our development then will implement this payload for send JPEG 2000 frames over RTP.

Finally, the client needs to decode JPEG 2000 codestreams, in this case by software. We have tested various JPEG 2000 decoders and analyzed its performance and we finally chose J2K-Codec [8]. It supports selective tile decoding, different decoding resolutions levels, and video decoding quality. Other decoders tested were Open JPEG [9] and JASPER [10], both open source, but its performance levels and functionalities are under J2K-Codec.

In the following, we introduce in more detail RTP streaming architecture performed for JPEG 2000 codestreams. Section 3 covers JPEG 2000 codestream in detail and its packetization with RTP. We conclude by summarizing in Section 4 the performance obtained while Section 5 includes the conclusions.

2 RTP Streaming Architecture for JPEG 2000

In Figure 1 we can see the two main systems of our architecture, the client and server. We can observe how in turn are divided into two subsystems. The control layer on the client is responsible for obtaining information from available streams in a server and asks it to start or stop sending a particular video signal. The control layer on server therefore will respond to such requests and communicate with the streaming subsystem to start or stop sending images to a particular client. Streaming subsystems will manage the compression/decompression of the JPEG 2000 frames as well as its transmission/reception through the network protocols described above.

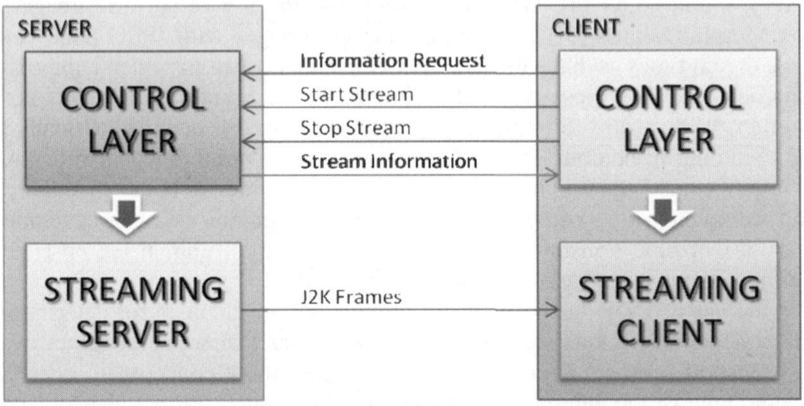

Fig. 1. Streaming Architecture for JPEG 2000

The main part of our architecture are the streaming subsystems, there are streaming server and streaming client. Now we will explain with more detail its operation mode.

2.1 JPEG 2000 Streaming Server

Streaming server basically works as follows. Single frames are acquired from camera and later transmitted to the compression module; once the frame has been compressed it is ready to send over network (see in Figure 2 (a)).

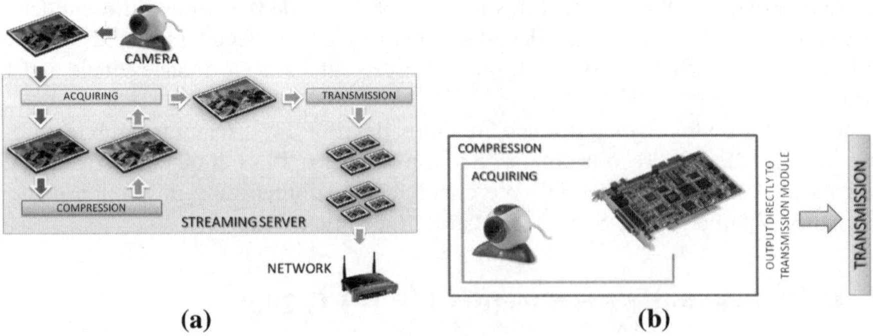

Fig. 2. (a) Streaming server architecture. **(b)** Acquiring and compression.

Acquiring and compression modules are performed using hardware devices as shown in Figure 2 (b), in our case Matrox Morphis boards. The output obtained from Matrox Morphis is sent directly to transmission module.

Transmission module has to analyze the JPEG 2000 codestream in real time once it has been received from Matrox Morphis board. This preprocess is required because transmission module has to perform RTP JPEG 2000 payloads in order to send it over RTP. These payloads are defined specially for make and intelligent packetization of the stream.

2.2 JPEG 2000 Streaming Client

Its architecture is similar to the Streaming server shown in Figure 2 (a), but instead of transmission, compression and acquiring modules it has receiving, decompression and displaying modules. JPEG 2000 streaming client running is shown in Figure 3.

Fig. 3. JPEG 2000 Streaming Client

Receiving module has to receive RTP packets over network and process it. Generally it consists in receive enough RTP packets to complete a tile, once a tile is completely received and identified it is sent to decompression module that will perform its decompression (in this case decompression is realized by software). Now when the tile is decompressed it is stored in a tile buffer until the rest of tiles of the current frame were ready too. When decompression module has the complete frame decompressed, it is ready for its displaying which is performed by displaying module.

3 JPEG 2000 Codestream

In JPEG2000 an image may be spatially divided into tiles, where each tile is independently coded.

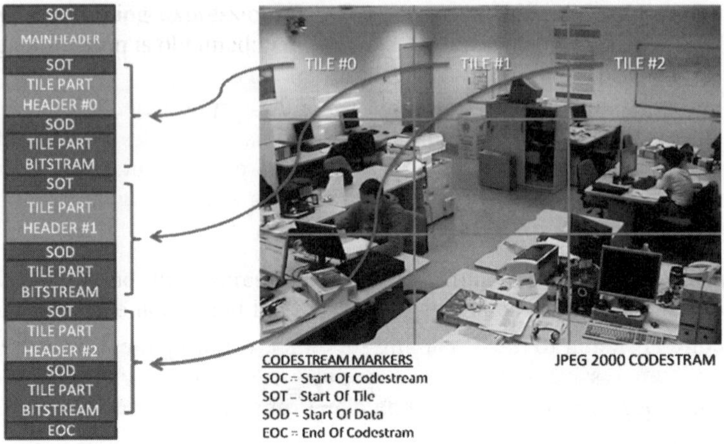

Fig. 4. JPEG 2000 Codestream

There are two types of headers in the codestream. The main header is at the beginning of the codestream. The tile-part header is found at the beginning of each tile-part, where tile-part is a portion of the codestream that makes up some or all of a tile. The main header provides information about uncompressed image such as width, height, width of a tile, height of a tile, number of component, bit-depth of each component, etc. The main header is followed by one or more tile-parts (each tile-part includes a tile-part header and the tile-part bitstream). Similar information can be included in the tile-part header. This organization of the codestream is shown in Figure 4.

Without one payload designed for JPEG 2000 codestreams we will have to do a "non-intelligent" packetization, for example, making separations of the codestream at arbitrary length (see Figure 5). This packetization is called "non-intelligent" because don't use the advantages of how JPEG 2000 codestream is structured.

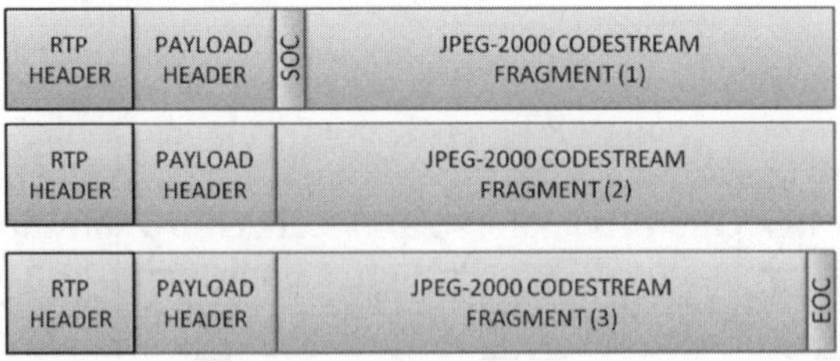

Fig. 5. Non-intelligent packetization of JPEG 2000 Codestream

With the JPEG 2000 RTP payload header [7] we can now make an intelligent packetization of the JPEG 2000 codestream. The packetization we have followed for each single frame is shown in Figure 6.

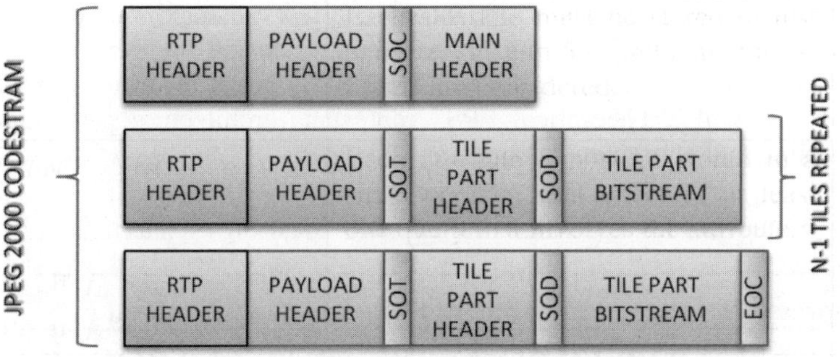

Fig. 6. Intelligent packetization of JPEG 2000 Codestream

It has many advantages from "non-intelligent" packetization as below:

- With the first RTP packet, the client receives main header of the codestream will receive. With the main header the client can know the image width, image height, number of tiles, number of components, etc., and can prepare buffers faster and know how many tiles have to receive to complete the frame.
- When a tile is received it can be decoded immediately without having to wait for the rest of tiles, because they are coded independently. Moreover while client is decoding one tile can continue receiving data, incrementing parallelism.
- The loss of one tile or packet doesn't suppose lose all information of the frame. Can be used the concept of main header recovery. In video streaming normally all frames uses to have the same format like with, height, components, etc. Then all their main headers will be the same too, and if one main header of a frame is lost, the client can use previous main header received for the current frame.

4 Performance of JPEG 2000 Live Streaming Architecture

All the results obtained in this section have been tested in our laboratory, using LAN at 100Mbps and CPU with two cores (this is important because client uses independent threads for decode each tile and performance may vary).

The results obtained in our tests consist basically in the maximum FPS (frames per second) of live video received in the client, that get us a great value of the video streaming performance. We have made various tests with different setups of compression, decompression, and architecture.

- Compression: JPEG 2000 supports lossy (some loss of original image information) and lossless (original image is not modified) compression. We have tested both modes of compression.
- Decompression: J2K-Codec supports two modes of decompression, one is "video on", which perform the decompression faster but losing some information of the codestream and "video off" that is slower but obtain all the codestream information.
- Architecture: In our tests we have seen that the time employed for image acquiring and compression performed by Matrox Morphis boards is not real "real-time", at

least for streaming purposes. Then, in order to test our maximum architecture performance without this hardware delay, we have tested the performance sending frames pre-acquired and compressed. “Real FPS” is the result using Matrox Morphis boards while “Theoretical FPS” represents architecture performance using pre-acquired and compressed frames.

Table 1. Performance of JPEG 2000 Streaming Architecture

Compression	Theoretical FPS		Real FPS	
LOSSY	**VIDEO OFF**	**VIDEO ON**	**VIDEO OFF**	**VIDEO ON**
768x576	20	32	13	13
384x288	48	62	13	13
LOSSLESS	**VIDEO OFF**	**VIDEO ON**	**VIDEO OFF**	**VIDEO ON**
768x576	7	8	N/A	N/A
384x288	21	52	13	13

Notice that the use of Matrox Morphis boards dont let our architecture obtain the maximun performance that we can see in “Theoretical FPS” in Table 1, due that Acquiring and JPEG 2000 compression is not real-time and a minimal lag in this part may vary performance drastically. Matrox Morphis boards only get us up to 13 FPS acquired and compressed in any compression mode or resolution. Anyway 13 FPS is just half of FPS of standard digital video (25 FPS) and still display the streaming normally.

5 Conclusions

In this paper, we developed a JPEG 2000 live video streaming architecture. In Table 1 we can see how with lossy compression we have obtained good values both for theoretical fps and real fps but in general for lossless compression these values are worse because lossless JPEG 2000 codestreams are about 4 times bigger than lossy codestreams. With these values and knowing that our final lag is about 100 to 200 ms with these FPS we can say that we have done a great job of JPEG 2000 live video streaming.

Acknowledgments. This work was supported in part by Projects MADRINET, TEC2005-07186-C03-02, SINPROB, TSI2005-07344-C02-02, CAM CCG06-UC3M/TIC-0781.

References

1. Information Technology – JPEG 2000 Image Coding System, ISO/IEC FDIS15444-1: 2000 (August 2000)
2. Koenen, R.: Overview of the MPEG-4 Standard. ISO/IEC JTC1/SC29/WG11 N2725 (March 1999), `http://www.cselt.it/mpeg/standards/mpeg-4/mpeg-4.htm`

3. Microsoft Corporation, Advanced systems format (ASF) specification, rev. 1.20.02 (June 2004), http://download.microsoft.com
4. Pfeiffer, S.: The Ogg Encapsulation Format Version 0. Request for Comments 3533 (2003)
5. http://www.matrox.com/imaging/products/morphis/home.cfm
6. Schulzrinne, H., Casner, S., Frederick, R., Jacobson, V.: RTP: A Transport Proto-col for Real-Time Applications, Request for Comments 3550, IETF (July 2003)
7. Edwards, E., Futemma, S., Tomita, N., Itakura, E.: Rtp payload format for jpeg 2000 video streams. Internet-Draft (June 2003)
8. http://j2k-codec.com/
9. http://www.openjpeg.org/
10. Adams, M.D., Ward, R.K.: JASPER: A portable flexible open-source software tool kit for image coding/processing (2004)
11. Deshpande, S., Zeng, W.: Scalable streaming of JPEG2000 images using hypertext transfer protocol (2001)
12. Apostolopoulos, J.G., Tan, W.-t., Wee, S.J.: Video Streaming: Concepts, Algorithms, and Systems (September 2002)

A Survey of Distributed and Data Intensive CBR Systems

Aitor Mata

Departamento Informática y Automática
Universidad de Salamanca
Plaza de la Merced s/n, 37008, Salamanca, Spain
aitor@usal.es

Abstract. Case-Based Reasoning is a methodology that uses information that has been considered as valid in previous situations to solve new problems. That use of the information allows CBR systems to be applied to different fields where the reutilization of past good solutions is a key factor. In this paper some of the most modern applications of the CBR methodology are revised in order to obtain a global vision of the techniques used to develop functional systems. In order to analyze the systems, the four main phases of the CBR cycled are considered as the key elements to organize an application based on CBR.

Keywords: Case-Based Reasoning, recover, reuse, revise, retain.

1 Introduction

Case-Based reasoning (CBR) is one of the paradigms developed by artificial intelligence in order to build intelligent knowledge-based systems. CBR has received a great deal of attention in recent years and has been successfully used in a great variety of areas.

CBR systems are normally structured following the classic CBR cycle. That structure is composed of four main phases: *retrieval, reuse, revision* and *retention.* Those four phases are a way of organize the process used to solve problems by human beings. This paper follows the CBR cycle structure with the aim of explain the modern approaches to the methodology used to face different problems, and concentrates in problems with big amount of data or distributed applications.

2 Case Based Reasoning Systems

CBR system requires efficient techniques for several important subtasks, such as organizing and maintaining the case base, retrieving cases (which are maximally similar to the problem) from the case base, and adapting stored cases to the problem at hand. The basic inference mechanism underlying CBR, where the concept of similarity plays a major role, is built upon the principle of instance based learning and nearest neighbor classification.

J.M. Corchado et al. (Eds.): DCAI 2008, ASC 50, pp. 582–586, 2009.
springerlink.com

2.1 Case Definition and Case Base Creation

The first steps in the design of a CBR application must consist in a transformation of the information available into a structure, into *cases*. In textual case bases, it is sometimes necessary to extract knowledge from the data before creating the case base [1]. Once the knowledge is obtained, it can be structured into the case base. Every new element is part of one or more of the *pieces of knowledge* previously identified, and then, the case is formed by the separated pieces that has inside it.

In medical applications, the case must include values referred to the patient, but also associated with the clinical evolution of the patient [2]. It is also interesting to include a *reputation* value, which is increased every time a case is recovered from the case base and used, every time the expert considers that the case is useful.

When the information to be transformed into cases contains a great amount of words, it is necessary to parse the original data [3] in order to obtain the list of terms used to create the cases. In some occasions, the information can be considered as hard to model, but after an analysis, it can be transformed into numerical variables [4] with what is quite easier to generate cases.

2.2 Recovering the Data from the Case Base

When a new problem appears, a selection of cases are recovered from the case based and will be used to solve that new problem. If the amount of variables is quite big, it is necessary to select which ones will be used to select the similar cases from the case base [2]. A two steps procedure occurs so first the interesting variables must be chosen, and then, the search in the case base of the most similar cases according to those variables. If different features are considered when defining the case base, they must all be considered when obtaining similar cases from the case base. In this kind of situations different metrics can be done to calculate the similarity of the different features [4], and then create a combined similarity metric that integrates all the metrics used.

In some circumstances, a previous search of context is done [5], to obtain a variety of cases that are used to perform a second and more specific search.

If the problem introduce in the system implies considering different scenarios, multiple retrievals can be done [6]. In this kind of situations the original problem introduced in the system defines the start point of the search and from that point and looking for in different directions, different sets of cases are recovered from the case base, in order to generate a complete perspective of the problem.

2.3 Adaptation of the Retrieved Cases

Transforming the recovered cases from the case base is the solution generator for the CBR systems. From the collection of cases retrieved from the cases base, a new solution must be generated in order to solve the proposed problem. Numeric situations, like those used in microarray problems, can be reused thru neural networks like Growing Cells Structures [7], where the aim is to cluster the retrieved information. Another way to use neural networks to adapt the retrieved information is to change the weight of the connection between the neurons depending on the retrieved cases [8]. Changing the weights allows the system to adapt the solution to the problem, as the retrieved cases will depend directly on the proposed problem.

If the problem to be solved may belong to more than one field of knowledge, and there may be more than one case base, a good solution can be to adapt the retrieved cases, from the different case bases, according to the characteristics of the problem [9]. In this case, neural networks were used to recover the data from the different case bases, and machine learning algorithms combined the retrieved cases in order to adapt those cases to the proposed problem.

When using genetic algorithms, the reuse may help to reduce convergence time if considering previously working solutions [10]. This approach may be applied to different fields where evolutionary algorithms are useful but slow.

2.4 Review of the Proposed Solution

When a solution is generated by a CBR system, it is necessary to validate the correction of that solution. This may be the most complicated phase of the cycle. In crucial fields, such as medical applications, it is normal to trust an expert in order to finally accept a solution [11]. Then, after being accepted by the corresponding expert, next time it will be considered as a better solution, being chosen from the case base with a higher probability.

Changing the values proposed by the system to others *similar* but not equal is a technique also used to revise the correction of a solution [12]. If the solution generated by the similar values is not better than the proposed one, then the chosen one is a good solution for the problem. Genetic algorithms are also used to revise the correction of the solutions [13]. After running those algorithms, the solutions can be accepted, and added to the case base.

2.5 Storage of the Solution and Maintenance of the Case Base

In most cases there a big amount of information is stored in the case base and it is not necessary to store every valid case, thus the information could be too redundant. In those situations a conditional retention is performed [14], keeping the new solution only if it is different enough to the closest existing case.

Even when the proposed solution is considered as an eventually good solution to be stored in the case base, the growth of the case base can be counterproductive. In some case, where the amount of stored information is huge and when there must be an economy of resources in order to manage a reasonable case base, case base editing is necessary [15].

When the case base grows to thousands of elements, it may be difficult to maintain it. Then dividing the case base in different parts with certain inner similarity [16] can help to structure the store information and also to make future retrievals. Another strategy used to control the growth of the case base is to group cases into prototypes that include the common characteristics of a series of cases with no plenty of variability.

3 Resume of Solutions

After analyzing the main phases of the CBR cycle, it is necessary to clearly resume the various techniques used to improve the capabilities of each one, depending, in most of the cases, on the problem to be solved or on the type of data managed.

Table 1. Resume of strategies used in the four main phases of the CBR cycle

Phase	Strategies
Case Definition	o Numerical list of parameters o List of abstract concepts extracted from textual information
Retrieval	o Metrics o Two steps search
Reuse	o Collaboration between algorithms o Transforming neural networks
Revision	o Experts o Alternative structures
Retention	o Eliminate redundant data o Division of the case base

Retrieval and reuse are the two phases where more improving efforts have been expended. Those are the phases where the quality of the information is checked. Table 1 shows a resume of the solutions used in every phase.

References

1. Mustafaraj, E.: Knowledge Extraction and Summarization for Textual Case-Based Reasoning, Philipps-Universitat Marburg (2007)
2. Montani, S.: Case-based Reasoning for managing non-compliance with clinical guidelines. In: International Conference on Case-Based Reasoning, ICCBR 2007, Proceedings (2007)
3. Patterson, D., Rooney, N., Dobrynin, V., Galushka, M.: Sophia: A novel approach for Textual Case-based Reasoning. In: Proceedings of the Eighteenth International Joint Conference on Artificial Intelligence (2005)
4. Ros, R., Veloso, M., de Mantaras, R.L., Sierra, C., et al.: Retrieving and Reusing Game Plays for Robot Soccer. In: Advances in Case-Based Reasoning, vol. 4106 (2006)
5. Spasic, I., Ananiadou, S., Tsujii, J.: MaSTerClass: a case-based reasoning system for the classification of biomedical terms. Bioinformatics 21(11), 2748–2758 (2005)
6. Aha, D.W., Molineaux, M., Ponsen, M.: Learning to Win: Case-Based Plan Selection in a Real-Time Strategy Game. In: Case-Based Reasoning Research and Development, pp. 5–20 (2005)
7. Diaz, F., Fdez-Riverola, F., Corchado, J.M.: Gene-CBR: A case-based reasoning tool for cancer diagnosis using microarray data sets. Computational Intelligence 22(3/4), 254–268 (2006)
8. Zhang, F., Ha, M.H., Wang, X.Z., Li, X.H.: Case adaptation using estimators of neural network. In: Proceedings of 2004 International Conference on Machine Learning and Cybernetics, p. 4 (2004)
9. Policastro, C.A., Carvalho, A.C., Delbem, A.C.B.: Automatic knowledge learning and case adaptation with a hybrid committee approach. Journal of Applied Logic 4(1), 26–38 (2006)
10. Pérez, E.I., Coello, C.A.C., Aguirre, A.H.: Extraction and reuse of design patterns from genetic algorithms using case-based reasoning. Soft Computing-A Fusion of Foundations, Methodologies and Applications 9(1), 44–53 (2005)
11. Chang, C.L.: Using case-based reasoning to diagnostic screening of children with developmental delay. Expert Systems With Applications 28(2), 237–247 (2005)

12. Li, H., Hu, D., Hao, T., Wenyin, L., et al.: Adaptation Rule Learning for Case-Based Reasoning. In: Third International Conference on Semantics, Knowledge and Grid, pp. 44–49 (2007)
13. Pavón, R., Díaz, F., Laza, R., Luzón, V.: Automatic parameter tuning with a Bayesian case-based reasoning system. A case of study, Expert Systems With Applications (2008)
14. Sharma, M., Holmes, M., Santamaria, J., Irani, A., et al.: Transfer learning in real-time strategy games using hybrid cbr/rl. In: Proceedings of the Twentieth International Joint Conference on Artificial Intelligence (2007)
15. Delany, S.J.: Using Case-Based Reasoning for Spam Filtering, Dublin Institute of Technology (2006)
16. Li, J.Y., Ni, Z.W., Liu, X., Liu, H.T.: Case-Base Maintenance Based on Multi-Layer Alternative-Covering Algorithm. In: International Conference on Machine Learning and Cybernetics, 2006, pp. 2035–2039 (2006)

QoS-Based Middleware Architecture for Distributed Control Systems

José L. Poza, Juan L. Posadas, and José E. Simó

Institute of Industrial Control Systems
Polytechnic University of Valencia
Camino de Vera s/n, 46022, Valencia, Spain
{jopolu,jposadas,jsimo}@ai2.upv.es

Abstract. This paper presents an implementation of a middleware architecture to control distributed systems. The main objective is providing a QoS level between the communications layer and the control layer. This architecture is based on the use of a hierarchical communications structure called "logical namespace tree" and a structured set of control processes interconnected, called "logical sensors graph". This architecture is named Frame Sensor Adapter Control (FSA-Ctrl). In this architecture both: communication layer and control layer can manage the QoS policies. The communication layer is based on the Data Distribution Service (DDS), a standard proposed by Object Management Group (OMG). Control layer is derived from the Sensor Web Enablement (SWE) model proposed by Open Geospatial Consortium (OGC). By means of QoS policies, control components can take important decisions about distributed questions, like components mobility or information redundancy detection.

Keywords: Distributed system, control architecture, quality of service.

1 Introduction

Usually quality of service (QoS) used in the communication layer provides simple temporal parameters like messages delay or message flow control like congestion control. When communication system has important requirements, like real-time support, information optimization or components hidden, then communication layer should use more QoS policies, with more complex parameters.

A middleware must control all communication parameters involved in message management like the message producers and consumers or all questions related to the use of message queues. Most of the communication paradigms are designed to improve message speed or hide the communications components to the application layer. Nevertheless, QoS covers some more aspects, like optimizing messages or metadata management. Publish/Subscribe (P/S) communications paradigm provides an appropriate environment for information distribution, and the messages queues provides a flow control extending middleware QoS policies.

Among the current communication architectures standards, the Data Distribution Service (DDS) provides a system based on publish-subscribe paradigm [1] with ability to manage a large amount of QoS policies [2]. Likewise, among the current control architecture standards, the Sensor Web Enablement (SWE) allows a simple intelligent

J.M. Corchado et al. (Eds.): DCAI 2008, ASC 50, pp. 587–595, 2009.
springerlink.com

control model based on services capable of managing complex sensor networks [3]. Joining DDS and SWE standards is interesting because it provides a unique environment to implement a solution to manage component-based distributed intelligent control system, based on the combination of several QoS policies.

This paper presents the model of an architecture called Framed Sensor Adapter Control (FSA-Ctrl) whose communication components are based on the DDS model and its control components are based on the SWE model. The QoS merges communications and control components. The communication layer is called Logical Namespace Tree (LNT) and it is a hierarchical abstraction of the real communications channels, like TCP/IP, EIB bus, and they adapters. Each information item in the LNT can be identified by means of a topic called Logical Data (LD). The control layer is known as Logical Sensor Graph (LSG) and it is a structured set of small control processes interconnected by means the LNT or internally. Each control process unit is called Logical Sensor (LS) and they are an abstraction of the control components.

The rest of the paper is organized as follows. Section 2 briefly describes the functional structure of DDS and SWE standards. Section 3 presents in detail the LNT and LSG components and the role of QoS in the system. Section 4 presents concluding remarks.

2 Fundamentals: DDS and SWE

Most of the communication systems that provide support to the distributed control architectures need a module that hides the details of the components connexions. Usually, when this module is separated from control components, is known as "middleware". The main responsibility of middleware is providing, to control components, the necessary services to increase efficiency of communication. Among the required services is outstanding identification of components, authentication, authorization, hierarchical structuring or components mobility.

Moreover, the underlying programming technology like object-oriented programming, component-based programming or service-based programming determines partially the resultant control architecture and its ability to provide more QoS features [4].

To develop a distributed system based on components, some components must adapt his technology to the communication interfaces. For example, if communication is based on CORBA [5], the distributed system must be implemented with the object-oriented programming technology. To avoid the use of a particular technology usually distributed systems use standardized protocols like FIPA [6]. To support distributed protocols, usually systems use a message-based service, like JMS [7], that offers some control to QoS.

2.1 Data Distribution Service

Data Distribution Service (DDS) provides a platform independent model that is aimed to real-time distributed systems. DDS is based on publish-subscribe communications paradigm. Publish-subscribe components connect information producers (publishers) and consumers (subscribers) and isolate publishers and subscribers in time, space and message flow [8]. To configure the communications, DDS uses QoS policies. A QoS

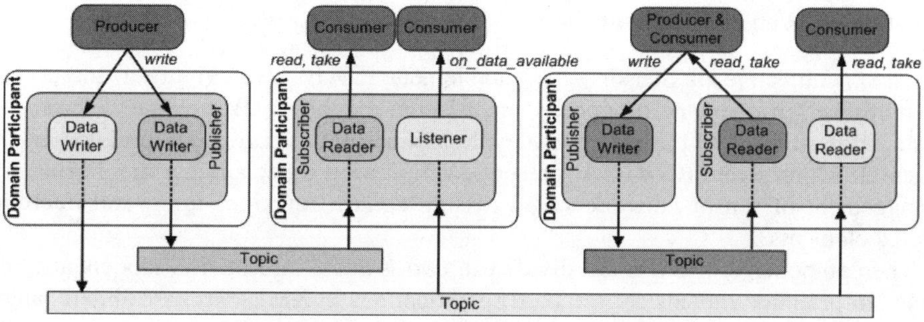

Fig. 1. Overview DCPS components from DDS model

policy describes the service behaviour according to a set of parameters defined by the system characteristics or by the administrator user. Consequently, service-oriented architectures are recommended to implement QoS in his communication modules.

DDS specifies two areas: Data-Centric Publish-Subscribe (DCPS) witch is responsible for data distribution and DLRL which is responsible for the data adaptation to the local applications level. The DLRL area is optional due to the DCPS components can work directly with the control objects without data translations.

DCPS has a lot of components, or classes in the case of the object-oriented (OO) model, in his formal model. However, there are mandatory components, presented at figure 1. When an producer (component, agent or application) wants to publish some information, should write it in a "Topic" by means of an component called "DataWriter" witch is managed by another component called "Publisher". Both components, DataWriter and Publisher, are included in another component called "DomainParticipant". On the other side of the communication, a Topic can be received by two kinds of components: "DataReaders" and "Listeners" by means a "Subscriber". A DataReader provides the messages to application when the application request-it, in lieu a "Listened" sends the messages without waiting for the application request. An application will use as many DataWriters, DataReader or Listeners as distributed topology requires. DataReaders, DataWriters and Listeners will be the elements responsible for maintaining message queues, and consequently responsible for implementing QoS policies.

2.2 Quality of Service

QoS is a concept that defines a set of parameters by which to evaluate a service offered. There are many definitions of quality of service. From the viewpoint of processing, QoS represents the set of both: quantitative and qualitative characteristics of a distributed system needed to achieve the functionality required by an application [9]. From the communication viewpoint, QoS is defined as all the requirements that a network must meet to message flow transport [10].

In DCPS model, all objects may have associated a set of policies QoS. At present the DCPS specification defines 22 different QoS policies that can be classified into four areas: times, flows, components and metadata management. All components of a DCPS based communication, can establish a set of QoS policies into them independently the others components.

2.3 Sensor Web Enablement

The main objective of Sensor Web Enablement (SWE) is providing a unique and revolutionary framework of open standards for exploiting Web-connected sensors systems of all types [11]. SWE was developed in 2004 as part of an initiative by the OpenGIS Consortium (OGC). At present SWE is used especially for monitoring and management of sensor networks. SWE assign control functions to several interconnected elements.

The components of SWE are divided in two groups: information models and services. Information models are standard specifications in XML, processes interchanges messages with these specifications. Services are control components that process the information models. Control processes are based on components interconnected, those receive information models from other components, and send the results to connected components.

From SWE viewpoint, a component is a particular physical process that transforms information. Simple examples of SWE components are sensors, effectors or physical process filters. Complex examples of SWE components are control kernels or sensor data fusion algorithms.

As shown in the Figure 2, a "Process Model" is a single component, used into a more complex structure, called "Process Chain". Moreover, a "Process Model" is based on a "Process Method" witch acts as a "Process Model" template. A "Process Method" specifies the interface and how to implement the "Process Model", also define inputs, outputs and the operating parameters. The model proposed by SWE is very interesting because allows to specify reusable process patterns. This scheme provides a highly scalable control system based on singles control kernels.

Anyway, it should take some precautions when using this scheme. The highly interconnected model increases redundant information because the model hides the data sources. Also, repetition in control patterns can lead to control actions repeated. Finally, the interconnection of control models can generate undesirable control cycles. Any SWE based architecture must prevent these aspects.

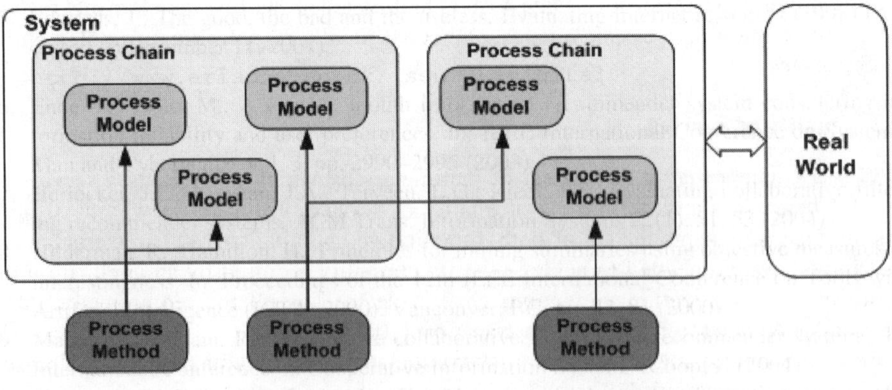

Fig. 2. SWE control architecture components overview

3 FSA-Ctrl Architecture

Frame Sensor Adapter to Control (FSA-Ctrl) is an architecture inspired by DDS and SWE models and is an evolution of an architecture developed by the research group called Frame Sensor Adapter (FSA) [12]. The architecture has two distinct areas: communication and control. QoS Policies connects both areas. Figure 3 shows the details of the architecture.

The "Frame" component of the FSA architecture takes the same role of the "DomainPaticipant" component of the DDS architecture. The "Adapter" component takes the role of both DDS components, "Publisher" and "Subscriber". A specialization of "Sensor" component takes the role of the "DataWriter", "DataReader" and "Listener" DDS components.

The function of the "Topic" DDS component is performed by the "LogicalData" component of the FSA-Ctrl architecture. The communication layer organizes the "LogicalData" in a hierarchical structure to hide any type of communication channel like the TCP/IP protocol or CAN bus. The structure is a symbolic tree called Logical Namespace Tree (LNT), details can be obtained from [13].

Control layer organizes the "Sensors" on a graph, called Logical Sensor Graph (LSG); details can be obtained from [14]. This model is based on SWE "Process Chain". The process units are known as "Logical Sensors". Some this "Logical Sensors" takes the role of some communication components. A "Logical Sensor" can receive, or send, messages from, or to, another "Logical Sensors".

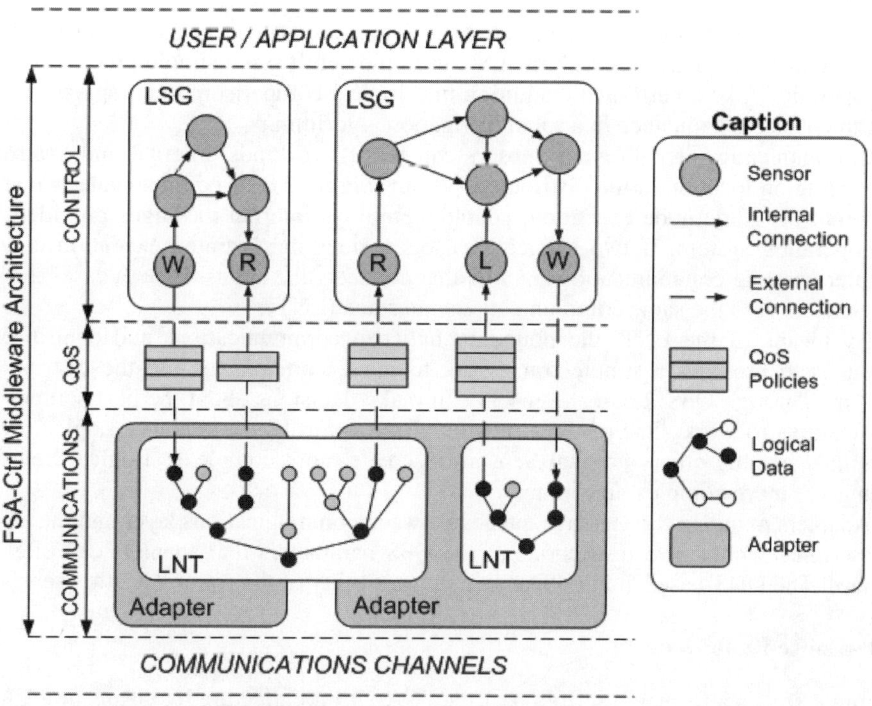

Fig. 3. FSA-Ctrl architecture overview

3.1 Communication: Adapters and LNT

Usually, communications platform use several methods to locate the components. Frequently the communications systems provide to control system a name or an address that represents the component. The name of the component can't offer more information, like type of component, location and other features. FSA-Ctrl architecture uses a layer, which hides the protocol-dependant location method and organizes the components depending on user -defined characteristics.

To connect the communications channels, like TCP/IP, EIB or CAN bus, FSA-Ctrl uses a type of component called adapter. Adapter hides details of the media and fit messages to a SWE message format standardized, like SensorML. Messages are dispatched to the control components that are interested in them.

To manage system components, FSA-Ctrl organizes the information in a tree structure called LNT. The LNT locates both main FSA-Ctrl components: adapters and sensors, by a concept named Logical Data (DL).

A logical data is a sequence of names separated by the symbol '/'. Every name is known as "logical node". Symbol '*' represents the sub-tree derivates form a node. This structure provides a common meta-information about the type of message or type of component involved in the communication. For example, is possible obtain all temperatures from a home automation system by a subscription to the logical data "root/home/sensors/temperature/*".

3.2 Control: Sensors and LSG

Components witch implements the control system are named Logical Sensors (LS) and contains the control algorithms. LS can implement from a simple process or single operation, like an arithmetic addition or a logical comparison, to complexes tasks like the obstacle avoidance in a robot navigation algorithm.

Communication into Logical Sensors can be of two kinds: internal and external, depending on location in the distributed system (Figure 3). To communicate two sensors into the same node execution, adapters employs internal messages provided by the operating system. If two Logical Sensors resides in separated execution nodes, adapter uses the communication channel that connects the nodes. In both cases, Logical Sensors uses the same communications interface (LNT).

By means of the LNT, the boundary between communications and control can change and provides a simple framework to move components into the distributed system. Through QoS, Logical Sensors can make decisions about the best sources or destinations to work. One of the strengths of the control system is that Logical Sensors may include others Logical Sensors to create more complexes Logical Sensors than gives more complex algorithms.

Adapters maintain the message queues between communications layer and the LNT. These queues only give restrictions of the QoS parameters that adapters can offer to control. The DDS-based QoS policies are responsibility of the communication sensors.

3.3 Conceptual Model

Figure 4 shows a formal description of the FSA-Ctrl architecture by means of a UML class diagram. "Entity" is the class base for all components, except for the QoS

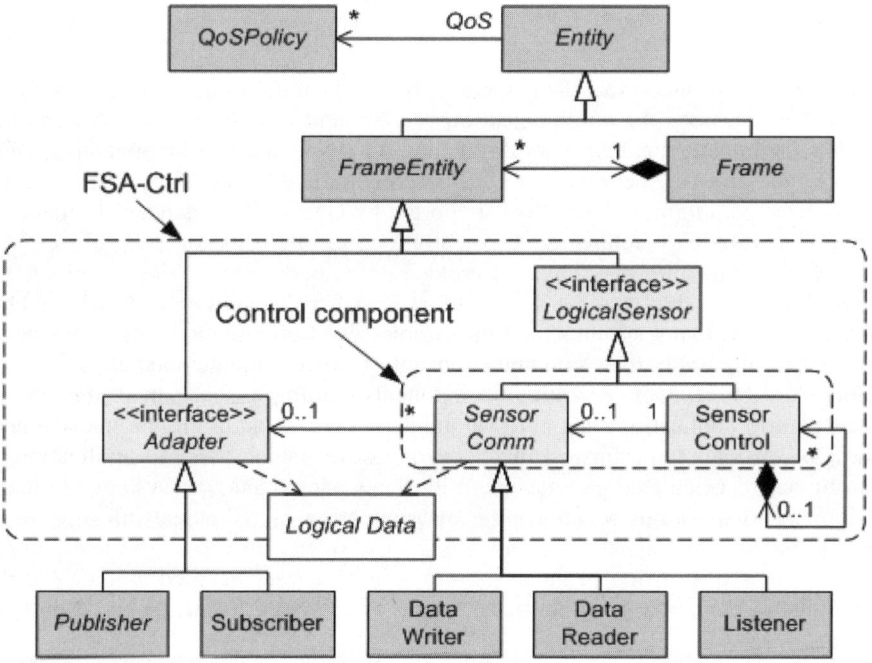

Fig. 4. UML-based class diagram of the FSA-Ctrl architecture

policy. Each component can have associated several QoS policies. This relationship is performed at the class base level to standardize the QoS to all components derived from "Entity" class.

The classes "Frame" and "FrameEntity" are derived from "Entity" class, these classes contain the elements of the architecture. "Frame" class represents the execution framework to sensors and adapters components and "FrameEntity" is the base class for "Adapter" class and "Sensor" class. "LogicalData" is a data model managed by the Adapter "class". A LogicalSensor object can be connected with some others LosgicalSensors, but only one LogicalSensor can be connected with an Adapter. Making a similarity with DCPS model of DDS standard, the "Frame" component of FSA-Ctrl takes the same role as the "DomainParticipant", "Adapter" is similar than "Publisher" and "Subscriber" and "Logical Sensors" makes the same role as the "DataWriter", "Datareader2 and "Listener" components. The role of a "LogicalData" is the same that "Topic" in DCPS. When a "Logical Sensor" does not have an associated "Adapter", then is a control component, and can be associated with others control components.

In FSA-Ctrl architecture, QoS is used by all components. "Adapters" and "Logical Sensors" specialized on communications uses the QoS policies to manage times and messages flow parameters, although "Logical Sensors" uses the QoS policies to manage the control action computation efficiency. DDS and QoS becomes a common interface between communications and control.

4 Conclusions

This article has presented the UML specification of a middleware with QoS support. The middleware hides the communications details and offer simple control components. The architecture, called FSA-Ctrl, is based in two standards architectures, DDS and SWE, and takes the benefits of a QoS-based communication. DDS, based on publish-subscribe paradigm, is a standard supported by OMG. SWE standard is endorsed by OGC. The FSA-Ctrl architecture especially focuses on the use of QoS policies.

FSA-Ctrl architecture may have different uses. A publish-subscribe-based system provides highly scalable systems. The fact that FSA-Ctrl is based on DDS will be able to adapt easily at many systems, and the simple set of communications components allows converting easily the communications interface system into another.

Moreover, due to communications components can implement part of the control system, control applications can perform another tasks. In addition, the management of the QoS policies is performed on the middleware, which gives to applications a large number of calculated parameters. With these parameters, the system can make decisions about questions regarding the discrimination of redundant information or taking decisions about agents' movement. The use of the LNT provides a structured and hierarchical abstraction of the system to control, a very detailed topology of the control algorithms and a great ability to tune the system performance based on QoS policies.

SWE-based control components allow the high-distributed control algorithms. Moreover, the logical sensor-based can be applied to evaluate the proper location for the components. Since certain control components with a set of QoS restrictions for each, the architecture could distribute automatically agents' control processes according to a set of parameters values.

System has some drawbacks, usually related to the use of a middleware. The main disadvantage is a slight speed reduction intrinsic to the use of queues and the LNT structure.

Currently, a beta version of the middleware, based on the FSA-Ctrl architecture, is being tested on a home automation project. System is used to obtain the most relevant QoS parameters from the basic algorithms, used on home automation control. The next step in the project is a real-time testing of the middleware to determine a set of QoS parameters to model performance aspects. The library has a set of adapters that can communicate an application with real-time CAN bus, the home automation bus EIB, TCP/IP services and a MySQL database service.

Middleware can be used in distributed systems to measure the communications QoS parameters. The obtained values can be used to distribute the components according to the communications performance, with the aim of achieve a system able to auto distribute its components according to some relevant changes in the environment.

Acknowledgments. The architecture described in this article is a part of the coordinated project KERTROL: Kernel control on embedded system strongly connected. Education and Science Department, Spanish Government. CICYT: DPI2005-09327-C02-01/02.

References

1. Matteucci, M.: Publish/Subscribe Middleware for Robotics: Requirements and State of the Art. Technical Report N 2003.3, Politecnico di Milano, Milano, Italy (2003)
2. OMG. Data Distribution Service for Real-Time Systems, v1.1. Document formal/2005-12-04 (2005)
3. Botts, M., Percivall, G., Reed, C., Davidson, J. (eds.): OGC. Sensor Web Enablement: Overview and High Level Architecture. OGC White Paper. OGC 06-050r2 (2006)
4. Coulouris, G., Dollimore, J., Kindberg, T.: Distributed systems, concepts and design, 3rd edn. Addison-Wesley, Reading (2001)
5. OMG. Real-Time Corba Specification version 1.1. Document formal /02-08-02 (2002)
6. FIPA. Specfication. Part 2, Agent Communication Language. Foundation for Intelligent Physical Agents (1997)
7. Hapner, M., Sharma, R., Fialli, J., Stout, K.: JMS specification, vol. 1.1. Sun Microsystems Inc., Santa Clara (2002)
8. Pardo-Castellote, G.: OMG Data-Distribution Service: architectural overview. In: Proceedings of 23rd International Conference on Distributed Computing Systems Workshops, Providence, USA, vol. 19-22, pp. 200–206 (2003)
9. Vogel, A., Kerherve, B., von Bochmann, G., Gecsei, J.: Distributed Multimedia and QoS: A Survey. IEEE Multimedia 2(2), 10–19 (1995)
10. Crawley, E., Nair, R., Rajagopalan, B.: RFC 2386: A Framework for QoS-based Routing in the Internet. IETF Internet Draft, pp. 1–37 (1998)
11. Botts, M., Percivall, G., Reed, C., Davidson, J.: OGC. Sensor Web Enablement: Overview and High Level Architecture. OpenGIS Consortium Inc. (2006)
12. Posadas, J.L., Perez, P., Simo, J.E., Benet, G., Blanes, F.: Communication structure for sensory data in mobile robots. Engineering Applications of Artificial Intelligence 15(3-4), 341–350 (2002)
13. Poza, J.L., Posadas, J.L., Simó, J.E., Benet, G.: Hierarchical communication system to manage maps in mobile robot navigation. In: Proceedings of International Conference on Automation, Control and Instrumentation, Valencia, Spain (2006)
14. Poza, J.L., Posadas, J.L., Simó, J.E.: Distributed agent specification to an Intelligent Control Architecture. In: 6th International Workshop on Practical Applications of Agents and Multiagent Systems, Salamanca, Spain (in press, 2007)

Distribution, Collaboration and Coevolution in Asynchronous Search

Camelia Chira, Anca Gog, and D. Dumitrescu

Department of Computer Science
Babes-Bolyai University
Kogalniceanu 1, 400084 Cluj-Napoca, Romania
{cchira,anca,ddumitr}@cs.ubbcluj.ro

Abstract. Spatial distribution of individuals in evolutionary search combined with agent-based interactions within a population formed of multiple societies can induce new powerful models for complex optimization problems. The proposed search model relies on the distribution of individuals in a spatial environment, the collaboration and coevolution of individuals able to act like agents. Asynchronous search process is facilitated through a gradual propagation of genetic material into the population. Recombination and mutation processes are guided by the population geometrical structure. The proposed model specifies three strategies for recombination corresponding to three subpopulations (societies of agents). Each individual (agent) in the population has the goal of optimizing its fitness and is able to communicate and select a mate for recombination. Numerical results indicate the performance of the proposed distributed asynchronous search model.

Keywords: evolutionary algorithms, distributed population topology, coevolution, asynchronous search, emergent behavior.

1 Introduction

Investigating connections between multi-agent systems, distributed networks and evolutionary models represents a novel approach to developing search models able to efficiently address complex real-world problems. The model proposed in this paper relies on a spatial distribution of individuals, coevolution and agent-based interactions within a population structured in societies with different strategies.

The proposed *Distributed Asynchronous Collaborative Evolutionary (DACE)* model uses a population of individuals distributed according to a predefined topological structure. Each individual acts like an agent being able to communicate and select a mate for recombination. The information exchanged between agents refers to specific environment characteristics such as the current fitness value. Furthermore, agents exchange request and inform type of messages for establishing recombination strategies.

Individuals belong to one of the following agent societies: *Local Interaction, Far Interaction* and *Global Interaction*. The society membership specifies the recombination strategy of individuals. The agent society of an offspring depends on the parents'

J.M. Corchado et al. (Eds.): DCAI 2008, ASC 50, pp. 596–604, 2009.
springerlink.com

agent society and a dominance probability between societies. The three societies of individuals form a complex system characterized by an emergent pattern of behavior and a phase transition interval.

Co-evolution of agent societies enables a useful balance between search diversification and intensification. Some individuals are specialized for local search facilitating exploitation while other individuals focus on global search.

The *DACE* model is a very general distributed search scheme admitting several instances able to cope with particular fitness landscapes (environments). A DACE instance using a discrete regulated population topology and a fitness-based distribution of individuals is investigated. All individuals are sorted according to their fitness and are distributed over concentric layers starting with the fittest individuals on the most inner layer. The proposed model uses an asynchronous search scheme. Individuals from a certain layer are updated through proactive recombination and are involved in forthcoming recombination and mutation processes within the same epoch.

Numerical experiments prove the efficiency of the proposed technique by comparing it with the results obtained by recent evolutionary algorithms for several difficult multimodal real-valued functions.

The structure of the paper is as follows: the DACE model and an instance of DACE based on a particular population topology are presented, numerical results and comparisons completed by a statistical analysis test are discussed and conclusions of research are briefly presented.

2 Distributed Asynchronous Search Model

The proposed *Distributed Asynchronous Collaborative Evolutionary (DACE)* model integrates evolutionary optimization with emergent behavior generated by agent-based interactions and recombination strategies facilitated by a certain population topology.

The DACE population is endowed with a topological structure guiding selection and search processes. Population spatial distribution is described by a topology T relying on a neighborhood system V. The population topology is associated to the asynchronous action of the search operators aiming to achieve a better balance between search intensification and diversification.

In order to ensure a flexible search process in solving very difficult problems, DACE individuals are identified with agents [5, 9] having the objective of optimizing a fitness function. This objective is pursued by communicating with other individuals, selecting a mate for recombination based on individual strategies and performing mutation. Individuals can exchange information regarding specific environment characteristics (such as the current fitness value) and specific messages (such as request and inform) for establishing recombination strategies.

The model allows a specialization of individuals in a population aiming to facilitate the search in all stages through emerging complex interactions. This aim is achieved by allowing individuals (agents) to belong to several well defined subpopulations (agent societies). These subpopulations are dynamic and their evolution reflects the search progress.

The proposed model implies three societies of agents (individuals) as follows:

1. *Local Interaction Agent (LIA) society:* LIA individuals select mates for recombination from their (local) neighbourhood.
2. *Far Interaction Agent (FIA) society:* FIA individuals select mates for recombination outside their neighbourhood.
3. *Global Interaction Agent (GIA) society:* GIA individuals select mates for recombination on a global basis from the entire population.

Population dynamics emerges through the recombination of individuals from different societies and a dominance principle – an offspring belongs to the class of the dominant parent.

LIA behaviour emphasizes local search while FIA individuals are able to guide the search towards unexplored regions. The GIA society focuses on the global exploration of the search space realizing the connection between the LIA and FIA societies.

Each individual invited to be a mate can accept or decline the proposal according to its own strategy. Individuals from LIA and FIA societies accept individuals from the same society or from the GIA society as mates. Individuals from GIA society accept any other individual as mate. Offspring are assigned to a certain society according to a dominance concept. If LIA is the dominant agent society then any combination of a GIA individual with a LIA individual results in an offspring belonging to LIA.

A dynamic dominance concept may allow complex interactions between the three societies. Let p be the probability of LIA (or FIA) dominating GIA. The dominance probability p may be viewed as the (probabilistic) membership degree of an offspring to the society LIA (FIA) when one of the parents is a GIA individual. Several assignment schemes for p are possible ranging from static to dynamic, including many intervals of probability.

Various instances of the proposed DACE model are obtained when a particular population topology is specified facilitating the emergence of complex interactions within the system.

3 An Instance of DACE Model

The proposed DACE model implemented using a simple population topology is presented.

3.1 Population Structure

Let $P(t)$ be the current population at iteration (epoch) t. The population size is fixed at n^2, where n is an even number. All individuals are sorted according to their fitness and are distributed over $n/2$ concentric layers starting with the fittest individuals on the most inner layer (see Figure 1). The number of individuals placed on layer i, $i=0,..., n/2-1$, is $4(n-2i-1)$. A corresponding grid is depicted in Figure 1. The neighborhood system of the considered topology is induced by the layers.

Let us denote the sorted population by $P(t) = x_1, x_2,..., x_{n^2}$ at iteration t, where x_1 is the fittest and x_{n^2} is the worst individual in the population. The most inner layer

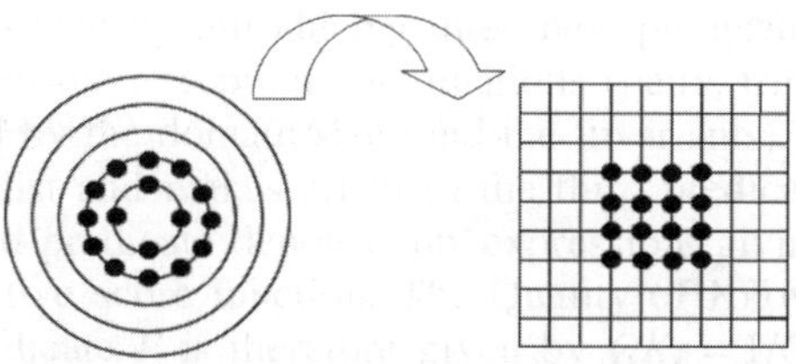

Fig. 1. A particular population topology

contains the first four individuals (x_1, x_2, x_3, x_4). The next layer holds 12 individuals ($x_5,..., x_{16}$) having the next best fitness values. It is assumed that the most outer layer is labeled by *0* and the label of the most inner layer has the highest value. For a population size of n^2, *n* even, there are *n/2* layers and the most inner layer is labeled *(n/2-1)*.

3.2 Search Progress

The proposed model uses an asynchronous search scheme. Individuals from a certain layer are updated through proactive recombination and are involved in forthcoming recombination and mutation processes within the same epoch.

Within the epoch *t* each individual from *P(t)* is considered as the first parent in a recombination process. The second parent involved in each recombination process is selected based on DACE agent society model. Moreover, an elitist scheme is considered by which the individuals from the most inner layer (the fittest individuals in the population) are copied in population *P(t+1)*.

LIA individuals from layer *c* address mating invitations to individuals from their neighborhood i.e. layer *(c+1)*, where *c = 0,…,n/2-2*. FIA individuals from layer *c* address mating invitations to individuals outside their neighborhood i.e. layer *(c+i)*, where *c = 0,…,n/2-3* and $i \geq 2$. Each FIA agent randomly generates the value of the index *i.* FIA individuals from layer *(n/2-2)* only invite individuals from the layer *(n/2-1)*. Individuals from the GIA society are more explorative. GIA individuals from a layer *c* may address mating invitations to any individuals except those belonging to the layer *c*.

The selection of a mate from a layer can be performed using any selection scheme. Tournament selection scheme is considered for all the experiments reported in this paper.

For each mating pair *(x, y)* one randomly chosen offspring *z* obtained by recombination is mutated. The best between *z* and *mut(z)* is compared to the first parent *x* and replaces *x* if it has a better quality. The useful genetic material collected from the entire population is propagated through the layers until it reaches the poorest individuals from the population. Furthermore, the co-existence of FIA and GIA individuals in the same population facilitates a more aggressive search space exploration.

4 Numerical Experiments

Numerical experiments reported in this section are based on the DACE instance presented in the previous section.

4.1 Test Functions

Numerical experiments are performed on a set of four multimodal benchmark functions (see for instance [8]):

1. *Shifted Rastrigin's Function*

$$f_1(x) = \sum_{i=1}^{D} (z_i^2 - 10\cos(2\pi z_i) + 10) + f\_bias_1,$$

$$z = x - x^*,\ x \in [-5,5]^D,\ f_1(x^*) = f\_bias_1 = -330.$$

2. *Shifted Rotated Rastrigin's Function*

$$f_2(x) = \sum_{i=1}^{D} (z_i^2 - 10\cos(2\pi z_i) + 10) + f\_bias_2,$$

$z = (x - x^*) * M$, *M*: linear transformation matrix, condition number=2,

$x \in [-5,5]^D,\ f_2(x^*) = f\_bias_2 = -330.$

3. *Shifted Rotated Weierstrass Function*

$$f_3(x) = \sum_{i=1}^{D} \left(\sum_{k=0}^{k\max} [a^k \cos(2\pi b^k (z_i + 0.5))] \right) - D \sum_{k=0}^{k\max} [a^k \cos(2\pi b^k * 0.5)] + f\_bias_3,$$

a=0.5, b=3, kmax=20,

$z = (x - x^*) * M$, *M*: linear transformation matrix, condition number=5,

$x \in [-0.5,0.5]^D,\ f_3(x^*) = f\_bias_3 = 90.$

4. *Schwefel's Problem*

$$f_4(x) = \sum_{i=1}^{D} (A_i - B_i(x))^2 + f\_bias_4,$$

$$A_i = \sum_{j=1}^{D} (a_{ij} \sin \alpha_j + b_{ij} \cos \alpha_j),$$

$$B_i = \sum_{j=1}^{D} (a_{ij} \sin x_j + b_{ij} \cos x_j),\ i = 1,\ldots,D,$$

a_{ij}, b_{ij} are integer random numbers in the range [-100,100],

α_j are random numbers in the range $[-\pi,\pi]$,

$x \in [-\pi,\pi]^D,\ f_4(x^*) = f\_bias_4 = -460.$

For all considered functions, x^* represents the global optimum.

4.2 Dominance Probability

In order to assign a meaningful dominance probability (between LIA and FIA societies), a set of preliminary experiments on several well-known functions has been performed.

Probabilistic dominance assignment suggests a potential emergence of complex population interactions. Emergent behavior emphasizing a geometrical phase transition of the system has been analyzed. Dynamics of the three agent societies in DACE have been investigated with respect to the dominance probability p.

A transition region of the system roughly corresponding to the interval *[0.45, 0.65]* has been detected. The emergence of a global pattern indicates that the simple system of three populations is merely a complex system of unpredictable behavior. These interesting results are used in the implementation of DACE model by setting an interval of dominance probability that matches the phase transition interval detected. We expect that the emergent complex behavior in the system societies produces the best results in this transition interval.

4.3 Parameter Setting and Search Progress

The search operators considered are simulated binary crossover for continuous search space [2], applied with a probability of 0.7 and one-position mutation applied with a probability of 0.2.

The population consists of 8*8 (64) individuals and the tournament size is ½ of the considered group of individuals.

The membership of individuals to agent societies is initially randomly generated based on uniform probability distribution.

Each individual from the population is involved in a recombination process by sending a mating invitation to several individuals according to the society recombination strategy. A potential mate accepts or declines the invitation. Mutation is applied for one randomly selected offspring generated by each mating pair.

4.4 Numerical Results and Comparisons

The DACE model has been compared to the following five evolutionary algorithms:

- Real-Coded Genetic Algorithm (RCGA) [3],
- Steady-State Real Parameter Genetic Algorithm (SPC-PNX) [1],
- Estimation of Distribution Algorithm (EDA) [10],
- Self-adaptive Differential Evolution Algorithm (SaDE) [7] and
- Evolutionary Strategy Algorithm (ESA) [6].

The error values $f(x)$-$f(x^*)$ – where x^* is the real optimum, are presented in Tables 1 and 2. Each column corresponds to a method used for comparison. The best and the average error values have been recorded after 1E+3 and 1E+4 function evaluations (FEs), after 25 runs of each algorithm for each function with dimension D=10.

The obtained standard deviations are also presented in these tables. The error values reported by the rival methods are bold in these tables if they are smaller than the error values recorded by our proposed method.

Smaller error values are reported by DACE in approximately 65% of the considered cases. For 1000 FEs DACE outperforms all the other algorithms for all considered functions. This indicates a better performance of the proposed model in the first stages of the algorithm.

Table 1. Error values achieved in 25 runs for functions $f_1 - f_4$ with D = 10 after 1E+3 FEs for DACE and five other methods

		DACE	RCGA	SPC-PNX	EDA	SaDE	ESA
	Best	7.25E+00	5.43E+01	5.52E+01	5.70E+01	3.69E+01	4.08E+01
f_1	Mean	2.30E+01	7.17E+01	7.49E+01	8.20E+01	5.44E+01	7.08E+01
	StdAvg	7.16E+00	9.36E+00	1.03E+01	9.96E+00	7.58E+00	1.63E+01
	Best	2.44E+01	6.24E+01	6.17E+01	8.03E+01	4.52E+01	5.80E+01
f_2	Mean	6.28E+01	8.99E+01	9.26E+01	1.01E+02	7.58E+01	9.37E+01
	StdAvg	1.53E+01	1.02E+01	1.51E+01	9.49E+00	1.17E+01	2.08E+01
	Best	6.05E+00	9.97E+00	8.07E+00	9.24E+00	8.94E+00	8.13E+00
f_3	Mean	1.01E+01	1.15E+01	1.11E+01	1.19E+01	1.14E+01	1.13E+01
	StdAvg	1.52E+00	6.42E-01	9.95E-01	8.18E-01	9.54E-01	1.20E+00
	Best	6.35E+03	1.83E+04	2.75E+04	3.32E+04	1.49E+04	2.34E+04
f_4	Mean	2.20E+04	4.63E+04	5.86E+04	7.93E+04	5.69E+04	5.39E+04
	StdAvg	9.94E+03	1.55E+04	1.77E+04	2.38E+04	1.85E+04	2.29E+04

Table 2. Error values achieved in 25 runs for functions $f_1 - f_4$ with D = 10 after 1E+4 FEs for DACE and five other methods

		DACE	RCGA	SPC-PNX	EDA	SaDE	ESA
	Best	2.35E-02	2.39E+00	2.29E+01	3.83E+01	3.87E+00	1.90E+01
f_1	Mean	1.21E+00	1.10E+01	3.18E+01	4.94E+01	6.69E+00	4.12E+01
	StdAvg	1.07E+00	7.17E+00	4.67E+00	5.26E+00	1.27E+00	1.60E+01
	Best	2.08E+01	1.68E+01	2.49E+01	2.69E+01	2.42E+01	1.84E+01
f_2	Mean	4.70E+01	3.41E+01	4.23E+01	4.79E+01	3.22E+01	4.14E+01
	StdAvg	2.50E+01	5.80E+00	5.72E+00	8.46E+00	5.41E+00	1.43E+01
	Best	5.96E+00	7.92E+00	2.62E+00	9.14E+00	5.78E+00	7.06E+00
f_3	Mean	8.84E+00	9.95E+00	5.26E+00	1.02E+01	8.02E+00	1.02E+01
	StdAvg	1.36E+00	7.50E-01	1.48E+00	5.97E-01	1.03E+00	1.39E+00
	Best	8.93E+02	9.82E+00	7.39E+02	1.68E+03	2.59E+03	7.89E+02
f_4	Mean	7.48E+03	7.61E+02	3.09E+03	1.02E+04	8.82E+03	5.03E+03
	StdAvg	6.08E+03	9.67E+02	1.58E+03	4.40E+03	2.80E+03	3.49E+03

4.5 Statistical Analysis

A statistical analysis is performed using the expected utility approach [4] to determine the most accurate algorithm. Let x be the percentage deviation of the solution given by the algorithm used and the best known solution on a given function:

$$x = \left| \frac{sol - opt}{opt} \right| \times 100 \cdot$$

The expected utility function can be:

$$euf = \gamma - \beta(1 - \bar{b}t)^{-\bar{c}},$$

where $\gamma = 500$, $\beta = 100$ and $t = 5.00E-17$ and

$$\bar{b} = \frac{s^2}{\bar{x}}, \bar{c} = \frac{(\bar{x})^2}{s^2}, \bar{x} = \frac{1}{4}\sum_{j=1}^{4} x_j, s^2 = \frac{1}{4}\sum_{j=1}^{4}(x_j - \bar{x})^2.$$

Table 3 presents the results of the statistical analysis test. It can be observed that DACE technique is ranked first for 1E+3 FEs.

Table 3. Statistical analysis for all considered algorithms on the average results obtained in 25 runs for functions $f_1 - f_4$ with $D=10$, after 1E+3 FEs

	DACE	RCGA	SPC-PNX	EDA	SaDE	ESA
euf	397.07	393.41	391.35	387.49	391.65	392.16
rank	1	2	5	6	4	3

The statistical results obtained for 1E+4 FEs indicate a slightly weaker performance of DACE. One reason for this behavior is the fact that, even if the global behavior favors DACE algorithm, each other algorithm obtains better results for several functions. Future work investigates mechanisms to improve the search particularly during the last stages of the algorithm.

5 Conclusions and Future Work

The distributed collaborative coevolutionary model proposed in this paper relies on a spatial distribution of individuals and agent-based interactions within a population that includes multiple societies. The coevolution and collaboration of individuals able to act like agents is a key feature of the proposed search model. Each individual (agent) in the population has the goal of optimizing its fitness and is able to communicate and select a mate for recombination. The coevolution of three different agent societies following different rules has the potential to induce a better balance between exploration and exploitation of the search space. An asynchronous search process is facilitated through a gradual propagation of genetic material into the population.

The proposed model has been implemented using a discrete regulated population topology and a regular fitness-based distribution of individuals. The performance of the obtained algorithm has been tested against several existing methods for optimizing difficult multimodal functions. Numerical results are encouraging and indicate a good performance of the proposed DACE model.

Future work explores other grid structures of the population aiming to enable even more complex interactions and emergent dynamics. Moreover, various strategies for recombination between agent societies will be investigated to allow a better exploitation in the late stages of the search process.

References

1. Ballester, P.J., Stephenson, J., Carter, J.N., Gallagher, K.: Real-parameter optimization performance study on the CEC 2005 benchmark with SPC-PNX. Congress on Evolutionary Computation, pp. 498–505 (2005)
2. Deb, K., Agrawal, R.B.: Simulated binary crossover for continuous search space. Complex Systems 9, 115–148 (1995)
3. García-Martínez, C., Lozano, M.: Hybrid Real-Coded Genetic Algorithms with Female and Male Differentiation. Congress on Evolutionary Computation, pp. 896–903 (2005)
4. Golden, B.L., Assad, A.A.: A decision-theoretic framework for comparing heuristics. European J. of Oper. Res. 18, 167–171 (1984)
5. Jennings, N.R.: On Agent-Based Software Engineering. Artificial Intelligence Journal 117(2), 277–296 (2000)
6. Posik, P.: Real Parameter Optimisation Using Mutation Step Co-evolution. Congress on Evolutionary Computation (2005)
7. Qin, A.K., Suganthan, P.N.: Self-adaptive differential evolution algorithm for numerical optimization. Congress on Evolutionary Computation, pp. 1785–1791 (2005)
8. Suganthan, P.N., Hansen, N., Liang, J.J., Deb, K., Chen, Y.-P., Auger, A., Tiwari, S.: Problem Definitions and Evaluation Criteria for the CEC 2005 Special Session on Real-Parameter Optimization, Technical Report, Nanyang Technological University, Singapore and KanGAL Report #2005005, IIT Kanpur, India (2005)
9. Wooldrige, M.: An Introduction to Multiagent Systems. Wiley & Sons, Chichester (2002)
10. Yuan, B., Gallagher, M.: Experimental results for the special session on real-parameter optimization at CEC 2005: a simple, continuous EDA. Congress on Evolutionary Computation, pp. 1792–1799 (2005)

Modeling the Nonlinear Nature of Response Time in the Vertical Fragmentation Design of Distributed Databases

Rodolfo A. Pazos R.[1], Graciela Vázquez A.[2], Joaquín Pérez O.[3], and José A. Martínez F.[1]

[1] Instituto Tecnológico de Cd. Madero, Cd. Madero, Mexico
r_pazos_r@yahoo.com.mx, jose.mtz@gmail.com
[2] ESIME, Instituto Politécnico Nacional, Mexico City, Mexico
gravazquez@hotmail.com
[3] Centro Nacional de Investigación y Desarrollo Tecnológico, Cuernavaca, Mexico
jperez@cenidet.edu.mx,

Abstract. The generalized use of the Internet has facilitated the implementation of distributed database (DDB) systems, which are becoming increasingly common-place. Unfortunately, though there exist many models for optimizing the design of DDBs (i.e., the distribution of data), they usually seek to optimize the transmission and processing costs of queries and overlook the delays incurred by their transmission and processing, which can be a major concern for Internet-based systems. In this paper a mathematical model is presented, which describes the behavior of a DDB with vertical fragmentation and permits to optimize its design taking into account the roundtrip response time (query transmission time, query processing time, and response transmission time).

Keywords: Distributed databases, vertical fragmentation, mathematical optimization.

1 Introduction

Nowadays, large organizations require software and hardware for carrying out their activities; thus, as a consequence of the growth and improvement of the communications technology (such as the Internet) and distributed database (DDB) management systems, and the reduction of computer costs, such organizations have opted for implementing DDB systems.

DDB systems managers have the responsibility of carrying out the fragmentation, allocation and monitoring of data at the network sites. In most DDB systems its design greatly affects the overall system performance, mainly because the time required for query transmission and processing, and the transmission of query responses, depends considerably on the sites where data resides, sites where queries are issued, data access frequencies, query processing capacity of database (DB) servers, and transmission speed of communication lines.

The purpose of this work is to develop a model that permits to carry out the vertical fragmentation and allocation of fragments such that the overall roundtrip response time (query transmission time, query processing time, and response transmission time) is minimized.

J.M. Corchado et al. (Eds.): DCAI 2008, ASC 50, pp. 605–612, 2009.
springerlink.com

As mentioned in Section 2, a survey of the specialized literature on vertical fragmentation for DDBs has revealed that previous works have used transmission, access and processing costs as optimization criteria. Unfortunately, these costs are directly proportional to query load. Therefore, minimizing only costs might result in concentrating loads on some DB servers or communication lines, which might almost reach or surpass servers/lines capacities, and therefore, incur unacceptable large response times.

It is important to point out that response time is not directly proportional to query load; in fact, it increases enormously as the load approaches the processing capacity of a DB server or the capacity of the line on which the load is transmitted; therefore, the importance of considering response time in a DDB optimization model.

2 Related Work

It has been recognized for many years the importance of considering response time in DDB modeling [2]. Unfortunately, a survey of the specialized literature revealed that only a few investigators have attempted to model response time, but only indirectly as shown in Table 1, which summarizes the most relevant works on vertical fragmentation of DDBs.

The second, third and fourth columns of Table 1 show that some works have addressed only the fragmentation problem, other works have only dealt with the fragment allocation problem and some others have integrally addressed both problems. The fifth column shows that all the previous works have considered transmission, access or processing costs; only the work reported in [4] for data warehouses has attempted to minimize response time, but only indirectly through the number of disk pages that must be accessed for processing a query.

Table 1. Related works on vertical fragmentation and allocation for DDBs

	Problems Addressed			**Value to Minimize**	
Works	**Fragmentation**	**Allocation**	**Integrated Fragmentation + Allocation**	**Transmission or processing costs**	**Response time**
Chakravarthy [2]	✓			✓	
Tamhankar [3]			✓	✓	
Golfarelli [4]	✓			✓	*
Pérez [5]			✓	✓	
Tolga [6]		✓		✓	
Pérez [7]			✓	✓	
Our work			✓		✓

* indirectly modeled.

In this last case, it is possible to minimize the response time by minimizing the number of disk pages accessed; unfortunately, this trick succeeds only when dealing with the fragmentation problem without including allocation, as in [4]. (Note: in this case, ignoring allocation is justified since all the fragments are stored in the same computer.)

3 Problem Definition and Mathematical Model

In this section a zero-one integer programming model is proposed for the vertical fragmentation and allocation problem for DDB design, which aims at minimizing response time taking into consideration that response time is not a linear function of query load. This problem can formally be defined as follows:

given

- a set of attributes (data-objects) $O = \{o_1, o_2, ..., o_L\}$,
- a computer communication network that consists of a set of sites $S = \{s_1, s_2, ..., s_I\}$,
- where a set of queries $Q = \{q_1, q_2, ..., q_K\}$ are executed, and
- the arrival rates f_{ki} of queries to the sites (where f_{ki} denotes the arrival rate of query k to source node i),

the problem consists of obtaining an allocation of data-objects into sites, such that the overall average response time is minimized.

The model proposed consists of an objective function and three groups of constraints. In this model the decision about allocating a data-object l to site i is represented by a binary variable x_{li}, where $x_{li} = 1$ if l is allocated to i, and $x_{li} = 0$ otherwise. At this point it is important to point out that the allocation of a subset of attributes to some site automatically implies the integration of a vertical fragment at the site; therefore, our model deals simultaneously with allocation and fragmentation.

3.1 Derivation of the Objective Function

The objective function is given by the following expression:

$$\min z = T_{TQ} + T_{PQ} + T_{TR} \tag{1}$$

where

$z =$ average response time,

$T_{TQ} =$ average transmission delay (waiting and transmission time) of queries,

$T_{PQ} =$ average processing delay (waiting and processing time) of queries,

$T_{TR} =$ average transmission delay (waiting and transmission time) of query responses.

Next, an expression for z in terms of the decision variables will be obtained; to this end, we introduce the following definitions:

$C_j =$ processing capacity of server j (expressed in queries/sec.),

$C_{ij} =$ transmission speed of the communication line from node i to node j (expressed in bits/sec.),

$1/\mu_Q =$ mean length of queries (expressed in bits/query),

$1/\mu_R =$ mean length of query responses (expressed in bits/response),

f_{ki} = arrival rate of query k to source node i (expressed in queries/sec.),
f = overall arrival rate of all the queries = $\sum_{ki} f_{ki}$,
γ_{ki} = emission rate of query k to source node i (= f_{ki}, $2f_{ki}$, $3f_{ki}$, etc.),
γ = overall emission rate of all the queries = $\sum_{ki} \gamma_{ki}$.
The average processing delay of queries is defined as follows:

$$T_{PQ} = \frac{1}{f}\sum_k \sum_i f_{ki} T_{ki} \tag{2}$$

where T_{ki} represents the average processing time of queries k issued from node i; but, given the difficulty for finding an exact expression for T_{PQ}, the following approximation will be used:

$$T^*_{PQ} = \frac{1}{\gamma}\sum_k \sum_i \gamma_{ki} T_{ki} \tag{3}$$

An easy way for calculating T^*_{PQ} is using Little's result [8], therefore

$$\gamma T^*_{PQ} = n = \sum_j n_j$$

where n represents the average number of queries that are in the DB servers (either waiting or being processed), and n_j represents the average number of queries in server j. Applying again Little's result to n_j, the following expression is obtained:

$$\gamma T^*_{PQ} = \sum_j \lambda_{*j} T_{PQj} \tag{4}$$

where λ_{*j} represents the arrival rate of queries to server j and T_{PQj} represents the average processing delay of queries at server j.

Now, the following simplifying assumption will made: query processing at the servers can be modeled as an M/M/1 queue. Consequently, the average processing delay of queries at server j can be approximated by the following expression [9]:

$$T_{PQj} = \frac{1}{\mathrm{e}_j - \lambda_{*j}} \tag{5}$$

Finally, substituting expression (5) for T_{PQj} into (4), and solving for T^*_{PQ}, the following expression is obtained:

$$T^*_{PQ} = \frac{1}{\gamma}\sum_j \frac{\lambda_{*j}}{\mathrm{e}_j - \lambda_{*j}}$$

$$= \frac{1}{\gamma}\sum_j \frac{1}{\mathrm{e}_j / \lambda_{*j} - 1} \tag{6}$$

On the other hand, the expression for λ_{*j} as a function of the arrival rates of queries to the source nodes and the location of attributes in the servers, can be calculated as follows:

$$\lambda_{*j} = \sum_k \sum_i f_{ki} y_{jk} \tag{7}$$

where y_{jk} is a dependent variable, such that $y_{jk} = 1$ if one or more attributes used by query k are stored in site j, and $y_{jk} = 0$ otherwise. (Note: y_{jk} is linked to the independent decision variables x_{li} by expression (21).)

Additionally, note that the sum of the emission rates of all the queries issued from the source nodes (γ) equals the sum of the arrival rates of all the queries that get to the servers; therefore,

$$\gamma = \sum_j \lambda_{*j} \tag{8}$$

Finally, substituting (7) and (8) into (6), the following expression for T^*_{PQ} is obtained as a function of the location of attributes in the servers:

$$T^*_{PQ} = \frac{1}{\sum_j \sum_k \sum_i f_{ki} y_{jk}} \sum_j \frac{1}{\dfrac{e_j}{\sum_k \sum_i f_{ki} y_{jk}} - 1} \tag{9}$$

Now a similar expression for the average transmission delay of queries will be developed, which is defined as follows:

$$T_{TQ} = \frac{1}{f} \sum_k \sum_i f_{ki} T_{ki} \tag{10}$$

Like the derivation for T_{PQ}, the following approximation for T_{TQ} will be used:

$$T^*_{TQ} = \frac{1}{\gamma} \sum_k \sum_i \gamma_{ki} T_{ki} \tag{11}$$

Using again Little's result, the following expression is obtained:

$$\gamma T^*_{TQ} - n - \sum_{ij} n_{ij}$$

where n represents the average number of queries on the transmission lines (either waiting or being transmitted), and n_{ij} represents the average number of queries on the line from source node i to server j. Applying again Little's result, the following expression is obtained:

$$\gamma T^*_{TQ} = \sum_{ij} \lambda_{ij} T_{TQij} \tag{12}$$

where λ_{ij} represents the arrival rate of queries to the line from i to j, and T_{TQij} represents the average transmission delay of the queries on the line from i to j.

On the other hand, considering an assumption similar to the one used for expression (5), then the average transmission delay of the queries on the line from i to j is given by

$$T_{TQij} = \frac{1}{\mu_Q C_{ij} - \lambda_{ij}} \tag{13}$$

Finally, substituting expression (13) for T_{TQij} into (12), and solving for T^*_{TQ}, the following expression is obtained:

$$T^*_{TQ} = \frac{1}{\gamma}\sum_{ij}\frac{\lambda_{ij}}{\mu_Q C_{ij} - \lambda_{ij}}$$

$$= \frac{1}{\gamma}\sum_{ij}\frac{1}{\mu_Q C_{ij} / \lambda_{ij} - 1} \tag{14}$$

On the other hand, the expression for λ_{ij} as a function of the arrival rates of the queries to the source nodes and the location of attributes in the servers, can be calculated as follows:

$$\lambda_{ij} = \sum_k f_{ki} y_{jk} \tag{15}$$

where y_{jk} is a dependent variable (defined previously).

Finally, substituting (7), (8) and (15) into (14) the following expression for T^*_{TQ} is obtained as a function of the location of attributes in the servers:

$$T^*_{TQ} = \frac{1}{\sum_j\sum_k\sum_i f_{ki} y_{jk}}\sum_{ij}\frac{1}{\frac{\mu_Q C_{ij}}{\sum_k f_{ki} y_{jk}} - 1} \tag{16}$$

Following a similar process, the following expression for T^*_{TR} can be derived:

$$T^*_{TR} = \frac{1}{\sum_j\sum_k\sum_i f_{ki} y_{jk}}\sum_{ij}\frac{1}{\frac{\mu_R C_{ij}}{\sum_k f_{ki} y_{jk}} - 1} \tag{17}$$

In the end, an expression can be obtained for the objective function in terms of the location of attributes, by substituting expressions (7), (16) and (17) into (1), which yields

$$\min z =$$

$$= \frac{1}{\sum_j\sum_k\sum_i f_{ki} y_{jk}}\left(\sum_{ij}\frac{1}{\frac{\mu_Q C_{ij}}{\sum_k f_{ki} y_{jk}} - 1} + \sum_j\frac{1}{\frac{C_j}{\sum_k\sum_i f_{ki} y_{jk}} - 1} + \sum_{ij}\frac{1}{\frac{\mu_R C_{ij}}{\sum_k f_{ki} y_{jk}} - 1}\right) \tag{18}$$

3.2 Description of Problem Constraints

This problem has three groups of constraints, which are formulated by expressions (19), (20) and (21) explained below.

$\sum_j x_{lj} = 1 \quad \forall l$	Each attribute must be stored in just one site, since in this problem replication is not considered.	(19)
$x_{li} \le \sum_k q_{kl}\varphi_{ki} \quad \forall l,i$ where $\varphi_{ki} = \begin{cases} 1, & \text{if } f_{ki} > 0 \\ 0, & \text{if } f_{ki} = 0 \end{cases}$	Each attribute l must be stored in a site (server) i that executes at least one query that involves the attribute.	(20)
$t\,y_{jk} - \sum_l q_{kl}x_{lj} \ge 0 \quad \forall j,k$ where t = number of attributes	This restriction links the values of the dependent variables y_{jk} to the values of the independent decision variables x_{lj}, by forcing the value of y_{jk} to 1 when some product $q_{kl}x_{lj}$ equals 1 (i.e., some attribute l used by query k is stored in server j) and induces y_{jk} to 0 otherwise.	(21)

4 Final Remarks and Future Work

This paper shows the possibility of modeling the nonlinear nature of query response time in the vertical fragmentation and allocation problem for DDB design. The optimal design problem was formulated as a zero-one integer programming model, which consists of an objective function (18) that aims at minimizing the overall roundtrip query response time, and three groups of constraints (19-21). This formulation is similar to the one described in [5], whose objective function was modified for taking response time into account.

One of the most important assumptions in the derivation of the objective function (specifically for expression (5)) was that query processing at the servers can be modeled as an M/M/1 queue. This assumption had to be made because obtaining an exact expression for the average processing delay of queries at servers is an extremely difficult task, as mentioned in [9], and if it could be obtained, the expression would be so complicated that it would not yield a practicable expression for optimization purposes.

Since it has been shown that the problem formulated in [5] is NP-hard, we conclude that the formulation proposed here is at least as hard as the first one. In order to get an idea of the complexity of both formulations, consider an algorithm that considers all the possible combinations of locating each attribute in each different site. The complexity of this algorithm would be $O(L^I)$ (where L stands for the number of attributes and I denotes the number of sites), which is obviously non-polynomial. Furthermore, the objective function (18) is much more complicated than the one presented in [5], which is linear.

Unfortunately, for many NP-hard problems only small instances can be solved by exact methods; therefore, in order to solve instances of real-world sizes, we are currently testing two metaheuristic algorithms on our model (the Threshold Accepting algorithm and Tabu Search), which have been successfully applied to similar problems [5]. These experiments will permit finding a good algorithm for solving our problem formulation and contrasting the solutions found for our formulation against the solutions to the formulation presented in [5], which will serve to assess the advantage of the new formulation.

Finally, the model proposed here can be improved in several ways:

1. Formulate a new objective function that adequately combines response time, and transmission, access and processing costs.
2. Include a new restriction that involves response time; thus, the objective function would involve only transmission, access and processing costs.
3. Include fragment replication in alternatives 1 and 2, including "write queries", since the model proposed here only considers "read queries".

References

1. Ozsu, M.T., Valduriez, P.: Principles of Distributed Database Systems. Prentice Hall, USA (1999)
2. Chakravarthy, S., Muthuraj, J., Varadarajan, R., et al.: An Objective Function for Vertically Partitioning Relations in Distributed Databases and its Analysis. Distributed and Parallel Databases 2(2), 183–207 (1994)
3. Tamhankar, A.M., Ram, S.: Database Fragmentation and Allocation: an Integrated Methodology and Case Study. IEEE Transactions Systems, Man and Cybernetics Part A 28, 288–305 (1998)
4. Golfarelli, M., Maio, D., Rizzi, S.: Vertical Fragmentation of Views in Relational Data Warehouses. In: Proc. Settimo Convegno Nazionale Sistemi Evoluti per Basi di Dati (SEBD 1999), Villa Olmo, Italy, pp. 23–25 (1999)
5. Pérez, J., Pazos, R., Frausto, J., et al.: Vertical Fragmentation and Allocation in Distributed Databases with Site Capacity Restrictions Using the Threshold Accepting Algorithm. In: Cairó, O., Cantú, F.J. (eds.) MICAI 2000. LNCS, vol. 1793, pp. 75–81. Springer, Heidelberg (2000)
6. Ulus, T., Uysal, M.: Heuristic Approach to Dynamic Data Allocation in Distributed Database Systems. Pakistan Journal of Information and Technology 2(3), 231–239 (2003)
7. Pérez, J., Pazos, R., Santaolaya, R., et al.: Data-Object Replication, Distribution, and Mobility in Network Environments. In: Broy, M., Zamulin, A.V. (eds.) PSI 2003. LNCS, vol. 2890, pp. 539–545. Springer, Heidelberg (2004)
8. Kleinrock, L.: Queueing Systems. Theory, vol. 1. Wiley-Interscience, USA (1975)
9. Kleinrock, L.: Communication Nets: Stochastic Message Flow and Delay. Dover Publications, USA (2007)

Discovering Good Sources for Recommender Systems

Silvana Aciar, Josep Lluis de la Rosa i Esteva, and Josefina López Herrera

Departamento Informática y Automática
University of Girona, Campus Montilivi, Edifici P4, 17071, Girona, Spain
saciar@eia.udg.edu, peplluis@eia.udg.edu, jlopez@eia.udg.edu

Abstract. Discovering user knowledge is a key issue in recommender systems and many algorithms and techniques have been used in the attempt. One of the most critical problems in recommender systems is the lack of information, referred to as Cold Start and Sparsity problems. Research works have shown how to take advantage of additional databases with information about users [1], but they do not solve the new problem that arises: which relevant database to use? This paper contributes to that solution with a novel method for selecting information sources in the belief that they will be relevant and will result in better recommendations. We describe a new approach to explore and discover relevant information sources in order to obtain reliable knowledge about users. The relation between the improvement of the recommendation results and the sources selected based on these characteristics is shown by experiments selecting source based on their relevance and trustworthiness.

Keywords: Recommender systems, Discovering User Knowledge.

1 Introduction

Recommender systems are programs that attempt to predict items that users are interested in. They use information from a user profile which is computational model containing information about tastes, preferences and user's behaviour. The success of the recommendations depends on the precision and reliability of the knowledge obtained to predict user preferences. Obtaining user knowledge is one of the most important tasks in recommender systems. If users are pleased with a first recommendation, they will probably interact with the system again [11]. One lifelong question in recommender systems is how we can guarantee that the knowledge is precise or reliable?

In this paper we present a way to "prospect" relevant information sources to guarantee that they contain relevant and reliable knowledge. Addressing this goal the main contributions presented in this work are:

1. A set of characteristics that allows knowing whether or not the source contains information necessary to discover user preferences.
2. A relevance measure based on these characteristics and a reliability measure obtained based on the result of the past use of the sources.
3. An algorithm to select sources based on their relevance and reliability has been defined to guarantee the suitability of the information from the sources.

The structure of the paper is as follows. Section 2 presents the set of characteristics defined to classify the sources. The relevance and reliability measures to select the most relevant sources from all those available are briefly explained in Section 3. The

J.M. Corchado et al. (Eds.): DCAI 2008, ASC 50, pp. 613–623, 2009.
springerlink.com

algorithm to select sources based on both measures is presented in Section 4. Section 5 presents the experiments that demonstrate the effectiveness of the proposed approach. Finally, the conclusions are discussed in Section 6.

2 Classification of Information Sources to Know the Information Contained on Them

Recommendation can be better if the recommender knows where is the suitable information to predict user's preferences to offer products. Sources that provide information that is timely, accurate and relevant are expected to be used more often than sources that provide irrelevant information. It is an intuitive idea but; *how to know which of the sources have to be selected to obtain better results?* Factors indicating if the sources can be used for recommender systems have to be study. There is much work made about the evaluation characteristic of web sources [2][5][10][16]. The characteristics as have been defined by the authors cannot be used for recommender systems. They use different sources which are those that allow obtaining information about user preferences. This information includes demographic information, product evaluations, products they have purchased in the past, information about the context and information about similar users [1] [3][15][6] and is generally contained in a database. Taking into consideration which information is necessary to supply recommenders, sources must contain the next characteristics: they must contain information about users that the system needs, demographic information, contain a lot of information, have updated information and contain the attributes needed to make the recommendations.

2.1 Defining the Characteristics to Classify Recommender Sources

We have defined measures that allow us to know if a source meets the information requirements for recommender systems.

2.1.1 Completeness

The aim of this work is to know how complete a source S with respect to the quantity of users for whom it can provide information. We define the completeness as:

Given the set U of users of a recommendation domain, the completeness of a source S is the quantity of users of U within S, known as $|C|$, divided by the quantity of users $|U|$.

$$Completeness(S) = \frac{|C|}{|U|} \tag{1}$$

2.1.2 Diversity

Dversity indexes provide information about the composition of the community (for example, the quantity of species), and also take into account the relative abundance of the various species. In this work, the diversity measure is used to represent, in a single value, the quantity of species (groups of users) in a source. Knowing information about the diversity of the source, the recommender systems can differentiate users one from the other and they are criterion to group users according to a relation or degree of similarity between them. The diversity of information sources is measured using the "index of diversity" defined by Shannon and Weber in biology [8].

$$Diversity(S) = \quad H = -\sum (p_i \log_2 p_i) \tag{2}$$

Adapted to the recommender systems each pi is calculated as follows:

$$p_i = \frac{n_i}{N} \tag{3}$$

Where n_i is the number of users included in the group i and N is the total quantity of users in source S. The users can be grouped according to gender, age, etc.

2.1.3 Frequency of Interactions

An analysis of the frequency of user's interactions stored in sources is the technique we use to determine the quantity of information about the users. Frequency is defined as the number of times that a client has made a purchase. While more purchase records for individual users found in the database, better their tastes for certain products can be known by analyzing the products they purchased in the past. Categories have been defined to obtain this measure. Each category represents a certain number of user interactions. For example, if we have a database containing information about the movies rented by the clients of a video club, each record in the database represents a rental. Each of these records contains the customer identifier, the movie rented and data about the movie. Let's suppose that the defined categories are:

Category f1: 1 - 10 interactions
Category f2: 11 - 25 interactions
Category f3: 26 - 50 interactions
Category f4: 51 - 100 interactions
Category f5: 101 - 200 interactions
Category f6: + 201 interactions

Category f1 includes the clients who have rented a movie from 1 to 10 times; category f2 includes the clients who have rented a movie from 11 to 25 times, and so on. It is clear that if the database contains more clients in category f6 there will be more information about them to obtain their preferences. This measurement is important because recommender systems choose sources from among those that can provide the most information to solve the problems of a lack of information. Mathematically, the frequency of interactions of a source S is the result of the weights w_i, given for each category f_i, multiplied by $|f_i|$ which represents the quantity of users within each category, divided by the quantity of users of S which is N.

$$Frequency(S) = \frac{\sum w_i * |f_i|}{N} \tag{4}$$

2.1.4 Timeliness

Generally the updating of the information is defined as the time that has passed since the last update of the data [10]. This way of measuring the "age" of the information is applied to resources on the Web, but cannot be applied to our problem. Our sources of information are databases containing information about users and their interactions. Application of this definition would provide the date of the client's last interaction but that does not mean that the database has been updated. The updating measure defined by Phillip Cykana [4] is more adapted to what we need to know about whether or not a source is more updated for our purpose. It measures update as the percentage of the data available within an interval of specified time (for example, day, hours or minutes). In order to apply this measure to sources containing information about users, the date of their interactions is analyzed. If more users have interacted recently, the information used to obtain their preferences and make recommendations to them will be more updated. Following the example mentioned in the measure of frequency, and using the database containing user purchase information, the more purchases made in the last months, the more updated will be the database and their preferences will be those of now and not those of 10 years ago. In order to obtain this measure time categories have been defined:

Category p1: 01/01/2001 - 31/12/2001
Category p2: 01/01/2002 - 31/12/2002
Category p3: 01/01/2003 - 31/12/2003
Category p4: 01/01/2004 - 31/12/2004
Category p5: 01/01/2005 - 31/12/2005
Category p6: 01/01/2006 - 31/12/2006

Category pl includes clients who rented movies from 01/01/2001 to 31/12/2001, category p2 includes clients who rented movies from 01/01/2002 to 31/12/2002 and so on continuously for each of the categories. Clearly if the data-base contains more clients in category p6, that information will be more updated. As the preferences and tastes of users change over time, it is important to take into account a measure of information updating when using this source.

The Timeliness of a source S is the sum of the weights w_i, given for each category p_i, multiplied by $|p_i|$, which represents the quantity of users within each category p_i, divided by the quantity of records of S, which is N.

$$Timeliness(S) = \frac{\sum w_i * |p_i|}{N} \tag{5}$$

2.1.5 Number of Relevant Attributes

This measurement is used to know of the existence of relevant information to make the recommendations. Continuing with the example of the video club, if a new movie is available for rent, it can be recommended to clients who like movies of the same genre (action, drama, etc). In order to recommend the new movie it is necessary to know information about the clients, the movie and the genre.

Given the set D of relevant attributes to make the recommendations, the quantity of relevant attributes of a source S is the quantity of attributes of D within S, $|B|$, divided by the quantity of attributes $|D|$.

$$Relevant\ Attributes(S) = \frac{|B|}{|D|} \tag{6}$$

3 Measuring the Relevance and Reliability of a Source

Finding a source that is more complete, more diverse, that has more interactions, is the most timeliness and has all the attributes necessary for the recommendations is unlikely. Some sources will have better characteristics than others and there will be still others with poorer characteristics than them. For that reason, when choosing a source, each of the characteristics must be considered and have a weight assigned to it according to how important it is for making the recommendations. For example, if the source required must be timeliness but its diversity does not matter as much, the "Timeliness" characteristic will have greater weight than the "Diversity" characteristic when the source is selected. We have define the Relevance (R) of a source S as the sum of the values w_j of each of the characteristics j multiplied by the weight c_j assigned to each of these characteristics divided by the quantity of characteristics $|N|$.

$$R(S) = \frac{w_j * c_j}{|N|} \tag{7}$$

Another parameter use to select a sources is the reliability that is based on the trust (T) of the source and is defined as the probability with which sources are evaluated to use their information. This trust value is obtained from observations of the past behaviour of the sources. Trust mechanisms have been applied in various fields such as e-commerce [12], recommender systems [13][9] and social networks [18][19]. In our work, trust is used to evaluate the reliability of the source (S) based on the record of successful or unsuccessful recommendations made with information from that particular source, and there is a trust value for each one of the sources. The information required to compute the degree of success of the recommendations is saved. This information is then used to evaluate recommendations made with information from a sourcc as "successful" or "not successful", indicating as the Result = 1 and Result = 0, respectively. The success of a recommendation is evaluated using one of the measures of evaluation of the recommendations [7].

With the information about the successful recommendations, the measure of trust defined by Jigar Patel [14] is applied. They define the value of trust in the interval between [0,1], 0 meaning an unreliable source and 1 a reliable source. The trust of a source S is computed as the expected value of a variable B_s given the parameters α and β is the probability that S has relevant information. This value is obtained using equation 8.

$$T(S) = E[B_s / \alpha\beta] \tag{8}$$

E is computed as follows:

$$E[B_s / \alpha\beta] = \frac{\alpha}{\alpha + \beta} \tag{9}$$

The parameters α and β are calculated as:

$$\alpha = m_s^{1:t} + 1 \qquad \beta = n_s^{1:t} + 1 \tag{10}$$

Where $n_s^{1:t}$ is the number of successful recommendations using source S and $n_s^{1:t}$ is the number of unsuccessful recommendations and t is the time of the interaction.

4 Selecting the Most Suitable Source

The most suitable and reliable sources are chosen to make the recommendations. A selection algorithm has been defined to make the choice automatically. The algorithm is composed by 3 elements: 1)A set (S) of candidate sources. 2)A selection function Selection(R(s) T(s)) to obtain the most relevant and reliable sources. This function uses the values of relevance R(s) and trust T(s) of the sources as parameters. 3)A solution set (F) containing the sources selected

$$(F \subset S).$$

With every step the algorithm chooses a source of S, let us call it s. Next it checks if the s ⊂ F can lead to a solution; if it cannot, it eliminates s from the set S, includes the source in F and goes back to choose another. If the sources run out, it has finished; if not, it continues.

Algorithm to select relevant and trustworthy source

```
Algorithm (S: Set of candidates sources)
F := Ø ;
while (S <> Ø) do
        if Selection(R(s),T(s)) > threshold then
                F := F∪ s;
                Eliminate (S, s)
                end if
end while
return F;
```

The selection function has as parameters the relevance R(s) and trust T(s). This function returns a value between 0 and 1, and is obtained through equation:

$$selection\ (R(s), T(s)) = R(s) * T(s) \tag{11}$$

5 Experimental Results

In this section, we will describe the recommendation domains choose to carry out our experiments. Suppose a scenario composed by four recommendation domains such as:

Table 1. Information of data set used in the experiment

Data set:	
Domains: Book, CD, DVD, Magazine	
Source S1: Information from books domain	**Source S3:** Information from DVDs domain
Rating = 732	Ratings = 225
Users = 124	Users = 45
Books = 699	DVDs = 212
Source S2: Information from CDs domain	**Source S4:** Information from Magazines domain
Rating = 188	Ratings = 72
Users = 40	Users = 35
CDs = 179	Magazines = 42

Book, Compact Disk (CD), Magazines and DVD. Each domain contains information about users, items and ratings. We have collected data from Amazon.com to obtain such information to build a data set for each domain. A popular feature of Amazon is the ability for users to submit reviews to the web page of each product. As part of their review, users must rate the product on a rating scale from one to five stars. Consumer's reviews about a product have been used to obtain information about users, the user's knowledge and experience with the product and a rating as valuation about the product. We have retrieved reviews about CDs, DVDs, Magazines and Books composing four information sources. The information collected is resumed in Table 1:

In order to evaluate the results of the recommendations, we divided the data set into a Training set and a Test set. 80% of the data were used to obtain the preferences of the users and the remaining 20% for Testing.

5.1 Experiment

The first step in selecting sources of information is obtaining the characteristics of each of them. Table 2 shows the values of each one of the characteristics obtained from applying the equations defined in Section 2.1. In this table the relevance of each one of the sources is also shown.

Table 2. Characteristic's values of each information source

	F1	F2	F3	F4
Completeness	1,00	0,64	0,68	0,24
Diversity	0,50	0,30	0,31	0,22
Frequency	0,63	0,37	0,27	0,32
Timeliness	0,79	0,33	0,26	0,24
Relevant Attrubutes	1,00	0,90	0,40	0,70
R(s)	**0,39**	**0,25**	**0,19**	**0,17**

Completeness was obtained taking into account the quantity of user of source F1 present in other sources. With the purpose of measuring diversity, the users were grouped according to the area where they lived in (this information exists on Amazon.com). To calculate frequency and timeliness the next categories were defined:

Frequency Category:
Category f1: 1 - 50 interactions
Category f2: 51 - 100 interactions
Category f3: 101 - 150 interactions
Category f4: + 151 interactions

Timeliness Category:
Category p1: 01/01/2000 - 31/12/2001
Category p2: 01/01/2002 - 31/12/2003
Category p3: 01/01/2004 - 31/12/2006

Fifty recommendations were made. Each one was made with information from sources selected according to their relevance value R and trust value T. The R values are shown in Table 2. These values indicate that F1 and F2 are the most relevant. Next the result of three of the 50 recommendations made with information from the selected sources is shown.

Three interactions performed to make recommendations			
	Iteration: 1		
	F2	F3	F4
Relevance	0.50	0.38	0.34
Trust 0.50	0.50	0.50	
Selected sources: F2, F3			
Precision of the recommendations: 0.58			
	Iteration: 2		
	F2	F3	F4
Relevance	0.50	0.38	0.34
Trust 0.66	0.66	0.50	
Selected sources: F2, F3			
Precision of the recommendations: 0.58			
	Iteration: 3		
	F2	F3	F4
Relevance	0.50	0.38	0.34
Trust 0.75	0.75	0.50	
Selected sources: F2, F3, F4			
Precision of the recommendations: 0.52			

The sources selected in the first recommendation were F2 and F3. The trust of each of them is 0.5. As this was the first time these sources had been used, there was no information about whether they are reliable or not. The recommendations were made with information from F1 plus information from F2 and F3.

$$Recommendations(F1 \leftarrow (F2 + F3))$$

Their precision was evaluated using the next equation [17] and the value obtained was 0.58.

$$\Pr ecision(\mathrm{Re}\, commendati\, on) = \frac{\Pr}{P} \tag{12}$$

Where Pr is the quantity of the successful recommendations. In our problem this parameter is calculated as the quantity of recommended products purchased by users. P is the total quantity of the recommendations made; in this work it is obtained as the quantity of recommended products. A value > 0,50 were considered successful in this study. With this result the trust value of F2 and F3 will be higher and therefore selected in the next recommendation. The results of the 50 recommendations made with information from the selected sources based on their relevance and trust are presented in Figure 1 (a). As can be seen, the recommendations made in all cases except the first two resulted in a precision value of 0.52. To evaluate the effectiveness of the results, recommendations were made with the same sources of information selected according to the following criteria:

1. Recommendations with the sources selected based on R obtained from the characteristics.
2. Recommendations with sources selected based on the measure of trust.
3. Recommendations only with source F1.
4. Recommendations with all the sources of information, F1, F2, F3 and F4.
5. Recommendations made with the optimal combination of sources.

The precision of recommendations made with sources selected based on their relevance value (R) are shown in Figure 1(b). Many of the recommendations made resulted in a precision value < 0, 5. Figure 1 (c) shows the precision of recommendations made with sources selected based on their trust value (T). The precision values indicate that more precise recommendations were made with sources selected according to T than with recommendations based on R.

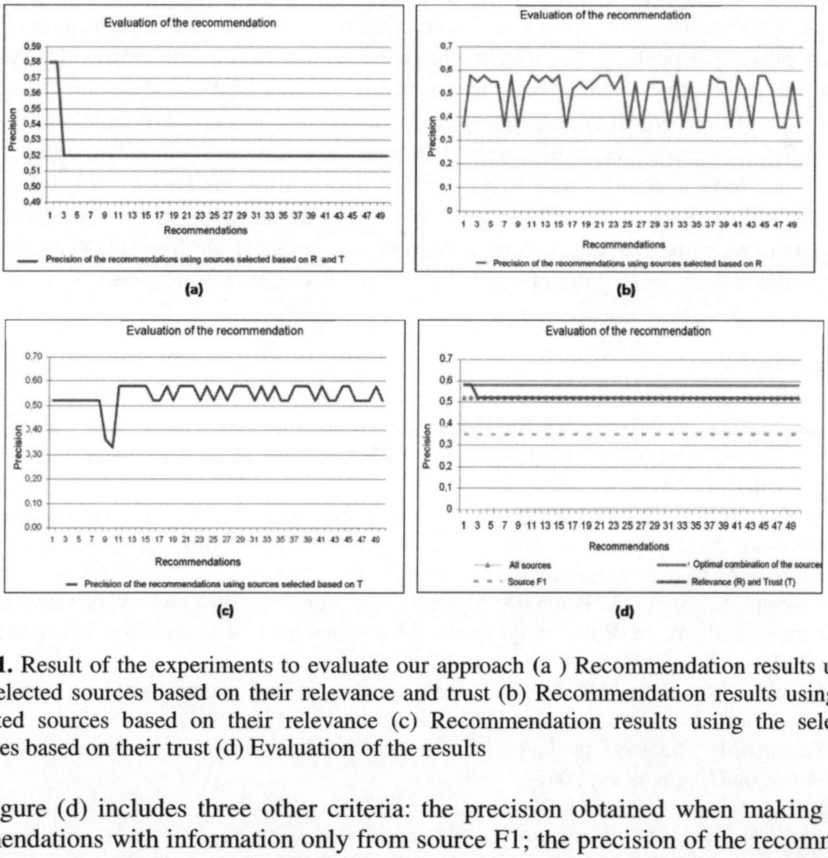

Fig. 1. Result of the experiments to evaluate our approach (a) Recommendation results using the selected sources based on their relevance and trust (b) Recommendation results using the selected sources based on their relevance (c) Recommendation results using the selected sources based on their trust (d) Evaluation of the results

Figure (d) includes three other criteria: the precision obtained when making recommendations with information only from source F1; the precision of the recommendations made with information from all the sources of information; and the precision achieved with the optimal combination of sources. The optimal combination of sources in each of the iterations was found making recommendations, adding information from the sources and taking the combination with the highest precision. As can be observed in the graph, recommendations made only with information from source F1 have a lower precision value than those made with the addition of information from other sources. However, adding information from all the sources is not optimal; the ideal would be to found the optimal combination of sources to make recommendations with better precision. In environments where the number of sources available is excessive, performing an exhaustive search for the optimal combination adds more complexity to the system.

6 Conclusions

The large amount of information available nowadays makes the process of detecting user preferences and selecting recommended products more and more difficult. Recommender techniques have been used to make this task easier, but the use of these techniques does not guarantee that the knowledge discovered to predict user preferences is reliable and that it results in better recommendations. Our approach proves to be such a technology that allows us discovering relevant and reliable user knowledge. We propose "prospecting" the information sources available to recommender systems and finding the most suitable ones to guarantee the reliability of the user knowledge acquired and used in recommender systems. Providing the retrieve suitable and reliable information sources could make the recommendations better. To prospect the sources we have defined a set of characteristics that allow us know if the source has relevant information or not. Observing the results of the experiments carried out in this paper, we note that the recommendations are better if we "prospect" to find the most suitable sources to guarantee the discovery of reliable knowledge.

References

1. Adomavicius, G.: Toward the Next Generation of Recommender Systems: A Survey of the State-of-the-Art and Possible Extensions. IEEE Transactions on Knowledge and Data Engineering 17 (2005)
2. Boklaschuk, K. and Caisse, K. Evaluation of educational web sites. resources (2006) (accessed on december 11), `http://www.usask.ca/education/coursework/802papers/bokcaisse/bokcaisse.htm`
3. Carenini, G., Smith, J., Poole, D.: Towards more conversational and collaborative recommender systems. In: Proceedings of the 7th International Conference on Intelligent User Interfaces. IUI (2003)
4. Cykana, P.A., Stern, M.: Dod guidelines on data quality management. In: Wang, R.Y. (ed.) Conference on Information Quality (IQ), pp. 154–171 (1996)
5. Edwards, J.: The good, the bad and the useless: Evaluating internet resources (1998) (Accessed on December 11, 2007), `http://www.ariadne.ac.uk/issue16/digital`
6. Endo, H., Noto, M.: A word-of-mouth information recommender system considering information reliability and user preferences. In: IEEE International Conference on Systems, Man and Cybernetics, vol. 3, pp. 2990–2995 (2003)
7. Herlocker, J.L., Konstan, J.A., Terveen, L.G., Riedl, J.T.: Evaluating collaborative filtering recommender systems. ACM Trans. Information Systems 22(1), 51–53 (2004)
8. Hilderman, R., Hamilton, H.: Principles for mining summaries using objective measures of interestingness. In: Proceedings of the 12th IEEE International Conference on Tools with Artificial Intelligence (ICTAI 2000), Vancouver, BC, pp. 72–81 (2000)
9. Massa, P., Avesani, P.: Trust-aware collaborative filtering for recommender systems. In: International Conference on Cooperative Information Systems (CoopIS) (2004)
10. Naumann, F., Freytag, J., Leser, U.: Completeness of integrated information sources. Information Systems 29, 583–615 (2004)

11. Nguyen, Q., Ricci, F.: User preferences initialization and integration in critiquebased mobile recommender systems. In: Proceedings of Workshop on Artificial Intelligence in Mobile Systems 2004. In conjunction with UbiComp 2004, Nottingham, UK (2004)
12. Noriega, P., Sierra, C., Rodriguez, J.A.: The fishmarket project. refleections on agent-mediated institutions for trustworthy e-commerce. In: Workshop on Agent Mediated Electronic Commerce (AMEC 1998), Seoul (1998)
13. O'Donovan, J., Smyth, B.: Trust in recommender systems. In: Proceedings of the 10th international conference on Intelligent user interfaces, San Diego, California, pp. 167–174 (2005)
14. Patel, J., Teacy, W.T.L., Jennings, N.R., Luck, M.: A probabilistic trust model for handling inaccurate reputation sources. In: Proceedings of Third International Conference on Trust Management, Rocquencourt, France (1998)
15. Pazzani, M.: A framework for collaborative, content-based and demographic filtering. Artificial Intelligence Review, 393–408 (1999)
16. Rieh, S.: Judgment of information quality and cognitive authority in the web. Journal of the American Society for Information Science and Technology 53(2), 145–161 (2002)
17. Salton, G., Buckley, V.: Term-weighting approaches in automatic text retrieval. Information Processing and Management 24(5), 513–523 (1988)
18. Yu, B., Singh, M.P.: Towards a probabilistic model of distributed reputation management. In: 4th Workshop on Deception, Fraud and Trust In Agent Societies, Montreal (2002)
19. Yu, B., Singh, P.: Searching social networks. In: Proceedings of Second International Joint Conference on Autonomous Agents and Multi-Agent Systems, pp. 65–72 (2003)

Quality of Information in the Context of Ambient Assisted Living

Luís Lima[1], Ricardo Costa[1], Paulo Novais[2], Cesar Analide[2], José Bulas Cruz[3], and José Neves[2]

[1] College of Management and Technology - Polytechnic of Porto, Rua do Curral – Casa do Curral, 4610-156 Felgueiras, Portugal
[2] Departamento de Informática/CCTC, Universidade do Minho, Campus de Gualtar, 4710-553 Braga, Portugal
[3] University of Trás-os-Montes e Alto Douro ,Vila Real, Portugal
lcl@estgf.ipp.pt, rfc@estgf.ipp.pt, pjon@di.uminho.pt, analide@di.uminho.pt, jneves@di.uminho.pt, jcruz@utad.pt

Abstract. With the use of new computational technologies and novel methodologies for problem solving, recurring to the use of Group Decision Support Systems, normally the problem of incomplete information is marginalized as if we were living in an ideal world. Common sense tells us that in the precise time a decision is make it is impossible to know all the information regarding to it, however decisions must be made. What we propose is, in the ambit of the VirtualECare project, is a possible solution to decision making, through the use of Group Decision Support Systems, aware of incomplete information but, even so, able to make decisions based in the quality of the information and its source.

Keywords: Incomplete information, knowledge representation, group decision support system, idea generation and argumentation.

1 Introduction

Imperfect information is ubiquitous; we take most of our decisions, if not all, of our day to day life based on incomplete, not precise and even uncertain information. Most information systems just ignore this characteristic of the information about the real world and build upon models were some idealisation expunges the inherent uncertainty [1]. The consequence is that one ends up with an elegant model which never gives correct answers, because it is not able to model exactly what is going on. Instead, one should deal with the uncertainty in the model itself, even at the cost of less simplicity. To implement useful information systems, in particular knowledge based ones, it is necessary to represent and reason with imperfect information.

Examples of such systems are Group Decision Support Systems (GDSS) based on agent perception that we try to associate with the healthcare practice and respective information systems (e-Health systems), in which lack of verification of the quality of information is a key omission [2].

The focus of this work is on a new class of systems from which VirtualECare [3], briefly described below, is an example. It represents an effort to find a new and cost effective way for health care delivery in the intersection of telemedicine, virtual

J.M. Corchado et al. (Eds.): DCAI 2008, ASC 50, pp. 624–633, 2009.
springerlink.com

healthcare teams and electronic medical records. One of the components of VirtualECare is a knowledge-based GDSS. As any system dealing with information and knowledge, it must encompass uncertainty. In this paper we define a method to evaluate the quality of knowledge involved in a GDSS, using VirtualECare as example, and present the foundations of a theory that permits to represent and reason with uncertain knowledge.

2 The VirtualECare Project

VirtualECare is an intelligent multi-agent system that will be able to monitor, interact and serve its customers, being those elderly people and/or their relatives. This system will be interconnected, not only to healthcare institutions, but also with leisure centers, training facilities, shops and relatives, just to name a few. The VirtualECare Architecture is a distributed one with different components interconnected through a network (e.g. LAN, MAN, WAN), each one with a different role (Fig. 1). A top-level description of the roles of the architecture components is presented below:

Supported User – Elderly with special health care needs, which are in constant supervision, thus allowing collecting of vital data information sent to the ***CallCareCenter*** and forwarded to the ***Group Decision Supported System***;

Home – ***SupportedUser*** natural ambient, under constant supervision, being the collected information sent to the ***Group Decision Supported System*** through the ***CallCareCenter*** in case of clinical one, or sent to the ***CallServiceCenter*** for all the remaining, allowing a better, more comfortable (Intelligent) Ambient;

Group Decision – There can be more than one, being responsible for all the decisions taken in the VirtualECare. Our work is centered in this key component;

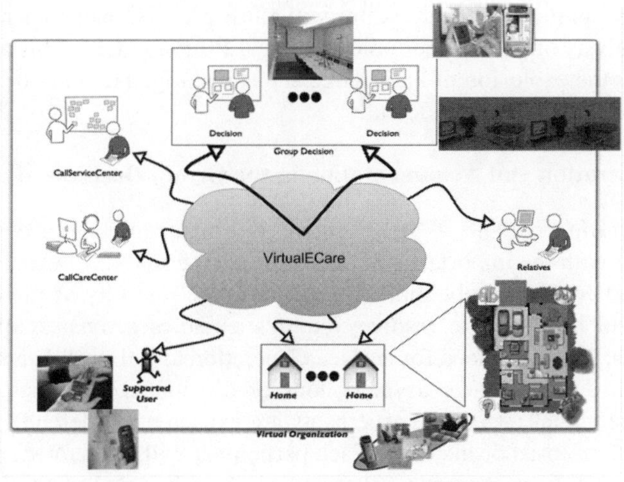

Fig. 1. The System Architecture, VirtualECare

Call Service Center – Entity with all the necessary computational and qualified personal resources, capable of receiving, analyze the diverse data, and take the necessary steps according to it;

Call Care Center – Entity with all the necessary computational and qualified personal resources (i.e., healthcare professionals and auxiliary), capable of receiving and analyze the diverse data and take the necessary steps according to it:

Relatives – ***SupportedUser*** relatives with an active role in the supervising task, giving complementary information and intervene, in some more specific crises (e.g. loneliness).

In order to the ***Group Decision Support System*** be able to correctly make their decisions, there is also the need to have a digital profile of the ***SupportedUser*** allowing a better understand of his/her special needs. In this profile we can have several different kinds of relevant information, from the patient Electronic Clinic Process to their own personal preferences.

3 Group Decision Support Systems

In the last years, we have assisted to a growing interest in combining the advances in information society - computing, telecommunications and presentation – in order to create Group Decision Support Systems (GDSS).

Decision Support Systems (DSS) are interactive computer-based systems aimed to help decision makers use communication technologies, information (structured or unstructured), knowledge and/or models to solve problems and make decisions.

DSS and particularly Group Decision Support Systems (GDSS) will benefit from progress in more basic research on behavioral topics in the areas of organizational decision making, behavioral decision theory and organizational behavior [4].

Our objective is to apply the above presented GDSS to a new sector. We believe the use of GDSS in the Healthcare sector will allow professionals to achieve better results in the analysis of one's Electronical Clinical Profile (ECP). This achievement is vital, regarding the explosion of knowledge and skills, together with the growing need to use limited resources more efficiently.

3.1 Idea Generation and Argumentation in the Group Decision Module

The ***Group Decision*** module, as stated above, is a major module of our system. This fact, associated with the importance of decision-making in today business activity and with the needed celerity in obtaining a decision in the majority of the cases that this key module will be defied to resolve, requires a real effectiveness of the decision making process. Thus, the need for an Idea Generation tool that will support the meetings, being those face-to-face, asynchronous or distributed, becomes crucial. After establishing individual ideas the participants are expected to "defend" those ideas in order to reach consensus or majority. Each participant will, therefore, and in a natural way, argue for the most interesting alternatives or against the worst alternatives, according to his/her preferences and/or skills [5].

3.2 Meeting Phases

In this work we will call *meeting* to all the phases necessary to the completion of a specific task, i.e., a meeting happens as an effect of the interaction between two or more individuals [6]. A meeting can be realized in one of the four scenarios: i) same time / same place; ii) same time / different places; iii) different times / same place; iv) different times / different places. Each one of these scenarios will require from the GDSS a different kind of action.

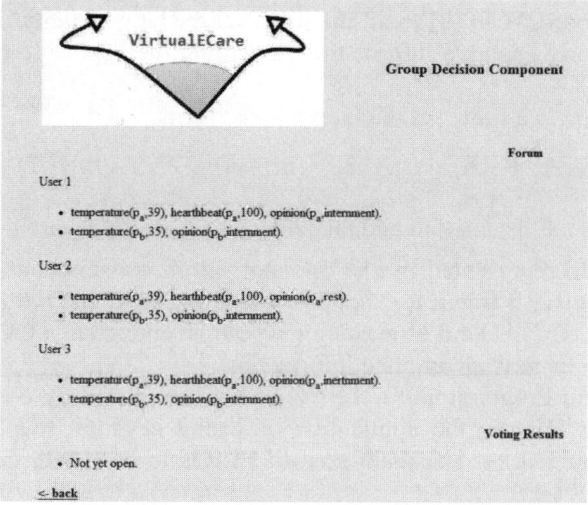

Fig. 2. Forum (Argumentation and Voting)

Besides the group members involved in the collaborative work process, it is very common to see a third element taking part in the course of action: the facilitator. The meeting facilitator is a person welcomed in the group, nonaligned, which arbitrate all the meeting phases [7]. The facilitator prepares the meeting, namely the group formation. The choice of the participants is a critical factor of success. In our model of the VirtualECare system, the GDSS assists the facilitator in this task, by providing quality metrics on the profile of each possible participant. In the In-Meeting phase, the participants will be working in order to accomplish the meeting goals and take de best decisions. In this, the participants use a knowledge database and exchange information. Again, the system must provide a measure of the quality of this knowledge and information. In the Post-Meeting phase it is important to evaluate the results achieved by the group, as well as how much each group member is acquit with the achieved results (satisfied/unsatisfied).

The VirtualECare GDSS is composed of several modules being the most important for this work the Argumentation and the Voting which are used to carry out the discussion, through systematic registering and voting of the statements supporting each participant position, reaching, in the end, the final decision (Fig. 2).

4 Knowledge Representation

A suitable representation of incomplete information is needed, one that supports non-monotonic reasoning. In a classical logical theory, the proof of a question results in a true or false value, or is made in terms of representing something about one could not be conclusive. In opposition, in a logic program, the answer to a question is only of two types: true or false. This is a consequence of the limitations of the knowledge representation in a logic program, because it is not allowed explicit representation of negative information. Additionally, the operational semantics applies the Closed-World Assumption (CWA) [8] to all the predicates. The generality of logic programs represents implicitly negative information, assuming the application of reasoning according to the CWA.

A logic program is a finite set of clauses in the form:

$$A_0 \leftarrow A_1 \wedge \ldots \wedge A_m \wedge not\ A_{m+1} \wedge \ldots \wedge not\ A_n \tag{1}$$

such as $\forall i \in N_0$, A is an atom and the terms A_i and $not\ A_i$ are literals.

Weak negation, represented by the operator *not* in conventional LP, is the ***negation-by-failure***: ***not A*** is true if it is not possible to prove ***A***, and ***not A*** is false when is possible to prove ***A***. This kind of reasoning would be enough in a CWA system, but is insufficient when there is incomplete information.

Extended Logic Programming (ELP) was first introduced by Neves [9] and Gelfond & Lifschitz [10] by the application of strong negation (explicit negation) to Logic Programming (LP). The main goal of ELP is to deal with the problem of incomplete information.

Extended logic programming introduces another kind of negation: strong negation, represented by the classical negation sign $\neg$. In most situations, it is useful to represent $\neg A$ as a literal, if it is possible to prove $\neg A$. In EPL, the expressions A and *not* A, being A a literal, are extended literals, while A or $\neg A$ are simple literals. Intuitively, *not p* is true whenever there is no reason to believe p, whereas $\neg p$ requires a proof of the negated literal.

An extended logic program is a finite collection of rules r of the form:

$$q \leftarrow p_1 \wedge \ldots \wedge p_m \wedge not\ p_{m+1} \wedge \ldots \wedge not\ p_{m+n} \tag{2}$$

where q and every p_i are literals, i.e. formulas like a or $\neg a$, being a an atom, for $m, n \in N_0$.

The objective is to provide expressive power for representing explicitly negative information, as well as directly describe the CWA for some predicates, also known as *predicate circumscription* [1]. Three types of conclusions for a question are then possible: *true*, *false* or, when there is no information to infer one or the other, the answer will be *unknown*. The representation of null values will be scoped by the ELP. In this work, we will consider two types of null values: the first will allow the representation of unknown values, not necessarily from a given set of values, and the second will represent unknown values from a given set of possible values.

We will show now some examples of how null values can be used to represent unknown information. In the following, consider the extensions of the predicates that represent some of the properties of the participants, as a measure of their "quality" for the decision process:

```
area_of_expertise: Entities x StrValue
role: Entities x StrValue
credible: Entities x Value
reputed: Entities x Value
```

The first argument denotes the participant and the second represents the value of the property (e.g. `credible(luis, 100)` means that the credibility of the participant `luis` has the value `100`).

```
credible(luis,100)
¬credible(E,V)← not credible(E,V)
```

Program 1. Extension of the predicate that describes the credibility of a participant

In Program 1, the symbol ¬ represents the strong negation, denoting what should be interpreted as false, and the term *not* designates negation-by-failure.

Let us now admit that the credibility of another possible participant *ricardo* has not, yet, been established. This will be denoted by a null value, of the type unknown, and represents the situation in Program 2: the participant is credible but it is not possible to be certain (affirmative) about its value. In the second clause of Program 2, the symbol ⊥ represents a null value of an undefined type. It is a representation that assumes any value as a viable solution, but without being given a clue to conclude about which value one is speaking about. It is not possible to compute, from the positive information, the value of the credibility of the participant *ricardo*. The fourth clause of Program 2 (the closure of predicate credibility) discards the possibility of being assumed as false any question on the specific value of credibility for participant *ricardo*.

```
credible(luis,100)
credible(ricardo,⊥)
¬credible(E,V)← not credible(E,V),
             not exception(credible(E,V))
exception(credible(E,V))← credible(E,⊥)
```

Program 2. Credibility about participant *ricardo*, with an unknown value

Let's now consider the case (Program 3) in which the value of the credibility of a participant is foreseen to be 60, with a margin of mistake of 15. It is not possible to be positive, concerning the credibility value. However, it is false that the participant has a credibility value of 80 or 100. This example suggests that the lack of knowledge may only be associated to a enumerated set of possible known values. As a different case, let's consider the credibility of the participant *paulo*, that is unknown, but one knows that it is specifically 30 or 50.

```
credible(luis,100)
credible(ricardo,⊥)
¬credible(E,V)←
        not credible(E,V), not exception(credible(E,V))
exception(credible(E,V))← credible(E,⊥)
exception(credible(carlos,V))← V ≥ 45 ∧ V ≤ 75
exception(credible(paulo,30))
exception(credible(paulo,50))
```

Program 3. Representation of the credibility of the participants *carlos* and *paulo*

Using ELP, a procedure given in terms of the extension of a predicate called demo is presented here. This predicate allows one to reason about the body of knowledge presented in a particular domain, set on the formalism previously referred to. Given a question, it returns a solution based on a set of assumptions.

This meta predicate is defined as: *Demo: Question x Answer*, Where Question indicates a theorem to be proved and Answer denotes a truth value (see Program 4): true (T), false (F) or unknown (U).

```
demo(Q,T)← Q
demo(Q,F)← ¬Q
demo(Q,U)← not Q ∧ not ¬Q
```

Program 4. Extension of meta-predicate `demo`

5 Quality of Knowledge

It is reasonable to argue that, in any decision making process, the decision is made without having all the information pertaining to the problem. How does a decision maker is confident about the reliability of the information at hand? In group decisions the situation is more complex - each participant must be confident on: The reliability of the computer support system; The other decision makers; The information rolling in and out of the system and the information exchanged between participants. The Group Decision of the VirtualECare system above operates in one such environment.We leave the first issue to others and concentrate in the last two, proposing a model for computing the quality of knowledge.

Let i ($i \in 1,\ldots, m$) represent the predicates whose extensions make an extended logic program that models the universe of discourse and j ($j \in 1,\ldots, n$) the attributes or those predicates. Let $x_j \in [min_j, max_j]$ be a value for attribute j. To each predicate is also associated a scoring function $V_{ij}[min_j, max_j] \rightarrow 0 \ldots 1$, that gives the score predicate i assigns to a value of attribute j in the range of its acceptable values, i.e., its domain (for simplicity, scores are kept in the interval $[0 \ldots 1]$), here given in the form: *all(attribute_exception_list, sub_expression, invariants).*

This denotes that *sub_expression* should hold for each combination of the exceptions of the extensions of the predicates that represent the attributes in the *attribute_exception_list* and the *invariants*. The invariants are integrity constraints in the form:

$$\leftarrow p_1 \wedge \ldots \wedge p_m \wedge not\ p_{m+1} \wedge \ldots \wedge not\ p_{m+n} \tag{3}$$

where all p_i are literals, i.e. formulas like a or $\neg a$, being a an atom, for $m, n \in N_0$.

This is further translated by introducing three new predicates. The first predicate creates a list of all possible exception combinations (pairs, triples, ..., n-tuples) as a list of sets determined by the domain size (and the invariants). The second predicate recurses through this list and makes a call to the third predicate for each exception combination. The third predicate denotes sub_expression, giving for each predicate, as a result, the respective score function. The Quality of Knowledge (QK) with respect to a generic predicate *P* is therefore given by $QK_P = 1/Card$, where *Card* denotes the cardinality of the exception set for *P*, if the exception set is not disjoint. If the exception set is disjoint, the quality of information is given by:

$$QK_P = \frac{1}{C_1^{Card} + \cdots + C_{Card}^{Card}} \tag{4}$$

where C_{Card}^{Card} is a card-combination subset, with *Card* elements.

The next element of the model to be considered is the relative importance that a predicate assigns to each of its attributes under observation: w_{ij} stands for the relevance of attribute *j* for predicate *i* (it is also assumed that the weights of all predicates are normalized, i.e.:

$$\forall i \sum_{j=1}^{n} w_{ij} = 1 \tag{5}$$

It is now possible to define a predicate's scoring function, i.e., for a value $x = (x_1, ..., n)$ in the multi dimensional space defined by the attributes domains, which is given in the form:

$$V_i(x) = \sum_{j=1}^{n} w_{ij} * V_{ij}(x_j) \tag{6}$$

It is now possible to measure the QK by posting the $V_i(x)$ values into a multi-dimensional space and projecting it onto a two dimensional one. Using this procedure, it is defined a circle, as the one given in Fig. 3.

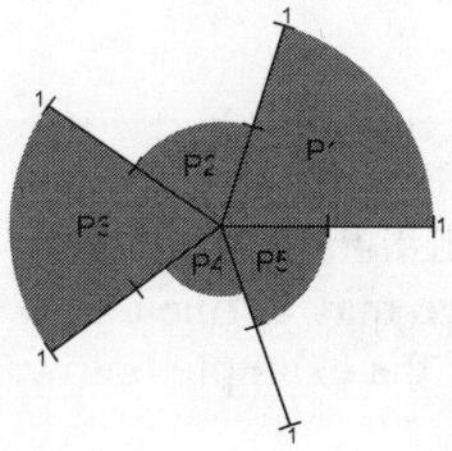

Fig. 3. A measure of the quality of knowledge for a logic program or theory P

Here, the dashed n-slices of the circle (in this example built on the extensions of five predicates, named as p_1 ... p_5) denote de QK that is associated with each of the predicate extensions that make the logic program. It is now possible to return to our case above and evaluate the global credibility of the system. Let us assume the logic Program 5.

As an example we represent the QK associated with participants *luis*, depicted in Fig. 4. In order to find the relationships among the extensions of these predicates, we evaluate the relevance of the QK, given in the form $V_{credible}$(luis) = 1; $V_{reputed}$(luis) = 0.785; V_{role}(luis) = 0. It is now possible to measure the QK associated to a logic program referred to above: the shaded n-slices (here n is equal to three) of the circle denote the QK for predicates *credible*, *reputed* and *role*.

```
¬credible(E,V)←
          not credible(E,V), not exception(credible(E,V))
exception(credible(E,V))← credible(E,⊥)
credible(luis,100)
credible(ricardo,⊥)
exception(credible(carlos,V))← V ≥ 45 ∧ V ≤ 75
exception(credible(paulo,30))
exception(credible(paulo,50))
role(luis,⊥)
role(ricardo,doctor)
exception(role(carlos,doctor))
exception(reputed(luis,80))
exception(reputed(luis,50))
exception(reputed(ricardo,40))
exception(reputed(ricardo,60))
reputed(carlos,100)
```

Program 5. Example of universe of discourse

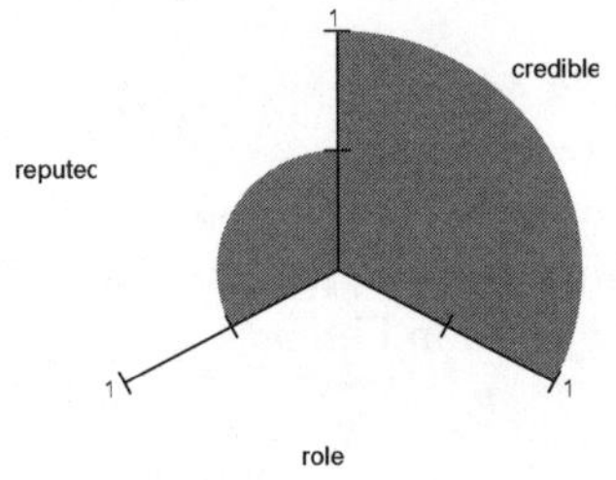

Fig. 4. A measure of quality of knowledge about participant luis

However, in order to accomplish the main goal of this work, we need to further extend the purpose of Fig. 4, i.e., we may define a new predicate, *reliance*; whose extension may be given in the form of the example below:

```
¬reliance(X,Y)←
            not reliance(X,Y), not exception(reliance(X,Y))
reliance(luis,((credible,1),(reputed,0.785)(role,0)))
reliance(ricardo,((credible,0),(reputed,0.785),(role,1)))
```

Program 6. Measuring the global quality

Besides being able to evaluate the quality of individual actors and individual pieces of information that flows in a group decision system, we aim to have an overall mechanism that allows one to measure the global quality of the system itself. The same mechanism used to evaluate individual parts of the system is consistently used to evaluate all the system, through an extension process.

6 Conclusion

Our agenda is to apply the above Knowledge Representation with the respective Quality of its Information to the VirtualEcare GDSS module. Thus, the suggestions/decisions presented by this module will consider the existence of incomplete information, and, even so, will present a possible way to try and, if possible, resolve the actual problem. Incomplete information may arise from several sources (e.g. unreachable sensors, incomplete Patient Electronic Clinical Profile) but what is important is to be able to measure the quality of the information we have access to and the quality of the ideas presented by the participants, based in factors like reputation, credibility, namely, in the discussion. However, we are certain, that some vital information, if incomplete, may even so, compromise any suggestion / decision but, in the majority of situations, we believe this will not be the case.

References

1. Parsons, S.: Current approaches to handling imperfect information in data andknowledge bases. IEEE Transactions on Knowledge and Data Engineering, 1996 8(3), 353–372 (1996)
2. Cruz-Correia, R.J., et al.: Reviewing the integration of patient data: how systems are evolving in practice to meet patient needs. BMC Medical Informatics and Decision Making 7(14) (2007)
3. Costa, R., et al.: Intelligent Mixed Reality for the Creation of Ambient Assisted Living. In: Neves, J., Santos, M.F., Machado, J.M. (eds.) EPIA 2007. LNCS (LNAI), vol. 4874, Springer, Heidelberg (2007)
4. Conklin, J.: Dialogue Mapping: Building Shared Understanding of Wicked Problems. Wiley, Chichester (2006)
5. Brito, L., Novais, P., Neves, J.: The logic behind negotiation: from pre-argument reasoning to argument-based negotiaion. In: Plekhanova, V. (ed.) Intelligent Agent Software Engineering, pp. 137–159. Idea Group Piblishing (2003)
6. Bostrom, R., Anson, R., Clawson, V.: Group facilitation and group support systems. In: Jessup, Valachic (eds.) Group Support Systems: New Perspectives. Macmillan, Basingstoke (2003)
7. Schwarz, R.M.: The Skilled Facilitator: Practical Wisdom for Developing Effective Groups. Jossey Bass (1994)
8. Hustadt, U.: Do we need the closed-world assumption in knowledge representation? In: Baader, Buchheit, Jeusfeld, Nutt (eds.) Working Notes of the KI 1994 Workshop, Saarbrüken, Germany (1994)
9. Neves, J.: A Logic Interpreter to Handle Time and Negation in Logic Data Bases. In: Proceedings of the ACM 1984, The Fifth Generation Challenge (1984)
10. Gelfond, M., Lifschitz, V.: Logic Programs with Classical Negation. In: Proceedings of the International Conference on Logic Programming (1990)

Towards Distributed Algorithm Portfolios

Matteo Gagliolo[1,2] and Jürgen Schmidhuber[1,2,3]

[1] IDSIA, Galleria 2, 6928 Manno (Lugano), Switzerland
[2] University of Lugano, Faculty of Informatics, Via Buffi 13, 6904 Lugano, Switzerland
[3] TU Munich, Boltzmannstr. 3, 85748 Garching, München, Germany
{matteo,juergen}@idsia.ch

Summary. In recent work we have developed an online algorithm selection technique, in which a model of algorithm performance is learned incrementally *while* being used. The resulting *exploration-exploitation* trade-off is solved as a bandit problem. The candidate solvers are run in parallel on a single machine, as an *algorithm portfolio*, and computation time is shared among them according to their expected performances. In this paper, we extend our technique to the more interesting and practical case of multiple CPUs.

1 Introduction

Existing parallel computing systems (e. g., Condor, Globus, JOpera), are aimed at improving speed and reliability of computation, but require the user to specify which computations should be carried out. In a more practical situation, a set of alternative algorithms, of unknown performance, is available for a given problem class, and one would like to automate the process of selecting which algorithm to use, independently for each problem instance.Such automatic *algorithm selection* is now a thirty years old field of artificial intelligence [19], even though most work in this area has been done in the last two decades. In the more general *algorithm portfolio* paradigm [16],the available algorithms are executed in parallel, and the aim of selection is to allocate computational resources among them. The problem of algorithm (portfolio) selection consists in deciding which experiments should be carried out, given a fixed amount of computational resources. In parallel computing, the set of experiments is fixed by the user, while the available computational resources may vary over time, due to failures, load fluctuations, node addition, etc. In both fields, the aim is obviously to minimize computation time, and to decrease the amount of human work and expertise required. With this paper, we intend to move a step towards the integration of these two orthogonal approaches, devising the blueprint of a more "intelligent" cluster front-end. In the following section we will briefly review related work. Section 3 describes our algorithm selection framework GAMBLETA. Section 4 discusses its extension to the case of multiple CPUs.After reporting experimental results (Section 5), we conclude the paper in Section 6.

2 Related Work

In general terms, algorithm selection can be defined as the process of allocating computational resources to a set of alternative algorithms, in order to improve some measure

J.M. Corchado et al. (Eds.): DCAI 2008, ASC 50, pp. 634–643, 2009.
springerlink.com

of performance on a set of problem instances. For *decision* or *search* problems, where a binary criterion for recognizing a solution is available (e. g., SAT [10]), the only meaningful measure of performance is runtime, and selection is aimed at minimizing it. The selection among different algorithms can be performed once for an entire set of problem instances (*per set* selection); or repeated for each instance (*per instance* selection). As in most practical cases there is no single "best" algorithm, per instance selection can only improve over per set selection. A further orthogonal distinction can be made among *static* algorithm selection, in which any decision on the allocation of resources precedes algorithm execution; and *dynamic* algorithm selection, in which the allocation can be adapted during algorithm execution. Selection based on a model of algorithm performance can be further distinguished as *offline*, if performance data is gathered during a preliminary training phase, after which the model is kept fixed; or *online*, if the model is updated at every instance solution.

A seminal paper in this field is [19], in which offline, per instance selection is first advocated. More recently, similar concepts have been proposed by the *Meta-Learning* community [7, 20]. *Parameter tuning* can also be modeled as an algorithm selection problem: in this case the algorithm set is composed of multiple copies of the same algorithm, differing only in the parameter values.

The foundation papers about algorithm portfolios [16, 14, 12] describe how to evaluate the runtime distribution of a portfolio, based on the runtime distributions of the algorithms.The RTD is used to evaluate mean and variance, and find the (per set optimal) *efficient frontier* of the portfolio, i.e., that subset of all possible allocations in which no element is dominated in both mean and variance. Another approach based on runtime distributions can be found in [4, 5], for parallel independent processes and shared resources respectively. The expected value of a cost function, accounting for both wall-clock time and resources usage, is minimized. In all these works the runtime distributions are assumed to be known *a priori*.

Further references on algorithm selection can be found in [8, 9]. Literature on parallel computing, grid computing, distributed computing [6, 1, 18] is focused on allocation of dynamically changing computational resources, in a transparent and fault tolerant manner. We are not aware of works in this field in which algorithm selection is considered. Performance modeling can be used to guide scheduling decisions, as in [1], where a simple estimate of expected runtime is performed, in order to be able to meet user-imposed deadlines.

3 Allocating a Single CPU

Online model-based algorithm selection consists in updating a model of performance *while* using it to guide selection. It is intuitive that this setting poses what is called an *exploration-exploitation* trade-off: on the one hand, collecting runtime data for improving the model can lead to better resource allocation in the future. On the other hand, running more experiments to improve a model which already allows to perform good selections can be a waste of machine time. A well known theoretical framework for dealing with such a dilemma is offered by the the *multi-armed bandit* problem [2], consisting of a sequence of trials against a K-armed slot machine. At each trial, the gambler chooses one of the available arms, whose losses are randomly generated from different unknown distributions, and incurs in the corresponding loss. The aim of the game is to minimize the *regret*, defined as the difference between the cumulative loss of the gambler, and the one of the best arm. A bandit problem solver (BPS) can be described as a mapping from the history of the observed losses l_k for each arm k, to a probability distribution $\mathbf{p} = (p_1, ..., p_K)$, from which the choice for the successive trial will be picked. Given the notion of regret, a straightforward application of a BPS to algorithm selection, in which "pick arm k" means "run algorithm a_k on next problem instance", would only allow to identify the per-set optimal algorithm [8]. Suppose instead we do have a model based method for *per-instance* algorithm selection, and we only want to know when the model is reliable enough. We can then play the game at an upper level, choosing, for each subsequent problem instance, between the model based selector and a simpler and more exploratory allocation strategy, such as a parallel portfolio of all available algorithms. Intuitively, the BPS will initially penalize the model-based allocator, but only until the model is good enough to outperform the exploratory allocator. Alternative allocation techniques can be easily added as additional "arms" of the bandit.

More precisely, consider a sequence $\mathcal{B} = \{b_1, \ldots, b_M\}$ of M instances of a decision problem, for which we want to minimize solution time; a set of K algorithms $\mathcal{A} = \{a_1, \ldots, a_K\}$; and a set of N *time allocators* (TA_j) [8]. Each TA_j can be an arbitrary function, mapping the current history of collected performance data for each a_k, to a share $\mathbf{s}^{(j)} \in [0, 1]^K$, with $\sum_{k=1}^{K} s_k = 1$. A TA is used to solve a given problem instance executing all algorithms in $\mathcal{A}$ in parallel, on a single machine, whose computational resources are allocated to each a_k proportionally to the corresponding s_k, such that for any portion of time spent t, $s_k t$ is used by a_k, as in a *static* algorithm portfolio [16]. The runtime before a solution is found is then $\min_k\{t_k/s_k\}$, t_k being the runtime of algorithm a_k when executed alone. A trivial example of an exploratory TA is the *uniform* time allocator, assigning a constant $\mathbf{s} = (1/K, ..., 1/K)$. Single algorithm selection can be represented in this framework by setting a single s_k to 1. Dynamic allocators will produce a time-varying share $\mathbf{s}(t)$. The TAs proposed in [8] are based on non-parametric models of the runtime distribution of the algorithms, which are used to optimize the share $\mathbf{s}$ according to different criteria (see next section).

At this higher level, one can use a BPS to select among different time allocators, $\text{TA}_j, \text{TA}_2 \ldots$, working on a same algorithm set $\mathcal{A}$. In this case, "pick arm j" means "use time allocator TA_j on $\mathcal{A}$ to solve next problem instance". In the long term, the BPS would allow to select, on a *per set* basis, the TA_j that is best at allocating time to algorithms in $\mathcal{A}$ on a *per instance* basis. The resulting "Gambling" Time Allocator (GAMBLETA) is described in Alg. 1.

One motivation for using a BPS is the guaranteed bound on regret. In [9], basing on [3], we introduced EXP3LIGHT-A, a BPS which guarantees a bound on regret when the maximum loss is unknown *a priori*. Note that any bound on the regret of the chosen BPS will determine a bound on the regret of GAMBLETA with respect to the best time allocator. Nothing can be said about the performance w.r.t. the best algorithm. In a worst-case setting, if none of the time allocator is effective, a bound can still be obtained by including the uniform share in the set of TAs. In practice, though, per-instance selection

Algorithm 1. GAMBLETA($\mathcal{A}$,$\mathcal{T}$,BPS) Gambling Time Allocator.

Algorithm set $\mathcal{A}$ with K algorithms;
A set $\mathcal{T}$ of N time allocators TA_j;
A bandit problem solver BPS
M problem instances.
initialize BPS(N, M)
for each problem $b_i, i = 1, \ldots, M$ **do**
  pick time allocator $I(i) = j$ with probability $p_j(i)$ from BPS.
  solve problem b_i using TA_I on $\mathcal{A}$
  incur loss $l_{I(i)} = \min_k\{t_k(i)/s_k^{(I)}(i)\}$
  update BPS
end for

can be much more efficient than uniform allocation, and the literature is full of examples of time allocators converging to a good performance.

4 Allocating Multiple CPUs

For simplicity, we assume a traditional cluster setup, with a front end controlling the allocation of jobs on different nodes. Consider again our K algorithms $\mathcal{A} = \{a_1, a_2, ..., a_K\}$. This time we need to allocate time on J CPUs, so the share will be represented as a $K \times J$ matrix, with columns summing to 1, $\mathbf{S} = \{s_{kj}\}, s_{kj} \in [0, 1], \sum_{k=1}^{K} s_{kj} = 1 \forall j \in \{0, 1, \ldots, J\}$. For a given share $\mathbf{S}$, the *survival function*[1] for the distributed portfolio can be evaluated as (compare with (12) from [8]):

$$S_{\mathcal{A},\mathbf{S}}(t) = \prod_{j=1}^{J} \prod_{k=1}^{K} S_k(s_{kj}t), \tag{1}$$

We will first see how to apply the model-based time allocators, introduced in [8, Sec. 4] for a single CPU, to a case in which $J > 1$ CPUs are used:

1. **Expected time.** The expected runtime value is minimized w.r.t. $\mathbf{S}$:

$$\mathbf{S}^* = \arg\min_{\mathbf{S}} \int_0^{+\infty} S_{\mathcal{A},\mathbf{S}}(t)dt. \tag{2}$$

2. **Contract.** This TA picks the $\mathbf{S}$ that maximizes the probability of solution within a *contract* time t_u , or, equivalently, minimizes the survival function at t_u:

$$\mathbf{S}^*(t_u) = \arg\min_{\mathbf{S}} S_{\mathcal{A},\mathbf{S}}(t_u). \tag{3}$$

3. **Quantile.** This TA minimizes a *quantile* α of (1):

$$\mathbf{S}^*(\alpha) = \arg\min_{\mathbf{S}} F^{-1}_{\mathcal{A},\mathbf{S}}(\alpha). \tag{4}$$

[1] $S(t) = 1 - F(t)$, F being the cumulative distribution function of the runtime distribution.

While in [8] the share s was found optimizing functions in a $(K-1)$ dimensional space, here the size of the search space grows also with the number of CPUs, as $J(K-1)$. Fortunately, and rather unexpectedly, we could prove that the optimal share for the contract and quantile allocators is *homogeneous*, i. e., the same on each CPU, such that the corresponding matrix has all its columns equal (see Appendix). This means that it can be found with a search in a $(K-1)$ dimensional space, regardless of the number of CPUs available.

The above criteria assume that one wants to use all available CPUs. This may or not be convenient, depending on the shape of the runtime distributions at play. One simple possibility for selecting the number of CPUs is to optimize the allocation for $1, 2, \ldots$ up to J CPUs, and then use the number j that gives the best result, considering that if a problem is solved in a time t using j CPUs, the total CPU time used will be jt.

A *dynamic* version of the above TAs can be implemented, periodically updating the model, for example conditioning on the time already spent as in [8], and re-evaluating the optimal share. Regarding the number of CPUs allocated, this should only be decreased, as increasing it would require to start the algorithms from scratch, with the unconditioned model: in this situation the optimal share would not be homogeneous, and we would have again a larger search space.

In the experiments described in the next section, we used a set of quantile allocators, with different parameter values for α, along with the uniform allocator, and considered the following heuristic: for each problem instance, the front end picks a time allocator, allowing it the use of all currently available CPUs J. The uniform allocator will take them all, and run all algorithms in parallel on each, obviously with a different random seed. The quantile allocators will instead evaluate the optimal share for $1, 2, \ldots, J$ CPUs, and pick a number j of them such that jt_α is minimized, $t_\alpha = F^{-1}(\alpha)$ being the quantile. The front-end will continue assigning the remaining $J - j$ CPUs for the next problem instance, until all CPUs are occupied. When a quantile TA dynamically updates its share, it re-evaluates it for $1, 2, \ldots, j$ CPUs, releasing some of the CPUs if necessary. Released CPUs are then reallocated by the front-end to solve the following problem instance. At the upper level, the BPS (EXP3LIGHT-A from [9]) is used as in GAMBLETA, using the *total* CPU time as a loss (i. e., jt if j CPUs are used for a wall-clock time t), in order to favor time allocators that do not use more CPUs than necessary.

Unfortunately, the bound on regret of the BPS does not hold in the case of multiple CPUs, as the proof assume a *sequential* game, in which the probability distribution over the arms is updated after each arm pull, based on the observed loss. In this multiple CPU version of GAMBLETA, instead, the choice of time allocators continues until there are CPUs available, and the feedback on the loss is received asynchronously, only when the corresponding problem instance is solved.

5 Experiments

The main objective of these preliminary experiments is to analyze the speedup (the ratio between runtime with 1 and $J > 1$ CPUs) and efficiency (the ratio between speedup and number of CPUs) of the proposed allocation method. Note that the notion of efficiency

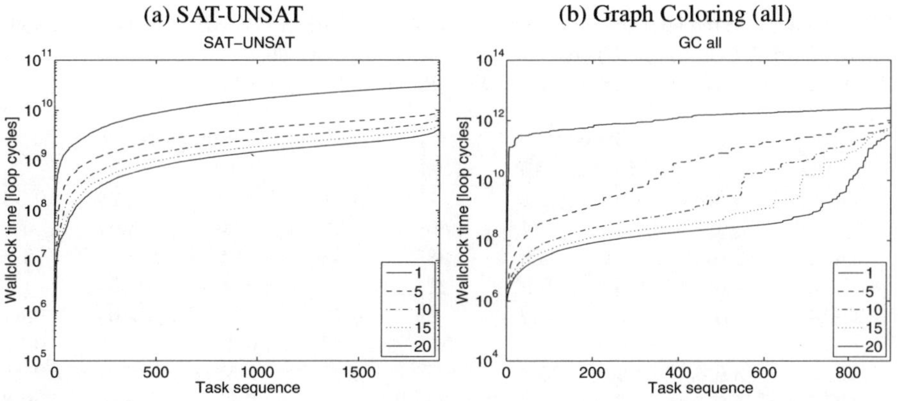

Fig. 1. (a): Wall-clock time for the SAT-UNSAT benchmark, for different numbers of CPUs ($10^9 \approx 1$ min.). (b): Wall-clock time for the Graph Coloring benchmark (all problems), for different numbers of CPUs ($10^9 \approx 1$ min.).

assumes a different connotation in the context of algorithm portfolios: traditionally one does not expect to achieve an efficiency larger than 1, as it is assumed that all computations performed on a single CPU have to be carried out on J CPUs as well. This is not the case for algorithm portfolios, as in this case we can stop the computation as soon as the fastest algorithm solves the problem, so we will see efficiencies greater than 1.

In the first experiment we apply GAMBLETA to a solve a set of satisfiable and unsatisfiable CNF3SAT problems (benchmarks `uf-*`, `uu-*` from [15], 1899 instances in total), using a small algorithm set composed of a local search and a complete SAT solver (respectively, G2-WSAT [17] and Satz-Rand [13]). Local search algorithms are more efficient on satisfiable instances, but cannot prove unsatisfiability, so are doomed to run forever on unsatisfiable instances; while complete solvers are guaranteed to terminate their execution on all instances, as they can also prove unsatisfiability. The set of time allocators includes the uniform one, and nine quantile allocators, with α ranging from 0.1 to 0.9, sharing the same conditional non-parametric model from [21]. As the clauses-to-variable ratio is fixed in this benchmark, only the number of variables, ranging from 20 to 250, was used to condition the model.

In a second set experiment we use Satz-Rand alone on 9 sets of structured graph-coloring (GC) problems [11], also from [15], each composed of 100 instances encoded in CNF 3 SAT format. This algorithm/benchmark combination is particularly interesting as the *heavy-tailed* behavior [13] of Satz-Rand differs for the various sets of instances[2] [11]. In this case, the time allocators decide only how many parallel copies of Satz-Rand

[2] A *heavy-tailed* runtime distribution $F(t)$ is characterized by a Pareto tail, i.e., $F(t) \rightarrow_{t\rightarrow\infty} 1 - Ct^{-\alpha}$. In practice, this means that most runs are relatively short, but the remaining few can take a very long time, with differences among runs of several orders of magnitude. In this situation, both restart strategies and parallel execution proved to be an effective way of reducing runtime [12].

Table 1. Speedup (on the left) and efficiency (on the right) for the SAT-UNSAT and the different subgroups of Graph Coloring (GC) benchmarks. Note the dramatic speed-up obtained on GC sets 1 and 2, the effect of the heavy-tailed RTD of Satz-Rand on these problem sets.

	Speedup				Efficiency			
#CPUs	5	10	15	20	5	10	15	20
GC 0	4.3	7.2	9.2	11	0.87	0.72	0.62	0.56
GC 1	1	280	84	86	0.2	28	5.6	4.3
GC 2	2.6	18	36	98	0.53	1.8	2.4	4.9
GC 3	2.6	6.6	9.8	21	0.52	0.66	0.65	1.1
GC 4	4	6.2	6.9	8.8	0.79	0.62	0.46	0.44
GC 5	3.4	3.5	7.7	7.8	0.67	0.35	0.51	0.39
GC 6	3.6	5.1	5.8	8.5	0.71	0.51	0.39	0.43
GC 7	5.9	8.3	13	12	1.2	0.83	0.84	0.6
GC 8	5.1	9	12	12	1	0.9	0.77	0.6
GC all	3.1	4.5	5.5	8.2	0.61	0.45	0.37	0.41
SAT-UNSAT	3.5	4.8	6	7.1	0.69	0.48	0.4	0.35

to run for each problem: as the share is 1 on each CPU, we allowed the TAs to dynamically shrink and also grow the number of CPUs used.

Both experiments were simulated in MATLAB, on a single machine, based on previously collected runtime data[3], for different numbers of CPUs $(1, 5, 10, 15, 20)$. Results reported are upper confidence bounds obtained from 20 runs, each time using fresh random seeds, and a different random reordering of the problem instances.

6 Conclusions

We presented a framework for distributed time allocation, aimed at minimizing solution time of decision problems. Preliminary results are encouraging, even though the efficiency sometimes decreases with the number of CPUs.

The difference with the "embarrassingly parallel" computation paradigm, in which all processes can be executed independently, is that in our case there is an indirect interaction among the algorithms, as resource allocation among elements of the algorithm set is periodically updated, based on runtime information (*dynamic* selection); and as soon as one algorithm solves a given problem, other algorithms working on the same problem are stopped. Note that, as our algorithm selection scheme has a very low computational overhead [8], it can be easily extended to a fully distributed situation, in which there is no front end, and all nodes are equal. In such case, runtime data can be broadcast, in order to allow each node to update a local copy of the RTD model; as the time allocation algorithm is pseudo-random, it can be reproduced deterministically, so each node can independently evaluate the same allocation, and execute the job(s)

[3] As we needed a common measure of time, and the CPU runtime measures are quite inaccurate, we modified the original code of the two algorithms adding a counter, that is incremented at every loop in the code. All runtimes reported for this benchmark are expressed in these loop cycles: on a 2.4 GHz machine, 10^9 cycles take about 1 minute.

assigned to itself. Existing distributed computing techniques can be used at a lower level, to deal with message losses and node failures.

Acknowledgments. This work was supported by the Hasler foundation with grant n. 2244.

References

1. Abramson, D., Giddy, J., Kotler, L.: High performance parametric modeling with Nimrod/G: Killer application for the global grid? In: Proceedings of the 14th International Parallel & Distributed Processing Symposium (IPDPS 2000), Cancun, Mexico, May 1-5, 2000, pp. 520–528 (2000)
2. Auer, P., Cesa-Bianchi, N., Freund, Y., Schapire, R.E.: The nonstochastic multiarmed bandit problem. SIAM J. Comput. 32(1), 48–77 (2002)
3. Cesa-Bianchi, N., Mansour, Y., Stoltz, G.: Improved second-order bounds for prediction with expert advice. In: Auer, P., Meir, R. (eds.) COLT 2005. LNCS (LNAI), vol. 3559, pp. 217–232. Springer, Heidelberg (2005)
4. Finkelstein, L., Markovitch, S., Rivlin, E.: Optimal schedules for parallelizing anytime algorithms: the case of independent processes. In: Eighteenth national conference on Artificial intelligence, pp. 719–724. AAAI Press, Menlo Park (2002)
5. Finkelstein, L., Markovitch, S., Rivlin, E.: Optimal schedules for parallelizing anytime algorithms: The case of shared resources. Journal of Artificial Intelligence Research 19, 73–138 (2003)
6. Foster, I., Kesselman, C.: Globus: A metacomputing infrastructure toolkit. The International Journal of Supercomputer Applications and High Performance Computing 11(2), 115–128 (1997)
7. Fürnkranz, J.: On-line bibliography on meta-learning, EU ESPRIT METAL Project (26.357): A Meta-Learning Assistant for Providing User Support in Machine Learning Mining (2001)
8. Gagliolo, M., Schmidhuber, J.: Learning dynamic algorithm portfolios. Annals of Mathematics and Artificial Intelligence 47(3–4), 295–328 (2006); AI&MATH 2006 Special Issue
9. Gagliolo, M., Schmidhuber, J.: Algorithm selection as a bandit problem with unbounded losses. Technical Report IDSIA-07-08, IDSIA (2008)
10. Gent, I., Walsh, T.: The search for satisfaction. Technical report, Dept. of Computer Science, University of Strathclyde (1999)
11. Gent, I., Hoos, H.H., Prosser, P., Walsh, T.: Morphing: Combining structure and randomness. In: Proc. of AAAI 1999, pp. 654–660 (1999)
12. Gomes, C.P., Selman, B.: Algorithm portfolios. Artificial Intelligence 126(1–2), 43–62 (2001)
13. Gomes, C.P., Selman, B., Crato, N., Kautz, H.: Heavy-tailed phenomena in satisfiability and constraint satisfaction problems. J. Autom. Reason. 24(1-2), 67–100 (2000)
14. Gomes, C.P., Selman, B., Kautz, H.: Boosting combinatorial search through randomization. In: AAAI 1998/IAAI 1998, USA, pp. 431–437. AAAI Press, Menlo Park (1998)
15. Hoos, H.H., Stützle, T.: SATLIB: An Online Resource for Research on SAT. In: Gent, I.P., et al. (eds.) SAT 2000, pp. 283–292 (2000), `http://www.satlib.org`
16. Huberman, B.A., Lukose, R.M., Hogg, T.: An economic approach to hard computational problems. Science 275, 51–54 (1997)
17. Li, C.M., Huang, W.: Diversification and determinism in local search for satisfiability. In: SAT 2005, pp. 158–172. Springer, Heidelberg (2005)
18. Pautasso, C., Bausch, W., Alonso, G.: Autonomic computing for virtual laboratories (2006)

19. Rice, J.R.: The algorithm selection problem. In: Rubinoff, M., Yovits, M.C. (eds.) Advances in computers, vol. 15, pp. 65–118. Academic Press, New York (1976)
20. Vilalta, R., Drissi, Y.: A perspective view and survey of meta-learning. Artif. Intell. Rev. 18(2), 77–95 (2002)
21. Wichert, L., Wilke, R.A.: Application of a simple nonparametric conditional quantile function estimator in unemployment duration analysis. Technical Report ZEW Discussion Paper No. 05-67, Centre for European Economic Research (2005)

Appendix

Definition 1 (Homogeneous share). *Be* $\mathbf{s}_j$ *the share for CPU* j *(i.e., the* j*-th column of* $\mathbf{S}$*). A share* $\mathbf{S}$ *is* homogeneous *iff* $s_{kj} = s_k \forall (k,j)$*, i.e.,* $\mathbf{s}_j = \mathbf{s} \forall j$*.*

It is easy to prove by examples that expected-value-optimal shares can be inhomogeneous. For contract and quantile optimal shares, the following theorems hold.

Theorem 1. $\forall$ *contract-optimal share* $\mathbf{S}^*(t_u)$ *there is an homogeneous equivalent* $\mathbf{S}^*_h(t_u)$ *such that* $S_{\mathcal{A},\mathbf{S}^*}(t_u) = S_{\mathcal{A},\mathbf{S}^*_h}(t_u)$*.*

Proof

$$\mathbf{S}^* = \arg\min_{\mathbf{S}} S_{\mathcal{A},\mathbf{S}}(t_u) = \arg\min_{\{\mathbf{s}_j\}} \prod_{j=1}^{J} S_{\mathcal{A},\mathbf{s}_j}(t_u) \tag{5}$$

Consider a non homogeneous contract-optimal share $\mathbf{S}^*(t_u)$, with $\mathbf{s}^*_j \neq \mathbf{s}^*_i$ for a pair of columns i, j. If $S_{\mathcal{A},\mathbf{s}_i}(t_u) > S_{\mathcal{A},\mathbf{s}_j}(t_u)$, then replacing $\mathbf{s}_i$ with $\mathbf{s}_j$ produces a better share, violating the hypothesis of optimality of $\mathbf{S}^*$. As this must hold for any i, j, then $S_{\mathcal{A},\mathbf{s}_i}(t_u)$ must be the same for all i. Setting all $\mathbf{s}_j$ to a same $\mathbf{s}_i$ will then produce a homogeneous optimal share $\mathbf{S}^*_h$.

Theorem 2. $\forall$ *quantile-optimal share* $\mathbf{S}^*(\alpha)$ *there is an homogeneous equivalent* $\mathbf{S}^*_h(\alpha)$ *such that* $F^{-1}_{\mathcal{A},\mathbf{S}^*}(\alpha) = F^{-1}_{\mathcal{A},\mathbf{S}^*_h}(\alpha)$*.*

Proof. From its definition, a quantile t_α is $t_\alpha = min\{t|F(t) = \alpha\} = min\{t|S(t) = (1-\alpha)\}$; this, together with the monotonicity of the survival function, and of the logarithm function, implies:

$$\ln(1-\alpha) = \ln S(t_\alpha) < \ln S(t) \;\; \forall \;\; t < t_\alpha. \tag{6}$$

Consider a non homogeneous optimal share $\mathbf{S}^*$, with $\mathbf{s}^*_j \neq \mathbf{s}^*_i$ for a pair $(i,j), i \neq j$. We can write:

$$\ln(1-\alpha) = \ln S_{\mathcal{A},\mathbf{S}}(t_\alpha) = \sum_{j=1}^{J} \ln S_{\mathcal{A},\mathbf{s}_j}(t_\alpha) \tag{7}$$

Pick now an arbitrary column $\mathbf{s}^*_i$ of $\mathbf{S}^*$, and set all other columns $\mathbf{s}^*_j$ to $\mathbf{s}^*_i$, obtaining a homogeneous share with quantile t_i

$$\ln(1-\alpha) = J \ln S_{\mathcal{A},\mathbf{s}_i}(t_i) \tag{8}$$

While $t_i < t_\alpha$ violates the hypothesis of optimality of $\mathbf{S}^*(\alpha)$, $t_\alpha < t_i$ would imply for (6) (with t_i in place of t_α):

$$\ln S_{\mathcal{A},\mathbf{s}_i}(t_i) = \frac{\ln(1-\alpha)}{J} < \ln S_{\mathcal{A},\mathbf{s}_i}(t_\alpha). \tag{9}$$

Summing over i gives a contradiction ($\ln(1-\alpha) < \ln(1-\alpha)$), so the only possibility left is $t_i = t_\alpha$. As this must hold for any i, then t_i must be constant $\forall i$. Setting all $\mathbf{s}_j$ to a same $\mathbf{s}_i$ will then produce a homogeneous optimal share $\mathbf{S}_h^*$. Note that, different from the contract, in this case $\mathbf{S}_h^*$ depends on J.

Learning and Comparing Trajectories with a GNG-Based Architecture

José García-Rodríguez, Francisco Flórez-Revuelta, and Juan Manuel García-Chamizo

Department of Computer Technology, University of Alicante, Apdo. 99. 03080 Alicante, Spain
{jgarcia,florez,juanma}@dtic.ua.es

Abstract. In this paper we use a kind of self-organising network, the Growing Neural Gas, as structure capable of characterizing hand posture, as well as its movement. Topology of a self-organizing neural network determines posture, whereas its adaptation dynamics throughout time determines gesture. This adaptive character of the network allows us to avoid the correspondence problem of other methods, so that the gestures are modelled by the movement of the neurons. Using this representation to an image sequence we are able to follow the evolution of the object along the sequence learning the trajectory that describes. Finally we use the Hausdorff distance among trajectories to compare and recognize them.

Keywords: Self-organising maps, topology preservation, topologic maps, trajectories, Delaunay Triangulation, Hausdorff distance.

1 Introduction

In this paper, we use a structure that is able to represent hand shape, as well as its movements. It is based on the behaviour of self-organizing networks that are capable of adapting their topology to an input manifold and, due to their dynamic character, to readapt it to new input patterns.

In this way, static gestures are modelled by the interconnection network between the neurons, whereas dynamic gestures are determined by the dynamics of the network throughout the image sequence of the gesture. That is, hand tracking is reduced to following the movement of the neurons. This avoids the correspondence problem of other methods to determine the location of the different parts of the hand in order to perform its tracking.

Self-organising neural networks, by means of a competitive learning, make an adaptation of the reference vectors of the neurons as well as the interconnection network among them; obtaining a mapping that tries to preserve the topology of an input space. Besides, they are capable of a continuous re-adaptation process even if new patterns are entered, with no need to reset the learning.

These capacities have been used for the representation of objects [1] (Figure 1) and their motion [2] by means of the Growing Neural Gas (GNG) [3] that has a learning process more flexible than other self-organising models, like Kohonen maps [4] and more flexible and faster than Topology Representing Networks [5].

We also introduce a new architecture based on GNG that uses the graph obtained from the learning algorithm of the net using as input an image sequence and representing the objects that appears in the images solving intrinsically the correspondence

J.M. Corchado et al. (Eds.): DCAI 2008, ASC 50, pp. 644–652, 2009.
springerlink.com

Fig. 1. Representation of two-dimensional objects with a self-organising network

problem following the dynamic of the net and using the neurons to predict and readjust the representation from frame to frame. The system is able to learn the trajectories that describe the objects that appear in the image following the trajectories that describe the neurons of the maps that represent the objects along the sequence. Comparing these trajectories with the Hausdorff distance we are able to recognize, for example, hand gestures previously learned and saved in a database.

The remainder of the paper is organized as follows: section 2 provides a detailed description of the topology learning algorithm GNG and in section 3 an explanation on how we can apply GNG to represent objects is given. Section 4 presents the GNG based architecture to represent image sequences called GNG-Seq and section 5 presents a technique to compare trajectories obtained with the system using the Hausdorff distance and shows some experimental results, followed by our major conclusions and further work.

2 Topology Learning

Self-organising maps are neural networks where input patterns are projected onto a network of neural units such that similar patterns are projected onto units adjacent in the network and vice versa. As a result of this mapping a representation of the input patterns is achieved that in post-processing stages allows one to exploit the similarity relations of the input patterns. Such models have been successfully used in applications such as speech processing [4], robotics [7,8] and image processing [9]. However, most common approaches are not able to provide good neighborhood and topology preservation if the logical structure of the input pattern is not known a priori. In fact, the most common approaches specify in advance the number of neurons in the network and a graph that represents topological relationships between them, for example, a two-dimensional grid, and seek the best match to the given input pattern manifold. When this is not the case the networks fail to provide good topology preserving as for example in the case of Kohonen's algorithm.

The approach presented in this paper is based on self-organising networks trained using the Growing Neural Gas learning method [3], an incremental training algorithm. The links between the units in the network are established through competitive hebbian learning [10]. As a result the algorithm can be used in cases where the topological structure of the input pattern is not known a priori and yields topology preserving maps of feature manifold [5].

2.1 Growing Neural Gas

With Growing Neural Gas (GNG) [3] a growth process takes place from minimal network size and new units are inserted successively using a particular type of vector quantisation [4]. To determine where to insert new units, local error measures are gathered during the adaptation process and each new unit is inserted near the unit which has the highest accumulated error. At each adaptation step a connection between the winner and the second-nearest unit is created as dictated by the competitive hebbian learning algorithm. This is continued until an ending condition is fulfilled, as for example evaluation of the optimal network topology based on some measure. Also the ending condition could it be the insertion of a predefined number of neurons or a temporal constrain. In addition, in GNG networks learning parameters are constant in time, in contrast to other methods whose learning is based on decaying parameters.

In the remaining of this Section we describe the growing neural gas algorithm. The network is specified as:

- A set N of nodes (neurons). Each neuron $c \in N$ has its associated reference vector $w_c \in R^d$. The reference vectors can be regarded as positions in the input space of their corresponding neurons.
- A set of edges (connections) between pairs of neurons. These connections are not weighted and its purpose is to define the topological structure. An *edge aging scheme* is used to remove connections that are invalid due to the motion of the neuron during the adaptation process.

The GNG learning algorithm to approach the network to the input manifold is as follows:

1. Start with two neurons a and b at random positions w_a and w_b in $\mathcal{R}^d$.
2. Generate a random input pattern ξ according to the data distribution $\mathcal{P}(\xi)$ of each input pattern. In our case since the input space is 2D, the input pattern is the (x, y) coordinate of the points belonging to the object. Typically, for the training of the network we generate 1000 to 10000 input patterns depending on the complexity of the input space.
3. Find the nearest neuron (winner neuron) s_1 and the second nearest s_2.
4. Increase the age of all the edges emanating from s_1.
5. Add the squared distance between the input signal and the winner neuron to a counter error of s_1 such as:

$$\Delta error(s_1) = \left\| w_{s_1} - \xi \right\|^2 \tag{1}$$

6. Move the winner neuron s_1 and its topological neighbours (neurons connected to s_1) towards ξ by a learning step ε_w and ε_n, respectively, of the total distance:

$$\Delta w_{s_1} = \varepsilon_w (\xi - w_{s_1}) \tag{2}$$

$$\Delta w_{s_n} = \varepsilon_n (\xi - w_{s_n}) \tag{3}$$

7. If s_1 and s_2 are connected by an edge, set the age of this edge to 0. If it does not exist, create it.

8. Remove the edges larger than a_{max}. If this results in isolated neurons (without emanating edges), remove them as well.
9. Every certain number λ of input signals generated, insert a new neuron as follows:
   - Determine the neuron q with the maximum accumulated error.
   - Insert a new neuron r between q and its further neighbour f:

$$w_r = 0.5(w_q + w_f) \tag{4}$$

   - Insert new edges connecting the neuron r with neurons q and f, removing the old edge between q and f.
   - Decrease the error variables of neurons q and f multiplying them with a constant α. Initialize the error variable of r with the new value of the error variable of q and f.
10. Decrease all error variables by multiplying them with a constant β.
11. If the stopping criterion is not yet achieved, go to step 2. (In our case the criterion is the number of neurons inserted)

3 Representation of 2D Objects with GNG

Given an image $I(x,y) \in \mathcal{R}$ we perform the transformation $\psi_{\mathcal{T}}(x,y) = \mathcal{T}(I(x,y))$ that associates to each one of the pixels its probability of belonging to the object, according to a property $\mathcal{T}$. For instance, in figure 2, this transformation is a threshold function.

Fig. 2. Silhouette extraction

If we consider $\xi = (x,y)$ and $\mathcal{P}(\xi) = \psi_{\mathcal{T}}(\xi)$, we can apply the learning algorithm of the GNG to the image I, so that the network adapts its topology to the object. This adaptive process is iterative, so the GNG represents the object during all the learning.

As a result of the GNG learning we obtain a graph, the Topology Preserving Graph $\mathcal{TPG} = \langle \mathcal{N}, \mathcal{C} \rangle$, with a vertex (neurons) set $\mathcal{N}$ and an edge set $\mathcal{C}$ that connect them (figure 1). This $\mathcal{TPG}$ establishes a Delaunay triangulation induced by the object [6].

4 GNG-Based Architecture to Represent Image Sequences

The technique used to represent objects and analyze the movement is based on the tracking throughout frames of some features of the object representation [11] that we obtain with the neural network, using the own neurons of the network like features to follow. For it, it is necessary to obtain a representation for each one of the instances, position and shape of the object, for any of the images in the sequence.

One of the most advantageous characteristics of the use of the graph obtained from the neural network to represent the present objects in any frame of a sequence of images that we have called GNG-Seq is which does not require to reset the learning of the network for each one of the images of the sequence, since we do not add or delete any neuron or edge we can use the representation obtained for the previous image whenever the speed of sampling is high. Other representation techniques were tested in [12] but GNG has better topology preservation and is faster than Kohonen maps [4] or Topology Preserving Networks [5].

In this way, using a prediction based on an historical information from previous frames and with a small readjustment of the network we will obtain the new representation in a very small time, which contributes to achieve a high processing speed.

The tracking of an object in each one of the images would be based on the following scheme: (figure 3)

- Calculation of the transformation function in order to segment the object from the background
- Prediction of the new positions of the neurons
- Re-adjustment of the neurons.

This way we will be able to segment the object in the new image, predict the new position and readjust the map based on the information from previous maps.

$$\psi_T(x, y, t) = T(I(x, y, t), TPG^{t-n} \quad (5)$$

Where n can represent the number of previous frames we use to help a prediction of segmentation and representation of the current frame. That means that we segment the image using information about colour saved in the neurons of TPG from previous maps.

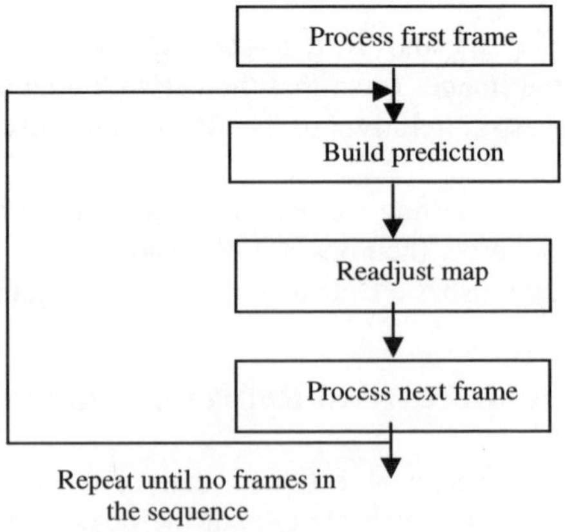

Fig. 3. GNG-Seq architecture

5 Measuring Trajectories with Hausdorff Distances

The movement of an object is interpreted like the trajectories followed by each one of the neurons of the GPT:

$$\mathcal{M} = \left[\mathcal{T}ray_i\right], \forall i \in \mathcal{A} \tag{6}$$

where the trajectory comes determined by the succession of positions for each one of the neurons throughout the map:

$$\mathcal{T}ray_i = \left\{w_{i_{t_0}}, \ldots, w_{i_{t_f}}\right\} \tag{7}$$

5.1 Hausdorff Distance

The use of direct measures of similarity between sets of points as it is the case of the *modified Hausdorff distance* [13] for the comparison of trajectories and the learning of semantic models of scenes [14] could provide good results.

Being the distance between two points a and b defined as the euclidean distance $d(a,b) = \|a-b\|$. The distance between a point a and a set of points $B = \{b_1, \ldots, b_{N_b}\}$ is defined as $d(a,B) = \min_{b \in B} \|a-b\|$. Different ways it exists to compute the direct distance between two sets of points $A = \{a_1, \ldots, a_{Na}\}$ and $B = \{b_1, \ldots, b_{N_b}\}$.

We now considered two direct measurements of distances of sets of points:

$$d(A,B) = \max_{a \in A} d(a,B) \tag{8}$$

$$d(A,B) = \frac{1}{N_a} \sum_{a \in A} d(a,B) \tag{9}$$

The direct measures between sets of points can be combined to obtain an indirect measurement with a great discriminatory power to compare sets of points that define trajectories:

$$f(d(A,B), d(B,A)) = \max(d(A,B), d(B,A)) \tag{10}$$

Combining equations 8 and 10 we obtain the Hausdorff distance and combining equations 9 and 10 we obtain the modified Hausdorff distance that has better capacities to match objects. Both measures obtain good results in the comparison of objects trajectories but a priori knowledge of behaviour or special features of the movement can improve the results, for example for hand or face gestures. In this case should it be interesting to normalize or parameterize the trajectories before applied the measures.

5.2 Experiments

We have tested the architecture to represent some video sequences of images. Figure 4 shows a set of gestures used to compare the discriminatory power of Hausdorff distance and extended Hausdorff distance when comparing the movement of a set of points, in our case the neurons that represent the hands.

Gesture I1 is the initial gesture for sequences that end with gestures G1 to G8 and I2 is the initial gesture for sequences that end with gestures G9 and G10.

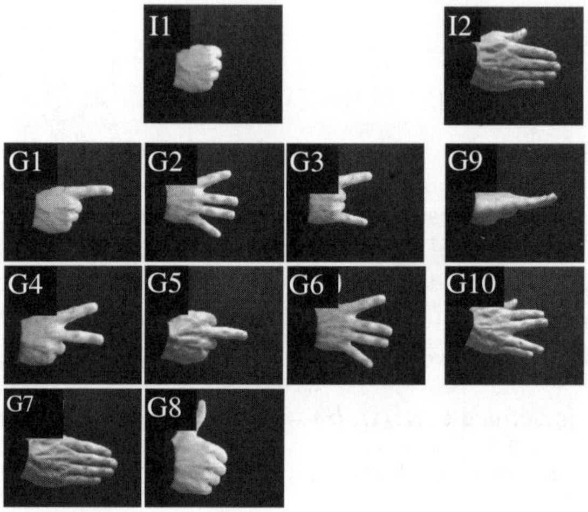

Fig. 4. Set of gestures used for the study of the trajectories

Table 1 and 2 show Hausdorff and Modified Hausdorff distances for gestures presented in figure 4. Any gesture has been compared with twenty versions of all the gestures made by different people and only the worst cases are shown in tables.

Table 1. Hausdorff distance for a set of gestures of the hand represented in figure 4

	G1	G2	G3	G4	G5	G6	G7	G8	G9	G10
G1	**445**	1058	847	955	635	1094	1344	998	1250	961
G2	1.397	**786**	787	771	1120	1011	1352	1370	2032	1981
G3	1.243	859	**534**	1020	1052	1021	1453	1298	1899	1699
G4	1.182	907	977	**521**	856	1143	1165	1267	1878	1762
G5	806	1068	894	694	**463**	1196	1123	1067	1498	1191
G6	2.122	1522	1765	1549	1654	**1932**	1659	2154	2861	2766
G7	1.059	859	858	1015	988	842	**786**	1081	1371	1359
G8	3.015	5984	3323	3488	2966	5898	4573	**3780**	3438	3580
G9	943	1110	982	1488	1285	1058	1461	1151	**636**	1076
G10	690	1174	739	1065	680	1157	1422	983	949	**517**

Table 2. Modified Hausdorff distance for a set of gestures of the hand represented in figure 4

	G1	G2	G3	G4	G5	G6	G7	G8	G9	G10
G1	**7,15**	19,60	21,04	17,27	12,04	25,46	23,47	15,39	6,47	18,87
G2	14,86	**8,33**	7,13	14,64	12,24	14,95	14,65	13,42	11,12	12,55
G3	23,62	15,15	**5,21**	19,82	22,34	25,90	27,37	18,18	22,34	28,16
G4	9,68	15,45	15,45	**6,16**	6,83	20,80	15,08	12,08	9,87	17,01
G5	17,21	20,54	23,52	13,15	**5,37**	28,05	18,75	14,63	12,40	26,32
G6	55,29	37,30	45,54	40,62	43,23	**50,08**	40,76	54,94	74,58	72,48
G7	12,30	12,10	12,02	12,01	9,82	11,75	**7,10**	8,46	7,62	10,34
G8	96,41	152,43	113,0	92,32	85,17	16,45	105,1	**74,35**	60,55	157,5
G9	12,35	12,18	13,80	19,84	20,23	14,91	17,97	20,75	**4,58**	18,01
G10	11,44	14,37	14,73	18,42	13,42	18,86	15,74	15,46	8,17	**10,05**

Distances calculated between any pair of gestures suggest that system allows to recognize most of them, obtaining the smallest distance values for the same gesture made by different users, even considering trajectories with different number of points. Both measures failed only in gestures G6 and G8 since some gesture trajectories without previous normalization are very similar as average but the hit-ratio for both measures is very high.

The system is able to work at a video rate speed since the re-adjustment for frames after the first one requires only one loop in the learning algorithm. GNG parameters used for the simulations are: $\lambda = 1000$, $\varepsilon_w = 0.1$, $\varepsilon_n = 0.001$, $\alpha = 0.5$, $\beta = 0.95$, $\alpha_{max} = 250$.

6 Conclusions and Further Work

In this paper, we have demonstrated the capacity of representation of bi-dimensional objects by a self-organising neural network. Establishing a suitable transformation function, the model is able to adapt its topology to images of the environment. Then, a simple, but very rich representation is obtained. One of the most advantageous characteristics of the use of the graph obtained from the neural network to represent the present objects in any frame of a sequence of images that we have called GNG-Seq is which does not require to reset the learning of the network for each one of the images of the sequence and we can use the representation obtained for the previous image, whenever the speed of sampling is high.

We have also consider a couple of distance measures among sets of points, the Hausdorff and extended Hausdorff distances, to compare trajectories obtained from the representation of the sequences with the GNG-Seq system testing the results to compare and recognize a set of hand gestures and obtaining good results without previous normalization. In future works some previous parameterization and normalization would be applied.

Acknowledgements

This work has been partially funded by the Spanish Government within the project DPI200509215C0201.

References

1. Flórez, F., García, J.M., García, J., Hernández, A.: Representation of 2D Objects with a Topology Preserving Network. In: Proceedings of the 2nd International Workshop on Pattern Recognition in Information Systems (PRIS 2002), Alicante, pp. 267–276. ICEIS Press (2001)
2. Flórez, F., García, J.M., García, J., Hernández, A.: Hand Gesture Recognition Following the Dynamics of a Topology-Preserving Network. In: Proc. of the 5th IEEE ICFGR, pp. 318–323 (2001)
3. Fritzke, B.: A Growing Neural Gas Network Learns Topologies. In: Tesauro, G., Touretzky, D.S., Leen, T.K. (eds.) Advances in Neural Information Processing Systems 7, pp. 625–632. MIT Press, Cambridge (1995)
4. Kohonen, T.: Self-Organising Maps. Springer, Heidelberg (1995)
5. Martinetz, T., Schulten, K.: Topology Representing Networks. Neural Networks 7(3), 507–522 (1994)
6. O'Rourke, J.: Computational Geometry in C. Cambridge University Press, Cambridge (2001)
7. Ritter, H., Schulten, K.: Topology conserving mappings for learning motor tasks. In: Neural Networks for Computing. AIP Conf. Proc. (1986)
8. Martinez, T., Ritter, H., Schulten, K.: Three dimensional neural net for learning visuomotor-condination of a robot arm. IEEE Transactions on Neural Networks 1, 131–136 (1990)
9. Nasrabati, M., Feng, Y.: Vector quantisation of images based upon kohonen self-organising feature maps. In: Proc. IEEE Int. Conf. Neural Networks, pp. 1101–1108 (1998)
10. Martinez, T.: Competitive hebbian learning rule forms perfectly topology preserving maps. In: ICANN (1993)
11. Cédras, C., Shah, M.: Motion-based recognition: a survey. Image and Vision Computing 13(2), 129–155 (1995)
12. Flórez, F., García, J.M., García, J., Hernández, A.: Representing 2D-objects. Comparison of several self-organizing networks. In: Advances in Neural Networks World, pp. 69–72. WSEAS Press (2002)
13. Dubbuison, M.P., Jain, A.K.: A Modified Hausdorff Distance for Object Matching. In: Proceedings of the International Conference on Pattern Recognition, Jerusalem, Israel, pp. 566–568 (1994)
14. Wang, X., Tieu, K., Grimson, E.: Learning Semantic Scene Models by Trajectory Analysis. MIT CSAIL Technical Report (2006)

Symbolic Summation of Polynomials in Linear Space and Quadratic Time

Jose Torres-Jimenez[1], Laura Cruz[2], and Nelson Rangel-Valdez[1]

[1] CINVESTAV-Tamaulipas, Cd. Victoria, Tamaulipas, México
{jtj,nrangel}@cinvestav.mx
http://www.tamps.cinvestav.mx/~jtj
[2] Tecnológico de Cd. Madero, Cd. Madero, Tamaulipas, México
lauracruzreyes@yahoo.com

Summary. Artificial Intelligence (AI) is the core of many current technologies with diverse applications like: a) systems that understand the spoken languages; b) intelligent tutors that assist in the process of learning new concepts; c) systems that detect patterns in huge amounts of data; etc. Also AI has originated many spin-off technologies that are seen as part of our daily lives, v.gr. a) the mouse; b) symbolic programming languages; c) symbolic computation systems like Macsyma. This work is related to the field of symbolic computation, specifically we present an optimized algorithm that is able to compute symbolic summation of polynomials. The algorithm is based on the solution of a system of simultaneous equations that delivers the coefficients of the resulting polynomial.

Keywords: Symbolic Summation, Algorithm Analysis.

1 Introduction

Artificial intelligence (AI) originated the field of symbolic computation that deals with the problem of computing solutions to mathematical questions through symbolic manipulation of mathematical objects, this work deals with the problem of computing symbolic summation of polynomials.

In the analysis of algorithms emerges the need to compute the symbolic summation of polynomials of the form $\sum_{k=1}^{n} \wp(k)$ where $\wp(k) = \sum_{j=0}^{J} C_j * k^j$, such polynomials satisfy the requirement to be a *hyper geometric* series. Some symbolic computation methods like [1, 2, 3] can be used to solve the symbolic summation $\sum_{k=1}^{n} \wp(k)$, but we present a very easy to implement and understand algorithm to accomplish this task.

The main goal of this paper is to present a linear space and quadratic time algorithm (*J-Algorithm*) to compute symbolic summation of polynomials like $\wp(k)$ (the algorithm was developed using well known ideas of undetermined coefficients). The rest of the paper is organized in 4 more sections, section 2 presents the analysis of one algorithm from which the need to compute symbolic summation of polynomials arises, in section 3 the derivation of the *J-Algorithm* is

J.M. Corchado et al. (Eds.): DCAI 2008, ASC 50, pp. 653–657, 2009.
springerlink.com

presented, section 4 presents the pseudocode of the *J-Algorithm* and its space and temporal analysis, finally some conclusions and future works are presented.

2 Algorithm Analysis Example

The theoretical determination of the amount of resources (such as time and storage) necessary to execute an algorithm is the field of work of the *Algorithm Analysis* area [4], in this area it is frequently required to express (as a polynomial) the number of times a particular (or a set of) instruction(s) is executed. In order to compute the previously mentioned polynomial, it is necessary to compute (maybe many times) a symbolic summation like $\sum_{k=a}^{n} \wp(k)$. The expression $\sum_{k=a}^{n} \wp(k)$ can be computed by: $\sum_{k=1}^{n} \wp(k) - \sum_{k=1}^{a-1} \wp(k)$, then it is only required a general procedure to compute summations like: $\sum_{j=1}^{m} \wp(j)$. Next the analysis of one algorithm is presented, illustrating the need of computing the symbolic summation of a polynomial. In algorithm 2.1 the method of Gaussian elimination is presented [4]. The number of times the statement **(i)** of the algorithm 2.1 is executed, and its simplification is presented in equation 1.

$$\begin{aligned}\sum_{i=1}^{n}\sum_{j=i+1}^{n}\sum_{k=i}^{n+1}(1) &= \sum_{i=1}^{n}\sum_{j=i+1}^{n}(n+2-i)\\ &= \sum_{i=1}^{n}(n^2+2n-i(n+2)+i^2)\\ &= \frac{n^3}{3}+\frac{n^2}{2}-\frac{5n}{6}\end{aligned} \tag{1}$$

Algorithm 2.1. GAUSSIANELIMINATION(M)

comment: $Input : M[n \times (n+1)]$
comment: $Output : M\ after\ Gaussian\ Elimination$
for $i \leftarrow 1$ **to** n
  do { **for** $j \leftarrow i+1$ **to** n
    do { **for** $k \leftarrow n+1$ **downto** i
      do { $M_{j,k} = M_{j,k} - M_{i,k} * M_{j,i}/M_{i,i}$ **(i)**

3 *J-Algorithm* Derivation

In the previous section an example in which the need to compute the symbolic summation of a polynomial was given, the problem can be stated in general form as computing the equation 2, where A is an array of constant coefficients, i.e. they do not depend on the index i.

$$\sum_{i=1}^{n}\left(\sum_{d=0}^{\delta} A_d i^d\right) = B_1 n + \cdots + B_\delta n^\delta + B_{\delta+1} n^{\delta+1} + \dots \tag{2}$$

The summation of (2) with upper limit equals to $n-1$ is given in (3).

$$\sum_{i=1}^{n-1}(\sum_{d=0}^{\delta} A_d i^d) = B_1(n-1) + \dots + B_\delta(n-1)^\delta + B_{\delta+1}(n-1)^{\delta+1} + \dots \tag{3}$$

Then subtracting (3) from (2), (4) is obtained.

$$\sum_{d=0}^{\delta} A_d n^d = B_1(n-(n-1)) + \dots + B_\delta(n^\delta - (n-1)^\delta) + B_{\delta+1}(n^{\delta+1} - (n-1)^{\delta+1}) + \dots \tag{4}$$

Simplifying the right side of (4), (5) is obtained.

$$\sum_{d=0}^{\delta} A_d n^d = B_1 + \dots + B_\delta \sum_{k=1}^{\delta}(-1)^{k+1}\binom{\delta}{k}n^{\delta-k} + B_{\delta+1}\sum_{k=1}^{\delta+1}(-1)^{k+1}\binom{\delta+1}{k}n^{\delta+1-k} + \dots \tag{5}$$

From equation 5 the set of simultaneous equations given in (6) is obtained.

$$\begin{bmatrix} \binom{1}{1} & -\binom{2}{2} & \binom{3}{3} & \dots & (-1)^{\delta+1}\binom{\delta}{\delta} & (-1)^{\delta+2}\binom{\delta+1}{\delta+1} \\ 0 & \binom{2}{1} & -\binom{3}{2} & \dots & (-1)^{\delta}\binom{\delta}{\delta-1} & (-1)^{\delta+1}\binom{\delta+1}{\delta} \\ 0 & 0 & \binom{3}{1} & \dots & (-1)^{\delta-1}\binom{\delta}{\delta-2} & (-1)^{\delta}\binom{\delta+1}{\delta-1} \\ \vdots & \vdots & \vdots & \ddots & \vdots & \vdots \\ 0 & 0 & 0 & \dots & \binom{\delta}{1} & -\binom{\delta+1}{2} \\ 0 & 0 & 0 & \dots & 0 & \binom{\delta+1}{1} \end{bmatrix} \begin{bmatrix} B_1 \\ B_2 \\ B_3 \\ \vdots \\ B_\delta \\ B_{\delta+1} \end{bmatrix} = \begin{bmatrix} A_0 \\ A_1 \\ A_2 \\ \vdots \\ A_{\delta-1} \\ A_\delta \end{bmatrix} \tag{6}$$

Given that the equation 6 is an upper triangular matrix, in order to compute the values of $B_1, \dots, B_{\delta+1}$, it is necessary to compute the next steps:

$$\begin{aligned}
(\text{step } \delta)\quad & A'_\delta = \frac{A_\delta}{\binom{\delta+1}{1}} \\
& A'_{\delta-1} = A_{\delta-1} - \binom{\delta+1}{2}(-A'_\delta), \dots \\
& A'_2 = A_2 - \binom{\delta+1}{\delta-1}((-1)^\delta A'_\delta), \\
& A'_1 = A_1 - \binom{\delta+1}{\delta}((-1)^{\delta+1} A'_\delta), \\
& A'_0 = A_0 - \binom{\delta+1}{\delta+1}((-1)^{\delta+2} A'_\delta) \\
(\text{step } \delta-1)\quad & A'^{(2)}_{\delta-1} = \frac{A'_{\delta-1}}{\binom{\delta}{1}}, \dots \\
& A'^{(2)}_2 = A'_2 - \binom{\delta}{\delta-2}((-1)^{\delta-1})A'^{(2)}_{\delta-1}), \\
& A'^{(2)}_1 = A'_1 - \binom{\delta}{\delta-1}((-1)^{\delta} A'^{(2)}_{\delta-1}), \\
& A'^{(2)}_0 = A'_0 - \binom{\delta}{\delta}((-1)^{\delta-1} A'^{(2)}_{\delta-1}) \\
& \vdots \\
(\text{step } 2)\quad & A'^{(\delta-1)}_2 = \frac{A'^{(\delta-2)}_2}{\binom{3}{1}}, \dots \\
& A'^{(\delta-1)}_1 = A'^{(\delta-2)}_1 + \binom{3}{2} A'^{(\delta-1)}_2, \\
& A'^{(\delta-1)}_0 = A'^{(\delta-2)}_0 - \binom{3}{3} A'^{(\delta-1)}_2
\end{aligned}$$

$$\text{(step 1)}\ A_1^{'(\delta)} = \frac{A_1^{'(\delta-1)}}{\binom{2}{1}}, \ldots$$
$$A_0^{'(\delta)} = A_0^{'(\delta-1)} + \binom{2}{2} A_1^{'}$$

The final values of $B_1, \ldots, B_{\delta+1}$ are stored in $A_0^{'\delta}, A_1^{'\delta}, A_2^{'\delta-1}, \ldots, A_{\delta-1}^{'(2)}, A_\delta^{'}$. As it can be inferred all the elements of array B are computed using only one array A, the super-indices of array A in the previous steps were used only to illustrate the order of the computations.

Also it can be seen that the matrix of the left side of (6) consists of binomial coefficients with alternating signs, then it is easy to compute one element using the previous one, and the fact that $\binom{x}{y+1} = \binom{x}{y}\frac{x-y}{y+1}$ (the sign change can be done multiplying by -1). Then it is not necessary to store the matrix of (6), the matrix elements are computed and accessed using one variable (α variable in the *J-Algorithm* in the next section).

4 *J-Algorithm* and Its Spatial and Temporal Analysis

The algorithm 4.1 presents the computation details of the *J-algorithm*, derived in the previous section.

The *J-Algorithm* only requires one *linear* array: $A[\delta+1]$ that contains the input polynomial, and at the end of the *J-Algorithm*, it contains the output polynomial; additionally it uses three variables: α, i, and j. Then the space requirements are: $(\delta+1)+3 = \delta+4$.

For the temporal complexity, only will be counted the number of times each relevant statement is executed, the statements **statement-1** and **statement-2** are executed δ times, and the statements **statement-3** and **statement-4** are executed $\frac{\delta(\delta+1)}{2}$.

Algorithm 4.1. J-ALGORITHM(A)

```
comment: Input : A[δ+1] coefficients of the input polynomial
comment: Output : A[δ+1] coefficients of the output polynomial
local α, i, j
for i ← δ downto 1
   do { α ← i + 1                          (statement-1)
        A[i] ← A[i] / α                    (statement-2)
        for j ← i − 1 downto 0
           do { α ← −α (j+1)/(i−j+1)       (statement-3)
                A[j] ← A[j] − αA[i]        (statement-4)
```

Then the claim of linear space($\delta+4$), and quadratic time ($\delta^2+3\delta$) is obtained.

5 Conclusions

In this paper a linear-space ($\delta + 4$) and quadratic-time($\delta^2 + 3\delta$) algorithm for computing symbolic summation of polynomials was presented. The algorithm requires the evaluation of a small number of instructions and conspicuously, it does not require a matrix to solve the associated system of linear equations.

Currently we are working in two extensions to the *J-Algorithm*: a) the case when the A vector contains polynomials with many variables; and b) the case when the limits of the summations are polynomials on many variables.

Acknowledgments. This research was partialy funded by the following projects CONACyT 58554-Calculo de Covering Arrays, 51623-Fondo Mixto CONACyT y Gobierno del Estado de Tamaulipas.

References

1. Fasenmyer, S.M.C.: Some generalized hypergeometric polynomials. PhD thesis, University of Michigan (1945)
2. Gosper Jr., R.W.: Indefinite hypergeometric sums in macsyma. In: Proceedings of the MACSYMA User's Conference Berkeley, pp. 237–251 (1977)
3. Graham, R., Knuth, D., Patashnik, O.: Concrete mathematics. Addison-Wesley, Reading (1989)
4. Cormen, T., Leiserson, C., Rivest, R., Stein, C.: Introduction to Algorithms, 2nd edn. MIT Press and McGraw-Hill (2001)

Solving the Oil Spill Problem Using a Combination of CBR and a Summarization of SOM Ensembles

Aitor Mata[1], Emilio Corchado[2], and Bruno Baruque[2]

[1] University of Salamanca, Spain
aitor@usal.es
[2] University of Burgos, Spain
escorchado@ubu.es, bbaruque@ubu.es

Abstract. In this paper, a forecasting system is presented. It predicts the presence of oil slicks in a certain area of the open sea after an oil spill using Case-Based Reasoning methodology. CBR systems are designed to generate solutions to a certain problem by analysing historical data where previous solutions are stored. The system explained includes a novel network for data classification and retrieval. Such network works as a summarization algorithm for the results of an ensemble of Self-Organizing Maps. This algorithm, called Weighted Voting Superposition (WeVoS), is aimed to achieve the lowest topographic error in the map. The WeVoS-CBR system has been able to precisely predict the presence of oil slicks in the open sea areas of the north west of the Galician coast.

Keywords: Case-Based Reasoning; oil spill; Self Organizing Memory; summarization; Radial Basis Function.

1 Introduction

After an oil spill, decisions must be taken fast and accurately in order to avoid natural disasters. Forecasting the probability of finding oil slicks after an oil spill in a certain area will offer a great support to take those critical decisions. In this paper a forecasting system is presented, which aim is to proportionate the probability of finding oil slicks in a certain area using the Case-Based Reasoning methodology and a new summarization algorithm for SOM ensembles.

The system explained here has been developed using historical data and the information obtained after the Prestige accident in November 2002. To obtain data about the oil slicks, like their positions and sizes, SAR (*Synthetic Aperture Radar*) satellite images [1] has been used. After presenting the oil spill problem, both the Case-Based Reasoning methodology and the Weighted Voting Summarization algorithm are explained. Then, the developed system is described, ending with the results and conclusions.

2 Oil Spill Risks Supervise

When an oil spill is produced, it is crucial to know the progress of the oil slicks generated. The evolution of those slicks should be overseen or even forecasted so that a prediction of the arrival of oil slicks to certain areas can be done. To obtain a precise

J.M. Corchado et al. (Eds.): DCAI 2008, ASC 50, pp. 658–662, 2009.
springerlink.com

prediction the behaviour of the oil slicks must be known. Working in an instable and highly changing environment such as the open sea increases the difficulty of providing good predictions.

Data about the oil slicks, like position, shape and size, must be obtained. The best way to acquire that information is by using satellite images, specially SAR satellite images. With those satellite images it is possible to discriminate between normal sea variability and slicks. Distinguish between oil slicks and similar look-alikes is also essential. There have been different approaches to estimate, analyze and forecast the critical situations after an oil spill. In this paper a CBR system is used to generate precise predictions using both historical data and information obtained from satellites.

3 CBR Methodology

Case-Based Reasoning methodology is developed from the evolution of knowledge based systems. Past situations are used by CBR systems as the key elements to solve new problems [2]. The *case* base is the main element in the CBR systems structure. The case base stores all the information used to generate new solutions. That information is organized so that similar elements can be easily recovered to solve new problems.

CBR systems learn from past experiences. The process of acquiring new knowledge from past data is divided in four main phases: *retrieval, reuse, revision* and *retention*. In the *retrieve* phase the most similar cases to the proposed problem are recovered from the case base. With those cases, the system must create a solution adapting them to the new problem, *reusing* them. Once the solution is generated by the system, it must be *revised* to check its correction. If the proposed solution is accepted then it can be eventually *retained* by the system, if there are not redundant information stored, and could serve as a solution to future problems.

4 WeVoS: Weighted Voting Summarization of SOM Ensembles

Case-Based Reasoning systems are highly dependent on stored information. The new algorithm presented here, *Weighted Voting Summarization of Som ensembles (WeVoS)* is used to organize the data that is accumulated in the case base. It is also used to recover the most similar cases to the proposed problem.

The main objective of the new fusion of an ensemble of topology preserving maps [3] algorithm presented here, WeVoS, is to generate a final map processed unit by unit. Instead of trying to obtain the best position for the units of a single map trained over a single dataset, it aims to generate several maps over different parts of the dataset. Then, it obtains a final summarized map by calculating by consensus which is the best set of characteristics vector for each unit position in the map. To do this calculation, first this meta-algorithm must obtain the "quality" [4] of every unit that composes each map, so that it can relay in some kind of informed resolution for the fusion of neurons.

The final map obtained is generated unit by unit. The units of the final map are first initialized by determining their centroids in the same position of the map grid in each of the trained maps. Afterwards, the final position of that unit is recalculated using data related with the unit in that same position in every of the maps of the ensemble. For each unit, a sort of voting process is carried out as shown in Eq. 1:

$$V(p,m)=\frac{|x_{p,m}|}{\sum_1^M |x_p|}\cdot\frac{q_{p,m}}{\sum_1^M q_p} \tag{1}$$

The final map is fed with the weights of the units as it is done with data inputs during the training phase of a SOM, considering the "homologous" unit in the final map as the BMU. The weights of the final unit will be updated towards the weights of the composing unit. The difference of the updating performed for each "homologous" unit in the composing maps depends on the quality measure calculated for each unit. The higher quality (or the lowest error) of the unit of the composing map, the stronger the unit of the summary map will be updated towards the weights of that neuron. The summarization algorithm will consider the weights of a composing unit "more suitable" to be the weights of the unit in the final map according to both the number of inputs recognized and the quality of adaptation of the unit (Eq. 1). With this new approach it is expected to obtain more faithful maps to the inner structure of the dataset.

5 WeVos-CBR

There have already been CBR systems created to solve maritime problems [5] in which different maritime variables have been used. In this occasion, the data used have been previously collected from different observations from satellites, and then pre-processed, and structured to create the case base. The created cases are the main elements to obtain the correct solutions to future problems, through the CBR system. The developed system determines the probability of finding oil slicks in a certain area after an oil spill has been produced.

5.1 Pre-processing and Retrieval

When the case base is created the WeVoS algorithm is used to structure it. The graphical capabilities of this novel algorithm are used in this occasion to create a model that represents the actual variability of the parameters stored in the cases. At the same time, the inner structure of the case base will make it easier to recover the most similar cases to the problem cases introduced in the system.

The WeVos algorithm is also used to recover the most similar cases to the problem introduced in the system. That process if performed once the case base is structured keeping the original distribution of the available variables.

5.2 Reuse

After recovering the most similar cases to the problem from the case base, those cases are used to obtain a solution. *Growing RBF networks* [6] are used to generate the predicted solution corresponding to the proposed problem. The selected cases are used to train the GRBF network. This adaptation of the RBF network lets the system grow during the training phase in a gradual way increasing the number of elements (prototypes) which work as the centres of the radial basis functions. The error definition for every pattern is shown below:

$$e_i = 1/_p \cdot \sum_{k=1}^{p} \| t_{ik} - y_{ik} \| \quad (2)$$

Where t_{ik} is the desired value of the kth output unit of the ith training pattern, y_{ik} the actual values ot the kth output unit of the ith training pattern. After the creation of the GRBF network, it is used to generate the solution to the introduced problem. The solution will be the output of the network using as input data the retrieved cases.

5.3 Revision and Retain

In order to verify the precision of the proposed solution, *Explanations* are used [7]. To justify and validate the given solution, the retrieved cases are used once again. The selected cases have their own future associated situation. Considering the case and its solution as two vectors, a distance between them can be measured by calculating the evolution of the situation in the considered conditions. If the distance between the proposed problem and the solution given smaller than the distances obtained from the selected cases, then the proposed solution considered as a good one.

Once the proposed prediction is accepted, it can be stored in the case base in order to serve to solve new problems. It will be used equally than the historical data previously stored in the case base. The *WeVoS* algorithm is used again to introduce new elements in the case base.

6 Results

The WeVoS-CBR system has been checked with a subset of the available data that has not been previously used in the training phase. The predicted situation was contrasted with the actual future situation as it was known (historical data was used to train the system and also to test its correction). The proposed solution was, in most of the variables, close to 90% of accuracy.

Table 1. Percentage of good predictions obtained with different techniques

Number of cases	RBF	CBR	RBF + CBR	WeVoS-CBR
500	46 %	41 %	44 %	47 %
1000	49 %	46 %	55 %	63 %
3000	58 %	57 %	66 %	79 %
5000	60 %	62 %	73 %	88 %

Table 1 shows a summary of the obtained results. In this table different techniques are compared. The evolution of the results is shown along with the augmentation of the number of cases stored in the case base. All the techniques analyzed improve their results at the same time the number of stored cases is increased. The solution proposed do not generate a trajectory, but a series of probabilities in different areas, what is far more similar to the real behaviour of the oil slicks.

References

[1] Palenzuela, J.M.T., Vilas, L.G., Cuadrado, M.S.: Use of ASAR images to study the evolution of the Prestige oil spill off the Galician coast. International Journal of Remote Sensing 27(10), 1931–1950 (2006)

[2] Aamodt, A.: A Knowledge-Intensive, Integrated Approach to Problem Solving and Sustained Learning, Knowledge Engineering and Image Processing Group. University of Trondheim (1991)

[3] Kohonen, T.: The self-organizing map. Neurocomputing 21(1-3), 1–6 (1998)

[4] Pölzlbauer, G.: Survey and Comparison of Quality Measures for Self-Organizing Maps. In: Rauber, J.P. (ed.) Fifth Workshop on Data Analysis (WDA 2004), pp. 67–82. Elfa Academic Press (2004)

[5] Corchado, J.M., Fdez-Riverola, F.: FSfRT: Forecasting System for Red Tides. Applied Intelligence 21, 251–264 (2004)

[6] Karayiannis, N.B., Mi, G.W.: Growing radial basis neural networks: merging supervised andunsupervised learning with network growth techniques. IEEE Transactions on Neural Networks 8(6), 1492–1506 (1997)

[7] Sørmo, F., Cassens, J., Aamodt, A.: Explanation in Case-Based Reasoning–Perspectives and Goals. Artificial Intelligence Review 24(2), 109–143 (2005)

A Symbiotic CHC Co-evolutionary Algorithm for Automatic RBF Neural Networks Design

Elisabet Parras-Gutierrez[1], Mª José del Jesus[1], Juan J. Merelo[2], and Víctor M. Rivas[1]

[1] Department of Computer Sciences
Campus Las Lagunillas s/n, 23071, Jaén, Spain
eparrasg@vrivas.es, mjjesus@ujaen.es, vrivas@vrivas.es
[2] Department of Computers Architecture and Technology
C/ Periodista Daniel Saucedo sn, 18071 Granada, Spain
jmerelo@geneura.ugr.es

Abstract. This paper introduces Symbiotic_CHC_RBF, a co-evolutionary algorithm intended to automatically establish the parameters needed to design models for classification problems. Co-evolution involves two populations, which evolve together by means of a symbiotic relationship. One of the populations is the method EvRBF, which provides the design of radial basis function neural nets by means of evolutionary algorithms. The second population evolves sets of parameters for the method EvRBF, being every individual of the population a configuration of parameters for the method. Results show that Symbiotic_CHC_RBF can be effectively used to obtain good models, while reducing significantly the number of parameters to be fixed by hand.

Keywords: neural networks, evolutionary algorithms, co-evolution, parameter estimation, CHC algorithm.

1 Introduction

Data mining methods have been proved to be very useful. Nevertheless, most of them need to be given a good parameter setting in order to make them work properly. Selecting good, or even the best, possible data mining method and finding a reasonable parameter setting are very time consuming tasks.

This paper introduces the Symbiotic_CHC_RBF algorithm. This method automatically designs models for classification using two populations that evolve together by means of co-evolution. First population is composed of many instances of the method EvRBF [14,15], which provides the design of radial basis function neural nets by means of evolutionary algorithms. The second population evolves sets of parameters for the method EvRBF, being every individual of the above mentioned population a configuration of parameters for it.

Thus, for every given problem, Symbiotic_CHC_RBF automatically searches the optimal configuration of parameters that the method EvRBF should use. Symbiosis shows then its ability to increase the usability of traditional methods, one of the key issues on future trends in data mining [7].

In Nature, co-evolution can be classified as competitive (parasitic), cooperative (symbiotic), and exploitation. Competitive evolution takes place when the organisms,

J.M. Corchado et al. (Eds.): DCAI 2008, ASC 50, pp. 663–671, 2009.
springerlink.com

belonging to the same species or not, compete for the same resources. On the other hand, exploitation appears when the presence of a species stimulates the development of another, but the presence of this one disables the development of the first one. And finally, the interactions among two species are cooperative when both (of them) benefit from this interaction. Symbiont organisms are extreme cases in which two species are so intimately integrated that they act as an alone organism. In any case, from a biological point of view, co-evolution can be defined as "the evolution of two or more species that do not cross among them but that possess a narrow ecological relation, across reciprocal selection pressures, the evolution of one of the species depends on the evolution of other one" [19]. The symbiotic co-evolution approach used in this work makes every population to contribute to the evolution of the other and vice versa.

The rest of the paper is organized as follows: section 2 introduces some of the papers found in literature and related to this research; section 3 shows the Symbiotic_CHC_RBF method, as well as EvRBF, the method it is based on. Section 4 shows the experimentation carried out and the result we obtained; finally, section 5 describes some conclusions and future works.

2 Related Work

One of the key issues in the study of artificial neural networks (ANN) is the search of method able to design optimal models [21], and both evolutionary and non-evolutionary algorithms have been developed to face this task. Most non-evolutionary algorithm can be classified into constructive and destructive methods. Constructive algorithm start with a minimal network (network with minimal number of hidden layers, nodes, and connections) and add new layers, nodes, and connections during training; destructive algorithms do the opposite, i.e., start with the maximal network and delete unnecessary layers, nodes, and connections during training.

Harpham et al. reviewed in [6] some of the best known methods that apply evolutionary algorithms to Radial Basis Function Neural Network (RBFNN) design. They concluded that, in general, methods tend to concentrate in only one aspect when designing this kind of nets. Nevertheless, there also exist methods intended to optimize the whole net, such as [9] for Learning Vector Quantization nets or [3] for multilayer perceptrons.

Chronologically, initial papers [4, 20] used binary codification and were constricted by the number of hidden neurons, that had to be set a priori. While Carse [4] works with a population of nets, Whitehead [20] works with a population of neurons (thus, only one RBFNN was build and evaluated) that compete and cooperate to find the optimal net.

Subsequent papers [2] presented algorithms based on real number representation. Once more, the imposition of a limit to the number of neurons, optimization of only the centers for hidden neurons, and a badly defined penalization term for the number of times an individual can reproduce, are the main drawbacks of this algorithm.

Recent applications try to overcome the disadvantages of the preceding methods. Rivera [16] made many neurons compete, modifying them by means of *fuzzy evolution*, i.e., using a fuzzy rules table that specifies the operator that must be applied to a neuron in order to improve its behavior.

Most recent methods still face single aspects when configuring a net. Thus, Ros et al.'s method [17] automatically initializes RBFNN, finding a good set of initial neurons, but relying in some other method to improve the net and to set its widths.

Different kinds of co-evolution have also been used to design neural nets, as can be found in literature. Paredis [10] competitively coevolved the weights of networks of fixed topology for a synthetic classification task. After this, Paredis [11] proposed a general framework for the use of co-evolution to boost the performance of genetic search, combining co-evolution with life-time fitness evaluation.

Potter and De Jong [13] proposed Cooperative Co-evolutionary Algorithms for the evolution of ANNs of cascade network topology. Co-evolution occurs by decomposing the problem into interacting subtasks.

Barbosa [1] employs a competitive co-evolutionary genetic algorithm for structural optimization problems using a game-theoretic approach. The first player, the designer of a mechanical structure, tries to minimize the Compliance (a measure of the overall deformability under a specific load). The second player, nature, challenges the designer's constructions by maximizing the compliance.

Schwaiger and Mayer [18] employed a genetic algorithm for the parallel selection of appropriate input patterns for the training datasets. After this, Mayer [7] investigated the use of symbiotic (cooperative) co-evolution, in order to generate coadapted networks and datasets without human intervention. In his approach, independent populations of ANNs and training datasets were evolved by a genetic algorithm, where the fitness of an ANN was equally credited to the dataset it had been trained with.

3 Method Overview

The optimal design of a net involves the establishment of its structure as well as the weights of the links. In this sense, RBFNN represent a special kind of NN since, once the structure has been fixed, the optimal set of weights can be computed. Symbiotic_CHC_RBF is based on EvRBF, an evolution method developed to automatically design asymmetric RBF.

The main goal of Symbiotic_CHC_RBF is to find a suitable configuration of parameters necessary for the method EvRBF, which is adapted automatically to every problem. In the section 3.1 and 3.2 both methods are described detailed.

3.1 The EvRBF Algorithm

EvRBF [14, 15] is a steady state evolutionary algorithm that includes elitism. It follows Pittsburgh scheme, in which each individual is a full RBFNN whose size can vary, while population size remains equal.

EvRBF codifies in a straightforward way the parameters that define each RBFN, using an object-oriented paradigm; for this reason, it includes two operators for recombination and four for mutation that directly deal with the neurons, centers and widths themselves. Recombination operators (X_FIX and X_MULTI) interchange information between individuals, trying to find the building blocks of the solution. In the other hand, mutation operators (centers and width modification: C_RANDOM and R_RANDOM) use randomness to increase diversity generating new individuals so that local minima can be avoided. Furthermore, EvRBF tries to determine the correct size of the hidden layer using the operators ADDER and DELETER to create and remove neurons, respectively.

The exact number of neurons affected by these operators (except ADDER) is determined by their internal application probabilities.

EvRBF incorporates tournament selection for reproduction. The *fitness function* measures the generalization capability of each individual as the percentage of samples it correctly classifies. When comparing two individuals, if and only if both two individuals have exactly the same error rate, the one with less neurons is said to be better. The LMS algorithm is used in order to train individuals, and to obtain the weights of links.

3.2 Description of Symbiotic_CHC_RBF

Symbiotic_CHC_RBF uses the CHC algorithm [5], an evolutionary approach which introduces an appropriate balance between diversity and convergence. Its main aim is to combine an elitist selection, preserving the best individuals who have appeared up to the moment, with an operator of crossing, which produces children very different from their parents. Therefore, it uses a high selective pressure based on an elitist scheme in combination with a highly disruptive crossover and a restart when the population is stagnated. Then, it is based on four components [5]:

- Elitist selection: it selects the best individuals to form the new population. The best individuals found up to the moment will remain in the current population.
- HUX crossover operator: it exchanges exactly the half of the individual's parts that they are different in the parents. As a result, it guarantees the most diverse offsprings from their parents. (The operator of mutation does not use).
- Incest prevention: two individuals are not crossed over if they are too similar, so this ensures diversity.
- Restart: it is restarted when the population reaches a stagnated state, keeping the best individual.

Every individual of the algorithm Symbiotic_CHC_RBF is a string composed of 0 and 1 representing a set of 8 parameters for the method EvRBF, as the size of the poppulation, the size of the tournament for the selection of the individuals or the maximum number of generations. This sequence of bits is organized as follows:

- 7 bits for the size of the population (a value between 2 and 100)
- 2 bits for the size of tournament size (between 2 and 5)
- 6 bits for the number of generations (between 0 and 50)
- 10 bits for the percentage of the initial limit of neurons (between 0.1 and 1.0.)
- 10 bits to represent the percentage of training patterns used for validation (between 0.1 and 0.5)
- 10 bits for the rate of replacement (between 0.1 and 1.0.)
- 10 bits for the rate of the cross over (between 0.0 and 1.0)
- 10 bits that will represent the rate of mutation (between 0.0 and 1.0.)

In order to set the fitness of an individual, its chromosome is decoded into the set of parameters it represents. After this, these parameters are used to perform a complete execution of EvRBF. Once EvRBF has finished, the percentage of training patterns correctly classified by the best net found is used as fitness for the individual. The skeleton of Symbiotic_CHC_RBF is showed in Figure 1.

Individuals of the first population are created randomly, although the algorithm is able to accept predefined individuals that can act as guide for the search of the solution.

1. Create, train, evaluate and set fitness of first generation.
2. Set threshold for cross over
2. Until stop condition is reached
   (a) Select and copy individuals from current population.
   (b) Modify, train, and set fitness of the copies.
   (c) Replace worst individuals by the copies.
   (d) Reduce the threshold if the population is the same
   (e) If threshold < 0 restart the population
3. Train last generation with training and validation data
4. Use test data set to obtain the generalization ability of each individual.

Fig. 1. General skeleton of Symbiotic_CHC_RBF

4 Experimental Results

The system Symbiotic_CHC_RBF has been evaluated across three data sets from UCI data set repository[1]: ionosphere (351 instances, 34 features, 2 classes), Sonar (208 instances, 60 features, 2 classes) and Wdbc (570 instances, 30 features, 2 classes). For every database, 7 different sets of data have been generated using various feature selection methods, so that Symbiotic_CHC_RBF has been tested over 21 different classification databases.

Feature selection algorithms used in this work included both filter and wrapper methods. Briefly, FS-LVF and FS-Focus are non-evolutionary filters; FS-Forward and FS-Backward algorithms are two classic, non-evolutionary wrapper greedy methods; FS-GGA generational genetic algorithm filter; FS-SSGA is a steady state genetic algorithm wrapper. Finally, the seventh dataset for every problem is composed of the original data. Table 1 shows the number of attributes present in every dataset.

In order to test the generalization capabilities of Symbiotic_CHC_RBF, a 10-crossfold validation method has been used, so that every dataset has been divided into 10 different sets of training-test patterns; for every training-set couple, the algorithm has been executed three times, and the results show the average of these 30 executions.

The performance of the method has been compared with two other evolutionary algorithms found in the literature: EvRBF [12, 14, 15], and CO^2RBFN [12][2]. Both of them are designed to automatically build RBF neural networks, the first one by means of a Pittsburg genetic algorithm, and the second using a cooperative-competitive evolutionary algorithm, i.e., another kind of co-evolutionary approach.

As CHC clearly defines the way individuals have to be selected, reproduce and substitute themselves, the parameters needed to run the experiments were the number of generations, set to 30, and the number of individuals, set to 50.

[1] http://www.ics.uci.edu/~mlearn/MLRepository.html

[2] Results yielded by EvRBF and CO^2RBFN have been taken from [12].

Table 1. Number of attributes for every problem with respect to the different feature selection methods

Preprocessing	Ionosphere	Sonar	WDBC
None	34	60	30
FS-GGA	7	14	7
FS-LFV	9	17	8
FS-SSGA	9	17	8
FS-BackwardLVO	29	39	27
FS-Focus	3	6	10
FS-ForwardLVO	4	4	5

Tables 2 to 4 show the results obtained by Symbiotic_RBF, EvRBF and CO^2RBFN in the above cited classification problems. The percentages of classification yielded by Symbiotic_CHC_RBF are similar or better to those obtained by EvRBF and CO^2RBFN, as can be seen. Thus, the overfitting problem does not seem to affect to this new algorithm, despite the fact that the training dataset is used more times in Symbiotic_CHC_RBF than in the rest of algorithms.

Table 2. Experimental results with Sonar database

Feature selection	Symbiotic_CHC_RBF		EvRBF		CO^2RBFN	
	RBF nodes	Test (%)	RBF nodes	Test (%)	RBF nodes	Test (%)
None	**6**	71,79	**6**	70,20	8	**75,27**
FS-GGA	15	**85,25**	**8**	80,49	**8**	70,47
FS-LFV	11	**82,81**	**8**	79,30	**8**	73,42
FS-SSGA	16	**84,70**	9	81,40	**8**	72,45
FS-BackwardLVO	15	**84,77**	10	81,34	8	73,68
FS-Focus	8	**85,52**	**6**	82,94	8	75,80
FS-ForwardLVO	6	**81,16**	**4**	80,90	8	74,51

Both EvRBF and Symbiotic_CHC_RBF do not fix neither imposes an upper limit to the size of the net. On the other hand CO^2RBFN is executed fixing the number of neurons by hand. Although this could turn into over-specified models, Symbiotic_CHC_RBF is able to build nets whose size is similar to those designed with EvRBF. The search for the effective number of neurons needed by the net is guided by means of the genetic operators used by EvRBF, and the probabilities used by these operators are set by the symbiotic method.

Table 2 to 4 can lead to conclude that Symbiotic_CHC_RBF can be effectively used to establish the parameters EvRBF needs, and can be easily extended to any other evolutionary algorithm. This property should be specially taken into account when statistical methods (like ANOVA) can not be considered because of the high

Table 3. Experimental results with Ionosphere database

Feature selection	Symbiotic_CHC_RBF		EvRBF		CO^2RBFN	
	RBF nodes	Test (%)	RBF nodes	Test (%)	RBF nodes	Test (%)
None	13	**97,45**	10	96,65	**8**	93,20
FS-GGA	18	**95,89**	12	93,55	**8**	83,39
FS-LFV	11	**94,03**	10	93,56	**8**	87,95
FS-SSGA	11	**95,83**	9	95,16	**8**	89,98
FS-BackwardLVO	19	**97,64**	14	96,34	**8**	91,61
FS-Focus	17	**94,90**	15	93,70	**8**	88,57
FS-ForwardLVO	10	94,87	9	95,05	**8**	92,49

Table 4. Experimental results WDBC database

Feature selection	Symbiotic_CHC_RBF		EvRBF		CO^2RBFN	
	RBF nodes	Test (%)	RBF nodes	Test (%)	RBF nodes	Test (%)
None	8	95,02	**7**	95,51	8	**96,27**
FS-GGA	10	**96,72**	12	96,42	**8**	95,85
FS-LFV	**7**	92,21	8	92,76	8	**94,51**
FS-SSGA	7	94,83	**6**	94,97	8	**95,53**
FS-BackwardLVO	**6**	94,79	8	95,45	8	**96,37**
FS-Focus	9	93,89	10	93,60	**8**	**94,76**
FS-ForwardLVO	**7**	95,14	9	95,40	8	**96,48**

number of parameters that must be evaluated. Actually, Symbiotic_CHC_RBF has been able to significantly reduce the number of parameters needed by EvRBF, so that only the parameters need to run the symbiotic method must be set (only number of generations and population size), and these ones can be set to standard values frequently used in literature.

5 Conclusions and Future Research

In this paper a system to automatically set the parameters of evolutionary algorithms using symbiosis is proposed. Thus, two populations co-evolve in cooperative way contributing one to the evolution of the other and vice versa.

One of the populations is the method EvRBF, which provides the design of radial basis function neural nets by means of evolutionary algorithms. The second population evolves sets of parameters for the method EvRBF, being every individual of the above mentioned population a configuration of parameters for the method. The main goal of the proposed system, Symbiotic_CHC_RBF, is to find a suitable configuration of parameters necessary for the method EvRBF by means of co-evolution.

The system Symbiotic_CHC_RBF has been evaluated across three data sets (Ionosphere, Sonar and WDBC, from UCI repository), but every one has been filtered with 7 different features selection methods, resulting into 21 different datasets, and it has been compared with two other systems (EvRBF and CoCoRBFN) in order to evaluate the performance of the system. The obtained results by Symbiotic_CHC_RBF are similar or even better than in the methods EvRBF and CO^2RBFN.

Future research will include the evaluation of Symbiotic_CHC_RBF on functional approximation problems and its extension in order to be used for time-series forecasting. In this case, Symbiotic_CHC_RBF will be able to both estimate the parameters for EvRBF and also select those intervals needed to perform the optimal forecasting.

Acknowledgements

This work has been partially supported by the CICYT Spanish TIN2005-04386-C05-03, and TIN2005-08386-C05-03 projects.

References

1. Barbosa, H.J.C.: A coevolutionary genetic algorithm for a game approach to structural optimization. In: Proceedings of the Seventh International Conference on Genetic Algorithms, San Francisco, California, pp. 545–552 (1997)
2. Burdsall, B., Giraud-Carrier, C.: GA-RBF: A Self- Optimising RBF Network. In: ICANNGA 1997, pp. 348–351. Springer, Heidelberg (1997)
3. Castillo, P.A., et al.: G-Prop: Global optimization of multilayer perceptrons using GAs. Neurocomputing 35, 149–163 (2000)
4. Carse, B., Fogarty, T.C.: Fast evolutionary learning of minimal radial basis function neural networks using a genetic algorithm. In: Evolutionary Computing, pp. 1–22. Springer, Heidelberg (1996)
5. Eshelman, L.J.: The CHC adptive search algorithm: How to have safe search when engaging in nontraditional genetic recombination. In: First Workshop on Foundations of Genetic Algorithms, pp. 265–283. Morgan Kaufmann, San Francisco (1991)
6. Harpham, C., et al.: A review of genetic algorithms applied to training radial basis function networks. Neural Computing & Applications 13(3), 193–201 (2004)
7. Kriegel, H., Borgwardt, K., Kroger, P., Pryakhin, A., Schubert, M., Zimek, A.: Future trends in data mining. Data Mining and Knowledge Discovery: An International Journal 15(1), 87–97 (2007)
8. Mayer, H.A.: Symbiotic Coevolution of Artificial Neural Networks and Training Data Sets. LNCS, pp. 511–520. Springer, Heidelberg (1998)
9. Merelo, J., Prieto, A.: G-LVQ, a combination of genetic algorithms and LVQ. In: Artificial Neural Nets and Genetic Algorithms, pp. 92–95. Springer, Heidelberg (1995)
10. Paredis, J.: Steps towards Co-evolutionary Classification Neural Networks. In: Brooks, R.L., Maes, P. (eds.) Proceedings Artificial Life IV, pp. 545–552. MIT Press / Bradford Books (1994)
11. Paredis, J.: Coevolutionary Computation. Journal Artificial Life, 355–375 (1995)
12. Perez-Godoy, M.D., Aguilera, J.J., Berlanga, F.J., Rivas, V.M., Rivera, A.J.: A preliminary study of the effect of feature selection in evolutionary RBFN design. In: Bouchon-Meunier, B., Yager, R.R. (eds.) Proceedings of Information Processing and Management of Uncertainty (IPMU 2008), Malaga, Spain (June 2008)

13. Potter, M.A., De Jong, K.A.: Evolving Neural Networkds with Collaborative Species. In: Proceedings of the 1995 Summer Computer Simulation Conference (1995)
14. Rivas, V.M., Merelo, J.J., Castillo, P.A., Arenas, M.G., Castellanos, J.G.: Evolving RBF neural networks for time-series forecasting with EvRBF. Information Sciences 165(3-4), 207–220 (2004)
15. Rivas, V.M., Garcia-Arenas, I., Merelo, J.J., Prieto, A.: EvRBF: Evolving RBF Neural Networks for Classification Problems. In: Proceedings of the International Conference on Applied Informatics and Communications (AIC 2007), pp. 100–106 (2007)
16. Rivera, J., Ortega, M., del Jesus, J.: Aproximación de funciones con evolución difusa mediante cooperación y competición de RBFs. In: AEB 2002, pp. 507–514 (2002)
17. Ros, F., Pintore, M., Deman, A., Chrétien, J.R.: Automatical initialization of RBF neural networks. Chemometrics and intelligent laboratory systems (2007)
18. Schwaiger, R., Mayer, H.A.: Genetic algorithms to create training data sets for artificial neural networks. In: Proceedings of the 3NWGA, Helsinki, Finland (1997)
19. Thompson, J.N.: The Geographic Mosaic of Coevolution. University of Chicago Press, Chicago (2005)
20. Whitehead, B.A., Choate, T.: Cooperative-Competitive Genetic Evolution of Radial Basis Function Centers and Widths for Time Series Prediction. IEEE Trans. on Neural Networks 7(4), 869–880 (1996)
21. Yao, X., Shi, Y.: A preliminary Study on Designing Artificial Neural Networks Using Co-Evolution. In: Proc. of the IEEE Singapore Intl Conf on Intelligent Control and Instrumentation, Singapore, pp. 149–154. IEEE, Los Alamitos (1995)

Modeling Processes of AOSE Methodologies by Means of a New Editor

Iván García-Magariño[1], Alma Gómez-Rodríguez[2], and Juan C. González-Moreno[2]

[1] Universidad Complutense de Madrid
ivan_gmg@fdi.ucm.es
[2] Universidade de Vigo
{alma,jcmoreno}@uvigo.es

Summary. This paper introduces a work in progress for definition of processes for AOSE methodologies. FIPA Methodology Technical Committee encourages the definition of software engineering process models for Multi-Agent Systems methodologies and recommends the use of Software Process Engineering metamodel (SPEM) defined by the Object Management Group. This paper follows those recommendations for obtaining a systematic method and an editor for describing software engineering process models for Multiagent Systems. The editor has been built by means of the Eclipse Modeling Framework and follows the SPEM standard. Several Agent Oriented processes have defined using the editor, in addition, the method for defining such processes is also addressed in this paper.

Keywords: Multiagent Systems(MAS), development process, INGENIAS, SPEM, Tools, Metamodel.

1 Introduction

Agent Oriented paradigm has shown its suitability in modeling and developing huge, complex and distributed systems. Therefore, Agent Oriented Software Engineering (AOSE) has become an important field for investigation in Systems Development. In this discipline, software quality assurance is based in methodology definition and usage. All these, jointly with the increasing interest in Multi-Agent Systems (MAS) have result in the definition of several Agent Oriented semiformal methodologies [20, 19, 9, 4, 18].

Nowadays, in the field of quality assurance, one of the more relevant lines of work is the study and improvement of processes for software development and maintenance. The foundations of this field lie on the direct relation between process quality and final product quality. This encourages the necessity of obtaining models of the development processes.

Nevertheless, for modeling development processes, it is necessary a tool which simplifies and automates the activity. The automation of process definition is addressed in this paper by proposing an editor tool which has been used in the specification of some development processes for INGENIAS methodology [20].

J.M. Corchado et al. (Eds.): DCAI 2008, ASC 50, pp. 672–681, 2009.
springerlink.com

Thus, the aim of this paper is twofold. On the one hand, it addresses the work done for obtaining a tool which facilitates methodology process definition. On the other hand, the paper introduces a systematic method for defining a development process using the tool. These objectives are part of the aim of two PhD works which are currently under development. Although the editor tool and mechanism have been used for defining agent-based development processes, since both of them are based on a general-purpose standard (SPEM [15]), they may be used in the definition of process models of any software development methodology.

The remainder of the paper is organized as follows. After making a brief explanation of other works related, section 3 introduces the editor tool constructed for defining process models. Next section addresses the steps to accomplish when defining a process for AOSE methodologies using the previous tool. Finally, section 5 presents conclusions and future work.

2 Related Work

Some authors have made a first approach in the definition of MAS processes introducing a taxonomy which depends on the corresponding AOSE methodology. Cernuzzi [7] provides a survey of the AOSE methodologies and classifies the most relevant ones according to their process models. Other works try to define a metamodel of the development process. For instance, Henderson-Sellers [17] compares four process metamodels and provides a new generic standard, trying to obtain a global development process, with independence of the chosen methodology. Other approach is introduced in [21] where processes and methodologies are described jointly.

Recent works have followed that latest approach for defining new methodologies and their associated processes. For instance, in 2008, a new MAS methodology, called MOBMAS [22], has been defined to combine MAS with ontologies. Furthermore, other authors introduce a deeper detail in existent MAS process/methodologies. That is the case of the work [14], which describes in depth the requirements analysis phase of the Tropos methodology. In addition, Brazier [5] incorporates the component-based software techniques to MAS processes. These works try, using different approaches, to improve, to describe and to adapt the MAS processes to the new software engineering trends. From this, we can conclude that the goal this paper addresses, the formalization of MAS processes, is a constant necessity.

Besides, until the moment there are still few tools which allow the definition of development processes. Among these tools APES [2], EPF [10] and Metameth [8] are the most relevant. Initially, we tried to model development processes using existent tools, in particular APES2 (part of APES for model definition), but it was not possible because it does not implement the full specification of SPEM. The other tools have also been taken into consideration for definition. For instance, Metameth has been constructed with a similar philosophy than the tool proposed in this paper, but there are few indications of how to use it for defining a new process. The results of the comparison of these tools with the editor proposed in this paper can be found in [13].

3 Editor of Process Models

The processes editor, based on SPEM, presented in this paper is built using the techniques of the *Domain-specific Model Languages* (DSML) [1] and the Model-driven Development (MDD) [3]. The use of these techniques makes the tool specially suitable for MAS modeling, because it allows savings in development costs and makes easier the tool adaptation to the changes in methodologies and/or development process. Following Model-driven Development (MDD) approach, first a language model (metamodel) is defined and, next, a framework generates the editor from the metamodel. In this way, the process of obtaining the editor is reduced to the proper definition of the metamodel.

At this moment, there are several metamodeling languages available. Some of them are MOF, EMOF, CMOF, SMOF, ECore, KM3, CORAL M3, Microsoft DSL language and GOPRR. In order to decide among so many languages, two facts are important to be aware of. First, the SPEM specification [15] uses MOF [16] language. Second, the ECore language is used by the most reliable tools and frameworks. In particular, ECore is supported by *Eclipse Modeling Framework (EMF)* [6, 11] and EMF can generate an editor automatically from an ECore metamodel. Therefore, we defined the SPEM metamodel with ECore, from its definition in MOF, so that we got the SPEM editor generated automatically by EMF. In addition, this editor serializes their SPEM models using XMI documents, which are widely supported.

To summarize, the editor proposed in this paper facilitates the definition of software process models with a graphical user interface. In next subsections, the editor is exposed. A more detailed explanation of the editor tool and of the process of definition can be found in [13].

3.1 Creation of Editor Following a MDD Approach

As it has been said before, process editor implementation has followed a MDD approach (see Fig. 1). This implementation has been divided in several phases: specification, design, implementation, testing and maintenance.

For the specification phase, the SPEM [15] notation provided by OMG is used, particularly in its version 1.1; although several characteristics of SPEM 2.0 have been taken into account. Because of the MDD approach utilized, tool can be easily adapted to variations requested in the different standards.

In the design phase, the selection of the metamodel language (metametamodel) and the framework for generating the editor are relevant. In [1], Amyot provides an evaluation of frameworks for developing tools for modeling languages. Following this work, the ECore model has been selected as metametamodel; while EMF is the chosen framework for editor generation. Furthermore, as SPEM is originally defined with MOF by OMG, in this paper the translation

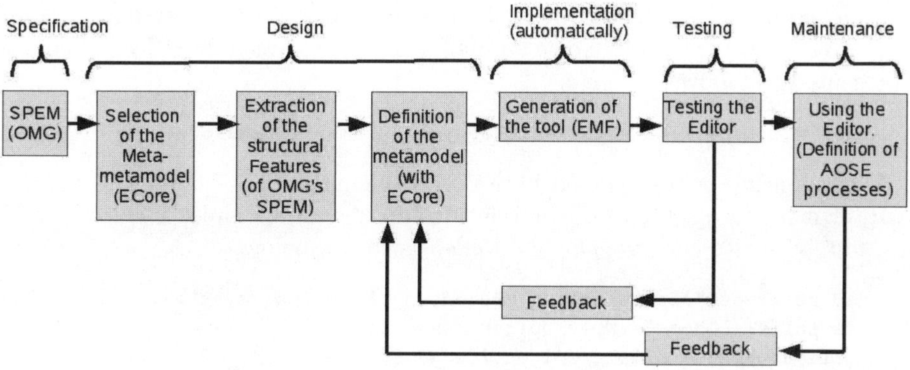

Fig. 1. Process of creation for the SPEM editor

from MOF to ECore is made providing support to the structural features of SPEM [13]. The way of defining the metamodel with ECore has effects on the generated editor. Thus, the stage of metamodel definition must consider the implications over the generated editor and, in some cases, the original metamodel must be adapted. In conclusion, the design phase (see Fig. 1) of the proposed editor includes three stages: the selection of the metametamodel, the extraction of the structural features of the OMG's SPEM, and the definition of SPEM with ECore.

The implementation phase has a low cost since it is computerized and is, therefore, achieved by the automatic generation provided by EMF.

If the results of the automatic generation are not the expected ones, feedback is provided to the design phase and a new iteration starts from the metamodel definition (see Fig. 1). For instance, in the first attempt errors in metamodel definition and problems of usability (too many entities available for definition of the model) were detected. Trying to overcome these problems the metamodel has been changed and the editor was generated again (in the problem of usability related previously, the entities were grouped, so the user only has available in a concrete moment the suitable group of entities).

When the editor is considered stable enough, the process reaches the maintenance phase and is provided to user. The user feedback (detected problems and suggested improvements) implied returning back to the design phase (see Fig. 1) and changing the metamodel definition. The process of generation started again from the metamodel definition.

3.2 The SPEM Editor Serialization

The SPEM editor generates XMI documents which describe the process models. The generation of XMI documents provides an standard way of information interchange with other tools, which will provide, for instance, the basis for customization of CASE tools for a particular development process. In the next paragraphs, a brief description of the generated XMI documents is introduced.

Each entity is represented by a XML document being its attributes represented by XML attributes. For example, a lifecycle named *INGENIAS* can be represented as follows.

```
<wdef xsi:type="plc_entities:Lifecycle" name="INGENIAS"/>
```

A relationship is represented by an XML element for its body and another XML element for each end of the relationship. As an example, a *precedes* relationship, with two ends *supplier* and *client*is shown below.

```
<dep xsi:type="dep_relations:Precedes" pKind="pk_fn_st">
  <supplier EDMod="//@plc/@plcent/@wdef.1"/>
  <client EDMod="//@plc/@plcent/@wdef.0"/>
</dep>
```

Each end of the relationship connects with another entity of the model. The entity is pointed using a path, for instance: "*//@plc/@plcent/@wdef.1*". *XPath* is the language used to represent this sort of paths. By the nature of this representation, the relationship can be n-ary and have attributes on both its body and its ends.

The root element of the XMI document is named *Specification* and contains elements for each package defined in the SPEM specification [15] (*Core, Actions, State_Machines, Activity_Graphs, Model_Management, BasicElements, Dependencies, Process Structure, Process Components* and *Process Lifecycle*). All the mentioned elements can contain entities or relationships and their tags for identification are their initials or abbreviations. In this way, the entities and relationships of a model are classified by the original SPEM packages.

4 Process Definition

Many works presented in Section 2 identify process and methodology, in the sense that a methodology introduces an inherent development process. Nevertheless, the proposal presented in this paper addresses this issue with a different perspective. It considers that a methodology can be applied using different development processes, so the selection of particular process will depend on several factors (such as kind of project, costs, development team, etc.). The process can even adapt to available human resources or include new phases or activities. But, in any case, aspects defined by the methodology like the entities of the model or the roles implied will not suffer any change. Taking into account this hypothesis, the process manager, using a particular methodology, may change dynamically the development process to readapt the development to new costs, time restrictions or new requirements.

This section introduces a framework for specification and modeling of developments processes for AOSE. The editor previously introduced has been used for applying the framework in the specification of different development processes. The framework is based on the development of three orthogonal views of the systems: the view of lifecycle process, the disciplines view and the guidance and suggestions view. This latest view introduces products, works and roles which constitute modeling elements of the other views. The first view describes the path which must be followed for obtaining as final product a MAS jointly with the activities that must be accomplished along the path. The second view establishes the disciplines of the process, the roles and participants and the tasks to perform, which will consume and produce a set of products.

The framework proposed has been used until the moment for fully modeling two different process for the INGENIAS methodology: an unified development process (UDP), named OpenUP [12, 13] and a SCRUM process.

4.1 Lifecycle

The process specification is started modeling the particular lifecycle selected for the development. This definition allows the analyst to forget about participants and products and focus in the definition of the activities to accomplish in each step. The development path is divided in consecutive phases and every phase in a different number of iterations.

One of the most relevant drawbacks in version 1.1 of SPEM is the lack of a formal language for defining restrictions about the admissible number of iterations. In order to overcome this problem, it is suggested to add in the iteration name the range for the number of iterations, that is, the maximum and minimum number of allowed iterations (for instance [1..3]). That will work if the number is always the same, in other case, it is mandatory to specify particularly each of the different iterations.

Using the editor described in previous section, the representation of the iterative and incremental lifecycle is made in several steps. In first place, the elements (phase, iteration and activities) which constitute the lifecycle are defined. Initially, a child of `Specification` of kind `PCLProcess Lifecycle` must be created. Afterward, the analyst must define an entity of `Lifecycle` type, and as much entities of type `Phase` as phases in the methodology and one entity of `Iteration` type, which will include in its name, as it was said before, the minimum and maximum number of iterations allowed in each phase. The following step is the definition of the existing temporal and membership relationships among these elements.

Once the previous activities are done, it is necessary to define for each iteration the phase it belongs to, and what the activities that must be performed are. Next step, after having specified the development process along with the activities to perform in each iteration, is defining the disciplines implied in these activities formalization.

4.2 Disciplines

Disciplines in SPEM determine process packages which take part in the process activities, related with a common *subject*. That is, disciplines represent a specialization of selected subactivities, where these new subactivities can not appear in other packages or disciplines. The definition of disciplines is made by creating inside of `Specification` a child of type `PC Ent` (*Process Component* Entities) and every discipline is included inside it by selecting new entities of `Discipline` type.

The definition of participants is made using as basis the different roles that each participant will play along the project and the products used. In INGENIAS, for instance, the roles implied are: *Analyst*, *Architect*, *Developer*, *Tester*, *Project Manager* and *FreeMan* and the products used are: *Use Case Diagram*, *Environment Model*, *Organization Model*, *Interaction Model*, *Agent Model*, *Object and Task Model*, *Glossary*, *NonFunctional Requirement* and *Stakeholder Vision*. In the editor, the mechanism for including roles and products is to create `PS Ent` entities with types `Proc Rol` and `Work Prod` respectively. Fig. 2(a) presents a snapshot of the editor after the definition of roles and products.

Once the participants, products, disciplines and process are defined, each activity must be divided inside its iteration depending on the task pertaining to each discipline. The activities must be performed by any of the specified participants

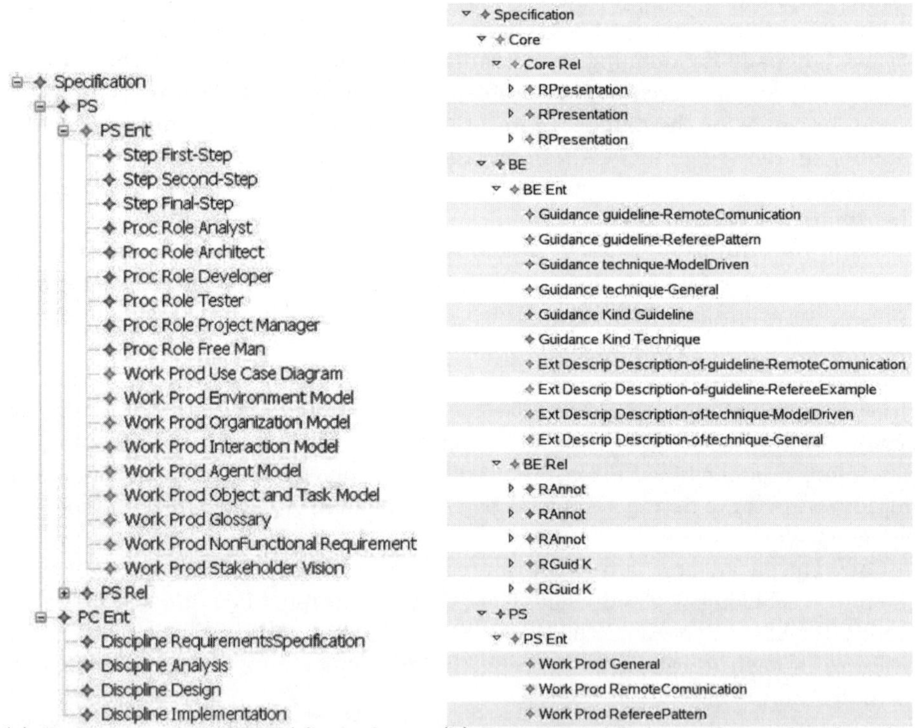

(a) Definition of INGENIAS products and roles using the editor.

(b) Entities and Relationships related to the Guidances. Entities are Guidances, Guidance Kinds, External Descriptions and Work Products. Relationships link these entities.

Fig. 2. Disciplines and Guidances

consuming and/or producing any of the existent products. The process for making up this step is similar to the previous ones, but modeling tasks like `Step` inside `PS Ent` and relations between activities and task like a `PS Rel` of type `RAc Step`. The relationships of making an activity by a participant will be defined using relations `PS Rel` of type `RAss`. In version 1.1 of SPEM, activities and consumed and/or produced products have the limitation that they can not be related directly; so this relationship is established using as intermediates the discipline where they are used and the role which manages them.

4.3 Guidances

The *Guidances* provide information about certain model elements. Each *Guidance* is associated to a *Guidance Kind.* SPEM provides a basic repertoire of Guidance Kinds, but SPEM is flexible about Guidance Kinds.

In the editor (Fig. 2(b)), the `Guidance` and `GuidanceKind` element to the `BE Ent` element (entities of the *Basic Elements* SPEM package) can be added. Each Guidance must be associated with its kind with the `RGuidK` relationship in the `BE Rel` element. Each Guidance must be associated with a `Model Element` , using the relationship `RAnnot` in the `BE Rel` element. In this example (Fig. 2(b)), all the guidances are linked with work products which are model elements. Each guidance has its own *External Description*, which contains a complete explanation of the guidance. The `Ext Descrip` element can be added to the `BE Ent` element to define an external description. The guidance can connect with its external description using the `RPresentation` element in the `Core Rel` element (relationships of the *Core* SPEM package).

In MAS, we recommend to use specially two kinds of guidance: *Technique* and *Guideline.* The *technique* provides an algorithm to create a work product. The *guideline* is a set of rules and recommendations about a work product organization.The most relevant INGENIAS Guidance can be found in [13].

5 Conclusions and Further Work

The definition of the software development processes of a MAS methodology makes it easier to learn and use a MAS methodology, in several ways. This paper presents a technique and an editor tool (based on SPEM and DSML techniques) which allow the definition of process models for agent-based development. In particular, the technique and the proposed editor have been used for the definition of the Unified Development Process for INGENIAS methodology and the SCRUM process for the same methodology. The application of the tool to a particular process definition proves its utility for the goal it was created for.

Furthermore, this work can guide a MAS developer through the steps of definition of a development process for MAS construction. The description of the steps to follow provided in the paper, can simplify the definition of processes for non expert engineers.

Next step in tool evolution will be to integrate the process editor with a tool for methodology support, so each software process engineering model will be available in the methodology support tool. In this way, the tool for development will be able to guide the user through the development steps, using the information provided by the XMI documents generated by the tool proposed in this paper. For instance, the defined INGENIAS software processes may be integrated with the INGENIAS Development Kit(IDK) tool.

In the future, several software process engineering models for different MAS methodologies can be defined with the presented editor. These process models can assist the MAS designer in selecting the appropriate methodology and process model for a specific MAS.

Acknowledgments

This work has been supported by the following projects: *Methods and tools for agent-based modeling* supported by Spanish Council for Science and Technology with grants TIN2005-08501-C03-01 and TIN2005-08501-C03-03 co-financed with FEDER funds and Grant for Research Group 910494 by the Region of Madrid (Comunidad de Madrid) and the Universidad Complutense Madrid.

References

1. Amyot, D., Farah, H., Roy, J.F.: Evaluation of Development Tools for Domain-Specific Modeling Languages. In: Proceedings of the 5th Workshop on System Analysis and Modelling (2006)
2. APES2: A Process Engineering Software, `http://apes2.berlios.de/en/links.html`
3. Atkinson, C., Kuhne, T.: Model-driven development: a metamodeling foundation. Software, IEEE 20(5), 36–41 (2003)
4. Bernon, C., Cossentino, M., Pavón, J.: Agent-oriented software engineering. Knowl. Eng. Rev. 20(2), 99–116 (2005)
5. Brazier, F.M.T., Jonker, C.M., Treur, J.: Principles of component-based design of intelligent agents. Data & Knowledge Engineering 41(1), 1–27 (2002)
6. Budinsky, F.: Eclipse Modelling Framework: Developer's Guide. Addison Wesley, Reading (2003)
7. Cernuzzi, L., Cossentino, M., Zambonelli, F.: Process models for agent-based development. Engineering Applications of Artificial Intelligence 18(2), 205–222 (2005)
8. Cossentino, M., Sabatucci, L., Seidita, V., Gaglio, S.: An Agent Oriented Tool for New Design Processes.. In: Proceedings of the Fourth European Workshop on Multi-Agent Systems (2006)
9. Cuesta, P., Gómez, A., González, J.C., Rodríguez, F.J.: The MESMA methodology for agent-oriented software engineering. In: Proceedings of First International Workshop on Practical Applications of Agents and Multiagent Systems (IWPAAMS 2002), pp. 87–98 (2002)
10. Eclipse: Eclipse Process Framework (EPF), `http://www.eclipse.org/epf/`
11. Moore, B., et al.: Eclipse Development using Graphical Editing Framework and the Eclipse Modelling Framework. IBM Redbooks (2004)

12. García-Magariño, I., Gómez-Rodríguez, A., González, J.C.: Modeling INGENIAS development process using EMF (In Spanish). In: 6th International Workshop on Practical Applications on Agents and Multi-agent Systems, IWPAAMS 2007, Salamanca Spain, November 12/13, 2007, pp. 369–378 (2007)
13. García-Magariño, I., Gómez-Rodríguez, A., González, J.C.: Definition of Process Models for Agent-based Development. In: 9th International Workshop on AOSE, Lisbon, Portugal, May 12/13 (2008)
14. Giorgini, P., Mylopoulos, J., Sebastiani, R.: Goal-oriented requirements analysis and reasoning in the Tropos methodology. Engineering Applications of Artificial Intelligence 18(2), 159–171 (2005)
15. Object Management Group. Software Process Engineering Metamodel Specification. Version 1.1, formal/05-01-06 (January 2005)
16. Object Management Group. Meta Object Facility (MOF) Specification. Version 2.0, formal/2006-01-01 (January 2006)
17. Henderson-Sellers, B., Gonzalez-Perez, C.: A comparison of four process metamodels and the creation of a new generic standard. Information and Software Technology 47(1), 49–65 (2005)
18. Mas, A.: Agentes Software y Sistemas Multi-Agentes. Pearson Prentice Hall (2004)
19. O'Malley, S.A., DeLoach, S.A.: Determining when to use an agent-oriented software engineering pradigm. In: Wooldridge, M.J., Weiß, G., Ciancarini, P. (eds.) AOSE 2001. LNCS, vol. 2222. Springer, Heidelberg (2002)
20. Pavón, J., Gómez-Sanz, J.: Agent Oriented Software Engineering with INGENIAS. Multi-Agent Systems and Applications III 2691, 394–403 (2003)
21. Penserini, L., Perini, A., Susi, A., Mylopoulos, J.: High variability design for software agents: Extending Tropos. ACM Transactions on Autonomous and Adaptative Systems 2(4), 16 (2007)
22. Quynh, G.L., Tran, N.N.: MOBMAS: A methodology for ontology-based multiagent systems development. Information and Software Technology 50(7-8), 697–722

An Agent-Based Architecture for the Decision Support Process

María A. Pellicer[1] and M. Lourdes Borrajo[2]

[1] Dept. de Ingeniería Civil, University of Burgos, Spain
[2] Dept. Informática, University of Vigo
lborrajo@uvigo.es

Abstract. Firms require a control mechanism that facilitate the analysis of work carried out in changing environments, such as finance. A tool for the decision support process has been developed based on a multi-agent system that incorporates a case-based reasoning system. The case-based reasoning system automates the process of case indexing and retrieval by means of a Maximum Likelihood Hebbian Learning-based method. The multi-agent system has been tested within ten small and medium companies and the results obtained have been very satisfactory.

Keywords: Agents Technology, Case Based Reasoning, Maximum Likelihood Hebbian Learning.

1 Introduction

All firms need to monitor their "modus operandi" and to analyse whether they are achieving their goals. As a consequence of this, it is necessary to construct models that facilitate the analysis of work carried out in changing environments, such as finance. Processes carried out inside any firm are grouped in Functions [1, 28]. A Function is a group of coordinated and related activities, which are necessary to reach the objectives of the firm and are carried out in a systematic and iterative way [2]. Purchases, Cash Management, Sales, Information Technology, Fixed Assets Management, Compliance to Legal Norms and Human Resources are the functions that are usually carried out in a company. In turn, each one of these functions is divided into a series of activities, which aim to achieve different objectives. For example, the function Information Technology is divided into the following activities: Computer Plan Development, Study of Systems, Installation of Systems, Treatment of Information Flows and Security Management.

Each activity is composed of a number of tasks. For example, the activity Computer Plan Development, belonging to the function Information Technology, can be divided in the following tasks:

1. Definition of the required investment in technology in the short and medium time of period.
2. Coordination of the technology investment plan and the development plan of the company.
3. Periodic evaluation of the established priorities on the technology investment plan to identify their relevance.

J.M. Corchado et al. (Eds.): DCAI 2008, ASC 50, pp. 682–691, 2009.
springerlink.com

4. Definition of a working group focused in the identification and control of the information technology policy.
5. Definition of a communication protocol in both directions: bottom-up and top-down, to involve the company's employees in the maintenance strategic plan.

Therefore, control procedures have to be established in the tasks to ensure that the objectives of the firm are achieved.

This paper presents a multiagent system which is able to analyse the activities of a firm and calculate its level of risk. The developed model is composed of four different agent types. The principal agent, whose objectives are: to identify the state or situation of each one of the activities of the company and to calculate the risk associated with this state, incorporates a case-based reasoning (CBR) system [4, 5, 6, 7]. The CBR system uses different problem solving techniques [8, 9]. Moreover, the CBR systems proposed in the framework of this research incorporate a Maximum Likelihood Hebbian Learning (MLHL) [12] based model to automate the process of case indexing and retrieval, which may be used in problems in which the cases are characterised, predominantly, by numerical information. One of the aims of this work is to improve the performance of the CBR system integrated within the principal agent by means of incorporating the MLHL into the CBR cycle stages. The ability of the Maximum Likelihood Hebbian Learning-based methods presented in this paper to cluster cases/instances and to associate cases to clusters can be used to successfully prune the case-base without losing valuable information.

This paper first presents the Maximum Likelihood Hebbian Learning based method and its theoretical background and then, the proposed multi-agent system is presented. Finally, the system results and conclusions are presented.

2 Maximum Likelihood Hebbian Learning Based Method

The use of Maximum Likelihood Hebbian Learning based method has been derived from the work of [13, 15, 16, 17], etc. in the field of pattern recognition as an extension of Principal Component Analysis (PCA) [10, 11].

PCA is a standard statistical technique for compressing data; it can be shown to give the best linear compression of the data in terms of least mean square error. There are several artificial neural networks which have been shown to perform PCA e.g. [10, 11]. We will apply a negative feedback implementation [18].

The basic PCA network is described by equations (1)-(3). Let us have an N-dimensional input vector at time t, x(t), and an M-dimensional output vector, y, with W_{ij} being the weight linking input j to output i. η is a learning rate. Then, the activation passing and learning is described by

Feedforward:

$$y_i = \sum_{j=1}^{N} W_{ij} x_j \ , \forall i \tag{1}$$

Feedback:

$$e_j = x_j - \sum_{i=1}^{M} W_{ij} y_i \tag{2}$$

Change weights:

$$\Delta W_{ij} = \eta e_j y_i \tag{3}$$

We can readily show that this algorithm is equivalent to Oja's Subspace Algorithm [10]:

$$\Delta W_{ij} = \eta e_j y_i = \eta (x_j - \sum_k W_{kj} y_k) y_i \tag{4}$$

The PCA network not only causes convergence of the weights but causes the weights to converge to span the subspace of the Principal Components of the input data. Exploratory Projection Pursuit (EPP) is a more recent statistical method aimed at solving the difficult problem of identifying structure in high dimensional data. It does this by projecting the data onto a low dimensional subspace in which we search for its structure by eye. However not all projections will reveal the data's structure equally well. We therefore define an index that measures how "interesting" a given projection is, and then represent the data in terms of projections that maximise that index. The first step in our exploratory projection pursuit is to define which indices represent interesting directions. Now "interesting" structure is usually defined with respect to the fact that most projections of high-dimensional data onto arbitrary lines through most multi-dimensional data give almost Gaussian distributions [19]. Therefore if we wish to identify "interesting" features in data, we should look for those directions onto which the data-projections are as far from the Gaussian as possible. It was shown in [0] that the use of a (non-linear) function creates an algorithm to find those values of W which maximize that function whose derivative is f() under the constraint that W is an orthonormal matrix. This was applied in [18] to the above network in the context of the network performing an Exploratory Projection Pursuit.

On the other hand, it has been shown [21] that the nonlinear PCA rule

$$\Delta W_{ij} = \eta \left(x_j f(y_i) - f(y_i) \sum_k W_{kj} f(y_k) \right) \tag{5}$$

can be derived as an approximation to the best non-linear compression of the data. Thus we may start with a cost function

$$J(W) = 1^T E\left\{ \left(\mathbf{x} - Wf\left(W^T \mathbf{x} \right) \right)^2 \right\} \tag{6}$$

which we minimise to get the rule (5). [22] used the residual in the linear version of (6) to define a cost function of the residual

$$J = f_1(\mathbf{e}) = f_1(\mathbf{x} - W\mathbf{y}) \tag{7}$$

where $f_1 = \|.\|^2$ is the (squared) Euclidean norm in the standard linear or nonlinear PCA rule. With this choice of $f_1(\ )$, the cost function is minimised with respect to any set of samples from the data set on the assumption that the residuals are chosen independently and identically distributed from a standard Gaussian distribution. We may show that the minimisation of J is equivalent to minimising the negative log probability of the residual, $\mathbf{e}.$, if e is Gaussian.

$$\text{Let } p(\mathbf{e}) = \frac{1}{Z} \exp(-\mathbf{e}^2) \tag{8}$$

Then, we can denote a general cost function associated with this network as

$$J = -\log\ p(\mathbf{e}) = (\mathbf{e})^2 + K \tag{9}$$

where K is a constant. Therefore performing gradient descent on J we have

$$\Delta W \propto -\frac{\partial J}{\partial W} = -\frac{\partial J}{\partial \mathbf{e}}\frac{\partial \mathbf{e}}{\partial W} \approx \mathbf{y}(2\mathbf{e})^T \tag{10}$$

where we have discarded a less important term. See [0] for details.

In general, the minimisation of such a cost function may be thought to make the probability of the residuals greater dependent on the probability density function (pdf) of the residuals [0] Thus, if the probability density function of the residuals is known, this knowledge could be used to determine the optimal cost function. [0] investigated this with the (one dimensional) function:

$$p(\mathbf{e}) = \frac{1}{2+\varepsilon}\exp\left(-|\mathbf{e}|_\varepsilon\right) \tag{11}$$

where

$$|e|_\varepsilon = \begin{cases} o & \forall |e| < \varepsilon \\ |e| - \varepsilon & otherwise \end{cases} \tag{12}$$

with ε being a small scalar ≥ 0.

Fyfe and MacDonald [0] described this in terms of noise in the data set. However, we feel that it is more appropriate to state that, with this model of the pdf of the residual, the optimal $f_1(\ )$ function is the ε-insensitive cost function:

$$f_1(\mathbf{e}) = |\mathbf{e}|_\varepsilon \tag{13}$$

In the case of the negative feedback network, the learning rule is

$$\Delta W \propto -\frac{\partial J}{\partial W} = -\frac{\partial f_1(\mathbf{e})}{\partial \mathbf{e}}\frac{\partial \mathbf{e}}{\partial W} \tag{14}$$

which gives:

$$\Delta W_{ij} = \begin{cases} o & if\,|e_j| < \varepsilon \\ otherwise & \eta y(sign(e)) \end{cases} \tag{15}$$

The difference with the common Hebb learning rule is that the sign of the residual is used instead the value of the residual. Because this learning rule is insensitive to the magnitude of the input vectors x, the rule is less sensitive to outliers than the usual rule based on mean squared error. This change from viewing the difference after feedback as simply a residual rather than an error permits us to consider a family of cost functions each member of which is optimal for a particular probability density function associated with the residual.

2.1 Applying Maximum Likelihood Hebbian Learning

The Maximum Likelihood Hebbian Learning algorithm is constructed now on the bases of the previously presented concepts as outlined here. Now the ε-insensitive learning rule is clearly only one of a possible family of learning rules which are suggested by the family of exponential distributions. This family was called an exponential family in [0] though statisticians use this term for a somewhat different family. Let the residual after feedback have probability density function

$$p(\mathbf{e}) = \frac{1}{Z}\exp(-|\mathbf{e}|^p) \tag{16}$$

Then we can denote a general cost function associated with this network as

$$J = E(-\log p(\mathbf{e})) = E(|\mathbf{e}|^p + K) \tag{17}$$

where K is a constant independent of W and the expectation is taken over the input data set. Therefore, performing gradient descent on J, we have

$$\Delta W \propto -\frac{\partial J}{\partial W}|_{W(t-1)} = -\frac{\partial J}{\partial \mathbf{e}}\frac{\partial \mathbf{e}}{\partial W}|_{W(t-1)} \approx E\{\mathbf{y}(p|\mathbf{e}|^{p-1}\, sign(\mathbf{e}))^T|_W \tag{18}$$

where T denotes the transpose of a vector and the operation of taking powers of the norm of e is on an element wise basis as it is derived from a derivative of a scalar with respect to a vector.

Computing the mean of a function of a data set (or even the sample averages) can be tedious, and we also wish to cater for the situation in which samples keep arriving as we investigate the data set and so we derive an online learning algorithm. If the conditions of stochastic approximation [0] are satisfied, we may approximate this with a difference equation. The function to be approximated is clearly sufficiently smooth and the learning rate can be made to satisfy $\eta_k \geq 0, \sum_k \eta_k = \infty, \sum_k \eta_k^2 < \infty$ and so we have the rule:

$$\Delta W_{ij} = \eta . y_i . sign(e_j)|e_j|^{p-1} \tag{19}$$

We would expect that for leptokurtotic residuals (more kurtotic than a Gaussian distribution), values of p<2 would be appropriate, while for platykurtotic residuals (less kurtotic than a Gaussian), values of p>2 would be appropriate. Researchers from the community investigating Independent Component Analysis [24, 26] have shown that it is less important to get exactly the correct distribution when searching for a specific source than it is to get an approximately correct distribution i.e. all supergaussian signals can be retrieved using a generic leptokurtotic distribution and all subgaussian signals can be retrieved using a generic platykutotic distribution. Our experiments will tend to support this to some extent but we often find accuracy and speed of convergence are improved when we are accurate in our choice of p. Therefore the network operation is:

Feedforward:

$$y_i = \sum_{j=1}^{N} W_{ij} x_j, \forall_i \tag{20}$$

Feedback:

$$e_j = x_j - \sum_{i=1}^{M} W_{ij} y_i \tag{21}$$

Weights change:

$$\Delta W_{ij} = \eta . y_i . sign(e_j) | e_j |^{p-1} \tag{22}$$

Fyfe and MacDonald [0] described their rule as performing a type of PCA, but this is not strictly true since only the original (Oja) ordinary Hebbian rule actually performs PCA. It might be more appropriate to link this family of learning rules to Principal Factor Analysis since PFA makes an assumption about the noise in a data set and then removes the assumed noise from the covariance structure of the data before performing a PCA. We are doing something similar here in that we are basing our PCA-type rule on the assumed distribution of the residual. By maximising the likelihood of the residual with respect to the actual distribution, we are matching the learning rule to the probability density function of the residual.

More importantly, we may also link the method to the standard statistical method of Exploratory Projection Pursuit: now the nature and quantification of the interest is in terms of how likely the residuals are under a particular model of the probability density function of the residuals.

3 Multi-agent System

This section describes the multi agent system in detail. Although the aim is to develop a generic model useful in any type of enterprise, the initial work has focused in small o medium firms to facilitate the research and its evaluation [3]. The model here presented may be extended or adapted for other sectors. Ten companies from the Northwest of Spain have collaborated in this research, working mainly for the Spanish market. After analyzing the data relative to the activities developed within a given firm, the constructed multi-agent system is able to determine the state of each of the activities and calculate the associated risk. A Firm agent has been assigned for each firm in order to collect new data and allow consults. The Expert agents help the auditors and business control experts that collaborate in the project to provide information and feedback to the multiagent system. These experts generate prototypical cases from their experience and they have help to develop the Store agent case-base.

The CBR-based agent incorporates a case-based reasoning system as reasoning mechanism. The cycle of operations of each case based reasoning system is based on the classic life cycle of a CBR system [4, 27]. This agent is communicated with the Store agent that stores the shared case base. A case represents the "shape" of a given activity developed in the company.

The CBR-based agent identifies the state or situation of each of the firm's activities and calculates the risk associated with this situation. The agent uses the data for the activity, introduced by the Firm agent, to construct the problem case. For each task making up the activity analyzed, the problem case is composed of the value of the realization state for that task, and its level of importance within the activity (according to the internal auditor).

In the retrieval step, the agent communicates with the Store agent to retrieve K cases – the most similar cases to the problem case; this is done with the Maximum Likelihood Hebbian Learning proposed method. Applying equations 20 to 22 to the case base, the MLHL algorithm groups the cases in clusters automatically. The proposed indexing mechanism classifies the cases/instances automatically, clustering together those of similar structure. One of the great advantages of this technique is that it is an unsupervised method so we do not need to have any information about of the data before hand. When a new problem case is presented to the CBR system, it is identified as belonging to a particular type by applying also equations 20 to 22 to it. This mechanism may be used as an universal retrieval and indexing mechanism to be applied to any problem similar to the presented here. Maximum Likelihood Hebbian Learning techniques are used because of the size of the database and the need to group the most similar cases together in order to help retrieve the cases that most resemble the given problem.

The re-use phase aims to obtain an initial estimation of the state of the activity analysed. In order to obtain this estimation, RBF networks are used [14, 28]. As in the previous phase, the number of attributes of the problem case depends on the activity analyzed. Therefore, it is necessary to establish an RBF network system, one for each of the activities to be analysed. The k cases retrieved in the previous phase are used by the RBF network as a training group that allows it to adapt its configuration to the new problem encountered before generating the initial estimation. The RBF network is characterized by its ability to adapt, to learn rapidly, and to generalize. Specifically, within this system the network acts as a mechanism capable of absorbing knowledge about a certain number of cases and generalizing from them.

The objective of the revision phase is to confirm or refute the initial solution proposed by the RBF network, thereby obtaining a final solution and calculating the control risk. In view of the initial estimation or solution generated by the RBF network, the internal auditor (through the Firm agent) will be responsible for deciding if the solution is accepted. For this it is based on the knowledge he/she retains, specifically, knowledge about the company with which he/she is working. If he/she considers that the estimation given is valid, the system will take the solution as the final solution and in the following phase of the CBR cycle, a new case will be stored in the Store agent case base consisting of the problem case and the final solution. The system will assign the case an initial reliability of 100%. If on the other hand, the internal auditor considers the solution given by the system to be invalid, he will give his own solution which the system will take as the final solution and which together with the problem case will form the new case to be stored by the Store agent in the following phase. This new case will be given a reliability of 30%. This value has been decided by various auditors which have considered to assign a reliability of 30% to the personal opinion of the internal auditor. From the final solution: state of activity, the agent calculates the control risk associated with the activity. Every activity developed in the business sector has a risk associated with it that indicates the negative influence that affects the good operation of the firm. In this study, the level of risk is valued at three levels: low, medium and high. The calculation of the level of control risk associated with an activity is based on the current state of the activity and its level of importance. This latter value was obtained after analysing data obtained from a series of questionnaires (98 in total) carried out by auditors throughout Spain. The level of control risk was then calculated from the level of importance given to the activity by the auditors and the final solution obtained after the revision phase. For this purpose, if-then rules are employed.

The last phase executed by the agent is the communication and incorporation of the system's memory managed by the Store agent of what has been learnt after resolving a new problem. Once the revision phase has been completed, after obtaining the final solution, a new case (problem + solution) is constructed, which is stored in the agent Store's memory. Apart from the overall knowledge update involving the insertion of a new case within the agent Store memory, the multi-agent system presented carries out a local adaptation of the knowledge structures that it uses. Maximum Likelihood Hebbian Learning technique contained within the prototypes related to the activity corresponding to the new case is reorganised in order to respond to the appearance of this new case, modifying its internal structure and adapting itself to the new knowledge available. In this way, the RBF network uses the new case to carry out a complete learning cycle, updating the position of its centres and modifying the value of the weightings that connect the hidden layer with the output layer.

4 Results and Conclusions

For a given company, each one of its activities was evaluated by the system, obtaining a level of risk. On the other hand, we request to six external and independent auditors that they analyzed the situation of each company. The mission of the auditors is to estimate the state of each activity, the same as the proposed system makes. Then, we compare the result of the evaluation obtained by the auditors with the result obtained by the system. The results obtained by the system are very similar to those obtained by the external auditors. In general, it could be said that these results demonstrate the suitability of the techniques used for their integration in the multiagent system.

This paper presents a multi-agent system that uses a CBR system employed as a basis for hybridization of a Maximum Likelihood Hebbian Learning technique, and a RBF net. The system is able to estimate or identify the state of the activities of the firm and their associated risk. Estimation in the environment of firms is difficult due to the complexity and the great dynamism of this environment. However, the developed model is able to estimate the state of the firm with precision. We have demonstrated a new technique for case indexing and retrieval, which could be used to construct case based reasoning systems. The basis of the method is a Maximum Likelihood Hebbian Learning algorithm. This method provides us with a very robust model for indexing the data and retrieving instances without any need of information about the structure of the data set.

References

1. Corchado, J.M., Borrajo, L., Pellicer, M.A., Yáñez, J.C.: Neuro-Symbolic System For Business Internal Control. In: Perner, P. (ed.) ICDM 2004. LNCS (LNAI), vol. 3275, pp. 1–10. Springer, Heidelberg (2004)
2. Mas, J., Ramió, C.: La Auditoría Operativa en la Práctica, Marcombo, Barcelona (1997)
3. Pellicer, M.A.: Arquitectura Multiagente para la Gestión Autónoma de Procesos Empresariales Phd Thesis, Universidade de Burgos, Spain (2008)

4. Aamodt, A., Plaza, E.: Case-Based Reasoning: Foundational Issues, Methodological Variations, and System Approaches. AI Communications 7(1) (1994)
5. Kolodner, J.: Case-Based Reasoning. Morgan Kaufmann, San Mateo (1993)
6. Lenz, M., Bartsch-Spörl, B., Burkhard, H.-D., Wess, S. (eds.): Case-Based Reasoning Technology. LNCS (LNAI), vol. 1400. Springer, Heidelberg (1998)
7. Watson, I.: Applying Case-Based Reasoning: Techniques for Enterprise Systems. Morgan Kaufmann, San Mateo (1997)
8. Hunt, J., Miles, R.: Hybrid Case-Based Reasoning. The Knowledge Engineering Review 9(4), 383–397 (1994)
9. Medsker, L.R.: Hybrid Intelligent Systems. Kluwer Academic Publishers, Dordrecht (1995)
10. Oja, E.: Neural Networks, Principal Components and Subspaces. International Journal of Neural Systems 1, 61–68 (1989)
11. Oja, E., Ogawa, H., Wangviwattana, J.: Principal Components Analysis by Homogeneous Neural Networks, Part 1, The Weighted Subspace Criterion. IEICE Transaction on Information and Systems E75D, 366–375 (1992)
12. Corchado, E., Fyfe, C.: Maximum and Minimum Likelihood Hebbian Rules as a Exploratory Method. In: Proc. ICNIP 2002, IEEE Catalog Number 02 EX575C (Suvisoft 2002); ISBN 981-04-7525-X
13. Corchado, E., Macdonald, D., Fyfe, C.: Optimal Projections Of High Dimensional Data. In: Proc. ICDM 2002, The 2002 IEEE International Conference on Data Mining. IEEE Computer Society, Los Alamitos (2002)
14. Fdez-Riverola, F., Corchado, J.M.: Fsfrt: Forecasting System For Red Tides. Applied Intelligence 21(3), 251–264 (2004)
15. Fyfe, C., Corchado, E.: Maximum Likelihood Hebbian Rules. In: Proc. ESANN 2002, Bruges, pp. 143–148 (2002)
16. Fyfe, C., Macdonald, D.: E-Insensitive Hebbian Learning. Neurocomputing 47(1-4), 35–57 (2002)
17. Fyfe, C., Corchado, E.: A New Neural Implementation of Exploratory Projection Pursuit. In: Proc. IDEAL 2002, Third International Conference on Intelligent Data Engineering and Automated Learning, Manchester, pp. 12–14 (2002)
18. Fyfe, C., Baddeley, R.: Non-Linear Data Structure Extraction Using Simple Hebbian Networks. Biological Cybernetics 72(6), 533–541 (1995)
19. Diaconis, P., Freedman, D.: Asymptotics of Graphical Projections. The Annals of Statistics 12(3), 793–815 (1984)
20. Karhunen, J., Joutsensalo, J.: Representation and Separation of Signals Using Non-Linear PCA Type Learning. Neural Networks 7, 113–127 (1994)
21. Xu, L.: Least Mean Square Error Reconstruction for Self-Organizing Nets. Neural Networks 6, 627–648 (1993)
22. Lai, P.L., Charles, D., Fyfe, C.: Seeking Independence Using Biologically Inspired Artificial Neural Networks. In: Girolami, M.A. (ed.) Developments in Artificial Neural Network Theory: Independent Component Analysis and Blind Source Separation. Springer, Heidelberg (2000)
23. Smola, A.J., Scholkopf, B.: A Tutorial on Support Vector Regression, Technical Report NC2-TR-1998-030, Neurocolt2 Technical Report Series (1998)
24. Hyvärinen, A., Karhunen, J., Oja, E.: Independent Component Analysis. Wiley, Chichester (2002)

25. Kashyap, R.L., Blaydon, C.C., Fu, K.S.: Stochastic Approximation. In: Mendel, J.M. (ed.) A Prelude to Neural Networks: Adaptive and Learning Systems. Prentice Hall, Englewood Cliffs (1994)
26. Hyvärinen, A.: Complexity Pursuit: Separating Interesting Components from Time Series. Neural Computation 13, 883–898 (2001)
27. Watson, I., Marir, F.: Case-Based Reasoning: A Review. The Knowledge Engineering Review 9(4), 355–381 (1994)
28. Fritzke, B.: Fast Learning With Incremental RBF Networks. Neural Processing Letters 1(1), 2–5 (1994)
29. Yañez, J.C.: Importancia del Sistema de Control Interno en la Auditoría Legal. Contrastes Empíricos. Phd Thesis. Universidade de Vigo (Spain) (2003)

Software Agents for Home Environment Automation

Ana Isabel Calvo Alcalde[1], Juan José Andrés Gutiérrez[1], Jesús Vegas Hernández[2], Valentín Cardeñoso Payo[2], and Esteban Pérez Castrejón[1]

[1] Telefónica R&D
Parque Tecnológico de Boecillo, 47151, Boecillo, Valladolid, Spain
aica@tid.es, jjangu@tid.es, esteban@tid.es
[2] Computer Science Department
University of Valladolid
Escuela Técnica Superior de Ingeniería Informática (E.T.S.I.I.), Campus Miguel Delibes s/n, 47011, Valladolid, Spain
jvegas@infor.uva.es, valen@infor.uva.es

Abstract. This article shows the *Software Agents for Home Environment Automation system.* This research aims to explore several device discovery technologies and ambient intelligence techniques in order to allow the user interaction at home as transparent as possible. Specifically, the main goal is to provide users suitable services according to their preferences and location. To achieve this a multi-agent system is proposed, focussing on the knowledge representation for the multi-agent system communication. This work is part of the activities related to "Digital Home" management, currently under development in Telefónica R&D in collaboration with Computer Science Department at the University of Valladolid.

Keywords: Ambient Intelligence, Digital Home, services discovery, UPnP, RfID, middleware, Artificial Intelligence, software intelligent agents, Multi-Agent System, ontology, context.

1 Introduction

Nowadays homes have successively more capabilities related to make easier the daily household tasks. However, in most cases, all these services operate in isolation and require users a high knowledge to take advantage of their capabilities. Currently, users are demanding greater simplicity. Due to this fact a new concept comes up, the *Ambient Intelligence* or AmI [1, 2] which is based on a natural and non-intrusive relation between people and the technological environment around them. Due to this reasons, home is one of the environments where the AmI concept gets more sense.

This is the main goal of this work, to provide an Ambient Intelligence solution to be applied at home by the design of a platform witch enables interoperability between users and home devices. For users, who must be automatically identified and located, the system will generate dynamically personalized interfaces that will let them to define the behavior, manage warnings and alarms at home based on the environment information and the available devices discovered automatically.

Among the available Knowledge Engineering methodologies appears CommonKADS [3] as the European standard for developing knowledge-based systems and it

J.M. Corchado et al. (Eds.): DCAI 2008, ASC 50, pp. 692–701, 2009.
springerlink.com

is the one that has been applied, specifically, to analyze the rule-based system Jess [4, 5]. It is worth to review some interesting initiatives that define languages, like SWRL, to share rules (OWL knowledge bases) and to infer new knowledge with different rule engines [6, 7].

Several approaches provide an overview of different agent oriented methodologies for multi-agent systems development (MAS) [8, 9]. The proposed context-sensitive MAS is analyzed by using MAS-CommonKADS methodology [10], which adds the relevant aspects for MAS [11] to CommonKADS.

Among several previous researches about ontology creation and its use in agent communication that have been reviewed [12, 13, 14] it is worth pointing out COBRA-ONT ontology [15]. For this work, the ontologies that are especially interesting are: agent, action, device, personal-device and location.

Some existing implementations, like CoBrA [16], have been taken into account before designing the architecture to improve agents reasoning abilities in relation to the knowledge-based system [17]. Some progresses have been made according to AMIGO project [18] researches about the use of UPnP devices for the automation of the home instead of handling multimedia aspects of UPnP A/V.

This paper is structured as follows. The second section presents a general overview of the scope. Then, third section presents an analysis of the proposed MAS and illustrates the system architecture. The scenario interactions between actors and the different components in order to achieve the objectives are detailed in fourth section. Finally, fifth section reports some conclusions and future works.

2 Multi-Agent System for Digital Home

The increasing new technological facilities at home create the necessity of an intelligent management. To accomplish this goals a MAS [11] has been designed composed of autonomous intelligent agents [19] that are able to communicate, to perceive the environment around them and to coordinate their activities in order to achieve the goals they were designed to.

Fig. 1. shows an illustrative scenario of our proposal. Home is composed of heterogeneous devices (WiFi, Bluetooth, etc.) such as sensors for user identification and location (RFID receivers [20]) and environmental sensors (weather station). To achieve the incorporation of these devices and new ones to be available at home is necessary to introduce a new concept: **services discovery**. These devices are integrated in a transparent way for the user, when they are connected to the network they are automatically *discovered and controlled* by the invocation of the services they offer.

This is possible using UPnP [21, 22] which provides, (1) automatic discovery of the devices connected to the Home network, (2) allows zero-configuration networking, (3) be independent of the kind of devices and networks, (4) use standard protocols and (5) easy system extensibility.

The system is in charge of scanning the home network, discovering devices and controlling them through the invocation of the offered services. This control ability

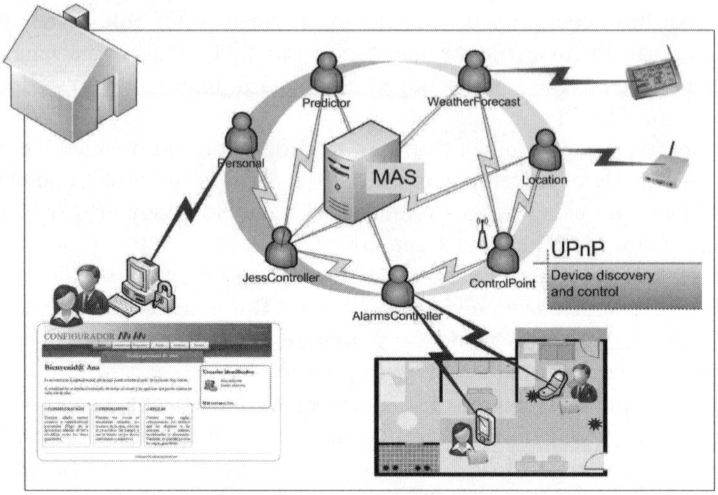

Fig. 1. Project general overview

added to the platform intelligence, allows to analyse events and context information, provided by sensors, and to perform actions according to this.

Using the environment information and the user's personal rules defined, the actuations that have to be performed can be dynamically decided and also who will be in charge of them. As well as conversations between them, there is a specific agent that will interact with the knowledge-based system Jess [4, 5]. The system allows users to ***create personal rules*** in order to define the system behaviour, that is, to associate several events with the actions which will be activated [23].

Next, the activities which are going to be performed at Home by the software agents and their behaviours are the following:

- *Home devices management*: UPnP devices [21, 22] will be discovered and incorporated into the platform in a transparent way for the user.
- *Home networks management*: agents keep the domotic networks information and perform actions according to their internal state and user profiles.
- *Alarms and integrations*: the knowledge-based system is able to find out event sequences that can activate alarms [23].
- *Location:* agents are aware of the users location at home (identification and location RfID [20]).
- *Personalization*: according to users preferences and users location agents can provide a personal ambient for everyone.

3 System Architecture Analysis and Design

The methodology MAS-CommonKADS [10] has been followed to design the system. This methodology extends CommonKADS models [3], a European standard for developing knowledge-based systems, adding the relevant aspects for MAS [11] and integrating object oriented techniques to make its application easier. Agents can carry out tasks that are detailed in the Task model using the CRC (Class Responsibility Collaboration) card modelling technique:

Table 1. CRC card for Personal Agent

Agent *Personal*	
Goal	*Personal rules management*
Plans	To show user´s information and inform to JessController
Knowledge	Connected user's identification
Collaborator	JessController
Services	Creation, modification and elimination of personal rules
Goal	*Users and tags Management*
Plans	To ask for information and show it to the user Send the information to JessController agent
Knowledge	Connected user's identification
Collaborator	JessController
Services	Creation, modification and elimination of users and tags

Table 2. CRC card for ControlPoint Agent

Agent *Control Point*	
Goal	*Devices Discovery*
Plans	To discover devices and to get the inf. to control them
Knowledge	Devices identification
Collaborator	WeatherForecast. Location. AlarmsController.
Services	New devices discovery to integrate them into the network

Table 3. CRC card for WeatherForecast Agent

Agent *WeatherForecast*	
Goal	*"Weather station" device control*
Plans	To keep information about the device state and to control it To know the weather forecast
Knowledge	Device identification and device state changes (events)
Collaborator	ControlPoint. JessController.
Services	To realize weather forecast changes and report them

Table 4. CRC card for Location Agent

Agent *Location*	
Goal	*"RfID receiver" device control*
Plans	To keep information about the device state and to control it To know users location
Knowledge	Device identification and device state changes (events)
Collaborator	ControlPoint. JessController.
Services	To realize users location changes and report them

Table 5. CRC card for AlarmsController Agent

Agent *AlarmsController*	
Goal	*To perform actions in the environment*
Plans	To keep environment information and control it
Knowledge	Devices identification
Collaborator	ControlPoint. JessController. Predictor
Services	To control the devices to perform actions in the environment

Table 6. CRC card for JessController Agent

Agent *JessController*	
Goal	*To update user's information*
Plans	To update Jess knowledge base
Knowledge	Personal rules, users and tags modifications
Collaborator	Personal
Services	To update user's information
Goal	*To infer knowledge according to user's personal rules*
Plans	To update Jess Knowledge Base and facts base
Knowledge	User's personal rules, events and inferred knowledge
Collaborator	Personal. WeatherForecast. Location. AlarmsController. Predictor
Services	To infer new knowledge according user's personal rules

Table 7. CRC card for Predictor Agent

Agent *Predictor*	
Goal	*To get user's behaviours at home predictions*
Plans	To consult past behaviours to predict future actions
Knowledge	Past behaviours and event
Collaborator	JessController. AlarmsController
Services	To predict user's behaviours at home

The architecture has been designed to ensure that the system will be as robust, flexible and decentralized as possible:

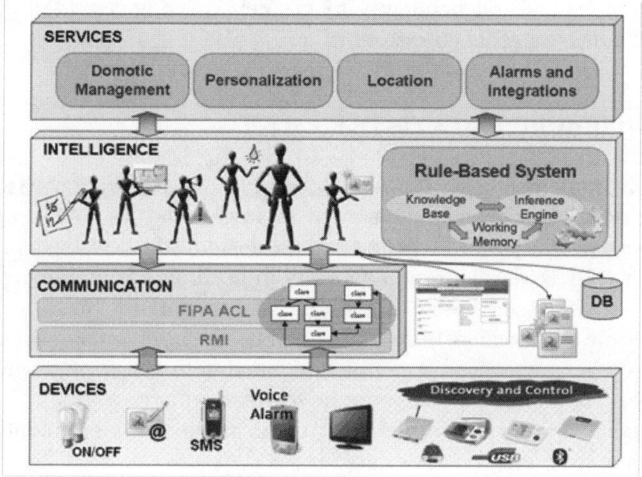

Fig. 2. Architecture design

To allow agents integration inside of an open and complex environment the **information representation** must be formalized and the semantic meaning must be the same for all the agents, that is, all of them must interpret the information in the same way. This consensus, needed in a particular domain, is reflected in the communication ontology [13, 14]. Furthermore, this is an open system because new agents can be added in the future **to increase the overall functionality**. These qualities are of special interest in the domain of application, the digital home, because it is an environment where heterogeneous devices and services are added frequently. The proposed architecture presents a four layers system:

- Devices layer. It supports direct interaction with the integrated devices. The discovery and control of them is carried out using UPnP [21, 22]. The system is able to respond dynamically to changing circumstances and adapt itself to user's profiles.
- Communication Layer. It supports agents conversations that allow them to synchronise tasks, to share knowledge by sending and receiving messages and solve conflicts. ACL messages are defined [25, 26] and their content is composed of the modelled ontology attributes [12].
- Intelligence layer. It provides the necessary abilities to interpret the environment information and to infer new knowledge. The **knowledge modeling** is essential to model and design the knowledge-based systems Jess [4, 5] and the agents knowledge which could require it.
- Services layer. It supports and exploits control home services like domotic control, alarms management, etc., using the lower layers defined.

This layer architecture allows (1) integrating new devices in a transparent way, (2) to make independent the communications between agents and devices, (3) to model the specific home domain for the knowledge-based system according to the context information sources and independently of the network devices, (4) to provide complex services thanks to agents cooperation.

4 Experimentation and Practice Cases

For the system implementation the JADE-Jess-Protégé integration has been chosen. Therefore the MAS has been implemented in Java using the JADE agent platform [27, 13], Protégé-3.3.1 [28] has been used to design and develop ontologies using BeanGenerator [29] as an integration gateway to generates Java files representing the ontology that can be used with the JADE environment. Its functionality has been proved in multiple solutions [30]. Jess has been used as rule engine because it is written entirely in Java language, so it can be integrated with JADE platform to supply the capacity to "reason" using knowledge in the form of declarative rules.

The main goal of the proposed MAS is to be applied in a real control home scenario through the following practice case:

"The user has available several devices that are automatically discovered and integrated into the home network when they are connected. User is able to configure his options through the system web interface. First of all, users and objects that he wants to identify at home must be pointed out. To achieve this, he has to assign a unique RfID tag [20] for each one of them. Once this information is introduced, the system will be able to know whether users are or not at home, and if they are it can know in which exactly room they are. With this available information users can create personal rules to associate events with the actions to be performed."

This figure shows the scenario with the implicated agents, their communications and the MAS organization:

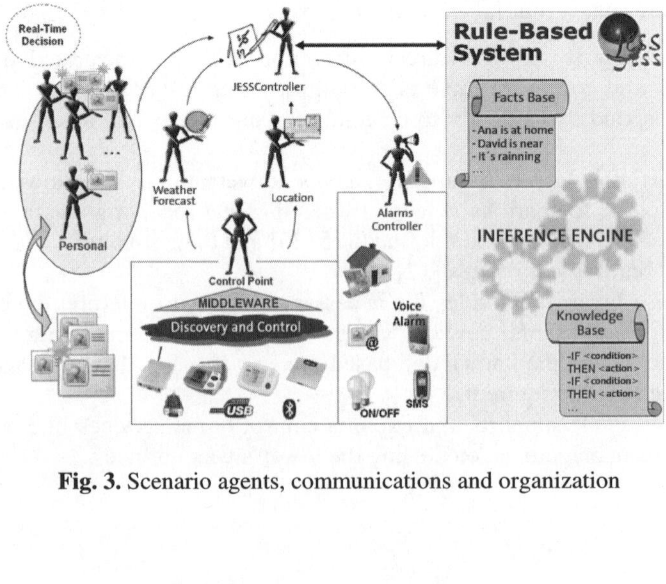

Fig. 3. Scenario agents, communications and organization

The exchanged messages sequence of the system running in JADE platform is shown in the next figure:

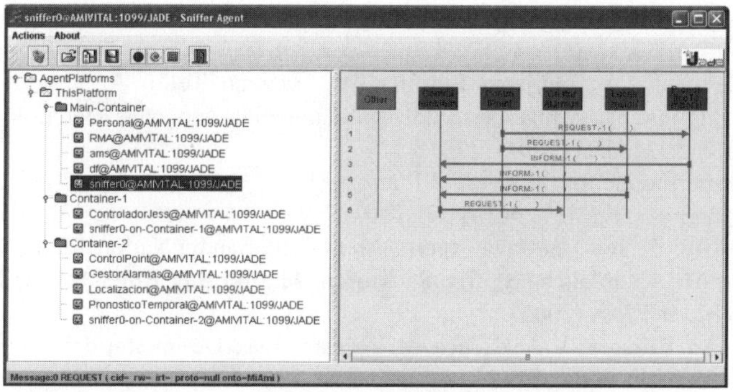

Fig. 4. Messages sequence in JADE platform

Finally an example of the services that can be provided is the following: user creates a personal rule to stop watering plants and to be informed with a SMS to his mobile phone when it starts raining and nobody is at home: If *It is raining* and *Nobody is at home* then *Stop watering plants* and *Send SMS*. Moreover, user creates another personal rule to be informed with a voice alarm to his PDA when he leaves home without his umbrella and the weather forecast is rainy: If *User leaves home (he)*, *User near (umbrella)* and *Weather forecast (rainy)* then *Send Voice alarm to the PDA*."

5 Conclusions and Future Works

An intelligent system has been designed for the digital home. To achieve this several technologies have been researched and used. The result is a system which let show how some existing technologies can work together to support users to carry out their everyday life activities in natural way using the environment information. Moreover, this work can be useful for the development and deployment of future projects that aim to increase home comfort.

The philosophy of this project is to provide a user-friendly system that can be easily used by users of any age and with or without technological knowledge. It is worth to point out the interaction between system and home surrounding and the application of technologies that allow to discovering and integrating different devices.

This work opens up the following future researches: (1) The incorporation of data mining techniques to infer user's behaviours, (2) Improvement of the platform learning system, (3) The integration of DLNA devices, (4) Development of an adaptive remote control for different devices, (5) To extend the intelligent system focused on home domain to a mobile scenario, (6) To carry out a usability study of the user interface.

References

1. Weber, W., Rabaey, J.M., Aarts, E.: Ambient Intelligence (2005)
2. Information Society Technologies Advisory Group, http://cordis.europa.eu/ist/istag.htm
3. Alonso Betanzos, A., Guijarro Berdiñas, B., Lozano Tello, A., Palma Méndez, J.T., Taboada Iglesias, M.J.: Ingeniería del Conocimiento. Aspectos Metodológicos. Pearson Education, S.A. Madrid (2004)
4. Jess, the Rule Engine for the JavaTM Platform, http://herzberg.ca.sandia.gov
5. Friedman-Hill, E.: Jess, the Java expert system shell. Sandia National Lab. (2000)
6. O'Connor, M., Knublauch, H., Tu, S., Musen, M.: Writing Rules for the Semantic Web Using SWRL and Jess (2005)
7. Guerrero, A., Villagrá, V.A., López de Vergara, J.E.: Definición del comportamiento de gestión de red con reglas SWRL en un marco de gestión basado en ontologías en OWL (2005)
8. Cernuzzi, L., Giret, A.: Methodological Aspects in the Design of a Multi-Agent System. In: Proceedings of Workshop on Agent Oriented Information Systems. VII National Conference on Artificial Intelligence (AAAI 2000), Austin, USA (2000)
9. Iglesias, C.A., Garijo, M., González, J.C.: A Survey of Agent-Oriented Methodologies. In: Proceedings of Fifth International Workshop on Agent Theories, Architectures and Languages (1998)
10. Iglesias Fernández, C.Á.: Definición de una metodología para el desarrollo de sistemas multiagente (1998)
11. Mas, A.: Agentes Software y Sistemas Multiagente: Conceptos, Arquitectura y Aplicaciones. Pearson-Prentice Hall (2005)
12. Van Aart, C., Pels, R., Caire, G., Bergenti, F.: Creating and Using Ontologies in Agent Communication. In: 2nd Workshop on Ontologies in Agent Systems (2002)
13. Caire, G.: Jade tutorial - application-defined content languages and ontologies (2002)
14. Krempels, K.-H., Nimis, J., Braubach, L., Herrler, R., Pokahr, A.: How words can tell what actions are doing. In: Challenges in Open Agent Systems 2003 Workshop (2003)
15. COBRA-ONT, http://daml.umbc.edu/ontologies/cobra
16. CoBrA Web Site, http://cobra.umbc.edu
17. Gua, T., Keng Punga, H., Qing Zhang, D.: A service-oriented middleware for building context-aware services (2004)
18. Amigo (Ambient intelligence for the networked home environment), http://www.hitechprojects.com/euprojects/amigo/
19. Stiglic, G., Verlic, M., Kokol, P.: Software Agents (2006)
20. RfID Journal, http://www.rfidjournal.com
21. UPnP Forum, http://www.upnp.org
22. Song, H., Kim, D., Lee, K., Sung, J.: UPnP-Based Sensor NetworkManagement Architecture. In: Second ICMU (2005)
23. Arias, F.J., Moreno, J., Ovalle, D.A.: Integración de Mecanismos de Razonamiento en Agentes de Software Inteligentes para la Negociación de Energía Eléctrica (2006)
24. Joo, I., Park, J., Paik, E.: Developing Ontology for Intelligent Home Service Framework. IEEE, Los Alamitos (2007)
25. Foundation for Intelligent Physical Agents (FIPA), http://www.fipa.org
26. Bellifemine, F., Poggi, A., Rimassa, G.: JADE – A FIPA compliant agent framework (1999)

27. Java Agent DEvelopment Framework (JADE), http://jade.tilab.com/
28. Protégé Home Page, http://protege.stanford.edu
29. BeanGenerator,
http://protege.cim3.net/cgi-bin/wiki.pl?OntologyBeanGenerator
30. Shreiber, G.: A Case Study in Using Protégé-2000 as a tool for CommonKADS. In: 12th Int. Conf. on Knowledge Engineering and Knowledge Management (EKAW) (2001)

An Indoor Location Method Based on a Fusion Map Using Bluetooth and WLAN Technologies*

Sofía Aparicio, Javier Pérez, Paula Tarrío, Ana M. Bernardos, and José R. Casar

ETSI Telecomunicación
Universidad Politécnica de Madrid
Madrid, Spain
saparicio@grpss.ssr.upm.es, jperez@grpss.ssr.upm.es,
paula@grpss.ssr.upm.es, abernardos@grpss.ssr.upm.es,
jramon@ grpss.ssr.upm.es

Abstract. This paper proposes a method for merging Bluetooth and WLAN technologies to face the problem of indoor positioning. The method consists in the construction of a fusion map based on calibrated WiFi RSS and simulated Bluetooth RSSI. On a recent work, we have presented a different approach for fusing both technologies. The performance of this method is tested experimentally and the comparison between the localization results obtained using both technologies and using only WiFi is presented.

Keywords: Context-awareness, Indoor Location, Fusion, WLAN, Bluetooth.

1 Introduction

Wireless technologies such as Bluetooth, WLAN (WiFi) or GPS have become worldwide known over the last few years. The proliferation of these technologies, concurrently with the mobile computing devices in which they are embedded, has triggered an increasing interest in context-aware systems and services. Several examples and a full taxonomy of these Location Based Services (LBS) are depicted in [1].

In this framework, indoor location methods come up as a key issue to be addressed. GPS-based systems offer good performances and are widely spread in outdoors environments. However, when the target is located indoors, GPS does not provide an acceptable accuracy due to its Line of Sight (LOS) dependent infrastructure. In recent years, indoor positioning has been studied by many researchers using different technologies. Most of them are developed based on a single technology.

The aim of this paper is to present an indoor location system that takes advantage of fusing Received Signal Strength (RSS) data gathered from multiple wireless technologies. Our goal is to develop a cost-effective system using easily accessible wireless devices. Today it is not uncommon for an average user to carry a mobile phone or a PDA with some of the technologies mentioned above. Several approaches have been made with RF [2], ultrasound [3] or RFID [4]; in this work we have selected IEEE

* This work has been financed by the Government of Madrid under grant S-0505/TIC-0255 (MADRINET) and the Spanish Ministry of Education and Science under grant TSI2005-07344 (COLOCAME).

J.M. Corchado et al. (Eds.): DCAI 2008, ASC 50, pp. 702–710, 2009.
springerlink.com

802.11 wireless LAN (WiFi) and Bluetooth. We will focus our efforts on emphasizing how the combination of multiple technologies improves the reliability of the location measurements with respect to a single one.

This paper is organized as follows. Section 2 reviews some existing methods for location estimation using a single technology or by means of fusing data from multiple sources. In section 3 we analyze the features of Bluetooth and WLAN in location systems. In section 4 we introduce the algorithm to fuse data from both sources based on RSS in WLAN and RSSI in Bluetooth. Section 5 gathers some real experiments. The comparison between the localization results of a real target object using the fusion of both technologies and using only WiFi is presented. Finally, Section 6 contains the conclusions and draws some guidelines for future research.

2 Background

There is a large amount of literature focused on indoor positioning systems using wireless technologies. In most of them, however, location estimation is addressed by handling a single data source. For example, the RADAR system [2] bases its position calculations on triangulation on RF Received Signal Strength (RSS); there are also ultrasound systems such as Active Bat [3] and Cricket [5]; Active Badge [6] uses Infrared, RFID is the key technology in LANDMARC [4] and SpotON [7], while Bluetooth is the foundation of ATLANTIS [8] and other projects [9], [10], [11].

Many other initiatives have also taken advantage of fusing multiple technologies. WLAN and sensors are used in [12] to improve location accuracy. A system based on RFID, WiFi and Vision is described in [13]. In [14] indoor location estimation based on Bluetooth and WLAN is presented. Most previous methods use RSSI and Link Quality information from Bluetooth devices.

Regarding the fusion methods, in most of the contributions above location is met independently with every single technology; then, results are compared or combined to obtain a final estimated position. An interesting algorithm to solve this shortcoming is presented in [15]. As we will describe in Section 4, our fusion procedure it is based on the construction of a fusion map consisting in calibrated WiFi RSS and simulated Bluetooth RSSI. Therefore, no weight has to be assigned to previous independent estimated locations.

3 WLAN and Bluetooth in Location Systems

3.1 WLAN

Location in indoor environments are challenging because of the reflections, absorptions and multi-path phenomenon suffered by the RF signals. Hence, LOS dependent parameters such as time of flight (time of arrival - TOA, time difference of arrival - TDOA) or angle (angle of arrival – AOA) based measurements used by other technologies are unsuitable. Moreover, most of these techniques need specialized hardware to extract the information [16].

One of the most well-known location techniques used with 802.11x wireless LAN [2] is based on fingerprinting building. It is based on the measurements of the RSS from the Access Points (APs) and infers an estimated position by means of non-geometrical algorithms. The actual RSS vector of values measured in the client (as many values as APs displayed in the covered area) is compared with the vectors of values of each point of the target map, previously measured and stored in a data base in an off-line phase. Finally an estimated location is calculated applying different distance algorithms (Euclidean, k-nearest Neighbors, etc.).

Specialized hardware is not needed this time to extract the RSS value; it can be directly read from the wireless card. However, this method has some limitations. For example, it is completely dependent on the target area that we want to cover. Any change in the distribution of the furniture, walls or even people walking along the area will substantially vary the RSS readings. Two kinds of problems are derived from these modifications. Firstly, sudden changes in the received RSS when continuous location is being carried out cause impossible deviations of several meters for the target device's location just in one second. These wrong locations are called outliers, and need to be suppressed to gain accuracy. Secondly, when variations in the environment are not occasional, former fingerprints in the database become obsolete causing steady errors in the location sensing. Systems that refresh the database are one of the solutions implemented to minimize this last problem.

With all its pros and cons, average precision with this technique has been reported to be around 3 m in RADAR. This accuracy may be improved using more APs or including fingerprints from more coordinates in the map, by means of a more dense data base.

3.2 Bluetooth

Bluetooth is a widely spread short range wireless technology used to connect computers, mobile/smart phones and peripherals. This proliferation has led many researchers, in the last few years, to investigate the use of this technology to implement location systems. In this section, we will describe the parameters selected to estimate the position in those systems.

RSSI and Link Quality in Bluetooth. RSSI in Bluetooth is obtained from the comparison of the received power level against two thresholds that determine the golden receive power range (GRPR), an interval of approximately 20 dB (it varies depending on the hardware) in which the RSSI value given is zero. Figure 1 shows the ideal correspondence between RSS and received power level, where GRPR is the horizontal line in the middle of the graphic. Assuming that receive power decreases proportionally to the square of distance to the transmitter, we can infer a relation between RSSI and distance to the transmitter.

On the other hand, Link Quality (LQ) is a discrete value varying from 0 to 255 which indicates the quality of the connection established between two Bluetooth devices. There is no specification that exactly defines how it varies; thereby it is strongly dependent on the vendor. Empirical tests show that the higher the link quality value, the better the link is. Therefore, we can also assume that LQ degrades with distance and use it for location estimation systems.

The majority of the works that have developed a Bluetooth-based location system have used RSSI to estimate the position [8], [10], as well as LQ [18]. Many works

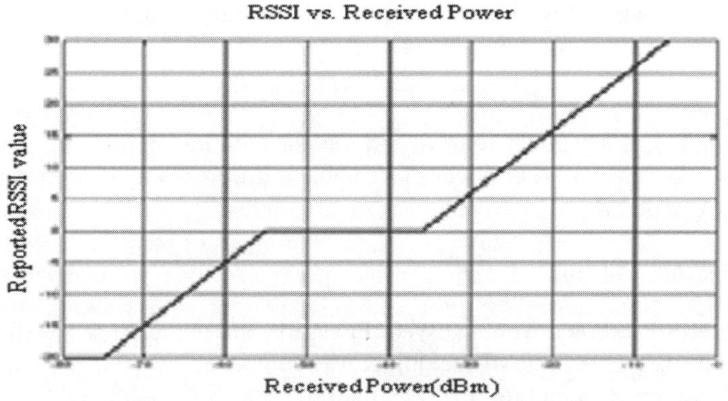

Fig. 1. Ideal Correspondence between Receive Power Level and RSSI in Bluetooth Devices [17]

[17], [11] have tackled the golden range. We opted for the construction of a fusion map based on calibrated WiFi RSS and simulated Bluetooth RSSI. A common drawback in both RSSI and LQ parameters is the time required to extract the value from the client. A piconet should be formed between the two Bluetooth devices, receiver and transmitter, which means an average time of 2,5 s, without an ensured maximum. This process should be repeated with each target client in the range, thus when more than three clients in the area needs to be located delays become unacceptable. We were able to configure the inquiry process for extracting the RSSI without connection. The basics of this method and the way it is merged with the WLAN location calculations are described in section 4.

4 The Fusion Algorithm

The proposed algorithm aims at determining the location of an object. It is based on the previous existence of a fusion WiFi RSS and Bluetooth RSSI map. Two distributions of access points are assumed: one consisting of Bluetooth stations and another one consisting of WiFi stations.

4.1 The Fusion Map

The first step is to build a fusion map based on WiFi RSS and Bluetooth RSSI. To construct the fusion map, we have used fingerprinting to measure the WiFi RSS received from each WiFi AP, and we have simulated the RSSI received from each Bluetooth AP for each point of the map.

To measure the WiFi RSS received from each WiFi AP, if we do not receive any RSS, a value of -92 dB is assigned, corresponding to the low limit of the RSS.

To simulate the Bluetooth RSSI received from each Bluetooth station for each point of the map, we have implemented a program using the Friis formula and empirical results. For that purpose, we formulate a propagation model that provides a relationship between the received power (P_{RX}), the transmitted power (P_{TX}) and the distance (d) between transmitter and receiver.

$$P_{RX} = P_{TX} + A - 10\eta\log(d) \tag{1}$$

In equation (1), A is a constant term and η is the path loss exponent.

We have also considered that walls introduce a different attenuation depending on what they are made of.

4.2 Localization Method

Knowing the RSS received from every WiFi station and the RSSI received from every Bluetooth station, the location of the object was computed.

We compare the RSS and the RSSI of the object with the RSS and the RSSI of the points inside the fusion map. To do that we use the minimax distance, i.e. the maximum absolute difference between the RSS or RSSI measurements, as they are related with the same station. For example, we assume that we have three stations and (p_1,p_2,p_3), $(p_{1,x,y},p_{2,x,y},p_{3,x,y})$ are the RSS or RSSI vectors of the target object and a point inside the fusion map respectively. Then, the minimax distance between these points is

$$d = \max\left(\left|p_1 - p_{1,x,y}\right|, \left|p_2 - p_{2,x,y}\right|, \left|p_3 - p_{3,x,y}\right|\right).$$

If we do not have a RSS or RSSI measure for an AP, a value of -92 dB is assigned, corresponding to the low limit of the RSS or RSSI.

Once we have defined a distance, the position of the object is obtained using the same method as in [19]. This method consists on averaging the 3 nearest neighbors with some weights as follows. If (x_1,y_1), (x_2,y_2), (x_3,y_3) are the coordinates of these 3 points and d_1, d_2 and d_3 are the respective distances to the target object, the coordinates considered for this point are

$$x_p = \frac{1}{2}\left(\frac{d_2+d_3}{d_1+d_2+d_3}x_1 + \frac{d_1+d_3}{d_1+d_2+d_3}x_2 + \frac{d_1+d_2}{d_1+d_2+d_3}x_3\right)$$
$$y_p = \frac{1}{2}\left(\frac{d_2+d_3}{d_1+d_2+d_3}y_1 + \frac{d_1+d_3}{d_1+d_2+d_3}y_2 + \frac{d_1+d_2}{d_1+d_2+d_3}y_3\right).$$

This method gives more weight to the points with a smaller distance.

5 Results

In this section the localization results in a real layout are presented. Some real experiments were done to locate a real target object using a fusion map based on calibrated WiFi RSS and simulated Bluetooth RSSI.

The results presented in this paper are based on an actual WiFi deployment at UPM Telecommunication Engineering School. The distribution of APs is shown in the map in Figure 2 for WiFi stations and in Figure 3 for Bluetooth stations. Black lines represent concrete walls and the green ones are glass and plastic walls.

The Bluetooth attenuation considered is 4 and 1.5 (RSSI) respectively.

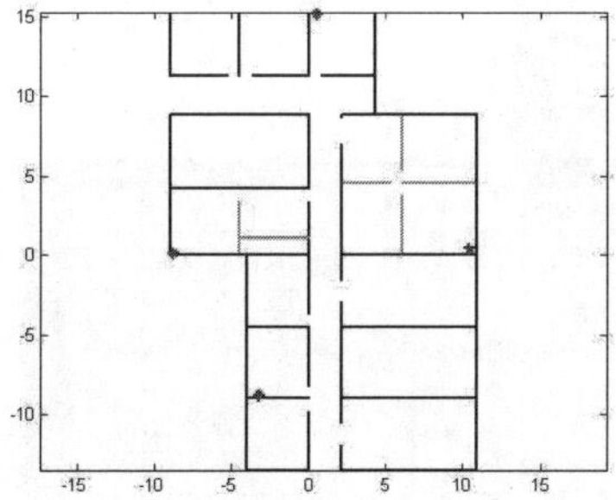

Fig. 2. Wifi access points distribution

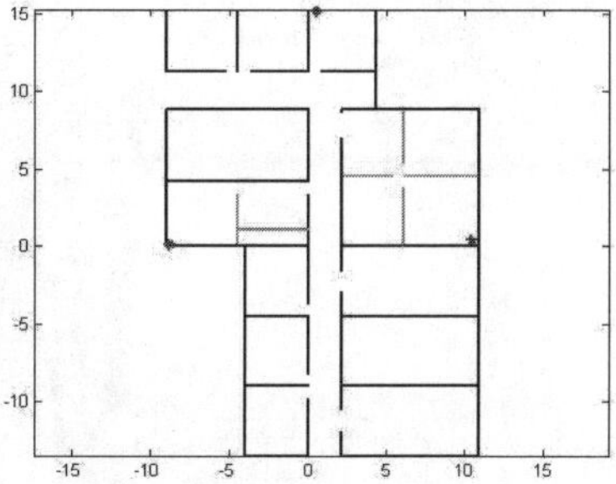

Fig. 3. Bluetooth access points distribution

Based on the concepts of [20], some experiments have been made to compute the path loss exponent η and the constant term A of the Friis formula for a given transmitted power. A Bluetooth AP and a PDA have been used to measure the RSSI of the received packets for different distances between transmitter and receiver.

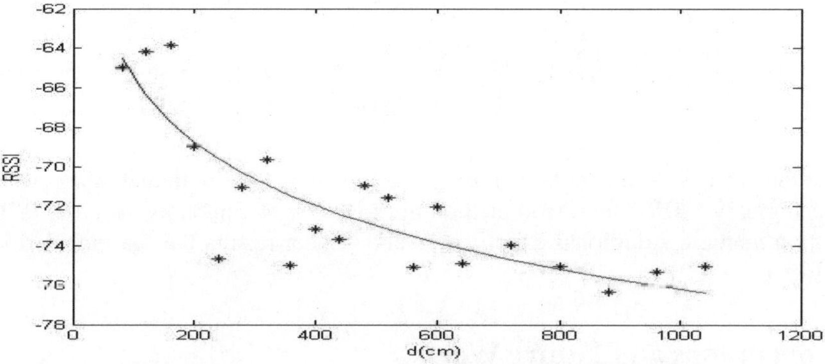

Fig. 4. RSSI vs. distance

The experimental data (blue dots) are fitted using equation (1) (solid red line), see Figure 4. From the fit in Figure 4, values for the path loss exponent η and for the constant term A were obtained: η=1.0680 and A=-97.5712.

Knowing the RSS received from every WiFi station and the RSSI received from every Bluetooth station, the computation of the position of a real target object was made. We computed the Bluetooth RSSI and the WiFi RSS received from every Bluetooth and WiFi station respectively for each real target object that we wanted to localize. We tested the localization method using both technologies and using only WiFi for several points around the map, Figure 5.

In Table 1 we summarize the average error obtained for the localization of several real target points using both technologies and using only WiFi. By average error we mean the average deviation for 100 measurements of each real target point.

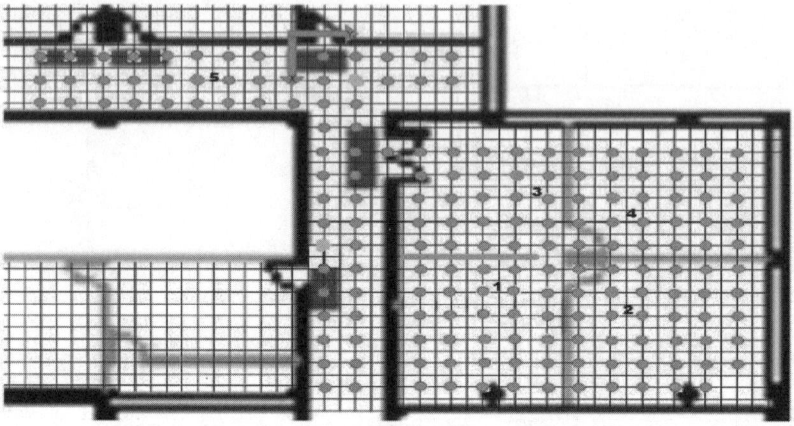

Fig. 5. Distribution of points where the algorithm has been tested

Table 1. Average error obtained for the localization of the target points using only WiFi and using Bluetooth-WiFi with 3 Bluetooth APs

Points	1	2	3	4	5	TOTAL
WIFI	2.39	1.34	3.28	3.21	3.45	2.73 m
BTWIFI	3.03	2.38	1.66	1.99	2.02	2.21 m

The localization results obtained using the fusion of Bluetooth and WiFi are better than using only WiFi. The fusion method improves by 50 cm on average the WiFi localization method. Additional experiments and further results will be included in the final paper.

6 Conclusions and Future Work

We presented in this paper the results obtained using the fusion of technologies in indoor location systems. We selected to fuse Bluetooth and WLAN, and we proposed a method based on the construction of a fusion map consisting of calibrated WiFi RSS and simulated Bluetooth RSSI.

In a previous paper [21], a different approach was presented for fusing both technologies. First, Bluetooth selects the region where the object is located (Cell-ID) and second WLAN is applied to that region for the precise determination of the object position.

In this paper some real experiments were done to locate a real target object. We show the localization results obtained using a fusion calibrated WiFi RSS and simulated Bluetooth RSSI map. Some Bluetooth real measurements were used for the computation of the simulated RSSI received from every Bluetooth station. The localization results were better using the fusion of Bluetooth and WiFi than using only WiFi.

For the simulated Bluetooth RSSI we formulated a propagation model taking into account some real measurements and the attenuation produced by walls. In a future work, it would be interesting to design accurate models including other important factors such as interferences due to obstacles present in the covered area, multiple reflections. This will lead us to avoid the fingerprinting and to obtain more accurate simulated maps.

Research aimed to improve fingerprint method will be developed. Monitor APs that refresh the fingerprint database to upgrade its robustness, or the deployment of a Bluetooth ad-hoc network to add mobility to the Bluetooth APs are also being studied and considered.

References

1. Bernardos, A., Tarrio, P., Casar, J.R.: A Taxonomy of Mobile Location-Based Services. In: International Conference on e-Business, Barcelona (2007)
2. Bahl, P., Padmanabhan, V.N.: RADAR: An In-Building RF-based User Location and Tracking System. In: INFOCOM 2000. Nineteenth Annual Joint Conference of the IEEE Computer and Communications Societies, Tel Aviv, vol. 2, pp. 775–784 (2000)
3. Harter, A., Hopper, A., Steggles, P., Ward, A., Webster, P.: The Anatomy of a Context-Aware Application. In: 5th Annual ACM/IEEE Int. Conference on Mobile Computing and Networking, Seattle USA, pp. 59–68 (1999)
4. Ni, L.M., Liu, Y., Lau, Y.C., Patil, A.P.: LANDMARC: Indoor Location Sensing Using Active RFID. In: Wireless Networks, pp. 701–710. Springer, Netherlands (2004)
5. Priyantha, N.B., Chakraborty, A., Balakrishnan, H.: The Cricket Location-Support System. In: 6th Int. Conference on Mobile Computing and Networking, Boston, pp. 32–43 (2000)
6. Want, R., Hopper, A., Falcao, V., Gibbons, J.: The Active Badge Location System. ACM Transactions on Information Systems 10(1), 91–102 (1992)
7. Hightower, J., Borriello, G., Want, R.: SpotON: An indoor 3D Location Sensing Technology Based on RF Signal Strength. In: UW CSE Technical Report #2000-02-02
8. Yipin Ye, J.: Atlantis: Location Based Services with Bluetooth. Master's thesis. Dept. of Computer Science, Brown University (2005)
9. Forno, F., Malnati, G., Portelli, G.: Design and Implementation of a Bluetooth ad hoc Network for Indoor Positioning. In: IEE Proceedings-Software Microsoft Research Cambridge. Workshop, Cambridge, vol. 152, pp. 223–228 (2005)
10. Feldmann, S., Kyamakya, K., Zapater, A., Lue, Z.: An Indoor Bluetooth-Based Positioning System: Concept, Implementation and Experimental Evaluation. In: International Conference on Wireless Networks, Las Vegas (2003)
11. Bandara, U., Hasegawa, M., Inoue, M., Morikawa, H.: Design and Implementation of a Bluetooth Signal Strength Based Location Sensing System. In: Radio and Wireless Conference, pp. 319–322. IEEE, Atlanta (2004)
12. Hii, P., Zaslavsky, A.: Improving Location Accuracy by Combining WLAN Positioning and Sensor Technology. Monash University, Melbourne (2005)
13. Matthieu, A., Crowley, J.L., Devin, V., Privat, G.: Localisation Intra-bâtiment Multi-technologies: RFID, WiFi et Vision. In: Proceedings of the 2nd French-speaking conference on Mobility and ubiquity computing, Grenoble, pp. 29–35 (2005)
14. Pandya, D., Jain, R., Lupu, E.: Indoor Location Estimation Using Multiple Wireless Technologies. In: The 14th IEEE 2003 Int. Symposium on Personal Indoor and Mobile Radio Communication Proceedings, Beijing, pp. 2208–2212 (2003)

15. Myllymaki, J., Edlund, S.: Location Aggregation from Multiple Sources. In: Proceedings of the Third Int. Conf. on Mobile Data Management, Singapore (2002)
16. Günther, A., Hoene, C.: Measuring round trip times to determine the distance between WLAN nodes. Technical Report TKN-04-016, Telecommunication Networks Group, Technische Universität Berlin (December 2004)
17. Bielawa, T.M.: Positioning Location of Remote Bluetooth Devices. Thesis of M. of Science in Electrical Engineering, Virginia Polytechnic Inst. And State University (2005)
18. Genco, A.: Three Step Bluetooth Positioning. Location and Context Awareness. LNCS, vol. 3479, pp. 52–62. Springer, Heidelberg (2005)
19. Collado, E.: Diseño y desarrollo de algoritmos de localización de dispositivos en redes inalámbricas para provisión de servicios móviles. Master´s thesis, ETSIT-UPM (2007)
20. Tarrío, P., Bernardos, A., Casar, J.R.: A Transmission Rate and Energy Design for Power Aware Localization in Ad Hoc and Sensor Networks. In: The 18th Annual IEEE International Symposium on Personal, Indoor and Mobile Radio Communications (PIMRC 2007), Athens (2007)
21. Pérez, J., Aparicio, S., Bernardos, A., Casar, J.R.: An Indoor Location System Based on Bluetooth and WLAN. In: Proceedings of the Workshop on User-Centric Technologies and Applications, Salamanca, pp. 77–86 (2007)

Design and Deployment of Context-Aware Services: A Prototyping Case-Study*

Ana M. Bernardos, Paula Tarrío, Josué Iglesias, and José R. Casar

Universidad Politécnica de Madrid - ETSI Telecomunicación
Ciudad Universitaria S/N, 28040 Madrid, Spain
{abernardos,paula,josue,jramon}@grpss.ssr.upm.es

Abstract. In this paper, we describe a design and deployment experience of context-aware services for a particular application environment. The design has been based on a conceptual framework, which identifies the main functional building blocks that a context-aware system should have. In practice, the context-aware system has been built on an application server as a central element, which interacts with different data bases where user and context information is stored. To enable positioning and communications, a hybrid infrastructure combining WiFi and Bluetooth has been deployed. The user experience analysis shows that the acceptance of the proposed context-aware services is related, firstly, to the stability of the communications and positioning infrastructure.

Keywords: Context-aware services, mobile services, ubiquitous architectures.

1 Introduction

The deployment of context-aware services depends on the development of logical architectures and physical systems capable of acquiring and handling "any information that can be used to characterize the situation of entities (i.e. whether a person, place or object)". By now, the existing literature gathers a significant number of architecture proposals, many of them claiming the need of decoupling the application development stage from the context provision infrastructures.

The primary objective of this contribution is to present a framework that gathers the main functionalities that a device-centric context-aware system should have. This functional model may make easier the prototyping of context-aware systems and may also increase the reusability of logical components. As secondary objective we present an implementation of the proposed functional model for a real application environment, namely an exhibition fairground.

2 Related Research

More than a decade after the first developments, context-aware systems have evolved from monolithic applications to different middleware infrastructures: based on widgets

* This work has been financed by the Spanish Ministry of Education and Science under grant TSI2005-07344 and by the Government of Madrid under grant S-0505/TIC-0255.

J.M. Corchado et al. (Eds.): DCAI 2008, ASC 50, pp. 711–719, 2009.
springerlink.com

(e.g. the Context Toolkit), centralized context servers (e.g. the CoBrA multiagent system), distributed architectures (e.g. Solar) or blackboard structures (e.g. CMF). In parallel, many areas of application for context-aware mobile services have been explored.

With respect to the research activity focused on context-aware systems or services for fairground or exhibition areas, one of the first trials was built on the well-known Context Toolkit platform. Among the various pilot applications that were developed on it, there was a Conference Assistant [1], a PC-based application that was configured to provide agenda information, downloads of slides when in a conference room and documentation recovery functionalities after the event. Other research developments can be the Hippie prototype [2], the mExpress Project [3] or the SAiMotion project [4], where user's preferences and positioning were used to facilitate context-awared services.

From a commercial viewpoint, in the last years some technology fairs have made some trials with mobile services. It's the case of CeBIT, that since 2000 has used several commercial context-aware applications to provide guiding and information services to its visitors (more information in [5][6]).

From a short review of previous references, it is possible to state that, although some pilot and trial experiences have been taken, commercial services are limited and mainly driven to "mobilize" the traditional catalogues, not taking real advantage from neither the possibilities of the connected environments nor the technology that visitors nowadays have in their pockets.

3 Analyzing the Functional Structure of Context-Awareness

Let us supposed a hypothetical scenario in which context-aware services may be offered; it may represent a ground level in an exhibition centre, an office floor, a house, an airport or even some open-air places. We consider that this space can be divided into symbolic geometric zones, with different shapes and sizes. When a user enters in a given zone a set of location-based and profile-adapted services will become active and offered to the user. For example, when the user enters in a conference room, a collection of services related with that place will appear in his/her device (i.e. next speaker's slides or a virtual cards exchange application). From the logical point of view, a zone will be able to contain smaller subzones.

In order to achieve this service provisioning, we can summarize the usual elements that will be find in a generic context-aware system for personal mobile devices: a communications infrastructure, probably heterogeneous (WiFi, Bluetooth, cellular, etc.), built on the connectivity features of the mobile devices; a number of sensors capable of acquiring context data (they may be part of the infrastructure or built-in in the users' devices); a positioning infrastructure, and its associated software, that may be configured depending on different architectural approaches (network-based, network-assisted or terminal-based); a logical system capable of reasoning on context data and users' preferences, in order to take decisions; and a bundle of context-aware services, which will determine their informational needs and possibly will provide feedback to the logical system on the quality of the received information.

From a functional view, the context-aware system should have some general capabilities, which will be implemented to a different extent depending on the final application needs. These functionalities and their relationships are represented in Figure 1 and explained below.

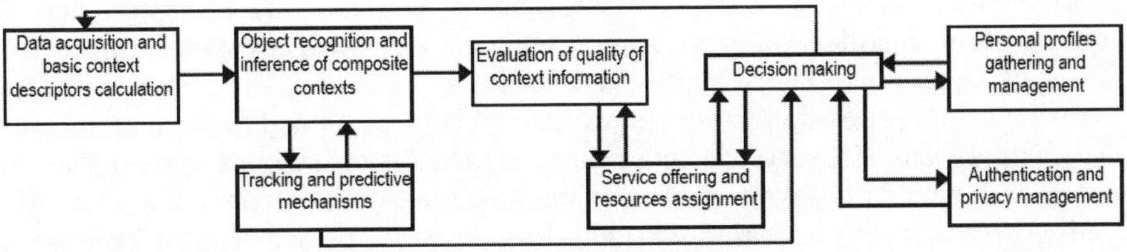

Fig. 1. Functional block diagram

• *Data acquisition and basic context descriptors calculation*. This feature includes sensor management, continuous data acquisition mechanisms and adaptive multisensor fusion to obtain appropriate first level descriptors from data streams acquired by different type of sensors. These first level descriptors will serve as a basis to infer activities or situations.
• *Personal profiles gathering and management*. Context interpretation will be also conditioned by personal descriptors that will shape user's profiles. These profiles will serve as a filter both to support context building, decision making and to anticipate to users' intentions, desires or needs.
• *Object recognition and inference of composite contexts*. Basic descriptors have to be associated to identities and histories, in order to be validated respecting to real entities. At a given moment, each entity will have knowledge of its own context, which will be composed by first level descriptors and reasoned inferences on these and other data. The composite context may include assumed relational linkages, which will be necessary for a number of applications.
• *Tracking and predictive mechanisms* to anticipate the entities' intentions, desires or needs. Designing anticipation mechanisms may both enhance data acquisition processes and support knowledge extraction on actions the user can take. This pre-activity knowledge will serve as a basis to configure a service offering. To achieve this objective, it is necessary to maintain certain knowledge about the user history.
• *Evaluation of quality of context information*. Acquiring context data is a costly procedure, so the system will need to optimize it, fulfilling the applications' requirements while minimizing complexity and cost. With this objective, the design of measurements of information quality will serve the system to take decisions about how to modify acquisition procedures and algorithms. Each application may have an indicator that will have to diagnose if the information received is updated, accurate, complete and correct for its needs, but also other elements in the system may be performing similar tasks.
• *Service offering and resources assignment*. In an environment where heterogeneous technologies are operative, efficient handling of communications may support the continuity of services and the service experience from the user point of view. On the other hand, the sensing infrastructure must be adapted to the diverse situations the system may be in.

• *Authentication and privacy management.* Securing the context-aware system and its users is a horizontal functionality that may be present in different stages of the process. Managing privacy is somehow related to a) adopting centralized or decentralized approaches when acquiring and processing sensors' data streams, b) using a convenient design of data collection, storing and exchange processes, c) establishing security policies, using pseudonym and camouflage methods.
• *Decision making.* Position, context, profiles, etc. are combined to form an informational picture that will serve as input data for a decision support system that will decide which actions to take regarding communications, service provision or user interaction needs. It will be based on controlling context, personalization parameters, user's history and predictions methods. It will also be in charge of updating users' histories, context parameters and knowledge data base. The final objective of a context-aware system is to provide its users with the information and services they expect to have or to optimize resources administration to fulfill a global mission.

It is important to underline that there is not any connection between the functionalities described above and a particular type of architecture: both centralized and distributed systems will have to consider how to implement them.

4 A Case Study: Context-Aware Services in an Exhibition Hall

Next we refer to a real experience of design and deployment of context-aware services in a rich content and dynamic environment, such as it can be a fairground or an exhibition hall. The application environment fits the generic description and contents in Section 3: a fixed sensing infrastructure composed by WiFi and Bluetooth access points; a WiFi communications infrastructure; a hybrid positioning system that calculates position basing on WiFi signal strength fingerprinting combined with Bluetooth propagation models [7][8]; a logical system, composed by an application server and several databases, to store, manage and reason on context descriptors in order to configure the offer of services for the final user; and a bundle of web-based context-aware applications specially developed for the fairground environment.

4.1 Characterizing the User's Context

In our case, the user's context has been described by the following personal, physical and social context parameters (Figure 2):

IDENTITY. The user must register in the system providing his personal data. The system will automatically retrieve information about his mobile device (WiFi and Bluetooth physical address), as it is needed to complete location procedures

ROLE. Each user will pertain to a meta-group of users. In a fairground, it is possible to identify at least three different roles: visitors, exhibitors and organizers. Users in the same meta-group will have access to some common services, that will be adapted to their personal configuration parameters

PROFILE. Each user is identified according to his main professional activity (student, researcher, professor, and practitioner)

PREFERENCES. The user's technology interests are gathered and used as a filter for information about events and related contents

SOCIAL INFORMATION. Users may belong to a private group. Special communication services are configured among the members

POSITION. It is obtained through a positioning system that integrates WiFi and Bluetooth and provides the approximate coordinates of the user's location, translating them into symbolic positions (zones)

TIME. Time filters facilitate browsing on the contents (for example, looking up in schedules)

Fig. 2. User's context parameters

4.2 Description of the Application Area

Deploying context-aware services in an exhibition hall is a challenging task. Due to the "boundary conditions" the system will have to work on a great heterogeneity of mobile devices, with different communication capabilities and applications installed; it will be necessary to adapt the positioning algorithms to minimize the deployment time while guaranteeing its correct stability and the accuracy of the results; as the system will deal with user's information, privacy issues will also have to be considered in the whole system.

Our testbed has been prepared for the Employment Fair that is annually held at the Telecommunications School of the Technical University of Madrid where around twenty exhibitors show their corporate activity to the community. Conferences and parallel recruitment activities are held simultaneously.

4.3 Service Offering Mapping

The service offer seeks to facilitate communications between different users of the system, optimize the time spent on the visit, improve the access to information of interest and allow its easy retrieval, and provide an efficient mechanism of control to the organizers during the event. With these objectives, and considering the three primary roles identified above, we have prepared the services offer shown in Figure 3. Services, which are grouped into four types, will be adapted to their users in the appropriate way using the context descriptors previously mentioned.

As position will enable services filtering, we have created and attached the services to physical spaces, in particular to four types of zones (Figure 4).

Exhibitor	Visitor	Organizer
INFORMATION SERVICES - Location-based advertising - Update of the exhibition geoblog - Presentations schedule - Finder - Same services as a visitor	- Exhibitors information (web or QR code-based) - Customized agenda - Maps and positioning service - Real time guide - Dynamic routes configuration - Finder	- Professional notifications agenda - Same services as a visitor - Ambient alerts
COMMUNICATION SERVICES - Ad-hoc communications - Sending alerts - Same services as a visitor	- vCards exchange - VoIP service - Fair geowiki - Geo-referenced messaging - Sending emails or notifications to the exhibitors	- VoIP service - Same services as a visitor
EXECUTION SERVICES - Same services as a visitor	- Direct access: notebook, recorder, QR reader, VoIP application	- Same services as a visitor
CONTROL SERVICES - Resources management - Presence notifications - Stand visits statistics	- Registration, profile management and preferences - Notifications about activities requiring personal data comm.	- Resource management - Tasks assignment - General statistics - Alerts on infrastructure problems

Fig. 3. Examples of user roles and services

GENERAL ZONE, in which information and communication services by default will be provided	**ENTRANCE ZONE**, where registration and welcome services will be offered
STAND ZONE, physical influence area of each of the exhibitors, for which they will be able to design their own services	**CONFERENCE ROOM ZONE**, where information and personal applications' activation will be enabled

Fig. 4. Contextual zone

Services are to be offered as browsing context-aware web applications, or as initiators of built-in applications, previously installed in the mobile devices. The context-aware client will show to the user a set of access icons that, when clicked, will f orward to the corresponding web service or launch the built-in applications.

5 From the Functional Structure to an Operative Architecture

The integrated system is based on a Context Provider, a logical element with reasoning and decision capabilities. This Context Provider configures the services offering that the Context Client installed in the devices shows to the final users. A Fusion Engine (providing location), an Applications Handler (containing the context-aware web applications) and several data bases are also considered. The External Contexts Generator facilitates the aggregation of new context parameters, making them available to the Context Provider through XML files. In the device-side a Positioning Client is also needed to make the fusion system more accurate. Applications such as a VoIP client or a QR reader are also installed in the device.

Let us now analyze our system under the view of the functional structure presented in Section 3.1. Figure 5 shows all the logical elements and the code of colours represents the functionalities implemented in each entity.

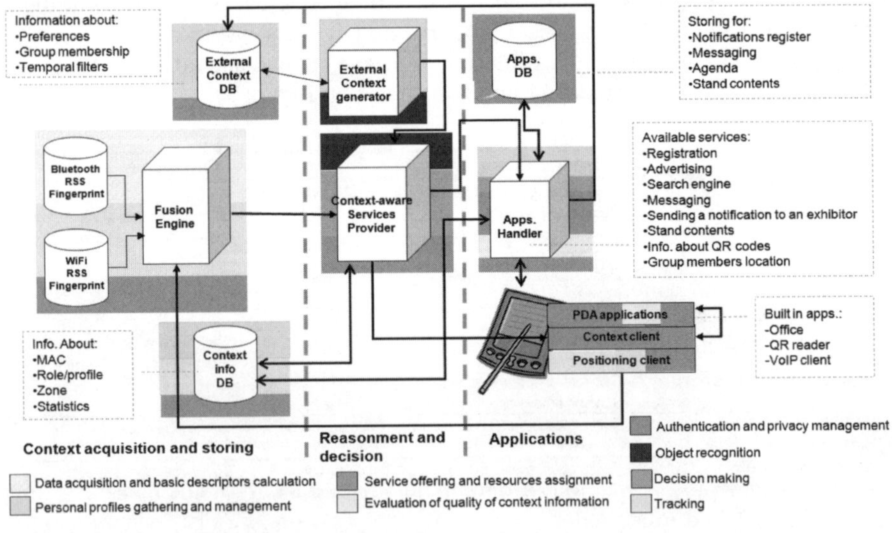

Fig. 5. A functional view of the system deployed

- *Data acquisition and basic context descriptors calculation.* Among the basic context descriptors that may act as filters, position is the one that need to be calculated from a sensing infrastructure. In our case, a fusion system combining WiFi and Bluetooth access points has been used to obtain it. The mobile devices need to run a small application to get WiFi RSS parameters, and together with the Fusion Engine support this functionality.

• *Personal profiles gathering and management.* Knowing the user's role and storing his preferences and group membership is needed to personalize the context-aware services. On the other hand, it is necessary to know the relationship between personal identities and devices, as it is a requirement of the positioning engine. For these reason, information about MAC addresses of Bluetooth and WiFi interfaces must be gathered. Web-based registration applications (running in the Applications Handler) and connected databases (Context DB and External Context DB) will be support the acquisition and storing process.
• *Object recognition and inference of composite contexts.* Once the basic descriptors have been calculated or gathered, the objective is to build an informational picture combining all the data about a given entity. For the fairground problem, identity, profile, preferences, group membership, location and time of the day are the filters that will enable service offering. On this information, the Context Provider will reason, supported by the External Context Generator.
• *Tracking and predictive mechanisms* to anticipate the entities' intentions, desires or needs. Real-time tracking is necessary for some of the services we are planning to offer, such as the statistic analysis available for exhibitors and organizers. This operation may be implemented as a feature of the Context Provider, and will be supported by the Fusion Engine, connected databases and applications.
• *Decision making.* Reasoning on the informational picture described above determines which services to offer to the user and also the need of interacting with the sensing and communications infrastructure. This functionality will be implemented in the Context Provider, which determines which services will appear in the user's device. It will be also present in the applications that may query for extra information and enhance the context description with external parameters.
• *Service offering and resources assignment.* The bundle of services previously described will be provided by the Applications Handler and its connected database, which will hand out the context-configurable services to the mobile devices under the order of the Context Provider.
• *Evaluation of quality of context information.* Different elements in the system may establish metrics to control the completeness and quality of context information. In our case, the applications themselves will be in charge of this task.
• *Authentication and privacy management.* The authentication process will be controlled from the Context Provider authentication module, which is connected to the Context Client running in the user's mobile devices. Regarding privacy, this is a horizontal function that may be built in different stages and blocks.

As the reader may notice, the different functionalities are neither linearly implemented nor attached to a single component of the system, but distributed among the different building blocks. Although the architecture follows a centralized approach, the client software also collaborates to acquire context-data, maintain the user's privacy or interact with the final user.

6 On the User's Experience

The design and deployment of context-aware services presents some challenges related to the integration complexity of acquisition systems and context management and reasoning tools. But, beyond technical issues, the design of this kind of services is strongly conditioned by the demanding requirements that users may impose to get a satisfactory experience.

To date, there are few works evaluating the user experience with respect to context-aware services. Below we present the results of a limited-scope evaluation we have carried out during the event. The experiment involved 10 fair visitors, who received a PDA with preinstalled applications and stated to have limited experience (on average, 6/10) with regard to the use of such devices. The usability questionnaire consisted of eight evaluation questions about the difficulty of use and usefulness of some of the services configured (Figure 6.1). Subsequently, users were asked about their overall user experience (Figure 6.2).

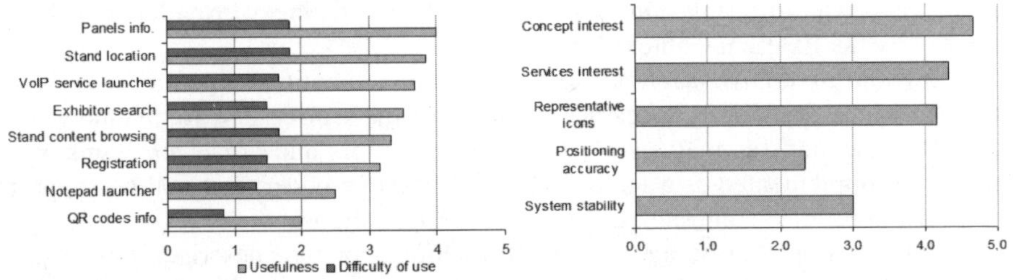

Fig. 6. 1) Service evaluation and 2) Overall evaluation of the user experience

As it can be seen in Figure 6.1, just one of the services provided (access to information via QR codes stuck on books and posters) received a lower valuation of usefulness, perhaps due to the lack of experience of the users with respect to 2D code readers. Information and communications services are among the most appreciated: an event agenda filtered according time and interests, a location-aware finder of stands and facilities and the simplification of launching VoIP clients. On the other hand, users showed their interest in location-based services: finders of members of your contact group (4.8/5), geo-referenced advertising (3.6/5), or tools to send messages linked to a given spatial address (3.6/5). It is also remarkable the interest in having VoIP communications services (4.6/5).

Users agreed that all services were highly functional and controlled them with minimal problems. Iconography also resulted to be intelligible and enlightening. Regarding the overall experience, users clearly ratified both the viability of the concept of offering context-aware services in this environment. The stability of the system obtains a middle score, fact attributable to the problems associated with the lack of accuracy of the positioning system. The correct operation of the positioning system is crucial to achieve a suitable level of tasks automation.

7 Lessons Learned and Conclusions

In this paper we have conceptually described an architectural design for mobile context-aware systems which covers from data acquisition processes to service deployment, following the functional model stated above. A real system has been built from scratch, combining proprietary and in-home developed modules, evaluating its technical performance in a real testbed.

From the deployment experience we can conclude that: 1) Functionalities identified in the theoretical model are explicitly present in the real system. They are finally built on different components or real building blocks, not attached to a single logical component. 2) The user expects the system to work in a predictable and solid way and does not tolerate irregularities or fall downs. This is influenced by the stability and accuracy of the positioning engine. So it is critical to find an effective positioning system in terms of capacity to deal with heterogeneous devices, sufficient accuracy and facility of deployment. 3) Using a centralized and externally managed system (both for positioning and context data acquisition) may have some advantages. But it is important to analyze the impact of this strategy on the feeling of control the user has over the process. A compromise between user interaction requests, automation and user control is needed.

In brief, the favorable analysis of the user experience let us think that device-centric context-aware services may be accepted if they are deployed on stable infrastructures (both regarding communications and positioning) and designed with an adequate automation level that guarantees the equilibrium between the feeling of control and the need of interaction.

References

1. Dey, A.K., et al.: The Conference Assistant: Combining Context-Awareness with Wearable Computing. ISWC 1999, pp. 21–28 (1999)
2. Oppermann, R., et al.: Hippie: A Nomadic Information System. In: Gellersen, H.-W. (ed.) HUC 1999. LNCS, vol. 1707. Springer, Heidelberg (1999)
3. Mathes, I.: Context-aware services in an exhibition environment - mexpress approach. In: Proc. of the eBusiness and eWork Conf., October 2002, pp. 685–692 (2002)
4. Schmidt-Belz, B., et al.: User Validation of a Nomadic Exhibition Guide. In: Brewster, S.A., Dunlop, M.D. (eds.) Mobile HCI 2004. LNCS, vol. 3160. Springer, Heidelberg (2004)
5. Bieber, G., Ide, R.: XyberScout: A Platform for the Efficient Construction of Mobile Location Aware Information Systems. In: HICSS 2003 (2003)
6. Kerer, C., et al.: XGuide - A Practical Guide to XML-Based Web Engineering. In: Networking Workshops 2002, pp. 104–117 (2002)
7. Collado, E., Bernardos, A.M., Tarrío, P., Besada, J.A., Casar, J.R.: Desarrollo y Despliegue de Sistemas de Localización en WLAN Basados en Huella de Potencia. In: Proc. of the 2nd Iberian Conf. on IST, Porto, Portugal, June 2007, vol. 2, pp. 453–464 (2007)
8. Pérez, J., Aparicio, S., Bernardos, A.M., Casar, J.R.: An Indoor Location System. Based on Bluetooth and WLAN. In: Proc. Workshop on User-Centric Technologies and Applications, Salamanca, Spain, pp. 77–86.

Object's Interaction Management by Means of a Fuzzy System within a Context-Based Tracking System

Ana M. Sánchez, Miguel A. Patricio, and J. García

Universidad Carlos III de Madrid, Computer Science Department, Applied Artificial Intelligence Group, Avda. Universidad Carlos III 22, 28270 Colmenarejo, Madrid
{amsmonte,mpatrici,jgherrer}@inf.uc3m.es

Summary. Tracking objects through interactions is a complex task, especially when it is important to be able to obtain the final trajectory followed by the object being track. This work proposes the use of a Context Layer to solve the problem of tracking through objects interactions, using a Fuzzy Reasoning System. Other authors have already used context information within a tracking system in order to improve its performance. The novelty of this work relies in that a Context Layer is created to reason over a general tracking system and thus improve the performance, instead of creating a context-based tracking algorithm. The experimentation shows how this Context Layer reasons over and improves a general tracking system.

1 Introduction

Video tracking is the problem of following objects moving across a video sequence automatically[7]. This is done typically by matching objects in consecutive frames using features such as points, lines or blobs. In a tracking scenario, an object can be defined as anything that is of interest for further analysis.

Tracking may be seen as a separate process, as a mean to prepare data for pose estimation, or as a mean to prepare data for recognition. If tracking prepares data for recognition, the task is usually to represent data in an appropriate manner. An example of this was published by Polana and Nelson [9] where flow information and down-sampling are used to represent image information in a compact manner which is processed by a classifier to recognize six different classes, e.g., walking and running.

Most trackers are usually designed for specific cases and particular conditions: For example, the MS (Mean Shift) tracker easily loses the object due to its intrinsic limitations of exploring local maximum, in particular, when the tracked object has quick movements. In addition the MS is hard to recover from a total occlusion [1]. MSPF (Particle filter based on MS weight)is restricted to applications with little changes of the target model[8]. Finally, CGA (Association by Canonical Genetic Algorithm) have a slow performance which can be a real problem when tracking several targets [6].

Even more when tracking multiple people in complex situations, many tracking algorithms fail. Complex situations can be described as those in which objects

J.M. Corchado et al. (Eds.): DCAI 2008, ASC 50, pp. 720–728, 2009.
springerlink.com

enter and / or leave the scene, objects interact with each other producing occlusions, crosses, unions, separations, etc [14] [15]. As a result of these situations problems such as tracks discontinuity, inconsistent track's labeling, inconsistent track's size, etc, may arise.

Tracks discontinuity occurs when a track disappears in the middle of a scene, and sometimes it reappears in another part of the scene. We call inconsistent track's labeling when an object has been labeled with different ids during the period it was in the scene. When an object goes behind, for example, a table, the tracker does not detect the entire object, it is able to see only the part of the object that is above the table, this is when inconsistent track's size occurs.

These problems add a complexity to other applications that make use of the tracking system's output. In order to solve these problems several authors have inserted context information into their tracking systems. The context notion is not new and has been explored in different areas such as linguistics, natural language processing and knowledge representation. Dey [3] defined contest as "*any information that can be used to characterize the situation of an entity*". An entity can be a person, place or object considered relevant to users and applications. The structure and representation of this information must be determined before being exploited by a specific application [2] Once a context model is built up and validated, it can be used to identified and correct erroneous or incomplete data coming from perceptual components. A string context can thus improve performance of detection and recognition algorithms [5][13].

Xu et al. [14], to treat both tracks discontinuity, inconsistent track's labeling , models the scene distinguishing between three types of static occlusions: Border occlusion (BO), Long-Term occlusion (LO), and Short-term occlusion (SO). C. Stauffer [11] considers that a tracking system with knowledge of the locations of doors, garages ... is capable of improving initialization of tracking sequences, tracks discontinuity and inconsistent track's labeling.

However, occlusions don't always occur due to static objects. Occlusions may happen because two targets get two close to each other, thus merging their tracks into one. Also a target can get occluded by another target. These type of occlusion will be referred to as dynamic occlusions for the rest of this work. In order to cope with these situations [12] detects merging and splitting events, using object tracking and segmentation results. They create a merging and splitting detection module, where the detected object are divided into four classes: existing objects, new objects, merge object and split object. The first two class objects will be directly used to update the tracker in the tracking management module. For the merge object, a group will be created which contains the trajectory and color feature of the objects in it. For the split object, the feature correspondence module is employed to assign a correct label to each split object. Even more [4], distinguish between tracks representing a person and tracks corresponding to groups of people, aiming to create the appropriate object's paths.

Recently the authors have proposed a novel architecture which aims to asses and improve a general tracking system by using context information [10]. This work centers on the inconsistent track's labeling problem due to dynamic occlusions and aims to solve this problem by using contextual information and a fuzzy reasoning system. Next section, gives an overview of a our two-lavel architecture Section 3 presents the fuzzy reasoning method used. Last Section 4 shows a performance evaluation and Section 5 some conclusions.

2 Overview

Authors have recently designed a novel framework to adapt a context reasoning layer to a general tracking system [10]. This architecture aims to be well structure and modular, in order to obtain a system easily reconfigurable.

The architecture is based on a two layer image-processing modules: General Tracking Layer (GTL) and Context Layer (CL). GTL describes a generic multipurpose tracking process for video-surveillance systems. CL is designed as a symbolic reasoning system that manages the symbolic interface data between GTL modules in order to asses a specific scenario and improves tracking.

GTL is arranged in a pipeline structure of several modules; it directly interfaces with the image stream coming from a camera and extracts the track information of the mobile objects in the current frame. The interface between adjacent modules in GTL is symbolic data and it is set up so that for each module different algorithms are interchangeable.

CL manages the symbolic interface data between GTL modules aiming to asses a specific tracking scenario. CL is designed as a symbolic reasoning system. One of the most employed reasoning systems in industry and services are knowledge-based systems (universally known as expert systems). Knowledge-based systems embed a large component of domain-specific knowledge, but differently from other heuristic-based systems, knowledge is represented in an identifiable separate part of the system rather than being dispersed throughout the whole program.

This architecture enables an easy adaptability of the system to any scenario, as well as the use of different implementations and algorithms for the general tracking system. This is done by the interface between the two layers.

3 Context Reasoning

This work centers on the reasoning needed to establish if a track being deleted by the GTL has left the scene, is behind an occlusion object or is grouped with another active track. In order to make a decision on how the CL must interpret what is happening in the scene, a fuzzy expert system is employed.

The first step to build this fuzzy system is the selection of adequate descriptions of inputs and rules relating them to the output, that is the confidence level for the possible decisions to be taken. The inputs are translated into linguistic variables. Using these concepts, for inputs h_i, a linguistic variable Lh_i, is introduced together with a set of values lh_{i1},lh_{i2},...,lh_{im}, whose cardinality is m_i. Each term lh_{ij} in the set, labels a fuzzy subset in the universe of discourse H_i, with membership function $u_{lh_{ij}}(h_i)$. A fuzzy relational algorithm (FRA) will store the knowledge required to obtain the final confidence level , CONF, for the output decision. It is composed of a finite set of fuzzy conditional statements of the form:

$$\text{IF } Lh_i \text{ is } lh_{ij} \text{ THEN LCONF is } l\alpha_k$$

where LCONF is a linguistic variable representing the decisions confidence level, with a set of possible values $l\alpha_1,\ldots, l\alpha_n$. The Mamdani implications has been chosen. Finally, α is the defuzzification of LCONF, and CONF represents its numerical domain (universe of discourse of LCONF). The adopted defuzzification process on LCONF will be the center of gravity procedure. The decision of what is happening in the scene will depend on the different confidence levels for each of the possible decisions.

The implemented system has four inputs:

- `DISTANCE`: Distance of the track to another element (static occlusion, another track or an exit or entrance).
- `PRED-DISTANCE`: Distance of the track's prediction to another element (static occlusion, another track or an exit or entrance).
- `CHANGE-SIZE`Change in track's sizes between two consecutive frames, in order to be able to detect if that track has grouped the eliminated track.
- `COLOR-SIMILARITY`: Color's different with other stored tracks.

and one output, the decision to be taken and it's confidence level (α). Each linguistic variable is defined with three fuzzy sets: small (S), medium (M) and large (L). The values defining the fuzzy sets depend on the context of the scene being analyzed.

Rules take into consideration the proximity to the occlusion objects, to other tracks and to exits or entrances, in order to make a decision on why the GTL deleted a track, or why it initialized a new track.

Two systems are created, one designed to solve problems when the GTL eliminates a track and the other one specific for when the GTL creates new tracks. Figure 1 shows the Fuzzy Systems for when eliminating tracks. It is composed of three inputs (h_i). This system must be evaluated for each track and static object that is in the same area as the track eliminated was last seen.

Figure 2 maps the confidence in respect to distance of the eliminated track and the object with the distance of the predicted position of the eliminated track with the object. As Figure 2 illustrates the importance of the distance to the last known position of the eliminated track is greater than the distance to the predicted position.

4 Experimentation

In this section, an experiment is discussed in order to demonstrate the performance of a knowledge-based tracking system at solving occlusion problems.

The video used for this experimentation belongs to a dataset recorded in our installations. It was filmed using a SONY EVI-D100P camera and Matrox

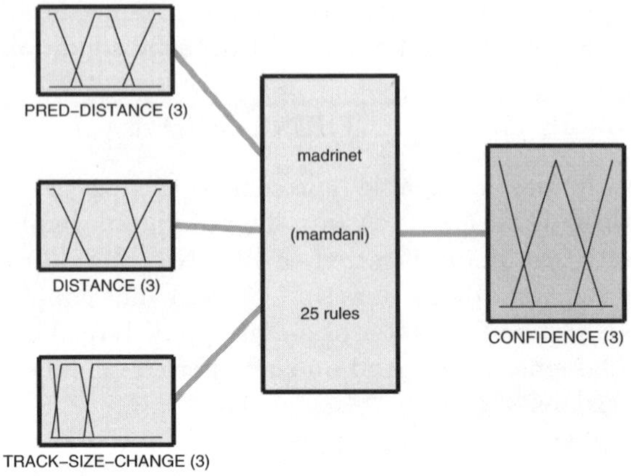

Fig. 1. Fuzzy reasoning system for when eliminating tracks

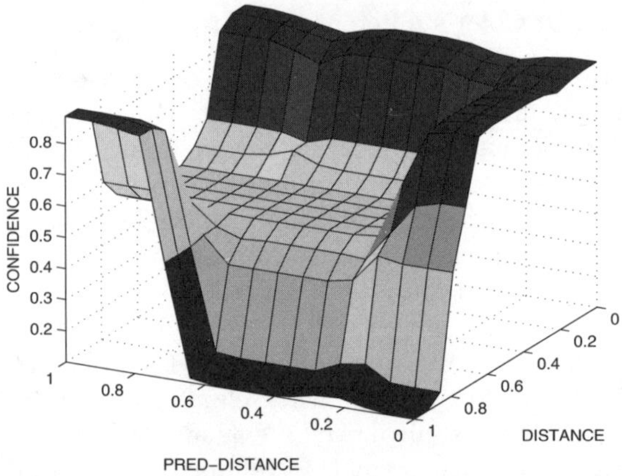

Fig. 2. Fuzzy surface map regarding `DISTANCE` and `PRED-DISTANCE`

Morphis frame grabber board. This video was recorded with a resolution of 768x576 pixels with 25 fps. The video presented in this paper contains 1172 frames. There is no noticeable change in the ambient lighting being an indoor scene, however the lighting source frequency (conventional fluorescent lamps) produces illumination changes between adjacent frames.

The context defined in this video are two entrances and exits, and one occlusion object. On Figure 3 the static objects are highlighted the right and left objects are entrances and exits, while the object in the middle is an occlusion object. During the video two people come into the scene, after that two new people also come into the scene and walk toward the other two people. Without CL when the people get together the GTL deletes two tracks and maintains a track for two people. Later when the groups separate, GTL assigns new ids to the people that separate from the other people.

Fig. 3. Static Objects defined in the scene

The goal of this experimentation is to show how the use of contextual information can help a general tracking system to generate the correct trajectory for an object that goes through dynamic occlusions. In order to generate the correct trajectory it is important that to identify an object with the same id at all times. Therefore, without the use of context information a general tracking system will assign a different id to a person after it has separated from another person.

CL reasons on the GTL symbolic interface data, in order to give a better understanding to the GTL output. In this sense, CL will be able to detected that the person with id 2 and the one with id 3 did not leave the scene and they are being grouped with other tracks and therefore it will be able to reassign the same id to the person once it comes out of the dynamic occlusion. This can be observed in Figure 4, where the objects maintains the same ids before and after the dynamic occlusion.

The contextual information will not change during the system's execution and therefore it will remain at all times within the knowledge base. Before analyzing the video, the system loads the knowledge base as well was the rules. The

Table 1. Initial and Static Facts

$$
\begin{array}{l}
(facts) \\
\qquad c_1 : (is\_a\_entrance\ (id\ e_1)) \\
\qquad c_2 : (is\_a\_entrance\ (id\ e_2)) \\
\qquad c_3 : (is\_a\_exit\ (id\ e_1)) \\
\qquad c_4 : (is\_a\_exit\ (id\ e_2)) \\
\qquad c_5 : (is\_a\_occlusion\ (id\ o_1))
\end{array}
$$

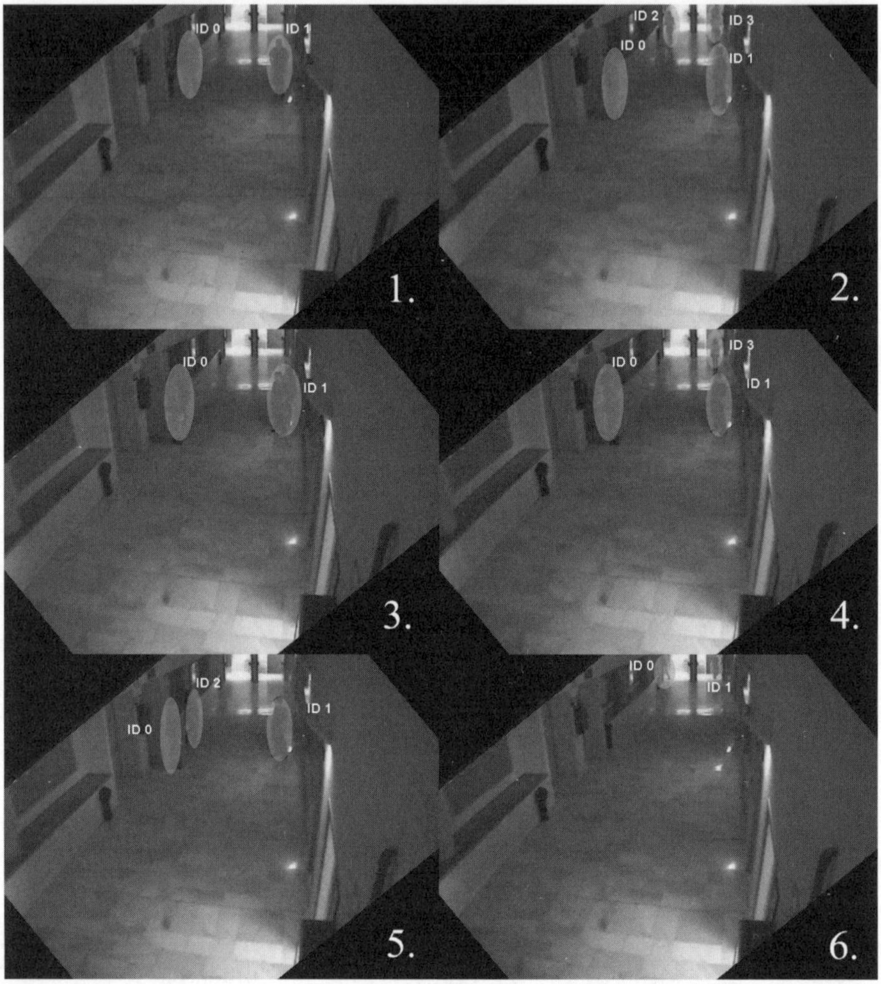

Fig. 4. Results obtained using the Context Layer. This is a sequence of frames where tracks id2 and id3 are going to be grouped with id0 and id1, respectively (frame 3). Then, in frame 4 track id3 leaves the group and is assigned the same identification (not new identification), and in frame 5 the track id2 leaves the group keeping the same identification.

facts shown in Table 1 are inserted at the knowledge base, these initial facts c_i represent the contextual information.

Where entrances, exits and occlusions are represented by polygons, that is height and width, and a position, as illustrated in Figure 3.

Once the initial facts, containing the contextual information, have been loaded into the system, the tracking system starts analyzing the video.

5 Conclusions

Tracking systems face a difficult task when required to track objects through dynamic occlusions. This work has presented a fuzzy reasoning within Context Layer that can be connected to most general tracking systems in order to reason on the tracking system's performance and improve it. This Context Layer can easily be adapted to any scenario, by just introducing a description of the scene. This description includes information about the location and size of important elements of the scene, such as entrances, exits and objects that may cause occlusions.

In order to cope with dynamic occlusions, every time a new object is detected or an existing object is lost by the general tracking system, the Context Layer analysis this information and the scenes information to decide if a dynamic occlusion has taken place.

The performance evaluation shows with an example the steps taken by the Context Layer to reason over the general tracking system's performance. From the example provided it is possible to see how a Context Layer can improve general tracking system's performance and the understanding of the output.

Acknowledgement. This work was supported in part by Projects MADRINET, TEC2005-07186-C03-02, SINPROB, TSI2005-07344-C02-02.

References

1. Tian, W., Zhang, B., Jin, Z.: Joint tracking algorithm using particle filter and mean shift with target model updating. Chinese Optics Letters 4(10), 569–572 (2006)
2. Brdiczka, O., Yuen, P.C., Zaidenberg, S., Reignier, P., Crowley, J.L.: Automatic acquisition of context models and its application to video surveillance. In: Proceedings of the 18th International Conference on Pattern Recognition (ICPR 2006) (2006)
3. Dey, A.K.: Understanding and using context. J. Personal and Ubiquitous Computing 5(1) (February 2001)
4. Bremond, F., Cupillard, F., Thonnat, M.: Tracking groups of people for video surveillance. In: Proc. of the 2nd European Workshop on Advanced Video-Based Surveillance System (2001)
5. Morency, L., Sidner, C., Lee, C., Darrell, T.: Contextual recognition of head gestures. In: Proc. of ICMI (2005)
6. Angel Patricio, M., Garcia, J., Berlanga, A., Manuel Molina, J.: Video tracking association problem using estimation of distribution algorithms in complex scenes. In: Mira, J., Álvarez, J.R. (eds.) IWINAC 2007. LNCS, vol. 4528, pp. 261–270. Springer, Heidelberg (2007)
7. Patricio, M.A., Castanedo, F., Berlanga, A., Perez, O., Garcia, J., Molina, J.M.: Computational Intelligence in Visual Sensor Networks: Improving Video Processing Systems. In: Computational Intelligence in Multimedia Processing: Recent Advances. Studies in Computational Intelligence, vol. 96, pp. 351–377. Springer, Heidelberg (2008)

8. Patricio, M.A., Garcia, J., Berlanga, A., Molina, J.M.: Solving video-association problem with explicit evaluation of hypothesis using edas. In: Proceedings of the 2008 Congress on Evolutionary Computation CEC 2008. IEEE Press, Los Alamitos (2008)
9. Polana, R., Nelson, R.: Low level recognition of human motion. In: Motion of Non-Rigid and Articulated Objects, pp. 77–82 (1994)
10. Sanchez, A.M., Patricio, M.A., Garcia, J., Molina, J.M.: Video tracking improvement using context-based information. In: The 10th International Conference on Information Fusion, Quebec (July 2007)
11. Stauffer, C.: Estimating tracking sources and sinks. In: Proc. IEEE Conf. Computer Vision and Pattern Recognition, Madison, WI, pp. 259–266 (July 2003)
12. Pan, Q., Yang, T., Li, S.Z., Li, J.: Real-time multiple objects tracking with occlusion handling in dynamic scenes. In: IEEE Computer Society Conference on Computer Vision and Pattern Recognition (CVPR 2005) (2005)
13. Torralba, A., Murphy, K.P., Freeman, W.T., Rubin, M.A.: Context-based vision system for place and object recognition. In: Proc. of ICCV (2003)
14. Xu, M., Ellis, T.: Augmented tracking with incomplete observation and probabilistic reasoning. Image and Vision Computing 24, 1202–1217 (2006)
15. Zhao, T., Nevatia, R.: Tracking multiple humans in complex situations. IEEE Transactions on Pattern Analysis and Machine Intelligence 26(9), 1208–1221 (2004)

On the Process of Designing an Activity Recognition System Using Symbolic and Subsymbolic Techniques

Rodrigo Cilla, Miguel A. Patricio, Antonio Berlanga, and Jose M. Molina

Universidad Carlos III de Madrid, Computer Science Department, Applied Artificial Intelligence Group, Avda. de la Universidad Carlos III, 22 28270 Colmenarejo (Madrid), Spain
rcilla@inf.uc3m.es

Summary. In this paper, we address the problem of human activity classification from videos, giving a special emphasis to feature extraction and good feature selection. Due to the cut down in cameras cost that have been in the last years, these kind of systems are becoming popular for their wide application area. Taking a video blob tracker output, a feature extraction process is defined to extract an extensive feature set, that is filtered in a later step to select the best features present. Three different type of classifiers are trained with the result feature set and results are shown.

Keywords: computer vision, activity recognition, feature selection.

1 Introduction

Human activity recognition from video sequences is an important research field in the computer vision community [14]. There are a lot of fields where Human activity recogition becomes useful such as Ambient Intelligence appliances[5], where the environment could interact with the user on different ways depending of the activity being performed; or video surveillance systems, that could automatically classify abnormal activities and show them on an operator's console.

Given an image sequence or video stream, where people are performing different kind of movements, the activity that are being performed must be stated. In the last years, it have been a great development of video tracking techniques [6] which provides the position of moving objects (people in our case). Taking tracker output as the input for activity recognition systems is straightforward and has been done in several works.

Works in activity recognition could be divided in two groups [13]: (1) those that are centered in small duration activities (i.e. walking, running,...); and (2) those that are in large duration activities (i.e. leaving place, going to living room,...). The former are centered in choosing good features for activity recognition, whereas the latter usually tackle the temporal structure in the classifier.

Regarding small duration activities, Ribeiro et al. [12] have shown how to perform feature extraction, feature selection and classification to solve the small

J.M. Corchado et al. (Eds.): DCAI 2008, ASC 50, pp. 729–738, 2009.
springerlink.com

duration activity recognition problem. In their approach, they use ground truth data from CAVIAR[9] dataset of people bounding boxes instead of a blob tracker output.

Robertson et al. [13] uses trajectory and velocity concatenated data for five frames, and blob optical flow, to decide what is the current activity being performed. Blob tracking is done using mean shift tracking. This small duration activity is then introduced in a Hidden Markov Model (HMM) to decide which is the current activity being performed.

Perez et al. [8] use different time averaged speed measures to classify the activities present in the CAVIAR dataset using HMM.

The main objective of this paper is to present a large feature set to model activities and a feature selection process to obtain a good feature set. Results are given using different classifiers trained from this feature set.

The article is organized as follows: section 2 gives an overview of the full system; the classification problem to be solved is defined on section 3; section 4 shows how to choose the most relevant features for activity classification;different activity classifier architectures are proposed in section 5; some results are shown in section 6 and finally some conclusions and future lines are given in section 7.

2 Overview

The proposed activity recognition system (see figure 1) takes as input video streaming. Before labeling activities, there is a preprocess step where, using a blob tracker, moving objects are tracked and their bounding box is extracted.

Blob tracker takes a sequence of video frames $I = \{I[0], I[1], \ldots, I[t]\{$, and places a bounding box around each object that is moving. To perform this process, it is necessary to maintain a background model to detect changes between frames, produced by moving targets[7]. The output from this first step is a binary foreground mask $F[t]$, containing blobs of moving objects. A blob is a set of connected pixels of moving objects. Using this output mask, object tracks are initialized, updated and deleted, and the temporal coherence of tracks is maintained. Readers are referred to [6] for more information about the tracking process. Output from the process is a temporal sequence of object bounding boxes $p = \{p_0[t], p_1[t], p_2[t], \ldots, p_k[t]\}$, being k the number of track in frame t.

Activity recognition process solves the correspondence

$$\max_{R[t]} f(R[t]) = P(p[t] \mid R[t], R[t-1], \ldots, R[1], p[0]) \quad (1)$$

where the matrix $R[t]$ is defined as $R_{ij}[t] = 1$ if track $p_i[t]$ is performing the action a_j; and $R_{ij}[t] = 0$ otherwise. So, the activity recognition decision will be which maximizes the likelihood of current activity recognition conditioned to available activity recognitions from previous frames, organized according to the chain of previous activity recognitions $R[t-1], R[t-2], \ldots, R[0]$.

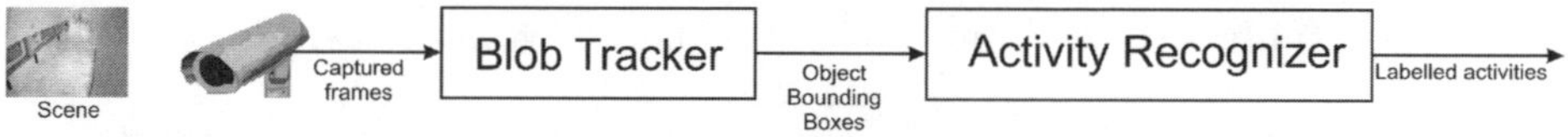

Fig. 1. Overview of the activity recognition process

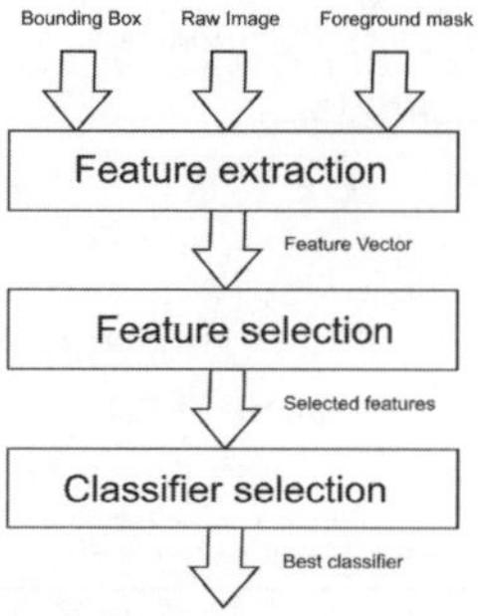

Fig. 2. The three steps to create an activity recognizer

Our approach to the activity recognition process has three different processes. First, a extensive feature set has to be extracted from the blob tracker output. Using this set, a feature selection process has to be performed to select good features, because to solve any recognition problem, it is compulsory to have a good set of features as input to the system. Then, the last problem to solve is to select a suitable classifier to perform activity recognition.

The activities that are going to be recognized in this paper are short term activities, such *stopped*, *walking*, *falling*, *lying down* or *rising*. The same architecture could be used to detect long term activities, but probably the type of classifiers used has to incorporate some type of temporal reasoning.

3 Feature Extraction

The objective of this section is to present a large set of features to characterize the activities to be classificated as complete as possible. The input to the feature extraction process includes the original frame $I(t,x,y)$, its foreground mask $F(t,x,y)$, and the blob bounding box $b(t)$ given by a blob tracker. Using these informations, different type of features are going to be derived. Most of the features presented here have been proposed on [12]. The first group of extracted features comes from the blob bounding box properties and its time evolution, taking first order derivatives and time averaged derivatives, using a temporal window of T frames, as show on table 1.

Second feature group consist of the seven Hu invariant moments[3] of the foreground mask enclosed by the blob bounding box. These 7 moments are shape descriptors invariant under translation, rotation and scaling. Each Hu moment is numbered from 26 to 32.

Table 1. First feature group: blob bounding box properties

#	Feature name	Definition
		Bounding box properties at time t
1	Position	$p(t) = (x(t), y(t))$
2	Size	$size(t) = (w(t), h(t))$
3	Velocity	$s(t) = \frac{\partial p(t)}{\partial t} = \left(\frac{\partial x(t)}{\partial t}, \frac{\partial y(t)}{\partial t}\right)$
4	Size derivative	$\frac{\partial size(t)}{\partial t} = \frac{\partial w(t)}{\partial t}, \frac{\partial h(t)}{\partial t}$
5	Speed	$s(t) = \sqrt{\left(\frac{\partial x(t)}{\partial t}\right)^2 + \left(\frac{\partial y(t)}{\partial t}\right)^2}$
6	Area	$s(t) = w(t) * h(t)$
		Properties averaged over T frames
7	Mean speed	$\bar{s}_T(t) = \frac{1}{T} \sum_{i=t-T+1}^{T} s(i)$
-	Mean velocity norm	$\Vert\bar{\mathbf{s}}_T(t)\Vert$ 3 different methods:
8	Averaging vectors	$\Vert\bar{\mathbf{s}}_T(t)\Vert_1 = \left\Vert \frac{1}{T} \sum_{i=t-T}^{t-1} \frac{\mathbf{p}(t) - \mathbf{p}(i)}{t-i} \right\Vert$
9	Mean vectors	$\Vert\bar{\mathbf{s}}_T(t)\Vert_2 = \left\Vert \frac{\mathbf{p(t)-p(t-T+1)}}{T} \right\Vert$
10	Linear fitting	$\Vert\bar{\mathbf{s}}_T(t)\Vert_3 =$ Linear Least Squares Fitting
11..13	speed/velocity ratio	$R_{\mathbf{s_T}}(i) = \frac{s_T(t)}{\Vert\mathbf{s}_T(t)\Vert_i}$ $i = 1, 2, 3$
		Second order moments
14	speed	$\sigma_v^2(t)_1 = \frac{1}{T-1} \sum_{i=t-T+1}^{t} s^2(i)$
15..17	speed (centered)	$\sigma_v^2(t)_{1+j} = \frac{1}{T-1} \sum_{i=t-T+1}^{t} (s(i) - \bar{s}_T(t))_j$ j=1,2,3
-	velocity	$\Sigma_{\mathbf{s}}(t)_1 = \frac{1}{T-1} \sum_{i=t-T+1}^{t} \mathbf{s}(i)\,\mathbf{s}(i)'$
-	velocity centered	$\Sigma_{\mathbf{s}}(t)_{1+j} = \frac{1}{T-1} \sum_{i=t-T+1}^{t} \left(\mathbf{s}(i) - \bar{\mathbf{s}}_T(t)_j\right)\left(\mathbf{s}(i) - \bar{\mathbf{s}}_T(t)_j\right)'$ j=1,2,3
18..21	trace	$tr_{\Sigma_{\mathbf{s}}}(t)_i = trace(\Sigma_{\mathbf{v}}(t))$ i=1,2,3,4
22..25	eigenvalues ratio	$R_{\Sigma_{\mathbf{s}}}(t)_i = \frac{\lambda_{min}}{\lambda_{max}}(\Sigma_{\mathbf{v}}(t))$ i=1,2,3,4

Table 2. Third and fourth group: Optical flow measures

# $I(t,x,y)$	# $F(t,x,y)$	Feature name	Definition
-	-	target pixels	$P(t) = \{(x, y) \in boundingbox(t)\}$
			Instantaneous
-	-	optical flow	$\mathbf{F}(t,x,y) = (f_x(t,x,y), f_y(t,x,y)), (x,y) \in P(t)$
-	-	mean flow	$\bar{\mathbf{F}}(t)_1 = \frac{1}{N} \sum_{(x,y)\in P(t)} \mathbf{F}(t,x,y)$
-	-	mean flow (centered)	$\bar{\mathbf{F}}(t)_2 = \mathbf{F}(t)_1 - \mathbf{v}_t$
-	-	flow norm	$f(t,x,y)_1 = \Vert\mathbf{F}(t,x,y)\Vert$
-	-	flow norm (centered)	$f(t,x,y)_2 = \Vert\mathbf{F}(t,x,y) - \mathbf{v}(t)\Vert$
			Spatial energy / 2nd order moments
33, 34	37, 38	motion energy	$f(t)_i = \frac{1}{N} \sum_{(x,y)\in P(t)} \mathbf{f}^2(t,x,y)_i$ $i = 1, 2$
-	-	flow cov.	$\Sigma_{\mathbf{F}}(t)_1 = \frac{1}{N} \sum_{(x,y)\in P(t)} \mathbf{F}(t,x,y)\,\mathbf{F}(t,x,y)'$
-	-	flow cov. centered	$\Sigma_{\mathbf{F}}(t)_2 = \frac{1}{N} \sum_{(x,y)\in P(t)} (\mathbf{F}(t,x,y) - \bar{\mathbf{F}}(t)_1)(\mathbf{F}(t,x,y) - \bar{\mathbf{F}}(t)_1)'$
35,36	39, 40	eigenvalues ratio	$R_{\Sigma_{\mathbf{F}}}(t)_i = \frac{\lambda_{min}}{\lambda_{max}}(\Sigma_{\mathbf{F}}(t))$ i=1,2
			Time averaged over T frames
-	-	mean flow	$\bar{\mathbf{F}}(t)_{i+2} = \bar{\mathbf{F}}(t)_i - \frac{1}{T} \sum_{j=t-T+1}^{t} \bar{\mathbf{F}}(j)_i$ $i = 1, 2$
41	42	motion energy	$f_T(T) = \frac{1}{T} \sum_{i=t-T+1}^{t} f(i)_1$
			Temporal energy / 2nd order moments
-	-	mean flow	$\Sigma_{\bar{\mathbf{F}}}(t)_i = \frac{1}{T} \sum_{j=t-T+1}^{t} \bar{\mathbf{F}}(j)_i\,\bar{\mathbf{F}}(j)'_i$ $i = 1, 2, 3, 4$
43...46	51...54	trace	$tr_{\Sigma_{\bar{\mathbf{F}}}}(t)_i = trace\left(\Sigma_{\bar{\mathbf{F}}}(t)_i\right)$ $i = 1, 2, 3, 4$
47 ...50	55...58	eigenvalues ratio	$R_{\Sigma_{\bar{\mathbf{F}}}}(t)_i = \frac{\lambda_{min}}{\lambda_{max}}\left(\Sigma_{\bar{\mathbf{F}}}(t)_i\right)$ $i = 1, 2, 3, 4$

Third and Fourth feature groups have been extracted from optical flow meassures from the enclosed blob bounding box of $I(t)$ and $F(t)$. Optical flow is a meassure of the motion of each one of the pixels of a given frame with respect to a previous one. In our work, it is obtained using the Lucas-Kanade method [4]. Again, some properties are taken from the time averaged meassures of different optical flow properties, using a temporal window of size T.

4 Feature Selection

Before training any classifier, good attributes have to be selected from the input dataset. Using the full dataset is usually not a good decision: as number of attributes increases, more computational time is needed to train classifiers and to perform classification in a later step. Also, some attributes do not provide so much information to classifiers and even, they can add noise degrading classifier performance. To avoid these problems, it's necessary to select a good feature vector as input to the classifier.

Feature selection is a difficult problem. Given an input feature set Y, a subset Z^* has to be chosen such a criteria function $J(X)$ is maximized:

$$Z^* = \arg\max_{Z \subseteq Y} J(Z) \tag{2}$$

To design a feature selection method, two decisions must be taken: What is the function $J(X)$? What optimization method is going to be used? The most straightforward way of answering the first question is to use the *wrapper*[10] method: for each feature subset to be evaluated, train a classifier and take as $J(X)$ the achieved classifier performance.

Another way of answering to the first question is to use *filter*[10] methods. These approaches choose a classifier independent measure as $J(X)$. Separability measures, information gain measures or correlation measures are suitable. Using these methods, a good solution to use with different classifiers could be achieved. Also, computational cost is usually better than using *wrapper* methods.

The problem is a discrete space state search problem. The trivial solution is to perform an exhaustive search, but the search space is usually too large, 2^n, n being the size of the feature set. To avoid this combinatorial explosion, more efficient search methods should be used, as traditional space state search methods such best-first search[10], or metaheuristics such genetic algorithms.

To extract the best features in our domain, the objective function $J(X)$ proposed on [2] have been chosen. It measures the degree of correlation that there is between features. As search method, best-first search is used.

5 Activity Classification

Once that the most discriminant features have been selected, a classifier has to be built to decide what is the target doing. In this work three different classifiers are going to be considered, a symbolic rule tree classifier like C4.5[11] and two sub-symbolic prototype classifiers, CSCA[1] and AIRS[15].

The first classifier to be considered is C4.5. This classifier is an extension of the previously released ID3 rule tree classifier to handle continuous attributes. Given a set of features, C4.5 finds the feature that best splits the dataset in the different objective classes using the concept of Information Gain. It applies this procedure recursively until the data contained on each leaf belongs to the same class.

CSCA and AIRS are two bioinspired prototype based classifiers. They belong to the family of the Artificial Immune System algorithms. They use a metaphor of the operation procedure of the Immune System of vertebrates to perform classification.

CSCA initializes a population of prototypes with N training instances randomly chosen. The population is evolved employing the concepts of somatic hypermutation and affinity maturation. The individuals with a highest classification score, calculated presenting all the training instances to each individual, are cloned and mutated and selected to be part of the next generation. This process is repeated until a stop criteria is satisfied. This criteria is usually a number of generations.

AIRS also employs the concepts of clonal selection and affinity maturation, but only presents each training individual to the classifier once. It initializes a population of prototypes and each training instance is iteratively presented to the population. On each iteration, the prototype with the highest affinity to the current instance is selected, and clonal selection and somatic hypermutation are applied. If the difference between the fitness ratio of the original and the best of the new individuals is greater than a threshold, the best individual replaces the selected one.

6 Results

Four video sequences were recorded at 25 fps on a corridor. Five activities were played by an actor on that videos: *walking*, *stopped*, *falling*, *lying down* and *rising*. Examples of this five activities can be seen on figure 3. All the features that were introduced on section 3 have been extracted using a temporal window from three to twenty-five, obtaining a total of attributes. The activity being performed on each frame has been manually annotated and 326 examples of each class were randomly chosen. As this set is too large, a feature extraction process has been applied using as $J(X)$ the function proposed on [2]. The search algorithm used has been best-first search. The 26 selected features are (see tables on section 3): 3, 4, 26,27,28,31,34,35_3,41_4,42_4,41_6,43_6, 42_8, 9_{10},42_{11},41_{14},

Table 3. Classification performance

C4.5	AIRS	CSCA
90.8589 %	74.4172 %	73.3742 %

Fig. 3. Example activities. First row: walking. Second row: stopped. Third row: falling. Fourth row:lying down. Fifth row: rising up.

Table 4. Training time comparative (in seconds)

C4.5	AIRS	CSCA
0.28	4.05	18.38

51_{14},45_{19},43_{20},14_{21},42_{21},11_{22}, 7_{25}, 41_{25}, 43_{25} and 45_{25}. Subscripts denotes the temporal window used to compute the feature value.

Activity recognition results using different classifiers can be seen on table 3. Also, confusion matrices of the classifiers could be seen on tables 5(a), 5(c) and 5(b). According to these matrices, the most difficult decision is to decide when a person is *walking,falling* or *rising.* C4.5 algorithm is the classifier that provides better results and requires less time to be trained (see table 4). 3

Table 5. Confusion matrices for different algorithms

(a) C4.5

class \ prediction	waking	stopped	falling	lied down	rising
walking	278	9	18	3	18
stopped	15	302	1	5	3
falling	17	2	297	2	8
lied down	4	4	1	311	6
rising	17	2	11	3	293

(b) AIRS

class \ prediction	waking	stopped	falling	lied down	rising
walking	215	2	41	2	66
stopped	18	253	6	47	2
falling	58	5	209	5	49
lied down	19	3	3	292	9
rising	45	7	12	18	244

(c) CSCA

class \ prediction	waking	stopped	falling	lied down	rising
walking	158	4	69	7	88
stopped	9	255	23	32	7
falling	20	1	263	0	42
lied down	6	36	3	271	10
rising	28	5	34	10	249

7 Conclusions

In this work a model to solve the activity recognition problem has been proposed. Features have been extracted from blob tracker output, and a procedure to select the most relevant has been employed, obtaining a good classification performance.

In future works longer durative activities have to be introduced in the system, to be able to know for example what is doing a person if he walks, stops and then starts walking again. Temporal reasoning techniques, like Hidden Markov Models, could be used to make this. Also, unsupervised learning could be used to discover activities that people do. Using that approach, more accurate models could be built, because it is not affected by possible errors in the pattern manually labeling process. Context information about the scene could be useful to provide priors about the occurrence of certain activities in different part of the scene, penalizing activity detection on places where they are not common.

Different feature searching strategies, that could provide different results, could be used to try to find more accurate feature sets. In particular, the use of multiobjective optimization for feature selection could be useful. It could be useful to obtain feature sets of different size extracted in a single feature selection run.

Activity recognition problem in multicamera environments have to be addressed. Having views of an activity from different perspectives could provide lot of information for classification.

Acknowledgement. This work was supported in part by Projects MADRINET, TEC2005-07186-C03-02, SINPROB, TSI2005-07344-C02-02.

References

1. Brownlee, J.: Clonal selection algorithm & CLONALG. The Clonal Selection Classification Algorithm(CSCA). Technical report, Swinburne University of Technology (January 2005)
2. Hall, M.A.: Correlation-based Feature Selection for Machine Learning. PhD thesis, The University of Waikato (1999)
3. Hu, M.K.: Visual pattern recognition by moment invariants. IEEE Transactions on Information Theory 8(2), 179–187 (1962)
4. Lucas, B.D., Kanade, T.: An iterative image registration technique with an application to stereo vision. In: International Joint Conference on Artificial Intelligence, vol. 81, pp. 674–679 (1981)
5. Pantic, M., Pentland, A., Nijholt, A., Huang, T.: Human computing and machine understanding of human behavior: a survey. In: ICMI 2006: Proceedings of the 8th international conference on Multimodal interfaces, pp. 239–248. ACM, New York (2006)
6. Patricio, M.A., Castanedo, F., Berlanga, A., Perez, O., Garcia, J., Molina, J.M.: Computational Intelligence in Visual Sensor Networks: Improving Video Processing Systems. In: Computational Intelligence in Multimedia Processing: Recent Advances (2008)
7. Pérez, O., Patricio, M.A., García, J., Molina, J.M.: Improving the segmentation stage of a pedestrian tracking video-based system by means of evolution strategies. In: 8th European Workshop on Evolutionary Computation in Image Analysis and Signal Processing. EvoIASP 2006, Budapest, Hungary (April 2006)
8. Perez, O., Piccardi, M., García, J., Patricio, M.A., Molina, J.M.: Comparison Between Genetic Algorithms and the Baum-Welch Algorithm in Learning HMMs for Human Activity Classification. In: Giacobini, M. (ed.) EvoWorkshops 2007. LNCS, vol. 4448, p. 399. Springer, Heidelberg (2007)
9. CAVIAR Project, `http://homepages.inf.ed.ac.uk/rbf/caviar/`
10. Pudil, P., Novovicova, J., Somol, P.: Recent Feature Selection Methods in Statistical Pattern Recognition. In: Pattern Recognition and String Matching. Springer, Heidelberg (2003)
11. Quinlan, J.R.: C4. 5: Programs for Machine Learning. Morgan Kaufmann, San Francisco (1993)
12. Ribeiro, P., Santos-Victor, J.: Human activity recognition from Video: modeling, feature selection and classification architecture. In: Proceedings of the International Workshop on Human Activity Recognition and Modelling 2005, vol. 1, pp. 61–78 (2005)

13. Robertson, N., Reid, I.: A general method for human activity recognition in video. Computer Vision and Image Understanding 104(2-3), 232–248 (2006)
14. Sebe, N.: Machine Learning in Computer Vision. Springer, Heidelberg (2005)
15. Watkins, A., Boggess, L.: A new classifier based on resource limited artificial immunesystems. In: Proceedings of the 2002 Congress on Evolutionary Computation, 2002. CEC 2002, vol. 2 (2002)

Building a Knowledge Based System for an Airport Case of Use*

Nayat Sánchez-Pi, Javier Carbó, and José Manuel Molina

Computer Science Department
Carlos III University of Madrid
Avda de la Universidad Carlos III, 22, 28270, Madrid, Spain
nayat.sanchez@uc3m.es, javier.carbo@uc3m.es, jose.molina@uc3m.es

Abstract. Knowledge based systems are advanced systems of complex problems representation. Its architecture and representation formalisms are the base of nowadays systems. The nature of the knowledge is usually derived from the experience in specific areas and its validation requires a different methodology of the one used in the conventional systems because the symbolic characteristic of the knowledge. This paper describes the design, definition and evaluation of a knowledge-based system using CommonKADS methodology in order to represent the contextual information in a formal way for a realistic environment: an airport case of use.

Keywords: Knowledge-based system, context-aware systems, real world applications.

1 Introduction

Computational resources are every time closer to all the people and that is possible because of the development of devices such as Personal Digital Assistants and mobile phones which have gained a lot of popularity in these days and are increasingly being networked. The use of this technology gave birth to new concepts like mobile computing and context aware computing. Mobile computing field is having an increasing attention as many systems are designed towards this direction. Most of them are desired to be context aware with the aim of optimizing and automating the distribution of their services in the right time and in the right place. Context aware computing was firstly defined by Shilit [1] some years ago where he claimed that the main components of context were: where you are, who you are with and where resources are nearby. This paradigm studies methods for modeling and utilizing contextual information. A more widely and used definition of what context is, was given by Dey [2] where he defines context as: "any information that characterizes a situation related to the interaction between humans, applications, and the surrounding environment." There are several approaches developing mobile and context aware systems such as platforms, frameworks and applications for offering context-aware services. The Context Toolkit proposed by Dey [2], assists for instance developers by providing them with abstractions enabling them to build context-aware applications. The Context Fusion Networks [3] was introduced by Chen and Kotz in 2004. It allows context aware applications to select distributed data sources and compose them with customized data

* Funded by projects MADRINET, TEC2005-07186-C03-02, SINPROB, TSI2005-07344-C02-02.

J.M. Corchado et al. (Eds.): DCAI 2008, ASC 50, pp. 739–747, 2009.
springerlink.com

fusion operators into an information fusion graph. The graph represents how an application computes high level understandings of its execution context from low-level sensory data. The Context Fabric [4] is another toolkit proposed by Hong and Landay in 2004 which facilitates the development of privacy-sensitive, ubiquitous computing applications.

Because of the increasing development of these kinds of systems and applications, the development of knowledge-based systems (KBS) are also being increased for being used in areas where the context aware systems failures can be costly because of the losses in services, property, etc... There are several methodologies for the development of KBS but the most known is KADS and it is the one we will use to represent, in a formal way, our case of use. KADS (Knowledge Acquisition and Document Structure) and its successor CommonKADS [5] (which is becoming the de facto European standard) is a knowledge engineering methodology. This paper provides the design, definition and validation of a KBS using CommonKADS in order to represent the contextual information and their behavior based on a predefined set of rules.

2 Domain, Inference and Task Layer

There are also several methodologies for the development of these systems but the most known is KADS and it is the one we will use in this section to represent, in a formal way, our case of use. KADS (Knowledge Acquisition and Document Structure) and its successor CommonKADS [5] (which is becoming the de facto European standard) is a knowledge engineering methodology.

In this section the representative knowledge of the domain is represented. It is here where the information and the static knowledge are described. *Concepts* in CommonKADS [6] are used to define objects collections with similar characteristics. In the case of the airport domain, the main concepts are identified as:

- Airport: It is the high level concept representing the domain on discourse.
- Location: Includes the airport location, their zones and the user location.
- User: Every person who has a role in the domain.
- Offering: A class containing a set of services grouped by categories.

These concepts can be represented in different ways. In Fig. 1 we can see the representation of the User concept using CML language where the characteristics are the

```
CONCEPT user ;
DESCRIPTION :
  "Every person who has
   a role in the domain"
ATTRIBUTES :
  Name: STRING;
  Address: STRING;
  Age: NATURAL;
  Role: value-role;
  ...
AXIOMS
  ...
END CONCEPT user;
```

```
VALUE-TYPE valor-role ;
  TYPE: ORDINAL;
  VALUE-LIST: (pilot,
      passenger, FTO,
      CBC, ASBG, ASIC,
      ASSV, ASCH )
END VALUE-TYPE valor-role;
```

Fig. 1. CML definition of the User Concept

represented by means of "attributes" unlike other approaches where as well as the characteristics, the functional information is included.

Another way of represent these concepts is the use of a domain ontology. The Domain Ontology is defined by means of three sets: a set of terms, a set that contains their definitions (typology) and a third set of relations between the terms (taxonomy). The ontology provides the explicit conceptualization of the domain terms as support of the knowledge base implementation which is prepared to be used by the applications and which solved different tasks [6].

CommonKADS has an extension of the initial phases for the development of ontologies that covers its entire life-cycle, from the feasibility study until the maintenance phase. Fig. 2 contains the representation of the domain knowledge for the airport.

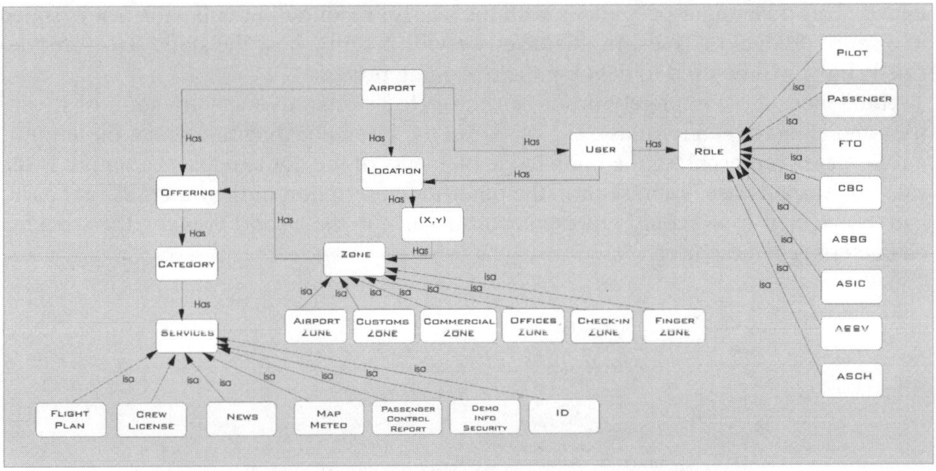

Fig. 2. Domain knowledge in the Airport KBS

Once we have the concepts we can establish relationships between them. In order to define the different type of rules in our domain schema, we need to represent a *previous/preamble* and a *consequent*. It also necessary to represent the *connection symbols* used to connect the preamble to the consequent.

The input/output curve represents a global relation or a group of partial I/O relations. The Domain Knowledge dynamics is gathered from these relations between input ranges and output ranges. For instance, if we know a passenger user of our domain is a check in zone the check-In service will be offered to him, so the rule in this case can be assessed as:

Rule Inference:

If a user with a role is into a zone at a time it implies some offering will be offered to him.

Rule Inference:

Given an offering, and category of the user, it implies that some services are provided to the user and others don't.

The CML specification and its graphic representation can be state as in Fig. 3:

```
BINARY-RELATION request ;
 ARGUMENT-1 :request-service;
 ARGUMENT-2 :user;
   CARDINALITY :ANY;
 ATTRIBUTES :
   time-request :ttime;
   zone:tzone;
   special-request:tspecial;
END BINARY-RELATION request;
```

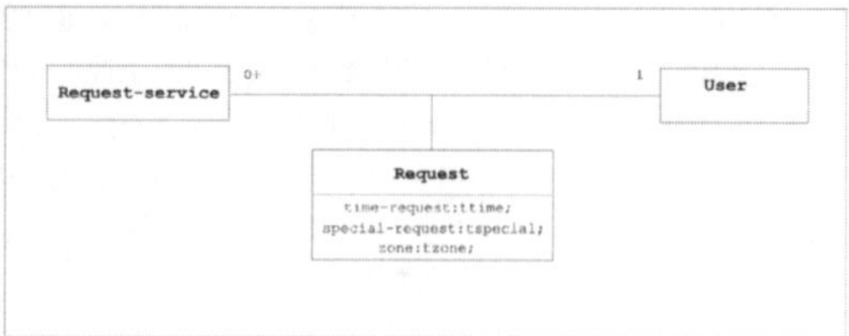

Fig. 3. CML definition of the User Concept

After getting the Domain Ontology, the domain dynamical component must be obtained. This dynamic aspect copes with the system input/output behavior that is stated as a set of production rules. In this layer we will describe how the static structures defined above will be used to develop the reasoning process.

Inferences are completely described through a declarative specification of its entries and exits (dynamic roles). Fig. 4 shows the inference diagram where the input is a role "case" representing the knowledge elements of the domain and the output is the role "abstracted-case" representing the qualified description of the input. Roles "case" and "abstracted-case" could represent, for instance in the airport domain, the smoking characteristics of the user.

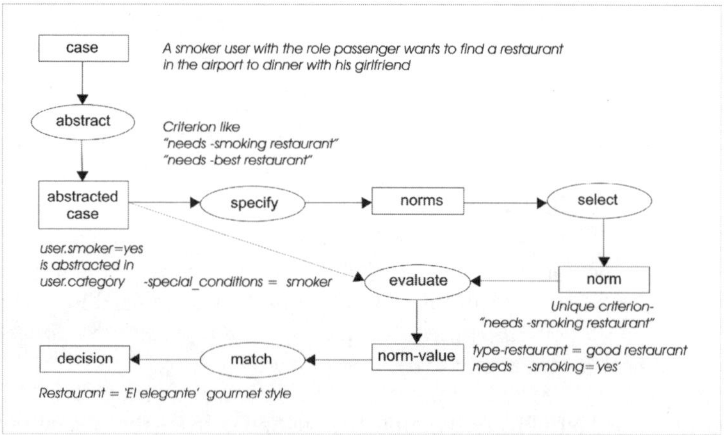

Fig. 4. Inference diagram for the assessment in the decision of choosing a restaurant in the Airport

When both Domain Ontology and Dynamics have been obtained, we have to group inference into tasks, where tasks are similar to traditional functions but the data manipulated by the task are described in a domain-independent way. They describe the input/output and how a task is realized through an ordered decomposition into sub-functions as: another task, inference and transfer function. They form a small program or algorithm that captures the reasoning strategy of the expert.

The particularity of CommonKADS is the partial reuse of Knowledge models in new applications, proposing a catalog of task templates comprised by: a tentative inference structure, a typical control structure and a typical domain schema from task point-of-view which are specific for a task type. Among the task types that CommonKADS suggests, we consider our problem as a particularization of an assessment task template; where an assessment problem consists of finding a decision category for a case based on domain specific norms.

3 Development of a KBS for an Airport Case of Use

In this section we will present the development of the KBS using a particular case of use. This case of use shows an approach to context aware services provisioning in the context of a traditional airport area and the services offered to the airline staff and passengers [7].

It is important to begin explaining we test our system in an HTC Touch terminal with a TI's OMAP™ 850, 200 MHz processor and Windows Mobile 6. We also deploy a WiFi positioning system in order to locate the user's terminals. Particularly, we used an Aruba WLAN deployment with a 2400 Mobility Controller with capacity of 48 access points and 512 users connected at the same time and several AP 60/ AP61 which are single/dual radio wireless access points. Aruba Networks[1] has a location tracking solution which uses an enterprise wide WLAN deployment to provide precise location tracking of any Wi-Fi device in the research facility. The RF Locate application can track and locate any Wi-Fi device within range of the Aruba mobility infrastructure. Using accurate deployment layouts and triangulation algorithms devices can be easily located with an accuracy of about 8 meters.[8] Although many alternatives of indoor positioning systems exists, we decided to use Aruba because its useful capability of configuring APs permanently in 'listening' mode in order to avoid missing transmissions. This is a very valuable capability because the "Air Monitors" (AMs: dedicated RF monitors) contribute not just to location accuracy but also by improving security coverage, detecting RF sources that may be security risks or interferers. The only drawback of using dedicated AMs is that they add to the capital costs of the network.

The case of use will be explained with two users of the system: Don and Donna who are boyfriends and are getting into the airport. Donna is a passenger who will take a flight to London and Don is pilot of Air France Airline. When they enter in the wireless network, it is evaluated the position of each user's device and it initiates the negotiation of the set of applications the user can access depending on his physical position and its role in the system. Our illustrative demo will describe the provisioning solution in the case users were at the commercial zone and later at the customs zone. It will illustrate the different categories of services users get depending on their role access to the system when the will be placed at the same zone. It will also illustrate how concepts, relationships and inferences, described in the dynamic domain knowledge, are represented in the way the provisioning of services to each user occurs and it is represented in the user interface.

---

[1] www.arubanetworks.com

4 Provisioning of Services into the Commercial Zone

While moving through the airport area, users with different roles in the system, receive different offerings of services. Our lead actors: Don and Donna were registered in the system and while they were moving they got different offerings of services (Fig. 5).

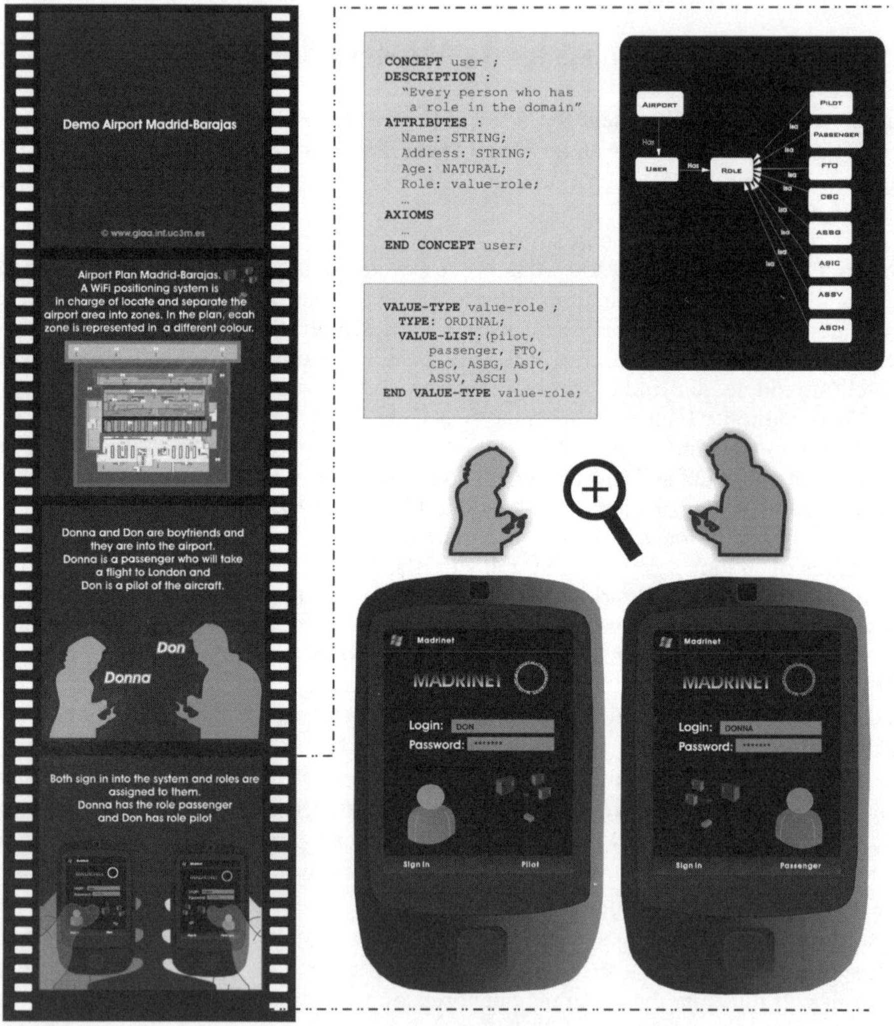

Fig. 5. Demo KBS validation. Frames 1-4

Context information in the system is used throughout the entire life-cycle of the service: selection based on the context profile, filtering of individual services, and enhancement of services at boot or runtime and the constant feed of context information

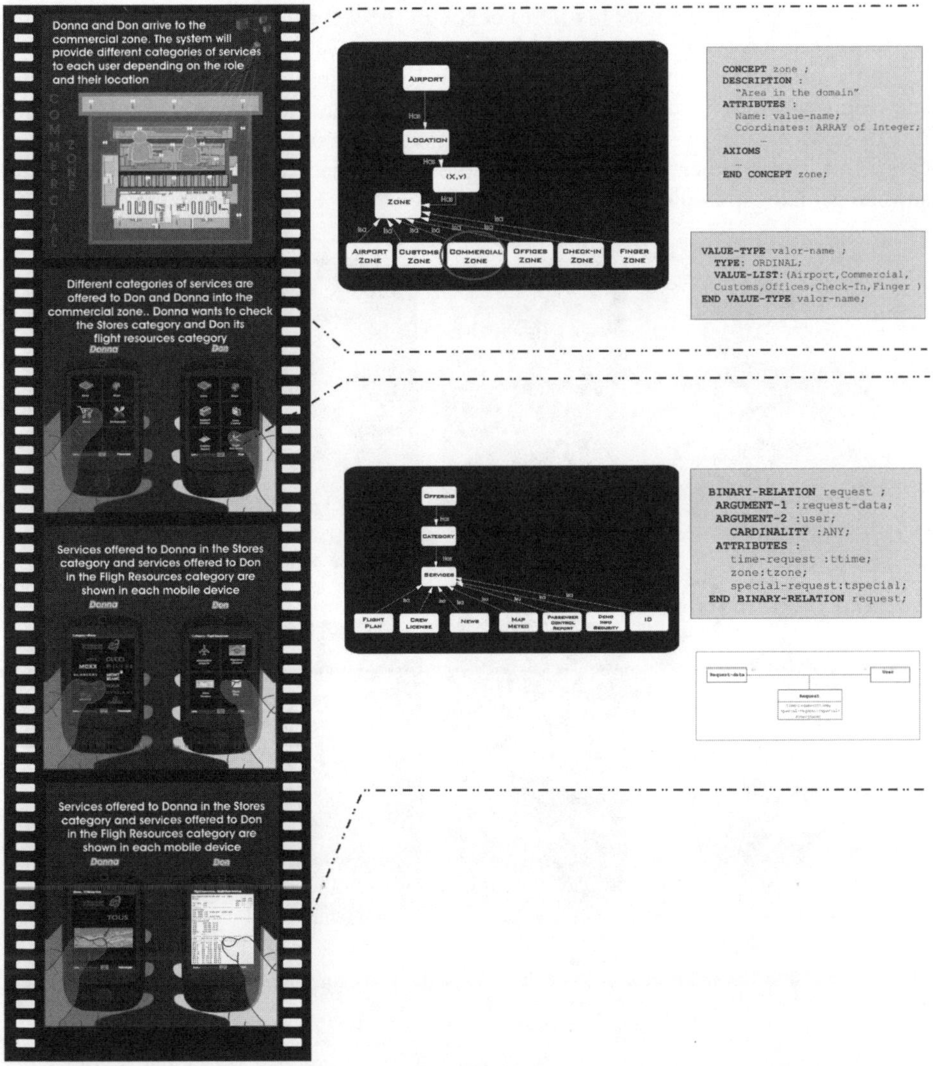

Fig. 6. Demo KBS validation. Frames 5-8

to services during execution to allow service adaptation depending on the zone the user is placed and the role in the system he may have.

Dona and Donna arrive to the commercial zone (Fig. 6) Fig.7shows how Don, who is a pilot, receives offerings of categories of services related to his role in the system: Maps, Control report, Load control, Support Control, News and Fight resources. Then he checks the Flight Resources category which has a group of services like: Alternative airports, Flight over license, Crew license and Flight plan. All of these services were designed as Java services, so if he decides then to check for instance, the Flight plan service, the system uses his ID to check in the server the information and return the request of information as a PDF file.

In the case of the other user: Donna, she is located in the same zone than Don, but she receives different offerings. News, Maps, Stores and Restaurants are the categories of services she was provided in the commercial zone. If she accesses the Stores categories, she can then choose among all the brands that has a store in the airport and she can get a location plan of the store.

Fig. 7. Demo KBS validation. Zoom of the UIs of the passenger and pilot user in the commercial zone

5 Conclusions

Our contribution in this paper is the design, definition and validation of a knowledge based system according to CommonKADS methodology and Conceptual Modeling Language (CML) in order to represent the contextual information in a formal way for a context aware system. Therefore we first defined the domain knowledge layer as a set of 22 concepts with their corresponding attributes and relationships between them. Additionally we model the set of production rules as a structure of inferences that copes dynamically with the system input/output behavior. In this inference layer we described how the concepts and relationships are assigned to declarative specifications of dynamic roles. Finally these inferences are fired in a sequential order defined by a control structure corresponding to one of the task templates that CommonKADS suggests. Particularly, we consider our problem as a particularization of an assessment task template; where an assessment problem consists of finding a decision category for a case based on domain specific norms. Therefore we have designed, implemented and validated a prototype that consists of a realistic context aware system in a formal way (through CommonKADS methodology).

References

[1] Schilit, B.N.: A System Architecture for Context-Aware Mobile Computing. Ph.D. Thesis, Columbia University, Department of Computer Science (May 1995)

[2] Dey, A., Abowd, G., Salber, D.: A Conceptual Framework and a Toolkit for Supporting the Rapid Prototyping of Context-Aware Applications. Human-Computer Interaction 16(2-4), 97–166 (2001)

[3] Chen, G., Li, M., Kotz, D.: Design and Implementation of a Large-Scale Context Fusion Network. In: MobiQuitous 2004, pp. 246–255 (2004)

[4] Hong, J.I.: The Context Fabric: An Infrastructure for Context-Aware Computing (doctoral consortium). In: Extended Abstracts of ACM Conference on Human Factors in Computing Systems (CHI 2002), pp. 554–555. ACM Press, Minneapolis (2002)

[5] Schreiber, G., Akkermans, H., Anjewierden, A., de Hoog, R., Shadbolt, N., Van de Velde, W., Wielinga, B.: Knowledge Engineering and Management, The CommonKADS Methodology. MIT Press, Cambridge (1999)

[6] Breuker, J.A., Van de Velde, W. (eds.): The Common Kads Library for Expertise Modelling. IOS Press, Amsterdam (1994)

[7] Sánchez-Pi, N., Carbó, J., Molina, J.M.: In: 8th Int. Conf. on Software Engineering, Artificial Intelligence, Networking, and Parallel/Distributed Computing (SNPD), Qingdao (2007)

[8] Sánchez-Pi, N., Carbó, J., Molina, J.M.: In: Workshop User-Centric technologies and Applications in Conjuction with CAEPIA 2007, Salamanca (2007)

Experimental Evaluation of Channel Modelling and Fingerprinting Localization Techniques for Sensor Networks*

Henar Martín, Paula Tarrío, Ana M. Bernardos, and José R. Casar

Universidad Politécnica de Madrid
ETSI Telecomunicación
Ciudad Universitaria S/N, 28040 Madrid, Spain
{hmartin,paula,abernardos,jramon}@grpss.ssr.upm.es

Abstract. In this paper we present the comparison of two localization systems where the Received Signal Strength (RSS) is measured to compute the position of a mobile node. We have implemented two approaches based on channel modelling and fingerprinting respectively. Our conclusions are supported by experimental results carried out in an indoor environment where a sensor network was deployed. The results show that in the environment of our experiments the precision of the fingerprinting method is better than the precision of the one based on channel modelling.

Keywords: Sensor network, RSS, channel model, fingerprinting, indoor.

1 Introduction

Nowadays localization in wireless sensor networks is becoming a very active research field. The applications that use the localization information in military, environmental and industrial areas are countless. It can also be applied in networks to improve routing and prevent cases of congestion.

There are different techniques in order to compute the position of a node, for instance: radiofrequency-based, optical, infrared, ultrasound techniques, etc. As sensor networks are limited by cost, battery and computational capabilities radiofrequency-based techniques are the most suitable for this purpose [1]. These techniques consist in measuring one parameter from the radio signal transmitted between two nodes [2]. In wireless sensor networks RSS techniques are commonly used since most of the commercial hardware includes the possibility of measuring this parameter. In this way, there is no need for developing new hardware, which reduces the cost and the size of the systems. The main drawback is that RSS is a very fluctuating parameter (especially indoors) due to effects on the radio signal like shadowing, multipath, etc. That is why achieving accurate localization based on RSS measurements is still a challenge. There are two main approaches to develop a localization method based on RSS in wireless sensor networks: channel modelling and fingerprinting.

* This work has been financed by the Government of Madrid under grant S-0505/TIC-0255 (MADRINET) and by the Spanish Ministry of Education and Science under grant TSI2005-07344 (COLOCAME).

J.M. Corchado et al. (Eds.): DCAI 2008, ASC 50, pp. 748–756, 2009.
springerlink.com

Our work consisted in developing two different indoor localization systems based on fingerprinting and a channel modelling respectively and carrying out some experiments in order to compare them and analyse the suitability of each of them depending on the application. The area of deployment was a laboratory environment. We worked with sensor networks based on the ZigBee standard.

This paper is organized as follows. In section 2 we survey related work concerning localization techniques based on RSS measurements. Section 3 presents the channel modelling localization method. Section 4 analyses the fingerprinting localization method. Finally, in section 5 we compare both methods and present our conclusions.

2 Related Work

As we mentioned before channel modelling and fingerprinting are the main methods when performing localization estimation based on RSS measurements. In both of them we must consider a network composed of both mobile and beacon nodes. Beacon nodes are nodes with known location. In localization based on fingerprinting the position of beacon nodes must be fixed too.

The first method is based on a channel model approximation [3-6]. A channel model is a function that provides a relation between the measured received signal strength and the distance between two nodes (a transmitter and a receiver, in our case a beacon node and a mobile node). A mobile node measures the received signal strength from a specific beacon node and estimates the distance between them using the channel model. It repeats this estimation for different beacon nodes and then it computes the position of the node using a triangulation algorithm.

The second one [6-7] can be divided into two different stages. The first one is an off-line stage which consists in measuring and storing in a database the signal strength that a node receives from the different beacon nodes in a number of known specific points throughout the coverage area of the network. We called signature a tuple which contains this information (x, y, RSS_1,...RSS_j), where x, y are the known coordinates of the point and RSS_j is the received signal strength from the beacon node whose identifier is j. From now on the database will be called map since it contains the information of the totality of the received signal strength information all over the coverage area.

The second stage is the on-line one. It is also composed of different phases. To begin with, the signature of the node whose location will be estimated must be measured. Then, it is compared with the signatures stored in the map and finally, the localization is computed using the coordinates of the signatures which are closer (in the RSS space) to the one of the mobile node.

Some groups have developed systems using these techniques. RADAR is a project from Microsoft Research based on both a channel modelling and a fingerprinting method using a WiFi network [6]. It achieves an average resolution around 3 meters using fingerprinting and a resolution of about 4.3 m with a channel model. Another project is MoteTrack from Harvard University. It proposes a method based on fingerprinting with a resolution of 2 and 3 meters with 50th and 80th percentile respectively [7]. In this project low-power, embedded wireless devices (such as the Berkeley "motes") are used. In the project WILMA a fingerprinting location system is developed. It presents an average positioning error of 2 meters using a WiFi network [8]. Alippi C. and Vanini G. propose a system based on channel models using Berkeley "motes" [9]. They achieve an average error of less than 3 meters outdoors when the node density is of about one node over 25 m^2 in an area of about $500m^2$. They also develop a map-based algorithm for indoor environments and they grant an error below 1.2 m with a 50% confidence [10]. Zemek R., Hara S., Yanahiara K., Kitayama K. propose a joint estimation of target location and channel model for sensor networks [11]. The location accuracy is about 2.5 meters.

The difference between these systems and our work is that, as we said before, we used in our networks sensors based on the ZigBee standard and that our main aim is to compare the performance of the two systems in a real environment where there are people moving and objects interfering with the signal.

3 Channel Model Localization Method

3.1 Localization Method

A channel model establishes a relation between the RSS and the distance between two nodes. In free propagation conditions RSS tends to diminish as the distance between the transmitter and the receiver increases. Some well-known channel models are: Nakagami fading model, Rayleigh fading, Ricean fading, lognormal shadowing path loss model, etc. The latter is the most commonly used in RSS-based localization due to its simplicity [12]. Using this model, the value of the received power (P_{RX}) can be calculated as:

$$P_{RX}(dBm) = P_{TX}(dBm) + A - 10 \cdot \eta \cdot \log\left(\frac{d}{d_0}\right) + N(0,\sigma) \tag{1}$$

where, A is a constant term which has to be experimentally determined, d is the distance between transmitter and receiver, P_{TX} is the transmitted power, η is the path loss exponent which typically ranges between 2 and 4 (depending on the environment), d_0 is a reference distance and N is a zero mean Gaussian noise where σ is its typical deviation.

The channel model is used to estimate the distance between the mobile node and a specific beacon node knowing the RSS measured by the mobile node from the latter. This estimation is repeated for all the beacon nodes in the area of coverage of the mobile node.

Then the position is estimated using a triangulation algorithm. Obviously, the mobile node must measure the RSS from at least three beacon nodes in order to be able to compute its location. One of the simplest positioning algorithms is the hyperbolic one which reduces the estimation problem to a linear least-squares problem [13]. The fundamentals of this algorithm are explained next. Consider (x,y) the coordinates of the position of a mobile node and (x_i,y_i) the coordinates of the ith beacon node. The square distance between a mobile and a beacon node can be expressed as follows:

$$d_i^2 = (x - x_i)^2 - (y - y_i)^2 \tag{2}$$

We consider the origin of the coordinates placed in the position of the i=1 beacon node ($x_1=0$, $y_1=0$). To compute the hyperbolic algorithm we operate the expression 2, include the equations for different beacon nodes (i>1) and express it in a matrix form. We obtain:

$$\begin{pmatrix} 2x_2 & 2x_2 \\ \vdots & \vdots \\ 2x_j & 2x_j \end{pmatrix} \cdot \begin{pmatrix} x \\ y \end{pmatrix} = \begin{pmatrix} x_2^2 + y_2^2 - d_2^2 + d_1^2 \\ \vdots \\ x_j^2 + y_j^2 - d_j^2 + d_1^2 \end{pmatrix} \quad (3)$$

Or

$$H \cdot \bar{x} = b \quad (4)$$

The least-squares solution of this equation is given by:

$$\bar{x} = \left(H^T \cdot H\right)^{-1} \cdot H^T \cdot \bar{b} \quad (5)$$

3.2 Experimental Results

In order to analyse the accuracy of this method some experiments were carried out. In the environment where we deployed the sensor network there are people whose movements can not be described using a fixed pattern. There also exist furniture and objects with fixed positions, as well as mobile ones that interfere with the propagation of the radio signal. We used MICAz nodes for the deployment of our system. These nodes work in the frequency band of 2.4 GHz and were programmed to transmit their maximum power (0 dBm). Our area of deployment is composed of four different rooms of dimensions 4x4m^2 as we show in Figure 1. In that figure the positions of the eight beacon nodes placed in the area are also shown. Their positions where chosen in order to get coverage of at least three beacon nodes in every point of the area.

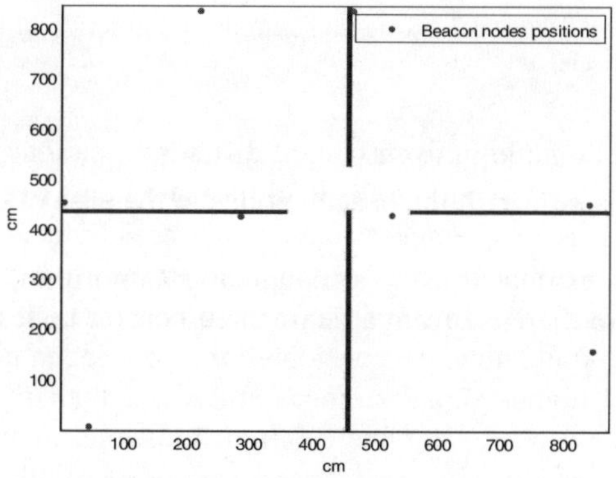

Fig. 1. Area of deployment with the positions of the beacon nodes

Firstly, we estimated the parameters of our channel model. We carried out some measurements between two nodes in a specific direction and at different distances in the area of deployment of the network and we fitted the lognormal model to the experimental data. These measurements were repeated in three different places of the deployment area. The parameters of the fittings corresponding to the three experiments are shown in Table 1. We used those fittings to estimate the distance between the beacon nodes and the mobile node in the localization phase. In Figure 2 we show the experimental data and the lognormal fitting for one of the experiments.

Table 1. Parameters of the lognormal fittings for the three experiments carried out in the deployment area

Fitting	1	2	3
A	-67.89	-73.27	-74.82
η	2.04	1.99	2.01

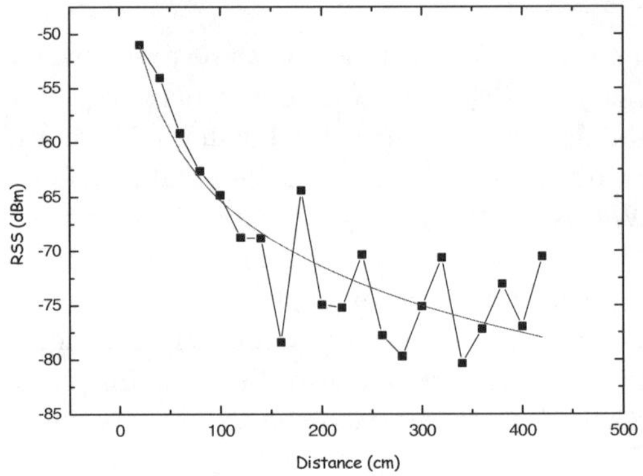

Fig. 2. Experimental data of the received signal strength as a function of the distance and fitting curve to the lognormal model

We placed the mobile node in twenty eight different locations in two of the rooms and, at each of these points, we stored the RSS measurements over a period of time (1 minute). This allowed us to average a number of measurements from the different beacon nodes so as to prevent an occasional change in the environment from having a big influence in the position estimation. In order to compute the physical coordinates of the position of the mobile node we considered the six beacon nodes from which the mobile node received higher signal strength and we solved the problem given by equation (5). This was repeated for the different locations of the mobile node that were considered and then we computed the average error of all of them. We did not consider the two farthest beacon nodes due to the fact that the estimation of their distances contained big errors and this would increase the error in the localization.

Table 2. Average localization error for the channel modelling localization system

Fitting	1	2	3
Average error (m)	2,80	2.74	2.78

In Table 2 we present the average localization error for the different fittings.

Analysing the results we observe that the average precision achieved in our deployment is less than 3 meters. An important factor is that in our case the channel model parameters have been estimated from measurements collected in a specific direction and in a specific place of the deployment area and we have generalised it for all the points. This introduces errors in the estimation of the distance between the mobile node and the beacon node that produce an error in the estimation of the position.

The accuracy of indoor localization methods based on a channel model depends on effects on the radio signal such as multipath, fading, shadowing, etc. Those effects produce variations of the radio signal which change its propagation in the environment. This leads to a bigger difference between the experimental measurements and the fitting curve. And this will introduce bigger errors in the estimation of the distances which will also involve bigger errors in the localization.

4 Fingerprinting Localization Method

4.1 Localization Method

As we explained in section 2 this method is divided in a calibration stage (off-line phase) and an on-line one. During the calibration stage we measure and store the signatures of different known positions throughout the deployment area. The measurement of the signature of the mobile node and the computation of the position is made in the on-line stage. To compute the position of the mobile node we must compare the signatures of the map and the mobile node to see which points are close (in the signal space). In this way, we must define a distance. The most common metrics are the Euclidian and the Manhattan distances. The Manhattan distance is more efficiently computed when the nodes have low computational capabilities. In our experiments we have proved that both of the distances produce similar results so we have used the Manhattan metric.

We must also consider how to compute the distance when there is a measurement from a beacon node that does not appear in the signature but it appears in the map or vice versa. We have penalized the distance by assigning a value of -100 dBm (similar to the sensibility of the nodes) to the RSS values that are missing.

Once we have computed the distances there are different possibilities to assign the location to the mobile node. We can simply opt to choose the coordinates of the closest point of the map, or to compute the average of the positions of the k nearest points of the map. Another possibility is establishing a threshold so the points which are closer than the threshold are averaged. Other option is computing a weighted average in which the closer points have more weight than the others.

4.2 Experimental Results

Some experiments were carried out to analyse the accuracy of this system. We used the same deployment as in the system based on channel modelling, using MICAz nodes. In this way, the position of the beacon nodes is the one shown in Figure 1.

Firstly we did the calibration stage, which involved the measurement of the signal map. We collected and stored the signatures of a lattice with 40 cm of distance between two adjacent points in two rooms of the deployment area.

Next, we stored the signature of the mobile node in twenty eight different test points of the area of deployment (averaging the RSS measured over a minute). These test points were the same we used for the evaluation of the channel modelling system. Then we estimated the location of the mobile node. In order to do this we only considered in the signature as well as in the map the six beacon nodes from which the mobile node measured higher signal strength. To assign the coordinates we considered the ones of the point of the map which was closest (in signal space) to the signature of the mobile node. We also considered the possibility of computing them averaging the three closest points of the map or the k closest points where k was the number of points whose distance is below a certain threshold. Finally, we computed the error of the estimation and we averaged the errors of the twenty eight points that were considered. The average localization errors achieved for different techniques are presented in Table 3.

Table 3. Average localization error using the fingerprinting system

Techniques	Nearest point	Average 3 closest points	Average k closest points
Average error (m)	2.28	2.02	2.02

We can see that the average error can be reduced by not considering only the closest point. Using a threshold to decide how many points are to be averaged reduces the average error too.

The accuracy of the system is around 2 meters. Due to the fact that we have used the same conditions in the experiments of the two localization systems we can compare the results of both of them. We observe that the system based on fingerprinting presents a lower error. This can be explained because the measurement of the signal map involves a calibration of the propagation of the signal in a huge amount of points in the deployment area. When we compute the location we apply the calibration that corresponds to the place of measurement. In contrast, the channel modelling system uses the same parameters for all the area, which is a bad approximation for the propagation of the signal.

5 Conclusions

To sum up, we have deployed a wireless sensor network using MICAz nodes and we have developed and tested two localization methods: one based on a channel model and one based on fingerprinting. With regard to the results we can state that the precision of both systems is less than 3 meters. Both cases allow us to give symbolic location of an object. As the area of deployment was a laboratory environment where there were people and fixed objects as well as mobile objects we expect that our results can be applied to similar environments.

The environment has a big influence over the fingerprinting localization system and over the channel modelling system as well. As a result of its change from the off-line stage to the on-line stage the localization results will be affected and the errors will increase. We must remark that in a laboratory environment as the one where we have carried out our experiments it is difficult to get static conditions so these errors will always be present.

On the other hand, fingerprinting can not be used in a deployment where there is no possibility to carry out the calibration stage. If we observe the results presented in sections 3 and 4 we note that the channel modelling localization system provides worst results than the fingerprinting system. But on the other hand it requires less effort, since the off-line stage of the fingerprinting system involves much more time than the estimation of the parameters of the channel model. In addition, we can state that, in environments similar to the one of our experiments, when it is not possible to measure the map the localization system based on channel modelling can be used although it will reduce the accuracy.

References

1. Akyildiz, F., Weilan, S., Sankarusubramaniam, Y., Cayirci, E.: A survey on Sensor Networks. IEEE Communications Magazine, 102–114 (August 2002)
2. Patwari, N., Ash, J.N., Kyperountas, S., Hero III, A.O., Moses, R.L., Correal, N.S.: Locating the nodes. Cooperative localization in wireless sensor networks. IEEE Signal Processing Magazine, 54–69 (July 2005)
3. Robinson, M., Psaromiligkos, I.: Received signal strength based location estimation of a wireless LAN client. In: Proceeding of the IEEE Conference on Wireless Communications and Networking, vol. 4, pp. 2350–2354 (March 2005)
4. Dogandzic, A., Amran, P.P.: Signal-strength based localization in wireless fading channels. In: Conference Record of the 38th Asilomar Conference on Signals, Systems and Computers, vol. 2, pp. 2160–2164 (2004)
5. MacDonald, J.T., Roberson, D.A., Ucci, D.R.: Location estimation of isotropic transmitters in wireless sensor networks. In: Military Communications Conference, pp. 1–5 (October 2006)
6. Bahl, P., Padmanabhan, V.N.: RADAR: an in-building RF-based user location and tracking system. In: Proc. IEEE Infocom, pp. 775–784 (March 2000)
7. Lorincz, K., Welsh, M.: Motetrack: a robust, decentralized approach to RF-based location tracking. In: Proc. of the International Workshop on Location and Context-Awareness at Pervasive 2005 (May 2005)
8. Brunato, M., Kiss Kalló, C.: Transparent Location Fingerprinting for Wireless Services. In: Proc. of Med-Hoc-Net, Mediterranean Wokshop on Ad-hoc Networks (2002)
9. Alippi, C., Vanini, G.: A RSSI-based and Calibrated centralized location technique for Wireless Sensor Networks. In: Proc. of the Fourth Annual IEEE Int. Conference on Pervasive Computing and Communications Workshops (PERCOMW 2006) (2006)

10. Alippi, C., Mottarella, A., Vanini, G.: A RF map-based localization algorithm for indoor environments. In: IEEE Int. Symposium on Circuits and Systems, vol. 1, pp. 652–655 (2005)
11. Zemek, R., Hara, S., Yanagihara, K., Kitayama, K.: A Joint estimation of target location and channel model parameters in an IEEE 805.15.4-based wireless sensor network. In: The 18th Annual IEEE International Symposium on Personal, Indoor and Mobile Radio Communications (PIMRC 2007) (2007)
12. Rappaport, T.S.: Wireless communications – Principles and practice. Prentice Hall PTR, Englewood Cliffs (1996)
13. Liu, B., Lin, K., Wu, J.: Analysis of hyperbolic and circular positioning algorithms using stationary signal-strength-difference measurements in wireless communications. IEEE Transactions on Vehicular Technology 55(2), 499–509 (2006)

Author Index

Printing: Krips bv, Meppel, The Netherlands
Binding: Stürtz, Würzburg, Germany